# Plant Pathology for Sustainable Agriculture

**Series Editors:**
Ravinder Kumar · Govind P. Rao

*Plant Pathology for Sustainable Agriculture* is a comprehensive book series that presents current research, tools, and practices that strengthen plant health and support sustainable crop production. The series brings together scientific knowledge, field experience, and technological advances to create a steady body of reliable and accessible resources. Each book adds to a growing collection that highlights practical solutions, modern diagnostics, and integrated approaches used in plant disease management across diverse crops and environments.

The series covers a wide range of topics, reflecting the depth of plant pathology research. It includes studies on viruses, bacteria, fungi, phytoplasmas, nematodes, seed-borne pathogens, point-of-care diagnostics, molecular and genomic tools, disease forecasting, climate-linked risks, crop resistance strategies, and sustainable management practices. It also includes emerging areas such as digital agriculture, remote sensing, and microbiome-based solutions.

The series serves researchers, faculty members, extension workers, and students who seek dependable scientific content. It also supports plant health professionals, diagnostic laboratories, breeding programs, policy groups, and anyone involved in crop protection and sustainable agricultural systems. It acts as an essential resource for readers who need clear, evidence-based information that supports plant disease research and practical decision-making.

Ravinder Kumar • Rahul Kumar Tiwari
Milan Kumar Lal
Virendra Kumar Baranwal • Brajesh Singh
Editors

# Vegetable Crops and Viral Pathogens Volume I

## Molecular Insights and Sustainable Control

Springer

*Editors*
Ravinder Kumar
Division of Plant Pathology
ICAR-Indian Agricultural Research Institute
New Delhi, Delhi, India

Rahul Kumar Tiwari
Division of Crop Protection
ICAR-Indian Institute of Sugarcane Research
Lucknow, Uttar Pradesh, India

Milan Kumar Lal
Division of Crop Physiology & Biochemistry
ICAR-National Rice Research Institute
Cuttack, Odisha, India

Virendra Kumar Baranwal
Division of Plant Pathology
ICAR-Indian Agricultural Research Institute
New Delhi, Delhi, India

Brajesh Singh
Physiology, Biochemistry & Post-Harvest Technology
ICAR-Central Potato Research Institute
Shimla, Himachal Pradesh, India

ISSN 3120-5976 ISSN 3120-5984 (electronic)
Plant Pathology for Sustainable Agriculture
ISBN 978-981-95-7264-9 ISBN 978-981-95-7265-6 (eBook)
https://doi.org/10.1007/978-981-95-7265-6



This Springer imprint is published by the registered company Springer Nature Singapore Pte Ltd.
The registered company address is: 152 Beach Road, #21-01/04 Gateway East, Singapore 189721, Singapore

# Preface

Vegetable crops constitute a cornerstone of global food and nutritional security, providing essential vitamins, minerals, dietary fiber, and bioactive compounds to human diets. However, their productivity and quality are persistently threatened by viral pathogens, which are among the most challenging constraints to sustainable vegetable production. The rapid emergence and re-emergence of vegetable-infecting viruses driven by climate change, globalized trade, evolving vector populations, and intensive agricultural practices have transformed plant virology from a largely descriptive discipline into a dynamic, multidisciplinary field requiring molecular, ecological, technological, and socio-regulatory perspectives. The book *Vegetable Crops and Viral Pathogens: Molecular Insights and Sustainable Control* is conceived to address this complexity. It brings together comprehensive, state-of-the-art knowledge on vegetable viruses, ranging from their global distribution and evolutionary dynamics to cutting-edge diagnostic, biotechnological, and management strategies. The volume is designed to serve as an authoritative reference for researchers, academicians, advanced students, plant breeders, extension specialists, and policy stakeholders engaged in plant health and sustainable agriculture. The opening chapters provide a global and environmental context, detailing the contemporary distribution of emerging vegetable viruses and elucidating how climate change influences virus evolution, host adaptation, and epidemic patterns. These perspectives are complemented by in-depth discussions on vector ecology, virus–host interactions, molecular evolution, and pathogenesis, offering readers a mechanistic understanding of how viral diseases establish, spread, and persist in vegetable cropping systems. A major strength of this book lies in its strong molecular and systems biology focus. Dedicated chapters explore viral suppressors of RNA silencing, altered nutrient dynamics during infection, and advanced proteomic and metabolomic approaches that reveal intricate host–virus interactions. The integration of bioinformatics tools and curated databases further equips readers with resources to analyze viral genomes, predict functional interactions, and accelerate discovery in vegetable virology. Recognizing that early detection is fundamental to disease management, the book devotes significant attention to modern diagnostic innovations. Chapters on CRISPR/Cas-based platforms (Cas9, Cas12, and Cas13), onsite diagnostic kits, nanotechnology-enabled diagnostics, and AI- and machine-learning–driven disease forecasting highlight the transformative potential of these technologies for real-time, field-level virus surveillance. Participatory plant health surveillance is also emphasized, underscoring the importance of farmer engagement and community-based monitoring in managing viral outbreaks. From a management and resistance standpoint, this volume offers an integrated view of sustainable control strategies. Topics such as seed-borne viruses, planting material certification, legislative frameworks, host plant resistance breeding, QTL mapping, RNA interference, dsRNA-based spray-induced gene silencing, peptide-based interventions, and advanced biotechnological engineering approaches are discussed in depth. Together, these chapters bridge fundamental research and practical application, aligning molecular insights with field-level disease management. The editors gratefully acknowledge the contributions of all authors, whose expertise spans plant virology, molecular biology, biotechnology, bioinformatics, ecology, and agricultural policy. Their collective efforts have resulted in a cohesive and forward-looking volume that not only synthesizes current knowledge but also identifies future research directions and translational opportunities. We hope that *Vegetable Crops and Viral Pathogens: Molecular Insights and Sustainable Control* will inspire innovative thinking, foster interdisciplinary collaboration, and support the development of resilient vegetable production systems capable of meeting the challenges of a changing global environment.

New Delhi, Delhi, India — Ravinder Kumar
Lucknow, Uttar Pradesh, India — Rahul Kumar Tiwari
Cuttack, Odisha, India — Milan Kumar Lal
New Delhi, Delhi, India — Virendra Kumar Baranwal
Shimla, Himachal Pradesh, India — Brajesh Singh

# Contents

# Editors and Contributors

## About the Editors

**Ravinder Kumar** is a senior scientist in Plant Pathology at ICAR-Indian Agricultural Research Institute, New Delhi. He earned his M.Sc. and Ph.D. in Plant Pathology from Sardar Vallabhbhai Patel University of Agriculture and Technology, Meerut, and previously served at the ICAR-Central Potato Research Institute, Shimla. His research focuses on host–pathogen interactions, plant virus epidemiology, and disease management, with major contributions in molecular characterization, genome sequencing, and diagnostics of plant viruses. He has developed advanced diagnostic tools, including multiplex RT-PCR, real-time PCR, LAMP, and RPA. Dr. Kumar has published over 150 research articles, edited 20 books with leading publishers, and serves on the editorial boards of several international journals.

**Rahul Kumar Tiwari** is currently working on melatonin-mediated mitigation of soil and tuber-borne diseases and physiological disorders in crop plants. He has also worked on the signaling mechanism of pathogens, pathogenomics, and the management of hemibiotrophic fungi. He has explored the genetic diversity of pathogenic microbes in field and horticultural crops. Additionally, his area of interest is on multiple aspects of pathophysiological alterations in food crops under the influence of diverse biotic and abiotic stresses. He has published more than 130 publications in reputed journals.

**Milan Kumar Lal** is a plant physiologist specializing in abiotic stress and nutritional aspects of potato and other starchy crops. His research focuses on heat, drought, salinity, and heavy metal stress, as well as plant physiological, biochemical, and molecular responses to biotic stresses caused by fungi, viruses, and bacteria. He also has expertise in nutrition and quality traits of starchy crops, including resistant starch, glycemic index, phytonutrients, bioactive compounds, functional fermented foods and beverages, and processing techniques to enhance nutritional value. His work has resulted in over 180 publications in international peer-reviewed journals.

**Virendra Kumar Baranwal** is a distinguished plant virologist. He earned his M.Sc. and Ph.D. in Botany (Virology) from Lucknow University and joined the Agricultural Research Service in 1985, and served at CPCRI and NCIPM before joining the Advanced Centre for Plant Virology, IARI, in 1999. He served as Professor (2016–2022) and ICAR–National Professor (2022–2025). He has guided 4 M.Sc. and 20 Ph.D. students and published over 200 peer-reviewed articles. A recipient of the IARI Best Teacher Award, he is a Fellow of NAAS and NASI. His international exposure includes IPM training in Indonesia, DBT–Overseas Associate tenure at the State University of New Jersey, USA, and academic visits to the University of Florida and the University of Queensland. Dr. Baranwal's major contributions include development of a microarray chip for the detection of over 1100 plant viruses, infectious clones of banana-infecting badnaviruses, studies on virus resistance in banana, HTS–based virus and viroid genomics, and rapid field-based diagnostics of plant viruses.

**Brajesh Singh** is a Doctor in Botany and currently serves as Director, ICAR–Central Potato Research Institute, Shimla, and Principal Scientist in the Crop Physiology, Biochemistry, and Post-Harvest Technology Division. His research interests focus on horticultural crop storage technologies, post-harvest processing, and nutritional quality assessment, with particular emphasis on improving the value and shelf life of potato crops. He is coordinating programs on genome editing of potato, post-harvest management, and nutritional enhancement, contributing to sustainable productivity and quality improvement. Dr. Singh has an extensive publication record with over 220 peer-reviewed research articles and is the author of four patents, reflecting his strong commitment to innovation, translational research, and the advancement of potato science in India.

## Contributors

**Kamal A. M. Abo-Elyousr** Department of Plant Pathology, Faculty of Agriculture, University of Assiut, Assiut, Egypt

Department of Agriculture, Faculty of Environmental Sciences, King Abdulaziz University, Jeddah, Saudi Arabia

**Munirah F. Aldayel** Department of Biological Sciences, College of Science, King Faisal University, Al-Ahsa, Saudi Arabia

**Mark Angelo O. Balendres** Department of Biology, College of Science, De La Salle University, Manila, Philippines

Plant and Soil Health Research Unit, Center for Natural Sciences and Environmental Research, College of Science, De La Salle University, Manila, Philippines

**Mritunjoy Barman** Department of Entomology, University of Nebraska-Lincoln, Lincoln, NE, USA

**Preetam Baruah** Department of Plant Pathology, Assam Agricultural University, Jorhat, Assam, India

**Laxmipriya Behera** ICAR-Central Rice Research Institute, Cuttack, Odisha, India

**Sonali Bhagat** Department of Plant Pathology, Sher-e-Kashmir University of Agricultural Sciences and Technology, Jammu, India

**Munmi Borah** Department of Plant Pathology, Assam Agricultural University, Jorhat, Assam, India

**Swapnil Borah** Department of Plant Pathology, Assam Agricultural University, Jorhat, Assam, India

**Archita Bordoloi** Department of Molecular Biology and Biotechnology, Tezpur University, Tezpur, Assam, India

**R. L. R. Cabral** Department of Agronomy, Plant Pathology Graduate Program, Federal Rural University of Pernambuco, Recife, PE, Brazil

**Sanjukta Chakraborty** Department of Plant Pathology, Assam Agricultural University, Jorhat, Assam, India

**Channakeshavaiah Chikkaputtaiah** Academy of Scientific and Innovative Research (AcSIR), Ghaziabad, India

Center for Biotechnology, Biological Sciences and Technology Division, CSIR-North East Institute of Science and Technology (CSIR-NEIST), Jorhat, Assam, India

**Devendra Kumar Choudhary** Department of Plant Pathology, College of Agriculture, Indira Gandhi Krishi Vishvavidyalaya, Raipur, Chhattisgarh, India

**Sneha Choudhary** Department of Plant Pathology, Sher-e-Kashmir University of Agricultural Sciences and Technology, Jammu, India

Department of Plant Pathology, School of Agriculture, Lovely Professional University, Phagwara, Punjab, India

**Debojit Debnath** Center for Infectious Diseases, Biological Sciences and Technology Division, CSIR-North East Institute of Science and Technology (CSIR-NEIST), Jorhat, Assam, India

**Puja Dey** Department of Plant Pathology, Assam Agricultural University, Jorhat, Assam, India

**Alfredo Diaz-Lara** Tecnologico de Monterrey, Campus Queretaro, Monterrey, Mexico

**Subham Dutta** School of Agriculture, Swami Vivekananda University, Barrackpore, West Bengal, India

Department of Plant Pathology, Bidhan Chandra Krishi Viswavidyalaya, Mohanpur, Nadia, West Bengal, India

**Leny Galvez** Philippine Fiber Industry Development Authority, Department of Agriculture, Quezon City, Philippines

**C. Gonzalez-Calquin** Center for Plant Molecular Biology, Faculty of Sciences, University of Chile, Ñuñoa, Santiago, Chile

**Darshana Goswami** Department of Molecular Biology and Biotechnology, Tezpur University, Tezpur, Assam, India

**Bernadetta Rina Hastilestari** National Research and Innovation Agency of Indonesia, Cibinong Bogor, Indonesia

**Agustín Hernández-Juárez** Department of Parasitology, Universidad Autónoma Agraria Antonio Narro, Saltillo, Coahuila, CP, Mexico

**Shridhar Shivakumar Hiremath** Center for Infectious Diseases, Biological Sciences and Technology Division, CSIR-North East Institute of Science and Technology (CSIR-NEIST), Jorhat, Assam, India

Academy of Scientific and Innovative Research (AcSIR), Ghaziabad, India

**Muhammad Imran** Department of Biological Engineering, Henan University of Technology, Zhengzhou, Henan, China

**Jose Isagani B. Janairo** Department of Biology, College of Science, De La Salle University, Manila, Philippines

Institute of Biological Control, College of Science, De La Salle University, Manila, Philippines

**P. Jayanthi** Department of Genetic Engineering, SRM Institute of Science and Technology, Kattankulattur, Tamil Nadu, India

**D. R. S. Jesus** Department of Agronomy, Genetic and Plant Breeding Graduate Program, Federal University of Espírito Santo, Alegre, ES, Brazil

**Priyam Kafle** Department of Plant Pathology, Assam Agricultural University, Jorhat, Assam, India

**Pronoy Kalita** Department of Plant Pathology, Assam Agricultural University, Jorhat, Assam, India

**Uchakankhi Kashyap** Department of Molecular Biology and Biotechnology, Tezpur University, Tezpur, Assam, India

**Ashulata Kaushal** Department of Plant Pathology, College of Agriculture, Indira Gandhi Krishi Vishvavidyalaya, Raipur, Chhattisgarh, India

**Purnama Isti Khaerani** National Research and Innovation Agency of Indonesia, Cibinong Bogor, Indonesia

**C. P. Khare** Department of Plant Pathology, College of Agriculture, Indira Gandhi Krishi Vishvavidyalaya, Raipur, Chhattisgarh, India

**A. S. Kotasthane** Department of Plant Pathology, College of Agriculture, Indira Gandhi Krishi Vishvavidyalaya, Raipur, Chhattisgarh, India

**P. J. Krishnapriya** College of Agriculture, Kerala Agricultural University, Trivandrum, Kerala, India

**S. Kruthika** ICAR-Central Rice Research Institute, Cuttack, Odisha, India

**Awadhesh Kumar** ICAR-Central Rice Research Institute, Cuttack, Odisha, India

**Bhumika Prasad Kumar** College of Agriculture, Kerala Agricultural University, Trivandrum, Kerala, India

**Ravinder Kumar** Department of Plant Pathology, ICAR-Indian Agricultural Research Institute, New Delhi, Delhi, India

**Milan Kumar Lal** ICAR-Central Rice Research Institute, Cuttack, Odisha, India

**B. W. S. Lima** Department of Agronomy, Plant Pathology Graduate Program, Federal Rural University of Pernambuco, Recife, PE, Brazil

**Soumya Majumder** Department of Tea Science, University of North Bengal, Siliguri, Darjeeling, West Bengal, India

Division of Biochemistry, Darjeeling Tea Research & Development Centre, Tea Board of India, Kurseong, West Bengal, India

**Juan Mayo-Hernandez** Department of Parasitology, Universidad Autónoma Agraria Antonio Narro, Saltillo, Coahuila, CP, Mexico

**Bandana Mohapatra** ICAR-Central Rice Research Institute, Cuttack, Odisha, India

**Nima D. Namsa** Department of Molecular Biology and Biotechnology, Tezpur University, Tezpur, Assam, India

**F. Neha Angelin** Department of Genetic Engineering, SRM Institute of Science and Technology, Kattankulattur, Tamil Nadu, India

**Adrij Pawan Neog** Department of Molecular Biology and Biotechnology, Tezpur University, Tezpur, Assam, India

**Mario Josafat Pacheco-Flores** Centro de Investigación en Alimentación y Desarrollo, Subsede Culiacán, Mexico

**Saswat Swarup Padhi** ICAR-Central Rice Research Institute, Cuttack, Odisha, India

**Moumita Panda** Department of Plant Pathology, Bidhan Chandra Krishi Viswavidyalaya, Mohanpur, Nadia, West Bengal, India

**Carla Frieda Pantouw** National Research and Innovation Agency of Indonesia, Cibinong Bogor, Indonesia

**Amitha Paul** College of Agriculture, Kerala Agricultural University, Trivandrum, Kerala, India

**Alejandro Payan-García** Centro de Investigación en Alimentación y Desarrollo, Subsede Culiacán, Mexico

**Varshini Pillai** Department of Genetic Engineering, SRM Institute of Science and Technology, Kattankulathur, Tamil Nadu, India

**Sebabrata Pramanik** Department of Computer Science and Engineering, School of Computing, Lovely Professional University, Phagwara, Punjab, India

**N. R. Rangare** Department of Genetics and Plant Breeding, College of Agriculture, Indira Gandhi Krishi Vishvavidyalaya, Raipur, Chhattisgarh, India

**Smita Bala Rangare** Department of Vegetable Science, College of Agriculture, Indira Gandhi Krishi Vishvavidyalaya, Raipur, Chhattisgarh, India

**Naropantul Appaji Rao** Department of Biochemistry, IISc, Bangalore, Karnataka, India

**Malini Ray** Department of Plant Pathology, School of Agriculture, Lovely Professional University, Phagwara, Punjab, India

**Mark Gabriel S. Sagarbarria** Department of Biology, College of Arts and Sciences, University of the Philippines Manila, Metro Manila, Philippines

**Anindita Saikia** Department of Molecular Biology and Biotechnology, Tezpur University, Tezpur, Assam, India

**Nilim Kumar Saikia** Center for Infectious Diseases, Biological Sciences and Technology Division, CSIR-North East Institute of Science and Technology (CSIR-NEIST), Jorhat, Assam, India

Academy of Scientific and Innovative Research (AcSIR), Ghaziabad, India

**Nashwa Sallam** Department of Plant Pathology, Faculty of Agriculture, University of Assiut, Assiut, Egypt

**B. S. Sandeep** College of Agriculture, Kerala Agricultural University, Trivandrum, Kerala, India

**Shravani Sanyal** School of Crop Health Biology Research NIBSM, Raipur, Chhattisgarh, India

**Devayani Sarmah** Department of Plant Pathology, Assam Agricultural University, Jorhat, Assam, India

Division of Plant Pathology, ICAR-Central Rice Research Institute, Cuttack, Odisha, India

**Handanahal Subbarao Savithri** Department of Biochemistry, IISc, Bangalore, Karnataka, India

**Mehdi Shahgolzari** Department of Medical Nanotechnology, School of Medicine, Zanjan University of Medical Sciences, Zan-jan, Iran

**K. P. Sindura** College of Agriculture, Kerala Agricultural University, Trivandrum, Kerala, India

**Zhongke Sun** Department of Biological Engineering, Henan University of Technology, Zhengzhou, Henan, China

**A. Swapna Geetanjali** Department of Genetic Engineering, SRM Institute of Science and Technology, Kattankulathur, Tamil Nadu, India

**Arpita Talukdar** Center for Infectious Diseases, Biological Sciences and Technology Division, CSIR-North East Institute of Science and Technology (CSIR-NEIST), Jorhat, Assam, India

Academy of Scientific and Innovative Research (AcSIR), Ghaziabad, India

**Jayanta Tarafdar** Department of Plant Pathology, Bidhan Chandra Krishi Viswavidyalaya, Mohanpur, Nadia, West Bengal, India

**Peter Immanuel Tenido** DLSU Innovation and Technology Office, De La Salle University, Manila, Philippines

**Rahul Kumar Tiwari** Division of Plant Protection, ICAR-Indian Sugarcane Research Institute, Lucknow, Uttar Pradesh, India

**Juan Manuel Tovar-Pedraza** Centro de Investigación en Alimentación y Desarrollo, Subsede Culiacán, Mexico

**Greeshma Varghese** Department of Plant Pathology, Assam Agricultural University, Jorhat, Assam, India

**Ubilfrido Vásquez-Gutierrez** Department of Parasitology, Universidad Autónoma Agraria Antonio Narro, Saltillo Coahuila, CP, Mexico

Department of Parasitology, Work Area of Plant Virology. Universidad Autónoma Agraria Antonio Narro, Saltillo Coahuila, CP, Mexico

**Vidya Venkat** Independent Researcher, Toronto, ON, Canada

**A. S. Xavier** Department of Agronomy, Plant Pathology Graduate Program, Federal Rural University of Pernambuco, Recife, PE, Brazil

# Abbreviations

| | |
|---|---|
| ABA | Abscisic acid |
| AgNPs | Silver nanoparticles |
| AGO | Argonaute protein |
| AI | Artificial intelligence |
| AmiRNA | Artificial microRNA |
| AMP | Antimicrobial peptide |
| AMV | Alfalfa mosaic virus |
| ANN | Artificial neural network |
| ATP | Adenosine triphosphate |
| AuNPs | Gold nanoparticles |
| AV1 | Begomovirus coat protein gene |
| BBTV | Banana bunchy top virus |
| BC1 | Begomovirus movement protein |
| BGMV | Bean golden mosaic virus |
| BiFC | Bimolecular fluorescence complementation |
| BLAST | Basic local alignment search tool |
| BNYVV | Beet necrotic yellow vein virus |
| BV1 | Nuclear shuttle protein |
| BWYV | Beet western yellows virus |
| BYMV | Barley yellow mosaic virus |
| BYMV | Bean yellow mosaic virus |
| BYV | Beet yellows virus |
| $Ca^{2+}$ | Calcium ion |
| CaATG8e | *Capsicum annuum* autophagy-related gene 8e |
| CamV | Cauliflower mosaic virus |
| Cas | CRISPR-associated protein |
| CGMMV | Cucumber green mottle mosaic virus |
| ChiLCV | Chilli leaf curl virus |
| CLE | CLAVATA3/ESR-related peptide |
| CMotV | Carrot mottle virus |
| CMV | Cucumber mosaic virus |
| CNN | Convolutional neural network |
| CNT | Carbon nanotube |
| CP | Capsid/coat protein |

| | |
|---|---|
| CPMV | Cowpea mosaic virus |
| CR | Common region |
| CRISPR | Clustered regularly interspaced short palindromic repeats |
| CRISPR/Cas9 | Clustered regularly interspaced short palindromic repeats/Cas9 |
| CRISPR-Dx | CRISPR-based diagnostics |
| CTAB | Cetyl trimethyl ammonium bromide |
| CTV | Citrus tristeza virus |
| CYSDV | Cucurbit yellow stunting disorder virus |
| DayMV | Dandelion yellow mosaic virus |
| DBIA | Dot blot immunobinding assay |
| DCL | Dicer-like protein |
| DCNN | Deep convolutional neural network |
| DETECTR | DNA endonuclease targeted CRISPR trans reporter |
| DI RNA | Defective interfering RNA |
| DNA | Deoxyribonucleic acid |
| dsDNA | Double-stranded DNA |
| dsRNA | Double-stranded RNA |
| DT | Decision Tree |
| DREB2C | Dehydration-responsive element-binding protein 2C |
| eIF4E | Eukaryotic initiation factor 4E |
| ELISA | Enzyme-linked immunosorbent assay |
| EMBL-EBI | European Molecular Biology Laboratory–European Bioinformatics Institute |
| ET | Ethylene |
| ETI | Effector-triggered immunity |
| FAM | Fluorescein amidite |
| FAO | Food and Agriculture Organization |
| FCN | Fully convolutional network |
| FFT | Fast Fourier transform |
| FQ | Fluorophore–quencher |
| GFLV | Grapevine fanleaf virus |
| GWAS | Genome-wide association study |
| HC-Pro | Helper component proteinase |
| HDA | Helicase-dependent amplification |
| HEPN | Higher eukaryotes and prokaryotes nucleotide-binding domain |
| HOLMES | One-hour low-cost multipurpose highly efficient system |
| HR | Hypersensitive response |
| HSP | Heat shock protein |
| HTS | High-throughput sequencing |
| ICTV | International Committee on Taxonomy of Viruses |
| IDM | Integrated disease management |
| INA | Isothermal nucleic acid amplification |
| IPM | Integrated Pest Management |
| iSCAN-OP | In vitro specific CRISPR-based assay (one pot) |
| iSCAN-TP | In vitro specific CRISPR-based assay (two pot) |

| | |
|---|---|
| ISTA | International Seed Testing Association |
| iTRAQ | Isobaric tags for relative and absolute quantification |
| IVT | In vitro transcription |
| JA | Jasmonic acid |
| KNN | K-nearest neighbor |
| LAMP | Loop-mediated isothermal amplification |
| LBVaV | Lettuce big-vein associated virus |
| LC-MS/MS | Liquid chromatography tandem mass spectrometry |
| LDH | Layered double hydroxide |
| LFA | Lateral flow assay |
| LFD | Lateral flow device |
| LFIA | Lateral flow immunoassay |
| LIYV | Lettuce infectious yellows virus |
| LNP | Lipid nanoparticle |
| LOC | Lab-on-a-chip |
| LoD | Limit of detection |
| LR | Logistic regression |
| LwCas13a | *Leptotrichia wadei* Cas13a |
| MAFFT | Multiple alignment using fast Fourier transform |
| MALDI-TOF | Matrix-assisted laser desorption/ionization time-of-flight |
| MAS | Marker-assisted selection |
| MCC | Mobile cloud computing |
| MED | Mediterranean cryptic species of *Bemisia tabaci* |
| MEAM1 | Middle East–Asia Minor 1 cryptic species of *Bemisia tabaci* |
| MEGA | Molecular evolutionary genetics analysis |
| miRNA | MicroRNA |
| ML | Machine learning |
| MP | Movement protein |
| mRNA | Messenger RNA |
| MSC | Multivariate scattering correction |
| MNSV | Melon necrotic spot virus |
| NASBA | Nucleic acid sequence-based amplification |
| NB | Naïve Bayes |
| NCBI | National Center for Biotechnology Information |
| NGS | Next-generation sequencing |
| NP | Nanoparticle |
| OCR | Optical character recognition |
| OECD | Organisation for Economic Co-operation and Development |
| ORF | Open reading frame |
| PAM | Protospacer adjacent motif |
| PAMP | Pathogen-associated molecular pattern |
| PAZ | Piwi–Argonaute–Zwille domain |
| PCR | Polymerase chain reaction |
| PDR | Pathogen-derived resistance |
| PEG | Polyethylene glycol |

| | |
|---|---|
| PLRV | Potato leafroll virus |
| PMMoV | Pepper mild mottle virus |
| PO | Peroxidase |
| POCT | Point-of-care testing |
| PPO | Polyphenol oxidase |
| PR | Pathogenesis-related protein |
| PRA | Pest risk analysis |
| PRSV | Papaya ringspot virus |
| PTGS | Post-transcriptional gene silencing |
| PTI | PAMP-triggered immunity |
| PVX | Potato virus X |
| PVY | Potato virus Y |
| qPCR | Quantitative polymerase chain reaction |
| qRT-PCR | Quantitative reverse transcription polymerase chain reaction |
| QTL | Quantitative trait locus |
| RCA | Rolling circle amplification |
| RdDM | RNA-directed DNA methylation |
| RdRP | RNA-dependent RNA polymerase |
| RefSeq | Reference sequence database |
| Rep | Replication initiator protein |
| R genes | Resistance genes |
| RISC | RNA-induced silencing complex |
| RNA | Ribonucleic acid |
| RNAi | RNA interference |
| RNA-seq | RNA sequencing |
| RNP | Ribonucleoprotein |
| ROS | Reactive oxygen species |
| RPA | Recombinase polymerase amplification |
| RT-LAMP | Reverse transcription loop-mediated isothermal amplification |
| RT-PCR | Reverse transcription polymerase chain reaction |
| RT-qPCR | Reverse transcription quantitative polymerase chain reaction |
| RT-RPA | Reverse transcription recombinase polymerase amplification |
| SA | Salicylic acid |
| SAR | Systemic acquired resistance |
| SavGol | Savitzky–Golay filter |
| SDA | Strand displacement amplification |
| sgRNA | Single-guide RNA |
| SHERLOCK | Specific high-sensitivity enzymatic reporter UnLOCKing |
| SIGS | Spray-induced gene silencing |
| siRNA | Small-interfering RNA |
| SNP | Single nucleotide polymorphism |
| SNV | Standard normal variate |
| SPVD | Sweet potato virus disease |
| ssDNA | Single-stranded DNA |
| ssRNA (+) | Single-stranded positive-sense RNA |

| | |
|---|---|
| SSR | Simple sequence repeat |
| SVM | Support vector machine |
| SVBT | Seed-borne and seed-transmitted virus |
| TBIA | Tissue blot immunoassay |
| TCDVd | Tomato chlorotic dwarf viroid |
| TEV | Tobacco etch virus |
| TGS | Transcriptional gene silencing |
| TMV | Tobacco mosaic virus |
| ToBRFV | Tomato brown rugose fruit virus |
| ToFBV | Tomato fruit blotch virus |
| ToLCNDV | Tomato leaf curl New Delhi virus |
| ToMV | Tomato mosaic virus |
| TRV | Tobacco rattle virus |
| TSWV | Tomato spotted wilt virus |
| TuMV | Turnip mosaic virus |
| TYLCD | Tomato yellow leaf curl disease |
| TYLCV | Tomato yellow leaf curl virus |
| TYMV | Turnip yellow mosaic virus |
| UAV | Unmanned aerial vehicle |
| VBT | Vegetable-borne virus |
| VCU | Value for cultivation and use |
| VIGS | Virus-induced gene silencing |
| ViPR | Virus pathogen resource |
| VSR | Viral suppressor of RNA silencing |
| WMV | Watermelon mosaic virus |
| WYMV | Wheat yellow mosaic virus |
| XGB | Extreme gradient boosting |
| YOLO | You only look once |
| ZIP | Zinc-regulated transporter/Iron-regulated transporter-like protein |
| ZYMV | Zucchini yellow mosaic virus |

# Contemporary Status of the Distribution of Emerging Viruses on a Global Scale

1

Juan Manuel Tovar-Pedraza, Alejandro Payan-Garcia, Mario Josafat Pacheco-Flores, and Alfredo Diaz-Lara

**Abstract**

Viruses are among the leading plant pathogens responsible for significant yield and quality losses in global crop production. The continuous expansion of international trade, climate change, and the intensification of agricultural systems have contributed to the emergence and reemergence of viral diseases in vegetable crops. Several newly identified or poorly characterized viruses have gained prominence in recent years due to their rapid spread, high transmission efficiency, and broad host range. This chapter provides a comprehensive and contemporary overview of the global distribution of key emerging plant viruses affecting vegetable crops, focusing on the genera *Begomovirus, Blunervirus, Crinivirus, Orthotospovirus, Potexvirus,* and *Tobamovirus.* These viruses have been associated with severe disease outbreaks that compromise plant development, fruit quality, and overall yield, posing a growing threat to food security and agricultural sustainability. Their detection in diverse agroecological zones reflects their adaptability and the increasing vulnerability of crop production systems. Understanding the current distribution patterns, transmission dynamics, and potential vectors of these pathogens is essential for designing integrated surveillance and diagnostic frameworks. By highlighting recent reports and trends in geographical expansion, this chapter aims to inform researchers and plant health authorities of the need to strengthen phytosanitary measures and invest in coordinated international efforts to manage emerging viral threats in vegetable crops.

J. M. Tovar-Pedraza (✉) · A. Payan-Garcia · M. J. Pacheco-Flores
Centro de Investigación en Alimentación y Desarrollo, Subsede Culiacán, Mexico
e-mail: juan.tovar@ciad.mx

A. Diaz-Lara
Tecnologico de Monterrey, Campus Queretaro, Monterrey, Mexico



R. Kumar et al. (eds.), *Vegetable Crops and Viral Pathogens Volume I*, Plant Pathology for Sustainable Agriculture, https://doi.org/10.1007/978-981-95-7265-6_1

**Keywords**

Begomovirus · Blunervirus · Crinivirus · Orthotospovirus · Potexvirus · Tobamovirus

## 1.1 Introduction

Emerging and re-emerging plant diseases are generally understood to correspond to those caused by viral agents that appear for the first time in a crop or region or that reappear after a period of apparent absence or control. These viruses often possess biological traits that favor rapid dissemination and the onset of epidemics under specific agro-climatic conditions. Both DNA and RNA plant viruses have been implicated in major plant disease outbreaks. Evolutionary processes such as recombination, mutation, and other forms of genetic exchange are considered fundamental drivers of plant virus diversification, contributing to the remarkable genetic variation within plant virus populations (Parrella et al. 2022). This chapter provides a broad and up-to-date overview of the global distribution of emerging viruses affecting economically important vegetable crops, including tomato, pepper, potato, cucurbits, lettuce, and sweet potato. These viruses have gained increasing relevance in recent years due to their ability to cause severe yield losses, which in extreme cases may result in the complete loss of crops (Kumar et al. 2022). The chapter places particular emphasis on the geographical distribution of these viral agents, while also incorporating information on their biological characteristics, symptomatology, host range, and transmission mechanisms, thereby offering a comprehensive framework to better understand their impact on modern agriculture. In recent decades, the frequency and intensity of viral disease outbreaks in plants have escalated, mainly because of global factors such as the expansion of international trade in agricultural products, climate change, and the insect vectors' adaptability to new environmental conditions. These factors have facilitated the establishment of viruses previously restricted to specific regions in new geographical areas, thereby broadening their distribution and posing an increasing threat to global agricultural production. As a result, farmers and agricultural specialists are continually confronted with challenges in managing these diseases, whose effective control is limited by the absence of direct curative treatments and the systemic nature of viral infections. Despite these constraints, recent progress in molecular and biotechnological methods has enabled the development of faster, more sensitive, and highly specific diagnostic tools, such as real-time PCR, digital PCR, high-throughput sequencing (HTS), and CRISPR-based assays. These methodologies have revolutionized plant virology by providing an early detection of viruses in plant tissues and their vectors, thereby contributing to epidemiological surveillance and the characterization of new viral species (Kumar et al. 2023; Paul et al. 2025). Furthermore, the increasing availability of these techniques in laboratories with varying levels of specialization has facilitated broader studies on the geographical distribution of emerging plant viruses across diverse production regions worldwide, enriching our understanding of their population dynamics and evolutionary patterns. Regarding viral diversity, vegetable crops host a broad spectrum of viruses from multiple families and taxonomic groups. Among RNA viruses of major agricultural importance are those that belong to the families *Virgaviridae* (*Tobamovirus*), *Bunyaviridae* (*Orthotospovirus*), *Alphaflexiviridae* (*Potexvirus*), and *Closteroviridae* (*Crinivirus*), which include species responsible for recurrent epidemics in different regions. In the case of DNA viruses, the most notable belong to the family *Geminiviridae*, particularly the genus *Begomovirus*, which has emerged as one of the most devastating viral groups in vegetable crops due to its extensive genetic diversity, high recombination capacity, and close association with insect vectors such as the whitefly (*Bemisia tabaci* cryptic species complex). Understanding the global distribution of emerging viruses in plants is essential not only to assess the magnitude of the problem but also to guide breeding programs, design integrated management strategies, and establish phytosanitary measures to prevent their dissemination. This chapter provides a detailed analysis of the distribution of the most relevant viruses that threaten crop production worldwide.

## 1.2 *Begomovirus coheni* (tomato yellow leaf curl virus, TYLCV)

*Begomovirus coheni,* formerly known as tomato yellow leaf curl virus (TYLCV), is the currently accepted species name for this widely distributed and economically significant *Begomovirus*. This virus includes multiple strains and variants, historically designated by geographic origin and phylogenetic differences. Its global distribution, high adaptability, and propensity for recombination have made it an important plant virus around the world (Yan et al. 2021). *Begomovirus coheni* comprises a genetically diverse collection of strains. These variants, such as TYLCV-IL (Kil et al. 2016), TYLCV-Mld (Ohnishi et al. 2016), TYLCV-Ge (Fazeli et al. 2009), TYLCV-Ker (Marchant et al. 2023), and TYLCV-Iran (Marchant et al. 2023), share a bipartite genomic organization; however, mutations have been reported in regions like the capsid protein (Gronenborn 2007), replication protein, V2, and C2 (Navot et al. 1991). These differences are reflected not only in their nucleotide sequences but also in their pathogenicity and geographic distribution (Juárez et al. 2014). TYLCV-IL, described in Israel, has been associated with severe symptoms like leaf curling, chlorosis, and floral abortion (Navot et al. 1991). In contrast, TYLCV-Mld is more frequently found in southern Europe, generally induces milder symptoms, and presents clear divergence in the CP and IR regions, distinguishing it phylogenetically from the TYLCV-IL lineage (Yan et al. 2021). Other variants, such as TYLCV-Iran and TYLCV-Ker, reported in Iran and India, respectively, display unique polymorphisms in genes like Rep and V2 that support their classification as independent regional lineages (Marchant et al. 2023). A key evolutionary feature of *B. coheni* is its high propensity for recombination. This has led to the appearance of variants, including TYLCV-IS176, TYLCV-IS141, and tomato yellow leaf curl Axarquia virus (TYLCAxV) (Urbino et al. 2020), that exhibit hybrid genomes derived from the Israel strain and TYLCSV, with recombination breakpoints in critical regions such as the CP, Rep, and Ori (García-Andrés et al. 2007; Papayiannis et al. 2011). Tomato yellow leaf curl Malaga virus (TYLCMalV) constitutes another well-characterized recombinant derived from TYLCV-Mld and TYLCSV, with mutations localized in the intergenic region and CP (Navas-Castillo et al. 2000). The phenological stage at which the virus infection occurs influences the severity of fruit-related symptoms. Early infections cause plants to lose vigor and completely halt fruit production, whereas later infections affect the development of new fruits but allow those already formed to ripen normally. Even though floral structures typically remain symptomless, flower abscission is frequently observed, contributing to reduced yields (Papayiannis et al. 2011). In tomato, TYLCV infection induces foliar symptoms like upward curling, interveinal chlorosis, and general dwarfing of the plant (Fig. 1.1). Infections at early developmental stages result in severe stunting, with erect terminal and axillary shoots, as well as abnormally shaped and reduced leaflets. Initially emerging leaves are often cupped downward, while those that appear later become distorted, with upward curved leaves and interveinal curling (Prasad et al. 2020). This virus is also capable of infecting other important crops, including common bean (*Phaseolus vulgaris*) and pepper, where it elicits a spectrum of foliar abnormalities (Lapidot 2002; Navas-Castillo et al. 1999). Affected plants commonly display leaf deformation, including crumpling, thickening, epinasty, and reduced leaf size, often accompanied by chlorosis or mosaic patterns. Leaflets may curl either upward or downward, and abnormal growth of both apical and lateral shoots is frequently observed. Symptom severity is linked with the cultivar and the host. In pepper, for example, some genotypes exhibit only mild symptomatology such as mottling, whereas others develop more pronounced distortion and growth suppression (Morilla et al. 2005).

*Begomovirus coheni* is transmitted exclusively by members of the *Bemisia tabaci* whitefly complex, under a persistent, non-replicative association with the vector (Rey et al. 2012). Transmission involves an acquisition access period of at least 15–30 min, followed by a latent period before it can be transmitted to another plant. Once acquired, *B. coheni* can persist in the insect for all its lifetime, significantly contributing to its dissemination (Czosnek et al. 2002). Populations of MED biotype (Q) have demonstrated greater transmission efficiency than other biotypes (Moriones and Navas-Castillo 2000; Li et al. 2010). Although tomato remains the primary host and the crop most severely affected, numerous reports have confirmed the susceptibility of other members of the family *Solanaceae*, including pepper, eggplant, tomatillo (*Physalis ixocarpa*), and tobacco (Kegler 1994; Reina et al. 1999; Gámez-Jiménez et al. 2009; Ahmed et al. 2023). Beyond the solanaceous group, the virus has also been identified infecting cucurbit species like cucumber (Al-Ali et al. 2016). This capacity for infection is not limited solely to economically important crops. There are many reports of infections in ornamental and medicinal plants, such as wild poinsettia (*Euphorbia heterophylla*) and Chinese mugwort (*Artemisia argyi*), and in weeds and ruderal plants, such as *Datura stramonium* and *Solanum nigrum*, which can harbor the virus without displaying visible symptoms (Kegler 1994; Jordá et al. 2001a; Lobin et al. 2023; Fan et al. 2025). The current

**Fig. 1.1** Symptoms induced by emerging *Begomovirus*, *Blunervirus,* and *Crinivirus* species in vegetable crops. (**a–c**) Symptoms induced by *Begomovirus coheni* in tomato leaves. (**d**) Symptoms caused by *Blunervirus solani* in tomato fruit. (**e**) Symptoms of yellowing and chlorotic mottling induced by *Crinivirus cucurbitae* in watermelon plants. (**f**) Symptoms of chlorosis of the tomato plant associated with *Crinivirus tomatochlorosis*

distribution of *B. coheni* reflects its rapid and extensive global spread (Fig. 1.2). Since the first identification in the Middle East (Navot et al. 1991), the virus has colonized key agricultural zones across America, Africa, Asia, and Europe, where tomato cultivation represents a vital economic activity. The first identification of *B. coheni* occurred in Israel during the 1960s, where it triggered severe tomato epidemics and was experimentally transmitted by whiteflies to diverse hosts (Czosnek

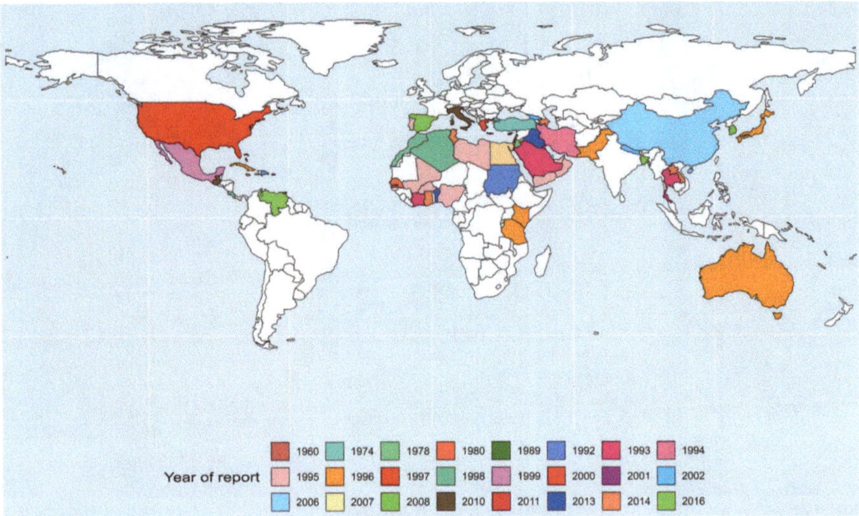


**Fig. 1.2** Worldwide distribution of *Begomovirus coheni* based on the year of first report in each country

et al. 1990). From there, its presence expanded throughout the eastern Mediterranean and North Africa. Molecular studies later confirmed its occurrence in Egypt, Jordan, and Cyprus during the 1980s and 1990s (Abdel-Salam 1991; Anfoka et al. 2005; Papayiannis et al. 2011), followed by records in Tunisia, Italy, Spain, and Turkey, frequently associated with novel variants and recombinant strains (Accotto et al. 2000, 2003; Pellegrin et al. 2008; Gul-Seker and Elibuyuk 2019). By the early 1990s, the virus had reached the Caribbean, where countries such as the Dominican Republic, Jamaica, and Cuba reported severe tomato outbreaks (McGlashan 1994). Its presence in South America was later confirmed in Venezuela in 2007 with natural field infections (Nava et al. 2006). *B. coheni* was first documented in Mexico in 1999 in Yucatán (Ascencio-Ibáñez et al. 1999) and subsequently in the state of Sinaloa (Gámez-Jiménez et al. 2009). More recent studies demonstrate its expansion into Baja California Sur and Sonora states, where it co-infects tomatoes and peppers alongside other begomoviruses, including *B. solanumlapazense* and *B. solanumseverugosi* (Cardenas-Conejo et al. 2010). Similarly, in the USA, isolates identified in Arizona and California showed high genomic similarity with strains from Mexico and the eastern Mediterranean, suggesting multiple independent introductions and the establishment of stable populations sustained by *B. tabaci* communities (Brown and Idris 2006; Rojas et al. 2007). In Asia, the virus became endemic in tomato-producing regions of China, India, Pakistan, and Iran. In China, the TYLCV-IL strain was reported in 2006 and rapidly spread across more than 11 provinces (Lefeuvre et al. 2010; Shamshiri et al. 2019). Other variants, such as TYLCV-Ker and TYLCV-Iran, emerged in India, driven by recombination and dense vector populations (Navas-Castillo et al. 2011). Though less frequent, occurrences in Oceania, including Australia and New Caledonia, have been associated with contaminated plants and vectors from Southeast Asia (Péréfarres et al. 2012).

## 1.3 *Begomovirus solanumdelhiense* (tomato leaf curl New Delhi virus, ToLCNDV)

*Begomovirus solanumdelhiense*, previously known as tomato leaf curl New Delhi virus (ToLCNDV) (Sarkar et al. 2024). This virus has geminate particles with icosahedral morphology, composed of two joined, tear-shaped capsids connected at a common base, measuring approximately 18–20 nm in width and 30 nm in length (Cai et al. 2023). Its genome is formed by two circular ssDNA (+), each measuring 2.7 kb in length (Zaidi et al. 2017). *B. solanumdelhiense* transmission is mediated by whitefly under a persistent, non-replicative association, which contributes to its rapid spread in agricultural environments with high host density (EPPO 2025a). Although it has also been detected in seeds in some cases, this transmission route is still considered of secondary importance (Moriones et al. 2017; Naganur et al. 2023). Epidemic significance is related to its high infectivity, with reported incidence levels exceeding 80%. In India, losses of 70% in melon and zucchini have been estimated (Panno et al. 2016). In Spain, during initial outbreaks in Murcia and Alicante, up to 80% of the incidences were recorded in greenhouse-grown zucchini. In contrast, tomato showed lower incidence but severe symptoms when mixed infections occurred (Juárez et al. 2014). *B. solanumdelhiense* host range is broad, including economically important crops, mainly from the Solanaceae and Cucurbitaceae, but also comprising species from Fabaceae, Malvaceae, Asteraceae, and Apiaceae (EPPO 2025a). Affected crops include tomato (Ruiz et al. 2015), zucchini (Venkataravanappa et al. 2025), cucumber (Ito et al. 2008), melon (C. melo) (Juárez et al. 2019), pumpkin (Phaneendra et al. 2012), watermelon (Venkataravanappa et al. 2020), chayote (*Sechium edule*) (Nagendran et al. 2017), potato (Lal et al. 2021), eggplant (Parrella et al. 2020), pepper (Khan et al. 2006), and common bean (Ji et al. 2012), as well as weeds including *Datura stramonium*, *Ecballium elaterium, Solanum nigrum*, and *Sonchus oleraceus* (Rabadan et al. 2019; Ansar et al. 2021). Symptoms induced by *B. solanumdelhiense* are host-related. In tomato, leaf curling, interveinal chlorosis, leaf size reduction, vein thickening, and fruit deformation are observed (Zaidi et al. 2017); in zucchini, melon, and cucumber, yellowing, chlorotic mosaic, shortened internodes, fruit deformation, and stunted growth have been reported (Panno et al. 2016; Espino de Paz et al. 2019); in chayote, marginal necrosis and leaf blade roughness have been observed (Nagendran et al. 2017); in potato, dwarfism, vein clearing, and foliage deformation are documented (Jeevalatha et al. 2023); and in chili pepper, mild mosaic, apical curling, and reduced flowering have been described (Khan et al. 2006). Since its initial detection (Cai et al. 2023), *B. solanumdelhiense* has exhibited remarkable geographical expansion (Fig. 1.3). Its presence is currently known in several Asian countries (Bangladesh, Iran, China) (Hussain et al. 2004; Maruthi et al. 2005;

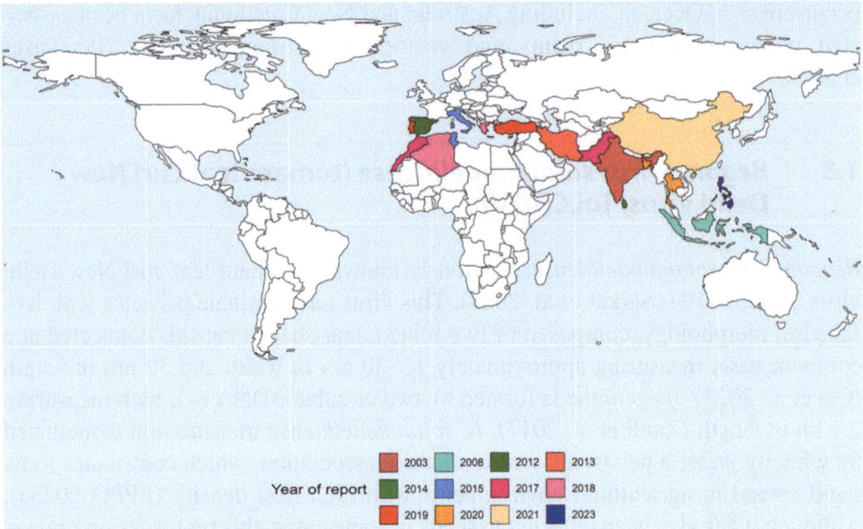


**Fig. 1.3** Global distribution of *Begomovirus solanumdelhiense*. The map shows the year of the first report in each country

Yazdani-Khameneh et al. 2013; Zeng et al. 2023), the Middle East (Oman and Saudi Arabia), North Africa (Sudan, Morocco, and Tunisia) (Mnari-Hattab et al. 2015; Radouane et al. 2018), and Europe (Spain, Italy, and Turkey) (Juárez et al. 2014; Panno et al. 2016; Fidan et al. 2023), causing epidemics, particularly in cucurbit crops. In Europe, the ToLCNDV-ES strain is predominant (Farina et al. 2023). This strain is genetically homogeneous and distinct from Asian isolates, sharing more than 98% nucleotide identity (Vo et al. 2022).

## 1.4 *Blunervirus solani* (tomato fruit blotch virus, ToFBV)

Symptoms caused by *Blunervirus solani* were first observed in Italy in 2018, causing blotchy, uneven tomato ripening and dimpling (Fig. 1.1) (Nakasu et al. 2022). Attempts to reproduce infection through mechanical inoculation were unsuccessful. Sequencing and phylogenetic analysis proved it to be a novel virus of the *Blunervirus* genus (Ciuffo et al. 2020). *B. solani* is an ssRNA (+) virus (ICTV 2025b). Its genome is segmented into four parts, and its virions have been reported to accumulate in the perinuclear space, the lumen of the endoplasmic reticulum of the epidermis, and peri- and mesocarp cells (Kitajima et al. 2023). As with other kitaviruses, transmission is dependent on mite vectors. Transmission of *B. solani* via the eriophyid, tomato russet mite (*Aculops lycopersici*) has been experimentally proven, as well as the virus's ability to cause systemic infections in tomato. Foliar symptoms include slight mosaic, chlorotic areas, discoloration, and stunting (Bertin et al. 2025), whereas severe fruit symptoms might include dimpling and uneven blotchy

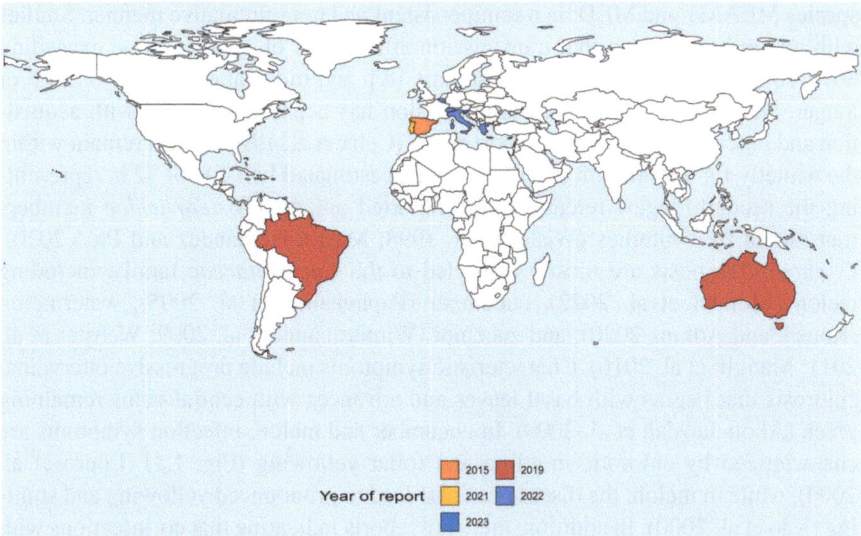


**Fig. 1.4** Geographical distribution of *Blunervirus solani*. The color scale represents the year of the first report in each country

ripening (Ciuffo et al. 2020). Aside from its first global report in Italy, *B. solani* distribution has reached several European countries, such as Belgium, France (EPPO 2025e), Greece (Beris et al. 2023), Malta (EPPO 2025c), Portugal (Maachi et al. 2021), Slovenia (Rivarez et al. 2023), Spain (Maachi et al. 2021), and Switzerland (Blouin et al. 2023). In America, it has only been reported in Brazil, and in Oceania, it has only been reported in Australia (Ciuffo et al. 2020) (Fig. 1.4). Interestingly, ToFBV has tomato as its only major host. Presumably, its erratic incidence could be attached to the fact that kitaviruses are closely related to arthropod-infecting viruses, and mutations that made it possible for *B. solani* to infect tomato might have conferred only partial adaptation (Bertin et al. 2025).

## 1.5 *Crinivirus cucurbitae* (cucurbit yellow stunting disorder virus, CYSDV)

*Crinivirus cucurbitae*, formerly known as *cucurbit yellow stunting disorder virus* (CYSDV), represents one of the most relevant viruses affecting cucurbits in warm and temperate regions (Marco and Aranda 2005; Hernández-Pérez et al. 2025). Its genome consists of two ssRNA (+), with approximate lengths of 9123 and 7976 nucleotides, respectively (Hernández-Pérez et al. 2025; Kiss et al. 2013). The virions are non-enveloped flexuous, ranging from 750 to 800 nm in length, specifically adapted to colonize the phloem, where the virus is restricted during infection (Aguilar et al. 2003; Marco and Aranda 2005). *C. cucurbitae* is transmitted exclusively by whitefly (*B. tabaci* cryptic species complex), particularly the cryptic species MEAM1 and MED, in a semipersistent and non-circulative manner. Studies with melon have shown that transmission efficiencies of *C. cucurbitae* exceeding 80% require acquisition periods of at least 18 h and inoculation periods of 24 h or longer. Nevertheless, successful transmission has been documented with acquisition and inoculation intervals as short as 2 h (Celix et al. 1996). It can remain within the whitefly for a minimum of 9 days, with an estimated half-life of 72 h, representing the second-longest retention time reported among *Closteroviridae* members transmitted by whiteflies (Wisler et al. 1998; Martín-Hernández and Picó 2020). *C. cucurbitae* hosts are mostly restricted to the *Cucurbitaceae* family, including melon (Mondal et al. 2022), cucumber (Papayiannis et al. 2005), watermelon (Kousik and Adkins 2020), and zucchini (Wintermantel et al. 2009; Webster et al. 2011; Manglli et al. 2016). Characteristic symptoms include progressive interveinal chlorosis that begins with basal leaves and advances with central veins remaining green (Abou-Jawdah et al. 2000). In cucumber and melon, infection symptoms are characterized by chlorotic mottling and foliar yellowing (Fig. 1.1) (Louro et al. 2000), while in melon, the disease can also lead to pronounced yellowing and stunting (Kao et al. 2000). In addition, there are reports indicating that co-infections with other viruses, such as *Begomovirus citrulli* (watermelon chlorotic stunt virus, WmCSV), may result in more severe symptom expressions in watermelon (Hernández-Pérez et al. 2025). Regarding its distribution (Fig. 1.5), *C. cucurbitae* was first reported in the Emirates in the early 1990s (Abou-Jawdah et al. 2000), but it was rapidly propagated across Mediterranean countries and parts of Western Asia (Mohammed et al. 2024; Nagendran et al. 2023), North Africa (Desbiez et al. 2000; Wisler et al. 1998), and later America. It is now widely distributed in countries such

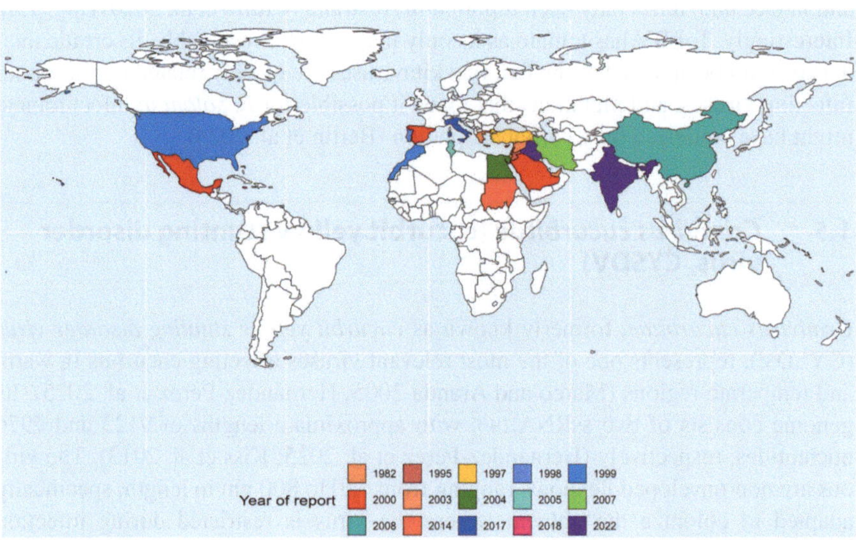


**Fig. 1.5** Geographical distribution of *Crinivirus cucurbitae*. The color scale represents the year of the first report in each country

as Spain (Celix et al. 1996), Lebanon (Abou-Jawdah et al. 2000), Morocco (Desbiez et al. 2000), Algeria (Kheireddine et al. 2020), Israel (Luria et al. 2019), Mexico (Hernández-Pérez et al. 2025), and the USA (Mondal et al. 2022), where it has become the most economically important virus affecting melon and cucumber in the southwestern production regions (Wintermantel et al. 2009; Pitter et al. 2024). Concerning incidence, field evaluations have revealed that the disease can reach variable values depending on the region, typically ranging from 40% to 100%, with associated yield losses of up to 50% under severe infections. In Mexico, it has been reported that incidence levels of 40–70% in watermelon (Hernández-Pérez et al. 2025), whereas in Jamaica, the disease has affected almost 100% of cucurbit crops (Pitter et al. 2024).

## 1.6 *Crinivirus ipomeae* (sweet potato chlorotic stunt virus, SPCSV)

*Crinivirus ipomeae*, renamed as sweet potato chlorotic stunt virus (SPCSV), constitutes an important etiological pathogen of sweet potato virus disease (SPVD), an important viral disorder of sweet potato (*Ipomoea batatas*) globally when infecting synergistically with *Potyvirus batataplumei* (*sweet potato feathery mottle virus*, SPFMV) (Sastry et al. 2019; Zhang et al. 2020). Its virions are flexuous filamentous particles measuring approximately 650–850 nm in length (ICTV 2025a). It has a bipartite genome, composed of (+) ssRNA (Kreuze et al. 2002). RNA1 (approximately 8.6 kb) encodes replication-associated proteins, including a polyprotein harboring an RdR polymerase and an RNase III implicated in RNA-silencing suppression, whereas RNA2 (approximately 8.1–8.2 kb) encodes movement proteins, a heat shock protein HSP70h, and the major and minor CPs (Kreuze et al. 2002; H. Liu et al. 2021). Phylogenetic analyses revealed the existence of two strains: the East African (EA) strain, reported in Uganda and Peru, characterized by the presence of the *p22* gene, and the West African (WA) strain, which lacks this gene and has a broader distribution worldwide (Clark et al. 2012; Tugume et al. 2013). *C. ipomeae* transmission is mediated by whiteflies in a non-circulative, semi-persistent mode, with *B. tabaci* as the primary vector (Karyeija et al. 2000). Besides vector-mediated spread, this virus is efficiently disseminated through stem cuttings and storage roots intended for planting (CABI 2021). Most of its known hosts are from the family *Convolvulaceae*, with sweet potato as the primary important cultivated host (Cheng et al. 2020), but ornamentals such as *Eustoma russellianum* and *Alcea rosea* serve as natural reservoirs, maintaining inoculum sources in agroecosystems (Cohen et al. 2001; Read et al. 2021). Single infections may cause generalized chlorosis, purpling of older leaves, reduced growth, and yield losses close to 50% (Kreuze et al. 2008). However, its most significant importance lies in its synergistic interactions with *Potyvirus batataplumei*, causing severe mosaic, leaf curling, plant stunting, and drastic yield reductions, with significant yield losses of 80–100% (Kreuze et al. 2008; Untiveros et al. 2007; Clark et al. 2012; Méndez-Navarro et al. 2025). *Crinivirus ipomeae* has a nearly worldwide distribution (Fig. 1.6), being

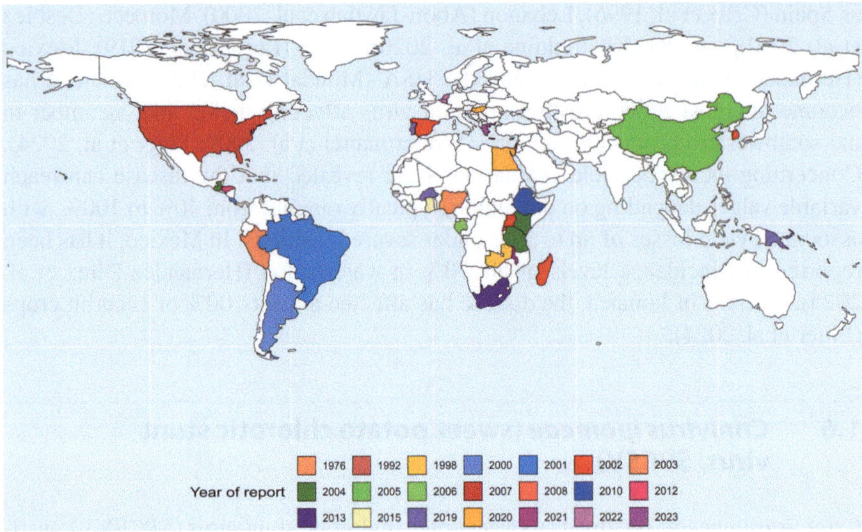


**Fig. 1.6** Reported global distribution of *Crinivirus ipomeae*. The map shows the year of the first report in each country

reported in America (Costa Rica, Peru, and Brazil) (Kroth et al. 2001; Gutiérrez et al. 2003; Valverde and Moreira 2004), throughout sub-Saharan Africa (Uganda, Tanzania, and Nigeria) (Tairo et al. 2004; Aritua and Adipala 2006), in Asia (China, Korea, Taiwan, and Israel) (Cohen et al. 2001; Qiao et al. 2011), and in some parts of Europe (Valverde et al. 2004; Orfanidou et al. 2022b).

## 1.7 *Crinivirus lactucaflavi* (lettuce infectious yellows virus, LIYV)

Lettuce infectious yellows virus (LIYV), currently designated as *Crinivirus lactucaflavi*, represents the type species of the genus and is the first to be molecularly characterized (Klaassen et al. 1995). Its genomic organization consists of two ssRNA (+), with RNA1 (approximately 8.1 kb) encoding the replication-associated polyproteins and RNA2 (approximately 7.2 kb) encoding a heat shock protein HPS70h and the major and minor CPs (Klaassen et al. 1995). Virions exhibit a flexuous, filamentous morphology; their length ranges from 700 to 800 nm and their width from 13 to 15 nm (Klaassen et al. 1996; Agrios 2005; Stewart et al. 2010). A distinctive characteristic of *C. lactucaflavi* is the capacity to consistently generate defective de novo RNAs from RNA2 (Rubio et al. 2002). Transmission of *C. lactucaflavi* occurs exclusively through *B. tabaci* cryptic species complex in a semipersistent, non-circulative manner. However, its transmission is limited to *B. tabaci* A biotype (New World) (Duffus 1986), where virions are retained in the foregut, whereas invasive biotypes MEAM1 and MED exhibit minimal or no retention, leading to inefficient propagation (Stewart et al. 2010; Chen et al. 2021). Consequently, *C. lactucaflavi* epidemics were historically restricted to regions of North America where the A biotype predominated, particularly in California and Arizona (Duffus 1986). The subsequent displacement of the A biotype by MEAM1 drastically reduced the incidence of LIYV. *Crinivirus lactucaflavi* was first described in California during the 1980s, causing serious disease outbreaks in melons and lettuce crops (Duffus 1986). Since then, its presence has been primarily confined to arid and semi-arid regions of the southwestern USA and northern Mexico, correlating with the prevalence of the A biotype vector (EPPO 2025d). Unlike many other criniviruses, *C. lactucaflavi* has not been reported to establish widespread distribution across multiple continents (Blua et al. 1994; Halliwell 1992). This virus infects lettuce, melon, zucchini, sugar beet, and cucumber (Duffus 1986; Halliwell 1992; Duffus et al. 1996). Plants affected by the virus typically develop pronounced leaf discoloration, often manifested as intense yellowing or reddening, accompanied by growth reduction, leaf rolling, vein clearing, and increased fragility (Brown 1986). Older foliage exhibits marked interveinal chlorosis in advanced stages, while the main veins characteristically retain their green coloration.

## 1.8 *Orthotospovirus tomatomaculae* (tomato spotted wilt virus, TSWV)

*Orthotospovirus tomatomaculae*, previously known as tomato spotted wilt virus (TSWV), is the type species of the genus *Orthotospovirus* (EPPO 2020a). Virions are pleomorphic, 80–120 nm in diameter, and are surrounded by a lipid envelope containing the Gn and Gc glycoproteins (Turina et al. 2016). Its genomic organization consists of ssRNA (−/ambisense) (Adkins 2000; Rotenberg et al. 2015), comprising an L ssRNA which encodes a RdR polymerase, an M ssRNA that encodes an envelope glycoprotein precursor (Gn and Gc) together with a cell-to-cell MP (NSm), and an S ssRNA that encodes the nucleocapsid (N) protein and the NSs protein, a key RNA-silencing suppressor (Oliver and Whitfield 2016; Pappu et al. 2009). The virus exhibits marked genetic diversity; the strains differ in virulence, host specificity, and their capacity to bypass resistance mechanisms such as the *Sw-5* resistance gene (Almási et al. 2020) and *Tsw* (Olaya et al. 2020; Wang et al. 2019). *Orthotospovirus tomatomaculae* transmission occurs exclusively through thrips in a circulative, propagative, and persistent manner, with *Frankliniella occidentalis* and *F. fusca* representing the most epidemiologically relevant vectors (Rotenberg and Whitfield 2018). Viral acquisition is restricted to larval stages, whereas adults remain infected and viruliferous throughout their life cycle (Rotenberg et al. 2015). Symptoms caused by this virus are genotype-related and are more severe in young plants, typically including foliar bronzing, concentric necrotic or chlorotic rings, chlorosis, mosaic, dwarfing, blotches, and deformation (Panno et al. 2021; García-Estrada et al. 2022). In peppers, symptoms usually appear as necrotic rings on fruits. In lettuce, symptoms tend to develop unilaterally

**Fig. 1.7** Symptoms caused by *Orthotospovirus tomatomaculae* in vegetable crops. (**a**, **b**) Necrosis and bronzing on tomato leaves. (**c**, **d**) Necrotic ring spots on tomato fruits. (**e**) Leaf deformation in bell pepper. (**f**) Ring spots on bell pepper fruit. (**g**, **h**) Necrotic spots on lettuce leaves

on the plant and include leaf yellowing, necrotic spots, and foliage twisting (Koike et al. 2007) (Fig. 1.7).

*Orthotospovirus tomatomaculae* infects more than 1000 species from approximately 85 botanical families, including commercial families such as Solanaceae (tomato, eggplant, tobacco, potato, pepper), Asteraceae, Fabaceae (peanut, common bean), and Rosaceae (strawberry) (Parrella et al. 2003; Sevik and Arli-Sokmen

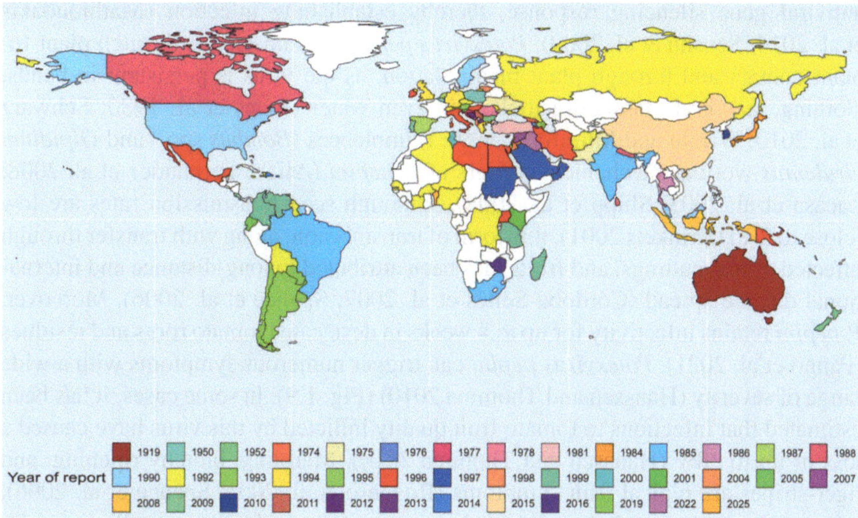


**Fig. 1.8** Worldwide distribution of *Orthotospovirus tomatomaculae* based on the year of first report in each country

2012; Batuman et al. 2020). The geographic distribution of *O. tomatomaculae* is nearly worldwide (Fig. 1.8). The first description was in Australia (Brittlebank 1919), and since then, it has spread extensively. In Europe, severe outbreaks have been documented in countries like Greece, Spain, Italy, and Turkey (Parrella et al. 2013; Deligoz et al. 2014). In America, including the USA, Mexico, Brazil, and Argentina (Ochoa et al. 1996; Alexandre et al. 2014; Salvalaggio et al. 2017; Rodriguez et al. 2024), throughout Asia, present in China, India, and Japan, affecting solanaceous and cucurbit crops (Maoka et al. 2013; Renukadevi et al. 2015; Zhang et al. 2021), and in several parts of Africa and the Middle East. Outbreaks caused by *O. tomatomaculae* frequently result in severe production losses. In Turkey, yield reductions of about 42% and declines of up to 95% in tomato marketability have been reported (Sevik and Arli-Sokmen 2012), while in the southeastern USA, the virus has generated cumulative economic damages in tomato, peanut, and tobacco valued at several hundred million dollars annually (Riley and Pappu 2000; Culbreath et al. 2003).

## 1.9 *Potexvirus pepini* (pepino mosaic virus, PepMV)

*Potexvirus pepini* was first reported in Peru, causing mosaic symptoms in *Solanum muricatum* (Jones et al. 1980). Nevertheless, for 20 years, no further reports of its presence were made until its reemergence in the UK and the Netherlands on tomato crops (EPPO 2000a, b). *P. pepini* has an ssRNA (+) genome, which is organized into five ORFs; ORF1-3 encode the TGB proteins (TGBp1–3) (Aguilar et al. 2002). Both TGBp1 and CP act as silencing suppressors, essential for subverting the plant's antiviral gene silencing response, thereby establishing infection (Mathioudakis et al. 2014; Senshu et al. 2009). *Potexvirus pepini* transmission is through plant-to-plant contact and through plant manipulation, as the virus is persistent in hands, clothing, standard glasshouse tools, and even water (Jones et al. 1980; Schwarz et al. 2010; Wright and Mumford 1999). Bumblebees (*Bombus* spp.) and *Olpidium virulentus* work as mechanical carriers of *P. pepini* (Alfaro-Fernández et al. 2008; Lacasa et al. 2003; Shipp et al. 2008). Although seed transmission rates are low (close to 2%) (Krinkels 2001), this form of transmission, along with transfer through infected grafts, cuttings, and fruits, has been attributed to long-distance and international disease spread (Córdoba-Sellés et al. 2007; Spence et al. 2006). Moreover, *P. pepini* retains infectivity for up to 4 weeks in desiccated tomato roots and residues (Panno et al. 2021). *Potexvirus pepini* can trigger numerous symptoms with a wide range of severity (Hanssen and Thomma 2010) (Fig. 1.9). In some cases, it has been estimated that infections to tomato fruit quality inflicted by this virus have caused a loss of up to 38% (Hanssen and Thomma 2010). Marbling, blotchy ripening, and tiger-stripes are typical fruit symptoms (Roggero et al. 2001; Spence et al. 2006), whereas the symptoms in leaves and plants range from leaf mosaics, yellow angular spotting, blisters, spiky or nettle-like heads, and dwarfing to systemic necrosis. The symptomatic expression of *P. pepini* depends on the virus isolate, virus accumulation level, and environmental factors, mainly light and temperature (García-Estrada et al. 2022; Jones et al. 1980; Sempere et al. 2016; Spence et al. 2006). Five *P. pepini* variants have been identified: the European (EU), Chilean (CH2), North American (US1/CH1), Peruvian (LP), and Southern Peruvian (PES) (Moreno-Pérez et al. 2014; Agüero et al. 2018; García-Estrada et al. 2022; Miranda-Campaña et al. 2024). Apart from its first report in Peru, and the reemergence in the Netherlands and the UK, in Europe it has also been reported in Austria (EPPO 2007), Bulgaria (EPPO 2004b), Croatia (EPPO 2011a), Cyprus (Papayiannis et al. 2012), Czechia (EPPO 2008), Denmark (EPPO 2002), Finland (EPPO 2011b), France (Cotillon et al. 2002), Germany (EPPO 2000a, b), Greece (Efthimiou et al. 2011), Hungary (Forray et al. 2004), Italy (Roggero et al. 2001), Lithuania (Sneideris et al. 2013), Norway, Poland, Portugal (EPPO 2002), Serbia (Stankovic et al. 2020), Spain (Jordá et al. 2001b), Slovakia (EPPO 2004a), Sweden (EPPO 2001), Switzerland (Stäubli 2005), and Türkiye (Özdemír, S. 2010). In America, reports of *P. pepini* have been made in Canada (French et al. 2001), Ecuador (Soler et al. 2005), Mexico (Ling and Zhang 2011), and the USA (French et al. 2001). Likewise, in Asia, there have been reports in China (Yaoliang and Zhongjian 2003), Israel (EPPO 2020b), Syria (Fakhro et al. 2010), and South Korea (I. S. Cho et al. 2023). As well as in African countries such as Egypt (Amer and Mahmoud 2020), South Africa (Carmichael et al. 2011), and in Oceania (Veerakone et al. 2024) (Fig. 1.10).

The two major hosts of *P. pepini* are *Solanum lycopersicum* and *Solanum muricatum* (Jones et al. 1980; van der Vlugt et al. 2000), although *Solanum melongena* has also been proven to be a viable host, either symptomless or showing mild symptoms during *P. pepini* infections (Blystad et al. 2015; Miranda-Campaña et al. 2024). Other plants belonging to the *Solanaceae* family that have been proven to host this virus either naturally or experimentally are as follows: *Datura innoxia* (Papayiannis

**Fig. 1.9** Symptoms caused by emerging species of *Potexvirus* and *Tobamovirus* in vegetable crops. (**a–c**) Mosaic, mottling, and leaf distortion in tomato leaves induced by *Potexvirus pepini*. (**d–f**) Tomato leaves showing distortion and mosaic patterns induced by *Tobamovirus fructirugosum*. (**g**, **h**) Tomato fruits exhibiting deformation and brown spots caused by *T. fructirugosum*. (**i**) Mosaic patterns on cucumber leaves caused by *T. viridimaculae*. (**j**) Mottling on cucumber fruits caused by *T. viridimaculae*

et al. 2012), *Datura metel*, *Datura stramonium* (Blystad et al. 2015), *Nicandra physalodes* (Jones et al. 1980), *Nicotiana debneyi* (Jones et al. 1980), *N. glauca* (Jordá et al. 2001c), *Nicotiana glutinosa, Nicotiana occidentalis* (Blystad et al. 2015; Miranda-Campaña et al. 2024), *Nicotiana quadrivalvis* (Jones et al. 1980), *N.*

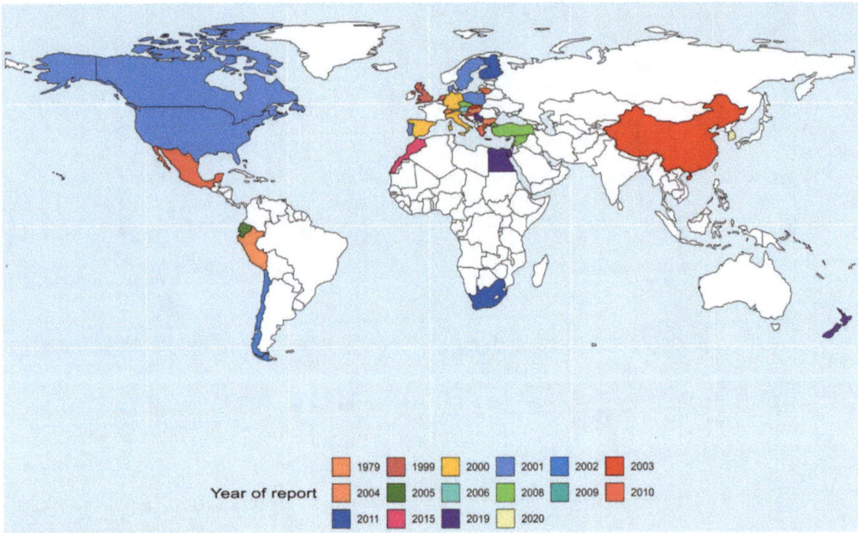


**Fig. 1.10** Worldwide distribution of *Potexvirus pepini* based on the year of first report in each country

*tabacum* (Blystad et al. 2015), *Physalis floridana, Physalis peruviana* (Jones et al. 1980), *Solanum americanum* (Stobbs et al. 2009), *Solanum boliviense, Solanum cardiophyllum* (Jones et al. 1980), *Solanum chilense, Solanum chmielewskii* (Soler et al. 2002), *Solanum curtilobum, Solanum demissum* (Jones et al. 1980), *Solanum dulcamara* (Stobbs et al. 2009), *Solanum microdontum, Solanum mochiquense* (Jones et al. 1980), *Solanum nigrum* (Papayiannis et al. 2012), *Solanum oxycarpum, Solanum palustre* (Jones et al. 1980), *Solanum peruvianum, Solanum pimpinellifolium* (Soler et al. 2002), *Solanum raphanifolium* (Jones et al. 1980), *Solanum sarrachoides* (Stobbs et al. 2009), *Solanum stenotomum and Solanum stoloniferum* (Jones et al. 1980; EPPO 2025f). As well as *Tetragonia tetragonoides* (Jones et al. 1980), *Amaranthus graecizans, Amaranthus retroflexus* (Papayiannis et al. 2012), *Amaranthus* sp. (Jordá et al. 2001c), *Amaranthus viridis* (Papayiannis et al. 2012), *Bassia scoparia, Chenopodiastrum murale, Taraxacum officinale* (Córdoba et al. 2004); *Allium sativum* (Fakhro et al. 2011); *Calendula arvensis* (Papayiannis et al. 2012), *Erigeron sumatrensis* (Córdoba et al. 2004), *Glebionis segetum, Onopordum cyprium* (Papayiannis et al. 2012), *Onopordum* sp. (Córdoba et al. 2004), *Sonchus asper* (Papayiannis et al. 2012), *Sonchus oleraceus* (Jordá et al. 2001c), *Sonchus tenerrimus* (Papayiannis et al. 2012); *Diplotaxis erucoides, Lepidium* sp., *Moricandia arvensis, Sisymbrium irio*; *Calystegia sepium* (Córdoba et al. 2004), and *Cucumis sativus* (Córdoba et al. 2004; Davino et al. 2009; Fakhro et al. 2011; Jones et al. 1980; Jordá et al. 2001c; Papayiannis et al. 2012). As with other seed-borne and mechanically transmitted plant pathogens, the cornerstone of its management is following strict hygiene measures, such as destroying plants that are source of inoculum, disinfecting mats, cleaning work tools and equipment, wearing clean clothing and allocating workers to specific areas within the greenhouse to avoid virus spread in combination with chemical seed treatments (Córdoba-Sellés et al. 2007; García-Estrada et al. 2022; Ling 2010).

## 1.10 *Tobamovirus fructirugosum* (tomato brown rugose fruit virus, ToBRFV)

Tomato (*Solanum lycopersicum*) infections by *Tobamovirus fructirugosum* (tomato brown rugose fruit virus, ToBRFV) were first recorded in Israel as a *Tobamovirus* capable of breaking the broadly used $Tm2^2$ (Luria et al. 2017). In 2015, infections spread through the Jordan Valley, affecting tomato plants grown in greenhouse conditions. Resistance through R-gene $Tm2^2$ relies on its interactions with *Tobamovirus* movement protein (MP) (Weber and Pfitzner 1998). Accumulated mutations in *T. fructirugosum* MP allow it to break such resistance (Hak and Spiegelman 2021; Maayan et al. 2018). *Tobamovirus fructirugosum* has been reported to spread rapidly and cause severe damage. Under experimental conditions, it was observed that starting with only two infected tomato plants in a greenhouse, the virus reached 100% incidence within 9 months (Panno et al. 2020). In contrast, in commercial greenhouses where cultural management was performed more often, thereby facilitating the spread of *T. fructirugosum*, the maximum incidence (100%) was achieved in just 4 months (González-Concha et al. 2021). Comparable to other Tobamoviruses, infections by *T. fructirugosum* are transmitted mechanically, this being the most effective form of *Tobamovirus* transmission, through plant-to-plant contact and plant manipulation as the virus remains active on shoes, machinery, tools, skin, and clothing (Broadbent 1976; Smith and Dombrovsky 2020). As *T. fructirugosum* is seedborne, contaminated fruit and seeds have effectively spread the virus over long distances (Smith and Dombrovsky 2020). Although *Tobamovirus fructirugosum* only persists on the coat of the seed (Salem et al. 2022a, b), seedling infection is still possible through seedling-seed contact during germination (Oladokun et al. 2019; Salem et al. 2023). Beneficial pollinators, such as the bumblebee (*Bombus terrestris*), can spread the infection during buzz pollination or by transferring crude sap, posing a challenge as bumblebees are extensively used in tomato production, especially in greenhouse conditions (Levitzky et al. 2019). In addition, *T. fructirugosum* virions are thermally stable, being able to survive on most materials used in greenhouses, for example, concrete (at least 7 days); glass, aluminum, and stainless steel (approximately 28 days); and polythene and rigid plastic (up to 180 days) (Skelton et al. 2023). A study on the longevity of *T. fructirugosum* virions in naturally contaminated soil showed that infectivity was still detectable 184 days after the previous growth cycle had ended (Molad et al. 2024). *T. fructirugosum* symptoms are influenced by abiotic factors like temperature, photoperiod, and growing system, as well as biotic factors such as the cultivar and the plant phenological stage (Salem et al. 2023). In tomato (Fig. 1.9), typical leaf symptoms include deformations, including leaf narrowing, mosaic, and mottling (Alkowni et al. 2019; Menzel et al. 2019; García-Estrada et al. 2022; González-Concha et al. 2023). Brown necrotic lesions may appear in the stem, calyx, and flowers (Chanda et al. 2021). Affected tomato fruits may display brown spots, skin wrinkling, deformities, and uneven ripening (EPPO 2020g; Salem et al. 2016; García-Estrada et al. 2022; González-Concha et al. 2023). More severe symptoms have been reported when co-infections occur between *T. fructirugosum* and other viruses, such as mixed infections with *Potexvirus pepini* (PepMV) (Klap et al. 2020) and *Orthotospovirus tomatomaculae* (TWSV) in tomato (Luria et al. 2017), as well as with *Tobamovirus paprikae* (PaMMV) in peppers (Luria et al. 2018). Since its emergence, *T. fructirugosum* has spread to most major tomato-producing areas and countries (Fig. 1.11), including China (Yan et al. 2019), India (Kavya et al. 2024), the USA (Dey et al. 2021), Italy (Panno et al. 2019), Spain (Alfaro-Fernández et al. 2021), Türkiye (Fidan et al. 2019), Iran (Ghorbani et al. 2021), and Mexico (Cambrón-Crisantos et al. 2018). Additionally, in Asia, there are reports in Iraq (Obaid and Adhab 2025), Lebanon (Abou-Kubaa et al. 2022), Palestine (Alkowni et al. 2019), Saudi Arabia (Sabra et al. 2022), Syria (Abou-Kubaa et al. 2022), and Uzbekistan (Bakhtiyorova et al. 2024). In America, it has also been reported in Argentina (Obregón et al. 2023), Canada (Sarkes et al. 2020), and Colombia (Bados et al. 2024), while in Europe it has also been reported in Albania (Orfanidou et al. 2022a), Austria, Belgium, Bulgaria (EPPO 2021a, b, g), Croatia (EPPO 2024c), Cyprus (EPPO 2020c), Czechia (EPPO 2020d), Estonia (EPPO 2021c), Finland (EPPO 2022), France (Skelton et al. 2022) Hungary (EPPO 2021d), Ireland (EPPO 2024d), Jersey (EPPO 2025b), Latvia (EPPO 2024e), Lithuania (EPPO 2024f), Malta (EPPO 2021e), the Netherlands (EPPO 2019), Norway (Hamborg and Blystad 2022), Poland (EPPO 2020f), Portugal (EPPO 2021f), Romania (EPPO 2024a),

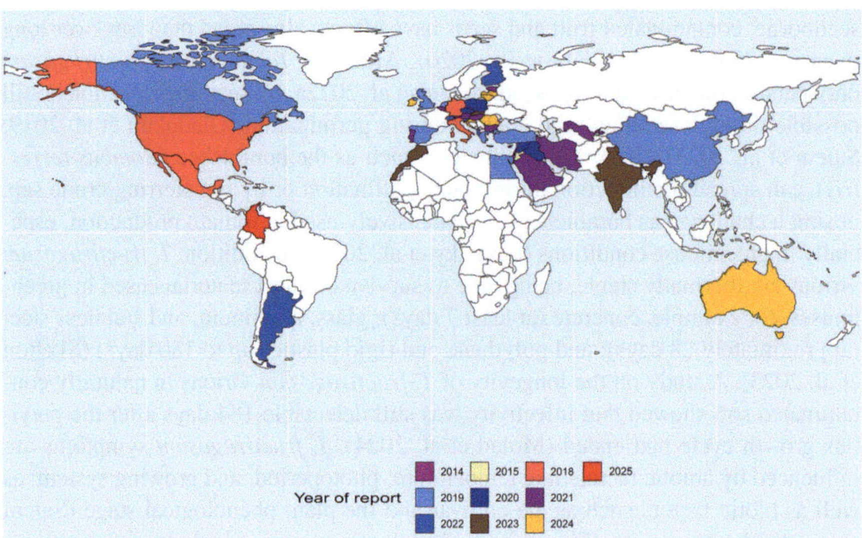


**Fig. 1.11** Worldwide distribution of *Tobamovirus fructirugosum* based on the year of first report in each country

Slovakia (EPPO 2023a), Slovenia (Vučurović et al. 2022), the UK (Skelton et al. 2019), and Switzerland (Mahillon et al. 2022). In Oceania, it has been reported in Australia (EPPO 2024b), and likewise in Africa, there have been reports of *T. fructirugosum* in Morocco (EPPO 2023b) and Egypt (Amer and Mahmoud 2020; EPPO 2020e), although later investigations by Egyptian authorities deemed the first report as unreliable (EPPO 2020h). Apart from tomato crops, the other major crops infected by *T. fructirugosum* are peppers (*Capsicum* spp.) (Cambrón-Crisantos et al. 2018; Salem et al. 2020). Different weeds belonging to ten different families also act as alternative hosts, comprising *Erigeron canadensis* (Salem et al. 2022a), *Polycarpon tetraphyllum*, *Convolvulus arvensis* (Cultrona et al. 2024), *Malva parviflora*, *Oxalis corniculate*, *Portulaca oleracea*, *Veronica syriaca*, *Solanum elaeagnifolium* (Salem et al. 2022a), *S. nigrum* (Luria et al. 2018), and *Corchorus olitorius* (Salem et al. 2022a).

Actions toward an integrated management strategy consist of disinfection of greenhouse surfaces, tools and equipment as washing hands is unreliable to reduce *T. fructigosum* inoculum, changing gloves and workers to different areas is successful (Skelton et al. 2023), sourcing seed which is virus free and/or disinfecting it with sodium hypochlorite solution has been proved to be efficient at reducing *Tobamovirus* infections although germination might be hindered by this method (Salem et al. 2022a; Smith and Dombrovsky 2020). Soil treatments such as chlorine-stabilized, triclosan-sodium, and chlorinated tri-sodium phosphate have been demonstrated to be effective against other *Tobamovirus* to inactive virions that might be present in the soil (Dombrovsky et al. 2022). Recent efforts to manage the virus have led to the design of novel resistance strategies, for example, by combining a locus on tomato 11 chromosome that confers virus tolerance with the *Tm*-1 resistance gene to obtain effective resistance (Zinger et al. 2025).

## 1.11 *Tobamovirus viridimaculae* (cucumber green mottle mosaic virus, CGGMV)

*Tobamovirus viridimaculae* has reached worldwide distribution, posing a significant threat to cucurbit farmers of economically important species, retailers, seedling nurseries, and breeding companies (Dombrovsky et al. 2017). Although *T. viridimaculae* was first reported as early as 1935 in England (Ainsworth 1935), its spread has accelerated notably over the last 20 years, possibly by multifactorial causes involving intensive cucurbit production and international seed trade, as this is a seedborne virus (Dombrovsky et al. 2022; Jones and Naidu 2019). *T. viridimaculae* has an ssRNA (+) genome with 4 ORFs encoding four distinct proteins (Ugaki et al. 1991). *Tobamovirus viridimaculae* is highly contagious, and virions can remain stable in infected soil for ~10 months (Mandal et al. 2008; Varveri et al. 2002). *Tobamovirus viridimaculae* principal transmission via is through plant-to-plant contact, either aerial or root-to-root, through irrigation water, or by contact with any other infected surface, such as greenhouse equipment, tools, workers' hands, clothing, and shoes (Dombrovsky et al. 2017; van Dorst 1988; Li et al. 2016). *T.*

*viridimaculae* is seedborne (Xinying et al. 2022). Nevertheless, in cucumber, seed transmission has been estimated at 12% (Liu et al. 2014). Under protected cropping conditions, honeybees (*Apis mellifera*) have been shown to contribute to the spread of *T. viridimaculae* (Darzi et al. 2018).

Typical symptoms caused by *T. viridimaculae* are leaf mottling and systemic green mosaics paired with severe fruit distortion (Fig. 1.9). However, these symptoms differ due to biotic (crop species, cultivar, virus strain, plant stage) and abiotic factors (temperature) (Antignus et al. 1990; Dombrovsky et al. 2017; Mandal et al. 2008). Additionally, in cucumber, it can cause stunt growth, leaf blistering, and flecks on fruit skin (Ainsworth 1935). In watermelon, it can cause leaf mosaic, fruit discoloration, and even rotting (Komuro 1971). In melon, infections can produce mosaic, mottling, and leaf blotching (Tian et al. 2014). Squash might be asymptomatic on leaves, but fruit flesh might be chlorotic or necrotic (Dombrovsky et al. 2017). Strains are determinant for expressed symptomatology, for instance, the UK isolate (CV3) only produces mosaic symptoms in cucumber (Ainsworth 1935), whereas the Spanish isolate (Sp) can produce mosaic and blistering (Celix 1996). The Greek (GC) isolate produces fruit pulp deterioration in watermelon (Varveri et al. 2002) and the Japanese (W) severe leaf mosaic and plant stunning (Komuro 1971). The plant stage is also crucial, as early infections can exacerbate symptoms (Reingold et al. 2016). During the first 50 years since the emergence of *T. viridimaculae* in England (Ainsworth 1935), reports were limited to Europe and Asia. It was not until 2007 that cases began to be reported in America, Africa, Oceania, and even Antarctica (Polischuk et al. 2007). In Asia, this virus has been detected in numerous countries, including China (Chen et al. 2006), India (Capoor and Varma 1948), Indonesia (Daryono et al. 2005), Iran (Nematollahi et al. 2014), Iraq (Obaid and Adhab 2025), Israel (Antignus et al. 1990), Japan (Komuro 1971), Korea (Lee et al. 1990), Lebanon (Faris-Mukhayyish and Makkouk 1983), Myanmar (Kim et al. 2010), Pakistan (Ali et al. 2004), Sri Lanka (Ariyaratne et al. 2005), Syria (Kassem and Halim 2005), Taiwan (Wang and Chen 1985), and Thailand (Noda et al. 1993). In Europe, it has spread through many countries such as in Austria (Webster and Jones 2018), Belgium (Ortega-Parra et al. 2024), Bulgaria (Orfanidou et al. 2019), Denmark (Blystad et al. 2015), Finland (Annikki 1966), France (Letschert et al. 2002), Georgia (Tarasashvili 1976), Germany (Brčák et al. 1962), Latvia (Dzirkale et al. 1990), Lithuania (Webster and Jones 2018), Moldova (Webster and Jones 2018), The Netherlands (Koot and Dorst 1959), Norway, Poland (Pospieszny et al. 1997), Romania (Pop and Jilaveanu 1985), Russia (Slavokhotova et al. 2007), Spain (Celix 1996), Sweden, Türkiye (Köklü and Yilmaz 2007), and Ukraine (Budzanivska et al. 2007). In America, it has been reported in the USA (Tian et al. 2014) and Canada (Ling et al. 2014). Its occurrence has also been documented in Nigeria (Ling et al. 2023) in Africa, and in Australia (Tesoriero et al. 2016) in Oceania (Fig. 1.12).

All major *T. viridimaculae* hosts belong to the *Cucurbitaceae* family including cucumbers (Hollings et al. 1975), gourds (*Benincasa hispida*) (Noda et al. 1993), *Lagenaria siceraria* (Boubourakas et al. 2004), *Luffa acutangula*, *L. cylindrica*, *Momordica charantia* (Noda et al. 1993), *Trichosanthes cucumerina* (Ariyaratne

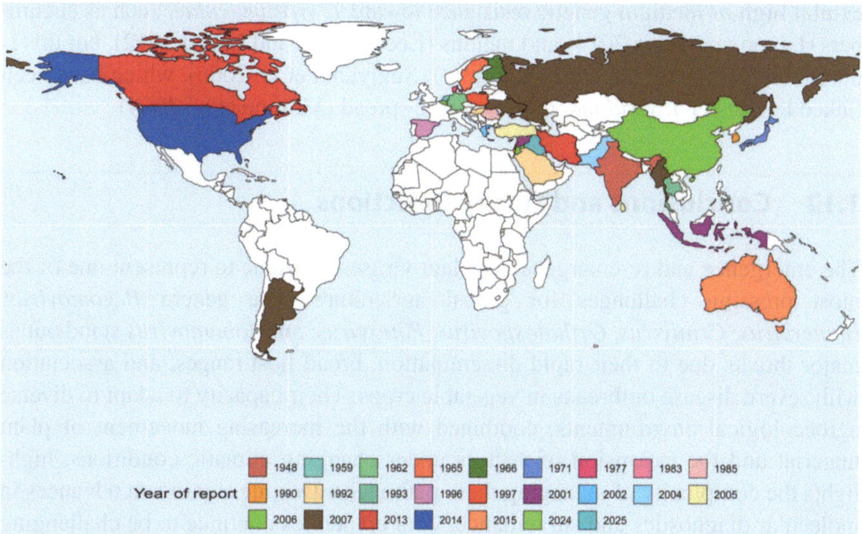


**Fig. 1.12** Global distribution of *Tobamovirus viridimaculae*. The map shows the year of the first report in each country

et al. 2005), melon (Tian et al. 2014), watermelon (Lee et al. 1990), and *Cucurbita pepo* (Ali et al. 2004). Other crops and weeds that host *T. viridimaculae* and belong to the same family are *Citrullus colocynthis* (Shargil et al. 2017) and *Ecballium elaterium* (Antignus et al. 1990). Alternative hosts as weed species from different families are *Amaranthus blitoides* (Boubourakas et al. 2004), *A. graecizans*, *A. muricatus* (Shargil et al. 2017), *A. retroflexus* (Boubourakas et al. 2004), *Heliotropium europaeum* (Boubourakas et al. 2004), *Chenopodium album* (Boubourakas et al. 2004), *Chrozophora tinctoria* (Shargil et al. 2017), *Moluccella laevis* (Shargil et al. 2017), *Portulaca oleracea* (Boubourakas et al. 2004), *Solanum nigrum*, and *Withania somnífera* (Shargil et al. 2017). Management of *T. viridimaculae* infections relies on cultural and phytosanitary measures to control spread. ELISA (Antignus et al. 1990), RT-PCR (Kim et al. 2003), and RT-qPCR (Ling et al. 2007) are validated techniques to detect possible infections in seeds or seedlings before they enter the farm. Seed heat treatment has been proven to be effective in inactivating *T. viridimaculae* in gourd seeds when heated at 75 °C for 72 h (Kim and Lee 2000). However, germination can be affected when treating older seeds. Chemical disinfectants applied to seeds have been proven to be insufficient to halt infection (Reingold et al. 2015); nevertheless, they have been proven beneficial to inactivate this virus when used in tools and greenhouse equipment (0.1–0.3% hypochlorite) (Reingold et al. 2016). The most useful practices to limit virus dissemination and to control the main sources of inoculum are the removal of plant debris, avoiding contaminated water sources, early monitoring for infections, and the adoption of worker allocation management strategies (always moving from younger crops to older crops) (Mandal et al. 2008; Reingold et al. 2016). There are cucurbit crops that exhibit high or medium genetic resistance toward *T. viridimaculae,* such as cucumbers (Lecoq and Katis 2014) and melons (Loebenstein and Lecoq 2012), but this is unreliable at higher temperatures (30 °C) (Sugiyama et al. 2006), which have been linked to a faster *T. viridimaculae* systemic spread (Moreno et al. 2004).

## 1.12 Conclusions and Future Directions

The emergence and re-emergence of plant viruses continue to represent one of the most pressing challenges for global agriculture. The genera *Begomovirus, Blunervirus, Crinivirus, Orthotospovirus, Potexvirus,* and *Tobamovirus* stand out as major threats due to their rapid dissemination, broad host ranges, and association with severe disease outbreaks in vegetable crops. Their capacity to adapt to diverse agroecological environments, combined with the increasing movement of plant material and the expansion of vectors under changing climatic conditions, highlights the complexity of managing these pathogens. Despite significant advances in molecular diagnostics and surveillance, viral epidemics continue to be challenging to predict and contain, underscoring the need for coordinated and proactive strategies. Future efforts must focus on strengthening integrated surveillance systems that combine classical field monitoring with advanced molecular and genomic tools. These tools will not only improve early detection but also facilitate real-time tracking of viral evolution and distribution. In parallel, the development of resistant crop varieties through breeding and biotechnological approaches should remain a cornerstone of long-term management strategies. A greater emphasis on vector ecology and control is also needed, as vectors play a pivotal role in the emergence and spread of viruses. Ultimately, addressing these challenges necessitates international collaboration among researchers, regulatory agencies, and growers to implement harmonized phytosanitary measures and facilitate rapid information exchange. Integrating technological innovations with global cooperation will mitigate the impact of emerging viral diseases and safeguard agricultural sustainability and food security.

## References


Abdel-Salam AM (1991) Tomato yellow leaf curl virus in Egypt: 1—characterization, partial purification and antiserum production. Bulletin of the Faculty of Agriculture, University of Cairo 42:507–520

Abou-Jawdah Y, Sobh H, Fayad A, Lecoq H, Delecolle B (2000) Cucurbit yellow stunting disorder virus. A new threat to cucurbits in Lebanon. J Plant Pathol 82(1):55–60

Abou-Kubaa R, Choueiri E, Heinoun K, Cillo F, Saponari M (2022) First report of tomato brown rugose fruit virus infecting sweet pepper in Syria and Lebanon. J Plant Pathol 104(1):425

Accotto GP, Navas-Castillo J, Noris E, Moriones E, Louro D (2000) Typing of tomato yellow leaf curl viruses in Europe. Eur J Plant Pathol 106(2):179–186

Accotto GP, Bragaloni M, Luison D, Davino S, Davino M (2003) First report of *Tomato yellow leaf curl virus* (TYLCV) in Italy. Plant Pathol 52(6):799

Adkins S (2000) Tomato spotted wilt virus-positive steps towards negative success. Mol Plant Pathol 1(3):151–157

Agrios GN (2005) Plant diseases caused by viruses. In: Plant pathology. Elsevier, pp 723–824

Agüero J, Gómez-Aix C, Sempere RN, García-Villalba J, García-Núñez J, Hernando Y, Aranda MA (2018) Stable and broad spectrum cross-protection against Pepino mosaic virus attained by mixed infection. Front Plant Sci 9:1810. https://doi.org/10.3389/fpls.2018.01810

Aguilar JM, Hernández-Gallardo MD, Cenis JL, Lacasa A, Aranda MA (2002) Complete sequence of the Pepino mosaic virus RNA genome. Arch Virol 147(10):2009–2015

Aguilar JM, Franco M, Marco CF, Berdiales B, Rodriguez-Cerezo E, Truniger V, Aranda MA (2003) Further variability within the genus Crinivirus, as revealed by determination of the complete RNA genome sequence of Cucurbit yellow stunting disorder virus. J Gen Virol 84(9):2555–2564

Ahmed N, Amin I, Mansoor S (2023) First report of Tomato yellow leaf curl virus and associated Tomato leaf curl betasatellite infecting eggplant (*Solanum melongena*) and in Pakistan. Aust Plant Dis Notes 18(1):15

Ainsworth GC (1935) Mosaic diseases of the cucumber. Ann Appl Biol 22(1):55–67

Al-Ali E, Al-Hashash H, Ben Heji A, Al-Aqeel H (2016) First report of *tomato yellow leaf curl virus* infecting cucumber in Kuwait. Plant Dis 100(3):656

Alexandre MAV, Duarte LML, Harakava R, Chaves ALR, Marque JM, Kitajima EW (2014) Amazon lily: a new natural host of *Tomato spotted wilt virus*. New Dis Rep 30(1):13

Alfaro-Fernández A, Cebrián MC, Córdoba-Sellés C, Herrera-Vásquez JA, Jordá C (2008) First report of the US1 strain of *Pepino mosaic virus* in tomato in the Canary Islands, Spain. Plant Dis 92(11):1590

Alfaro-Fernández A, Castillo P, Sanahuja E, Rodríguez-Salido MC, Font MI (2021) First report of tomato brown rugose fruit virus in tomato in Spain. Plant Dis 105(2):515

Ali A, Natsuaki T, Okuda S (2004) Identification and molecular characterization of viruses infecting cucurbits in Pakistan. J Phytopathol 152(11–12):677–682

Alkowni R, Alabdallah O, Fadda Z (2019) Molecular identification of tomato brown rugose fruit virus in tomato in Palestine. J Plant Pathol 101(3):719–723

Almási A, Nemes K, Salánki K (2020) Increasing diversity of resistance breaking pepper strains of Tomato spotted wilt virus in the Mediterranean region. Phytopathol Mediterr 59(2):385–391

Amer MA, Mahmoud SY (2020) First report of *Tomato brown rugose fruit virus* on tomato in Egypt. New Dis Rep 41(1):24

Anfoka G, Abhary M, Nakhla M (2005) Molecular identification of species of the tomato yellow leaf curl virus complex in Jordan. J Plant Pathol 87(1):65

Annikki L (1966) Virus diseases of cucumber in Finland and characteristics of their causal agents cucumber mosaic and cucumber green mottle mosaic viruses. Ann Agric Fenn 5(4):305–323

Ansar M, Agnihotri AK, Ranjan T, Karn M, Srinivasaraghavan A, Kumar RR, Bhagat AP (2021) Nightshade (*Solanum nigrum*), an intermediate host between tomato and cucurbits of Tomato leaf curl New Delhi virus. Phytopathol Mediterr 60(3):409–419

Antignus Y, Pearlsman M, Ben-Yoseph R, Cohen S (1990) Occurrence of a variant of cucumber green mottle mosaic virus in Israel. Phytoparasitica 18(1):50–56

Aritua V, Adipala E (2006) Characteristics and diversity in sweetpotato-infecting viruses in Africa. Acta Hortic 703:175–182

Ariyaratne I, Weeraratne W, Ranatunge R (2005) Identification of a new mosaic virus disease of snake gourd in Sri Lanka. Ann Sri Lanka Dep Agric 7:13–21

Ascencio-Ibáñez JT, Diaz-Plaza R, Méndez-Lozano J, Monsalve-Fonnegra ZI, Argüello-Astorga GR, Rivera-Bustamante RF (1999) First report of tomato yellow leaf curl geminivirus in Yucatán, México. Plant Dis 83(12):1178

Bados JP, Sanchez J, Davino S, Panno S, Marín M (2024) Detección por RT-LAMP del tomato brown rugose fruit virus (ToBRFV, *Tobamovirus fructirugosum*) en semillas de tomate en Colombia. Acta Agronomica 73(1):55–65

Bakhtiyorova M, Norov T, Khodjaeva S, Botirova N, Cillo F, Abou Kubaa R (2024) First report of tomato brown rugose fruit virus on tomato (*Solanum lycopersicum* L.) in Uzbekistan. J Plant Pathol 106(2):779

Batuman O, Turini TA, LeStrange M, Stoddard S, Miyao G, Aegerter BJ, Chen L-F, McRoberts N, Ullman DE, Gilbertson RL (2020) Development of an IPM strategy for thrips and tomato spotted wilt virus in processing tomatoes in the Central Valley of California. Pathogens 9(8):636

Beris D, Galeou A, Kektsidou O, Varveri C (2023) First report of *Tomato fruit blotch fruit virus* infecting tomato in Greece. New Dis Rep 48(2):e12219

Bertin S, Sybilska A, Luigi M, Tarchi F, Goggioli D, Taglienti A, Luison D, Faggioli F, Simoni S, Lewandowski M, Tiberini A (2025) Transmission of tomato fruit blotch virus by the tomato russet mite: epidemiological implications for an emerging/re-emerging tomato disease. Sci Rep 15(1):12079

Blouin AG, Dubuis N, BrodarD J, Apothéloz-Perret-Gentil L, Altenbach D, Schump P O (2023) Symptomatic, widespread, and inconspicuous: new detection of tomato fruit blotch virus. Phytopathol Mediterr 62(3):371–376

Blua MJ, Perring TM, Nuessly GS, Duffus JE, Toscano NC (1994) Seasonal cropping pattern effects on abundance of *Bemisia ta*baci (Homoptera: Aleyrodidae) and incidence of lettuce infectious yellows virus. Environ Entomol 23(6):1422–1427

Blystad D-R, van der Vlugt R, Alfaro-Fernández A, del Carmen Córdoba M, Bese G, Hristova D, Pospieszny H, Mehle N, Ravnikar M, Tomassoli L, Varveri C, Nielsen SL (2015) Host range and symptomatology of Pepino mosaic virus strains occurring in Europe. Eur J Plant Pathol 143(1):43–56

Boubourakas IN, Hatziloukas E, Antignus Y, Katis NI (2004) Etiology of leaf chlorosis and deterioration of the fruit interior of watermelon plants. J Phytopathol 152(10):580–588

Brčák J, Ulrychová M, Čech M (1962) Infection of tobacco and some *Chenopodium* species by the cucumber virus 4 (and 3) and by its nucleic acid. Virology 16(2):105–114

Brittlebank C (1919) Tomato diseases. J Dep Agric Victoria 17:1348–1352

Broadbent L (1976) Epidemiology and control of tomato mosaic virus. Annu Rev Phytopathol 14(1):75–96

Brown JK (1986) Whitefly-borne viruses of melons and lettuce in Arizona. Phytopathology 76(2):236

Brown JK, Idris AM (2006) Introduction of the exotic monopartite *tomato yellow leaf curl virus* into West Coast Mexico. Plant Dis 90(10):1360

Budzanivska IG, Rudneva TO, Shevchenko TP, Boubriak I, Polischuk VP (2007) Investigation of Ukrainian isolates of cucumber green mottle mosaic virus. Arch Phytopathol Plant Protect 40(5):376–380

CABI (2021) Sweet potato chlorotic stunt virus. CABI Compendium. https://doi.org/10.1079/cabicompendium.18605

Cai L, Mei Y, Ye R, Deng Y, Zhang X, Hu Z, Zhou X, Zhang M, Yang J (2023) Tomato leaf curl New Delhi virus: an emerging plant begomovirus threatening cucurbit production. aBIOTECH 4(3):257–266

Cambrón-Crisantos JM, Rodríguez-Mendoza J, Valencia-Luna JB, Alcasio-Rangel S, García-Ávila CDJ, López-Buenfil JA, Ochoa-Martínez DL (2018) Primer reporte de Tomato brown rugose fruit virus (ToBRFV) en Michoacán, México. Revista Mexicana de Fitopatología 37(1):185–192

Capoor SP, Varma PM (1948) A mosaic disease of *Lagenaria vulgaris* in the BOMBAY Province. Curr Sci 17(9):274–275

Cardenas-Conejo Y, Arguello-Astorga G, Poghosyan A, Hernandez-Gonzalez J, Lebsky V, Holguin-Peña J, Medina-Hernandez D, Vega-Peña S (2010) First report of *Tomato yellow leaf curl virus* co-infecting pepper with *Tomato chino La Paz virus* in Baja California Sur, Mexico. Plant Dis 94(10):1266

Carmichael DJ, Rey MEC, Naidoo S, Cook G, van Heerden SW (2011) First report of *Pepino mosaic virus* infecting tomato in South Africa. Plant Dis 95(6):767

Celix A (1996) First report of cucumber green mottle mosaic tobamovirus infecting greenhouse-grown cucumber in Spain. Plant Dis 80:1303

Celix A, Lopez-Sese A, Almarza N, Gomez-Guillamon ML, Rodriguez-Cerezo E (1996) Characterization of cucurbit yellow stunting disorder virus, a *Bemisia tabaci*-transmitted closterovirus. Phytopathology 86(12):1370–1376

Chanda B, Gilliard A, Jaiswal N, Ling K-S (2021) Comparative analysis of host range, ability to infect tomato cultivars with *Tm-2 2* gene, and Real-Time Reverse Transcription PCR detection of tomato brown rugose fruit virus. Plant Dis 105(11):3643–3652

Chen H, Bai J, Zhu S, Wen-jun Z, Li M (2006) Sequence analysis of the CP gene and its 3′ non-coding region of a watermelon Liaoning isolate of cucumber green mottle mosaic virus. Virol Sin 21(4):516–518

Chen AYS, Zhou JS, Liu J-X, Ng JCK (2021) Nuances of whitefly vector–Crinivirus interactions revealed in the foregut retention and transmission of Lettuce chlorosis virus by two *Bemisia tabaci* cryptic species. Viruses 13(8):1578

Cheng YH, Wang YC, Wang LY, Huang LH, Chen TC (2020) First report of sweet potato chlorotic stunt virus infecting sweetpotato in Taiwan. Plant Dis 104(9):2535

Cho IS, Chung BN, Yoon JY, Hammond J, Lim HS (2023) First report of Pepino mosaic virus infecting tomato in South Korea. Plant Dis 107(3):971

Ciuffo M, Kinoti WM, Tiberini A, Forgia M, Tomassoli L, Constable FE, Turina M (2020) A new blunervirus infects tomato crops in Italy and Australia. Arch Virol 165(10):2379–2384

Clark CA, Davis JA, Abad JA, Cuellar WJ, Fuentes S, Kreuze JF, Gibson RW, Mukasa SB, Tugume AK, Tairo FD, Valkonen JPT (2012) Sweetpotato viruses: 15 years of progress on understanding and managing complex diseases. Plant Dis 96(2):168–185

Cohen J, Lapidot M, Loebenstein G, Gera A (2001) First report of sweet potato sunken vein virus occurring in Lisianthus. Plant Dis 85(6):679

Córdoba MC, Martínez-Priego L, Jordá C (2004) New natural hosts of *Pepino mosaic virus* in Spain. Plant Dis 88(8):906

Córdoba-Sellés M d C, García-Rández A, Alfaro-Fernández A, Jordá-Gutiérrez C (2007) Seed transmission of *Pepino mosaic virus* and efficacy of tomato seed disinfection treatments. Plant Dis 91(10):1250–1254

Cotillon A-C, Girard M, Ducouret S (2002) Complete nucleotide sequence of the genomic RNA of a French isolate of Pepino mosaic virus (PepMV). Arch Virol 147(11):2231–2238

Culbreath AK, Todd JW, Brown SL (2003) Epidemiology and management of tomato spotted wilt in peanut. Annu Rev Phytopathol 41(1):53–75

Cultrona M, Bonini N, Pacifico D, Tessitori M (2024) First report of *Convolvulus arvensis* and *Polycarpon tetraphyllum* as natural hosts of tomato brown rugose fruit virus. Plant Dis 108(3):827

Czosnek B, Czosnek H, Navot N, Laterrot H (1990) Geographical distribution of Tomato yellow leaf curl virus. A first survey using a specific DNA probe. Phytopathol Mediterr 29(1):1–6

Czosnek H, Ghanim M, Ghanim M (2002) The circulative pathway of begomoviruses in the whitefly vector *Bemisia tabaci*-insights from studies with Tomato yellow leaf curl virus. Ann Appl Biol 140(3):215–231

Daryono BS, Somowiyarjo S, Natsuaki KT (2005) Biological and molecular characterization of melon-infecting kyuri green mottle mosaic virus in Indonesia. J Phytopathol 153(10):588–595

Darzi E, Smith E, Shargil D, Lachman O, Ganot L, Dombrovsky A (2018) The honeybee *Apis mellifera* contributes to *Cucumber green mottle mosaic virus* spread via pollination. Plant Pathol 67(1):244–251

Davino S, Accotto GP, Masenga V, Torta L, Davino M (2009) Basil (*Ocimum basilicum*), a new host of *Pepino mosaic virus*. Plant Pathol 58(2):407

Deligoz I, Sokmen MA, Sari S (2014) First report of resistance breaking strain of *Tomato spotted wilt virus* (Tospovirus; Bunyaviridae) on resistant sweetpepper cultivars in Turkey. New Dis Rep 30(1):26

Desbiez C, Lecoq H, Aboulama S, Peterschmitt M (2000) First report of *Cucurbit yellow stunting disorder virus* in Morocco. Plant Dis 84(5):596

Dey KK, Velez-Climent M, Soria P, Batuman O, Mavrodieva V, Wei G, Zhou J, Adkins S, McVay J (2021) First report of *Tomato brown rugose fruit virus* infecting tomato in Florida, USA. New Dis Rep 44(1):e12028

Dombrovsky A, Tran-Nguyen LTT, Jones RAC (2017) *Cucumber green mottle mosaic virus*: rapidly increasing global distribution, etiology, epidemiology, and management. Annu Rev Phytopathol 55(1):231–256

Dombrovsky A, Mor N, Gantz S, Lachman O, Smith E (2022) Disinfection efficacy of tobamovirus-contaminated soil in greenhouse-grown crops. Horticulturae 8(7):563

Duffus JE (1986) Lettuce infectious yellows virus—new type of whitefly-transmitted virus. Phytopathology 76(1):97

Duffus JE, Liu H-Y, Wisler GC, Li R (1996) Lettuce chlorosis virus–A new whitefly-transmitted closterovirus. Eur J Plant Pathol 102(6):591–596

Dzirkale L, Ikaunietse L, Putnaerglis E, Ignash Y (1990) Agents causing cucumber virus diseases in Latvia. Virusy Rastenii i Nasekomykh Elgava, pp 3–11

Efthimiou KE, Gatsios AP, Aretakis KC, Papayiannis LC, Katis NI (2011) First report of *Pepino mosaic virus* infecting greenhouse cherry tomatoes in Greece. Plant Dis 95(1):78

EPPO (2000a) Addition of Pepino mosaic potexvirus to the EPPO alert list EPPO reporting service: Eur. Mediterr. *Plant Prot. Organ., Paris; 2000/003*. https://gd.eppo.int/reporting/article-3033

EPPO (2000b) Situation of several quarantine pests in Germany in 1999 and 2000 EPPO reporting service: Eur Mediterr Plant Prot Organ, Paris; 2000/171

EPPO (2001) First report of Pepino mosaic potexvirus in Sweden EPPO reporting service: EPPO reporting service Eur. Mediterr. Plant Prot. Organ., Paris; 2001/157

EPPO (2002) Update on the situation of Pepino mosaic potexvirus in the EPPO region EPPO reporting service: EPPO reporting service Eur. Mediterr. Plant Prot. Organ., Paris; 2002/092

EPPO (2004a) First finding of Pepino mosaic potexvirus in Slovakia EPPO reporting service: EPPO reporting service Eur. Mediterr. Plant Prot. Organ., Paris; 2004/008

EPPO (2004b) First report of Pepino mosaic potexvirus in Bulgaria EPPO reporting service Eur. Mediterr. Plant Prot. Organ., Paris; 2004/076

EPPO (2007) First report of Pepino mosaic virus in Austria EPPO reporting service: EPPO reporting service Eur. Mediterr. Plant Prot. Organ., Paris; 2007/006

EPPO (2008) First outbreak of Pepino mosaic virus in The Czech Republic EPPO reporting service: EPPO reporting service Eur. Mediterr. Plant Prot. Organ., Paris; 2008/144

EPPO (2011a) First report of Pepino mosaic virus in Croatia EPPO reporting service: EPPO reporting service Eur. Mediterr. Plant Prot. Organ., Paris; 2011/245. https://gd.eppo.int/reporting/article-1814

EPPO (2011b) Outbreak of Pepino mosaic virus in Finland EPPO reporting service: EPPO reporting service Eur. Mediterr. Plant Prot. Organ., Paris; 2011/201. https://gd.eppo.int/reporting/article-1770

EPPO (2019) First report of tomato brown rugose fruit virus in The Netherlands EPPO reporting service: Eur. Mediterr. Plant Prot. Organ., Paris; 2019/209. https://gd.eppo.int/reporting/article-6639

EPPO (2020a) PM 7/139 (1) Tospoviruses (genus *Orthotospovirus*). EPPO Bull 50(2):217–240. https://doi.org/10.1111/epp.12676

EPPO (2020b) First report of Pepino mosaic virus in Israel EPPO reporting service: EPPO reporting service Eur. Mediterr. Plant Prot. Organ., Paris; 2020/104. https://gd.eppo.int/reporting/article-6782

EPPO (2020c) First report of tomato brown rugose fruit virus in Cyprus EPPO reporting service: Eur. Mediterr. Plant Prot. Organ., Paris; 2020/173: https://gd.eppo.int/reporting/article-6851

EPPO (2020d) First report of tomato brown rugose fruit virus in Czechia EPPO reporting service: Eur. Mediterr. Plant Prot. Organ., Paris; 2020/223: https://gd.eppo.int/reporting/article-6901

EPPO (2020e) First report of tomato brown rugose fruit virus in Egypt EPPO reporting service: Eur. Mediterr. Plant Prot. Organ., Paris; 2020/102. https://gd.eppo.int/reporting/article-6780

EPPO (2020f) First report of tomato brown rugose fruit virus in Poland EPPO reporting service: Eur. Mediterr. Plant Prot. Organ., Paris; 2020/122. https://gd.eppo.int/reporting/article-6800

EPPO (2020g) Tomato brown rugose fruit virus. EPPO Bull 50(3):529–534. https://doi.org/10.1111/epp.12711

EPPO (2020h) Tomato brown rugose fruit virus does not occur in Egypt EPPO reporting service: Eur Mediterr Plant Prot Organ Paris; 2020/125. https://gd.eppo.int/reporting/article-6803

EPPO (2021a) First report of tomato brown rugose fruit virus in Belgium EPPO reporting service: Eur. Mediterr. Plant Prot. Organ., Paris; 2021/017: https://gd.eppo.int/reporting/article-6956

EPPO (2021b) First report of tomato brown rugose fruit virus in Bulgaria EPPO reporting service: Eur. Mediterr. Plant Prot. Organ., Paris; 2021/135: https://gd.eppo.int/reporting/article-7074

EPPO (2021c) First report of tomato brown rugose fruit virus in Estonia EPPO reporting service: Eur. Mediterr. Plant Prot. Organ., Paris; 2021/176: https://gd.eppo.int/reporting/article-7115

EPPO (2021d) First report of tomato brown rugose fruit virus in Hungary EPPO reporting service: Eur. Mediterr. Plant Prot. Organ., Paris; 2021/134: https://gd.eppo.int/reporting/article-7073

EPPO (2021e) First report of tomato brown rugose fruit virus in Malta EPPO reporting service: Eur. Mediterr. Plant Prot. Organ., Paris; 2021/106. https://gd.eppo.int/reporting/article-7045

EPPO (2021f) First report of tomato brown rugose fruit virus in Portugal EPPO reporting service: Eur. Mediterr. Plant Prot. Organ., Paris; 2021/196. https://gd.eppo.int/reporting/article-7135

EPPO (2021g) First report of tomato brown rugouse fruit virus in Austria EPPO reporting service: Eur. Mediterr. Plant Prot. Organ., Paris; 2021/159: https://gd.eppo.int/reporting/article-7098

EPPO (2022) First report of tomato brown rugose fruit virus in Finland EPPO reporting service: Eur. Mediterr. Plant Prot. Organ., Paris; 2022/167: https://gd.eppo.int/reporting/article-7399

EPPO (2023a) First record of tomato brown rugose fruit virus in Slovakia EPPO reporting service: Eur. Mediterr. Plant Prot. Organ., Paris; 2023/049. https://gd.eppo.int/reporting/article-7531

EPPO (2023b) First report of tomato brown rugose fruit virus in Morocco EPPO reporting service: Eur Mediterr Plant Prot Organ, Paris; 2023/235. https://gd.eppo.int/reporting/article-7717

EPPO (2024a) First report and eradication of tomato brown rugose fruit virus in Romania EPPO reporting service: Eur. *Mediterr.* Plant Prot. Organ., Paris; 2024/133. https://gd.eppo.int/reporting/article-7883

EPPO (2024b) First report of tomato brown rugose fruit virus in Australia EPPO reporting service: Eur. Mediterr. Plant Prot. Organ., Paris; 2024/172. https://gd.eppo.int/reporting/article-7922

EPPO (2024c) First report of tomato brown rugose fruit virus in Croatia EPPO reporting service: Eur. Mediterr. Plant Prot. Organ., Paris; 2024/225: https://gd.eppo.int/reporting/article-7975

EPPO (2024d) First report of tomato brown rugose fruit virus in Ireland EPPO reporting service: Eur. Mediterr. Plant Prot. Organ., Paris; 2024/087: https://gd.eppo.int/reporting/article-7837

EPPO (2024e) First report of tomato brown rugose fruit virus in Latvia EPPO reporting service: Eur. Mediterr. Plant Prot. Organ., Paris; 2024/132. https://gd.eppo.int/reporting/article-7882

EPPO (2024f) First report of tomato brown rugose fruit virus in Lithuania EPPO reporting service: Eur. Mediterr. Plant Prot. Organ., Paris; 2024/111. https://gd.eppo.int/reporting/article-7861

EPPO (2025a) Begomovirus solanumdelhiense. EPPO datasheets on pests recommended for regulation. https://gd.eppo.int

EPPO (2025b) First report of tomato brown rugose fruit virus in Jersey EPPO reporting service: Eur. Mediterr. Plant Prot. Organ., Paris; 2025/017: https://gd.eppo.int/reporting/article-8028

EPPO (2025c) First report of tomato fruit blotch virus in Malta EPPO reporting service: Eur Mediterr Plant Prot Organ, Paris; 2025/094

EPPO (2025d) Crinivirus lactucaflavi. EPPO datasheets on pests recommended for regulation. https://gd.eppo.int

EPPO (2025e) First reports of tomato fruit blotch virus in France and in Belgium EPPO reporting service: Eur. Mediterr. Plant Prot. Organ., Paris; 2025/095

EPPO (2025f) Potexvirus pepini. EPPO datasheets on pests recommended for regulation. https://gd.eppo.int

Espino de Paz AI, Botella-Guillén M, Otazo-González HC, Alfaro-Fernández A, Font-San-Ambrosio I, Galipienso L, Rubio L (2019) First report of *Tomato leaf curl New Delhi virus* infecting cucurbits in the Canary Islands. Plant Dis 103(7):1798

Fakhro A, Von Bargen S, Bandte M, Büttner C (2010) Pepino mosaic virus, a first report of a virus infecting tomato in Syria. Phytopathol Mediterr 49:99–101

Fakhro A, von Bargen S, Bandte M, Büttner C, Franken P, Schwarz D (2011) Susceptibility of different plant species and tomato cultivars to two isolates of Pepino mosaic virus. Eur J Plant Pathol 129(4):579–590

Fan XY, Liu Y, Cheng XF, Wu X (2025) First report of tomato yellow leaf curl virus on Chinese mugwort in China. New Dis Rep 51(1):e70018

Farina A, Rapisarda C, Fiallo-Olivé E, Navas-Castillo J (2023) Tomato leaf curl New Delhi virus Spain strain is not transmitted by Trialeurodes vaporariorum and is inefficiently transmitted by Bemisia tabaci Mediterranean between Zucchini and the Wild Cucurbit Ecballium elaterium. Insects 14(4):384. https://doi.org/10.3390/insects14040384

Faris-Mukhayyish S, Makkouk KM (1983) Detection of four seed-borne plant viruses by the Enzyme-Linked Immunosorbent Assay (ELISA). J Phytopathol 106(2):108–114

Fazeli R, Heydarnejad J, Massumi H, Shaabanian M, Varsani A (2009) Genetic diversity and distribution of tomato-infecting begomoviruses in Iran. Virus Genes 38(2):311–319

Fidan H, Sarikaya P, Calis O (2019) First report of *Tomato brown rugose fruit virus* on tomato in Turkey. New Dis Rep 39(1):18

Fidan H, Yildiz K, Sarikaya P, Calis O (2023) First report of Tomato leaf curl New Delhi virus in Türkiye. New Dis Rep 47(2):e12180

Forray A, Tuske M, Gáborjányi R (2004) First report on occurrence of Pepino mosaic virus in Hungary. Novenyvedelem 40:471–473

French CJ, Bouthillier M, Bernardy M, Ferguson G, Sabourin M, Johnson RC, Masters C, Godkin S, Mumford R (2001) First report of *Pepino mosaic virus* in Canada and the United States. Plant Dis 85(10):1121

Gámez-Jiménez C, Romero-Romero JL, Santos-Cervantes ME, Leyva-López NE, Méndez-Lozano J (2009) Tomatillo (*Physalis ixocarpa*) as a natural new host for *Tomato yellow leaf curl virus* in Sinaloa, Mexico. Plant Dis 93(5):545

García-Andrés S, Accotto GP, Navas-Castillo J, Moriones E (2007) Founder effect, plant host, and recombination shape the emergent population of begomoviruses that cause the tomato yellow leaf curl disease in the Mediterranean basin. Virology 359(2):302–312

García-Estrada RS, Diaz-Lara A, Aguilar-Molina VH, Tovar-Pedraza JM (2022) Viruses of economic impact on tomato crops in Mexico: from diagnosis to management–a review. Viruses 14(6):1251

Ghorbani A, Rostami M, Seifi S, Izadpanah K (2021) First report of *Tomato brown rugose fruit virus* in greenhouse tomato in Iran. New Dis Rep 44(2):e12040

González-Concha LF, Ramírez-Gil JG, García-Estrada RS, Rebollar-Alviter Á, Tovar-Pedraza JM (2021) Spatiotemporal analyses of tomato brown rugose fruit virus in commercial tomato greenhouses. Agronomy 11(7):1268

González-Concha LF, Ramírez-Gil JG, Mora-Romero GA, García-Estrada RS, Carrillo-Fasio JA, Tovar-Pedraza JM (2023) Development of a scale for assessment of disease severity and impact of tomato brown rugose fruit virus on tomato yield. Eur J Plant Pathol 165:579–592

Gronenborn B (2007) The tomato yellow leaf curl virus genome and function of its proteins. In: Tomato yellow leaf curl virus disease. Springer Netherlands, pp 67–84

Gul-Seker M, Elibuyuk IO (2019) Occurrence of Tomato yellow leaf curl virus and Tomato chlorosis virus mixed infections in protected tomato plants, Antalya, Turkey. Phytoparasitica 47(3):441–449

Gutiérrez DL, Fuentes S, Salazar LF (2003) Sweetpotato virus disease (SPVD): distribution, incidence, and effect on sweetpotato yield in Peru. Plant Dis 87(3):297–302

Hak H, Spiegelman Z (2021) The tomato brown rugose fruit virus movement protein overcomes *Tm-2 2* resistance in tomato while attenuating viral transport. Mol Plant-Microbe Interact 34(9):1024–1032

Halliwell RS (1992) Lettuce infectious yellows virus infecting watermelon, cantaloupe, honey dew melon, squash, and cushaw in Texas. Plant Dis 76(6):643

Hamborg Z, Blystad D-R (2022) First report of tomato brown rugose fruit virus in tomato in Norway. Plant Dis 106(7):2004

Hanssen IM, Thomma BPHJ (2010) *Pepino mosaic virus*: a successful pathogen that rapidly evolved from emerging to endemic in tomato crops. Mol Plant Pathol 11(2):179–189. https://doi.org/10.1111/j.1364-3703.2009.00600.x

Hernández-Pérez DMJ, Obregón-Flores V, Diaz-Lara A, López-Cordova JP, Tovar-Pedraza JM (2025) Occurrence of mixed infections of cucurbit yellow stunting disorder virus (*Crinivirus cucurbitae*) and watermelon chlorotic stunt virus (*Begomovirus citrulli*) in watermelon (*Citrullus lanatus*) in Jalisco, Mexico. Plant Health Progr 26(3):367–369

Hollings M, Komuro Y, Tochihara H (1975) Descriptions of plant viruses no. 154. AAB, Wellesbourne, 4

Hussain M, Mansoor S, Iram S, Zafar Y, Briddon RW (2004) First report of *Tomato leaf curl New Delhi virus* affecting chilli pepper in Pakistan. Plant Pathol 53(6):794

ICTV (2025a) Family: Closteroviridae. ICTV report. https://ictv.global/report/chapter/closteroviridae/closteroviridae

ICTV (2025b) Family: Kitaviridae (interim report). ICTV report. https://ictv.global/report/chapter/kitaviridae/kitaviridae

Ito T, Sharma P, Kittipakorn K, Ikegami M (2008) Complete nucleotide sequence of a new isolate of tomato leaf curl New Delhi virus infecting cucumber, bottle gourd and muskmelon in Thailand. Arch Virol 153(3):611–613

Jeevalatha, A.; Siddappa, S.; Kumar, R*.; Tiwari, R.K.; Lal, M. K.; Sharma, S.; Chakrabarty, S.K.; Singh, B.P. 2023. RNA-seq analysis reveals an early defense response to tomato leaf curl New Delhi virus in potato cultivar Kufri Bahar. Funct Integr Genomics, 23, 215. https://doi.org/10.1007/s10142-023-01138-5

Ji YH, Cai ZD, Zhou XW, Liu YM, Xiong RY, Zhao TM, Yu WG, Tao XR, Zhou YJ (2012) First report of *Tomato yellow leaf curl virus* infecting common bean in China. Plant Dis 96(8):1229

Jones RAC, Naidu RA (2019) Global dimensions of plant virus diseases: current status and future perspectives. Annu Rev Virol 6(1):387–409

Jones RAC, Koenig R, Lesemann DE (1980) Pepino mosaic virus, a new potexvirus from Pepino (*Solanum muricatum*). Ann Appl Biol 94(1):61–68

Jordá C, Font I, Martínez P, Juarez M, Ortega A, Lacasa A (2001a) Current status and new natural hosts of *Tomato yellow leaf curl virus* (TYLCV) in Spain. Plant Dis 85(4):445

Jordá C, Pérez AL, Martínez-Culebras P, Abad P, Lacasa A, Guerrero MM (2001b) First report of *Pepino mosaic virus* on tomato in Spain. Plant Dis 85(12):1292

Jordá C, Pérez AL, Martínez Culebras PV, Lacasa A (2001c) First report of *Pepino mosaic virus* on natural hosts. Plant Dis 85(12):1292

Juárez M, Tovar R, Fiallo-Olivé E, Aranda MA, Gosálvez B, Castillo P, Moriones E, Navas-Castillo J (2014) First detection of *Tomato leaf curl New Delhi virus* infecting zucchini in Spain. Plant Dis 98(6):857

Juárez M, Rabadán MP, Martínez LD, Tayahi M, Grande-Pérez A, Gómez P (2019) Natural hosts and genetic diversity of the emerging tomato leaf curl New Delhi virus in Spain. Front Microbiol 10:140

Kao J, Jia L, Tian T, Rubio L, Falk BW (2000) First report of *Cucurbit yellow stunting disorder virus* (genus *Crinivirus*) in North America. Plant Dis 84(1):101

Karyeija RF, Kreuze JF, Gibson RW, Valkonen JPT (2000) Synergistic interactions of a Potyvirus and a phloem-limited Crinivirus in sweet potato plants. Virology 269(1):26–36

Kassem A, Halim K (2005) The most important of viruses affecting cucurbits in Syria. Arab J Plant Prot 23:1–6

Kavya SS, Mahantesha V, Chowdappa A, Mantesh M, Pooja PS, Venkataravanappa V, Reddy CNL (2024) Tomato brown rugose fruit virus associated with leaf mosaic, mottling and brown rugose patches on fruits of tomato in India. Aust Plant Dis Notes 19(1):9

Kegler H (1994) Incidence, properties and control of tomato yellow leaf curl virus—a review. Arch Phytopathol Plant Protect 29(2):119–132

Khan MS, Raj SK, Singh R (2006) First report of *Tomato leaf curl New Delhi virus* infecting chilli in India. Plant Pathol 55(2):289

Kheireddine A, Sáez C, Sifres A, Picó B, López C (2020) First report of Cucurbit chlorotic yellows virus infecting cucumber and zucchini in Algeria. Plant Dis 104(4):1264

Kil EJ, Kim S, Lee YJ, Byun HS, Park J, Seo H, Kim CS, Shim JK, Lee JH, Kim JK, Lee KY, Choi HS, Lee S (2016) Tomato yellow leaf curl virus (TYLCV-IL): a seed-transmissible geminivirus in tomatoes. Sci Rep 6:19013

Kim DH, Lee JM (2000) Seed treatment for cucumber green mottle mosaic virus (CGMMV) in gourd (*Lagenaria siceraria*) seeds and its detection. J Korean Soc Hortic Sci 41(1):1

Kim SM, Nam SH, Lee JM, Yim KO, Kim KH (2003) Destruction of Cucumber green mottle mosaic virus by heat treatment and rapid detection of virus inactivation by RT-PCR. Mol Cells 16(3):338–342

Kim O, Mizutani T, Natsuaki KT, Lee K, Soe K (2010) First report and the genetic variability of *Cucumber green mottle mosaic virus* occurring on bottle gourd in Myanmar. J Phytopathol 158:572–575

Kiss ZA, Medina V, Falk BW (2013) Crinivirus replication and host interactions. Front Microbiol 4:99. https://doi.org/10.3389/fmicb.2013.00099

Kitajima EW, Nakasu EYT, Inoue-Nagata AK, Salaroli RB, Ramos-González PL (2023) Tomato fruit blotch virus cytopathology strengthens evolutionary links between plant blunerviruses and insect negeviruses. Sci Agric 80:e20220045

Klaassen VA, Boeshore ML, Koonin EV, Tian T, Falk BW (1995) Genome structure and phylogenetic analysis of lettuce infectious yellows virus, a whitefly-transmitted, Bipartite Closterovirus. Virology 208(1):99–110. https://doi.org/10.1006/viro.1995.1133

Klaassen VA, Mayhew D, Fisher D, Falk BW (1996) In vitrotranscripts from cloned cDNAs of the Lettuce infectious yellows closterovirus bipartite genomic RNAs are competent for replication in *Nicotiana benthamiana* protoplasts. Virology 222(1):169–175

Klap C, Luria N, Smith E, Hadad L, Bakelman E, Sela N, Belausov E, Lachman O, Leibman D, Dombrovsky A (2020) Tomato brown rugose fruit virus contributes to enhanced Pepino mosaic virus titers in tomato plants. Viruses 12(8):879

Koike ST, Gladders P, Paulus AO (2007) Vegetable diseases: a colour handbook. Manson Publishing, London, p 448

Köklü G, Yilmaz Ö (2007) Occurrence of cucurbit viruses on field-grown melon and watermelon in the Thrace region of Turkey. Phytoprotection 87(3):123–130

Komuro Y (1971) Cucumber green mottle mosaic virus on cucumber and watermelon and melon necrotic spot virus on muskmelon. Jpn Agric Res Q 6(1):41–45

Koot Y, Dorst HJM (1959) Virusziekten van de komkommer in Nederland. Tijdschr PlZiekt 65(6):257–271

Kousik CS, Adkins S (2020) Detection of Cucurbit yellow stunting disorder virus infecting watermelon in South Carolina. Plant Health Prog 21(2):133–134

Kreuze JF, Savenkov EI, Valkonen JPT (2002) Complete genome sequence and analyses of the subgenomic RNAs of *Sweet potato chlorotic stunt virus* reveal several new features for the genus *Crinivirus*. J Virol 76(18):9260–9270

Kreuze JF, Klein IS, Lazaro MU, Chuquiyuri WJC, Morgan GL, Mejía PGC, Ghislain M, Valkonen JPT (2008) RNA silencing-mediated resistance to a Crinivirus (Closteroviridae) in cultivated sweetpotato (*Ipomoea batatas* L.) and development of sweetpotato virus disease following co-infection with a potyvirus. Mol Plant Pathol 9(5):589–598

Krinkels M (2001) Pepino mosaic virus causes sticky problem. Prophyta 30(3): 30-33.

Kroth L, Fuentes S, Salazar LF, Daniels J (2001) Serological virus detection (NCM-ELISA) in sweet potato fields at Rio Grande do Sul State, Brazil. Rev Bras Agrocienc 7(2):117–119

Kumar R, Tiwari RK, Sundaresha S, Kaundal P, Raigond B (2022) Potato viruses and their management. In: Chakrabarti SK, Sharma S, Shah MA (eds) Sustainable management of potato pests and diseases. Springer, Singapore, pp 309–335. https://doi.org/10.1007/978-981-16-7695-6_12

Kumar R, Kaundal P, Tiwari RK, Lal MK, Kumari H, Kumar R, Naga KC, Kumar A, Singh B, Sagar V, Sharma S (2023) Development of reverse transcription recombinase polymerase amplification (RT-RPA): A methodology for quick diagnosis of potato Leafroll viral disease in potato. Int J Mol Sci 24:2511. https://doi.org/10.3390/ijms24032511

Lacasa A, Guerrero M, Hita I, Martinez M, Jorda C, Bielza P, Contreras J, Alcazar A, Cano A (2003) Implicaciones de los abejorros (*Bombus* spp.) en la dispersión del virus del mosaico del pepino dulce (Pepino Mosaci Virus) en cultivos de tomate. Boletin de Sanidad Vegetal Plagas 3er Trimestre 29(3):393–403

Lal MK, Tiwari RK, Kumar R, Naga KC, Kumar A, Singh B, Raigond P, Dutt S, Chourasia KN, Kumar D, Parmar V, Changan SS (2021) Effect of potato apical leaf curl disease on glycemic index and resistant starch of potato (*Solanum tuberosum* L.) tubers. Food Chem 359:129939

Lapidot M (2002) Screening common bean (*Phaseolus vulgaris*) for resistance to *Tomato yellow leaf curl virus*. Plant Dis 86(4):429–432

Lecoq H, Katis N (2014) Control of cucurbit viruses. Adv Virus Res 90:255–296

Lee K, Lee B, Park H, Lee Y (1990) Occurrence of cucumber green mottle mosaic virus disease of watermelon in Korea. Korean J Plant Pathol 6(2):250–255

Lefeuvre P, Martin DP, Harkins G, Lemey P, Gray AJA, Meredith S, Lakay F, Monjane A, Lett J-M, Varsani A, Heydarnejad J (2010) The spread of tomato yellow leaf curl virus from the Middle East to the World. PLoS Pathog 6(10):e1001164

Letschert B, Adam G, Lesemann D-E, Willingmann P, Heinze C (2002) Detection and differentiation of serologically cross-reacting tobamoviruses of economical importance by RT-PCR and RT-PCR-RFLP. J Virol Methods 106(1):1–10

Levitzky N, Smith E, Lachman O, Luria N, Mizrahi Y, Bakelman H, Sela N, Laskar O, Milrot E, Dombrovsky A (2019) The bumblebee *Bombus terrestris* carries a primary inoculum of Tomato brown rugose fruit virus contributing to disease spread in tomatoes. PLoS One 14(1):e0210871

Li M, Hu J, Xu FC, Liu SS (2010) Transmission of *Tomato yellow leaf curl virus* by two invasive biotypes and a Chinese indigenous biotype of the whitefly *Bemisia tabaci*. Int J Pest Manag 56(3):275–280

Li J, Liu S, Gu Q (2016) Transmission efficiency of *Cucumber green mottle mosaic virus* via seeds, soil, pruning and irrigation water. J Phytopathol 164(5):300–309

Ling KS (2010) Effectiveness of chemo- and thermotherapeutic treatments on *Pepino mosaic virus* in tomato seed. Plant Dis 94(3):325–328

Ling KS, Zhang W (2011) First report of *Pepino mosaic virus* infecting tomato in Mexico. Plant Dis 95(8):1035

Ling KS, Wechter WP, Jordan R (2007) Development of a one-step immunocapture real-time TaqMan RT-PCR assay for the broad spectrum detection of Pepino mosaic virus. J Virol Methods 144(1–2):65–72

Ling KS, Li R, Zhang W (2014) First report of *Cucumber green mottle mosaic virus* infecting greenhouse cucumber in Canada. Plant Dis 98(5):701

Ling KS, Agindotan B, Thomas S (2023) Cucumber green mottle mosaic virus. Emerging Viruses in Cucurbits Working Group. https://doi.org/10.1094/GROW-CUC-02-23-011

Liu HW, Luo LX, Li JQ, Liu PF, Chen XY, Hao JJ (2014) Pollen and seed transmission of *Cucumber green mottle mosaic virus* in cucumber. Plant Pathol 63(1):72–77

Liu H, Zhao F, Qiao Q, Zhang D, Wang Y, Wang S, Tian Y, Zhang Z (2021) Complete genome sequence of a divergent sweet potato chlorotic stunt virus isolate infecting *Calystegia hederacea* in China. Arch Virol 166(7):2037–2040

Lobin KK, Booluck K, Ramessur I, Taleb-Hossenkhan N (2023) First report of tomato yellow leaf curl virus (TYLCV) infecting *Solanum nigrum* plants in Mauritius. J Plant Pathol 105(2):591–592

Loebenstein G, Lecoq H (2012) Viruses and virus diseases of vegetables in the Mediterranean Basin. Adv Virus Res 84:570

Louro D, Vicente M, Vaira AM, Accotto GP, Nolasco G (2000) *Cucurbit yellow stunting disorder virus* (genus *Crinivirus*) associated with the yellowing disease of cucurbit crops in Portugal. Plant Dis 84(10):1156

Luria N, Smith E, Reingold V, Bekelman I, Lapidot M, Levin I, Elad N, Tam Y, Sela N, Abu-Ras A, Ezra N, Haberman A, Yitzhak L, Lachman O, Dombrovsky A (2017) A new Israeli Tobamovirus isolate infects tomato plants harboring Tm-$2^2$ resistance genes. PLoS One 12(1):e0170429

Luria N, Smith E, Sela N, Lachman O, Bekelman I, Koren A, Dombrovsky A (2018) A local strain of Paprika mild mottle virus breaks L3 resistance in peppers and is accelerated in Tomato brown rugose fruit virus-infected Tm-$2^2$-resistant tomatoes. Virus Genes 54(2):280–289

Luria N, Smith E, Sela N, Koren A, Lachman O, Dombrovsky A (2019) Insights into a watermelon virome contribute to monitoring distribution of whitefly-borne viruses. Phytobiomes J 3(1):61–70

Maachi A, Torre C, Sempere RN, Hernando Y, Aranda MA, Donaire L (2021) Use of high-throughput sequencing and two RNA input methods to identify viruses infecting tomato crops. Microorganisms 9(5):1043

Maayan Y, Pandaranayaka EPJ, Srivastava DA, Lapidot M, Levin I, Dombrovsky A, Harel A (2018) Using genomic analysis to identify tomato Tm-2 resistance-breaking mutations and their underlying evolutionary path in a new and emerging tobamovirus. Arch Virol 163(7):1863–1875

Mahillon M, Kellenberger I, Dubuis N, Brodard J, Bunter M, Weibel J, Sandrini F, Schumpp O (2022) First report of *Tomato brown rugose fruit virus* in tomato in Switzerland. New Dis Rep 45(1):e12065

Mandal S, Mandal B, Varma A (2008) Properties, diagnosis and management of Cucumber green mottle mosaic virus. Plant Viruses 2(1):25–34

Manglli A, Murenu M, Sitzia M, Tomassoli L (2016) First report of *Cucurbit yellow stunting disorder virus* infecting cucurbits in Italy. New Dis Rep 34(1):23

Maoka T, Nakayama T, Taniguchi M, Kano Y, Suzuki A, Sato M, Hataya T, Koizumi E, Noguchi K (2013) Multivirus detection from Japanese landraces of potato by Reverse Transcription–Polymerase Chain Reaction–microplate hybridization. Potato Res 56(2):147–156

Marchant WG, Mugerwa H, Gautam S, Al-Aqeel H, Polston JE, Rennberger G, Smith H, Turechek B, Adkins S, Brown JK, Srinivasan R (2023) Phylogenomic and population genetics analyses of extant tomato yellow leaf curl virus strains on a global scale. Front Virol 3:221156

Marco CF, Aranda MA (2005) Genetic diversity of a natural population of Cucurbit yellow stunting disorder virus. J Gen Virol 86(3):815–822

Martín-Hernández AM, Picó B (2020) Natural resistances to viruses in cucurbits. Agronomy 11(1):23

Maruthi MN, Rekha AR, Cork A, Colvin J, Alam SN, Kader KA (2005) First report of *tomato leaf curl New Delhi virus* infecting tomato in Bangladesh. Plant Dis 89(9):1011

Mathioudakis MM, Rodríguez-Moreno L, Sempere RN, Aranda MA, Livieratos I (2014) Multifaceted capsid proteins: multiple interactions suggest multiple roles for *Pepino mosaic virus* capsid protein. Mol Plant-Microbe Interact 27(12):1356–1369

McGlashan D (1994) Tomato yellow leaf curl geminivirus in Jamaica. Plant Dis 78:1219C

Méndez-Navarro D, Chacón-Cerdas R, Alvarado-Ulloa C, Camacho-Montero JR, Sánchez-Navarro JÁ, Alvarado-Marchena L (2025) Characterization of virus species associated with sweet potato virus disease in Costa Rica and riboprobe development for its rapid detection. Plant Pathol 74(2):413–421

Menzel W, Knierim D, Winter S, Hamacher J, Heupel M (2019) First report of *Tomato brown rugose fruit virus* infecting tomato in Germany. New Dis Rep 39(1):1–1

Miranda-Campaña OA, Diaz-Lara A, García-Estrada RS, Carrillo-Fasio JA, Tovar-Pedraza JM (2024) Molecular and biological characterization of pepino mosaic virus isolates occurring in the main tomatoproducing areas in Mexico. Physiol Mol Plant Pathol 131:102269

Mnari-Hattab M, Zammouri S, Belkadhi MS, Doña DB, Nahia E b, Hajlaoui MR (2015) First report of *Tomato leaf curl New Delhi virus* infecting cucurbits in Tunisia. New Dis Rep 31(1):21

Mohammed MS, Lahuf AA, Jeddoa ZMA, Al-Taey DKA (2024) First detection of cucurbit yellow stunting disorder virus in Iraq. Plant Health Progress 25(1):95–97

Molad O, Smith E, Luria N, Bakelman E, Lachman O, Reches M, Dombrovsky A (2024) Studying tomato brown rugose fruit virus longevity in soil and virion susceptibility to pH treatments helped improve virus control by soil disinfection. Plant Soil 505:543–558

Mondal S, Jenkins Hladky L, Melanson RA, Singh R, Sikora EJ, Wintermantel WM (2022) First report of Cucurbit yellow stunting disorder virus and Cucurbit chlorotic yellows virus in cucurbit crops in Alabama. Plant Dis 106(2):776

Moreno IM, Thompson JR, García-Arenal F (2004) Analysis of the systemic colonization of cucumber plants by Cucumber green mottle mosaic virus. J Gen Virol 85(3):749–759

Moreno-Pérez MG, Pagán I, Aragón-Caballero L, Cáceres F, Fraile A, García-Arenal F (2014) Ecological and genetic determinants of Pepino mosaic virus emergence. J Virol 88(6):3359–3368

Morilla G, Janssen D, García-Andrés S, Moriones E, Cuadrado IM, Bejarano ER (2005) Pepper (*Capsicum annuum*) is a dead-end host for *Tomato yellow leaf curl virus*. Phytopathology 95(9):1089–1097

Moriones E, Navas-Castillo J (2000) Tomato yellow leaf curl virus, an emerging virus complex causing epidemics worldwide. Virus Res 71(1–2):123–134

Moriones E, Praveen S, Chakraborty S (2017) Tomato leaf curl New Delhi virus: an emerging virus complex threatening vegetable and fiber crops. Viruses 9(10):264

Naganur P, Shankarappa KS, Mesta RK, Rao CD, Venkataravanappa V, Maruthi MN, Reddy LRCN (2023) Detecting tomato leaf curl New Delhi virus causing ridge gourd yellow mosaic disease, and other Begomoviruses by antibody-based methods. Plants 12(3):490

Nagendran K, Mohankumar S, Mohammed Faisal P, Bagewadi B, Karthikeyan G (2017) Molecular evidence for the occurrence of tomato leaf curl New Delhi virus on chayote (*Sechium edule*) in southern India. Virusdisease 28(4):425–429

Nagendran K, Kumari S, Pandey S, Karmakar P, Chaubey T, Kumar R, Vignesh S, Karthikeyan G, Behera TK (2023) Emergence of yellowing disease in cucurbitaceous vegetables caused by Crinivirus and Polerovirus in India. Virology 587:109876

Nakasu EYT, Nagata T, Inoue-Nagata AK (2022) First report of tomato fruit blotch virus infecting tomatoes in Brazil. Plant Dis 106(8):2271. https://doi.org/10.1094/PDIS-07-21-1392-PDN

Nava AR, Patte CP, Hiebert E, Polston JE (2006) Detection and variability of Begomoviruses in tomato from the Andean States of Venezuela. Plant Dis 90(1):61–66

Navas-Castillo J, Sánchez-Campos S, Díaz JA, Sáez-Alonso E, Moriones E (1999) Tomato yellow leaf curl virus-is causes a novel disease of common bean and severe epidemics in tomato in Spain. Plant Dis 83(1):29–32

Navas-Castillo J, Sánchez-Campos S, Noris E, Louro D, Accotto GP, Moriones E (2000) Natural recombination between Tomato yellow leaf curl virus-Is and Tomato leaf curl virus. J Gen Virol 81(11):2797–2801

Navas-Castillo J, Fiallo-Olivé E, Sánchez-Campos S (2011) Emerging virus diseases transmitted by whiteflies. Annu Rev Phytopathol 49(1):219–248

Navot N, Pichersky E, Zeidan M, Zamir D, Czosnek H (1991) Tomato yellow leaf curl virus: a whitefly-transmitted geminivirus with a single genomic component. Virology 185(1):151–161

Nematollahi S, Haghtaghi E, Koolivand D, Hajizadeh M (2014) Molecular detection of *Cucumber green mottle mosaic virus* variants from cucurbits fields in Iran. Arch Phytopathol Plant Protect 47(11):1303–1310

Noda C, Kittipakorn K, Inchan P, Wanapee L, Deema N (1993) Distribution of cucurbit viruses and reactions of some cucurbit species to certain viruses. Proceedings of the 31th Kasetsart University Annual Conference 7, p 341–347

Obaid HK, Adhab M (2025) Outbreak of tobamoviruses and potexviruses associated with disease epidemics in tomato production area of Iraq. Iraqi J Agric Sci 56:237–246

Obregón VG, Ibañez JM, Lattar TE, Juszczak S, Groth-Helms D (2023) First report of *Tomato brown rugose fruit virus* in tomato in Argentina. New Dis Rep 48(1):e12203

Ochoa MDL, Zavaleta-Mejía E, Johansen NRM, Herrera GA, Soriano EC (1996) Tospoviruses, weeds and thrips associated with chrysanthemum *(Dendranthema grandiflora* Tzvelev cv. Polaris). Int J Pest Manag 42(3):157–159

Ohnishi J, Yamaguchi H, Saito A (2016) Analysis of the Mild strain of tomato yellow leaf curl virus, which overcomes Ty-2 gene–mediated resistance in tomato line H24. Arch Virol 161(8):2207–2217

Oladokun JO, Halabi MH, Barua P, Nath PD (2019) Tomato brown rugose fruit disease: current distribution, knowledge and future prospects. Plant Pathol 68(9):1579–1586

Olaya C, Fletcher SJ, Zhai Y, Peters J, Margaria P, Winter S, Mitter N, Pappu HR (2020) The Tomato spotted wilt virus (TSWV) genome is differentially targeted in TSWV-infected tomato (*Solanum lycopersicum*) with or without Sw-5 Gene. Viruses 12(4):363

Oliver JE, Whitfield AE (2016) The Genus *Tospovirus*: emerging Bunyaviruses that threaten food security. Annu Rev Virol 3(1):101–124

Orfanidou CG, Tsvetkov I, Efthimiou KE, Maliogka VI, Katis NI (2019) First report of cucumber green mottle mosaic virus in cucumber greenhouse crops in Bulgaria. J Plant Pathol 101(2):423

Orfanidou CG, Cara M, Merkuri J, Papadimitriou K, Katis NI, Maliogka VI (2022a) First report of tomato brown rugose fruit virus in tomato in Albania. J Plant Pathol 104(2):855–855

Orfanidou CG, Efthimiou K, Katis NI, Maliogka VI (2022b) Elucidating the sweet potato virome in Greece with the aid of high-throughput sequencing technology. Plant Pathol 71(9):1880–1891

Ortega-Parra N, Zisi Z, Vogel E, Vos C, Hanssen I (2024) First report of infections with isolates of the Asian genotype of cucumber green mottle mosaic virus in cucumber crops in Belgium. Plant Dis 108(4):1121

Özdemír, S. (2010) First report of Pepino mosaic virus in tomato in Turkey. J Plant Pathol 92:107

Panno S, Iacono G, Davino M, Marchione S, Zappardo V, Bella P, Tomassoli L, Accotto GP, Davino S (2016) First report of *Tomato leaf curl New Delhi virus* affecting zucchini squash in an important horticultural area of southern Italy. New Dis Rep 33(1):6

Panno S, Caruso AG, Davino S (2019) First report of tomato brown rugose fruit virus on tomato crops in Italy. Plant Dis 103(6):1443

Panno S, Caruso AG, Barone S, Lo Bosco G, Rangel EA, Davino S (2020) Spread of tomato brown rugose fruit virus in Sicily and evaluation of the spatiotemporal dispersion in experimental conditions. Agronomy 10(6):834

Panno S, Davino S, Caruso AG, Bertacca S, Crnogorac A, Mandić A, Noris E, Matić S (2021) A review of the most common and economically important diseases that undermine the cultivation of tomato crop in the Mediterranean Basin. Agronomy 11(11):2188

Papayiannis LC, Ioannou N, Boubourakas IN, Dovas CI, Katis NI, Falk BW (2005) Incidence of viruses infecting cucurbits in Cyprus. J Phytopathol 153(9):530–535

Papayiannis LC, Katis NI, Idris AM, Brown JK (2011) Identification of weed hosts of *Tomato yellow leaf curl virus* in Cyprus. Plant Dis 95(2):120–125

Papayiannis LC, Kokkinos CD, Alfaro-Fernández A (2012) Detection, characterization and host range studies of Pepino mosaic virus in Cyprus. Eur J Plant Pathol 132(1):1–7

Pappu HR, Jones RAC, Jain RK (2009) Global status of tospovirus epidemics in diverse cropping systems: successes achieved and challenges ahead. Virus Res 141(2):219–236

Parrella G, Gognalons P, Gebre-Selassiè K, Vovlas C, Marchoux G (2003) An update of the host range of Tomato spotted wilt virus. J Plant Pathol 85:227–264

Parrella G, Greco B, Cavicchi L, Bellardi MG (2013) Dieback and wilting caused by *Tomato spotted wilt virus* in *Arctotis* × *hybrida* in Italy. Plant Dis 97(10):1387

Parrella G, Troiano E, Lee S, Kil E-J (2020) Tomato leaf curl New Delhi virus found associated with eggplant yellowing disease in Italy. Plant Dis 104(7):2034

Parrella G, Elbeaino T, Guy PL (2022) Emerging and reemerging plant viruses in a context of global change. Front Plant Sci 13:1108211

Paul S, Emmadi V, Saxena S, Sarkar M, Mandal B, Kumar R, Sinha P, Roy A (2025) RPA-assisted CRISPR-Cas12a-enabled point-of-care diagnostic platform for chili leaf curl virus with fluorescent and colorimetric readouts. Front Microbiol 16:1644322. https://doi.org/10.3389/fmicb.2025.1644322

Pellegrin F, Mnari Hattab M, Tahiri A, Dalleau Clouet C, Peterschmitt M, Bonato O (2008) First report of simultaneous presence of Tomato yellow leaf curl Sardinia virus and Tomato yellow leaf curl Israel virus infecting crops and weeds in Tunisia. J Plant Pathol 90:145

Péréfarres F, De Bruyn A, Kraberger S, Hoareau M, Barjon F, Lefeuvre P, Pellegrin F, Caplong P, Varsani A, Lett J-M (2012) Occurrence of the Israel strain of *Tomato yellow leaf curl virus* in New Caledonia and Loyalty Islands. New Dis Rep 25(1):6

Phaneendra C, Rao KRSS, Jain RK, Mandal B (2012) Tomato leaf curl New Delhi virus is associated with pumpkin leaf curl: a new disease in Northern India. Indian J Virol 23(1):42–45

Pitter PL, Mondal S, Chang PG, Morgan LM, Aikman S, Wintermantel WM, Tennant PF (2024) First report of Cucurbit Yellow stunting disorder virus infecting cucurbit crops in Jamaica. Plant Dis 108(4):1120

Polischuk V, Budzanivska I, Shevchenko T, Oliynik S (2007) Evidence for plant viruses in the region of Argentina Islands, Antarctica. FEMS Microbiol Ecol 59(2):409–417

Pop I, Jilaveanu A (1985) Identification of cucumber green mottle virus in Romania. An Inst Cercet Pentru Protectia Plant 18:43–47

Pospieszny H, Zielinska L, Cajza M (1997) New plant viruses in Poland. Prog Plant Prot 36(1):244–249

Prasad A, Sharma N, Hari-Gowthem G, Muthamilarasan M, Prasad M (2020) Tomato yellow leaf curl virus: impact, challenges, and management. Trends Plant Sci 25(9):897–911

Qiao Q, Zhang ZC, Qin YH, Zhang DS, Tian YT, Wang YJ (2011) First report of *Sweet potato chlorotic stunt virus* infecting sweet potato in China. Plant Dis 95(3):356

Rabadan M, Aranda M, Gomez P, Juarez M (2019) El virus de rizado del tomate de Nueva Delhi (ToLCNDV) en cultivos y malas hierbas. Phytoma-España 306:18–25

Radouane N, Tahiri A, El Ghadraoui L, Al Figuigui J, Lahlali R (2018) First report of *tomato leaf curl New Delhi virus* in Morocco. New Dis Rep 37(1):2

Read DA, Roberts R, Thompson GD (2021) Orchid fleck virus and a novel strain of sweet potato chlorotic stunt virus associated with an ornamental cultivar of *Alcea rosea* L. in South Africa. Eur J Plant Pathol 160(1):227–232

Reina J, Morilla G, Bejarano ER, Rodríguez MD, Janssen D (1999) First report of *Capsicum annuum* plants infected by tomato yellow leaf curl virus. Plant Dis 83(12):1176

Reingold V, Lachman O, Blaosov E, Dombrovsky A (2015) Seed disinfection treatments do not sufficiently eliminate the infectivity of *Cucumber green mottle mosaic virus* (CGMMV) on cucurbit seeds. Plant Pathol 64(2):245–255

Reingold V, Lachman O, Belausov E, Koren A, Mor N, Dombrovsky A (2016) Epidemiological study of *Cucumber green mottle mosaic virus* in greenhouses enables reduction of disease damage in cucurbit production. Ann Appl Biol 168(1):29–40

Renukadevi P, Nagendran K, Nakkeeran S, Karthikeyan G, Jawaharlal M, Alice D, Malathi VG, Pappu HR (2015) First report of *Tomato spotted wilt virus* infection of chrysanthemum in India. Plant Dis 99(8):1190

Rey MEC, Ndunguru J, Berrie LC, Paximadis M, Berry S, Cossa N, Nuaila VN, Mabasa KG, Abraham N, Rybicki EP, Martin D, Pietersen G, Esterhuizen LL (2012) Diversity of dicotyledenous-infecting Geminiviruses and their associated DNA molecules in Southern Africa, including the South-West Indian Ocean Islands. Viruses 4(9):1753–1791

Riley DG, Pappu HR (2000) Evaluation of tactics for management of thrips-vectored *Tomato spotted wilt virus* in tomato. Plant Dis 84(8):847–852

Rivarez MPS, Pecman A, Bačnik K, Maksimović O, Vučurović A, Seljak G, Mehle N, Gutiérrez-Aguirre I, Ravnikar M, Kutnjak D (2023) In-depth study of tomato and weed viromes reveals undiscovered plant virus diversity in an agroecosystem. Microbiome 11(1):60

Rodriguez SD, Allan C, Duarte SD, Matzkin LM, Palumbo J, Carrière Y (2024) First report of tomato spotted wilt virus infecting lettuce in Yuma. Ariz Plant Dis 108(3):823

Roggero P, Masenga V, Lenzi R, Coghe F, Ena S, Winter S (2001) First report of *Pepino mosaic virus* in tomato in Italy. Plant Pathol 50(6):798

Rojas MR, Kon T, Natwick ET, Polston JE, Akad F, Gilbertson RL (2007) First report of *tomato yellow leaf curl virus* associated with tomato yellow leaf curl disease in California. Plant Dis 91(8):1056

Rotenberg D, Whitfield AE (2018) Molecular interactions between tospoviruses and thrips vectors. Curr Opin Virol 33:191–197. https://doi.org/10.1016/j.coviro.2018.11.007

Rotenberg D, Jacobson AL, Schneweis DJ, Whitfield AE (2015) Thrips transmission of tospoviruses. Curr Opin Virol 15:80–89

Rubio L, Tian T, Yeh H, Livieratos Y, Falk BW (2002) *De novo* generation of *lettuce infectious yellows virus* defective RNAs in protoplasts. Mol Plant Pathol 3(5):321–327

Ruiz ML, Simón A, Velasco L, García MC, Janssen D (2015) First report of *Tomato leaf curl New Delhi virus* infecting tomato in Spain. Plant Dis 99(6):894

Sabra A, Al-Saleh MA, Al-Shahwan IM, Amer MA (2022) First report of tomato brown rugose fruit virus infecting the tomato crop in Saudi Arabia. Plant Dis 106(4):1310

Salem N, Mansour A, Ciuffo M, Falk BW, Turina M (2016) A new tobamovirus infecting tomato crops in Jordan. Arch Virol 161(2):503–506

Salem NM, Cao MJ, Odeh S, Turina M, Tahzima R (2020) First report of tobacco mild green mosaic virus and tomato brown rugose fruit virus infecting *Capsicum annuum* in Jordan. Plant Dis 104(2):601

Salem NM, Abumuslem M, Turina M, Samarah N, Sulaiman A, Abu-Irmaileh B, Ata Y (2022a) New weed hosts for tomato brown rugose fruit virus in wild Mediterranean vegetation. Plants 11(17):2287

Salem NM, Sulaiman A, Samarah N, Turina M, Vallino M (2022b) Localization and mechanical transmission of tomato brown rugose fruit virus in tomato seeds. Plant Dis 106(1):275–281

Salem NM, Jewehan A, Aranda MA, Fox A (2023) Tomato brown rugose fruit virus pandemic. Annu Rev Phytopathol 61(1):137–164

Salvalaggio AE, López Lambertini PM, Cendoya G, Huarte MA (2017) Temporal and spatial dynamics of *Tomato spotted wilt virus* and its vector in a potato crop in Argentina. Ann Appl Biol 171(1):5–14

Sarkar M, Gupta D, Singh OW, Paul S, Kumar R, Mandal B, Roy A (2024) Identification of pathogenicity determinants in ToLCNDV and their RNAi-based knockdown for disease management in *Nicotiana benthamiana* and tomato plants. Front Microbiol 15:1481523. https://doi.org/10.3389/fmicb.2024.1481523

Sarkes A, Fu H, Feindel D, Harding M, Feng J (2020) Development and evaluation of a loop-mediated isothermal amplification (LAMP) assay for the detection of Tomato brown rugose fruit virus (ToBRFV). PLoS One 15(6):e0230403

Sastry KS, Mandal B, Hammond J, Scott SW, Briddon RW (2019) Encyclopedia of plant viruses and viroids. Springer India. https://doi.org/10.1007/978-81-322-3912-3

Schwarz D, Beuch U, Bandte M, Fakhro A, Büttner C, Obermeier C (2010) Spread and interaction of *Pepino mosaic virus* (PepMV) and *Pythium aphanidermatum* in a closed nutrient solution recirculation system: effects on tomato growth and yield. Plant Pathol 59(3):443–452

Sempere RN, Gómez-Aix C, Ruíz-Ramón F, Gómez P, Hasiów-Jaroszewska B, Sánchez-Pina MA, Aranda MA (2016) *Pepino mosaic virus* RNA-dependent RNA polymerase POL domain is a hypersensitive response-like elicitor shared by necrotic and mild isolates. Phytopathology 106(4):395–406

Senshu H, Ozeki J, Komatsu K, Hashimoto M, Hatada K, Aoyama M, Kagiwada S, Yamaji Y, Namba S (2009) Variability in the level of RNA silencing suppression caused by triple gene block protein 1 (TGBp1) from various potexviruses during infection. J Gen Virol 90(4):1014–1024

Sevik MA, Arli-Sokmen M (2012) Estimation of the effect of Tomato spotted wilt virus (TSWV) infection on some yield components of tomato. Phytoparasitica 40(1):87–93

Shamshiri M, Heydarnejad J, Kamali M, Pouramini N, Massumi H (2019) Identification of wild hosts of tomato yellow leaf curl virus in South-Eastern Iran. Arch Phytopathol Plant Protect 52(9–10):917–929

Shargil D, Smith E, Lachman O, Reingold V, Darzi E, Tam Y, Dombrovsky A (2017) New weed hosts for cucumber green mottle mosaic virus in wild Mediterranean vegetation. Eur J Plant Pathol 148(2):473–480

Shipp JL, Buitenhuis R, Stobbs L, Wang K, Kim WS, Ferguson G (2008) Vectoring of *Pepino mosaic virus* by bumble-bees in tomato greenhouses. Ann Appl Biol 153(2):149–155

Skelton A, Buxton-Kirk A, Ward R, Harju V, Frew L, Fowkes A, Long M, Negus A, Forde S, Adams IP, Pufal H, McGreig S, Weekes R, Fox A (2019) First report of *Tomato brown rugose fruit virus* in tomato in the United Kingdom. New Dis Rep 40(1):12

Skelton A, Gentit P, Porcher L, Visage M, Fowkes A, Adams IP, Harju V, Webster G, Pufal H, McGreig S, Ward R, Fox A (2022) First report of *Tomato brown rugose fruit virus* in tomato in France. New Dis Rep 45(1):e12061
Skelton A, Frew L, Ward R, Hodgson R, Forde S, McDonough S, Webster G, Chisnall K, Mynett M, Buxton-Kirk A, Fowkes AR, Weekes R, Fox A (2023) Tomato brown rugose fruit virus: survival and disinfection efficacy on common glasshouse surfaces. Viruses 15(10):2076
Slavokhotova AA, Andreeva EN, Shijan AN, Odintsova TI, Pukhalskij VA (2007) Specifics of the coat protein gene in Russian strains of the cucumber green mottle mosaic virus. Russ J Genet 43(11):1221–1226
Smith E, Dombrovsky A (2020) Aspects in *Tobamovirus* management in intensive agriculture. In: Plant diseases-current threats and management trends. IntechOpen
Sneideris D, Zizyte-Eidetiene M, Zitikaite I, Urbanavičienė L, Staniulis J (2013) First report of two distinct strains of Pepino mosaic virus infecting tomatoes in greenhouses in Lithuania. J Plant Pathol 95(1):217
Soler S, Prohens J, Díez MJ, Nuez F (2002) Natural occurrence of *Pepino mosaic virus* in *Lycopersicon* species in Central and Southern Peru. J Phytopathol 150(2):49–53
Soler S, López C, Nuez F (2005) Natural occurrence of viruses in *Lycopersicon* spp. in Ecuador. Plant Dis 89(11):1244
Spence NJ, Basham J, Mumford RA, Hayman G, Edmondson R, Jones DR (2006) Effect of *Pepino mosaic virus* on the yield and quality of glasshouse-grown tomatoes in the UK. Plant Pathol 55(5):595–606
Stankovic I, Vucurovic A, Zecevic K, Petrovic B, Ristic D, Vucurovic I, Krstic B (2020) Pepino mosaic virus, a new threat for Serbia's tomatoes. Span J Agric Res 18(4):e10SC05
Stäubli A (2005) Nouvelle virose sur tomate. Rev Suisse Vitic Arboric Hortic 37(3):150
Stewart LR, Medina V, Tian T, Turina M, Falk BW, Ng JCK (2010) A mutation in the *lettuce infectious yellows virus* minor coat protein disrupts whitefly transmission but not *in planta* systemic movement. J Virol 84(23):12165–12173
Stobbs LW, Greig N, Weaver S, Shipp L, Ferguson G (2009) The potential role of native weed species and bumble bees (*Bombus impatiens*) on the epidemiology of *Pepino mosaic virus*. Can J Plant Pathol 31(2):254–261
Sugiyama M, Ohara T, Sakata Y (2006) A new source of resistance to cucumber green mottle mosaic virus in melon. J Jpn Soc Hortic Sci 75(6):469–475
Tairo F, Kullaya A, Valkonen JPT (2004) Incidence of viruses infecting sweetpotato in Tanzania. Plant Dis 88(9):916–920
Tarasashvili L (1976) Study on the spread and harmfulness of cucumber ordinary mosaic virus in Georgia and on the resistance of varieties to the disease. Trudy Nauchno-Issled Inst Zashch Rast Gruz SSR 4:28–31
Tesoriero LA, Chambers G, Srivastava M, Smith S, Conde B, Tran-Nguyen LTT (2016) First report of cucumber green mottle mosaic virus in Australia. Aust Plant Dis Notes 11(1):1
Tian T, Posis K, Maroon-Lango CJ, Mavrodieva V, Haymes S, Pitman TL, Falk BW (2014) First report of *cucumber green mottle mosaic virus* on melon in the United States. Plant Dis 98(8):1163
Tugume AK, Amayo R, Weinheimer I, Mukasa SB, Rubaihayo PR, Valkonen JPT (2013) Genetic variability and evolutionary implications of RNA silencing suppressor genes in RNA1 of sweet potato chlorotic stunt virus isolates infecting sweet potato and related wild species. PLoS One 8(11):e81479
Turina M, Kormelink R, Resende RO (2016) Resistance to tospoviruses in vegetable crops: epidemiological and molecular aspects. Annu Rev Phytopathol 54(1):347–371
Ugaki M, Tomiyama M, Kakutani T, Hidaka S, Kiguchi T, Nagata R, Sato T, Motoyoshi F, Nishiguchi M (1991) The complete nucleotide sequence of cucumber green mottle mosaic virus (SH strain) genomic RNA. J Gen Virol 72(7):1487–1495
Untiveros M, Fuentes S, Salazar LF (2007) Synergistic interaction of *sweet potato chlorotic stunt virus* (*Crinivirus*) with Carla-, Cucumo-, Ipomo-, and Potyviruses infecting sweet potato. Plant Dis 91(6):669–676

Urbino C, Regragui ZF, Granier M, Peterschmitt M (2020) Fitness advantage of inter-species TYLCV recombinants induced by beneficial intra-genomic interactions rather than by specific mutations. Virology 542:20–27

Valverde R, Moreira MA (2004) Identificación de virus en el cultivo de camote (*Ipomoea batatas* L.) en Costa Rica. Agron Mesoam 15(1):1–7

Valverde RA, Lozano G, Navas-Castillo J, Ramos A, Valdés F (2004) First report of *Sweet potato chlorotic stunt virus* and *Sweet potato feathery mottle virus* infecting sweet potato in Spain. Plant Dis 88(4):428

van der Vlugt RAA, Stijger CCMM, Verhoeven JTJ, Lesemann D-E (2000) First report of *Pepino mosaic virus* on tomato. Plant Dis 84(1):103–103

van Dorst HJM (1988) Surface water as source in the spread of cucumber green mottle mosaic virus. Neth J Agric Sci 36(3):291–299

Varveri C, Vassilakos N, Bem F (2002) Characterization and detection of Cucumber green mottle mosaic virus in Greece. Phytoparasitica 30(5):493–501

Veerakone S, Waite DW, Delmiglio C, Kanchiraopally D, Kelly M, Khan S, Liefting L, Lilly ST, Perez-Egusquiza Z, Tang J, Yan J, Tomiczek L, Thompson JR (2024) Detection, characterization, and distribution of the first case of Pepino mosaic virus in Aotearoa New Zealand. Plant Dis 108(2):291–295

Venkataravanappa V, Ashwathappa KV, Reddy CNL, Shankarappa KS, Reddy MK (2020) Characterization of Tomato leaf curl New Delhi virus associated with leaf curl and yellowing disease of watermelon and development of LAMP assay for its detection. 3 Biotech 10(6):282

Venkataravanappa V, Madhu GS, Muralidhara BM, Reddy MK, Reddy CNL (2025) Tomato leaf curl New Delhi virus (*Begomovirus solanumdelhiense*) causes yellow mosaic disease in white pattypan squash (*Cucurbita pepo* L. ssp. pepo cv. Patissonina) in India. Indian Phytopathol 78(3):749–755. https://doi.org/10.1007/s42360-025-00887-6

Vo TTB, Troiano E, Lal A, Hoang PT, Kil E-J, Lee S, Parrella G (2022) ToLCNDV-ES infection in tomato is enhanced by TYLCV: evidence from field survey and agroinoculation. Front Microbiol 13:954460

Vučurović A, Brodarič J, Jakomin T, Pecman A, Benko Beloglavec A, Mehle N (2022) First report of *Tomato brown rugose fruit virus* in tomato in Slovenia. New Dis Rep 45(2):e12079

Wang S, Chen M (1985) New strain of cucumber mottle mosaic virus causing mosaic symptom. Plant Prot Bull 27:105–110

Wang D, Vinson R, Holmes M, Seibel G, Bechar A, Nof S, Tao Y (2019) Early detection of tomato spotted wilt virus by hyperspectral imaging and outlier removal auxiliary classifier generative adversarial nets (OR-AC-GAN). Sci Rep 9(1):4377

Weber H, Pfitzner AJP (1998) *Tm-2 2* Resistance in tomato requires recognition of the carboxy terminus of the movement protein of tomato mosaic virus. Mol Plant-Microbe Interact 11(6):498–503

Webster C, Jones R (2018) Cucumber green mottle mosaic virus. CABI Compendium, Wallingford

Webster CG, Kousik CS, Roberts PD, Rosskopf EN, Turechek WW, Adkins S (2011) *Cucurbit yellow stunting disorder virus* detected in Pigweed in Florida. Plant Dis 95(3):360

Wintermantel WM, Hladky LL, Cortez AA, Natwick ET (2009) A new expanded host range of *cucurbit yellow stunting disorder virus* includes three agricultural crops. Plant Dis 93(7):685–690

Wisler GC, Duffus JE, Liu H-Y, Li RH (1998) Ecology and epidemiology of whitefly-transmitted closteroviruses. Plant Dis 82(3):270–280

Wright D, Mumford R (1999) Pepino mosaic Potexvirus (PepMV) first records in tomato in the United Kingdom. Plant Dis 89:400

Xinying Y, Xin L, Lili Y, Qiuyue Z, Yongzhe P, Jijuan C (2022) Detection of cucumber green mottle mosaic virus in low-concentration virus-infected seeds by improved one-step pre-amplification RT-qPCR. Plant Methods 18(1):70. https://doi.org/10.1186/s13007-022-00901-2

Yan ZY, Ma HY, Han SL, Geng C, Tian YP, Li XD (2019) First report of tomato brown rugose fruit virus infecting tomato in China. Plant Dis 103(11):2973

Yan Z, Wolters A-MA, Navas-Castillo J, Bai Y (2021) The global dimension of tomato yellow leaf curl disease: current status and breeding perspectives. Microorganisms 9(4):740

Yaoliang Z, Zhongjian S (2003) Preliminary characterization of Pepino mosaic virus Shanghai isolate (PepMV-Sh) and its detection with ELISA. Acta Agric Shanghai 19(3):90–92

Yazdani-Khameneh S, Golnaraghi AR, Rakhshandehroo F (2013) Report of a new begomovirus on melon in Iran. New Dis Rep 28(1):17–17

Zaidi SS, Martin DP, Amin I, Farooq M, Mansoor S (2017) *Tomato leaf curl New Delhi virus*: a widespread bipartite begomovirus in the territory of monopartite begomoviruses. Mol Plant Pathol 18(7):901–911

Zeng R, Gu H, Fan J, Xu L, Gao S, Song Z, Gao P, Zhang K, Zhang C, Zhu P, Dai F (2023) Occurrence of tomato leaf curl New Delhi virus in cucurbit plants in China. Plant Dis 107(8):2561

Zhang K, Lu H, Wan C, Tang D, Zhao Y, Luo K, Li S, Wang J (2020) The spread and transmission of sweet potato virus disease (SPVD) and its effect on the gene expression profile in sweet potato. Plants 9(4):492

Zhang Z, Zheng K, Zhao L, Su X, Zheng X, Wang T (2021) Occurrence, distribution, evolutionary relationships, epidemiology, and management of Orthotospoviruses in China. Front Microbiol 12:686025

Zinger A, Doron-Faigenboim A, Gelbart D, Levin I, Lapidot M (2025) Contribution of the tobamovirus resistance gene Tm-1 to control of tomato brown rugose fruit virus (ToBRFV) resistance in tomato. PLoS Genet 21(5):e1011725

# Impact of Climate Change on Viral Pathogens in Vegetables

2

Ashulata Kaushal, Devendra Kumar Choudhary, C. P. Khare, A. S. Kotasthane, and N. R. Rangare

**Abstract**

Climate changes are transforming ecological and molecular determinants of the risk of viral diseases in vegetable crops. The interactions among increasing temperatures, altered rainfall, elevated $CO_2$ and ozone, and increased extreme weather events, along with the biology of vectors, host physiology, and viral evolution, result in the development of disease pressure. The presence of warmer temperatures and less severe winters increases the range and activity of the most important insect vectors, such as aphids, whiteflies, and thrips, shortens their development cycles, and multiplies the number of generations. There are also disturbances like droughts, floods, and storms, which further alter the availability of hosts and the movement of vectors, exposing the crops to vulnerable vectors with the virulence factor that are in excess. These environmental changes are paralleled by the non-linear behavioral modifications in vectors, such as a change in probing and feeding behavior due to virus-mediated changes in plant signals. Direct effects on the virus replication process include temperature stress, which alters stability and mutation rates and provides opportunities for recombination and the production of variants able to overcome the initial resistance. The alteration of host-specific plant hormone signaling, RNA silencing, proteostasis, and microbiome structure brought by climate change can either facilitate vulnerability or offer conditional tolerance. The case studies of TYLCV, TSWV, ToBRFV, and the recent cucurbit and solanaceous viruses demonstrate high rates of disease outbreak in relation to the expansion of vectors, collapse of resistance, and insufficient sanitation. Addressing these compound risks will necessitate system-level adjustments, which consist of an integrated pest and vector management, clean seed systems, enhanced diagnostics, genomic surveillance, and breeding

A. Kaushal (✉) · D. K. Choudhary · C. P. Khare · A. S. Kotasthane · N. R. Rangare
Department of Plant Pathology, College of Agriculture, Indira Gandhi Krishi Vishvavidyalaya, Raipur, Chhattisgarh, India



R. Kumar et al. (eds.), *Vegetable Crops and Viral Pathogens Volume I*, Plant Pathology for Sustainable Agriculture, https://doi.org/10.1007/978-981-95-7265-6_2

approaches that incorporate thermotolerance with broad-spectrum resistance. Further development requires mechanistic climate-pathogen-vector research, predictive modeling, and socio-economic studies to aid in the implementation of robust practices.

**Keywords**

Climate Change · Insect vector · Viral pathogens · Aphids · Vegetables

## 2.1 Introduction

Climate change and the emerging infectious diseases are closely connected threats that degrade health, economies, and environmental stability in every part of the world (Whitmee et al. 2015). Higher average temperatures, changed rainfall, and increased frequency of extremes all lead to new ecological opportunities for pathogens and their vectors, speeding up the evolution of pathogens as well as the intensity of transmission pathways, which is caused by climate change. Simultaneously, changes in land use and intensive farming elevate the rate of host contact and decrease the resilience of the ecosystems, forming a positive feedback loop, which increases the risk of outbreaks. The intersection point of these pressures is in vegetable production. Vegetables are the foundation of nutrition and livelihood, particularly in smallholder households, and also one of the most climate-sensitive crops. Climatic changes have an impact on plant physiology, development, and nutrient levels; extreme events cause direct harm; changing climatic conditions alter the distribution, seasonality, and behavior of insect vectors that transmit a large number of viral pathogens (Jones and Ejeta 2015; Patil et al. 2014; Kumar et al. 2023). The net impact is increased risk of disease, decreased output, lower quality, and increased susceptibility to the poor producers. Several mechanisms are associated with climate and disease. Warming may extend the range of vectors, extend the generation time, increase transmission pressure; high $CO_2$ and drought stress alter plant chemistry and immunity and can be more susceptible or less depending on the system; heat waves and storms create wounds and facilitate mechanical spread; and environmental change may facilitate viral mutation, recombinations, and resistance breakdown (Oladokun et al. 2019; Prasad et al. 2022). The transmission of the seed and the anthropogenic distribution of the plant material also allow the dissemination of the viral problem over a long distance and its establishment. Socio-economic consequences are significant. Outbreaks lower yields and nutritional value, strike marketable vegetables, and raise production costs as farmers take emergency steps (changed planting dates, use of chemicals, investment in sheltered buildings). Such adaptation reactions tend to benefit more resource-endowed farms, which increase inequities and may lead to a decline in ecosystems due to increased pesticide application or monoculture (Ratnadass et al. 2012). The solutions to these overlapping issues involve a cross-sectoral response intervention. In practice, the strategies are mixed resistant and tolerant, seed and quarantine, integrated control of vectors, better diagnostics and surveillance, and climate agronomy (water-efficient irrigation, balanced nutrition, measures to maintain the health of the soil). No less significant are institutional measures: the system of strengthened extension and participation, financing schemes of access to protective infrastructure, insurance associated with speedy reporting, and international collaboration regarding the phytosanitary standards and the exchange of germplasm (Patz et al. 2004; Jenkins et al. 2018). The chapter is a synthesis of the existing evidence on the effects of climate drivers on the emergence, spread, and impact of the viral pathogens in vegetable systems. It considers mechanisms—vector ecology, viral biology, host physiology, and extreme events—and evaluates the impacts on yield, quality, and livelihoods. Lastly, it identifies adaptation and mitigation strategies: breeding to multiple abiotic and biotic resilience, IPM to adapt to changing vector regimes, sanitation and seed systems, surveillance and decision support, and policy priorities to ensure the protection of smallholders and food security. Its objective is to synthesize in practice: to educate researchers, policymakers, and growers on the source of risk increase and the most promising resilient and equitable responses (Oladokun et al. 2019; Jones and Ejeta 2015).

## 2.2 Fundamentals of Plant Viruses and Their Vectors

### 2.2.1 What Is a Plant Virus?

Plant viruses are obligate intracellular parasites—they are not able to replicate or perform metabolism outside a living cell (Wendimu and Gurmu 2024). Beyond the host, they are inert particles: nucleic acid in a protein coat, and additional proteins are present in some instances. Most plant viruses have no means of movement or reproduction and rely upon vectors (primarily insect vectors) to infect one plant and leave another healthy (Chavan et al. 2025; Jeevalatha et al. 2021). Such reliance is reflected in the strategy of virus life: viruses need to enter cells of the host, hijack the host machineries, relocate to different plants, and count on vectors or human activities to cross between plants (Wendimu and Gurmu 2024).

### 2.2.2 The Critical Role of Vectors

Vectors are not merely carriers of the virus particles, but they also belong to the epidemiological system. Viruses and vectors co-evolve: viruses develop features to be easier to obtain or transmit, vectors develop strategies to behave or physiologically, and plants develop resistance to both virus behavior and interaction with vectors. Some systems even cause viruses to alter plant characteristics (volatiles, phloem composition, visual cues) to be more or less attractive to vectors and hence change the transmission probabilities (Chandi et al. 2018). Such manipulations may be so indirect that there is a high probability of the virus getting to new hosts. There are a lot of insect groups that propagate various vegetable viruses. Many begomoviruses, criniviruses, and associated groups are transmitted by whiteflies (*Bemisia tabaci*) that have caused massive losses in tomato and other vegetables (Wang et al. 2020). Aphids host a significant proportion of economically significant viruses, including representatives of *Potyviridae* and *Cucumovirus*, such as cucumber mosaic virus (CMV), since aphids are mobile, polyphagous, and probe plants frequently (Pidikiti et al. 2023). The illustrative examples of the real effects include tomato yellow leaf curl virus (TYLCV) transmitted by whitefly, potato leafroll virus (PLRV) transmitted by aphids, and cucumber mosaic virus (CMV), which can change the interactions between plants and vectors and has an extremely wide range of hosts (Wang et al. 2020; Naga et al. 2020, 2021; Pidikiti et al. 2023). The vast majority of incidences of plant virus transmission are carried out by arthropods; insects are dominant. The ecology and management issues of various vectors are different. Aphids are effective due to their piercing-sucking mouthparts and quick piercing-probing behavior, which enable them to acquire and inoculate fast, thus supporting non-persistent and persistent data transfer. Each of the whiteflies, thrips, leafhoppers, planthoppers, mealybugs, and mites carries its specific viruses and requires a particular method of control (Pidikiti et al. 2023).

### 2.2.3 Transmission Pathways: Modes of Viral Spread

The modes of transmission define the dynamics of outbreaks and the control choices. In general, they include persistent (virus stays in the insect and might last throughout life), semi-persistent (virus stays in foregut or cuticular structures and can last hours to days), non-persistent or stylet-borne (attached briefly to the stylet and passed during short probes), mechanical (contaminated tools or hands), and vertical (transmitted by seeds or vegetative growth). The transmission may need hours to days due to persistent spreads; the virus may be transmitted over a long duration and on a large scale by salivary glands that may be passed through gut barriers. Non-persistent transmission is rapid and short-lived, and hence the rapid-acting insecticides tend to work inadequately to prevent the propagation since aphids can infect before they are killed (Shi et al. 2021; Falla and Cunniffe 2024). Of particular importance are mechanical transmission and infected seed or planting material, particularly in nurseries, greenhouses, and clonally propagated crops (potato, banana, sugarcane). Hygiene and adequate care of tools, certification and utilization of virus-free planting material and bedsores, and effective use of the crop (which is good hygiene) are the key practical solutions to avoid the long-term durability and geographical transmission of most plant viruses (Shi et al. 2021; Wendimu and Gurmu 2024).

### 2.2.4 Climate Parameters Studied in Plant Virus-Vector Research

Climate has a strong influence on the dynamics of plant viruses by modulating host physiology, population of vectors and behaviors, and replication of the virus. The parameters to be studied include temperature, $CO_2$, water availability, humidity, oxidants (ozone/reactive species), and light, among others. Signs of temperature change (warming or extremes) can move vector activity and virus frequency and may undermine or destroy host defenses or occasionally decrease the viral titres based on the system (Roos et al. 2011; Kim et al. 2019). High $CO_2$ alters vegetation nutrition and protective mechanisms and alters the feeding of vectors and the concentration of viruses in complicated, occasionally inconsistent manners (Guo et al. 2021; Peñalver-Cruz et al. 2020). Drought and shifted precipitation alter the water status of plants and probing by vectors: transmission is promoted by drought by modulating feeding, whereas in other cases, reduced transmission is caused by drought (van Munster et al. 2017; Bergès et al. 2018). The humidity is interacting with temperature to influence the survival and fecundity of vectors, thereby indirectly changing the transmission. Reactive species and ozone modify defense signaling and are in clinical trials to activate immunity/sanitation against viruses. The intensity of light also changes host metabolism and immunity, which affects viral replication and transmission of seeds (Montes and Pagán 2019). Climate impacts in general are extremely context-dependent, usually non-additive ($CO_2$ × temperature), and demand system-specific experiments and models to forecast the effects of increased risk of viruses.

## 2.3 Impact of Climate Change on Viral Pathogens in Vegetables

Climate change changes the ecological and molecular background of the interaction between vegetable crops, insect vectors, and the viruses they are infected by. Higher temperatures, changes in rainfall, increased $CO_2$, changing ozone concentrations, and increasing extreme events create a web of both direct and indirect effects that alter the location of virus occurrence, the rate of virus replication, the behavior of vectors, and the resistance of plants to the infection. These impacts are usually species-specific, non-linear, and thus difficult to predict and control (Canto et al. 2008; Jones 2009; Prasad et al. 2022).

**Changeable Disease Risk Responses** Vectors—primarily ectothermic insects like whiteflies, aphids, and thrips—are very vulnerable to climatic changes. Three outcomes are specifically important. Shifts in ranges and new introductions. The warmer winters and warmer conditions in the higher latitudes and heights allow the poleward and upslope expansions of most vectors, e.g., *Bemisia tabaci* is predicted to increase in range and become more dangerous to begomoviruses in open-field tomato and other vegetables as climatic conditions get warmer. Another factor that allows non-native vectors to find new habitats is human movement of plant materials and irrigated microhabitats (greenhouses, oases) (Ramos et al. 2018).

**Phenology and Population Dynamics** Alterations in season duration, bringing about offspring or delayed autumn, have a tendency to raise the general number of generations per year in insects that are vectored. In certain areas, aphids are now capable of working longer hours, which makes the maintenance of virus-free planting material more difficult, and the total number of virus transmission events will continue to grow (Kim and Kwon 2019). Local reactions are dependent on rainfall, availability of hosts, and communities of natural enemies.

**Physiological and Behavioral Changes** Insect feeding, movement, and probing frequency, which are major determinants of virus acquisition and inoculation, are affected by temperature and humidity. Higher temperatures tend to promote higher feeding rates and movement, thereby increasing the probability of transmission. The alterations in the phloem composition and secondary metabolites due to elevated $CO_2$ and changes in the host physiology can also modify the performance of vectors and their feeding behavior in a complex manner (Peñalver-Cruz et al. 2020; Moreno-Delafuente et al. 2020).

**Management Implications** Warming increases the rate of vector turnover and tends to favor the evolution of insecticide resistance and may compromise biological control, compelling growers to increase pesticide application and providing selection pressure to produce so-called super vectors. Combined methods that cut down the frequency of sprays and focus on ecological management are hence imperative (Kenyon et al. 2014; Gilbertson et al. 2015).

**Direct Actions on Viral Biology** Viruses are directly affected by temperature and other climate variables that affect viral replication, stability, and evolution. Dynamic stability and replication. Rates of viral replication, capsid stability, and polymerase fidelity are temperature sensitive and virus specific. Warming that causes local conditions to approach the thermal optimum of a virus can moderate warming and have an effect of causing within-plant titres and more severe symptoms; the extreme heat can instead reduce infectivity by denaturing proteins. Multi-pathogen viruses, such as seed, tools, and humans, can take advantage of the appropriate environmental opportunities to distribute in a short period of time; a current-day example of a high-risk virus in the horticulture sector is tomato brown rugose fruit virus (ToBRFV) (Oladokun et al. 2019).

**Evolutionary Acceleration** An increase in replication, co-infection, and more widespread activity of vectors increases the chances of mutation, recombination, and the development of new variants. RNA viruses are particularly liable to produce genetic diversity; sometimes new variants have various virulence, host range, or compatibility with vectors, which might dilute resistance bred into vegetable cultivars (Pyott et al. 2016; Cai et al. 2023).

**Host-Mediated Pathways and Plant Physiological Pathways** Plant physiological and plant–microbe interactions changes that are caused by climate change may increase or decrease viral susceptibility. These changes involve immunity, microbiomes, and metabolism. Shifts in heat, drought, and $CO_2$ restructure photosynthesis, hormone signaling, and secondary metabolism. Stress may inhibit important antiviral pathways—salicylic acid signaling and RNA silencing—or disrupt useful microbiomes that aid in host defenses against pathogens (Sundaresha et al. 2022, 2024). On the other hand, in other environments, viruses can admittedly assist hosts in withstanding abiotic stress (such as water relations), generating multifaceted, contextualized consequences in managing diseases (Prasch and Sonnewald 2013; Shteinberg et al. 2021).

**Effects of Water Stress and Humidity** Both host and vector biology are affected by drought and changed precipitation. In certain systems, drought enhances feeding on the remaining green tissue by vectors and thus transmission; in some systems, drought lowers the abundance of the vectors, hence reducing transmission. The non-vector transmission risks are also subject to change by the moisture and wetness of roots surrounding the leaf surfaces and the farming equipment. It has been modeled that patterns of seasonal vector movement and alternate host availability are often dictated by precipitation regimes (van Munster et al. 2017; Jones 2016). Hidden reservoirs and seed transmission. The seed-borne infection is likely to occur under environmental conditions during reproductive stages. The seed infection under stress can form silent reservoirs that overcome quarantine and enable long-range movement of viruses by seed trade—a key route of many vegetable pathogens (Montes and Pagán 2019; Gutierrez-Sanchez et al. 2023).

**Triggers of Outbreaks and Extreme Events** Hail, mechanical damage, and storms can cause wounds and spread infected tissue and vectors on the surfaces, causing a sudden outbreak. Heat waves have the temporary effect of reducing the immunity of plants and altering the behavior of vectors, transforming the latent infection into an overt epidemic. Therefore, climate extremes are like pulses, which elevate background risk into acute epidemics (Miralles et al. 2018).

## 2.4 Case Studies: Climate Change and the Emergence of Viral Diseases in Vegetable Crops

Climate change is reshaping the landscape of plant–virus interactions by altering host physiology, vector ecology, and the environmental envelope that determines where pathogens and their carriers can survive and reproduce. Rising temperatures, changing precipitation patterns (including increased frequency of drought), elevated atmospheric $CO_2$, and altered atmospheric chemistry (for example, elevated ozone) interact with intensive agricultural practices to create new opportunities for viruses to emerge, spread, and cause severe losses in vegetable production (Table 2.1).

**Table 2.1** Key virus-vector pathosystems studied

| Virus name and vector name | Impact of climate change/ stress | Example of research finding | Reference |
|---|---|---|---|
| Cauliflower mosaic virus (CaMV) (*caulimovirus*) and aphids | Enhanced vector transmission under water deficit; variable host responses across genotypes | The rate of aphid transmission was significantly *increased* from water-deprived source plants (by 34%). Water deficit had contrasting effects on CaMV transmission rate and virulence among different *Arabidopsis thaliana* accessions. Viral infection decreased plant performance, and these effects were increased by water deficit | van Munster et al. (2017), Bergès et al. (2021) |
| *Barley yellow dwarf virus species PAV (BYDV-PAV) (luteovirus/polerovirus)* and bird cherry-oat aphid (*Rhopalosiphum padi*) | Complex interactions under combined elevated $CO_2$ and temperature | Combined elevated $CO_2$ (800 ppm) and temperature (22 °C) increased wheat growth and C:N ratio, decreasing *R. padi* fecundity; however, virus infection reversed these plant traits, increasing *R. padi* fecundity. Earlier and more distinct virus symptomatology was highlighted under the future predicted scenario | Moreno-Delafuente et al. (2020), Finlay and Luck (2011) |
| Cereal yellow dwarf virus (CYDV-RPV) (*polerovirus*) and aphids | Enhanced stability and transmission under thermal stress | A naturally occurring, single-amino-acid substitution (P247S) in the P0 protein of CYDV-RPV enhances its protein stability under increasing thermal stress, contributing to increased transmission by aphid vectors and higher disease prevalence | Cai et al. (2023), Cai et al. (2023) |
| Turnip yellows virus (TuYV) (*polerovirus*) and Green peach aphid (*Myzus persicae*) | Increased infection risk due to milder seasons | Milder autumns provide prolonged active periods for green peach aphids (*Myzus persicae*), increasing the risk for infection of winter oilseed rape (OSR) with TuYV | Puthanveed et al. (2023) |
| Beet curly top virus (BCTV) (*begomovirus*) and beet leafhopper (*Circulifer tenellus*) | Earlier migration leads to outbreaks (phenology shift) | Early spring migration of the beet leafhopper is the primary factor contributing to BCTV outbreaks, a phenomenon associated with vegetation greenness metrics | Lee et al. (2025) |

(continued)

**Table 2.1** (continued)

| Virus name and vector name | Impact of climate change/ stress | Example of research finding | Reference |
|---|---|---|---|
| Tomato mosaic virus (ToMV) (*tobamovirus*) and mechanical/seed (not insect vector) | Reduced virus severity under elevated $CO_2$ | Elevated $CO_2$ significantly reduced the disease index and viral RNA expression of ToMV in tomato plants | Marino et al. (2025) |
| Pepper mild mottle virus (PMMoV) (*tobamovirus*) and contaminated water/ feces (not insect vector, used as environmental tracer) | Increased importance due to water reuse in the context of climate change | PMMoV is identified as a reliable domestic wastewater tracer and an enteric virus surrogate, necessary as wastewater reuse becomes more prevalent due to climate change. Detected in non-native winter crane flies in Antarctica, suggesting spread potential due to climate change and human activity | Symonds et al. (2018) |
| Zucchini yellow mosaic virus (ZYMV) (*potyvirus*) and aphids | Increased incidence associated with warming temperatures and vector movement | New ZYMV isolates in Korea are linked to warming temperatures and vector movement caused by climate change. ZYMV, often found with other potyviruses like Moroccan watermelon mosaic virus (MWMV), poses an expanding threat facilitated by aphid vectors | Kim et al. (2019), Jagunic et al. (2025) |

### 2.4.1 Tomato Yellow Leaf Curl Virus (TYLCV) in Tomato

Tomato yellow leaf curl virus (TYLCV), an economically devastating begomovirus transmitted exclusively by the whitefly *Bemisia tabaci*, exemplifies how climate-driven changes to both host and vector biology can accelerate disease impact. Several complementary mechanisms have been proposed and documented. First, rising temperatures threaten the durability of host resistance: resistance conferred by the classical Ty-1 locus in tomato can be compromised under high-temperature regimes, undermining an important component of integrated TYLCV management (Koeda et al. 2024). Second, TYLCV infection itself interacts antagonistically with the plant's heat-stress response. Experimental studies show that infection down-regulates key heat shock transcription factors, notably HSFA2 and HSFB1, and their downstream targets; viral proteins (including coat protein candidates) can sequester HSFA2 in nuclei, thereby reducing the plant's ability to mount protective heat-responsive transcriptional programs and worsening symptom severity under heat stress (Ghandi et al. 2016). Third, vector ecology responds sensitively to temperature change: climate models and distributional projections indicate a substantial expansion of high-risk areas for *B. tabaci* biotypes (for example, projections to

2050 show up to ~180% greater area at high risk in some tomato-growing regions), increasing the geographic footprint where TYLCV can establish (Ramos et al. 2018, 2019). Fourth, interactions among abiotic stresses and viral infection further amplify risk. Elevated ozone and TYLCV infection together can change plant nutritional and defense chemistry, increasing free amino acids while weakening the jasmonic-acid-dependent defense pathway, and thereby raising whitefly fitness and fecundity on infected plants (Cui et al. 2019). Paradoxically, some begomovirus–host interactions can also enhance host tolerance to particular abiotic stresses: TYLCV and closely related begomoviruses (for example, TYLCSV) have been reported to stabilize osmoprotectants (soluble sugars, amino acids, and protective proteins) under drought, delaying wilting and enabling faster recovery after rehydration in infected tomato plants (Shteinberg et al. 2021; Botto et al. 2023). Taken together, these studies underscore a complex picture in which warming both increases the probability of virus spread via vectors and interferes with host resistance and stress responses, while viral infection can modify host physiology in ways that sometimes reduce, and sometimes exacerbate, crop losses, depending on the specific stress and management context (Koeda et al. 2024; Ghandi et al. 2016; Ramos et al. 2018, 2019; Cui et al. 2019; Shteinberg et al. 2021; Botto et al. 2023).

### 2.4.2 Tomato Spotted Wilt Virus (TSWV) in Tomato

Tomato spotted wilt virus (TSWV), a bunyavirus transmitted by thrips (principally Frankliniella occidentalis), is a salient example of an arthropod-borne pathogen whose epidemiology is strongly shaped by the movement and phenology of its insect vectors. Rising temperatures and altered precipitation regimes facilitate vector migration and establishment in previously unsuitable regions, making TSWV an increasingly global threat (Walter and Barr 2011). Beyond vector movements, high-temperature stress in the plant root zone can predispose tomato to more severe and earlier expression of TSWV symptoms independent of vector pressure: experiments that raised root zone temperatures (for example, conditions created under black plastic mulch) reported reduced plant growth and yield and earlier symptom onset compared with plants at milder root zone temperatures, suggesting that heat stress can reduce host tolerance or accelerate viral symptom development (Díaz-Pérez et al. 2007). At the same time, climate-sensitivity analyses for F. occidentalis show that temperature and precipitation are central determinants of the thrips' potential geographic distribution and seasonal abundance; projected shifts in these climatic drivers point to expanded areas with high suitability for thrips, which in turn increases the risk of TSWV outbreaks in both open-field and protected production systems (Wang et al. 2023). Thus, TSWV dynamics under climate change are driven by a combination of vector range shifts, altered vector abundance and phenology, and stress-mediated changes in host susceptibility and symptom expression (Walter and Barr 2011; Díaz-Pérez et al. 2007; Wang et al. 2023).

### 2.4.3 Tomato Brown Rugose Fruit Virus (ToBRFV) in Tomato and Pepper

Tomato brown rugose fruit virus (ToBRFV), a *tobamovirus*, has emerged rapidly as a major threat to tomato and pepper production worldwide. Climatic change is thought to be one of several interacting drivers that have facilitated the virus's establishment in regions previously considered unfavorable, alongside trade, movement of contaminated plant material, and changing agronomic practices (Oladokun et al. 2019; Rodríguez-Román et al. 2023). In contemporary intensive systems, the epidemiology of ToBRFV is further complicated by the mechanics of modern protected cultivation: the virus can be transmitted via recirculating hydroponic nutrient solution and other mechanized contact routes common in greenhouse production, meaning that environmental instability and system design become critical in determining outbreak severity. Promising mitigation research has identified cold-plasma ozone treatment as an effective viral inactivation method for water used in recirculating systems, suggesting an actionable control point for producers who face heightened risk from both changing external climates and the infection-friendly conditions of intensive cultivation (Zhou et al. 2024). The ToBRFV case stresses the lesson that climate change interacts with globalization and technology: where environmental shifts expand the window for establishment, modern production practices can then amplify transmission unless specific sanitation and treatment measures are adopted (Oladokun et al. 2019; Rodríguez-Román et al. 2023).

### 2.4.4 Viral Threats in Cucurbit Crops (Cucumber, Melon, Squash)

Cucurbit production is increasingly affected by an expanding suite of viral pathogens, whose distribution and impact are influenced by climate-driven vector movement, altered host physiology under elevated $CO_2$, and compounded abiotic stresses. The recent detection of Moroccan watermelon mosaic virus (MWMV) in northern Europe (for example, a first report from Croatia), identified through high-throughput sequencing of symptomatic cucurbits, highlights how previously tropical or subtropical potyviruses can appear at higher latitudes, a pattern consistent with climate-mediated northward range shifts of aphid vectors and changing suitability of new environments (Jagunic et al. 2025). Conversely, not all climate factors uniformly worsen viral infection: controlled environment experiments show that elevated atmospheric $CO_2$ (750 ppm vs. ~390 ppm) can reduce accumulation and severity of cucumber mosaic virus (CMV) in tobacco (a common proxy host for vegetable pathosystems); the mechanism involves $CO_2$-driven increases in the host regulator rgs-CaM, which interferes with the CMV silencing suppressor VSR-2b, thereby weakening viral suppression of host RNA interference pathways and lowering infection efficiency (Guo et al. 2021). Yet, combined stresses remain problematic: in melon, simultaneous infection by CMV and cucurbit aphid-borne yellows virus (CABYV), when coupled with drought, produced synergistic reductions in marketable fruit yield, demonstrating that the net effect of multiple simultaneous stresses can be more severe than any single factor alone (Jimenez et al. 2024). These cases show that climate change effects are virus- and context-specific, sometimes facilitating emergence and spread, sometimes altering host resistance or tolerance in unexpected directions, and that multi-factor experiments are essential to predict crop outcomes under future climates (Jagunic et al. 2025; Guo et al. 2021; Jimenez et al. 2024).

### 2.4.5 Emerging Viral Problems in Potato and Brassicas

Potato and brassica crops illustrate how elevated temperatures and seasonal warming patterns can shift both host susceptibility and vector phenology with large epidemiological consequences. For potato virus Y (PVY), higher temperatures (for instance, 28 °C vs. 22 °C in controlled comparisons) significantly increase viral accumulation and host susceptibility in susceptible potato cultivars; warming suppresses the expression of host defense genes in cultivars like "Chicago," leading to higher virus titres and greater potential yield loss (Makarova et al. 2018; Glushkevich et al. 2022). Long-term monitoring of aphid vectors in Korea revealed that spring migrations are beginning earlier and total aphid populations have risen substantially over 40 years, correlating with a mean annual temperature increase (Kim and Kwon 2019); such phenological shifts increase the window of exposure for potato to vector-borne viruses. In brassica systems, turnip yellows virus (TuYV) incidence has been linked to milder autumns that prolong periods of vector activity (notably Myzuspersicae), producing very high infection levels in winter oilseed rape and other brassicas, surveys in parts of Sweden found TuYV incidences averaging around 75%, a striking signal that seasonal warming can substantially raise epidemic risk (Puthanveed et al. 2023). Together, these studies emphasize that warming can both weaken plant defenses and reorganize vector life cycles, with consequences that are amplified in cropping systems where vectors are abundant and crop windows overlap with prolonged vector activity (Makarova et al. 2018; Glushkevich et al. 2022; Kim and Kwon 2019; Puthanveed et al. 2023).

## 2.5 Future Outlook and Conclusion

The issue of whether research, policy, and practice can be integrated into workable systems to forecast change and deliver knowledge into action on the spot will be the limiting factor to the health of vegetable crops in a warming world. Climate change alters the time and space of vectors, alters the susceptibility of the host, and transforms the social and economic environment in which farmers decide on what to grow and the ways to protect it. The fact implies that it is necessary to shift to proactive, combined plant-health models that connect climate-sensitive systems with operational advisory services in such a manner that the intervention can be made in a timely and context-dependent manner. That shift revolves around the working surveillance, diagnostics, and early warning. The models are only useful when reliable data is fed in time. Practical surveillance should comprise a combination of sentinel field surveillance, low-cost diagnostics (lateral-flow, LAMP), and regional lab support with clearly defined data pipelines. With simple bioinformatics and standards, it is possible to identify strains very quickly by using portable sequencing. Farmer reporting and trap networks are both important; machine learning is capable of identifying suspicious trends, but results need to be translated into simple, practical advice to growers. Remote sensing and drones can offer landscape-scale images of stress and hot spots, but will have to be locally validated and linked with ground-level verification. Technical solutions are evaluated in social, economic, governance, and ethical contexts on whether they work on the ground or not. The adoption should be researched under realistic conditions of experimentation, longitudinal follow-up, incentives, training, and institutions (cooperatives, seed certification) that promote long-term adoption. Gender issues: women tend to be at the center of the vegetable production process but lack access to input and information equally; the interventions should not contribute to the aggravation of the differences. One of the policies to consider is control of emerging technologies, ownership of data in common surveillance mechanisms, and the law of collective measures in certain areas. Privacy, ownership of data, and fair reporting are ethical concerns. There is a need to invest and to build up capacity. Labs and extension services ought to be more prepared; staff educated in unsophisticated digital instruments; and data portals opened enough to speed up investigation with data protected by metadata principles and regional information trusts. The applied priorities involve modular models that relate the characteristics of vectors, crop responses, and landscape maps; integrated surveillance that integrates traps, rapid diagnostics, remote sensing, and farmer reporting; multi-site field trials to test interventions in realistic climates; international consortia that share data and germplasm; and socio-economic trials that develop workable incentives while addressing equity concerns. Climate change is already transforming vegetable production by altering the interaction of the crops, pathogens, and vectors and by raising the susceptibility of the seed systems and livelihoods. The warmer temperatures, the changes in the rain, high levels of $CO_2$, extreme events, intercontinental trade, and the adjusted agricultural use enhance the risk of spreading the virus, mixing, and sudden outbreaks. The tools available are integrated surveillance, climate-conscious models explicitly addressing uncertainty, actionable early warnings, sustainable genetic resilience, and concerted landscape actions. It is an urgent investment in integrated science, policy, and practice that will be steered by adaptive governance and by focusing on equity, which will define whether vegetable systems can adapt and whether risks to yield, nutrition, and livelihoods are mitigated and not intensified.

## References

Bergès SE, Vile D, Vazquez-Rovere C, Blanc S, Yvon M, Bédiée A, Rolland G, Dauzat M, van Munster M (2018) Interactions between drought and plant genotype change epidemiological traits of Cauliflower mosaic virus. Front Plant Sci 9:703. https://doi.org/10.3389/fpls.2018.00703

Bergès SE, Vile D, Yvon M, Masclef D, Dauzat M, van Munster M (2021) Water deficit changes the relationships between epidemiological traits of Cauliflower mosaic virusacross diverse *Arabidopsis thaliana* accessions. Sci Rep 11(1):24103. https://doi.org/10.1038/s41598-021-03462-x

Botto CS, Matic S, Moine A, Chitarra W, Nerva L, D'Errico C, Pagliarani C, Noris E (2023) Tomato yellow leaf curl Sardinia virus increases drought tolerance of tomato. Int J Mol Sci 24(3):2893. https://doi.org/10.3390/ijms24032893

Cai LN, Dang MQ, Yang YW, Mei RX, Li F, Tao XR, Palukaitis P, Beckett R, Miller WA, Gray SM, Xu Y (2023) Naturally occurring substitution of an amino acid in a plant virus gene-silencing suppressor enhances viral adaptation to increasing thermal stress. PLoS Pathog 19(4):e1011301. https://doi.org/10.1371/journal.ppat.1011301

Canto T, Aranda MA, Fereres A (2008) Climate change effects on physiology and population processes of hosts and vectors that influence the spread of hemipteranborne plant viruses. Wiley. https://doi.org/10.1111/j.1365-2486.2008.01820.x

Chandi RS, Kataria SK, Kaur J (2018) Arthropods as vector of plant pathogens viz-a-viz their management. Int J Curr Microbiol App Sci. https://doi.org/10.20546/ijcmas.2018.708.415

Chavan DD, Sarkar M, Majumdar A, Mondal F, Babu YM, Lal SK, Mandal B, Kumar R, Roy A (2025) Enhanced expression and interaction of GmRDR1 and GmSGS3 proteins in resistant soybean cultivars synergistically regulate antiviral defense against mungbean yellow mosaic India virus. Plant Biol. https://doi.org/10.1111/plb.70078

Cui HY, Sun YC, Zhao ZH, Zhang YJ (2019) The combined effect of elevated O3 levels and TYLCV infection increases the fitness of *Bemisia tabaci* Mediterranean on tomato plants. Environ Entomol 48(6):1425–1433. https://doi.org/10.1093/ee/nvz113

Díaz-Pérez JC, Gitaitis R, Mandal B (2007) Effects of plastic mulches on root zone temperature and on the manifestation of tomato spotted wilt symptoms and yield of tomato. Sci Hortic 114(2):90–95. https://doi.org/10.1016/j.scienta.2007.05.013

Falla EK, Cunniffe NJ (2024) Why aphid virus retention needs more attention: modelling aphid behaviour and virus manipulation in non-persistent plant virus transmission. Public Libr Sci. https://doi.org/10.1371/journal.pcbi.1012479

Finlay KJ, Luck JE (2011) Response of the bird cherry-oat aphid (*Rhopalosiphumpadi*) to climate change in relation to its pest status, vectoring potential and function in a crop-vector-virus pathosystem. Agric Ecosyst Environ 144(1):405–421. https://doi.org/10.1016/j.agee.2011.08.011

Ghandi A, Adi M, Lilia F, Linoy A, Or R, Mikhail K, Mouhammad Z, Henryk C, Rena G (2016) Tomato yellow leaf curl virus infection mitigates the heat stress response of plants grown at high temperatures. Sci Rep 6:19715. https://doi.org/10.1038/srep19715

Gilbertson RL, Batuman O, Webster CG, Adkins S (2015) Role of the insect supervectors Bemisia tabaci and Frankliniella occidentalis in the emergence and global spread of plant viruses. Ann Rev Virol 2(1):67–93

Glushkevich A, Spechenkova N, Fesenko I, Knyazev A, Samarskaya V, Kalinina NO, Taliansky M, Love AJ (2022) Transcriptomic reprogramming, alternative splicing and RNA methylation in potato (*Solanum tuberosum* L.) plants in response to Potato virus Y infection. Plants 11(5):635. https://doi.org/10.3390/plants11050635

Guo HJ, Ge PP, Tong JH, Zhang YJ, Peng XH, Zhao ZH, Ge F, Sun YC (2021) Elevated carbon dioxide levels decreases Cucumber mosaic virus accumulation in correlation with greater accumulation of rgs-CaM, an inhibitor of a viral suppressor of RNAi. Plants 10(1):59. https://doi.org/10.3390/plants10010059

Gutierrez-Sanchez A, Cobos A, Lopez-Herranz M, Canto T, Pagan I (2023) Environmental conditions modulate plant virus vertical transmission and survival of infected seeds. Phytopathology 113(9):1773–1787. https://doi.org/10.1094/PHYTO-11-22-0448-V

Jagunic M, Grbin D, Marohnic M, Novak A, Cajkulic AM, Skoric D (2025) Severely symptomatic cucurbits in Croatia dominantly harbor a complex of potyviruses including the emerging Moroccan watermelon mosaic virus. Agronomy 15(7):1613. https://doi.org/10.3390/agronomy15071613

Jeevalatha A, Vanishree G, Siddappa S, Kumar R, Kaundal P, Kumar A, Chakrabarti SK (2021) Molecular characterization and infectivity analysis of tomato leaf curl New Delhi virus isolates infecting potato. 3 Biotech 11:203. https://doi.org/10.1007/s13205-021-02752-5

Jenkins A, Capon A, Negin J, Marais BJ, Sorrell TC, Parkes MW, Horwitz P (2018) Watersheds in planetary health research and action. Elsevier BV. https://doi.org/10.1016/s2542-5196(18)30228-6

Jimenez J, Sadras VO, Espaillat N, Moreno A, Fereres A (2024) Interplay between drought and plant viruses co-infecting melon plants. Sci Rep 14(1):15833. https://doi.org/10.1038/s41598-024-66344-y

Jones RAC (2009) Plant virus emergence and evolution: origins, new encounter scenarios, factors driving emergence, effects of changing world conditions, and prospects for control. Virus Res 141(2):113–130. https://doi.org/10.1016/j.virusres.2008.07.028

Jones RAC (2016) Future scenarios for plant virus pathogens as climate change progresses. Adv Virus Res 95:87–147. https://doi.org/10.1016/bs.aivir.2016.02.004

Jones AD, Ejeta G (2015) A new global agenda for nutrition and health: the importance of agriculture and food systems. World Health Organization. https://doi.org/10.2471/blt.15.164509

Kenyon L, Kumar S, Tsai WS, Hughes JD (2014) Virus diseases of peppers (Capsicum spp.) and their control. In: Control of plant virus diseases — seed-propagated crops, vol 90, pp 297–354. https://doi.org/10.1016/B978-0-12-801246-8.00006-8

Kim J, Kwon M (2019) Population dynamics of aphid species in Korean seed potato cultivation area over four decades. Entomol Res 49(4):179–184. https://doi.org/10.1111/1748-5967.12348

Kim B, Cho IS, Kim IH, Choi GW, Ju HK, Hu WX, Oh JP, Kim JK, Seo E, Domier LL, Hammond J, Lim HS (2019) Length of poly(A) tail affects transcript infectivity of three ZYMV symptom variants differing at only five amino acid positions. J Plant Pathol 101(4):1187–1193. https://doi.org/10.1007/s42161-019-00316-4

Koeda S, Fortes IM, Rodrguez-Lpez M, Fernndez-Muoz R, Moriones E (2024) Resistance to the insect vector *Bemisia tabaci* enhances the robustness and durability of tomato yellow leaf curl virus resistance conferred by ty-1. Plant Dis. https://doi.org/10.1094/PDIS-06-24-1281-RE

Kumar R, Lal MK, Tiwari RK, Kumar A, Behera B, Nayak L, Dash GK, Sahoo SK, Jena J, Lal P, Altaf MA (2023) Traditional and emerging climate-resilient agricultural practices for enhancing food production and nutritional quality. In: Aftab T (ed) New Frontiers in plant-environment interactions: innovative technologies and developments. Environmental Science and Engineering. Springer, Cham, pp 551–570. https://doi.org/10.1007/978-3-031-43729-8_21

Lee H, Wintermantel WM, Trumble JT, Nansen C (2025) Timing matters: remotely sensed vegetation greenness can predict insect vector migration and therefore outbreaks of curly top disease. J Pest Sci 98(1):607–617. https://doi.org/10.1007/s10340-024-01771-4

Makarova S, Makhotenko A, Spechenkova N, Love AJ, Kalinina NO, Taliansky M (2018) Interactive responses of potato (*Solanum tuberosum* L.) plants to heat stress and infection with potato virus Y. Front Microbiol 9:2582. https://doi.org/10.3389/fmicb.2018.02582

Marino G, Carli A, Raschi A, Centritto M, Noris E, D'Errico C, Matić S (2025) Elevated Atmospheric $CO_2$ Concentrations Reduce Tomato Mosaic Virus Severity in Tomato Plants. Plants 14(5):811

Miralles DG, Gentine P, Seneviratne SI, Teuling AJ (2018) Land–atmospheric feedbacks during droughts and heatwaves: state of the science and current challenges. Ann N Y Acad Sci 1436(1):19–35

Montes N, Pagán I (2019) Light intensity modulates the efficiency of virus seed transmission through modifications of plant tolerance. Plants 8(9):304. https://doi.org/10.3390/plants8090304

Moreno-Delafuente A, Viñuela E, Fereres A, Medina P, Trebicki P (2020) Simultaneous increase in $CO_2$ and temperature alters wheat growth and aphid performance differently depending on virus infection. Insects 11(8):459. https://doi.org/10.3390/insects11080459

Naga KC, Buckseth T, Subhash S, Bairwa A, Verma G, Kumar R, Malik K, Sharma S, Chakrabarti SK (2020) Transmission efficiency of potato leafroll virus (PLRV) by potato aphid Aulacorthum solani and green peach aphid Myzus persicae. Indian J Entomol 71:68–71. https://doi.org/10.5958/0974-8172.2020.00017.6

Naga KC, Sundaresha S, Kumar R, Tiwari RK, Subhash S, Verma G, Buckseth T, Bairwa A, Sharma S, Katare S, Srivastava RM (2021) A new record of Asia II 5 genetic group of Bemisia tabaci (Gennadius) in the major potato growing areas of India and its relationship with tomato leaf curl New Delhi virus infecting potato. 3 Biotech 11:421. https://doi.org/10.1007/s13205-021-02966-7

Oladokun JO, Halabi MH, Barua P, Nath PD (2019) Tomato brown rugose fruit disease: current distribution, knowledge and future prospects. Plant Pathol 68(9):1579–1586. https://doi.org/10.1111/ppa.13096

Patil B, Crosby K, Byrne D, Hirschi K (2014) The intersection of plant breeding, human health, and nutritional security: lessons learned and future perspectives. None. https://doi.org/10.21273/HORTSCI.49.2.116

Patz JA, Daszak P, Tabor GM, Aguirre AA, Pearl MC, Epstein J, Wolfe N, Kilpatrick AM, Foufopoulos J, Molyneux D, Bradley DJ (2004) Unhealthy landscapes: policy recommendations on land use change and infectious disease emergence. National Institute of Environmental Health Sciences. https://doi.org/10.1289/ehp.6877

Peñalver-Cruz A, Garzo E, Prieto-Ruiz I, Diaz-Carro M, Winters A, Moreno A, Fereres A (2020) Feeding behavior, life history, and virus transmission ability of *Bemisiatabaci* Mediterranean species (Hemiptera: Aleyrodidae) under elevated CO2. Insect Sci 27(3):558–570. https://doi.org/10.1111/1744-7917.12661

Pidikiti P, Sudhakar CS, Singh H, Kumar A, Meshram S (2023) A review on molecular aspects of virus-vector relationship to the aphids. Applied and Natural Science Foundation. https://doi.org/10.31018/jans.v15i2.4391

Prasad A, Sett S, Prasad M (2022) Plant-virus-abiotic stress interactions: a complex interplay. Environ Exp Bot 199:104869. https://doi.org/10.1016/j.envexpbot.2022.104869

Prasch CM, Sonnewald U (2013) Simultaneous application of heat, drought, and virus to *Arabidopsis* plants reveals significant shifts in signaling networks. Plant Physiol 162(4):1849–1866. https://doi.org/10.1104/pp.113.221044

Puthanveed V, Singh K, Poimenopoulou E, Pettersson J, Siddique A, Kvarnheden A (2023) Milder autumns may increase risk for infection of crops with turnip yellows virus. Phytopathology 113(9):1788–1798. https://doi.org/10.1094/PHYTO-11-22-0446-V

Pyott DE, Sheehan E, Molnar A (2016) Engineering of CRISPR/Cas9-mediated potyvirus resistance in transgene-free Arabidopsis plants. Mol Plant Pathol 17(8):1276–1288

Ramos RS, Kumar L, Shabani F, Picanço MC (2018) Mapping global risk levels of *Bemisia tabaci* in areas of suitability for open field tomato cultivation under current and future climates. PLoS One 13(6):e0198925. https://doi.org/10.1371/journal.pone.0198925

Ramos RS, Kumar L, Shabani F, Picanço MC (2019) Risk of spread of tomato yellow leaf curl virus (TYLCV) in tomato crops under various climate change scenarios. Agric Syst 173:524–535. https://doi.org/10.1016/j.agsy.2019.03.020

Ratnadass A, Fernandes P, Avelino J, Habib R (2012) Plant species diversity for sustainable management of crop pests and diseases in agroecosystems: a review. Agron Sustain Dev 32(1):273–303

Rodríguez-Román E, León Y, Perez Y, Amaya P, Mejías A, Montilla JO, Ortega R, Zambrano K, Olivares BO, Marys E (2023) Peppers under siege: revealing the prevalence of viruses and discovery of a novel potyvirus species in Venezuela. Sustainability 15(20):14825. https://doi.org/10.3390/su152014825

Roos J, Hopkins R, Kvarnheden A, Dixelius C (2011) The impact of global warming on plant diseases and insect vectors in Sweden. Eur J Plant Pathol 129(1):9–19. https://doi.org/10.1007/s10658-010-9692-z

Shi X, Zhang Z, Zhang C, Zhou X, Zhang D, Liu Y (2021) The molecular mechanism of efficient transmission of plant viruses in variable virus-vector-plant interactions. KeAi. https://doi.org/10.1016/j.hpj.2021.04.006

Shteinberg M, Mishra R, Anfoka G, Altaleb M, Brotman Y, Moshelion M, Gorovits R, Czosnek H (2021) Tomato yellow leaf curl virus (TYLCV) promotes plant tolerance to drought. Cells 10(11):2875. https://doi.org/10.3390/cells10112875

Sundaresha S, Jeevalatha A, Kumar R, Sood S, Sharma S, Bhardwaj V, Singh BP, Chakrabarti SK (2022) RNA interference: a versatile tool to augment plant protection strategies in potato. In: Chakrabarti SK, Sharma S, Shah MA (eds) Sustainable management of potato pests and diseases. Springer, Singapore, pp 369–388. https://doi.org/10.1007/978-981-16-7695-6_15

Sundaresha S, Tiwari JK, Bhardwaj V, Sharma S, Jeevalatha A, Kumar R, Sagar V, Singh BP, Chakrabarti SK, Kumar M (2024) Genetic engineering in potato: status and future perspectives. In: Kalia P (ed) Genetic engineering of vegetable crops. CABI, Nosworthy Way, Wallingford, pp 133–140. https://doi.org/10.1079/9781800627123.0008

Symonds EM, Nguyen KH, Harwood VJ, Breitbart M (2018) Pepper mild mottle virus: a plant pathogen with a greater purpose in (waste)water treatment development and public health management. Water Res 144:1–12. https://doi.org/10.1016/j.watres.2018.06.066

van Munster M, Yvon M, Vile D, Dader B, Fereres A, Blanc S (2017) Water deficit enhances the transmission of plant viruses by insect vectors. PLoS One 12(5):e0174398. https://doi.org/10.1371/journal.pone.0174398

Walter CT, Barr JN (2011) Recent advances in the molecular and cellular biology of bunyaviruses. J Gen Virol 92:2467–2484. https://doi.org/10.1099/vir.0.035105-0

Wang Y, He Y, Liu S, Wang X (2020) Mechanisms of plant virus transmission by the whitefly Bemisia tabaci. None. https://doi.org/10.1360/tb-2019-0854

Wang ZL, Xu DP, Liao WK, Xu Y, Zhuo ZH (2023) Predicting the current and future distributions of *Frankliniella occidentalis* (Pergande) based on the MaxEnt species distribution model. Insects 14(5):458. https://doi.org/10.3390/insects14050458

Wendimu GY, Gurmu AK (2024) Insect vectors of plant viruses: host interactions, their effects, and future opportunities. Adv Agric. https://doi.org/10.1155/aia/6006985

Whitmee S, Haines A, Beyrer C, Boltz F, Capon A, Dias BFDS, Ezeh A, Frumkin H, Gong P, Head P, Horton R, Mace GM, Marten R, Myers SS, Nishtar S, Osofsky SA, Pattanayak SK, Pongsiri MJ, Romanelli C et al (2015) Safeguarding human health in the Anthropocene epoch: report of the Rockefeller Foundation-Lancet Commission on planetary health. Elsevier BV. https://doi.org/10.1016/s0140-6736(15)60901-1

Zhou J, Gilliard A, Ling KS (2024) Tomato brown rugose fruit virus is transmissible through a greenhouse hydroponic system but may be inactivated by cold plasma ozone treatment. Horticulturae 10(4):416. https://doi.org/10.3390/horticulturae10040416

# Molecular Evolution of Vegetable-Infecting Viruses and Implications for Durable Resistance

3

Carla Frieda Pantouw, Purnama Isti Khaerani, and Bernadetta Rina Hastilestari

**Abstract**

Vegetable-infecting viruses represent a persistent, notorious, and fast-evolving threat to global crop production, driven by their high genetic variability and rapid adaptation to host defences. The molecular evolution of these viruses, mainly by mutation, recombination, and host–vector co-adaptation, has caused an emergence of variants that can break resistance, undermining the durability of resistance incorporated into widely cultivated vegetables such as tomato, pepper, and cucurbits. This chapter synthesizes current insights into the evolutionary dynamics of key viral families, including *Potyviridae*, *Geminiviridae*, *Tospoviridae*, and *Tobamoviridae*, with a focus on mechanisms that underlie viral diversification and resistance evasion. Additionally, we examine plant-defence mechanisms ranging from classical R gene-mediated recognition, RNA silencing pathways, and novel genome-editing approaches. By integrating evolutionary insights with advances in breeding and biotechnology, we emphasize options for creating persistent virus resistance, such as gene pyramiding, targeting conserved viral components, and predictive resistance deployment guided by virus monitoring and population genomics. Understanding the interplay between viral evolution and plant immunity is essential for designing robust plant-defence strategies to encounter emerging viral challenges and agroecosystem pressures.

**Keywords**

Vegetable · Virus · Crop protection · Durable resistance

C. F. Pantouw · P. I. Khaerani · B. R. Hastilestari (✉)
National Research and Innovation Agency of Indonesia, Cibinong Bogor, Indonesia
e-mail: bern003@brin.go.id



R. Kumar et al. (eds.), *Vegetable Crops and Viral Pathogens Volume I*, Plant Pathology for Sustainable Agriculture, https://doi.org/10.1007/978-981-95-7265-6_3

## 3.1 Introduction

The global demand for vegetable commodities has been steadily increasing in line with population growth and the rising need for healthy dietary components, including vitamins, minerals, and calories (Devirgiliis et al. 2024; Glabska et al. 2020). Various plant organs are consumed as vegetables, such as flower buds, fruits, leaves, stems, and tubers (Thomas et al. 2024). Vegetables consist of various genera and species. Some legume crops like cowpea (*Vigna unguiculata*), soybean (*Glycine* sp.), and common bean (Phaseolus) are categorized as vegetables as they are consumed at the immature stage, including the leaves and green pods (Singh et al. 2020). The most widely cultivated vegetables are allium, chillies, cucurbits, and tomatoes (Schreinemachers et al. 2018). According to Statistics of Agriculture (Statista 2023), among those vegetables, tomatoes represent the most widely produced vegetable worldwide, with an estimated production of 192.32 million tonnes. Onions follow in second place with about 111 million tonnes, while cucumbers and gherkins rank third at 98 million tonnes, indicating strong demand for both fresh consumption and pickling. Cabbages, including varieties such as Chinese cabbage, contribute approximately 74 million tonnes, and eggplants (aubergines) are positioned among the five main vegetables produced (FAO 2024). Global vegetable producers are mainly from China, Turkey, India, and the United States (Thomas et al. 2024). Global vegetable production is negatively affected by abiotic and biotic stresses (Savary et al. 2017; Tabasum et al. 2025). Among various biological stresses, viral diseases remain one of the most critical constraints. Unlike fungi and bacteria, plant viruses cannot be effectively managed through conventional physical or chemical strategies due to their biological characteristics. Furthermore, the management of virus-transmitting vectors such as insects, fungi, and nematodes is often limited and unsustainable when relying only on chemical approaches. Consequently, the creation and utilization of virus-free cultivars represent a highly practical, durable, and environmentally responsible strategy for viral disease control. Tomato plants, the most cultivated vegetable, are vulnerable to a broad spectrum of viral pathogens. Recent studies estimate over 180 viral species infecting tomatoes, including genera such as *Begomovirus*, *Orthotospovirus*, *Tobamovirus*, *Potyvirus*, and *Crinivirus* (Shahriari et al. 2023). Besides being transmitted by vectors, seeds are also media for virus transmission (Dall et al. 2023). Therefore, global seed trades have induced a higher risk of virus and vector spread in new plantations. New environmental conditions may alter the virus or host genome and subsequently enhance the severity of viral pathogenicity. Upon virus infection, the plant activates its defence strategies to counter the attacks and ensure plant survival (Khan et al. 2025). Plants can recognize pathogen invasion and activate a cascade of defence responses, extending from physical and biochemical defences through activation of complex signalling networks to restrict virus proliferation and movement (Kaur et al. 2022). In addition, climate change phenomena have affected virus transmission in plants by altering both the physiology of hosts and the biology of vectors (Hastilestari 2021; Hanssen et al. 2010; Pantouw et al. 2022). Elevated temperatures can increase host insect reproduction, survival, and biting frequency, which subsequently accelerates virus spread (Hossain et al. 2024). Drought may affect plant susceptibility by altering biochemical activities, weakening defences, and promoting viral transmission (Van Munster 2020; Bahar et al. 2021). Meanwhile, increased levels of atmospheric $CO_2$ and fluctuating thermal patterns influence plant growth and shift the geographical and seasonal ranges of both hosts and vectors (Lahlali et al. 2024). A combination of environmental conditions may affect virus spread, such as the appearance of viral disease in new geographic regions, posing greater management challenges unless strategies adapt accordingly (Hossain et al. 2024). Considering that vegetables are enriched with health-promoting phytochemicals, their widespread production and increased quality are beneficial to consumers. Pests and diseases deteriorate crop yield and can be further aggravated by abiotic factors that are unfavourable for plant growth. The improvement of defeating common diseases in vegetables has been reviewed by Thomas et al. (2024). However, recognizing the complexity of managing viral diseases underscores the need for further exploration of recent advances in viral disease management. This chapter recaptures vegetable viral disease, defence strategies to combat virus infection and proliferation, and strategies to achieve durable resistance. Finally, anticipation and resistance breakdown management, as there is increased worldwide seed trade, are also proposed.

## 3.2 Vegetable-Infecting Viruses

A wide diversity of viruses infects vegetables, with some of the most economically important belonging to distinct taxonomic groups. *Geminiviridae* and *Potyviridae* are the largest families of RNA viruses. *Geminiviridae* includes genera of *Curtovirus* and *Begomovirus*, which cause diseases in tomato, beet, and bean that are transmitted by whiteflies. *Potyviridae* comprises the *potyvirus* genus, causing diseases in beet, bean, and tomato. Other plant viruses are classified into several major families (Table 3.1). Classifying vegetable-infecting viruses not only provides insight into their evolutionary relationships but also helps in understanding their epidemiology, host adaptation, and vector interactions. Such frameworks are important in developing effective detection tools, generating resistant plant lines, and integrating disease management practices.

## 3.3 Plant-Defence System

Plants have two main defence systems due to pathogen infection: (i) through structural barriers to limit pathogen infection and spread within the plant, and (ii) through biochemical responses that can inhibit the pathogen replication and movement (Kaur et al. 2022). Physical supports are formed in plant cell walls, accumulated with lignin, callose, and suberin, as well as hydroxyproline-rich glycoproteins that strengthen the cell wall and limit the invading pathogen (Al-Khayri et al. 2023; Pinto et al. 2025). From a cellular perspective, plants employ innate immunity

**Table 3.1** Primary plant virus families

| Family | Genus | Virus | Host | References |
|---|---|---|---|---|
| *Geminiviridae* | *Curtovirus* | Beet curly top virus (BCTV) | *Beta vulgaris* | Strausbaugh et al. (2017), Creamer (2020) |
| | *Begomovirus* | Bean golden mosaic virus (BGMV)<br>Tomato yellow leaf curl virus (TYLCV) | *Phaseolus vulgaris*<br>*S. lycopersicum* | Bonfim et al. (2007) |
| *Potyviridae* | *Potyvirus* | Potato virus Y (PVY) | Potato, tomato | Al-Kuwaiti et al. (2016) |
| | | Bean yellow mosaic virus (BYMV) | *Vicia faba* | |
| | | Zucchini yellow mosaic virus (ZYMV) | Zucchini, cucumber | |
| *Partiviridae* | *Varicosavirus* | Lettuce big-vein associated virus (LBVaV) | (*Lactuca sativa*) | Verbeek et al. (2013) |
| *Rhabdoviridae* | *Cytorhabdovirus* | Lettuce necrotic yellow virus (LNYV) | Lettuce (*Lactuca sativa*) | Dietzgen et al. (2006) |
| | *Nucleorhabdovirus* | Potato yellow dwarf virus (PYDV) | Potato (*Solanum tuberosum*) | Bandyopadhyay et al. (2010) |
| *Bunyaviridae* | *Tospovirus* | Tomato spotted wilt virus (TSWV) | Tomato (*S. lycopersicum*) | Guo et al. (2017) |
| *Bromoviridae* | *Cucumovirus* | Cucumber mosaic virus (CMV) | Cucumber (*Cucumis* spp.) | Roossinck (2002) |
| | *Alfamovirus* | Alfalfa mosaic virus (AMV) | (*Phaseolus vulgaris*) | Mueller et al. (2012) |
| *Closteroviridae* | *Closterovirus* | Beet yellows virus (BYV) | (*Beta vulgaris*), (*Spinacia oleracea*) | Dolja (2003) |
| | *Crinivirus* | Lettuce infectious yellow (LIYV) | Lettuce (*Lactuca sativa*) | Zhang et al. (2020) |
| *Comoviridae* | *Comovirus* | Cowpea mosaic virus (CPMV) | Cowpea (*Vigna unguiculata*) | Martins et al. (2020) |
| *Luteoviridae* | *Enamovirus* | Pea enation mosaic virus (PEMV) | Pea (*Pisum sativum*) | Chaudhary et al. (2025), Doumayrou et al. (2016) |
| *Sequiviridae* | *Sequivirus* | Lettuce mottle virus (LeMov) | Lettuce (*Lactuca sativa*) | Jadao et al. (2007) |
| | | Dandelion yellow mosaic virus (DayMV) | | |
| *Tombusviridae* | *Gammacarmovirus* | Melon necrotic spot virus (MNSV) | Cucumber (*Cucumis* sp.) | Kwak et al. (2015) |
| | *Umbravirus* | Carrot mottle virus (CMotV) | Carrot (*Daucus carota L.*) | Adams et al. (2014) |
| | *Tymovirus* | Turnip yellow mosaic (TYMV) | Cabbage | Schirawski et al. (2000), Maskell et al. (1999) |

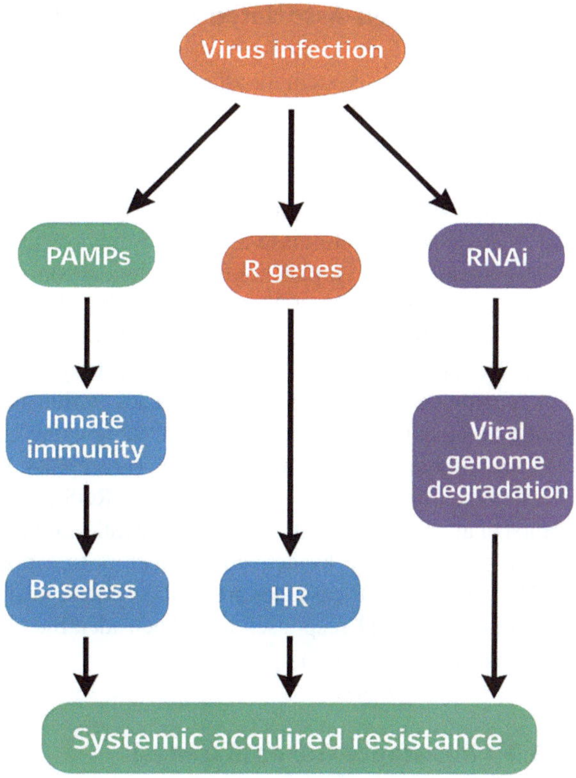


**Fig. 3.1** Plants' defence system upon virus infection

through recognition of viral components as pathogen-associated molecular patterns, known as PAMPs, that are recognized by receptor-like kinases to trigger basal defences (Niehl and Heinlein 2019). A more specific layer of defence is mediated by resistance (R) genes that detect viral effectors and activate localized hypersensitive responses (HR). These processes often determine systemic acquired resistance (SAR) (Han et al. 2024). Furthermore, RNA silencing, a small RNA-based mechanism, affects an important function in defending against viruses by degrading viral genomes or transcripts (Cao et al. 2020; Prasad et al. 2022; Chavan et al. 2025). Furthermore, pathways signalled by hormones, particularly ethylene, jasmonic acid, and salicylic acid, regulate and fine-tune antiviral responses (Wang et al. 2021; Wu et al. 2023). Besides that, immune priming mechanisms, such as those mediated by rgs-CaM, strengthen SAR by counteracting viral suppressors of RNA silencing (Li et al. 2017). Collectively, these interconnected defence strategies form a dynamic antiviral immune system, enabling plants to cope with continuous viral pressure (Fig. 3.1).

### 3.3.1 Pathogen-Associated Molecular Patterns (PAMPs)

The plant immune system is affected by pattern recognition receptors (PRRs). These receptors will recognize PAMPs, which subsequently activate PAMP-triggered immunity (PTI) to limit pathogen entry and replication (DeFalco and Zipfel 2021). Activation of pathogenesis-related (PR) proteins, including proteinase inhibitors, will deactivate pathogen proteolytic enzymes by secreting enzymes like β-1,3-glucanase and chitinase that destroy microbial cell walls (Jain and Khurana 2018), synthesizing isoflavonoids and phytoalexin to block the pathogen growth (Dos Santos and Franco 2023). Finally, the invaded cells activate programmed cell death in which necrotic lesions are formed at the infection site (Jain and Khurana 2018; DeFalco and Zipfel 2021). This necrosis system isolates the pathogen, decelerates pathogen development, and spreads within the plant. In Arabidopsis, for example, receptor FLS 2 is capable of recognizing bacterial flagellin (flg22) and subsequently develops complexes with co-receptors, e.g. BAK1, to initiate downstream signalling features through kinases, ROS production, and MAPK cascades (Hurst et al. 2023; Ryu et al. 2025). Meanwhile, at the pathogen's site, they counteract the plant defence by activating effectors to suppress or avoid PTI (Yu et al. 2024). Antiviral PRR cascades are well described in animals (Majzoub et al. 2019), but comparable PTI responses have been identified only recently (Hudson et al. 2024; Naveed et al. 2020).

### 3.3.2 Resistance Genes (R Genes)

The relation between pathogens and host plants is determined by pathogen–host compatibility. Pathogens, such as viruses, carry avirulence (Avr) genes, which encode avirulence proteins (effectors) during pathogen infection (Rahman et al. 2021). These effectors function to assist the pathogen to infect the host as well as to trigger plant-defence signals in plants (Zhang et al. 2022a, b). Meanwhile, in plants, there are resistance (R) genes encoding a protein to specifically recognize the corresponding pathogen Avr protein (Huang 2021). The recognition can take place directly or indirectly via receptor proteins, e.g. NLR proteins (Chen et al. 2022). The gene-for-gene model explains compatibility by predicting that resistance arises only when a pathogen possesses a dominant allele at its avirulence (Avr) locus, as well as the host plant carries a corresponding dominant allele at its resistance (R) locus (Fig. 3.2). Any alternative pairing of dominant and recessive alleles leads to a compatible interaction that permits successful infection. Conversely, when the pathogen's dominant avirulence genes correspond to the plants' dominant resistance genes, the plant initiates a hypersensitive response.

Plant receptors can be various. Nucleotide-binding leucine-rich repeat (NLR) proteins are pivotal for immunity, serving as intracellular receptors that detect the effectors of pathogens and trigger immune responses. Structurally, plant NLRs typically comprise N-terminal coiled-coil (CC) or Toll/interleukin-1 receptor (TIR) domain, and central nucleotide-binding (NB) domain, as well as C-terminal

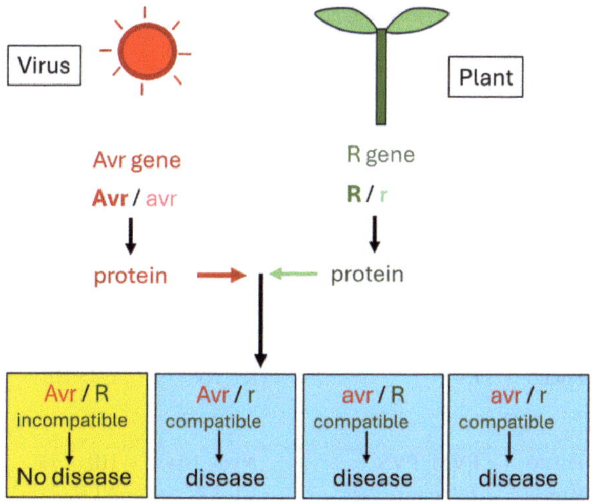


**Fig. 3.2** Virus–host compatibility

leucine-rich repeat (LRR) region, collectively termed CC-NB-LRR or TIR-NB-LRR proteins. Upon pathogen recognition, NLR activation initiates a cascade of defence responses, such as the hypersensitive response (HR), characterized by localized programmed cell death at the infection site to limit pathogen spread. Currently, studies have shed light on the NLR proteins' dynamic localization within plant cells, with certain NLRs accumulating in the endoplasmic reticulum (ER) during infection, indicating a function for the ER in NLR activation and signalling. Additionally, the specificity of NLR-mediated immunity is often allele-dependent, with different NLR alleles conferring resistance to distinct pathogen strains, reflecting their evolutionary adaptation to diverse pathogenic challenges. R genes and their receptors, due to virus infection, have been reported as depicted in Table 3.2.

### 3.3.3 RNA Silencing (RNA Interference, RNAi)

RNA interference (RNAi) represents a highly conserved gene-regulatory mechanism in eukaryotes, in which double-stranded RNA (dsRNA) mediates the sequence-specific suppression of target genes (Svoboda 2020). During this process, long perfect or imperfect dsRNA precursors are enzymatically cleaved into small RNA (sRNA) molecules of approximately 21–26 nucleotides by the Dicer enzyme in animals—including mammals and insects—or by Dicer-like (DCL) proteins in plants (Zapletal et al. 2023). One strand of the resulting sRNA is incorporated into the RNA-induced silencing complex (RISC), enabling the complex, via the core Argonaute (AGO) protein, to recognize and degrade, or inhibit the translation of, complementary RNA sequences (Iwakawa and Tomari 2022). In recent years, RNA silencing has emerged as a central antiviral defence mechanism in both plants and insects. During viral infection, double-stranded RNA (dsRNA) replication intermediates or structured viral RNA regions are processed into primary small interfering

**Table 3.2** Plant virus R genes

| Vegetable | Virus | R Gene(s) | Type/pathway | References |
|---|---|---|---|---|
| *Solanaceae* | | | | |
| Tomato | TMV/ToMV | Tm-1, Tm-2, Tm-22 | NLR, ER/HR | Lanfermeijer et al. (2003), Yamakawa and Nagata (1975) |
| Tomato | TYLCV | Ty-1 to Ty-6 | NLR & other mechanisms | Anbinder et al. (2009), Verlaan et al. (2013) |
| Tomato | TSWV | Sw-5 | CC-NB-LRR | Dianese et al. (2010), Sharma et al. (2022) |
| Potato | PVX | Rx1, Rx2 | NLR, ER | Bendahmane et al. (1999) |
| Potato | PVY | Ry, Ny | ER / HR | Li et al. (2022), Valkonen et al. (1994), Song et al. (2005) |
| Potato | PVM, PVS | Nvm, Nsv | HR | Valkonen et al. (1994) |
| Pepper | Tobamoviruses | L1–L4 | NLR, allele-specific | Kang et al. (2005) |
| Pepper | Potyviruses (PVY, TEV) | pvr1, pvr2, pvr6 | Recessive (eIF4E) | Ruffel et al. (2005) |
| *Cucurbitaceae* | | | | |
| Melon | Cucumber mosaic virus (CMV) | cmv1 | VPS41 protein | Giner et al. (2017), Nieto et al. (2006) |
| Zucchini | Zucchini yellow mosaic virus (ZYMV) | *Zym-0*, *Zym-1*, *Zym-2* | NBS-LRR, MLO-like | Lecoq and Desbiez (2012), Pérez-Moro et al. (2025) |
| *Brassicaceae* | | | | |
| Cabbage | TuMV | TuRB01, retr01 | NLR / recessive | Walsh et al. (2002) |

RNAs (siRNAs). In plants, these primary siRNAs additionally act as primers for RNA-dependent RNA polymerases (RDRs) (Sanan-Mishra et al. 2021). RDR enzymes then convert aberrant target RNAs into new long dsRNA molecules, which are subsequently cleaved into secondary siRNAs (Iwakawa and Tomari 2022). This amplification step strengthens the silencing response and extends siRNA production along the full length of the target RNA, a process known as transitive silencing (Sanan-Mishra et al. 2021). Such siRNA amplification is critical for establishing a strong antiviral RNAi response. In *Arabidopsis*, RDR1, RDR2, and RDR6—along with their orthologs in other species—play pivotal roles in this amplification pathway, and mutants lacking these genes exhibit enhanced susceptibility to numerous plant viruses (Willmann et al. 2011; Hashimoto et al. 2016).

## 3.4 Virus Evolution

Plant viruses generate genetic diversity through mutation, recombination, and reassortment, with RNA viruses showing particularly high variability due to error-prone replication. While DNA viruses also display diversity, their origins are less clear.

The degree of variation differs across virus species, even though mutation rates appear similar. Understanding viral evolution requires consideration of the key factors shaping it.

### 3.4.1 Intrinsic Mutation

Virus genetic variation is determined by recombination, mutation, and reassortment (Butković and González 2022). Mutation rates are influenced by genetic characteristics. A virus with single-stranded RNA (ssRNA) and reverse-transcribing (RT) has a greater mutation rate than a virus with DNA (Sanjuan and Domingo-Calap 2016; Duffy 2018). Faster mutation of RNA or RT viruses because they have no proofreading function during replication, which depends on RNA-dependent RNA polymerases and reverse transcriptases (Marinez et al. 2011). Besides, interaction with host antiviral enzymes and contact with chemicals also influence the mutation (Duffy 2018). RNA viruses, which represent a large portion of plant-infecting viruses, exhibit elevated mutation frequencies resulting from the low fidelity of RNA-dependent RNA polymerases. This leads to the quasispecies structure, fostering genetic diversity that enables viral escape from host immune responses as well as adjustment to new host organisms, for example, TYLCV and CMV (Scholthof 2023).

### 3.4.2 Recombination and Reassortment

The evolution of viruses can be induced by recombination and reassortment. Recombination is a process where two different viruses switch their genome and combine their genetic material. This may develop stronger virulence, wider host range and ability to attack the host defence system (Scholthof 2023). This phenomenon occurs frequently in positive-sense RNA viruses, single-stranded DNA (ssDNA) viruses, and reverse-transcribing viruses, yet it is comparatively uncommon among negative-sense RNA viruses (Martin et al. 2011; Pérez-Losada et al. 2015). Reassortment happens when viruses split their genome into separate pieces instead of one long strand. If there is an infection in the same cell, two segmented viruses can get mixed and assembled into new virus particles (Pérez-Losada et al. 2015). Recombination rarely takes place in segmented RNA viruses (Butković and González 2022). Both processes can increase genetic diversity, although they may generate reduced fitness. The diversity may increase due to environmental factors and trade-offs between replication speed and stability (Manrubia and Lazaro 2006).

### 3.4.3 Host Adaptation

Selective pressures from resistant cultivars and environmental factors drive viral adaptation. Viruses can cope with the harmful effects of mutation through epistasis

(Lalić and Elena 2012). Epistasis refers to interactions between mutations in which the effect of one genetic change depends on the presence of another. This interaction influences fitness and adaptive mechanisms across hosts (Lalić and Elena 2013). Therefore, viruses can cross host barriers, which provoke new diseases (Bernet and Elena 2015). The capacity of viruses and viroids to undergo rapid genetic change poses major challenges for disease management. Beyond migration from other regions, variability within pathogen populations provides the foundation for host shifts, resistance breaking (RB), and alterations in symptom expression and virulence. The rise and decline of particular genotypes are often shaped by natural selection, with variants that produce more viable progeny increasing in frequency. Understanding these evolutionary dynamics is critical for designing durable control strategies, as resistance-breaking pathotypes frequently arise. Recent advances in nucleotide sequencing, along with the generation of genetically engineered plants that resist viruses, have further stimulated interest in viral diversity and evolution. The following section will explore strategies aimed at achieving durable resistance in vegetable crops to secure both yield and quality.

## 3.5 Strategies for Achieving Durable Resistance

As mentioned in the previous part, plant viral disease is not easy to control; it requires management of germplasm as well as disease control. Durable resistance needs true immunity in which plants do not support the multiplication of the infected virus inside the plant system (Jeger 2023). This virus resistance can work either by preventing virus replication within the plant or the virus vector outside the plant. Conventional farmers still have difficulties in detecting and controlling the viral diseases, especially in the early stages, which can significantly reduce yields. This is even worse for tropical regions where plant viral diseases remain prevalent because of abundant active vectors all year (Khakimov et al. 2022; Tatineni and Hein 2023). Accordingly, viral disease vectors have been controlled by synthetic chemical pesticides. Imprudent application of chemical pesticides may lead to disease resistance, pesticide residues in vegetables and the environment and endanger worker health (Pathak et al. 2022; Ul Haq et al. 2024). Application of pesticides to eradicate viral diseases in plants is not effective, as the virus is hardly eradicated with this chemical. To reduce viral infection, the use of virus-resistant cultivars can provide a simple and relatively cheap approach to reducing plant damage due to plant viruses (Anikina et al. 2023; Tatineni and Hein 2023). Therefore, breeding to get a virus-free plant is important. However, traditional breeding for virus-resistant lines through crossing takes time and is a relatively expensive process. The developed lines should be durably resistant because it makes it easier to protect their plants throughout the commercial life of the cultivars. Plant virus populations have a wide range of pathogenicity. Variation in overcoming resistance determines how the virus manages to infect plants and break plant resistance (Rubio et al. 2020; Jeger 2023). On one hand, plants carry resistance (R) genes which either already exist in the plant or emerge through selection after breeding, or widespread

**Table 3.3** Virus-associated markers

| Crop | Virus | Gene/QTL | Ch. No | Markers | Reference |
|---|---|---|---|---|---|
| Tomato | TYLCV | Ty-3 | – | ACY (Indel) | Nevame et al. (2018) |
| | TYLCV | Ty-2 and Ty-4 | 11 and 3 | CAPS marker | Prasanna et al. (2015) |
| | Geminivirus | Ty-2, Ty-3, and Ty-5 | | SNP | Perdanaputra (2017) |
| | TSWV | Sw-5b | | Specific markers used in MAS breeding | Tong et al. (2023) |
| | ToMV | Tm2 | | SNP/SCAR markers | Asadi et al. (2025) |
| Cucumber | CMV | cmv6.1 | 6 | SSR11 | Shi et al. (2018) |
| Pea | Pea common mosaic virus | Mo | – | RFLP | Dirlewanger et al. (1994) |
| Pepper (Capsicum spp.) | *Tobamovirus* | L1–L4 | | | Djian-Caporalino et al. (2007) |

cultivation of the resistant lines. On the other hand, a virus can break plant resistance due to mutation in viral genes, so the plant's defence system cannot be activated, recombination between virus strains like geminiviruses and potyviruses, and virus adaptation. Therefore, to develop virus-resistant vegetable varieties that have higher yield, molecular breeding tools, in addition to conventional breeding, become strategies to achieve durable sustainability. Those strategies are (1) Harnessing wild relatives and landraces gene pyramiding: stacking multiple R genes; (2) Targeting conserved viral regions: via RNAi, CRISPR, or breeding; (3) Marker-assisted and genomic selection; (4) Exploiting wild relatives and landraces; (5) Surveillance and prediction: using virus genome databases and molecular epidemiology; and (6) Engineering virus-resistance pathways: synthetic biology approaches.

### 3.5.1 Harnessing Wild Relatives and Landraces

Breeding is a means to develop a virus-resistant cultivar in which resistance (R) genes are valuable for allele mining and marker development for this program (Asadi et al. 2025). Some of the R genes are mentioned in Table 3.3. However, domestication and subsequent selective breeding have reduced the genetic diversity within widely cultivated varieties (Smýkal et al. 2018). Wild crops usually grew in mixed species ecosystems in which plant pathogens coexisted and coevolved with them (Vavilov, 1992). However, modern domesticated crops are typically cultivated as genetically similar, dense populations, a practice that reduces genetic buffering capacity and increases vulnerability to disease in comparison to the native ones

(Renzi et al. 2022). The diversity reduction constitutes challenges for crop improvement, particularly under changing environmental conditions. Domesticated crops are more susceptible to disease and pests compared to the wild ones. Some domesticated lines have lost their R genes; therefore, crop wild lines may function as a critical source for crop adaptation and evolution (Zhang et al. 2017; Curtin et al. 2022). Wild Solanum lines, like *Solanum peruvianum, S. chilense*, *S. habrochaites*, and *S. pimpinellifolium,* have been reported to have R genes (Scott and Gardner 2007; Laterrot 2000; Foolad and Sharma 2005). *S. chilense* has been identified to exhibit R genes to tomato yellow leaf curl virus (TYLCV). Resistance in *S. chilense* is identified by allelic variants of Ty-1, Ty-4, and Ty-6 (Butterbach et al. 2014; Gill et al. 2019). Some accessions retain resistance despite Ty-1 silencing, indicating the involvement of other loci beyond Ty-1/Ty-3 (Verlaan et al. 2013; Caro et al. 2015; Yan et al. 2018).

### 3.5.2 Molecular Markers: A Bridge Between Wild Relatives, Landraces, and Modern Breeding

Crop wild relatives and landraces harbour reservoirs of genetic pool, for example, resistance to biological and physical stresses (Kapazoglou et al. 2023). Interestingly, direct application of genetic resources in the breeding system is often restricted by some factors, such as linkage drag, long generation times, and complex inheritance patterns. Therefore, markers can be applied to tag or indicate specific traits (Hastilestari et al. 2013; Amiteye 2021). Markers can be characterized into four categories: Morphological, biochemical, cytological and molecular (Nadeem et al. 2018; Adhikari et al. 2017). The first three markers belong to classical markers (Feng et al. 2018), while the molecular marker is a marker that applies nucleotide sequences estimated by polymorphism (Amom and Nongdam 2017). Morphological markers are applied to characterize phenotypic traits by visual characters such as shape, colour, growth habit, and disease response (Khadka et al. 2020). These markers enable the assessment and evaluation of genetic variability among the population and diversity based on a single phenotypic difference (Govindaraj et al. 2015). Biochemical markers are markers in the form of isozymes that are applied based on the electrophoretic mobility of protein, in which these markers are encoded by different genes with the same functions (Singh et al. 2022). These markers show protein and amino acid banding motifs, which are applied to determine the gene frequency, diversity, gene flow, and structure (Sun et al. 2024). Biochemical markers have been used to analyse the genetic difference between domesticated and wild counterparts (Hassan et al. 2021; Simko et al. 2021). Cytological markers can be applied to detect diversity associated with the chromosome morphology, such as chromosomal number, size, shape, order, and sequence specificity, to identify mapping and linkage groups (Feng et al. 2018; Nadeem et al. 2018).

Molecular marker approaches are broadly divided into two classes: (1) non–PCR-based markers and (2) PCR-based markers. Non–PCR-based methods primarily include restriction fragment length polymorphism (RFLP) (Begna 2020; Sun et al. 2024). PCR-based systems, in contrast, comprise techniques such as random amplified polymorphic DNA (RAPD) (Atienzar and Jha 2006), sequence-characterized amplified regions (SCARs) (Michelmore et al. 1991), and amplified fragment length polymorphism (AFLP) (Vos et al. 1995). These markers are fundamental tools for assessing genetic diversity in germplasm collections (Ruiz and Martinez 2005) and are widely employed in DNA fingerprinting, phylogenetic and evolutionary analyses, heterotic grouping, gene tagging, and marker-assisted selection (MAS) (Ansari 2015; Mehra et al. 2024). This marker technique has been applied to identify vegetable varieties and co-dominant markers that are heterozygous for a resistance gene, as mentioned in Table 3.3. These can detect heterozygosity for a resistance gene, so they enable introgression of resistance genes in a shorter time of selection (Naveed et .al. 2020; Sharma et al. 2018). Markers can also be developed by the integration of several resistance genes into genotypes to achieve stable, long-term resistance (Haque et al. 2021) and also a stepping point for cloning the genes to identify their action in vegetables (Peters et al. 2003).

### 3.5.3 Gene Pyramiding: Stacking Multiple R Genes

Resistance cultivars carrying R genes are effective in controlling yield losses due to virus infection. As a virus can rapidly mutate and break such resistance, a single R gene is often not durable in resistance (Jeger 2023; Tatineni and Hein 2023). Plant R genes should be analysed regarding their molecular and biological processes in plant-virus interaction. These genes have been categorized as non-host and host resistance. Non-host resistance refers to plant genotypes that are divided into two major categories (Nohtani et al. 2025): Non-host resistance, unlike host resistance, describes a condition in which every genotype within a plant species inherently blocks infection by a particular virus. Mechanistically, it reflects the absence of any susceptibility alleles or molecular pathways that the virus can exploit, indicating that no genetic polymorphism enabling infection exists within that plant taxon. (Taliansky et al. 2021). For example, in tomato, some cultivars are preferred by customers due to their morphological characteristics such as shape, size, flavour, and skin colour (Athinodorou et al. 2021). Cultivated varieties have undergone extensive domestication, during which many defensive genes and traits against pests and diseases have been lost (Khush 2001; Whitehead et al. 2017). Through artificial selection, breeders have understandably favoured plant genotypes with reduced toxicity and lower levels of secondary metabolites that are harmful to humans (Jadaun et al. 2023), as well as those with desirable fruit characteristics such as improved colour, shape, flavour, or enhanced nutritional quality (Alkanan et al. 2024; Goraya et al. 2025). Concurrently, the intensification of agricultural systems and widespread pesticide use have further shaped plant–pest interactions (Marcelino et al. 2024). Most of these processes are believed to have weakened core plant-defence machinery, producing cultivars with reduced innate immunity and heightened vulnerability to viruses such as ToMV, TYLCV (Pandey and Prasad 2024; Biswas and Mandal 2025), TMV (Yamakawa and Nagata 1975), TSWV,

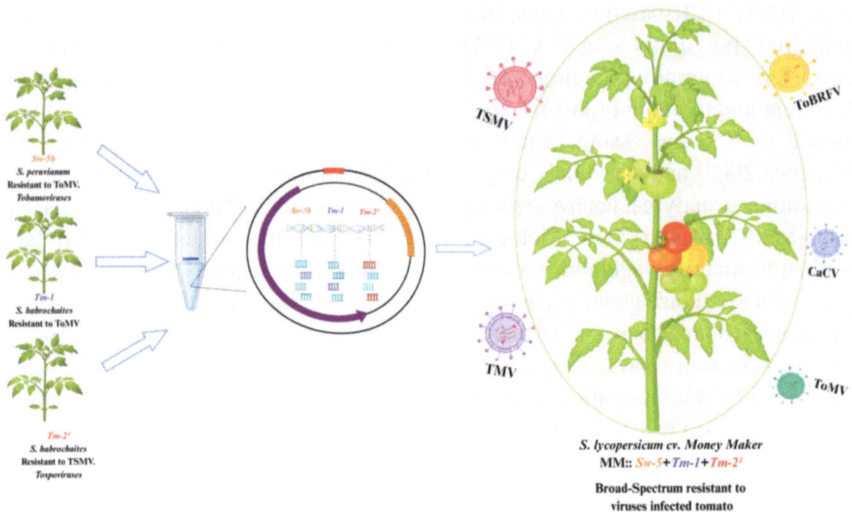


**Fig. 3.3** Pyramiding R genes in *S. Lycopersicum* cv. Money Maker

Capsicum chlorosis virus (Kunkalikar et al. 2018), and ToBRFV (Zhang et al. 2022a, b). Wild Solanum cultivars such as *S. peruvianum* are considered to offer some resistance genes. Of interest in the resistance genes, this species has been identified as an accession with resistance to biological and physical pressures and used as a female parent in breeding due to its unilateral incompatibility (Hogenboom 1972). Many R genes have been detected for some diseases, such as the *Sw-5* gene for TSWV (Picó et al. 2002; Stevens et al. 1992; Paterson and Queitsch 2024), *Ty-5* gene for TYLCD (Anbinder et al. 2009; Picó et al. 2002; Pilowsky and Cohen 2000; Muniyappa et al. 1991), Tm-2 gene for TMV (Yamakawa and Nagata 1975). Another wild tomato, *Solanum habrochaites* has been reported to have a genetic resource for resistance to ToMV, most notably Tm-1, Tm-2, and Tm-22 (Ishibashi et al. 2012**;** Mahfouze et al. 2022). Although those wild potato cultivars are resistant to virus diseases, those cultivars are not favoured by customers (Bhardwaj and Nain 2021). On the other hand, commercial tomato cultivars have a narrow genetic base due to domestication bottlenecks, breeding trade-offs, and uniform cultivation, which reduced their natural virus resistance. Therefore, pyramiding those R genes of wild cultivars offers an opportunity for developing broad-spectrum virus resistance in commercial ones (Fig. 3.3).

### 3.5.4 Targeting Conserved Viral Regions: Via RNAi, CRISPR, or Breeding

Besides the approaches mentioned above, molecular strategies such as RNA interference (RNAi) and CRISPR/Cas genome editing can also be applied to target conserved viral genome regions. RNAi suppresses viral transcripts required for replication or movement, whereas CRISPR can be programmed to disrupt host components indispensable for viral propagation. When integrated with conventional breeding, these tools add an extra defensive layer that reinforces R-gene-mediated resistance (Allemailem et al. 2024; Kalinina et al. 2025). Genome editing enables precise, site-specific DNA alterations. Early systems such as zinc finger nucleases (ZFNs) represented the first major advance in programmable nuclease technologies (Osakabe and Osakabe 2015; Costa et al. 2017), followed by TALENs, which broadened their utility. However, the CRISPR/Cas platform has become the dominant approach because of its simple design, high editing efficiency, and capacity to generate transgene-free crops (He and Zhao 2020; Bhattacharjee et al. 2023). CRISPR-Cas9 now exceeds ZFNs and TALENs in use and has become the leading tool for targeted genome manipulation in plants (Bhardwaj and Nain 2021; Shamshirgaran et al. 2022). The combination of gene pyramiding breeding with targeting of conserved regions creates a layered resistance system that is more difficult for viruses to break, while simultaneously leveraging wild germplasm diversity and precision molecular methods (Mishra et al. 2023; Kaushal et al. 2024). Therefore, vegetable productivity will be enhanced. In field application, vegetable lines carrying gene pyramiding and/or molecular technology should be monitored, so long-term field data on resistance durability can be gathered.

## 3.6 Conclusion

Virus has become a limiting factor in increasing vegetable production and quality. Molecular evolution in vegetable-infecting viruses presents a significant challenge to crop protection. Viral diseases are difficult to manage because they are easily evolving and the ineffectiveness of applying chemicals to eradicate viral diseases, vectors and host adaptation. Therefore, understanding viral evolution is critical for both improving management and prophylactic control of plant diseases and advancing insights into evolutionary dynamics. Advances in genomics, CRISPR gene editing, and molecular breeding provide new opportunities for durable resistance. Continuous surveillance of viral populations and evolution will be critical to adapting resistance strategies. Integrating knowledge of viral evolution with innovative breeding and management strategies will be essential for achieving durable resistance in vegetable crops.

## References

Adams IP, Skelton A, Macarthur R, Hodges T, Hinds H, Flint L et al (2014) Carrot yellow leaf virus is associated with carrot internal necrosis. PLoS One 9(11):e109125

Adhikari S, Saha S, Biswas A, Rana TS, Bandyopadhyay TK, Ghosh P (2017) Application of molecular markers in plant genome analysis: a review. Nucleus 60(3):283–297

Alkanan ZT, Altemimi AB, Hassan NA, Didar Z, Hesarinejad MA, Rahman Salama NA et al (2024) Effects of microwave utilization on the color properties of food: a review. ChemBioEng Rev 11(3):483–494

Al-Khayri JM, Rashmi R, Toppo V, Chole PB, Banadka A, Sudheer WN et al (2023) Plant secondary metabolites: the weapons for biotic stress management. Meta 13(6):716

Al-Kuwaiti NA, Maruthi MN, Seal SE (2016) Molecular characterization of potyviruses infecting potato and vegetables in Iraq. J Plant Pathol 98:603–606

Allemailem KS, Almatroudi A, Rahmani AH, Alrumaihi F, Alradhi AE, Alsubaiyel AM et al (2024) Recent updates of the CRISPR/Cas9 genome editing system: novel approaches to regulate its spatiotemporal control by genetic and physicochemical strategies. Int J Nanomedicine:5335–5363

Amiteye S (2021) Basic concepts and methodologies of DNA marker systems in plant molecular breeding. Heliyon 7(10):e08093

Amom T, Nongdam P (2017) The use of molecular marker methods in plants: a review. Int J Curr Res Rev 9(17):1–7

Anbinder I, Reuveni M, Azari R, Paran I, Nahon S, Shlomo H et al (2009) Molecular dissection of Tomato leaf curl virus resistance in tomato line TY172 derived from *Solanum peruvianum*. Theor Appl Genet 119(3):519–530

Anikina I, Kamarova A, Issayeva K, Issakhanova S, Mustafayeva N, Insebayeva M et al (2023) Plant protection from virus: a review of different approaches. Front Plant Sci 14:1163270

Ansari AM (2015) Molecular markers in vegetable improvement. Horticult Biotechnol Res 1:5–10

Asadi Z, Savadroudbari NS, Amini F, Ramshini H (2025) Marker-assisted selection in segregating populations of tomatoes for resistance to TYLCV, ToMV, and Fusarium wilt. Mol Biol Rep 52(1):107

Athinodorou F, Foukas P, Tsaniklidis G, Kotsiras A, Chrysargyris A, Delis C et al (2021) Morphological diversity, genetic characterization, and phytochemical assessment of the cypriot tomato germplasm. Plants 10(8):1698

Atienzar FA, Jha AN (2006) The random amplified polymorphic DNA (RAPD) assay and related techniques applied to genotoxicity and carcinogenesis studies: a critical review. Mutat Res 613(2–3):76–102

Bahar T, Qureshi AM, Qurashi F, Abid M, Zahra MB, Haider MS (2021) Changes in phytochemical status upon viral infections in plant: A critical review. Phyton 90(1):75

Bandyopadhyay A, Kopperud K, Anderson G, Martin K, Goodin M (2010) An integrated protein localization and interaction map for Potato yellow dwarf virus, type species of the genus Nucleorhabdovirus. Virology 402(1):61–71

Begna T (2020) Challenges and progresses of molecular plant breeding in crop improvement. Int J Res Agric Sci 7(2):110–123

Bendahmane A, Kanyuka K, Baulcombe DC (1999) The Rx gene from potato controls separate virus resistance and cell death responses. Plant Cell 11(5):781–791

Bernet GP, Elena SF (2015) Distribution of mutational fitness effects and of epistasis in the 5′ untranslated region of a plant RNA virus. BMC Evol Biol 15:274. https://doi.org/10.1186/s12862-015-0555-2

Bhardwaj A, Nain V (2021) TALENs—an indispensable tool in the era of CRISPR: a mini review. J Genet Eng Biotechnol 19(1):125

Bhattacharjee S, Bhowmick R, Kant L, Paul K (2023) Strategic transgene-free approaches of CRISPR-based genome editing in plants. Mol Gen Genomics 298(3):507–520

Biswas D, Mandal A (2025) Whitefly transmitted begomovirus infection in plant and host response: a review. Physiol Mol Biol Plants:1–11

Bonfim K, Faria JC, Nogueira EO, Mendes ÉA, Aragão FJ (2007) RNAi-mediated resistance to Bean golden mosaic virus in genetically engineered common bean (*Phaseolus vulgaris*). Mol Plant-Microbe Interact 20(6):717–726

Butković A, González R (2022) A brief view of factors that affect plant virus evolution. Front Virol 2:994057

Butterbach P, Verlaan MG, Dullemans A, Lohuis D, Visser RG, Bai Y, Kormelink R (2014) Tomato yellow leaf curl virus resistance by Ty-1 involves increased cytosine methylation of viral genomes and is compromised by cucumber mosaic virus infection. Proc Natl Acad Sci 111(35):12942–12947

Cao M, Du P, Wang X, Yu YQ, Qiu YH, Li W et al (2020) Virus-derived small interfering RNAs play a role in antiviral defense in plants. Phytopathol Res 2:7

Caro M, Verlaan MG, Julián O, Finkers R, Wolters AMA, Hutton SF, Scott JW, Kormelink R, Visser RG, Díez MJ, Pérez-de-Castro A (2015) Assessing the genetic variation of Ty-1 and Ty-3 alleles conferring resistance to tomato yellow leaf curl virus in a broad tomato germplasm. Mol Breed 35(6):132

Chaudhary A, Naik A, Mistry S (2025) Pea enation mosaic virus. In: Compendium of Phytopathogenic microbes in agro-ecology: Vol. 2 viruses and Viroids. Springer Nature Switzerland, Cham, pp 221–242

Chavan DD, Sarkar M, Majumdar A, Mondal F, Babu YM, Lal SK, Mandal B, Kumar R, Roy A (2025) Enhanced expression and interaction of GmRDR1 and GmSGS3 proteins in resistant soybean cultivars synergistically regulate antiviral defense against mungbean yellow mosaic India virus. Plant Biol. https://doi.org/10.1111/plb.70078

Chen J, Zhang X, Rathjen JP, Dodds PN (2022) Direct recognition of pathogen effectors by plant NLR immune receptors and downstream signalling. Essays Biochem 66(5):471–483

Costa JR, Bejcek BE, McGee JE (2017, November 20) Genome editing using engineered nucleases and their use in genomic screening. In: Markossian S, Grossman A, Baskir H et al (eds) Assay guidance manual [Internet]. Eli Lilly & Company and the National Center for Advancing Translational Sciences, Bethesda

Creamer R (2020) Beet curly top virus transmission, epidemiology, and management. In: Applied plant virology. Academic Press, pp 521–527

Curtin S, Qi Y, Peres LE, Fernie AR, Zsögön A (2022) Pathways to de novo domestication of crop wild relatives. Plant Physiol 188(4):1746–1756

Dall DJ, Lovelock DA, Penrose LD, Constable FE (2023) Prevalences of tobamovirus contamination in seed lots of tomato and capsicum. Viruses 15(4):883

DeFalco TA, Zipfel C (2021) Molecular mechanisms of early plant pattern-triggered immune signaling. Mol Cell 81(17):3449–3467

Devirgiliis C, Guberti E, Mistura L, Raffo A (2024) Effect of fruit and vegetable consumption on human health: an update of the literature. Foods 13(19):3149

Dianese EC, de Fonseca MEN, Goldbach R, Kormelink R, Inoue-Nagata AK, Resende RO, Boiteux LS (2010) Development of a locus-specific, co-dominant SCAR marker for assisted-selection of the Sw-5 (Tospovirus resistance) gene cluster in a wide range of tomato accessions. Mol Breed 25(1):133–142

Dietzgen RG, Callaghan B, Wetzel T, Dale JL (2006) Completion of the genome sequence of Lettuce necrotic yellows virus, type species of the genus Cytorhabdovirus. Virus Res 118(1–2):16–22

Dirlewanger E, Isaac PG, Ranade S, Belajouza M, Cousin R, De Vienne D (1994) Restriction fragment length polymorphism analysis of loci associated with disease resistance genes and developmental traits in *Pisum sativum* L. Theor Appl Genet 88(1):17–27

Djian-Caporalino C, Fazari A, Arguel MJ, Vernié T, VandeCasteele C, Faure I et al (2007) Root-knot nematode (Meloidogyne spp.) Me resistance genes in pepper (Capsicum annuum L.) are clustered on the P9 chromosome. Theor Appl Genet 114(3):473–486

Dolja VV (2003) Beet yellows virus: the importance of being different. Mol Plant Pathol 4(2):91

Dos Santos C, Franco OL (2023) Pathogenesis-related proteins (PRs) with enzyme activity activating plant defense responses. Plants 12(11):2226

Doumayrou J, Sheber M, Bonning BC, Miller WA (2016) Role of pea enation mosaic virus coat protein in the host plant and aphid vector. Viruses 8(11):312

Feng S, Zhu Y, Yu C, Jiao K, Jiang M, Lu J et al (2018) Development of species-specific SCAR markers, based on a SCoT analysis, to authenticate Physalis (Solanaceae) species. Front Genet 9:192

Food and Agriculture Organization of the United Nations (2024) The state of food and agriculture 2024: value-driven transformation of agrifood systems. FAO. https://www.fao.org/publications/fao-flagship-publications/the-state-of-food-and-agriculture/en

Foolad M, Sharma A (2005) Molecular markers as selection tools in tomato breeding. Acta Hortic 695:225–240

Gill U, Scott JW, Shekasteband R, Ogundiwin E, Schuit C, Francis DM et al (2019) Ty-6, a major begomovirus resistance gene on chromosome 10, is effective against Tomato yellow leaf curl virus and Tomato mottle virus. Theor Appl Genet 132(5):1543–1554

Giner A, Pascual L, Bourgeois M, Gyetvai G, Rios P, Picó B et al (2017) A mutation in the melon Vacuolar Protein Sorting 41prevents systemic infection of Cucumber mosaic virus. Sci Rep 7(1):10471

Glabska D, Guzek D, Groele B, Gutkowska K (2020) Fruit and vegetables intake in adolescents and mental health: a systematic review. Rocz Panstw Zakl Hig 71(1):15

Goraya RK, Singla M, Kaura R, Singh CB, Singh A (2025) Exploring the impact of high pressure processing on the characteristics of processed fruit and vegetable products: a comprehensive review. Crit Rev Food Sci Nutr 65(20):3856–3879

Govindaraj M, Vetriventhan M, Srinivasan M (2015) Importance of genetic diversity assessment in crop plants and its recent advances: an overview of its analytical perspectives. Genet Res Int 2015(1):431487

Guo Y, Liu B, Ding Z, Li G, Liu M, Zhu D et al (2017) Distinct mechanism for the formation of the ribonucleoprotein complex of tomato spotted wilt virus. J Virol 91(23):10–1128

Han J, Ding SW, Zhang X (2024) RNA silencing is activated by N gene-mediated hypersensitive response and contributes to systemic acquired resistance against tobacco mosaic virus. Phytopathol Res 6:27

Hanssen IM, Lapidot M, Thomma BP (2010) Emerging viral diseases of tomato crops. Mol Plant-Microbe Interact 23(5):539–548

Haque MA, Rafii MY, Yusoff MM, Ali NS, Yusuff O, Datta DR et al (2021) Recent advances in rice varietal development for durable resistance to biotic and abiotic stresses through marker-assisted gene pyramiding. Sustainability 13(19):10806

Hashimoto M, Neriya Y, Yamaji Y, Namba S (2016) Recessive resistance to plant viruses: potential resistance genes beyond translation initiation factors. Front Microbiol 7:1695

Hassan MN, Mekkawy SA, Mahdy M, Salem KF, Tawfik E (2021) Recent molecular and breeding strategies in lettuce (*Lactuca spp.*). Genet Resour Crop Evol 68(8):3055–3079

Hastilestari BR (2021) Analysis of potential genes for the development of potatoes (Solanum tuberosum) with heat stress-tolerance. Nusantara Science and Technology Proceedings, 8–14

Hastilestari BR, Mudersbach M, Tomala F, Vogt H, Biskupek-Korell B, Van Damme P et al (2013) *Euphorbia tirucalli L.*–Comprehensive characterization of a drought tolerant plant with a potential as biofuel source. PLoS One 8(5):e63501

He Y, Zhao Y (2020) Technological breakthroughs in generating transgene-free and genetically stable CRISPR-edited plants. aBIOTECH 1(1):88–96

Hogenboom NG (1972) Breaking breeding barriers in Lycopersicon. Euphytica 21(2):244–260

Hossain MM, Sultana F, Mostafa M, Ferdus H, Rahman M, Rana JA et al (2024) Plant disease dynamics in a changing climate: impacts, molecular mechanisms, and climate-informed strategies for sustainable management. Discov Agric 2(1):132

Huang C (2021) From player to pawn: viral avirulence factors involved in plant immunity. Viruses 13(4):688

Hudson A, Mullens A, Hind S, Jamann T, Balint-Kurti P (2024) Natural variation in the pattern-triggered immunity response in plants: investigations, implications and applications. Mol Plant Pathol 25(3):e13445

Hurst CH, Turnbull D, Xhelilaj K, Myles S, Pflughaupt RL, Kopischke M et al (2023) S-acylation stabilizes ligand-induced receptor kinase complex formation during plant pattern-triggered immune signaling. Curr Biol 33(8):1588–1596

Ishibashi K, Masuda K, Naito S, Meshi T, Ishikawa M (2012) An inhibitor of viral RNA replication is encoded by a plant resistance gene. Proc Natl Acad Sci 109(9):3731–3736

Iwakawa HO, Tomari Y (2022) Life of RISC: formation, action, and degradation of RNA-induced silencing complex. Mol Cell 82(1):30–43

Jadao AS, Krause-Sakate R, Liberti D, Pavan MA, Echer MM, Svanella-Dumas L et al (2007) Further characterization of two sequiviruses infecting lettuce and development of specific RT-PCR primers. Arch Virol 152(5):999–1007

Jadaun JS, Yadav R, Yadav N, Bansal S, Sangwan NS (2023) Influence of genetics on the secondary metabolites of plants. In: Natural secondary metabolites: from nature, through Science, to industry. Springer International Publishing, Cham, pp 403–433

Jain D, Khurana JP (2018) Role of pathogenesis-related (PR) proteins in plant defense mechanism. In: Molecular aspects of plant-pathogen interaction. Springer Singapore, Singapore, pp 265–281

Jeger MJ (2023) Tolerance of plant virus disease: its genetic, physiological, and epidemiological significance. Food Energy Secur 12(6):e440

Kalinina NO, Spechenkova NA, Taliansky ME (2025) Biotechnological approaches to plant antiviral resistance: CRISPR-Cas or RNA interference? Biochem Mosc 90(6):804–817

Kang BC, Yeam I, Jahn MM (2005) Genetics of plant virus resistance. Annu Rev Phytopathol 43(1):581–621

Kapazoglou A, Gerakari M, Lazaridi E, Kleftogianni K, Sarri E, Tani E, Bebeli PJ (2023) Crop wild relatives: A valuable source of tolerance to various abiotic stresses. Plants 12(2):328

Kaur S, Samota MK, Choudhary M, Choudhary M, Pandey AK, Sharma A, Thakur J (2022) How do plants defend themselves against pathogens-Biochemical mechanisms and genetic interventions. Physiol Mol Biol Plants 28(2):485–504

Kaushal A, Sadashiva AT, Ravishankar KV (2024) Genomics and molecular breeding of major vegetable crops. In: Genetic engineering of vegetable crops. CABI, GB, pp 258–309

Khadka K, Earl HJ, Raizada MN, Navabi A (2020) A physio-morphological trait-based approach for breeding drought tolerant wheat. Front Plant Sci 11:715

Khakimov A, Salakhutdinov I, Omolikov A, Utaganov S (2022) Traditional and current-prospective methods of agricultural plant diseases detection: A review. In: IOP Conference series: earth and environmental science, vol 951, No. 1. IOP Publishing, p 012002

Khan MHU, Muhammad A, Din I, Ullah N, Zheng S, Ahmar S et al (2025) Apoplast engineering unveiled: multi-target strategies to disarm pathogens and elevate global food security. Plant Cell Environ

Khush GS (2001) Green revolution: the way forward. Nat Rev Genet 2(10):815–822

Kwak HR, Kim JS, Cho JD, Lee JH, Kim TS, Kim MK, Choi HS (2015) Characterization of Melon necrotic spot virus occurring on watermelon in Korea. Plant Pathol J 31(4):379

Lahlali R, Taoussi M, Laasli SE, Gachara G, Ezzouggari R, Belabess Z et al (2024) Effects of climate change on plant pathogens and host-pathogen interactions. Crop Environ 3(3):159–170

Lalić J, Elena SF (2012) Magnitude and sign epistasis among deleterious mutations in a positive-sense plant RNA virus. Heredity 109(2):71–77

Lalić J, Elena SF (2013) Epistasis between mutations is host-dependent for an RNA virus. Biol Lett 9(1):20120396

Lanfermeijer FC, Dijkhuis J, Sturre MJ, de Haan P, Hille J (2003) Cloning and characterization of the durable tomato mosaic virus resistance gene tm-22 from Lycopersicon esculentum. Plant Mol Biol 52(5):1039–1051

Laterrot H (2000) Disease resistance in tomato: practical situation. Acta Physiol Plant 22:328–331

Lecoq H, Desbiez C (2012) Viruses of cucurbit crops in the Mediterranean region: an ever-changing picture. In: Advances in virus research, vol 84. Academic Press, pp 67–126

Li F, Zhao N, Wu G (2017) rgs-CaM detects and counteracts viral RNA silencing suppressors in plant immune priming. J Virol 91(19):e00761–e00717

Li G, Shao J, Wang Y, Liu T, Tong Y, Jansky S et al (2022) Rychc confers extreme resistance to potato virus Y in potato. Cells 11(16):2577

Mahfouze HA, Mahfouze SA, Ottai MES (2022) Molecular characterization of markers linked to Tomato spotted wilt virus and Tomato mosaic virus resistance loci in tomato. J Appl Biol Biotechnol:135–144

Majzoub K, Wrensch F, Baumert TF (2019) The innate antiviral response in animals: an evolutionary perspective from flagellates to humans. Viruses 11(8):758

Manrubia S, Lazaro E (2006) Viral evolution. Phys Life Rev 3:65–92. https://doi.org/10.1016/j.plrev.2005.11.002

Marcelino SM, Gaspar PD, do Paço A, Lima TM, Monteiro A, Franco JC et al (2024) Agricultural practices for biodiversity enhancement: evidence and recommendations for the viticultural sector. AgriEngineering 6(2):1175–1194

Martin DP, Biagini P, Lefeuvre P, Golden M, Roumagnac P, Varsani A (2011) Recombination in eukaryotic single stranded DNA viruses. Viruses 3(9):1699–1738

Martins TF, Souza PF, Alves MS, Silva FDA, Arantes MR, Vasconcelos IM, Oliveira JT (2020) Identification, characterization, and expression analysis of cowpea (Vigna unguiculata [L.] Walp.) miRNAs in response to cowpea severe mosaic virus (CPSMV) challenge. Plant Cell Rep 39(8):1061–1078

Maskell LC, Raybould AF, Cooper JI, Edwards ML, Gray AJ (1999) Effects of turnip mosaic virus and turnip yellow mosaic virus on the survival, growth and reproduction of wild cabbage (*Brassica oleracea*). Ann Appl Biol 135(1):401–407

Mehra P, Patel S, Pansare UD (2024) Application of molecular markers in vegetable improvement: a review. Plant Arch 24(1):50–63

Mishra MG, Singh PK, Pandey P, Sohi MA (2023) Breeding approaches for vegetable crops. Golden Leaf PublishersTM, Uttar Pradesh

Mueller EE, Groves RL, Gratton C (2012) Crop and non-crop plants as potential reservoir hosts of alfalfa mosaic virus and cucumber mosaic virus for spread to commercial snap bean. Plant Dis 96(4):506–514

Muniyappa V, Jalikop SH, Saikia AK, Raman R (1991) Reaction of tomato genotypes to tomato leaf curl virus. Euphytica 56(1):37–41

Nadeem MA, Nawaz MA, Shahid MQ, Doğan Y, Comertpay G, Yıldız M et al (2018) DNA molecular markers in plant breeding: current status and recent advancements in genomic selection and genome editing. Biotechnol Biotechnol Equip 32(2):261–285

Naveed ZA, Wei X, Chen J, Mubeen H, Ali GS (2020) The PTI to ETI continuum in Phytophthora-plant interactions. Front Plant Sci 11:593905

Nevame AYM, Emon RM, Malek MA, Hasan MM, Alam MA, Muharam FM et al (2018) Relationship between high temperature and formation of chalkiness and their effects on quality of rice. Biomed Res Int 2018(1):1653721

Niehl A, Heinlein M (2019) Virus perception at the cell surface: revisiting the roles of receptor-like kinases as viral pattern recognition receptors. Mol Plant Pathol 20(9):1183–1192

Nieto C, Morales M, Orjeda G, Clepet C, Monfort A, Sturbois B et al (2006) An eIF4E allele confers resistance to an uncapped and non-polyadenylated RNA virus in melon. Plant J 48(3):452–462

Nohtani H, Solouki M, Aghnoum R, Fakheri BA, Mahdinezhad N, Bakhshi B (2025) Genome-wide association study reveals novel marker-trait associations for agronomic traits and covered smut resistance in barley (*Hordeum vulgare* L.). Euphytica 221(8):127

Osakabe Y, Osakabe K (2015) Genome editing with engineered nucleases in plants. Plant Cell Physiol 56(3):389–400

Pandey S, Prasad M (2024) Recent advances in tomato genetic engineering for virus resistance: achievements and limitations. Genetic Engineering of Vegetable Crops, 1–22

Pantouw CF, Hapsari BW, Hastilestari BR (2022) Pengaruh peningkatan suhu pada fase pembentukan umbi tanaman kentang (*Solanum tuberosum*) cv. Granola. J Agro 9(1):147–161

Paterson AH, Queitsch C (2024) Genome organization and botanical diversity. Plant Cell 36(5):1186–1204

Pathak VM, Verma VK, Rawat BS, Kaur B, Babu N, Sharma A et al (2022) Current status of pesticide effects on environment, human health and it's eco-friendly management as bioremediation: A comprehensive review. Front Microbiol 13:962619

Perdanaputra F (2017) Ketahanan Pangan di Kampung Adat dan Non-Kampung Adat. Kampung Ciptagelar dan Kampung Tonjong II, Provinsi Jawa Barat, Kasus

Pérez-Losada M, Arenas M, Galán JC, Palero F, González-Candelas F (2015) Recombination in viruses: mechanisms, methods of study, and evolutionary consequences. Infect Genet Evol 30:296–307. https://doi.org/10.1016/j.meegid.2014.12.022

Pérez-Moro C, D'Esposito D, Capuozzo C, Guadagno A, Pérez-de-Castro A, Ercolano MR (2025) Discovery of variation in genes related to agronomic traits by sequencing the genome of Cucurbita pepo varieties. BMC Genomics 26(1):335

Peters JL, Cnudde F, Gerats T (2003) Forward genetics and map-based cloning approaches. Trends Plant Sci 8(10):484–491

Picó B, Díez MJ, Nuez F (2002) Viral diseases causing the greatest economic losses to the tomato crop. II. The tomato yellow leaf curl virus—a review. Sci Hortic 92(2):73–113

Pilowsky M, Cohen S (2000) Inheritance of resistance to tomato yellow leaf curl virus in tomatoes. Phytopathology 90(5):574–578

Pinto L, Soler-López L, Serrano A, Sánchez-Rodríguez C (2025) Between host and invaders: the subcellular cell wall dynamics at the plant–pathogen interface. Annu Rev Plant Biol 76(1):255–284

Prasad A, Sharma N, Tewari AK (2022) Transgene silencing, RNA interference, and the antiviral defense mechanism directed by small interfering RNAs. Phytopathology 112(10):1945–1956

Prasanna HC, Kashyap SP, Krishna R, Sinha DP, Reddy S, Malathi VG (2015) Marker assisted selection of Ty-2 and Ty-3 carrying tomato lines and their implications in breeding tomato leaf curl disease resistant hybrids. Euphytica 204(2):407–418

Rahman A, Sinha KV, Sopory SK, Sanan-Mishra N (2021) Influence of virus–host interactions on plant response to abiotic stress. Plant Cell Rep 40(11):2225–2245

Renzi JP, Coyne CJ, Berger J, von Wettberg E, Nelson M, Ureta S et al (2022) How could the use of crop wild relatives in breeding increase the adaptation of crops to marginal environments? Front Plant Sci 13:886162

Roossinck MJ (2002) Evolutionary history of cucumber mosaic virus deduced by phylogenetic analyses. J Virol 76(7):3382–3387

Rubio L, Galipienso L, Ferriol I (2020) Detection of plant viruses and disease management: relevance of genetic diversity and evolution. Front Plant Sci 11:1092

Ruffel S, Gallois JL, Lesage ML, Caranta C (2005) The recessive potyvirus resistance gene pot-1 is the tomato orthologue of the pepper pvr2-eIF4E gene. Mol Gen Genomics 274(4):346–353

Ryu H, Choi S, Cheng M, Koo BK, Kim EY, Lee HS, Lee DH (2025) Flagellin sensing, signaling, and immune responses in plants. Plant Commun 6(7):101383

Sanjuán R, Domingo-Calap P (2016) Mechanisms of viral mutation. Cellular and molecular life sciences 73(23):4433–4448

Sanan-Mishra N, Abdul Kader Jailani A, Mandal B, Mukherjee SK (2021) Secondary siRNAs in plants: biosynthesis, various functions, and applications in virology. Front Plant Sci 12:610283

Savary S, Bregaglio S, Willocquet L, Gustafson D, Mason D'Croz D, Sparks A, Castilla N, Djurle N, Allinne C, Sharma M, Rossi V, Amorim L, Bergamin A, Yuen J, Esker P, McRoberts N, Avelino J, Duveiller E, Koo J, Garrett K (2017) Crop health and its global impacts on the components of food security. Food Secur 9:311–327

Schirawski J, Voyatzakis A, Zaccomer B, Bernardi F, Haenni AL (2000) Identification and functional analysis of the turnip yellow mosaic tymovirus subgenomic promoter. J Virol 74(23):11073–11080

Scholthof KBG (2023) The past is present: coevolution of viruses and host resistance within geographic centers of plant diversity. Annu Rev Phytopathol 61(1):119–136

Schreinemachers P, Simmons EB, Wopereis MC (2018) Tapping the economic and nutritional power of vegetables. Glob Food Sec 16:36–45

Scott J, Gardner R (2007) Breeding for resistance to fungal pathogens. In: Publ Science (ed) Genetic improvement of Solanaceous crops, Enfield, pp 421–456

Shahriari Z, Su X, Zheng K, Zhang Z (2023) Advances and prospects of virus-resistant breeding in tomatoes. Int J Mol Sci 24(20):15448

Shamshirgaran Y, Liu J, Sumer H, Verma PJ, Taheri-Ghahfarokhi A (2022) Tools for efficient genome editing; ZFN, TALEN, and CRISPR. In: Applications of genome modulation and editing, pp 29–46

Sharma M, Tarafdar A, Chandran US, Chobe DR, Ghosh R (2018) Molecular markers as tools for identification and introgression of virus resistant genes. *Genes,* genetics and transgenic for virus resistance in plants. Caister Academic Press, UK, pp 87–102

Sharma N, Prasad A, Prasad M (2022) Role of the Sw5 gene cluster in the fight against plant viruses. J Virol 96(5):e02084–e02021

Shi L, Yang Y, Xie Q, Miao H, Bo K, Song Z et al (2018) Inheritance and QTL mapping of cucumber mosaic virus resistance in cucumber (*Cucumis Sativus* L.). PLoS One 13(7):e0200571

Simko I, Jia M, Venkatesh J, Kang BC, Weng Y, Barcaccia G et al (2021) Genomics and marker-assisted improvement of vegetable crops. Crit Rev Plant Sci 40(4):303–365

Singh G, Dukariya G, Kumar A (2020) Distribution, importance and diseases of soybean and common bean: A review. Biotechnol J Int 24(6):86–98

Singh HP, Raigar OP, Chahota RK (2022) Estimation of genetic diversity and its exploitation in plant breeding. Bot Rev 88(3):413–435

Smýkal P, Nelson MN, Berger JD, Von Wettberg EJ (2018) The impact of genetic changes during crop domestication. Agronomy 8(7):119

Song YS, Hepting L, Schweizer G, Hartl L, Wenzel G, Schwarzfischer A (2005) Mapping of extreme resistance to PVY (Ry sto) on chromosome XII using anther-culture-derived primary dihaploid potato lines. Theor Appl Genet 111(5):879–887

Statista (2023) Global production of vegetables in 2022, by type. Statista. https://www.statista.com/statistics/264065/global-production-of-vegetables-by-type/

Stevens MA, Scott SJ, Gergerich RC (1992) Inheritance of a gene for resistance to Tomato spotted wilt virus (TSWV) from *Lycopersicon peruvianum.* Euphytica 59(1):9–17

Strausbaugh CA, Eujayl IA, Wintermantel WM (2017) Beet curly top virus strains associated with sugar beet in Idaho, Oregon, and a Western US Collection. Plant Dis 101(8):1373–1382

Sun L, Lai M, Ghouri F, Nawaz MA, Ali F, Baloch FS et al (2024) Modern plant breeding techniques in crop improvement and genetic diversity: from molecular markers and gene editing to artificial intelligence—a critical review. Plants 13(19):2676

Svoboda P (2020) Key mechanistic principles and considerations concerning RNA interference. Front Plant Sci 11:1237

Tabasum S, Nazir G, Hussain K, Gautam A, Mushtaq F, Nabi A, Ahsan S (2025) Climate smart breeding of vegetable crops: A review. Plant Arch 25(1):3298–3307

Taliansky M, Samarskaya V, Zavriev SK, Fesenko I, Kalinina NO, Love AJ (2021) RNA-based technologies for engineering plant virus resistance. Plants 10(1):82

Tatineni S, Hein GL (2023) Plant viruses of agricultural importance: current and future perspectives of virus disease management strategies. Phytopathology 113(2):117–141

Thomas WJ, Amas JC, Dolatabadian A, Huang S, Zhang F, Zandberg JD et al (2024) Recent advances in the improvement of genetic resistance against disease in vegetable crops. Plant Physiol 196(1):32–46

Tong C, Huang S, Shi Y, Wu Q, Shangguan L, Yu H et al (2023) Development of novel specific molecular markers for the Sw-5b gene to assist with tomato spotted wilt virus-resistant tomato breeding. Phytopathol Res 5(1):59

Ul Haq I, Khan M, Khan I (2024) Phytopathological management through bacteriophages: enhancing food security amidst climate change. J Ind Microbiol Biotechnol 51:kuae031

Valkonen JPT, Xu YS, Pulli S, Pehu E, Rokka VM (1994) Transfer of resistance to potato leafroll virus, potato virus Y and potato virus X from *Solarium brevidens* to *S. tuberosum* through symmetric and designed asymmetric somatic hybridisation. Ann Appl Biol 124(2):351–362

Van Munster M (2020) Impact of abiotic stresses on plant virus transmission by aphids. Viruses 12:216

Verbeek M, Dullemans AM, Van Bekkum PJ, Van Der Vlugt R (2013) Evidence for Lettuce big-vein associated virus as the causal agent of a syndrome of necrotic rings and spots in lettuce. Plant Pathol 62(2):444–451

Verlaan MG, Hutton SF, Ibrahem RM, Kormelink R, Visser RG, Scott JW et al (2013) The tomato yellow leaf curl virus resistance genes Ty-1 and Ty-3 are allelic and code for DFDGD-class RNA–dependent RNA polymerases. PLoS Genet 9(3):e1003399

Vos P, Hogers R, Bleeker M, Reijans M, Lee TVD, Hornes M et al (1995) AFLP: a new technique for DNA fingerprinting. Nucleic Acids Res 23(21):4407–4414
Walsh JA, Rusholme RL, Hughes SL, Jenner CE, Bambridge JM, Lydiate DJ, Green SK (2002) Different classes of resistance to turnip mosaic virus in Brassica rapa. Eur J Plant Pathol 108(1):15–20
Wang Y, Mostafa S, Zeng W, Jin B (2021) Function and mechanism of jasmonic acid in plant responses to abiotic and biotic stresses. Int J Mol Sci 22(16):8568
Whitehead SR, Turcotte MM, Poveda K (2017) Domestication impacts on plant–herbivore interactions: a meta-analysis. Philos Trans R Soc B 372(1712):20160034
Willmann MR, Endres MW, Cook RT, Gregory BD (2011) The functions of RNA-dependent RNA polymerases in Arabidopsis. Arabidopsis Book 9:e0146
Wu J, Yang Z, Wang Y, Zheng L, Ye R, Ji Y et al (2023) Salicylic acid and RNA interference mediate antiviral immunity of plant stem cells. Proc Natl Acad Sci 120(22):e2302069120
Yamakawa S, Nagata T (1975) Resistance of wild tomato species to tobacco mosaic virus. Jpn J Phytopathol 41(1):12–20
Yan Z, Pérez-de-Castro A, Díez MJ, Hutton SF, Visser RG, Wolters AMA et al (2018) Resistance to tomato yellow leaf curl virus in tomato germplasm. Front Plant Sci 9:1198
Yu XQ, Niu HQ, Liu C, Wang HL, Yin W, Xia X (2024) PTI-ETI synergistic signal mechanisms in plant immunity. Plant Biotechnol J 22(8):2113–2128
Zapletal D, Kubicek K, Svoboda P, Stefl R (2023) Dicer structure and function: conserved and evolving features. EMBO Rep 24(7):e57215
Zhang H, Mittal N, Leamy LJ, Barazani O, Song BH (2017) Back into the wild—apply untapped genetic diversity of wild relatives for crop improvement. Evol Appl 10(1):5–24
Zhang Y, Xie Z, Fletcher JD, Wang Y, Wang R, Guo Z, He Y (2020) Rapid and sensitive detection of lettuce necrotic yellows virus and cucumber mosaic virus infecting lettuce (*Lactuca sativa* L.) by reverse transcription loop-mediated isothermal amplification. Plant Pathol J 36(1):76
Zhang S, Griffiths JS, Marchand G, Bernards MA, Wang A (2022a) Tomato brown rugose fruit virus: an emerging and rapidly spreading plant RNA virus that threatens tomato production worldwide. Mol Plant Pathol 23(9):1262–1277
Zhang S, Li C, Si J, Han Z, Chen D (2022b) Action mechanisms of effectors in plant-pathogen interaction. Int J Mol Sci 23(12):6758

# Role of Vector Ecology and Climate Adaptation in Virus Spread in Vegetable Crops

4

Ashulata Kaushal, Devendra Kumar Choudhary,
Smita Bala Rangare, Shravani Sanyal, and N. R. Rangare

**Abstract**

Vegetables, being a good source of nutrients, are often attacked by many viral diseases, which are mainly transmitted by specific vectors such as aphids, whiteflies, and thrips. The wide distribution of the viral diseases is not merely due to the presence of virus particles in the host but is rather the net result of the tripartite interactions of viruses within the host plants, their vectors, and the environmental factors. Each vector species has specific requirements and adaptability to cope with the adverse conditions. Availability and activity of vectors and their adaptation to climatic conditions can influence the range of distribution of the vector populations in an area and the amount of virus spread in a season. Temperature and wind are the main environmental factors that affect the activity of aerial vectors. Temperature may strongly influence the activity, incidence, and reproductive rates of the vectors. Water not only promotes plant growth but may also attract vectors and help the virus to perpetuate. Irrigation and erratic rainfall provide the most conducive conditions for rapid buildup of the vector population in dry seasons, which are otherwise challenging to the vectors. Additionally, the polyphagous nature of *Bemisia tabaci*, wider distribution of *Thrips tabaci* and

A. Kaushal (✉) · D. K. Choudhary
Department of Plant Pathology, College of Agriculture, Indira Gandhi Krishi Vishvavidyalaya, Raipur, Chhattisgarh, India

S. B. Rangare
Department of Vegetable Science, College of Agriculture, Indira Gandhi Krishi Vishvavidyalaya, Raipur, Chhattisgarh, India

S. Sanyal
School of Crop Health Biology Research NIBSM, Raipur, Chhattisgarh, India

N. R. Rangare
Department of Genetics and Plant Breeding, College of Agriculture, Indira Gandhi Krishi Vishvavidyalaya, Raipur, Chhattisgarh, India

 
R. Kumar et al. (eds.), *Vegetable Crops and Viral Pathogens Volume I*, Plant Pathology for Sustainable Agriculture, https://doi.org/10.1007/978-981-95-7265-6_4

*Frankliniella schultzei* in temperate as well as tropical regions, and the transmission ability of aphids to transmit viruses in semi-persistent, persistent, and non-persistent manner open wide opportunities for plant virus spread by these vectors. An insight into the interrelationships of vector populations with the environment can help to manage the devastating viral diseases, and strategies based on vector management, regular monitoring, early detection, and climate-smart agriculture can help to mitigate the rapid buildup of vector populations and virus spread.

**Keywords**

Plant viruses · Vegetable crops · Virus epidemiology · Insect vectors · Virus transmission · Climate change · Host–vector–virus interactions · Integrated disease management · Vector management

## 4.1 Introduction

Plant viruses cause a significant impact on agriculture economically, resulting in significant yield losses and financial burdens on farmers and consumers alike. Historically, viruses like curly top in sugar beets and swollen shoot in cacao have had disastrous consequences, decimating entire industries and causing widespread damage. With over 1000 known virus species affecting crops globally, the threat to food security is palpable. Significant yield loss due to plant viral diseases is a matter of great concern not only to the farmers but also to the scientists and the policy makers. Plant viruses can reduce crop productivity, increase market prices, and affect the livelihoods of those dependent on agriculture. The severity of yield losses varies depending on factors like the duration of infection and the type of virus, with some viruses having a more devastating impact than others. Some of the most notable viruses causing significant yield losses include tobamo, tospo, begomo, cucumo, and poty viruses, which can have far-reaching consequences for agricultural economies. Losses due to plant viruses are due to a reduction in the crop yield potential (Waterworth and Hadidi 1998; Scholthof et al. 2011; Sastry and Zitter 2014; Mandal et al. 2017). Crop yield reduction leads to heavy infections, which result in heightened market prices; hence, disease losses affect farmers, middlemen, and consumers. Worldwide viral epidemics have been reported across the globe in various environmental conditions. Sugar beet yellows virus causes serious epidemics in temperate regions. In tropical and subtropical areas, especially in Africa, cassava mosaic viruses, major potato viruses (PVS, PVX, PVY, and PLRV), Tomato spotted wilt virus in tomato and capsicum, begomoviruses in chili, tomato, and okra, and Cucumber mosaic virus in many horticultural crops cause heavy yield losses and are among the most important viral threats worldwide. (Kumar et al. 2020, 2025; Lal et al. 2023; Mazumder et al. 2024). Annual and perennial vegetable crops are highly susceptible to various plant viruses, resulting in substantial yield losses. Perennial crops face significant immediate losses and additional costs for replacing infected plants, making it a pressing concern. Asexually propagated horticultural crops like cassava, potatoes, and sweet potatoes are particularly vulnerable to viral infections, leading to considerable yield reductions (Kumar et al. 2022). The widespread adoption of intensive monoculture farming practices increases the risk of viral disease epidemics. Furthermore, rapid asexual methods of propagation and global trading of plant materials exacerbate the issue. Looking at this, effective management of plant viral diseases is very crucial for the benefit of the farmers. It is estimated that plant viruses cause over $30 billion in annual yield losses worldwide, impacting both quantity and quality of agricultural products (Sastry and Zitter 2014).

## 4.2 Viral Threats to Vegetables

Potato crops are severely impacted by over 30 viral diseases, with mosaic and leaf roll being two of the most prevalent diseases. Mixed infections, such as PVX + PVY and PVX + PVA, can exacerbate yield losses (Table 4.1). In potato, Potato leaf roll virus and potato virus Y can cause yield reductions of up to 80% (Khurana 2000; Mukherjee et al. 2003), while attenuated strains like PVX and PVS can decrease yields by 10–30% (Gupta and Thind 2006). Viral diseases are a major restriction to potato production worldwide, with estimated losses ranging from 40% to 85% due to PVY (Nagaich and Agrawal 1969; Khurana and Garg 2000). Other viruses, such as GBNV, also badly affected the yield of potato up to 29% (Singh et al. 1997). Tomato crops affected by leaf curl disease result in considerable yield losses as well as fruit quality. TLCV causes yield losses of 78–90% in Karnataka, India (Saikia and Muniyappa 1989). Another significant threat to tomato crops is tomato leaf curl New Delhi virus (ToLCNDV), causing yield losses of 17.6–99.7% depending on the disease severity and time of infection. The severity of yield losses varies depending on the infection stage, with losses ranging from 28.9% to 92.3% (Sastry and Singh 1973; Butter and Rataul 1981; Kalloo 1996). Significant losses are also caused by other viruses, such as tomato leaf curl virus, tobacco mosaic virus, tomato bushy stunt virus, and TYLCV. TYLCV causes yield losses of up to 100% in some regions, with estimated losses of $10–50 and million (Gilbertson et al. 2007; Polston and Anderson 1997; Petershmitt et al. 1999). Tomato mottle virus causes estimated losses of $140 million (Schuster 1992).

Chili leaf curl disease is a significant constraint to chili production, causing substantial yield losses and impacting fruit quality. This resulted in yield reductions of more than 90% when infection occurs within 3–4 weeks after transplantation. Recent studies have highlighted the severity of the issue, with estimated losses varying from 20% to 80% in India (Saha et al. 2005; Paul et al. 2025). Other viruses, such as CMV, TMV, PVY, and TEV, also contribute to significant losses in crops of the chili family worldwide (Table 4.1). The economic impact of viral diseases on chili yields is substantial, with estimated losses of 1.0%, 1.0%, 0.5%, and 0.5% annually in the United States due to TMV, CMV, TEV, and PVY, respectively (USDA 1965; Lamptey and Bonsi 1977), 40–80% in susceptible pepper cultivars in Argentina (Feldman et al. 1969). Studies have also shown that the timing of infection significantly impacts yield losses. For example, infection within 3–6 weeks

**Table 4.1** Losses due to important viral diseases in vegetables

| S. no. | Crop name | Virus | Country | Losses due to disease | Reference |
|---|---|---|---|---|---|
| 1 | Potato | PLRV | Washington State (1959) | $3–4 million | Hoyman (1963) |
| | | | India | 50–60% | Khurana (2000) and Mukherjee et al. (2003) |
| | | | India, Egypt, Canada, Bangladesh, and Scotland | 20.0–98.0% | Loughnane (1941), Nagaich and Agrawal (1969), Allam et al. (1974), Nelson and Tarfason (1974), Killick (1979), and Hossain et al. (1989) |
| | | Potato virus A and X | Brazil, India, and Australia | 20–89% and 75% | Cupertino et al. (1973), Dedic (1975), Bald and Norris (1941), and Khurana and Singh (1986) |
| | | GBNV | India | 29% | Singh et al. (1997) |
| | | Potato virus A | Brazil | 25% | Cupertino et al. (1973) |
| | | PVY | India | 80% | Khurana and Garg (2000) |
| 2 | Tomato | ToMoV, Tom GV1, and Tom GV2 | – | $140 million | Schuster (1992) |
| | | | – | 100% | Polston and Anderson (1997) |
| | | TYLCV | Dominican Republic | $10 million | Gilbertson et al. (2007) |
| | | | Malawi, Kenya, Tanzania, Sudan | 40% 50% 75% 100% | CIAT (1998) |
| 3. | Chili/ pepper | ChLCV | India | 20–80% | Saha et al. (2005) |
| 4 | Brinjal | PVY | – | 9.51–27.03% | Singh and Singh (1975) |
| | | Tobacco ring spot virus | – | 65.2–70.3% | Sastry and Nayudu (1978) |
| 5 | Okra | YVMV | India | Up to 90% | Chellaiah and Sellammal Murugesan (1976) |
| | | Okra leaf curl virus (OLCV) | India and Pakistan | 30–70% | Lana (1976) |
| | | Enation leaf curl virus (ELPV) | India | 93.10% | Singh (1990) |
| | | BYVMV | Sri Lanka, Pakistan, China, and Thailand | 94–100% | Tsai et al. (2013) and Nath et al. (1992) |

(continued)

**Table 4.1** (continued)

| S. no. | Crop name | Virus | Country | Losses due to disease | Reference |
|---|---|---|---|---|---|
| 6 | Cucurbits | Watermelon mosaic virus in Summer squash | United States | 9–43% | Demski and Chalkley (1972) |
| | | Cucumber tobacco ringspot virus (TRSV) | – | 2% | USDA (1965) |
| | | Buttercup squash, golden Hubbard squash, and pumpkin WMV-2 | – | 63%, 53%, and 49% | Wayne (1971) |

after transplanting can result in yield losses of up to 47% (Fernandez and Goborjanyi 1978). Pre-flowering infection with TMV can lead to more significant yield reductions. The severity of leaf curl disease in chili crops underscores the need for effective management strategies to mitigate the impact of viral diseases on chili production. Brinjal Mosaic is caused by nine different viruses; losses may go up to 9.51–27.03% (Basu and Giri 1992; Singh and Singh 1975). Another viral disease that can cause serious losses is TRSV (Sastry and Nayudu 1978).

The yellow vein mosaic virus (YVMV) is a highly destructive disease affecting the ladyfinger (*okra*) crop, particularly when infection occurs during early plant growth stages. Studies have shown that YVMV infection can lead to significant loss (Chellaiah and Sellammal Murugesan 1976). The impact of YVMV on okra yield is influenced by the timing of infection, with losses ranging from 86% to 93.10% depending on the stage of infection (Sinha and Chakrabarthi 1978; Singh 1990). Other viruses, such as okra leaf curl virus (OLCV) and enation leaf curl virus (ELCV), also cause significant losses in okra production. OLCV has been reported to prime financial losses of 30–70%, while ELCV can result in yield losses of up to 100% (Lana 1976; Sastry and Singh 1974). Tobacco stunt virus is another major okra virus, causing reported yield damage of 63% (Krishnareddy et al. 2003). The severity of yield losses due to viral diseases in okra underscores the need for effective management strategies to mitigate the impact of these diseases on okra production. Mosaic disease is a prevalent issue affecting various cucurbitaceous crops worldwide, resulting in significant yield losses. Cucumber mosaic virus, cucumber green mottle mosaic virus, watermelon mosaic virus, and tobacco ringspot virus (TRSV) are among the viruses responsible for these losses. Studies have quantified the economic impact of viral diseases on cucurbit yields, with estimated losses ranging from 2% to 100% depending on the virus and crop stage. Watermelon bud necrosis virus causes estimated losses of 37.4–100% depending on the stage of infection (Anon 2011; Krishnareddy and Singh 1993).

## 4.3 Vector–Virus Relationships

### 4.3.1 Aphids

Most of the aphids transmitting plant viruses are polyphagous, with a piercing and sucking type of mouthparts, wingless to winged, which help to migrate and spread into new areas (Table 4.2). Reproduction is parthenogenic and may also be viviparous. Aphids are the largest group of vectors that transmit plant viruses in all three categories of virus transmission, i.e., non-persistent, persistent, and semi-persistent. Non-persistent virus transmission occurs when the viruses are acquired by the vectors during probing and feeding the host cells in a few seconds and is acquired from a few periods to a few minutes. Vector becomes infective immediately without any period of latency and remains infective up to 40 h. However, the infectivity is lost during molting (bean yellow mosaic virus). Persistent virus transmission is associated with a highly specific virus–vector relationship, requiring a little more time in acquiring the virus, as the virus often is deep-seated in the phloem tissues, and the vector takes at least 60 s or more to probe into these tissues. However, it takes about 5–15 min to acquire the viruses. Virus gets into the alimentary canal, passes through the gut wall, circulates in the body fluid, and finally contaminates the saliva before transmission. Hence, this transmission is also referred to as circulative, and the time taken to circulate inside the vector is called the latent period. Virus remains inside the vector even after molting (potato leaf roll virus). Many times, the virus even multiplies inside the vector, and hence this type of transmission is referred to as circulative-propagative transmission, which may be carried in the eggs of the infected vector (Naga et al. 2020; Modgill et al. 2025). In semi-persistent transmission, the uptake of virus by the vector is from the deep-seated tissues, takes a long time to acquire, as in the persistent transmission, but becomes infective immediately with no latent period. Virus remains for 3–4 days, and it may accumulate along the lining of the anterior portion of the foregut. The virus is transmitted when the vector probes into the new site for feeding (Cauliflower mosaic virus, Beet yellows virus). Bimodal transmission is also reported in aphids, where few viruses are transmitted both non-persistently and semi-persistently. Chalfant and Chafman (1962) reported bimodal transmission of CaMV in *Brevicoryne brassicae*. Bimodal transmission is also reported in pea seed–borne mosaic virus (PSbMV) by Lim and Hadedorn (1974). Transmission of some viruses depends upon the presence of another virus, which is known as dependent transmission. Clinch et al. (1936) first noticed that potato aucuba mosaic could not be transmitted by aphids unless the same potato plant was infected with PVA also. Similar studies carried out by Kassanis and Govier (1971a) found the concept of helper component. Based on his observations, it acts as the connecting bridge between aphids and virus particles, helping in the attachment of the virus particle to adhere to a specific site in the aphid for further transmission (Basu and Giri 1992).

**Table 4.2** Vectors responsible for spreading major viral diseases in vegetable crops

| S. no. | Vector name | Country | Virus | Virus group | Reference |
|---|---|---|---|---|---|
| | Whitefly | India | Tomato leaf curl virus | *Begomovirus* | Padidam et al. (1995), Chakraborty et al. (2003), Punam Kumari et al. (2010), Ranjan et al. (2021), Kumar et al. (2008), Kumari et al. (2011), Dokka et al. (2021), and Venkataravanappa et al. (2016) |
| | | India | Mungbean yellow mosaic virus | | |
| | | Worldwide | Pepper leafroll virus | | Martinez-Ayala et al. (2014) |
| | | Worldwide | Pepper yellow leaf curl Aceh virus | | Kesumawati et al. (2019) |
| | | Worldwide | Tomato chlorotic leaf curl virus | | Quadros et al. (2019) |
| | | Worldwide | Tomato chlorotic leaf distortion virus | | Zambrano et al. (2011) |
| | | Worldwide | Tomato dwarf leaf virus | | Vaghi Medina and López (2012) |
| | | Worldwide | Tomato interveinal chlorosis virus and tomato interveinal chlorosis virus-2 | | Albuquerque et al. (2012) and Rego-Machado et al. (2019) |
| | | Worldwide | Tomato leaf curl purple | | Macedo et al. (2018) |
| | | Worldwide | Tomato mottle wrinkle virus | | Vaghi Medina et al. (2015) |
| | | Worldwide | Tomato yellow margin leaf curl virus | | Nava et al. (2013) |
| 2. | Aphids | India | Potato virus Y | *Potyvirus* | Chaudhary et al. (2023) |
| | | | Cucumber mosaic virus | *Cucumovirus* | |
| 3. | Thrips | India | Capsicum chlorosis virus | *Tospovirus* | Kunkalikar et al. (2007) |
| | | | Tomato spotted wilt virus | | Aravintharaj et al. (2023) |

### 4.3.2 Whitefly

Three well-known species of whitefly, which have the inherent ability to transmit plant viruses, are *Bemisia tabaci, Trialeurodes vaporarium, and T. abutilonea. Bemisia tabaci* is a highly opportunistic and polyphagous (colonizes many cultivated plants and weeds) and builds up high population in a very short time, under favorable conditions, may have overlapping generations, and has a worldwide distribution (Table 4.2) (Ghahari et al. 2013; CABI 2017). It is also known to cause direct injury to plants due to its dense population and feeding. It is known to transmit plant viruses in a persistent manner; however, semi-persistent transmission is also reported. Transmitted viruses are not sap transmissible with some exceptions. A unique phenomenon of periodic transmission of TYLCV is also reported. Low humidity and high temperature (more than 32 °C) are vulnerable to its immature nymphal stages, but microclimate conditions in the field might favor its population buildup (Gerling and Mayer 1996). Warm, humid conditions (July–October) are conducive to its rapid multiplication. Heavy rains and strong winds are detrimental to the adult population. The minimum period from egg to adult stage requires 11–14 days in summer, while in winter it takes about 42–84 days.

### 4.3.3 Thrips

These are minute, slender insects having a complex feeding mechanism, narrow wings, an absence of right mandible, an elongated left mandible, living in flowers, leaf sheaths, and other parts of the plant. They do not feed on the vascular tissues and are restricted to epidermal cells or up to deeper mesophyll cells. Five species of thrips, viz., *Frankliniella occidentalis, F. schultzei, F. fusca, Thrips tabaci,* and *Scirtothrips dorsalis* (chili thrips), are reported to transmit TSWV in a persistent manner; however, mechanical transmission of the virus is also reported. An interesting fact is that TSWV has to be acquired during larval stages to be transmitted by the adults. Thrips can retain the virus for weeks and may retain it lifelong (Table 4.2).

## 4.4 Virus–Vector–Host–Environmental Interactions

Vector ecology is the study of the interactions between organisms that transmit pathogens and their environment. It explores vector behavior, vector population dynamics, and environmental influences. Understanding vector ecology helps us to predict and prevent disease outbreaks, develop effective control strategies, and mitigate changes in disease transmission.

### 4.4.1 Reservoirs of Virus

A mere presence of a susceptible host and an efficient vector, but no source of virus, then there would not be any virus problem (Basu and Giri 1992). Viral sources in and around the field are the major source that may lead to the early onset of viral disease, especially for the non-persistent aphid-borne viruses. For the annual crops, volunteer plants and weeds are the major source of many viruses, such as CMV. There are a number of examples where unrelated plants often become the source of a virus. Gladiolus may act as a reservoir of bean yellow mosaic or cucumber mosaic virus. Not only this, but weeds play a prominent role in the ecology of plant viruses and their spread (Duffus 1971). In the ecology of crop viruses, there has been a growing realization of the importance of wild vegetation (Thresh 1981). They not only serve as a vector host but also as perpetuating hosts of the vectors and/or viruses, or may be responsible for the selection of new virus strains.

### 4.4.2 Availability and Activity of Vectors

Secondly, vector availability and their activity during the season greatly influence virus spread. Vector abundance depends upon several factors, such as the presence of natural enemies, temperature, precipitation, and available vegetation. Even though the vector population is abundant, their activity includes feeding, dispersal, and host selection. Virus infection can also alter changes in the host plant, which may attract vectors to feed on it. For example, infection with certain viruses can result in the accumulation of carbohydrates in the leaves, which mostly turn red or yellow, which attracts the vector aphids, leading to their multiplication and further spread (Basu and Giri 1992). Plant viruses are known to manipulate the biology of the vector. Some workers have reported on the changing feeding behavior of whitefly and the mutually beneficial virus–vector relationship. After acquiring the virus (tomato yellow leaf curl China virus), whiteflies showed more frequent feeding on phloem tissues as compared to non-acquired ones (Moreno-Delafuente et al. 2013).

### 4.4.3 Vector–Virus Relationship

Reservoirs of propagative viruses are the vector itself, and the transovarial transmission of the virus subtracts one step of acquisition from the diseased plants. Insecticidal treatments are more effective in preventing the spread of persistently transmitted virus through the vectors as compared to the non-persistently transmitted virus spread, as in non-persistent transmission, vectors acquire the virus in brief probing periods only and can further transmit it in a very short period of probing to healthy plants. However, semi-persistent and persistent transmission can lead to long-distance spread of the acquired virus by the vector. Vector specificity and differences in transmissibility may account for the degree of disease development from one area to another as well as from one year to the next (Basu and Giri 1992). For instance, from the mixed infections of CMV and PVY, *Myzus ascalonicus* is able to transmit only CMV, but *Myzus persicae* transmits both the viruses.

### 4.4.4 Host Response

The host response to virus infection may differ. There may be three types of responses: either total immunity (non-host resistance), subliminal infection (viruses can replicate in the initially infected cell but cannot spread to adjacent cells), or non-permissive infection (local or systemic infection). Hypersensitivity responses in TMV Samsun NN tobacco were reported at 20–24 °C, while higher temperatures showed no resistant responses (Hull 2009).

## 4.5 Changing Climatic Conditions and Their Effect on the Distribution and Abundance of the Vector Population

Climatic conditions, particularly temperature, strongly influence the incidence and activity of the vectors. Temperature and wind are important factors influencing vector activity. Weak fliers are assisted by the wind for long-distance movement. Vectors have specific requirements and adaptability to cope with the changing environmental influences. Environmental factors such as temperature, moisture, and drought have a substantial impact on plant viruses. For example, warm temperatures boost the reproduction and flying activity of vector insects, increasing virus transmission (Gutiérrez-Sánchez et al. 2023). Drought stress also caused plants to feed more frequently and for longer durations by the vectors, making them more vulnerable to viruses due to poor defenses (Van Munster et al. 2017). This makes viruses easier to establish and spread. This can raise viral loads in plants, making them more susceptible to transmission by insects (Van Munster et al. 2017; Gutiérrez-Sánchez et al. 2023). Moreover, host preferences also influence the vector activity. Feeding preference is influenced not only by visual or olfactory cues but also by temperature, humidity, wind, photoperiod, etc. For instance, aphid *Brevicoryne brassicae* is much more numerous on green cultivars of cabbage than reddish ones. Virus-infected plants, such as sugarbeet yellows, become more attractive to aphids than virus-free leaves. The aster leafhopper showed a marked preference for aster yellows-affected celery or carrot over healthy plants (Basu and Giri 1992). Aphids are very sensitive to even small temperature deviations from the mean. Change in the climatic conditions can have significant shifts in the spread, distribution, and abundance of the vector population during the cropping seasons, especially in the temperate regions, where a small increase in temperature (2 °C) may lead to more aphid generations per year. Similarly, an increase in mean winter temperature can even displace the existing vector population, e.g., *B. tabaci*, displacing the *T. vaporariorum* population, which subsequently increases the begomovirus spread in cooler regions (Jones and Barbetti 2012; Jones 2014). Drizzling and early showers during the dry season may provide plants with new growth; they may act as breeding sites, attracting the early buildup of the vector population in the season. However, heavy rainfall becomes unfavorable for some vectors, such as the beet leaf hopper, a vector of curly top virus. As already discussed, wind helps in long-distance migration and the spread of the aerial vectors. Hence, the extent of invasion in the next season depends upon the environmental conditions prevailing in the previous season and the distance travelled by the viruliferous vector population. (Basu and Giri 1992).

## 4.6 Mitigation Strategies

### 4.6.1 Plant Virus Epidemiology

The field of plant virus epidemiology calls for the precise understanding of the ecological mechanisms underlying vector and virus population dynamics and behavioral changes in the vectors. Tripartite interactions of the host, vector, and virus, along with the environment, leading to successful transmission, are the key factor in plant virus epidemiology (Jeger 2020). Major gaps in this study is to identify the consequences of climate change and its effects on virus and vector spread and distribution, to plan improved methods of control measures according to the prevailing conditions, to solve the complex tripartite interactions between virus, vector and host with environment which can be achieved by studying the type of virus transmission and how it effects the vector behavior and also, to deepen our understanding on whether virus can manipulate the host, vector and its natural enemies and to know whether the virus–vector interaction is synergistic, neutral, or antagonistic. Field observation on vector population and disease outbreaks should be correlated to epidemiological parameters to determine the possible epidemic outcomes.

### 4.6.2 Cultural Control

Elimination of the inoculum sources, such as volunteer crop plants and weed hosts, must be removed from the field, as they harbor viruses and their vectors. Moreover, certified virus-free vegetative stock should be used to avoid primary inocula. Moreover, increasing the distance between virus sources and the crop, to minimize the chances of the crop getting exposed to viruliferous vectors, especially for non-persistent viruses, is practiced. Virus-free potato tubers for seed are grown in isolated, cool, and windy places. Virus incidence can be reduced by early planting before the peak period of vector population, as plants are mostly susceptible to virus infection during the earlier stage, as compared to older plants, due to increased resistance. Crop-free period to break the cycle is very effective in onion (onion yellow dwarf virus), crucifers (cauliflower mosaic virus), and lettuce (lettuce mosaic). Virus-free potato tubers in India are grown on the northern plains during the aphid-free period (i.e., October–December), and by the time of the advent of peak vector population (January), the crop is harvested. Modifying the vector flight and settling behavior by increasing the crop density (as closely spaced plants are visited by aerial vectors less frequently than widely spaced ones; diseased plants can be removed by thinning and roguing out), by increasing the field area (as aerial vectors mostly visit along the edges of the field), cover crops (should be non-susceptible to the virus/es to be protected), and other barriers (rows of maize around chili crops), use of reflective mulches (which manipulate the behavior of whiteflies), and installation of yellow sticky traps (especially to control aphid-borne viruses). The cheapest and most effective way to avoid viral epidemics is to cultivate resistant varieties. In plant virus, six categories of resistance were given by Russell (1978) where immunity (no infection possible) resistance to virus infection (host susceptible but may escape infection), resistance to virus spread (plant highly sensitive to virus infection may show hypersensitivity), resistance to virus multiplication (concentration of virus particles remains low), tolerance (no apparent loss in yield), vector resistance (non-preference, antibiosis, and tolerance to vectors).

## 4.7 Conclusion

Plant viruses cause a significant impact on agriculture economically, resulting in significant yield losses and financial burdens on farmers and consumers alike. An insight into the interrelationships of vector populations with the environment and strategies based on vector management, regular monitoring, early detection, and climate-smart agriculture can help to mitigate the rapid buildup of vector populations and virus spread, and can help to manage the devastating viral diseases. Not a single method measures, but a judicious combination of different measures, depending upon the ecology of a particular virus and its vector, understanding the agro-ecosystem, socio-economic background of the area, for an effective and economic control of plant viral diseases should be practiced.

## References


Albuquerque LC, Varsani A, Fernandes FR, Pinheiro B, Martin DP, de Tarso Oliveira Ferreira P, Inoue-Nagata AK (2012) Further characterization of tomato-infecting begomoviruses in Brazil. Archives of Virology 157(4):747–752

Aravintharaj R, Asokan R, Prasad Babu K, Manamohan M, Nagendran K (2023) Molecular characterization of the Indian isolate (Ka-To) of tomato spotted wilt virus (TSWV) infecting tomato (Solanum lycopersicum L.). 3 Biotech 13(6):169

Allam EK, Omar RA, Gamal El-Din AS (1974) Studies on potato leafroll virus (PLRV). I. Comparative study of methods diagnosis. Egypt J Phytopathol 6:1–10

Anon (2011) Four decades of research accomplishments. Indian Institute of Horticultural "Research" Hessaraghatta, Bangalore, p 73

Bald JG, Norris DO (1941) Obtaining virus-free potatoes. J Coun Sci Res Aust 14:187–190

Basu AN, Giri BK (1992) The essentials of viruses, vectors and plant diseases. Wiley Eatern Limited, New Delhi

Butter NS, Rataul HS (1981) Nature and extent of losses in tomatoes due to tomato leaf curl virus (TLCV) transmitted by whitefly, Bemisia tabaci Gen. (Hemiptera, Aleyrodidae). Indian J Ecol 8:299–300

CABI (2017) *Bemisia tabaci* (tobacco whitefly). Center for Agriculture and Biosciences International, Invasive Species Compendium

Chalfant RB, Chapman KR (1962) Transmission of cabbage viruses A and B by the cabbage aphid and the green peach aphid. Journal of Economic Entomology 55(5):584–590

Chakraborty S, Pandey PK, Banerjee MK, Kalloo G, Fauquet CM (2003) Tomato leaf curl Gujarat virus, a new begomovirus species causing a severe leaf curl disease of tomato in Varanasi, India. Phytopathology 93(12):1485–1495

Chaudhary P, Kaur A, Singh B, Kumar S, Hallan V, & Nagpal AK (2023) First report of tomato chlorosis virus (ToCV) and detection of other viruses in field-grown tomatoes in North-Western region of India. VirusDisease 34(1):56–75

Chellaiah S, Sellammal Murugesan S (1976) Seasonal incidence of bhendi yellow vein mosaic. Annamalai Univ Agric Res Ann 6:167–168

CIAT (Centro Internacional de Agricultura Tropical) (1998) Sustainable integrated management of whiteflies as pests and vectors of plant viruses in the tropics, Progress Report. CIAT, Cali, 131 pp

Clinch PE, Loughnane JB, Murphy PA (1936) A study of the aucuba or yellow mosaics of the potato

Cupertino FP, Botelho J, Costa AS (1973) Estimation of losses in potato production caused by viruses. II virus A. Avaliacao das perdascansadaspor virus, naproducao da batata. II virus A. Bragantia 32:1–7

Dedic P (1975) Effect of virus A on yield in some varieties of potato. Vliv a virurambornavynos u nekterychodrudbramboru. Ochrana Rostl 11:127–133

Demski JW, Chalkley JH (1972) Effect of watermelon mosaic viruson yield and marketability of summer squash. Plant Dis Rep 56:146–150

Dokka N, Mahajan MM, Sahu B, Marathe A, Singh HK, Sivalingam PN (2021) Molecular analysis, infectivity and host range of Tomato leaf curl Karnataka virus associated with Corchorus yellow vein mosaic betasatellite. Virus Research 303:198521

Duffus JE (1971) Role of weeds in the incidence of virus diseases. Annual Review of Phytopathology 9(1):319–340

Feldman JM, Gracia O, Pontis RE, Boninsegna J (1969) Effect of tobacco mosaic virus on pepper yield. Plant Dis Rep 53:541–543

Fernandez FT, Goborjanyi R (1978) Reduction in capsicum and tomato yield caused by tobacco etch virus (TEV). Agrotech Cuba 10(2):45–50

Gerling, D., Mayer, R.T. (1996). *Bemisia* taxonomy, biology, damage, control, and management. Intercept, Andover

Ghahari H, Abd-Rabou S, Zahradnik J, Ostovan H (2013) Annotated catalogue of whiteflies (Hemiptera: Sternorrhyncha: Aleyrodidae) from Arasbaran, Northwestern Iran. Int J Nematol Entomol 1:42–52

Gilbertson, R., Rojas, M., Kon, T., Jaquez, J. (2007). Introduction of tomato yellow leaf curl virus into The Dominican Republic: the development of a successful integrated pest management strategy. Springer, Dordrecht

Gupta S.K. and Thind T.S. (2006). Disease problems in vegetable production. Pawan Kumar Scientific publisher, Jodhpur

Gutiérrez-Sánchez Á, Cobos A, López-Herranz M, Canto T, Pagán I (2023) Environmental conditions modulate plant virus vertical transmission and survival of infected seeds. Phytopathology 113(9):1773–1787

Hossain M, Ali MS, Khan AL, Mia SA (1989) Effect of PLRV on the growth and yield of potato. Bangladesh J Bot 18(1):15–21

Hoyman WG (1963) Summary of cooperative potato research programs: plant diseases. Wash State Univ Inst Agric Sci:7–9

Hull R (2009) Comparative plant virology, 2nd edn. Academic

Jeger MJ (2020) The epidemiology of plant virus disease: Towards a new synthesis. Plants, 9(12):1768

Jones RAC (2014) Plant virus ecology and epidemiology: historical perspectives, recent progress and future prospects. Ann Appl Biol 164:320–347

Jones RAC, Barbetti MJ (2012) Influence of climate change on plant disease infections and epidemics caused by viruses and bacteria. CAB Rev 7(22):1–32

Kalloo G (1996) Leaf curl virus of tomato and chilli in India. In: Proceeding of the phase I final workshop of the south asian vegetable research network, Kathmandu, Nepal, January 23–28, pp 229–234

Kassanis B, Govier DA (1971) The role of the helper virus in aphid transmission of potato aucuba mosaic virus and potato virus C. Journal of General Virology 13(2):221–228

Kesumawati E, Okabe S, Homma K, Fujiwara I, Zakaria S, Kanzaki S, Koeda S (2019) Pepper yellow leaf curl Aceh virus: A novel bipartite begomovirus isolated from chili pepper, tomato, and tobacco plants in Indonesia. Archives of Virology 164(9):2379–2383

Khurana SMP (2000) Potato viruses: detection and management. In: Adipala E, Nampata P, Osiru M (eds) African potato association conference proceedings, vol 5, pp 257–269
Khurana SMP, Garg ID (2000) Potato mosaics. In: Khurana SMP (ed) Diseases and pests of potato-a manual. CPR, Shimla, pp 36–40
Khurana SMP, Singh MN (1986) Viral and mycoplasma diseases of potato. Rev Tropical Plant Pathol 3:123–184
Killick RJ (1979) The effect of infection with potato leaf roll virus (PLRV) on yield and some of its components in a variety of potato (Solanum tuberosum). Ann Appl Biol 91:67–74
Krishnareddy M, Singh SJ (1993) Immunology and molecular diagnosis of tospovirus infecting watermelon. In: Golden Jubilee symposium on horticultural research: changing scenario. Indian Institute of Horticultural Research, Bangalore, pp 247–248
Krishnareddy M, Salil J, Samuel DK (2003) Fruit distortion mosaic diseases of okra in India. Plant Dis 87:1395
Kumar R, Jeevalatha A, Baswaraj R, Tiwari RK (2020) Viral and viroid diseases of potato and their management. In: Singh AK, Chakrabarti SK, Singh B, Sharma J, Dua VK (eds) Potato science & technology for sub tropics. New India Publishing Agency, New Delhi, pp 267–292. ISBN: 978-93-89571-93-6
Kumar PL, Rao RDVJP, Reddy AS, Madhavi KJ, Anitha K, Waliyar F (2008) Emergence and spread of tobacco streak virus menace in India and control strategies. Indian Journal of Plant Protection 36(1):1–8
Kumari P, Singh AK, Chattopadhyay B, Chakraborty S (2010) Molecular characterization of a new species of begomovirus and betasatellite causing leaf curl disease of tomato in India. Virus Research 152(1–2):19–29
Kumari P, Singh AK, Sharma VK, Chattopadhyay B, Chakraborty S (2011) A novel recombinant tomato-infecting begomovirus capable of transcomplementing heterologous DNA-B components. Archives of Virology 156(5):769–783
Kumar R, Tiwari RK, Sundaresha S, Kaundal P, Raigond B (2022) Potato viruses and their management. In: Chakrabarti SK, Sharma S, Shah MA (eds) Sustainable management of potato pests and diseases. Springer, Singapore, pp 309–335. (Hardcover ISBN 978-981-16-7694-9; Softcover ISBN978-981-16-7697-0; eBook ISBN978-981-16-7695-6). https://doi.org/10.1007/978-981-16-7695-6_12
Kumar R, Tiwari RK, Lal MK, Majumdar SG, Swaraj K, Mall AK, Tripathi MK, Singh D, Singh G (2025) Contemporary status of occurrence, spread, and yield loss due to tomato leaf curl New Delhi virus in global crop cultivation. In: Tomato leaf curl New Delhi virus (ToLCNDV) insights into virome dynamics and management. Springer Nature Singapore, Singapore, pp 1–12. https://doi.org/10.1007/978-981-95-2852-3_1
Kunkalikar S, Poojari S, Rajagopalan P, Zehr UB, Naidu RA, & Kankanallu RS (2007) First report of Capsicum chlorosis virus in tomato in India. Plant Health Progress 8(1):37
Lal P, Tiwari RK, Kumar A, Altaf MA, Alsahli AA, Lal MK, Kumar R (2023) Bibliometric analysis of real-time PCR-based pathogen detection in plant protection research: a comprehensive study. Front Plant Sci 14:1129714. https://doi.org/10.3389/fpls.2023.1129714
Lamptey PN, Bonsi C (1977) The effects of pepper veinal mosaic virus in relation to the time of inoculation on the symptom expression, flower initiation and yield of pepper (Capsicum frutescens L.). Acta Hort 53:227–234
Lana AF (1976) Mosaic virus and leaf curl diseases of okra in Nigeria. PANS 22:474–478
Loughnane JB (1941) The susceptibility to leaf roll of certain potato varieties and its effect on their yield. J Dept Agric Ir Free St 40:291–298
Macedo MA, Albuquerque LC, Maliano MR, Souza JO, Rojas MR, Inoue-Nagata AK, Gilbertson RL (2018) Characterization of tomato leaf curl purple vein virus, a new monopartite New World begomovirus infecting tomato in Northeast Brazil. Archives of Virology, 163(3):737–743
Mandal B, Rao GP, Baranwal VK, Jain RK (2017) A century of plant virology in India. Springer

Martínez-Ayala A, Sánchez-Campos S, Cáceres F, Aragón-Caballero L, Navas-Castillo J, Moriones E (2024) Characterisation and genetic diversity of pepper leafroll virus a new bipartite begomovirus infecting pepper bean and tomato in Peru. Ann Appl Biol 164(1):62–72. https://doi.org/10.1111/aab.2014.164.issue-110.1111/aab.12074

Mazumder P, Mondal F, Sarkar M, Majumdar A, Biswas KK, Sharma SK, Lal MK, Tiwari RK, Kumar R, Roy A (2024) Biological and molecular characterization of a new isolate of tomato mottle mosaic virus causing severe shoestring and fruit deformities in tomato plants in India. Plants 13:2811. https://doi.org/10.3390/plants13192811

Modgill N, Naga KC, Firake DM, Tiwari RK, Shivaramu S, Kumar R, Bhatnagar A, Buckseth T, Bairwa A, Maharana C, Sharma S, Singh B (2025) Flupyradifurone application: a promising approach to curtail tomato leaf curl New Delhi virus (ToLCNDV) transmission vectored by Bemisia tabaci in potato farming. Crop Prot 187:106951. https://doi.org/10.1016/j.cropro.2024.106951

Moreno-Delafuente A, Garzo E, Moreno A, Fereres A (2013) A plant virus manipulates the behavior of its whitefly vector to enhance its transmission. PLoS One 8:e61543

Mukherjee K, Verma Y, Chakrabarti SK, Singh MN, Khurana SMP (2003) Cloning and sequencing of coat protein gene of an Indian potato leaf roll virus (PLRV) isolate and its similarity with other members of luteoviridae. Virus Genes 26:247–253

Naga KC, Buckseth T, Subhash S, Bairwa A, Verma G, Kumar R, Malik K, Sharma S, Chakrabarti SK (2020) Transmission efficiency of potato leafroll virus (PLRV) by potato aphid Aulacorthum solani and green peach aphid Myzus persicae. Indian J Entomol 71:68–71. https://doi.org/10.5958/0974-8172.2020.00017.6

Nagaich BB, Agrawal HO (1969) Research on potato viruses in India. Indian J Agric Sci 39:286–296

Nath P, Gupta MK, Bura P (1992) Influence of sowing time on the incidence of yellow vein mosaic and whitefly population on okra. Indian J Virol 8:45–48

Nava A, Londoño A, Polston JE (2013) Characterization and distribution of tomato yellow margin leaf curl virus, a begomovirus from Venezuela. Archives of Virology 158(2):399–406

Nelson GA, Tarfason FW (1974) Association effects of leaf roll virus and ring rot on disease expression and yield of potatoes. Am Potato J 51:12–15

Padidam M, Beachy RN, & Fauquet CM (1995) Classification and identification of geminiviruses using sequence comparisons. Journal of General Virology 76(2):249–263

Paul S, Emmadi V, Saxena S, Sarkar M, Mandal B, Kumar R, Sinha P, Roy A (2025) RPA-assisted CRISPR-Cas12a-enabled point-of-care diagnostic platform for chili leaf curl virus with fluorescent and colorimetric readouts. Front Microbiol 16:1644322. https://doi.org/10.3389/fmicb.2025.1644322

Petershmitt P, Granier M, Aboulama S (1999) First report of Tomatoyellow leaf curl geminivirus in Morocco. Plant Dis 83:1074

Polston JE, Anderson PK (1997) The emergence of whitefly transmitted gemini viruses in tomato in Western hemisphere. Plant Dis 81:1359–1369

Quadros AF, Silva JP, Xavier CA, Zerbini FM, Boari AJ (2019) Two new begomoviruses infecting tomato and Hibiscus sp. in the Amazon region of Brazil. Archives of Virology 164(7)

Ranjan P, Kumar RV, Bhatt BS, Chahwala FD, Yadav BK, Patel S, Singh AK (2021) Two distinct monopartite begomovirus–betasatellite complexes in western India cause tomato leaf curl disease. Virus Research 295:198319

Rego-Machado CM, Nakasu EY, Blawid R, Nagata T, Inoue-Nagata AK (2019) Complete genome sequence of a new bipartite begomovirus infecting tomato in Brazil. Archives of Virology, 164(11)

Russell GE (1978) Plant breeding for pest and disease resistance. Butterworths

Saha AK, Nath PS, De BK (2005) Assessment of yield loss in chilli dueto chilli leaf curl virus (CLCV) disease. Res Crop 6(2):259–260

Saikia AK, Muniyappa V (1989) Epidemiology and control of tomatoleaf curl virus in southern India. Trop Agric (Trinidad) 66:350–354

Sastry KS, Nayudu MV (1978) Studies on tobacco ringspot virus from brinjal (Solanum melongena L) with particular reference to purification and assessment of losses. Curr Sci 47:846–848

Sastry KSM, Singh SJ (1974) Effect of yellow-vein mosaic virus infection on growth and yield of okra crop. Indian Phytopathol 27(3):294–297

Sastry KS, Zitter TA (2014) Plant virus and viroid diseases in the tropics: volume 2: epidemiology and management. Springer Science & Business Media, Dordrecht

Scholthof KB, Adkins S, Czosnek H, Palukaitis P, Jacquot E, Hohn T (2011) Top 10 plant viruses in molecular plant pathology. Mol Plant Pathol 12(9):938–954

Schuster DJ (1992) Report. Bemisia (Work group on Bemisia tabaci, Newsletter). 5:1–3

Singh SJ (1990) Etiology and epidemiology of whitefly-transmitted virus diseases of okra in India. Plant Dis Res 5:64–70

Singh BR, Singh DR (1975) Brinjal mosaic incidence and effect on yield. Indian J Farm Sci 3:118–119

Singh RB, Srivastava KK, Khurana SMP, Pandey SK (1997) Assessment of yield losses due to potato stem necrosis disease. Indian J Virol 3:135–137

Sinha SN, Chakrabarthi AK (1978) Effect of yellow vein mosaic virus infection on okra seed production. Seed Res 6:67–70

Singh SJ, Sastry KSM, Sastry KS (1973) Effect of oil spray on the control of tomato leaf-curl virus in field. Indian Journal of Agricultural Sciences 43:669–672

Thresh JM (1982) Plant virus disease epidemiology. Oxford University Press.

Tsai WS, Shih SL, Lee LM, Wang JT, Duangsong U, Kenyon L (2013) First report of Bhendi yellow vein mosaic virus associated with yellow vein mosaic of okra (Abelmoschus esculentus) in Thailand. Plant Dis 97(2):291

USDA (1965) Losses in agriculture, Agric Res Serv, Agric Handbook. 291. USDA, Washington, DC, 120 pp

Vaghi Medina CG, Martin DP, López Lambertini PM (2015) Tomato mottle wrinkle virus, a recombinant begomovirus infecting tomato in Argentina. Archives of Virology 160(2):581–585

Vaghi Medina CG, López Lambertini PM (2012) Tomato dwarf leaf virus, a New World begomovirus infecting tomato in Argentina. Archives of Virology 157(10):1975–1980

Van Munster M, Yvon D, Vile B, Dader A, Fereres, Blanc S (2017) Water deficits enhance the transmission of plant viruses by insect vectors. PLoS One 12(5):e0174398

Venkataravanappa V, Swarnalatha P, Reddy CNL, Chauhan N, & Reddy MK (2016) Association of recombinant chilli leaf curl virus with enation leaf curl disease of tomato: A new host for chilli begomovirus in India. Phytoparasitica 44(2):213–223

Waterworth HE, Hadidi A (1998) Economic losses due to plant viruses. In A. Hadidi, R. K. Khetarpal, H. Koganezawa (Eds.), Plant virus disease control. APS Press

Wayne T (1971) The incidence and economic importance of watermelon mosaic virus. NZJ Agric Res 14:242–247

WL, Lim DJ, Hagedorn Aphids As Virus Vectors BIMODAL TRANSMISSION OF PLANT VIRUSES Elsevier 237–251

Zambrano K, Geraud-Pouey F, Chirinos D, Romay G, & Marys E (2011) Tomato chlorotic leaf distortion virus, a new bipartite begomovirus infecting Solanum lycopersicum and Capsicum chinense in Venezuela. Archives of Virology 156(12):2263–2266

# Understanding the Virus Infection Biology and Pathogenesis Mechanisms in Vegetable Crops

5

Vidya Venkat, Mehdi Shahgolzari, Handanahal Subbarao Savithri, and Naropantul Appaji Rao

**Abstract**

Plant viruses, being obligately intracellular parasites, cause widespread infections in plants. They are dependent on the biochemical mechanisms of the host plants for their replication and dissemination, causing several symptoms and resulting in huge economic losses. Viral diseases in vegetable crops seriously impact the growth and yield of these crops, thereby threatening food security. Pathogenesis, the process by which these viruses cause diseases in crops, is associated with a complex interplay between the respective virus and the host. This entails virus entry into the plant cell, replication, intercellular movement, systemic spread, and the resultant host responses that in turn manifest as specific disease symptoms. The defense mechanisms of the plants and the strategies devised by the viruses to evade the plant defenses are critical aspects of this process. Given the enormous diversity of these viruses, the variety of vegetable crops that are affected, and the insect vectors involved in the transmission of plant viruses, there is a compelling need to examine the molecular biology and pathogenesis of these viruses in order to devise integrated management approaches against viral infection and spread. This chapter presents the recent advancements made over the past 5–10 years regarding mechanisms of infection biology and pathogenesis of major plant viruses infecting vegetable crops, as

V. Venkat (✉)
Independent Researcher, Toronto, ON, Canada

M. Shahgolzari
Department of Medical Nanotechnology, School of Medicine, Zanjan University of Medical Sciences, Zan-jan, Iran

H. S. Savithri · N. A. Rao
Department of Biochemistry, IISc, Bangalore, India



R. Kumar et al. (eds.), *Vegetable Crops and Viral Pathogens Volume I*, Plant Pathology for Sustainable Agriculture, https://doi.org/10.1007/978-981-95-7265-6_5

well as their complex interactions with their hosts. Our chapter will serve as an insightful theoretical source for the latest information on viral infection and pathogenesis mechanisms in vegetable crops.

**Keywords**

Virus infection · Pathogenesis · Obligate parasite · Plant virus · Vegetable

## 5.1 Introduction

Plant viruses negatively impact agriculture on a global scale, leading to severe economic damage by compromising crop quality and yield, threatening food security, and impacting the sustenance of farmers, particularly in developing countries. Vegetable crops are necessary for human nutrition as they provide vitamins, antioxidants, fiber, and minerals required for the good health of the global populace by improving digestion, reducing the risk of heart disease, preventing certain cancers, and regulating blood pressure (Abobatta 2021). They combat hidden hunger by providing essential micronutrients, such as vitamin A, iron, iodine, and zinc, vital for growth and cognitive advancement, particularly in children. Vegetable crops undergo several biotic and abiotic threats. The most serious among these are plant viruses impacting the growth, development, and productivity of vegetable crops (Jiang and Zhou 2023). More than 200 virus species have been found to infect these crops, among which begomoviruses, tospoviruses, and potyviruses are the most significant (Venkataravanappa et al. 2024). Plant viruses undergo transmission via several vectors, seeds, and mechanical means (Borah et al. 2023). They elicit widespread infections by inducing morphological, physiological, and anatomical symptoms in their hosts. Symptoms of viral infections can range from moderate to high and, at times, symptomless depending on the type and variant of the viruses, hosts they infect, and environmental conditions or a combination of these factors (Lacomme and Jacquot 2017). These viruses can spread across large areas within a short time period, augmenting the scale and frequency of disease epidemics (Trebicki 2020).

Plant viruses are obligate plant pathogens, replicating in cells by utilizing host components for their own propagation (Gomaa et al. 2024). They are composed of DNA or RNA (ssRNA or dsRNA) genomes encapsidated inside a capsid made up of the capsid protein (Tatineni and Hein 2023). Plant viral capsids display diverse morphologies such as spherical, bacilliform, filamentous, and rod-shaped structures (Evtushenko et al. 2020). Infection pathogenesis starts with the entry of the virus, uncoating of the genome, translation and replication, following which they are transported from cell-to-cell via plasmodesmata, finally spreading systemically within the host plant via the phloem (Morozov and Solovyev 2024; Wu et al. 2024). The lifecycle of the virus is deemed complete upon successful transmission to a new host. Owing to their circumscribed coding capacity, they are dependent on the cellular machinery of the host to complete their infection cycle (Tatineni and Hein 2023). Exploring the infection and spread of plant viruses allows for the design of powerful defense measures to ensure food security. Deciphering how viruses undergo transmission helps in the development of strategies to preclude their dissemination and practice vector control. Discerning the molecular fundamentals of antiviral immune responses of plants enables the genetic engineering and cultivation of crops having augmented resistance. This is particularly important as traditional chemical treatments are increasingly becoming ineffective against several viral infections. Insights into viral infection mechanisms help in the formulation of rapid and precise diagnostic tools. Early detection of plant viral infection is critical for precluding outbreaks and exercising timely virus control measures aimed toward decreasing economic losses due to compromised crop yield and quality. In the complex arms race between plant hosts and viruses, knowledge of viral pathogenesis enables the discernment of antiviral host defenses, such as gene silencing, and how viruses generate silencing suppressors to impede host defenses. Furthermore, studying viral infection mechanisms facilitates the exploration of the cellular machinery of plants such as hormonal signaling pathways, protein synthesis and gene expression. Research into plant virus infection biology helps discern how environmental factors affect the development of symptoms and induce disease severity, affording insights into the management of viral epidemics. There has been no recent review on the infection biology and pathogenesis of plant viruses affecting vegetable crops. In this context, this chapter addresses current literature regarding the basic biochemistry and pathogenesis of plant viruses affecting vegetable crops wherein five major representative viruses, namely the most destructive geminivirus, the tomato leaf curl New Delhi virus (ToLCNDV) as well as the tomato spotted wilt virus (TSWV), potato virus X (PVX), potato virus Y (PVY), and cucumber mosaic virus (CMV), are discussed in detail. Through this chapter, we provide a good theoretical source presenting a bird's eye view of the infection biology and pathogenesis of viruses impacting vegetable crops in addition highlighting recent developments in this field.

## 5.2 CMV

Cucumber mosaic virus (CMV) is among the most destructive viruses invading several cultivated vegetable crops worldwide, causing severe losses. It belongs to the superfamily of alpha-like viruses, family *Bromoviridae*, and genus *Cucumovirus* (Goldbach 1987; Kenyon et al. 2014; Sinha and Samad 2021).

### 5.2.1 Host Range

CMV has a significantly broad host range and infects more than 1200 species of plants across over 100 families, including a variety of vegetables, weeds, and ornamentals. Typical vegetable hosts include cucurbits (such as cucumbers, melons, and squash), carrots, beans, celery, lettuce, peppers, spinach, onion, and tomatoes (Mrkvová et al. 2024).

### 5.2.2 Symptoms

In celery, CMV causes downward curling of immature petioles, conferring plants with a flattened appearance. Clearing of the veins and mosaic symptoms develop on the leaves, and the interveinal areas could become thick and dark green, making the leaves appear crinkled. In cucurbits, CMV causes severe stunting, malformation of leaves and the plants, bright yellow foliar mosaic, and stunting owing to a drastic decrease in stem internodes and leaf size. Flowers show abnormalities in addition to greenish petals. Infected fruits are frequently discolored and distorted, stay small, and, upon severe infection, generate a minute amount of seeds. CMV-infected peppers develop necrotic and mosaic symptoms. Upon systemic infection, leaves remain small and narrow, with stunting of the plants caused by shortened internodes. Infected plants often generate smaller and fewer fruits than uninfected plants. Fruits display a bumpy, wrinkled appearance with pale to yellowish green color, irregular ripening, and sunken lesions having necrotic centers.

### 5.2.3 Disease Cycle

CMV dissemination occurs through different means, such as seeds, mechanical inoculation, aphid transmission, and contaminated tools. Various aphid species like *Pentalonia nigronervosa*, *Myzus persicae,* and *Aphis gossypii* mediate most of the transmission (Palukaitis et al. 1992). The primary source of CMV infection is from CMV-infected weeds or host crops bordering fields. More than 75 aphid species vector CMV, wherein the CMV exists in a non-persistent mode.

### 5.2.4 Genome Organization and Molecular Biology of CMV

The virus has icosahedral morphology with a diameter of ~28 nm (Francki et al. 1966). It is composed of a tripartite genome with linear, positive-sense single-stranded RNA named RNA1, RNA2, and RNA3 in addition to two subgenomic RNAs called RNA4 and RNA4a (Fig. 5.1). Each of the RNA segments has a 5′ terminal methylated nucleotide cap and a 3′ terminal tRNA-like structure found to be implicated in translation, replication, and stability of the RNA (Dreher 1999; Dreher 2009).

### 5.2.5 CMV Replication and Movement

The replication of CMV is an intricate process implicating both host factors and viral proteins (Huh et al. 2011; Seo et al. 2009; Cillo et al. 2002). The CMV 1a and 2a proteins are necessary for the assembly of the active replication complex and are primarily localized on the tonoplasts/vacuolar membranes. The virus interacts with Tsip1, a host-encoded zinc finger protein that is involved in the regulation of

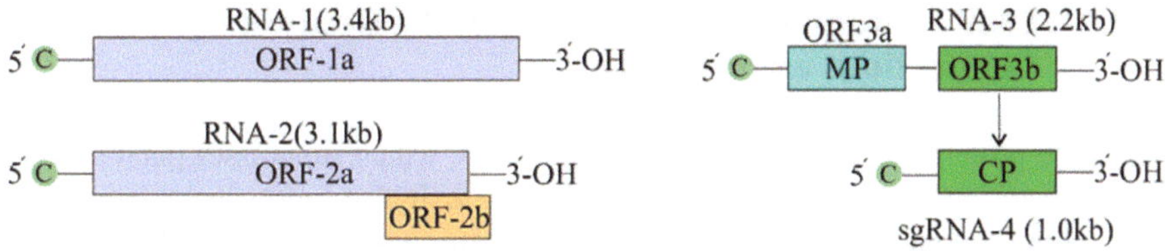


**Fig. 5.1** Genome organization of CMV: RNA1 codes for a single, highly conserved 1a protein of 110 kDa. The N-terminus of this protein harbors a methyltransferase domain implicated in RNA capping of the viral genomic and subgenomic RNAs, while the C-terminus functions as a helicase with NTPase/helicase motifs that enable viral RNA unwinding (Rozanov et al. 1992; Palukaitis and García-Arenal 2003). RNA2 codes for 1a and 2a proteins that are implicated in the generation of the replicase complex that enables virus replication (Palukaitis and García-Arenal 2003), wherein the 2a protein acts as the RNA-dependent RNA polymerase. The sgRNA4a originated from the 3′ region of RNA2 codes for the 2b protein that acts as a suppressor of the RNA silencing mechanism of the host (González et al. 2010; Guo and Ding 2002), in addition to acting as a pathogenicity determinant (Ding et al. 1996; Soards et al. 2002). It also facilitates aphid transmission as well as long-distance transport of CMV (Ziebell et al. 2011). RNA3 encodes the movement protein (MP) that is implicated in virus movement (Suzuki et al. 1991), and its subgenomic RNA called sgRNA4 encodes the capsid protein (CP), which performs genomic RNA encapsidation into virus particles, mediates cell-to-cell movement and systemic dissemination, influences expression of symptoms, and enables aphid transmission (Chen and Francki 1990; Shintaku et al. 1992; Mochizuki and Ohki 2012). CP is the sole protein found in the virus particles and is the singular determinant of aphid transmission

replication (Huh et al. 2011; Seo et al. 2009). The Tsip1 protein interacts with both the CMV 1a and 2a proteins, wherein these interactions are essential for assembly and subcellular localization of the replicase complex, besides regulating RNA replication (Huh et al. 2011; Seo et al. 2019). Upon entering the plant, CMV invades the epidermal cells where it replicates and spreads cell-to-cell through the plasmodesmata, finally reaching the plant vascular system to enable systemic infection. The MP encoded by RNA3 interacts with the cytoskeleton of the plant and increases the plasmodesmata size exclusion limit (Blackman et al. 1998). The replication of CMV is a dynamic process implicating the interactions of viral proteins, host factors and the cellular environment. Discerning these interactions is essential for formulating effective strategies to contain CMV infection.

### 5.2.6 CMV Assembly

The assembly of CMV involves the generation of icosahedral virions that measure 29 nm in diameter, constituted by 180 copies of CP, the single capsid protein, and three segments of CMV genomic RNAs. The capsid is made of the CP that envelopes the viral RNAs, wherein the subunits of CMV self-assemble to surround the viral genomic RNAs, leading to the formation of mature virions. The process of assembly implicates interactions between the viral RNAs and the CP via the basic N-terminal arm of the CP. Mutational analysis has revealed the role of specific amino acid residues in aphid transmission and assembly.

### 5.2.7 CMV Pathogenesis and Impacts of the Occurrence of CMV Coinfections with satRNAs

CMV 2b protein is a major factor in pathogenesis, which suppresses plant defense mechanisms and helps in establishing the infection. It also causes hypomethylation of the host genome, leading to up- or down-regulation of host genes. CMV MP can suppress the production of ROS. Isolates of both CMV I and II groups enable satRNA replication as an additional genomic segment. These short non-coding satRNA molecules are encapsidated within the virions (Palukaitis 2016) and, as a result, spread along with the associated isolate of CMV (Alonso-Prados et al. 1998).

### 5.2.8 Symptom Expression Is Influenced by Coinfections of CMV with Satellite RNAs

CMV satRNAs can variously impact the expression of symptoms based on the host plant. For instance, D-satRNA elicits necrosis in tomato plants infected with CMV while attenuating symptoms in tobacco plants (Garcia-Arenal and Palukaitis 1999). Similarly, CMV-PepY-satRNA occurring in pepper elicits chlorosis in hot pepper while not showing such symptoms in tomato or tobacco (Choi et al. 2011). Strain K8-satRNA caused attenuation of symptoms in tomato and tobacco but did not affect symptom development in squash.

### 5.2.9 satRNA Effects on CMV Accretion and Pathogenicity

Usually, the occurrence of satRNAs associated with CMV decreases the helper virus titer (Escriu et al. 2000; Palukaitis and García-Arenal 2003; Liao et al. 2007). Some studies propose that this effect of the satRNA could correlate with the replication competition between the satRNA and its helper virus (Wu and Kaper 1995). Symptom attenuation observed in plants infected with CMV was surmised to result from the lowered accretion of CMV genomic RNAs (Liao et al. 2007). However, the decreased accumulation of CMV due to the presence of its satRNAs does not always correlate with satRNA-enabled attenuation of symptoms, as satRNAs that cause necrosis in tomatoes also diminish CMV accretion (Escriu et al. 2000). satRNAs could function as RNA silencing targets, thus contributing to the protection of helper RNAs, particularly that of the subgenomic RNA4A (Hou et al. 2011).

## 5.3 Tomato Leaf Curl New Delhi Virus (ToLCNDV)

Geminiviruses are the largest group of plant viruses that infect diverse crop varieties, including vegetables. There are about 243 species of the genus *begomovirus* (family *Geminiviridae*) prevalent mainly in the subtropical and tropical regions, which impact the cultivation of vegetable crops and ultimately lead to significant crop losses. They are classified based on the number of genomic components (that could be monopartite or bipartite), the host they infect and their geographical origins. The tomato leaf curl New Delhi virus (ToLCNDV) of the genus *begomovirus* poses a major impediment to tomato cultivation as it is responsible for causing the foremost economically significant disease infecting tomato crops and other commercially important crops in the Indian subcontinent (Moriones et al. 2017).

### 5.3.1 Symptoms of ToLCNDV Infection

ToLCNDV elicits distinctive symptoms on several host plants. In tomato plants, it causes curling, yellowing, and mosaic patterns on leaves with stunted growth, frequently leading to poor setting of fruits (Chakraborty et al. 2008). In cucurbits like pumpkins, zucchini, and melons, classic symptoms such as vein clearing, yellow mosaic patterns, and blistering of leaves appear, accompanied by deformed or malformed fruits. In sponge gourd plants, symptoms such as severe mottling and yellow mosaic appear on the leaves. In cucumber and watermelon, upward curling and yellowing of leaves occur (Juárez et al. 2019; Venkataravanappa et al. 2020). These typical symptoms of varying intensity reveal that the ToLCNDV has serious, diverse impacts on its host plants that compromise crop quality as well as yield potential.

### 5.3.2 Wide Host Range

ToLCNDV is regarded as among the most significant and dangerous begomoviruses due to its pathogenicity. It has a diverse host range wherein nearly 58 species of plants have been recognized as natural hosts. Initially, this virus was identified in major crops such as eggplant, pepper, potato, and tomato within the Indian subcontinent. Later on, it spreads into cucurbit species like cucumber, melon, zucchini, and pumpkin within the Mediterranean region (Health et al. 2020; Padidam et al. 1995; Parrella et al. 2020; Phaneendra et al. 2012; Pratap et al. 2011; Ruiz et al. 2015; Usharani et al. 2004; Zaidi et al. 2017; Kumar et al. 2025). Many other species of cucurbits, including gourds (spine gourd, ridge gourd, bottle gourd, sponge gourd, and wax gourd), chayote, luffa, and watermelon, are also reported as ToLCNDV hosts specifically in Asia (Anwar et al. 2020; Espino De Paz et al. 2019; Nagendran et al. 2017; Troiano and Parrella 2023; Venkataravanappa et al. 2018b, 2019, 2020; Wilisiani et al. 2019). Besides major families (Cucurbitaceae and Solanaceae), crops in the families Apiaceae (carrot), Malvaceae (kenaf, okra, cotton), Fabaceae (soybean, mung bean), and Caricaceae (papaya) have also been found to be susceptible to ToLCNDV infection wherein however, no major losses have been identified as yet (Jamil et al. 2017; Sivalingam et al. 2011; Venkataravanappa et al. 2018a; Zaidi et al. 2016).

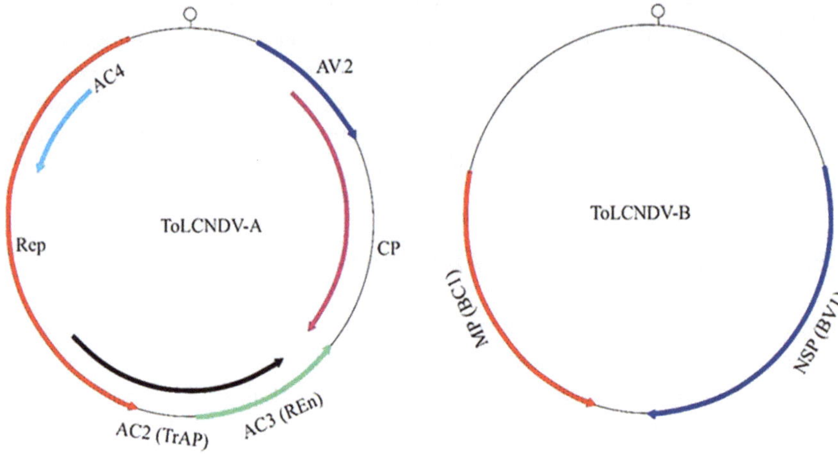


**Fig. 5.2** The DNA-A segment harbors six ORFs named as AV1, AV2, AC1, AC2, AC3, and AC4, which are implicated in crucial roles such as viral replication, encapsidation, and pathogenesis (Vo et al. 2025). The AC1 ORF gene product functions as a replication initiator protein called Rep, which initiates rolling circle replication via its nickase and ligase activities. The AC1 gene also harbors the AC4 ORF contained within its sequence. The AC4-encoded protein acts as a determinant of pathogenicity as well as an RNA silencing suppressor (RSS). The AC2 ORF codes for a 15 kDa polypeptide that acts as both an RSS and a pathogenicity factor. The AC3 generates a replication enhancer protein of size 15.2 kDa. The AV1 ORF encodes the capsid protein, while the AV2 ORF, which overlaps with the AV1 ORF, codes for a pre-coat protein with probable additional function as an RSS. The DNA-B segment enables movement of the virus by encoding two proteins, BV1 and BC1, expressed from its ORFs (Hanley-Bowdoin et al. 2013). The DNA-B segment has the ORFs BV1 and BC1, which respectively code for a nuclear shuttle protein (NSP) and a movement protein (MP). The NSP is unique to members of bipartite begomoviruses and plays a role as an avirulence determinant and a symptom determinant. MP facilitates cell-to-cell movement of the virus. Analogous to other bipartite begomoviruses, both MP and NSP are essential in promoting systemic infection

### 5.3.3 Genome Organization of ToLCNDV

ToLCNDV has a bipartite genome characterized by circular single-stranded DNA-A and DNA-B genomic components of size 2.7 and 2.6 kb, respectively (Fig. 5.2).

DNA-A is capable of autonomous replication (Panno et al. 2019). However, DNA-B is unable to replicate without the occurrence of DNA-A. The DNA-B is vital for the systemic movement of the virus, nuclear localization, systemic infection, and symptom expression (Sanderfoot and Lazarowitz 1996).

### 5.3.4 ToLCNDV Replication

Within the DNA-A and DNA-B segments, the Open Reading Frames (ORFs) are arranged as two divergent clusters divided by an intergenic region. This intergenic region harbors conserved sequences called the common region (CR) occurring in both DNA components (Chatterji et al. 1999). Within the CR sequence, there occurs the origin of replication (ori) that is necessary for replication initiation. The ori is characterized by a conserved hairpin structure and a Rep protein interaction site placed upstream of the hairpin (Fontes et al. 1994a, b). The Rep protein is indispensable for virus replication and possesses a nickase-ligase activity. Rolling circle replication is initiated by a site-specific cleavage occurring within the loop of the TAATATTAC, the conserved nonamer sequence (Heyraud-Nitschke et al. 1995; Laufs et al. 1995a, b; Orozco and Hanley-Bowdoin 1996). The binding site of the Rep protein is situated between the Rep gene transcription start site and the TATA box. This functions as the origin recognition sequence as well as a negative regulatory sequence for the transcription of the Rep gene (Eagle et al. 1994; Eagle and Hanley-Bowdoin 1997; Fontes et al. 1994b). The Rep protein interacts with the CR, specifically with the iterons in the CR sequence, wherein it causes a nick at a specific site in the DNA. This nicking generates a 3′-OH primer for the DNA polymerase of the host that then generates a complementary DNA strand, which displaces the original virus strand. The displaced DNA single strand acts as a template for synthesizing new molecules of viral DNA. The replication occurs unidirectionally, proceeding in a single direction around the DNA circle. ToLCNDV DNA and proteins associated with replication are localized within the infected host cell nuclei (Prasad and Prasad 2022). Replication of ToLCNDV is affected by host factors such as the FKBP-type peptidyl-prolyl cis-trans isomerase, which is involved in interaction with the virus MP and influences viral replication (Chang et al. 2022). The ToLCNDV can occur along with other viruses and even satellites wherein these combined infections influence host range and replication of the virus.

### 5.3.5 Transmission and Disease Cycle

ToLCNDV, like all other begomoviruses, undergoes transmission primarily by whiteflies, *Bemisia tabaci*, in a persistent and circulative fashion (Butter and Rataul 1977). Even one whitefly is capable of virus transmission, wherein females transmit the virus more efficiently than their male counterparts. Whiteflies need a 30-min minimum acquisition access time to enable transmission of the virus. Extending the acquisition and inoculation access times leads to greater transmission efficacies. Ghanim and Czosnek (2016) report that intimate cellular and molecular interactions occur between the proteins and genes of both the begomovirus and the whitefly and that critical protein receptors within the B. tabaci gut allow the entry of the virus into the hemolymph of the insect (Ghanim and Czosnek 2016). However, ToLCNDV can also be transmitted via the pollen and seeds in some cases. The disease cycle is associated with whitefly transmission, replication of the virus in the host plant, and probable seed transmission. The spread of the virus can be rampant, particularly in areas with abundant populations of whiteflies and vulnerable host plants.

### 5.3.6 ToLCNDV Pathogenesis

The ToLCNDV is conceivably the most destructive among bipartite begomoviruses within the Indian subcontinent (Moriones et al. 2017). The successful establishment of the virus within the host is due to attenuation of the defense responses of the host, followed by efficacious replication, encapsidation, movement, and spread. Plant defense response suppression is vital for viral infection. In tomato plants, genetic resistance mediated by Ty resistance loci for controlling viruses associated with the tomato yellow leaf curl disease has been recruited to control losses incurred due to ToLCNDV infection (Prasanna et al. 2015). Some among these loci have been functionally characterized, suggesting their involvement in RNAi-mediated host defense (Butterbach et al. 2014; Verlaan et al. 2013). To counteract this host defense, begomoviruses are armed with proteins having RSS activity that control pathogenicity. Viral proteins expressed by the ORFs BV1, AV2, AC2, and AC4 play roles in establishing pathogenicity, likely through the regulation of RNAi that constitutes a basal host defense mechanism.

### 5.3.7 Virus Recombination as a Diversification Mechanism

At the field level, isolates of ToLCNDV have been recognized in mixed infections accompanied by isolates of many other bipartite begomoviruses, including cotton leaf curl virus, chili leaf curl virus, bhendi yellow vein mosaic virus, betasatellites, and numerous tomato-begomoviruses (Chauhan et al. 2023; Sarkar et al. 2025; Mazumder et al. 2025; Venkataravanappa et al. 2015; Singh et al. 2016; Zaidi et al. 2016). Genetic recombination enables parental viruses occurring in mixed infections to interchange genetic information and transmit it to their progeny viruses in a manner of parasexual reproduction. This is a vital process involved in the evolution of several virus families and has been widely reported for geminiviruses (Lefeuvre and Moriones 2015). Recombination-based exchanges at the interspecies and intraspecies levels are often observed in begomoviruses. Studies of sequences of ToLCNDV have propounded that recombination is the driving factor in the evolution of this virus. Therefore, as reported by Fortes et al. (2016), the origin of ToLCNDV isolates differing from the other isolates at the level of the strain threshold has been attributed to recombination (Fortes et al. 2016). Recombination sites within the ToLCNDV DNA-A indicate that sequences having the ORFs AV2, AC2, and AC4 are more susceptible to recombination. Intriguingly, viral proteins expressed by these ORFs play roles in RSS activity and pathogenicity.

### 5.3.8 Biology and Association of Plant Viral Satellites with ToLCNDV

Analogous to the monopartite begomoviruses of the Old World, it has been found that ToLCNDV is associated with satellite DNAs at the field level (Kumar et al.

2025; Zaidi et al. 2016). These satellites can be betasatellites and alphasatellites (Health et al. 2020). In Pakistan, an alphasatellite is involved in the tomato leaf curl disease (Briddon et al. 2004). Another alphasatellite involved in cotton leaf curl disease has been obtained from crops having tomato leaf curl disease in Varanasi, India. The ToLCNDV association with both the betasatellite and alphasatellite simultaneously has been further reported (Anwar 2017). The involvement of the chili leaf curl betasatellite (ChLCB) with the DNA-B component of ToLCNDV and isolates of a bipartite begomovirus, the pepper leaf curl Lahore virus, was first identified from plants affected by chili leaf curl disease in Pakistan (Shafiq et al. 2010). Furthermore, Jyothsna et al. (2013) showed the presence of ToLCNDV occurring in mixed infections in association with isolates of tomato yellow leaf curl Thailand betasatellite (TYLCTHB), papaya leaf curl betasatellite from tomato plants, and cotton leaf curl Multan betasatellite from tomato plants (Jyothsna et al. 2013).

ToLCNDV DNA-A promotes the trans-replication of various betasatellites in tomato and *N. benthamiana* plants, wherein host-specific interactions could be implicated in this process (Ranjan et al. 2014; Jeevalatha et al. 2021). The combination of betasatellite and the DNA-B causes acute disease damage compared to that caused by the disease complex that lacks the betasatellite (Singh et al. 2016). Betasatellite association with the ToLCNDV DNA-A when the DNA-B was absent further demonstrates that the betasatellite is able to substitute the DNA-B component movement function. Investigations showed the presence of betasatellite molecules in most of these infections, propounding the involvement of betasatellites in widening the host range of this virus. ToLCNDV association with the betasatellite enables the long-distance dissemination of the virus within the host plant and augments the accumulation of the virus. Singh et al. (2012) reported that the association between the ToLCNDV and its betasatellites, including the croton yellow vein mosaic betasatellite and radish leaf curl betasatellite isolates, led to leaf curl disease in radish and tomato, mirroring the role played by the betasatellite in influencing the host range of ToLCNDV (Singh et al. 2012). Overall, these findings reiterate the observation that ToLCNDV uses mixed infections with DNA betasatellites efficiently as a way to adapt to many plants that were previously non-hosts. In a pioneering study, Pratap et al. (2011) reported a new ToLCNDV variant infecting eggplants and eliciting yellow mosaic disease in India (Pratap et al. 2011). The ToLCNDV has been by and large identified to be seedborne, however, not in melon plants (Fortes et al. 2023). The ToLCNDV host range reportedly extends to over 30 different vegetable crop species, including Cucurbitaceae, Malvaceae, Fabaceae, and Solanaceae (Reddy et al. 2024).

## 5.4 Tomato Spotted Wilt Virus (TSWV)

Tospoviruses, the only plant-infecting viruses within the family *Bunyaviridae*, can affect vegetable plants significantly. Among them, Impatiens necrotic spot virus (INSV), tomato spotted wilt virus (TSWV), and groundnut bud necrosis virus (GBNV) (Mandal et al. 2012) have been shown to elicit disease in a diverse array of vegetable crops in varied geographical locations, resulting in decreased crop quality and yield. These viruses undergo transmission by thrips in a circulative and propagative manner, causing a drastic impact with symptoms such as stunting, leaf spotting, uneven ripening, and distorted fruit (Turina et al. 2016).

### 5.4.1 Host Range of TSWV

Among plant viruses, TSWV has one of the broadest host ranges. This pathogen infects over 1000 species of plants across 85 families. It infects vegetable crops, such as eggplants, lettuce, peppers, potatoes, onions, spinach, tomatoes, and squashes, in addition to plants like peppermint, tobacco, floriculture plants, and weeds. https://ipm.ucanr.edu/agriculture/tomato/tomato-spotted-wilt/#gsc.tab=0

### 5.4.2 Symptomatology of TSWV

Tomato plants infected with TSWV show a bronzed appearance with small brown lesions. Peppers, upon infection, develop chlorotic ring spots and/or tiny brown necrotic spots on the plant foliage. In both peppers and tomatoes, infected foliage becomes riddled with several spots that turn the entire leaf brown with rolling and wilting symptoms. Infection at the early stages leads to complete loss of fruit. Established plants upon infection show stunting with diminished yields. Brown necrotic sunken areas and yellow ringspots emerge in the fruits of both plants (Andy Cooper 2024). TSWV infection in potatoes leads to the growth of smaller tubers, reducing crop yield.

### 5.4.3 Disease Cycle and Epidemiology

TSWV undergoes transmission by thrips belonging to the order Thysanoptera. At least ten distinct species of thrips have been found to function as TSWV vectors, such as *Frankliniella occidentalis* (the Western flower thrips), *Thrips tabaci* (the onion thrips), *Frankliniella fusca* (tobacco thrips), and *Scirtothrips dorsalis* (chili thrips) (Andy Cooper 2024). The thrips vector is also infected by TSWV, and nymphs pick up the virus through feeding on the plants infected with TSWV. Upon laying eggs on TSWV-infected plants, the larvae that hatch from these eggs can acquire TSWV since they feed by puncturing the surface cells and extracting their contents. To enable the uptake of TSWV, the larvae and adults have to feed for over 15 min. Once the larvae undergo maturity into adults, they fly and feed on other crops, transmitting TSWV in the process. There is no effective cure once a plant becomes infected. There exist only strategies for minimizing the future impact of this virus. Plants remain asymptomatic for some time following inoculation, and virus symptoms may not be detectable until weeks or days once transmission takes place.

### 5.4.4 Molecular Biology of Tomato Spotted Wilt Virus

#### 5.4.4.1 Genome Organization, Cytopathology, and Replication of TSWV

The *Tospovirus* particles are of spherical morphology and are enclosed within a host-derived membrane having an 80–120 nm diameter. The membrane is composed of two glycoproteins, Gc and Gn, that generate spikes on the virus surface and are essential for the acquisition and transmittance of the virus by thrips vectors (Turina et al. 2016). The virus particle core harbors ribonucleoproteins (RNPs) composed of RNA-dependent RNA polymerase (RdRp/L protein) and three segments of genomic RNA that are tightly encapsidated within the nucleocapsid protein (N). The RNPs appear to be pseudocircular due to the complementarity of sequence in the genomic RNA termini. The *Tospovirus* genome is of ambisense/negative polarity and is composed of three segments of RNA called large (L, ~8.9 kb), small (S, ~2.9 kb), and medium (M, ~4.8 kb), based on their sizes (Turina et al. 2016) (Fig. 5.3). In all three RNA segments, the termini contain a highly conserved, complementary (5-AGAGCAAU and 3-UCUGCUUA) sequence within their first eight nucleotides that is common to all negative-strand segmented RNA viruses, and function as essential promoters of replication and transcription.

Tospoviruses undergo replication in the cytoplasm, wherein electron-dense areas are commonly referred to as viroplasms. In the process of replication, genomic RNA molecules of the virus concurrently associate with the RdRp and N proteins to form infectious progeny RNPs. During the initial stages of replication of the virus, RNPs of the infectious progeny further bind with NSm to enable their transportation via modified plasmodesmata guided by tubules. However, in the case of GBNV, the NSm was shown to bind to ER through the C-terminal coiled domain to form vesicles that could be responsible for the transport of RNPs (Singh and Savithri 2015). In later stages, these progeny RNPs traverse toward the Golgi complex and the ER, dependent on actins, following which they assemble and concentrate at loci harboring the Gn and Gc glycoproteins of the virus. These Gn–Gc RNP complexes move from the ER to the Golgi complex, where they are enwrapped within a Golgi stack, to generate double-enveloped virus particles (DEVs).

### 5.4.5 Pathogenesis of TSWV: Host Factors and Impacted Molecular Pathways

Among the viral encoded proteins, NSS, acting as a silencing suppressor, and NSM are crucial for pathogenesis. Homologs of DNA-J-protein from tomato, *Arabidopsis*, and *Nicotiana tabacum* are among the first plant proteins recognized to associate with the NSM for cell-to-cell transportation of the *Tospovirus*, which indicates that infectious RNP movement likely implicates a mechanism requiring Hsp-70 (Soellick et al. 2000; Von Bargen et al. 2001). These studies also revealed associations with At-4/1 chaperone protein that shares homology with the alpha-helical domains of kinesin-, ankyrin-, and myosin-like proteins (Von Bargen et al. 2001), wherein these

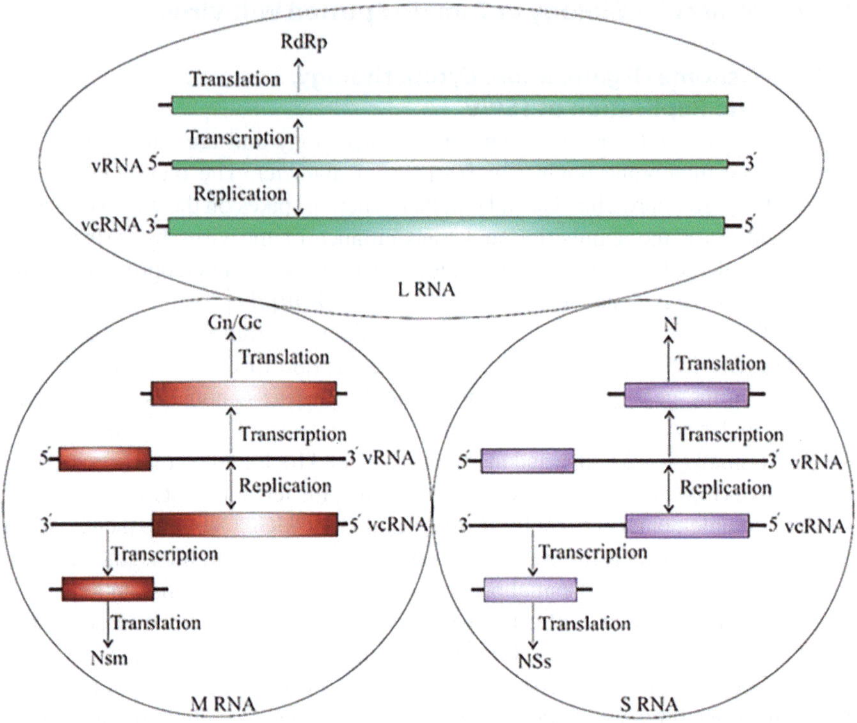


**Fig. 5.3** The L RNA has a single large ORF occurring on the viral complementary (vc) strand coding for the ~330 kDa RdRp (Mitter et al. 2013). The S and M RNAs both display an ambisense gene arrangement coding for a couple of nonoverlapping ORFs on each complementary strand (the vcRNA and vRNA). These ORFs are divided by an intergenic region (IR) in between and do not overlap. The IR sequence has stretches of A- and U-rich regions predicted to form a stable, folded, stem-loop/hairpin structure. The M RNA encodes the nonstructural protein involved in cell-to-cell transport (NSM), and this coding is derived from the viral (vRNA) strand, while the vcRNA strand encodes the protein precursor of the Gn and Gc glycoproteins (n and c refer to the amino-terminal and carboxy-terminal positions of the precursor proteins, respectively). The S RNA encodes the second nonstructural (NSS) protein that acts as a suppressor of RNA silencing on the vRNA strand, while the vcRNA strand codes for the N protein (Turina et al. 2016). Upon entering the cytoplasm of the host cell, Tospoviruses undergo transcription to mRNAs that are close to genome-length for the gene encoding the RdRp generated from the L RNA and are notably smaller than corresponding genome segments specific for all genes encoded by the ambisense S and M RNA segments (Kormelink et al. 1992a). Viral RNAs contain a cap structure but do not have a typical eukaryotic poly(A)-tail. Due to the absence of methytransferase activity, the RdRp of the virus is unable to provide a cap structure for the viral mRNAs. In lieu, cap structures are contributed by the cap snatching mechanism, wherein the viral RdRp protein snatches capped leader RNA sequences from the host mRNA cytoplasmic pool (Duijsings et al. 2001; Van Knippenberg et al. 2005). This snatched cap leader sequence in turn primes the transcription of the viral genomic RNA template (Kormelink et al. 1992b). GBNV NSS was also shown to possess helicase activity (Bhushan et al. 2015)

proteins are implicated in viral extra- and intracellular movement (Morozov et al. 2014; Paape et al. 2006). Intracellular dissemination of the N-RNPs, dependent on actin, needs a myosin XI K motor protein (Feng et al. 2013; Ribeiro et al. 2013). The host factor, eukaryotic translation elongation factor 1A (eEF1A) isolated from evacuolated BY-2 tobacco protoplast extracts, was identified to be essential for TSWV transcription and replication (Komoda et al. 2014).

## 5.5 Potato Virus X (PVX)

### 5.5.1 Infection Biology and Pathogenesis

Potato virus X is considered the type member of the *Potexvirus* genus in the *Alphaflexiviridae* family and is among the earliest identified potato viruses (Loughnane and Murphy 1938). The PVX virion measures 470–580 nm in length, with a diameter of 12–13 nm, and particles are filamentous, flexuous, and non-enveloped (Dutta et al. 2024). The viral RNA genome is linear and (+) sense and functions as the messenger RNA. It is encapsidated within a protein coat containing 1300 capsid protein (CP) subunits assembled as left-handed helical particles (Grinzato et al. 2020).

### 5.5.2 Genome Organization

The PVX genome measures 6.4 kb in size, having a 5′ methylated cap in addition to a 3′ poly-A tail (Fig. 5.4) (Sonenberg et al. 1978). The viral genome harbors five ORFs (Huisman et al. 1988). ORF 1 codes for the replicase (165 kDa) generated from the genomic RNA. The replicase is the sole protein found to be necessary for PVX replication (Longstaff et al. 1993; Doronin and Hemenway 1996; Plante et al. 2000).

The ORFs 2, 3, and 4 code for a set of three genes that are overlapping and together addressed as the triple block genes (TGBs). These ORFs generate three proteins called TGBp1, TGBp2, and TGBp3. The TGBp2 and TGBp1 proteins are generated from two distinct sgRNAs, and TGBp3 is synthesized by leaky scanning of the subgenomic RNA encoding TGBp2. The p25 TGBp1 protein is implicated in the translation and transport of the viral RNA in addition to its involvement as a suppressor of antiviral silencing (Voinnet et al. 2000). This protein also functions as

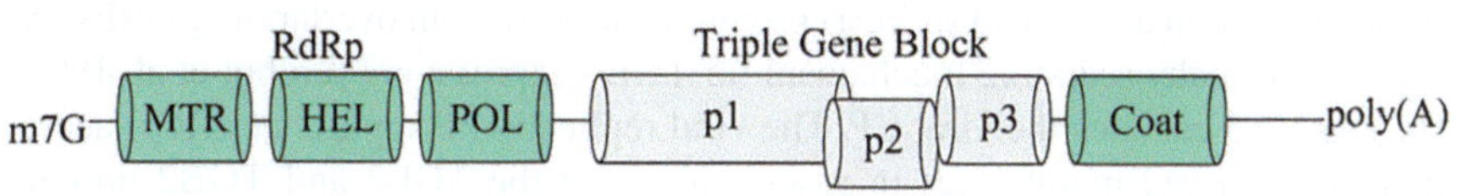


**Fig. 5.4** Genome organization of PVX: MTR stands for methyltransferase, HEL for helicase, POL for polymerase; p1, p2, and p3 are TGB proteins, coat stands for coat protein. (Adapted from Verchot J. 2022)

the major pathogenicity determinant responsible for the development of symptoms and the synergistic interaction between potex and potyviruses (Aguilar et al. 2015). TGBp3 and TGBp2 are associated with the ER membrane and possess transmembrane domains (Krishnamurthy et al. 2003; Mitra et al. 2003). The TGBp2 stimulates the formation of vesicles derived from the ER membrane and these vesicles are necessary for viral movement (Ju et al. 2007). The last ORF, ORF5, codes for the 25 kDa CP that is also generated from the subgenomic RNA promoter (Huisman et al. 1988; Lee et al. 2000). The CP enables virion formation, cell-to-cell transportation, and systemic vascular transport. It also functions as an inducer of the R-gene-associated plant resistance (Rathjen and Moffett 2003). The 5′ untranslated part of the PVX genome harbors the origin of assembly sequence, which overlaps with the sequences essential for virus replication as well as movement (Sit et al. 1994; Kwon et al. 2005; Lough et al. 2006). Only the PVX gRNA is encapsidated within the protein shell of the viral particles (Dutta et al. 2024). The PVX RNA coding sequence contains an 84-nucleotide UTR (untranslated region) at its 5′ end (Smirnyagina et al. 1991) and a 74-nucleotide UTR at its 3′ end (Pillai-Nair et al. 2003). These sequences are cis-acting and highly conserved between potexviruses and are vital for the movement (Lough et al. 2006) and replication (Kim and Hemenway 1996) of the virus.

### 5.5.3 Replication and Movement

Upon entry into the cells, virions of PVX unpack via phosphorylation of the exposed CP N-terminus using cellular enzymes. Following this, the exposed RNA genome undergoes translation to synthesize the viral RdRp, which enables viral replication and synthesis of the subgenomic RNAs. Replication takes place inside structures referred to as X-bodies that assemble close to the nucleus (Tilsner and Oparka 2012). The viral RdRp is the singular protein encoded by the PVX genomic RNA, and the rest of the viral proteins are expressed via subgenomic (sg)RNAs that are 3′ coterminal (Huisman et al. 1988; Ravin et al. 2008). The PVX RdRp is nearly 166 kDa in size and harbors an N-terminal conserved methyltransferase motif essential for the addition of the 5′ methyl guanosine cap, along with the catalytic motifs for a central helicase and an RNA polymerase at the C-terminus. The TGB proteins are generated from a conserved set of three ORFs that code for the virus movement proteins via two sgRNAs. The sgRNA1 codes for the 25 kDa TGB1 protein and can also express all the other TGB proteins as well through the ribosome leaky scanning mechanism (Fujimoto et al. 2021). The sgRNA2 codes for the TGB2 (12 kDa) protein and TGB3 (8 kDa) protein generated from overlapping ORFs that also use the leaky scanning mechanism for TGB3 expression (Verchot et al. 1998). The sgRNA3 encodes the viral CP. The viral replication complexes (VRCs) accrue along ER-derived membranes in association with the TGB3 and TGB2 proteins (Bamunusinghe et al. 2009). VRCs occur as membrane-bound bodies that aggregate in proximity to the nucleus or near plasmodesmata, wherein the adjacent plasmodesmata particularly orient the viral RNA toward cell-to-cell movement. Viral genome recruitment from the VRCs to the plasmodesmata entrance is referred to as "co-replicational insertion" (Linnik et al. 2013; Tilsner et al. 2013). The TGB3 and TGB2 proteins assign the CP and TGB1 to transport viral RNA present inside the plasmodesmata into neighboring cells. The CP and TGB proteins are localized within specific subdomains inside the plasmodesmata. The TGB3 and TGB2 proteins are located in the cortical ER close to the entrance and exit of the pores of plasmodesmata (Tilsner et al. 2013). The TGB1 has multiple activities, such as RNA binding, RNA helicase, and ATPase activities, and harbors a canonical helicase/NTPase domain (Leshchiner et al. 2006; Morozov and Solovyev 2015). TGB1 enhances the plasmodesmata size exclusion limit to facilitate the intercellular movement of viral RNA (Howard et al. 2004; Parker et al. 2002). The TGB1 interacts with the virion CP to enable the translation of the viral RNA (Mukhamedzhanova et al. 2009; Zayakina et al. 2008). Moreover, TGB1 functions as a posttranscriptional gene silencing suppressor through the inactivation of SGS3, which is an RDR6 cofactor and is essential for secondary viral siRNA synthesis that performs systemic RNA silencing. The TGB1 acts as an RNA silencing suppressor by interacting with AGO4, AGO3, AGO2, and AGO1proteins and drives their degradation through the proteasome (Voinnet et al. 2000).

### 5.5.4 PVX Assembly

The flexuous rod-shaped PVX particle contains 1270 CP subunits, having 8.9 subunits per turn, which forms a 3.45-nm helical pitch (Röder et al. 2019; Tollin et al. 1980). The CP C-terminus occurs inside the assembled virus particle while the N-terminus is exposed externally (Sõber et al. 1988; Baratova et al. 1992; Nemykh et al. 2008). PVX CP subunits do not assemble into virus-like filamentous particles without the genomic RNA in vitro or in vivo. This could likely be attributed to the specific recognition of the viral genomic RNA by the CP subunits. Within the viral RNA 5′ region there occurs the origin of assembly sequence consisting of stem loop 1 (Miller et al. 1998; Cho et al. 2012). Assembly starts upon interaction of the CP subunits with this sequence in the viral RNA 5′-terminus. The PVX RNA 5′ cap structure functions as an assembly initiation signal by inducing a distinct conformation within the 5′-proximal site of the RNA (Petrova et al. 2015).

### 5.5.5 Host Range

PVX exhibits a broad host range, infecting 62 species of plants of 27 families (Edwardson and Christie 1997). The *Solanaceae* family of plants is the major host of PVX. This virus is significantly predominant in potato plants (Beemster and De Bokx 1987) throughout the world, while in several countries, important solanaceous crops such as peppers (Paulus et al. 1960; Steepy et al. 1967; Fribourg and Fernandez-Northcote 1972; Makkouk and Gumpf 1974) and tomato (Vanterpool 1925; Ainsworth 1934) are impacted by PVX infection. Despite weeds being natural hosts for PVX, the virus also naturally occurs in other species of vegetable crops, such as turnip (Brassica rapa) in India (Sangar et al. 1980; Samad et al. 1991), red clover (Trifolium pratense) in Belarus (Ambrosov and Bltoskaya 1978), and artichoke (Cynara scolymus) in Italy (Chabbouh et al. 1990; Giunchedi 1973).

### 5.5.6 Transmission of PVX

PVX is widely distributed and readily transmitted through seeds, pollen, or between plants via contact with infected roots or foliage. Predominantly, the virus undergoes mechanical transmission via contact between plants when friction due to machines, humans, animals, wind, etc. occurs. PVX has no invertebrate vector hitherto identified.

### 5.5.7 Symptoms Caused by PVX in Host Plants

PVX elicits a mild mosaic disease in potatoes, tomatoes, tobacco, and other mainly solanaceous plant species. PVX can induce severe rugosity, mosaic or crinkling of the leaves, necrotic streaks, dwarfing, or considerable necrosis of tubers and tops (Beemster and De Bokx 1987; Brunt and Loebenstein 2001; Stevenson et al. 2001). More characteristic symptoms appear in older leaves, wherein prominent greenish veins are visible while the rest of the leaf becomes yellow (Beemster and De Bokx 1987). The tuber size and number can be somewhat reduced (Beemster and De Bokx 1987). PVX coinfections with potato leafroll virus (PLRV) and the potyviruses of potato, namely, PVA or PVY, cause synergistic effects leading to a crop production loss of 40–70% due to PVX-PLRV coinfection (Loebenstein 2008) and 50–80% upon PVX-PVA or -PVY coinfections (Kreuze et al. 2020).

### 5.5.8 Disease Cycle

The disease cycle of PVX starts with mechanical transmission to a new plant host upon contact. Once inside the host cell, the PVX virions are uncoated by the action of cellular enzymes and the N-terminal phosphorylation of the coat protein. The exposed viral RNA undergoes translation to synthesize the viral proteins, including the RdRp. The RdRp enables viral replication within X-bodies, and new virus particles are synthesized, which subsequently spread inside the plant as well as to other host plants, thus completing the infection cycle. The newly synthesized PVX RNA and the CP undergo assembly to form new, filamentous, helical virions.

### 5.5.9 PVX–Host Interactions

Three R genes conferring PVX resistance, specifically the $R_x$, $N_x$, and $N_b$ genes, have been characterized in potato plants (Liu et al. 2021; Cockerham 1970; Solomon-Blackburn and Barker 2001). The $R_x$ gene codes for a receptor having a coiled-coil nucleotide-binding domain with leucine-rich repeat (CC-NLR) that particularly recognizes the PVX CP and bestows extreme resistance (ER), a distinct resistance form bereft of eliciting identifiable local necrotic lesions or hypersensitive reactions (Bendahmane et al. 1995, 1997, 1999; Moffett et al. 2002). The $N_b$ and $N_x$ elicit necrotic lesions at the sites of infection through recognition of viral TGBp1 and CP, respectively (Tommiska et al. 1998; Malcuit et al. 1999; Marano et al. 2002). Additionally, many atypical dominant viral resistance proteins (ADVRPs) from PVX-infected host plants have been described such as the *Arabidopsis thaliana* Jacalin-type lectin required potexvirus resistance 1 (JAX1) protein, the *Nicotiana benthamiana* putative methyltransferase 1 (PNbMTS1) protein, and potato Remorins (REMs). Both PNbMTS1 and JAX1 block virus replication through targeting the RdRp of the virus, while REMs interfere with TGBp1 function in the cell-to-cell transport of the virus (Cheng et al. 2009; Yamaji et al. 2012; Sugawara et al. 2013; Yoshida et al. 2019). Furthermore, the PVX P25 protein acts as the major pathogenicity determinant, inducing systemic necrosis during synergistic infections of PVX and other viruses (Aguilar et al. 2015).

## 5.6 Potato Virus Y (PVY)

Potato virus Y belongs to the potyvirus genus and the *Potyviridae* family. *Potyviridae* is the next largest family of plant viruses following *Geminiviridae* and encompasses nearly 30% of all characterized plant viruses (Lacomme et al. 2017; Inoue-Nagata et al. 2022). *Potyvirus* is among the largest plant virus genus, having 162 species of viruses identified (Inoue-Nagata et al. 2022).

### 5.6.1 PVY Genome Organization

PVY is characterized by flexuous, rod-shaped particles measuring 700 nm in length, with a diameter of 11–13 nm and a helix pitch of 3.4–3.5 nm (Lacomme et al. 2017). The virus particles encapsidate the viral genomic RNA inside a capsid shell made of 2000 units of a single capsid protein (CP) of size 30 kDa. The PVY genome contains a (+)-sense single-stranded RNA of length nearly 9700 nucleotides, in which a viral protein genome-linked (VPg) is linked to the 5′ end of the RNA in addition to having a poly(A) tail at the RNA 3′ end. The genomic RNA of PVY harbors a single ORF that is translated to generate a large polyprotein (340–370 kDa). This polyprotein then undergoes cleavage by three virus-encoded proteases to yield ten multifunctional proteins, namely, P1, P3, 6 K1, HC-Pro, 6 K2, CI, NIa, Nib, CP, and VpG (Danci et al. 2009). An additional protein called P3N-PIPO has been identified to be encoded from the Pretty Interesting Potyvirus ORF (PIPO), wherein this protein is obtained either by a ribosomal slippage causing a +2 frameshift inside the ORF of P3 (Wei et al. 2010) or through the incorporation of an extra nucleotide by slippage of the PVY RdRp at the G1-2A6-7 motif, which is highly conserved at the PIPO sequence 5′ end (Olspert et al. 2015; Rodamilans et al. 2015; White 2015). All of the above proteins exhibit a notable degree of functional multiplicity and are frequently associated with distinct subcellular compartments, including the ER, Golgi, and chloroplasts (Fig. 5.5). These viral proteins interact with several other virus-encoded proteins as well as with host proteins to perform all the vital functions of the virus and complete its life cycle. The association between the NIa-Pro and VpG of PVBV affects the protease activity. The region of the NIa-Pro containing Trp 143 is most probably involved in the interaction of the protease with VPg (Mathur and Savithri 2012). Furthermore, the Ser129 residue of the protease domain undergoes phosphorylation by a host cell kinase. Such VPg-NIa-Pro interaction or Ser129 phosphorylation influences the orientation of the amino acids occurring at the active site, thus affecting catalysis. The PVBV VpG protein binds with the NIa-Pro domain and regulates the protease kinetics, leading to the induction of ATPase activity (Mathur and Savithri 2012).

### 5.6.2 PVY Assembly

The assembly of PVY is a complex process in which the viral ssRNA genome is encapsidated by helical capsid proteins (CPs) to form a virion. Viral assembly is an essential part of the replication cycle of this virus, wherein the CP begins assembly by initially interacting with the 5′ terminus of the viral RNA and then proceeding toward the RNA 3′ end, forming a left-handed helical structure of nearly 8.8 CP subunits per turn (KeŽar et al. 2019). The CP generates a protective layer enclosing the viral RNA, forming a complete virion having a diameter of 130 Å. The CP interacts with the viral ssRNA at a site within a groove located between the C-terminal and core regions of the CP that harbors five nucleotides. The mainly conserved CP amino acid residues Arg157, Ser125, and Asp201 are involved in tight interaction with the genomic RNA. The CP N-terminal region is crucial for assembly, while its C-terminal region is dispensable for this process. However, the binding of the CP C-terminal sequence with the PVY RNA is necessary for the successful execution of viral replication, whereas the CP N-terminus is essential for assembly and virus movement. Both regions of the CP are obligately required to enable systemic infection. The PVBV, a distinct species in the *potyvirus* genus was also shown to require the surface-exposed N—and C-terminal regions for the assembly of virus-like particles (Anindya and Savithri 2003). The CP is multifunctional, acting as a shield for the viral genome, enabling aphid transmission, genome replication, and translation, and local and systemic trafficking of the virus. RNA replication must take place successfully before the stable assembly of the virions. Despite not being a

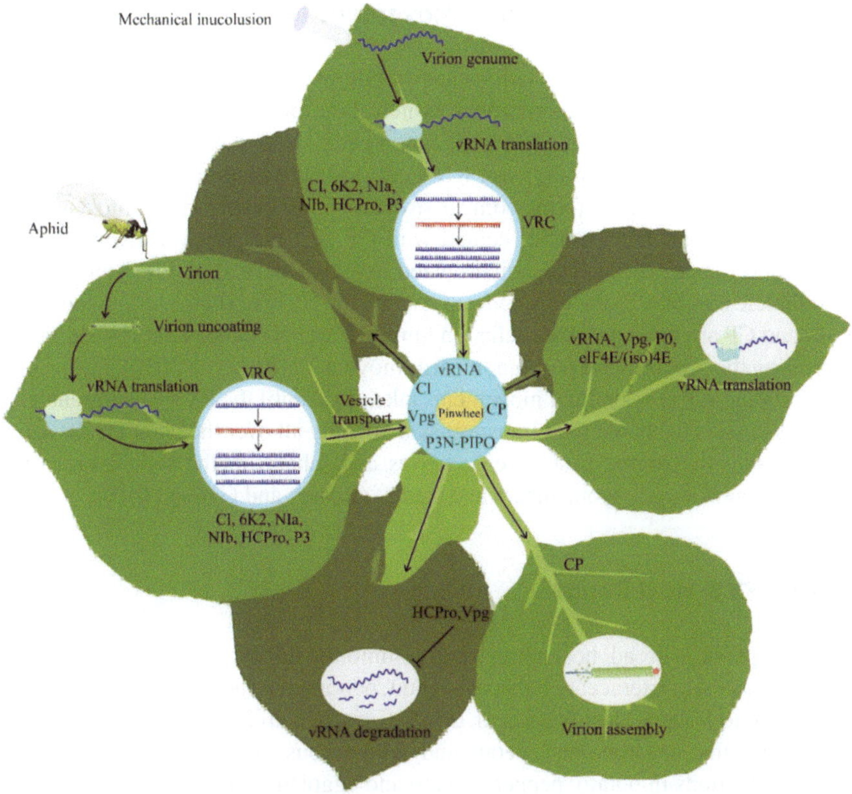


**Fig. 5.5** Replication cycle of PVY. Upon infection of the plant cell, the virus undergoes uncoating, which exposes the viral genomic RNA. Following this, the translational machinery of the host (inclusive of the ribosomes, eIF(iso)4E, etc.) recruits the viral RNA to express viral proteins (Ivanov et al. 2014). Subsequently, the proteins get processed and undergo maturation, whereupon viral replication complexes (VRCs) are formed. These VRCs associate with the host organelle membranes (that of the Golgi, ER, and vesicles). Replication begins through the synthesis of the (−) ssRNA using the (+) sense genomic RNA as the template by the viral replicase. The newly generated (−) ssRNA in turn acts as the template for the replicase to synthesize several (+) ssRNA, which then undergo encapsidation to assemble into new virions or alternatively are engaged by the viral proteins and the host proteins to establish ribonucleic complexes. These ribonucleic complexes, along with virions, engage the cytoskeleton of the host, following which they are transmitted into neighboring cells via plasmodesmata by the coordinated activity of the host and viral proteins

component of the virion structure, viral proteins, inclusive of the HC-Pro and cylindrical inclusion protein (CI), enable proper assembly of the virus. The CI protein creates structures funneling virus particles across the plasmodesmata, facilitating cell-to-cell movement, which suggests a close association between virus assembly and movement (Xue et al. 2023).

### 5.6.3 PVY Systemic and Local Movement

After encapsidation, the PVY virions are mobilized within the first infected cell of the plant to enter the plasmodesmata, which they traverse to gain entry into the neighboring cells and begin a new replication cycle. Both the CP and the CI (RNA helicase) are implicated in the viral RNA cell-to-cell transport across the plasmodesmata, wherein the CI forms conical structures that enable virion movement or that of the ribonucleic particles coated with CP past the cell wall (Roberts et al. 1998; Sorel et al. 2014). The synchronized activity of the P3N-PIPO anchors the ribonucleic complex to the plasmodesmata, and the HC-Pro (helper component protein) and CP augment the size exclusion limit of these plasmodesmata in addition to recruiting the host proteins to promote virus movement (Wei et al. 2010). Succeeding intracellular and extracellular movements take place until the PVY virions enter the phloem vessels to enable their transport systemically all over the plant via the sap to spread from the source tissues (e.g., leaves and tubers) to the sink tissues (e.g., emerging new leaves, roots, tubers and stems) (Revers and García 2015).

### 5.6.4 PVY–Host Range

PVY displays a broad host range and commonly infects plants in the families Amaranthaceae, Asteraceae, Fabaceae, and Chenopodiaceae, and especially plants belonging to Solanaceae (Wani et al. 2021; Kreuze et al. 2020). It also infects several weeds that are non-solanaceous and solanaceous. Among vegetable crops, it causes infections in potato, pepper, tomato and eggplant. (Quenouille et al. 2013).

### 5.6.5 PVY Transmission

PVY is naturally transmitted by aphids and other sap-sucking insects. PVY undergoes non-persistent transmission enabled by aphids (Lacomme et al. 2017). The viral HC-Pro facilitates virus particle binding to the stylets of aphids during short spells of probing on a virus-infected leaf. Consequently, the virus is transmitted mechanically into a new host at the time of further feeding. These non-persistently transferred virus particles are acquired in under a minute's time, do not require any latency period, and are able to spread immediately following acquisition (Torrance and Talianksy 2020). PVY can also disseminate into plants via contaminated farming equipment or tools, or simply by contact with healthy plants following the handling of infected plants. The green peach aphid and the bird cherry-oat aphid are efficacious PVY vectors.

### 5.6.6 Symptoms Elicited by PVY Infection

PVY elicits disease in several types of vegetable crops, such as potato, pepper, tomato, and eggplant. In potato plants, the most typical symptoms are curled and crinkled leaves having light green or yellowish mottling, wherein mottles vary in size and are not defined by leaf veins. Frequently, necrotic spots emerge on the infected leaves. Plants become stunted with lower petioles, and additionally, the leaves develop yellowing, and dead leaves remain clinging to the stem. Such a symptom is referred to as leaf drop. Plants infected with PVY typically die early. Infected potato plants develop smaller and fewer tubers compared to healthy plants. Vein darkening, otherwise termed as vein banding, also develops due to PVY infection.

### 5.6.7 PVY–Host Interactions

The viral VpG recruits the eIF4E protein of the host to enable initiation of viral protein translation. This association is a vital and essential requirement for establishing infection (Lebedeva et al. 2024). The viral VpG also uses the eIF4E host protein for virus replication as well as systemic transport (Lebedeva et al. 2024). The viral 6 K1 protein coded for by PVY blocks the interaction between the host defense proteins in the process of viral infection.

## 5.7 Conclusion

Thus far, plant viral infections have emerged as the primary cause of infections in vegetable crops, posing significant challenges to agriculture on a global scale. Despite scrupulous prophylactic and disease management measures, viral diseases continue to persist, fueled by the emergence of recombinant virus variants, which complicate virus detection and containment. Plant viral populations continue to evolve, underscoring the interaction complexity between the viruses, their hosts, vectors, and the prevailing environmental conditions. The economic outcomes of these infections can be devastating, resulting in global food insecurity. Plant virologists are continuously researching to evolve mitigation strategies through the generation of crop varieties resistant to these viruses, in addition to enhanced disease management practices. However, not all of the interactions between viruses and plants lead to disease and crop destruction. Viruses can co-exist with their host plants without displaying any symptoms, a process called asymptomatic infection. Furthermore, some virus–plant interactions can turn out to be mutually beneficial. For instance, certain viruses can promote plant tolerance to environmental stressors such as extreme temperatures (Anfoka et al. 2016) or drought (Prakash et al. 2024). Effectual mitigation requires multifaceted scientific research across key areas such as the characterization of viral pathogenicity and diversity to design targeted intervention strategies, identification of host genes that confer virus resistance and evaluation of their long-term endurance, investigating virus epidemiology across multiple environments to discern their transmission dynamics, development of sensitive and rapid diagnostic tools to enable early detection of the viruses and finally, the implementation of integrated agricultural management measures tailored to the specific prevalent local conditions. Recent progress in molecular biology and genomics has transformed plant virology. Now, researchers can sequence and characterize entire plant virus genomes that provide significant insights into their infection mechanisms, genetic composition, and evolution. Such information is pivotal in the development of diagnostic tools to enable rapid detection of plant viral infections and to design targeted disease control strategies. Furthermore, the invention of technologies like high-throughput/next-generation sequencing has allowed virologists to identify and explicate new plant viruses at a higher speed and precision. This has been specifically beneficial in the discernment of emerging plant viral diseases and the development of strategies to control virus spread. As we become increasingly dependent on plants for sustenance of life, environmental stability, and economic security, the investigation of plant virology, in particular the characterization of the virus infection biology and pathogenesis, remains of paramount significance. Discerning these miniature saboteurs is vital to preserving our food supply, protecting our ecosystems, and exploiting the potential of research in biotechnology to confront these global challenges. In the dynamic field of science, the complex dance between viruses and plants continues to fascinate researchers, demonstrating the potential of discoveries that could determine the future of plant virology and global agriculture. In the above context, several key scientific queries remain: why do certain strains of plant viruses predominate in specific regions? What promotes the increasing genetic variability of these viruses, and how does it affect virulence and adaptation to the host? Which molecular mechanisms determine virus-host-vector interactions and influence transmission outcomes? How does virus epidemiology vary across wide ecological niches, and which environmental parameters shape infection patterns? Addressing the above questions necessitates interdisciplinary and coordinated scientific research. Through a combination of epidemiological, genetic, and ecological data, scientists can generate predictive models to forecast virus outbreaks and devise timely intervention measures. Finally, advancing our discernment of plant virus infection biology and epidemiology is vital to protecting vegetable crops, enhancing their productivity, and ensuring global food security in the context of emerging viral threats.

## References

Abobatta W (2021) The potential health benefits of vegetable crops. J Food Nutr Health 2:2

Aguilar E, Almendral D, Allende L, Pacheco R, Chung BN, Canto T, Tenllado F (2015) The P25 protein of potato virus X (PVX) is the main pathogenicity determinant responsible for systemic necrosis in PVX-associated synergisms. J Virol 89:2090–2103

Ainsworth G (1934) A comparison of certain English and Canadian potato viruses found infecting tomatoes. Ann Appl Biol 21:581

Alonso-Prados JL, Aranda MA, Malpica JM, García-Arenal F, Fraile A (1998) Satellite RNA of cucumber mosaic cucumovirus spreads epidemically in natural populations of its helper virus. Phytopathology 88:520–524

Ambrosov A, Bltoskaya ZV (1978) Study of strains of potato virus X and their natural sources under Byelorussian conditions

Andy Cooper IM (2024) Tomato spotted wilt virus on tomato and pepper

Anfoka G, Moshe A, Fridman L, Amrani L, Rotem O, Kolot M, Zeidan M, Czosnek H, Gorovits R (2016) Tomato yellow leaf curl virus infection mitigates the heat stress response of plants grown at high temperatures. Sci Rep 6:19715

Anindya R, Savithri H (2003) Surface-exposed amino-and carboxy-terminal residues are crucial for the initiation of assembly in pepper vein banding virus: a flexuous rod-shaped virus. Virology 316:325–336

Anwar S (2017) Distinct association of an alphasatellite and a betasatellite with tomato leaf curl New Delhi virus in field-infected cucurbit. J Gen Plant Pathol 83:185–188

Anwar I, Bukhari HA, Nahid N, Rashid K, Amin I, Shaheen S, Hussain K, Mansoor S (2020) Association of cotton leaf curl Multan betasatellite and Ageratum conyzoides symptomless alphasatellite with tomato leaf curl New Delhi virus in Luffa cylindrica in Pakistan. Australas Plant Pathol 49:25–29

Bamunusinghe D, Hemenway CL, Nelson RS, Sanderfoot AA, Ye CM, Silva MA, Payton M, Verchot-Lubicz J (2009) Analysis of potato virus X replicase and TGBp3 subcellular locations. Virology 393:272–285

Baratova L, Grebenshchikov N, Dobrov E, Gedrovich A, Kashirin I, Shishkov A, Efimov A, Järvekulg L, Radavsky YL, Saarma M (1992) The organization of potato virus X coat proteins in virus particles studied by tritium planigraphy and model building. Virology 188:175–180

Beemster A, De Bokx J (1987) Survey of properties and symptoms. In: Viruses of potatoes and seed potato production. Pudoc, Wageningen, pp 84–113

Bendahmane A, Köhm BA, Dedi C, Baulcombe DC (1995) The coat protein of potato virus X is a strain-specific elicitor of Rx1-mediated virus resistance in potato. Plant J 8:933–941

Bendahmane A, Kanyuka K, Baulcombe D (1997) High-resolution genetical and physical mapping of the Rx gene for extreme resistance to potato virus X in tetraploid potato. Theor Appl Genet 95:153–162

Bendahmane A, Kanyuka K, Baulcombe DC (1999) The Rx gene from potato controls separate virus resistance and cell death responses. Plant Cell 11:781–791

Bhushan L, Abraham A, Choudhury NR, Rana VS, Mukherjee SK, Savithri HS (2015) Demonstration of helicase activity in the nonstructural protein, NSs, of the negative-sense RNA virus, groundnut bud necrosis virus. Arch Virol 160:959–967

Blackman LM, Boevink P, Cruz SS, Palukaitis P, Oparka KJ (1998) The movement protein of cucumber mosaic virus traffics into sieve elements in minor veins of Nicotiana clevelandii. Plant Cell 10:525–537

Borah A, Bhausaheb MA, Rana M, Srivastava S (2023) Mechanism of arthropod-mediated transmission of plant viruses-a review. J Adv Zool 44:38

Briddon RW, Bull SE, Amin I, Mansoor S, Bedford ID, Rishi N, Siwatch SS, Zafar Y, Abdel-Salam AM, Markham PG (2004) Diversity of DNA 1: a satellite-like molecule associated with monopartite begomovirus–DNA β complexes. Virology 324:462–474

Brunt A, Loebenstein G (2001) The main viruses infecting potato crops. In: Virus and virus-like diseases of potatoes and production of seedpotatoes. Kluwer Academic Publishers, pp 65–67

Butter N, Rataul H (1977) The virus-vector relationship of the tomato leafcurl virus (TLCV) and its vector, Bemisia tabaci Gennadius (Hemiptera: Aleyrodidae). Phytoparasitica 5:173–186

Butterbach P, Verlaan MG, Dullemans A, Lohuis D, Visser RG, Bai Y, Kormelink R (2014) Tomato yellow leaf curl virus resistance by Ty-1 involves increased cytosine methylation of viral genomes and is compromised by cucumber mosaic virus infection. Proc Natl Acad Sci 111:12942–12947

Chabbouh N, Cherif C, Martelli G (1990) Natural infections of artichoke by potato virus X in Tunisia. J Phytopathol 129:257–260

Chakraborty S, Vanitharani R, Chattopadhyay B, Fauquet C (2008) Supervirulent pseudorecombination and asymmetric synergism between genomic components of two distinct species of begomovirus associated with severe tomato leaf curl disease in India. J Gen Virol 89:818–828

Chang HH, Lee CH, Chang CJ, Jan FJ (2022) FKBP-type peptidyl-prolyl cis-trans isomerase interacts with the movement protein of tomato leaf curl New Delhi virus and impacts viral replication in Nicotiana benthamiana. Mol Plant Pathol 23:561–575

Chatterji A, Padidam M, Beachy RN, Fauquet CM (1999) Identification of replication specificity determinants in two strains of tomato leaf curl virus from New Delhi. J Virol 73:5481–5489

Chauhan PK, Mehta N, Chauhan RS, Kumar A, Singh H, Lal MK, Tiwari RK, Kumar R (2023) Utilization of primary and secondary biochemical compounds in cotton as diagnostic markers for measuring resistance to cotton leaf curl virus (CLCuV). Front Plant Sci 14:1185337. https://doi.org/10.3389/fpls.2023.1185337

Chen B, Francki R (1990) Cucumovirus transmission by the aphid Myzus persicae is determined solely by the viral coat protein. J Gen Virol 71:939–944

Cheng C-W, Hsiao Y-Y, Wu H-C, Chuang C-M, Chen J-S, Tsai C-H, Hsu Y-H, Wu Y-C, Lee C-C, Meng M (2009) Suppression of Bamboo mosaic virus accumulation by a putative methyltransferase in Nicotiana benthamiana. J Virol 83:5796–5805

Cho S-Y, Cho WK, Choi H-S, Kim K-H (2012) Cis-acting element (SL1) of potato virus X controls viral movement by interacting with the NbMPB2Cb and viral proteins. Virology 427:166–176

Choi S-K, Jeon Y-W, Yoon J-Y, Choi J-K (2011) Characterisation of a satellite RNA of Cucumber mosaic virus that induces chlorosis in Capsicum annuum. Virus Genes 43:111–119

Cillo F, Roberts IM, Palukaitis P (2002) In situ localization and tissue distribution of the replication-associated proteins of Cucumber mosaic virus in tobacco and cucumber. J Virol 76:10654–10664

Cockerham G (1970) Genetical studies on resistance to potato viruses X and Y. Heredity 25:309

Danci O, Ziegler A, Torrance L, Gasemi S, Danci M (2009) Potyviridae family-short review

Ding B, Kwon MO, Warnberg L (1996) Evidence that actin filaments are involved in controlling the permeability of plasmodesmata in tobacco mesophyll. Plant J 10:157–164

Doronin SV, Hemenway C (1996) Synthesis of potato virus X RNAs by membrane-containing extracts. J Virol 70:4795–4799

Dreher TW (1999) Functions of the 3′-untranslated regions of positive strand RNA viral genomes. Annu Rev Phytopathol 37:151–174

Dreher TW (2009) Role of tRNA-like structures in controlling plant virus replication. Virus Res 139:217–229

Duijsings D, Kormelink R, Goldbach R (2001) In vivo analysis of the TSWV cap-snatching mechanism: single base complementarity and primer length requirements. EMBO J 20:2545

Dutta P, Lõhmus A, Ahola T, Mäkinen K (2024) The Replicase protein of potato virus X is able to recognize and trans-replicate its RNA component. Viruses 16:1611

Eagle PA, Hanley-Bowdoin L (1997) cis elements that contribute to geminivirus transcriptional regulation and the efficiency of DNA replication. J Virol 71:6947–6955

Eagle PA, Orozco BM, Hanley-Bowdoin L (1994) A DNA sequence required for geminivirus replication also mediates transcriptional regulation. Plant Cell 6:1157–1170

Edwardson JR, Christie RG (1997) Viruses infecting peppers and other solanaceous crops. Volume 1

Escriu F, Perry KL, García-Arenal F (2000) Transmissibility of Cucumber mosaic virus by Aphis gossypii correlates with viral accumulation and is affected by the presence of its satellite RNA. Phytopathology 90:1068–1072

Espino De Paz A, Botella-Guillen M, Otazo-González H, Alfaro-Fernández A, Font-San-Ambrosio I, Galipienso L, Rubio L (2019) First report of tomato leaf curl New Delhi virus infecting cucurbits in the Canary Islands. Plant Dis 103:1798–1798

Evtushenko EA, Ryabchevskaya EM, Nikitin NA, Atabekov JG, Karpova OV (2020) Plant virus particles with various shapes as potential adjuvants. Sci Rep 10:10365

Feng Z, Chen X, Bao Y, Dong J, Zhang Z, Tao X (2013) Nucleocapsid of tomato spotted wilt tospovirus forms mobile particles that traffic on an actin/endoplasmic reticulum network driven by myosin XI-K. New Phytol 200:1212–1224

Fontes E, Eagle PA, Sipe PS, Luckow VA, Hanley-Bowdoin L (1994a) Interaction between a geminivirus replication protein and origin DNA is essential for viral replication. J Biol Chem 269:8459–8465

Fontes E, Gladfelter HJ, Schaffer RL, Petty I, Hanley-Bowdoin L (1994b) Geminivirus replication origins have a modular organization. Plant Cell 6:405–416

Fortes IM, Sánchez-Campos S, Fiallo-Olive E, Díaz-PendóN JA, Navas-Castillo J, Moriones E (2016) A novel strain of tomato leaf curl New Delhi virus has spread to the Mediterranean basin. Viruses 8:307

Fortes IM, Perez-Padilla V, Romero-Rodríguez B, Fernández-Muñoz R, Moyano C, Castillo AG, De LeóN L, Moriones E (2023) Begomovirus tomato leaf curl New Delhi virus is seedborne but not seed transmitted in melon. Plant Dis 107:473–479

Francki R, Randles J, Chambers T, Wilson S (1966) Some properties of purified cucumber mosaic virus (Q strain). Virology 28:729–741

Fribourg C, Fernandez-Northcote E (1972) Potato virus X and tobacco mosaic in Capsicum species grown in Peru

Fujimoto Y, Keima T, Hashimoto M, Hagiwara-Komoda Y, Hosoe N, Nishida S, Nijo T, Oshima K, Verchot J, Namba S (2021) Short 5′ UTR enables optimal translation of plant virus tricistronic RNA via leaky scanning. bioRxiv, 2021.05. 14.444105

Garcia-Arenal F, Palukaitis P (1999) Structure and functional relationships of satellite RNAs of cucumber mosaic virus. In: Satellites and defective viral RNAs. Springer, Berlin, pp 37–63

Ghanim M, Czosnek H (2016) Interactions between the whitefly Bemisia tabaci and begomoviruses: biological and genomic perspectives. In: Management of insect pests to agriculture: lessons learned from deciphering their genome, transcriptome and proteome. Springer

Giunchedi L (1973) Virus X della patata isolato da radici e foglie di vite. Rivista di Patologia vegetale:73–77

Goldbach R (1987) Genome similarities between plant and animal RNA viruses. Microbiol Sci 4:197–202

Gomaa AE, El Mounadi K, Parperides E, Garcia-Ruiz H (2024) Cell fractionation and the identification of host proteins involved in plant–virus interactions. Pathogens 13:53

González I, Martínez L, Rakitina DV, Lewsey MG, Atencio FA, Llave C, Kalinina NO, Carr JP, Palukaitis P, Canto T (2010) Cucumber mosaic virus 2b protein subcellular targets and interactions: their significance to RNA silencing suppressor activity. Mol Plant-Microbe Interact 23:294–303

Grinzato A, Kandiah E, Lico C, Betti C, Baschieri S, Zanotti G (2020) Atomic structure of potato virus X, the prototype of the Alphaflexiviridae family. Nat Chem Biol 16:564–569

Guo HS, Ding SW (2002) A viral protein inhibits the long range signaling activity of the gene silencing signal. EMBO J 21:398

Hanley-Bowdoin L, Bejarano ER, Robertson D, Mansoor S (2013) Geminiviruses: masters at redirecting and reprogramming plant processes. Nat Rev Microbiol 11:777–788

Health EPOP, Bragard C, Dehnen-Schmutz K, Di Serio F, Gonthier P, Jacques MA, Jaques Miret JA, Justesen AF, Macleod A, Magnusson CS (2020) Pest categorisation of tomato leaf curl New Delhi virus. EFSA J 18:e06179

Heyraud-Nitschke F, Schumacher S, Laufs J, Schaefer S, Schell J, Gronenborn B (1995) Determination of the origin cleavage and joining domain of geminivirus rep proteins. Nucleic Acids Res 23:910–916

Hou WN, Duan CG, Fang RX, Zhou XY, Guo HS (2011) Satellite RNA reduces expression of the 2b suppressor protein resulting in the attenuation of symptoms caused by cucumber mosaic virus infection. Mol Plant Pathol 12:595–605

Howard AR, Heppler ML, Ju H-J, Krishnamurthy K, Payton ME, Verchot-Lubicz J (2004) Potato virus X TGBp1 induces plasmodesmata gating and moves between cells in several host species whereas CP moves only in N. benthamiana leaves. Virology 328:185–197

Huh SU, Kim MJ, Ham BK, Paek KH (2011) A zinc finger protein Tsip1 controls cucumber mosaic virus infection by interacting with the replication complex on vacuolar membranes of the tobacco plant. New Phytol 191:746–762

Huisman MJ, Linthorst HJ, Bol JF, Cornelissen BJ (1988) The complete nucleotide sequence of potato virus X and its homologies at the amino acid level with various plus-stranded RNA viruses. J Gen Virol 69:1789–1798

Inoue-Nagata AK, Jordan R, Kreuze J, Li F, López-Moya JJ, Mäkinen K, Ohshima K, Wylie SJ, Consortium IR (2022) ICTV virus taxonomy profile: Potyviridae 2022. J Gen Virol 103:001738
Ivanov KI, Eskelin K, Lõhmus A, Mäkinen K (2014) Molecular and cellular mechanisms underlying potyvirus infection. J Gen Virol 95:1415–1429
Jamil N, Rehman A, Hamza M, Hafeez A, Ismail H, Zubair M, Mansoor S, Amin I (2017) First report of tomato leaf curl New Delhi virus, a bipartite begomovirus, infecting soybean (Glycine max). Plant Dis 101:845–845
Jeevalatha A, Vanishree G, Siddappa S, Kumar R, Kaundal P, Kumar A, Chakrabarti SK (2021) Molecular characterization and infectivity analysis of tomato leaf curl New Delhi virus isolates infecting potato. 3 Biotech 11:203. https://doi.org/10.1007/s13205-021-02752-5
Jiang T, Zhou T (2023) Unraveling the mechanisms of virus-induced symptom development in plants. Plants 12:2830
Ju H-J, Brown JE, Ye C-M, Verchot-Lubicz J (2007) Mutations in the central domain of potato virus X TGBp2 eliminate granular vesicles and virus cell-to-cell trafficking. J Virol 81:1899–1911
Juárez M, RabadáN MP, Martínez LD, Tayahi M, Grande-Perez A, Gómez P (2019) Natural hosts and genetic diversity of the emerging tomato leaf curl New Delhi virus in Spain. Front Microbiol 10:140
Jyothsna P, Haq Q, Singh P, Sumiya K, Praveen S, Rawat R, Briddon RW, Malathi V (2013) Infection of tomato leaf curl New Delhi virus (ToLCNDV), a bipartite begomovirus with betasatellites, results in enhanced level of helper virus components and antagonistic interaction between DNA B and betasatellites. Appl Microbiol Biotechnol 97:5457–5471
Kenyon L, Kumar S, Tsai W-S, Hughes JDA (2014) Virus diseases of peppers (Capsicum spp.) and their control. Adv Virus Res. Elsevier 90:297–354
KeŽar A, Kavčič L, PoláK M, Nováček J, Gutierrez-Aguirre I, Žnidarič MT, Coll A, Stare K, Gruden K, Ravnikar M, Pahovnik D, Žagar E, Merzel F, Anderluh G, Podobnik M (2019) Structural basis for the multitasking nature of the potato virus Y coat protein. Sci Adv 5:eaaw3808
Kim K-H, Hemenway C (1996) The 5′ nontranslated region of potato virus X RNA affects both genomic and subgenomic RNA synthesis. J Virol 70:5533–5540
Komoda K, Ishibashi K, Kawamura-Nagaya K, Ishikawa M (2014) Possible involvement of eEF1A in tomato spotted wilt virus RNA synthesis. Virology 468:81–87
Kormelink R, De Haan P, Meurs C, Peters D, Goldbach R (1992a) The nucleotide sequence of the M RNA segment of tomato spotted wilt virus, a bunyavirus with two ambisense RNA segments. J Gen Virol 73:2795–2804
Kormelink R, Van Poelwijk F, Peters D, Goldbach R (1992b) Non-viral heterogeneous sequences at the 5′ ends of tomato spotted wilt virus mRNAs. J Gen Virol 73:2125–2128
Kreuze JF, Souza-Dias J, Jeevalatha A, Figueira A, Valkonen J, Jones R (2020) Viral diseases in potato. In: The potato crop: its agricultural, nutritional and social contribution to humankind. Springer International Publishing, Cham
Krishnamurthy K, Heppler M, Mitra R, Blancaflor E, Payton M, Nelson RS, Verchot-Lubicz J (2003) The potato virus X TGBp3 protein associates with the ER network for virus cell-to-cell movement. Virology 309:135–151
Kumar R, Tiwari RK, Lal MK, Majumdar SG, Swaraj K, Mall AK, Tripathi MK, Singh D, Singh G (2025) Contemporary status of occurrence, spread, and yield loss due to tomato leaf curl New Delhi virus in global crop cultivation. In: Tomato leaf curl New Delhi virus (ToLCNDV) insights into virome dynamics and management. Springer Nature Singapore, Singapore, pp 1–12. https://doi.org/10.1007/978-981-95-2852-3_1
Kwon S-J, Park M-R, Kim K-W, Plante CA, Hemenway CL, Kim K-H (2005) cis-acting sequences required for coat protein binding and in vitro assembly of potato virus X. Virology 334:83–97
Lacomme C, Jacquot E (2017) General characteristics of potato virus Y (PVY) and its impact on potato production: an overview. In: Potato virus Y: biodiversity, pathogenicity, epidemiology and management. Springer, Cham, pp 1–19
Lacomme C, Glais L, Bellstedt DU, Dupuis B, Karasev AV, Jacquot E (2017) Potato virus Y: biodiversity, pathogenicity, epidemiology and management. Springer

Laufs J, Schumacher S, Geisler N, Jupin I, Gronenborn B (1995a) Identification of the nicking tyrosine of geminivirus Rep protein. FEBS Lett 377:258–262

Laufs J, Traut W, Heyraud F, Matzeit V, Rogers SG, Schell J, Gronenborn B (1995b) In vitro cleavage and joining at the viral origin of replication by the replication initiator protein of tomato yellow leaf curl virus. Proc Natl Acad Sci 92:3879–3883

Lebedeva M, Nikonova E, Babakov A, Kolesnikova V, Razhina O, Zlobin N, Taranov V, Nikonov O (2024) Interaction of Solanum tuberosum L. translation initiation factors eIF4E with potato virus Y VPg: apprehend and avoid. Biochimie 219:1–11

Lee Y-S, Hsu Y-H, Lin N-S (2000) Generation of subgenomic RNA directed by a satellite RNA associated with bamboo mosaic potexvirus: analyses of potexvirus subgenomic RNA promoter. J Virol 74:10341–10348

Lefeuvre P, Moriones E (2015) Recombination as a motor of host switches and virus emergence: geminiviruses as case studies. Curr Opin Virol 10:14–19

Leshchiner AD, Solovyev AG, Morozov SY, Kalinina NO (2006) A minimal region in the NTPase/helicase domain of the TGBp1 plant virus movement protein is responsible for ATPase activity and cooperative RNA binding. J Gen Virol 87:3087–3095

Liao Q, Zhu L, Du Z, Zeng R, Feng J, Chen J (2007) Satellite RNA-mediated reduction of cucumber mosaic virus genomic RNAs accumulation in Nicotiana tabacum. Acta Biochim Biophys Sin 39:217–223

Linnik O, Liesche J, Tilsner J, Oparka KJ (2013) Unraveling the structure of viral replication complexes at super-resolution. Front Plant Sci 4:6

Liu J, Liu Y, Fang Y, Zhang L, Yu K, Wu X, Cheng X (2021) Evaluation of potato virus X resistance in potato cultivars and identification of an innate immunity-independent resistance phenotype. Phytopathol Res 3:21

Loebenstein G (2008) Plant virus diseases: economic aspects

Longstaff M, Brigneti G, Boccard F, Chapman S, Baulcombe D (1993) Extreme resistance to potato virus X infection in plants expressing a modified component of the putative viral replicase. EMBO J 12:379–386

Lough TJ, Lee RH, Emerson SJ, Forster RL, Lucas WJ (2006) Functional analysis of the 5′ untranslated region of potexvirus RNA reveals a role in viral replication and cell-to-cell movement. Virology 351:455–465

Loughnane J, Murphy PA (1938) Mode of dissemination of potato virus X. Nature 141:120–121

Makkouk K, Gumpf D (1974) Further identification of naturally occurring virus diseases of pepper in California

Malcuit I, Marano MR, Kavanagh TA, De Jong W, Forsyth A, Baulcombe DC (1999) The 25-kDa movement protein of PVX elicits Nb-mediated hypersensitive cell death in potato. Mol Plant-Microbe Interact 12:536–543

Mandal B, Jain R, Krishnareddy M, Krishna Kumar N, Ravi K, Pappu H (2012) Emerging problems of tospoviruses (Bunyaviridae) and their management in the Indian subcontinent. Plant Dis 96:468–479

Marano M, Malcuit I, De Jong W, Baulcombe D (2002) High-resolution genetic map of Nb, a gene that confers hypersensitive resistance to potato virus X in Solanum tuberosum. Theor Appl Genet 105:192–200

Mathur C, Savithri HS (2012) Novel ATPase activity of the polyprotein intermediate, viral protein genome-linked-nuclear inclusion-a protease, of pepper vein banding potyvirus. Biochem Biophys Res Commun 427:113–118

Mazumder P, Majumdar A, Roy B, Mohanbabu Y, Valli NS, Sarkar M, Mondal F, Verma N, Kumar R, Roy A (2025) RNA vaccination in plants: a non-transgenic approach to combat plant viruses with special reference to tomato leaf curl New Delhi virus. In: Tomato leaf curl New Delhi virus (ToLCNDV) insights into virome dynamics and management. Springer Nature Singapore, Singapore, pp 339–352. https://doi.org/10.1007/978-981-95-2852-3_17

Miller ED, Plante CA, Kim K-H, Brown JW, Hemenway C (1998) Stem-loop structure in the 5′ region of potato virus X genome required for plus-strand RNA accumulation. J Mol Biol 284:591–608

Mitter N, Koundal V, Williams S, Pappu H (2013) Differential Expression of Tomato Spotted Wilt Virus-Derived Viral Small RNAs in Infected Commercial and Experimental Host Plants. PLOS ONE 8(10): e76276. https://doi.org/10.1371/journal.pone.0076276

Mitra R, Krishnamurthy K, Blancaflor E, Payton M, Nelson RS, Verchot-Lubicz J (2003) The potato virus X TGBp2 protein association with the endoplasmic reticulum plays a role in but is not sufficient for viral cell-to-cell movement. Virology 312:35–48

Mochizuki T, Ohki ST (2012) Cucumber mosaic virus: viral genes as virulence determinants. Mol Plant Pathol 13:217–225

Moffett P, Farnham G, Peart J, Baulcombe DC (2002) Interaction between domains of a plant NBS–LRR protein in disease resistance-related cell death. EMBO J 21:4511

Moriones E, Praveen S, Chakraborty S (2017) Tomato leaf curl New Delhi virus: an emerging virus complex threatening vegetable and fiber crops. Viruses 9:264

Morozov SY, Solovyev AG (2015) Phylogenetic relationship of some "accessory" helicases of plant positive-stranded RNA viruses: toward understanding the evolution of triple gene block. Front Microbiol 6:508

Morozov SY, Solovyev AG (2024) Mechanisms of plant virus cell-to-cell transport: new lessons from complementation studies. Front Plant Sci 15:1453464

Morozov SY, Makarova SS, Erokhina TN, Kopertekh L, Schiemann J, Owens RA, Solovyev AG (2014) Plant 4/1 protein: potential player in intracellular, cell-to-cell and long-distance signaling. Front Plant Sci 5:26

Mrkvová M, Kemenczeiová J, Hlebová M, Glasa M (2024) Weeds and wild plants as natural hosts of cucumber mosaic virus: a case study from Western Slovakia. J Microbiol Biotechnol Food Sci 14:e11521–e11521

Mukhamedzhanova A, Karpova O, Rodionova N, Atabekov I (2009) Nonspecific activation of translation of encapsidated potexviral RNA with involvement of potato virus X movement protein TGB1. Doklady Biochemistry & Biophysics 428:239

Nagendran K, Mohankumar S, Mohammed Faisal P, Bagewadi B, Karthikeyan G (2017) Molecular evidence for the occurrence of tomato leaf curl New Delhi virus on chayote (Sechium edule) in southern India. Virus 28:425–429

Nemykh MA, Efimov AV, Novikov VK, Orlov VN, Arutyunyan AM, Drachev VA, Lukashina EV, Baratova LA, Dobrov EN (2008) One more probable structural transition in potato virus X virions and a revised model of the virus coat protein structure. Virology 373:61–71

Olspert A, Chung BY, Atkins JF, Carr JP, Firth AE (2015) Transcriptional slippage in the positive-sense RNA virus family Potyviridae. EMBO Rep 16:995–1004

Orozco B, Hanley-Bowdoin L (1996) A DNA structure is required for geminivirus replication origin function. J Virol 70:148–158

Paape M, Solovyev AG, Erokhina TN, Minina EA, Schepetilnikov MV, Lesemann D-E, Schiemann J, Morozov SY, Kellmann J-W (2006) At-4/1, an interactor of the tomato spotted wilt virus movement protein, belongs to a new family of plant proteins capable of directed intra-and intercellular trafficking. Mol Plant-Microbe Interact 19:874–883

Padidam M, Beachy RN, Fauquet CM (1995) Tomato leaf curl geminivirus from India has a bipartite genome and coat protein is not essential for infectivity. J Gen Virol 76:25–35

Palukaitis P (2016) Satellite RNAs and satellite viruses. Mol Plant-Microbe Interact 29:181–186

Palukaitis P, García-Arenal F (2003) Cucumoviruses. Adv Virus Res 62:241–323

Palukaitis P, Roossinck MJ, Dietzgen RG, Francki RI (1992) Cucumber mosaic virus. Adv Virus Res 41:281–348

Panno S, Caruso A, Troiano E, Luigi M, Manglli A, Vatrano T, Iacono G, Marchione S, Bertin S, Tomassoli L (2019) Emergence of tomato leaf curl New Delhi virus in Italy: estimation of incidence and genetic diversity. Plant Pathol 68:601–608

Parker L, Kendall A, Stubbs G (2002) Surface features of potato virus X from fiber diffraction. Virology 300:291–295

Parrella G, Troiano E, Lee S, Kil E-J (2020) Tomato leaf curl New Delhi virus found associated with eggplant yellowing disease in Italy. Plant Dis 104(10):1094

Paulus A, Kendrick J, Desjardins P (1960) Natural occurrence of potato virus X in field-grown peppers in California

Petrova E, Nikitin N, Trifonova E, Protopopova A, Karpova O, Atabekov J (2015) The 5′-proximal region of potato virus X RNA involves the potential cap-dependent "conformational element" for encapsidation. Biochimie 115:116–119

Phaneendra C, Rao KS, Jain R, Mandal B (2012) Tomato leaf curl New Delhi virus is associated with pumpkin leaf curl: a new disease in northern India. Indian J Virol 23:42–45

Pillai-Nair N, Kim K-H, Hemenway C (2003) Cis-acting regulatory elements in the potato virus X 3′ non-translated region differentially affect minus-strand and plus-strand RNA accumulation. J Mol Biol 326:701–720

Plante CA, Kim K-H, Pillai-Nair N, Osman TA, Buck KW, Hemenway CL (2000) Soluble, template-dependent extracts from Nicotiana benthamiana plants infected with potato virus X transcribe both plus-and minus-strand RNA templates. Virology 275:444–451

Prakash V, Sharma V, Devendran R, Prajapati R, Ahmad B, Kumar R (2024) A transition from enemies to allies: how viruses improve drought resilience in plants. Stress Biol 4:33

Prasad A, Prasad M (2022) Interaction of ToLCNDV TrAP with SlATG8f marks it susceptible to degradation by autophagy. Cell Mol Life Sci 79:241

Prasanna H, Sinha D, Rai G, Krishna R, Kashyap SP, Singh N, Singh M, Malathi V (2015) Pyramiding T y-2 and T y-3 genes for resistance to monopartite and bipartite tomato leaf curl viruses of I ndia. Plant Pathol 64:256–264

Pratap D, Kashikar AR, Mukherjee SK (2011) Molecular characterization and infectivity of a tomato leaf curl New Delhi virus variant associated with newly emerging yellow mosaic disease of eggplant in India. Virol J 8:305

Quenouille J, Vassilakos N, Moury B (2013) Potato virus Y: a major crop pathogen that has provided major insights into the evolution of viral pathogenicity. Mol Plant Pathol 14:439–452

Ranjan P, Singh A, Kumar RV, Basu S, Chakraborty S (2014) Host-specific adaptation of diverse betasatellites associated with distinct Indian tomato-infecting begomoviruses. Virus Genes 48:334–342

Rathjen JP, Moffett P (2003) Early signal transduction events in specific plant disease resistance. Curr Opin Plant Biol 6:300–306

Ravin N, Mardanova E, Kotlyarov RY, Novikov V, Atabekov J, Skryabin K (2008) Complete sequencing of potato virus X new strain genome and construction of viral vector for production of target proteins in plants. Biochem Mosc 73:44–49

Reddy MK, Nagendran K, Kumari S, Pradeep RM, Karthikeyan G (2024) Global scenario of begomovirus diseases in vegetable crops. Veg Sci 51:43–53

Revers F, García JA (2015) Molecular biology of potyviruses. Adv Virus Res 92:101–199

Ribeiro D, Jung M, Moling S, Borst JW, Goldbach R, Kormelink R (2013) The cytosolic nucleoprotein of the plant-infecting bunyavirus tomato spotted wilt recruits endoplasmic reticulum–resident proteins to endoplasmic reticulum export sites. Plant Cell 25:3602–3614

Roberts IM, Wang D, Findlay K, Maule AJ (1998) Ultrastructural and temporal observations of the potyvirus cylindrical inclusions (Cls) show that the Cl protein acts transiently in aiding virus movement. Virology 245:173–181

Rodamilans B, Valli A, Mingot A, San León D, Baulcombe D, López-Moya JJ, García JA (2015) RNA polymerase slippage as a mechanism for the production of frameshift gene products in plant viruses of the potyviridae family. J Virol 89:6965–6967

Röder J, Dickmeis C, Commandeur U (2019) Small, smaller, nano: new applications for potato virus X in nanotechnology. Front Plant Sci 10:158

Rozanov MN, Koonin EV, Gorbalenya AE (1992) Conservation of the putative methyltransferase domain: a hallmark of the 'Sindbis-like' supergroup of positive-strand RNA viruses. J Gen Virol 73:2129–2134

Ruiz M, Simón A, Velasco L, García M, Janssen D (2015) First report of tomato leaf curl New Delhi virus infecting tomato in Spain. Plant Dis 99:894–894

Samad A, Naqvi Q, Garg I, Twhouvenel JC (1991) Potato virus X—a new report from turnip (Brassica rapa L.) in India. J Phytopathol 133:65–75

Sanderfoot AA, Lazarowitz SG (1996) Getting it together in plant virus movement: cooperative interactions between bipartite geminivirus movement proteins. Trends Cell Biol 6:353–358

Sangar R, Lal S, Singh R (1980) Occurrence of potato virus X on winter cherry. Potato J 7:95–97

Sarkar M, Mondal F, Majumdar A, Valli NS, Roy B, Verma N, Mohanbabu Y, Kumar R, Roy A (2025) Molecular arsenal of tomato leaf curl New Delhi virus: viral suppressors of RNA silencing and their roles in pathogenicity. In: Tomato leaf curl New Delhi virus (ToLCNDV) insights into Virome dynamics and management. Springer Nature Singapore, Singapore, pp 325–337. https://doi.org/10.1007/978-981-95-2852-3_16

Seo J-K, Kwon S-J, Choi H-S, Kim K-H (2009) Evidence for alternate states of cucumber mosaic virus replicase assembly in positive-and negative-strand RNA synthesis. Virology 383:248–260

Seo J-K, Kwon S-J, Kim K-H (2019) Chapter 11: Replication and the replicase. In: Cucumber mosaic virus. Am Phytopath Society

Shafiq M, Asad S, Zafar Y, Briddon RW, Mansoor S (2010) Pepper leaf curl Lahore virus requires the DNA B component of tomato leaf curl New Delhi virus to cause leaf curl symptoms. Virol J 7:367

Shintaku MH, Zhang L, Palukaitis P (1992) A single amino acid substitution in the coat protein of cucumber mosaic virus induces chlorosis in tobacco. Plant Cell 4:751–757

Singh P, Savithri H (2015) GBNV encoded movement protein (NSm) remodels ER network via C-terminal coiled coil domain. Virology 482:133–146

Singh A, Chattopadhyay B, Chakraborty S (2012) Biology and interactions of two distinct monopartite begomoviruses and betasatellites associated with radish leaf curl disease in India. Virol J 9:43

Singh AK, Kushwaha N, Chakraborty S (2016) Synergistic interaction among begomoviruses leads to the suppression of host defense-related gene expression and breakdown of resistance in chilli. Appl Microbiol Biotechnol 100:4035–4049

Sinha S, Samad A (2021) Characterization and evolutionary analysis of Cucumber mosaic virus isolate infecting Salvia sclarea in India. 3 Biotech 11:468

Sit TL, Leclerc D, Abou-Haidar MO (1994) The minimal 5′ sequence for in vitro initiation of papaya mosaic potexvirus assembly. Virology 199:238–242

Sivalingam P, Sumiya K, Malathi V (2011) Carrot as a new host for a begomovirus: yellow mosaic disease of carrot reported in India. New Dis Rep 23:2044-0588.2011

Smirnyagina E, Morozov SY, Rodionova N, Miroshnichenko N, Solovyev A, Fedorkin O, Atabekov J (1991) Translational efficiency and competitive ability of mRNAs with 5′-untranslated αβ-leader of potato virus X RNA. Biochimie 73:587–598

Soards AJ, Murphy AM, Palukaitis P, Carr JP (2002) Virulence and differential local and systemic spread of cucumber mosaic virus in tobacco are affected by the CMV 2b protein. Mol Plant-Microbe Interact 15:647–653

Sõber J, Järvekulg L, Toots I, Radavsky J, Villems R, Saarma M (1988) Antigenic characterization of potato virus X with monoclonal antibodies. J Gen Virol 69:1799–1807

Soellick T-R, Uhrig J, Bucher G, Kellmann J-W, Schreier P (2000) The movement protein NSm of tomato spotted wilt tospovirus (TSWV): RNA binding, interaction with the TSWV N protein, and identification of interacting plant proteins. Proc Natl Acad Sci 97:2373–2378

Solomon-Blackburn RM, Barker H (2001) A review of host major-gene resistance to potato viruses X, Y, a and V in potato: genes, genetics and mapped locations. Heredity 86:8–16

Sonenberg N, Shatkin AJ, Ricciardi RP, Rubin M, Goodman RM (1978) Analysis of terminal structures of RNA from potato virus X. Nucleic Acids Res 5:2501–2512

Sorel M, Garcia JA, German-Retana S (2014) The Potyviridae cylindrical inclusion helicase: a key multipartner and multifunctional protein. Mol Plant-Microbe Interact 27:215–226

Steepy T, Lewis G, Varney E (1967) Identification of the viruses affecting commercial varieties of pepper (Capsicum annuum) in New Jersey

Stevenson, W., Loria, R., Franc, G. & Weingartner, D. 2001. Compendium of potato diseases; American Phytopathological Society (APS) Press: St. Paul, 72–77

Sugawara K, Shiraishi T, Yoshida T, Fujita N, Netsu O, Yamaji Y, Namba S (2013) A replicase of potato virus X acts as the resistance-breaking determinant for JAX1-mediated resistance. Mol Plant-Microbe Interact 26:1106–1112

Suzuki M, Kuwata S, Kataoka J, Masuta C, Nitta N, Takanami Y (1991) Functional analysis of deletion mutants of cucumber mosaic virus RNA3 using an in vitro transcription system. Virology 183:106–113

Tatineni S, Hein GL (2023) Plant viruses of agricultural importance: current and future perspectives of virus disease management strategies. Phytopathology 113:117–141

Tilsner J, Oparka KJ (2012) Missing links?—the connection between replication and movement of plant RNA viruses. Curr Opin Virol 2:705–711

Tilsner J, Linnik O, Louveaux M, Roberts IM, Chapman SN, Oparka KJ (2013) Replication and trafficking of a plant virus are coupled at the entrances of plasmodesmata. J Cell Biol 201:981–995

Tollin P, Wilson H, Bancroft J (1980) Further observations on the structure of particles of potato virus X. J Gen Virol 49:407–410

Tommiska T, HäMäLäinen J, Watanabe K, Valkonen J (1998) Mapping of the gene Nxphu that controls hypersensitive resistance to potato virus X in Solanum phureja IvP35. Theor Appl Genet 96:840–843

Torrance L, Talianksy ME (2020) Potato virus Y emergence and evolution from the Andes of South America to become a major destructive pathogen of potato and other Solanaceous crops worldwide. Viruses 12

Trebicki P (2020) Climate change and plant virus epidemiology. Virus Res 286:198059

Troiano E, Parrella G (2023) First report of tomato leaf curl New Delhi virus in Lagenaria siceraria var. longissima in Italy. Phytopathol Mediterr 62:17–24

Turina M, Kormelink R, Resende RO (2016) Resistance to tospoviruses in vegetable crops: epidemiological and molecular aspects. Annu Rev Phytopathol 54:347–371

Usharani K, Surendranath B, Paul-Khurana S, Garg I, Malathi V (2004) Potato leaf curl-a new disease of potato in northern India caused by a strain of tomato leaf curl New Delhi virus. Plant Pathol 53:235

Van Knippenberg I, Lamine M, Goldbach R, Kormelink R (2005) Tomato spotted wilt virus transcriptase in vitro displays a preference for cap donors with multiple base complementarity to the viral template. Virology 335:122–130

Vanterpool TC (1925) Streak or winter blight of tomato in Quebec. McGill University (Canada)

Venkataravanappa V, Reddy CL, Jalali S, Reddy MK (2015) Association of tomato leaf curl New Delhi virus DNA-B with bhendi yellow vein mosaic virus in okra showing yellow vein mosaic disease symptoms. Acta Virol 59:125–139

Venkataravanappa V, Reddy CL, Saha S, Reddy MK (2018a) Recombinant tomato leaf curl New Delhi virus is associated with yellow vein mosaic disease of okra in India. Physiol Mol Plant Pathol 104:108–118

Venkataravanappa V, Reddy LRCN, Saha S, Subbanna SK, Manem KR (2018b) Detection and characterization of tomato leaf curl New Delhi virus association with mosaic disease of ivy gourd (Coccinia grandis (L.) Voigt) in North India. Arch Biol Sci 70:339–347

Venkataravanappa V, Reddy CL, Shankarappa K, Reddy MK (2019) Association of Tomato leaf curl New Delhi virus, betasatellite, and alphasatellite with mosaic disease of spine gourd (Momordica dioica Roxb. Willd) in India. Iran J Biotechnol 17:e2134

Venkataravanappa V, Ashwathappa K, Reddy CL, Shankarappa K, Reddy MK (2020) Characterization of Tomato leaf curl New Delhi virus associated with leaf curl and yellowing disease of Watermelon and development of LAMP assay for its detection. 3 Biotech 10:282

Venkataravanappa V, Madhu GS, Manjunatha L, Rakshith YP, Reddy CNL (2024) Viral disease of vegetable crops: current status and management strategies. In: Souvenir book on Golden Jubilee National Seminar on "technological innovations in vegetable production under changing climate regime". Indian Society of Vegetable Science, pp 63–78

Verchot J (2022) Potato virus X: a global potato-infecting virus and type member of the Potexvirus genus. Mol Plant Pathol 23:315–320

Verchot J, Angell SM, Baulcombe DC (1998) In vivo translation of the triple gene block of potato virus X requires two subgenomic mRNAs. J Virol 72:8316–8320

Verlaan MG, Hutton SF, Ibrahem RM, Kormelink R, Visser RGF, et al. (2013) The Tomato Yellow Leaf Curl Virus Resistance Genes Ty-1 and Ty-3 Are Allelic and Code for DFDGD-Class RNA–Dependent RNA Polymerases. PLOS Genetics 9(3): e1003399. https://doi.org/10.1371/journal.pgen.1003399

Vo TTB, Tabassum M, Nattanong B, Qureshi MA, Im H-J, Parrella G, Kil E-J, Lee S (2025) The insidious threat: assessing the dangers and spread of tomato leaf curl New Delhi virus. Plant Pathol J 41:1

Voinnet O, Lederer C, Baulcombe DC (2000) A viral movement protein prevents spread of the gene silencing signal in Nicotiana benthamiana. Cell 103:157–167

Von Bargen S, Salchert K, Paape M, Piechulla B, Kellmann J-W (2001) Interactions between the tomato spotted wilt virus movement protein and plant proteins showing homologies to myosin, kinesin and DnaJ-like chaperones. Plant Physiol Biochem 39:1083–1093

Wani S, Saleem S, Nabi SU, Ali G, Paddar BA, Hamid A (2021) Distribution and molecular characterization of potato virus Y (PVY) strains infecting potato (Solanum tuberosum) crop in Kashmir (India). Virus 32:784–788

Wei T, Zhang C, Hong J, Xiong R, Kasschau KD, Zhou X, Carrington JC, Wang A (2010) Formation of complexes at plasmodesmata for potyvirus intercellular movement is mediated by the viral protein P3N-PIPO. PLoS Pathog 6:e1000962

White KA (2015) The polymerase slips and PIPO exists. EMBO Rep 16:885–886

Wilisiani F, Neriya Y, Tagami M, Kaneko M, Hartono S, Nishigawa H, Natsuaki T (2019) Complete genome sequence of tomato leaf curl New Delhi virus from luffa in Indonesia. Microbiol Resour Announc 8. https://doi.org/10.1128/mra.01605-18

Wu G, Kaper J (1995) Competition of viral and satellite RNAs of cucumber mosaic virus for replication in vitro by viral RNA-dependent RNA polymerase. Res Virol 146:61–67

Wu J, Zhang Y, Li F, Zhang X, Ye J, Wei T, Li Z, Tao X, Cui F, Wang X (2024) Plant virology in the 21st century in China: recent advances and future directions. J Integr Plant Biol 66:579–622

Xue M, Arvy N, German-Retana S (2023) The mystery remains: how do potyviruses move within and between cells? Mol Plant Pathol 24:1560–1574

Yamaji Y, Maejima K, Komatsu K, Shiraishi T, Okano Y, Himeno M, Sugawara K, Neriya Y, Minato N, Miura C (2012) Lectin-mediated resistance impairs plant virus infection at the cellular level. Plant Cell 24:778–793

Yoshida T, Shiraishi T, Hagiwara-Komoda Y, Komatsu K, Maejima K, Okano Y, Fujimoto Y, Yusa A, Yamaji Y, Namba S (2019) The plant noncanonical antiviral resistance protein JAX1 inhibits potexviral replication by targeting the viral RNA-dependent RNA polymerase. J Virol 93. https://doi.org/10.1128/jvi.01506-18

Zaidi SS-E-A, Shafiq M, Amin I, Scheffler BE, Scheffler JA, Briddon RW, Mansoor S (2016) Frequent occurrence of tomato leaf curl New Delhi virus in cotton leaf curl disease affected cotton in Pakistan. PLoS One 11:e0155520

Zaidi SSEA, Martin DP, Amin I, Farooq M, Mansoor S (2017) Tomato leaf curl New Delhi virus: a widespread bipartite begomovirus in the territory of monopartite begomoviruses. Mol Plant Pathol 18:901–911

Zayakina O, Arkhipenko M, Kozlovsky S, Nikitin N, Smirnov A, Susi P, Rodionova N, Karpova O, Atabekov J (2008) Mutagenic analysis of potato virus X movement protein (TGBp1) and the coat protein (CP): in vitro TGBp1–CP binding and viral RNA translation activation. Mol Plant Pathol 9:37–44

Ziebell H, Murphy AM, Groen SC, Tungadi T, Westwood JH, Lewsey MG, Moulin M, Kleczkowski A, Smith AG, Stevens M (2011) Cucumber mosaic virus and its 2b RNA silencing suppressor modify plant-aphid interactions in tobacco. Sci Rep 1:187

# Viral Suppressors of RNA Silencing in Vegetable Viruses: Mechanism and Diversity

6

Uchakankhi Kashyap, Anindita Saikia, Darshana Goswami, Archita Bordoloi, and Nima D. Namsa

**Abstract**

RNA silencing is a fundamental antiviral defense in plants, comprising processes that detect viral double-stranded RNA (dsRNA), process it into small interfering RNAs (siRNAs) via Dicer-like enzymes, and direct degradation or translational repression of viral mRNAs through Argonaute (AGO)-containing effector complexes. Vegetable-infecting viruses encounter the host defense system, and many have evolved viral suppressors of RNA silencing (VSRs) to circumvent RNA silencing at various levels. In this chapter, we examine the diversity of VSRs in vegetable viruses (e.g., those infecting Solanaceae, Cucurbitaceae, Brassicaceae) and dissect their molecular mechanisms. VSRs operate via multiple strategies: sequestering dsRNA or siRNAs to prevent their detection or loading into RISC; degrading or destabilizing key components such as AGO proteins; interfering with the amplification of the silencing signal (e.g., through RNA-dependent RNA polymerases (RdRPs) and Suppressor of Gene Silencing 3 (SGS3); blocking systemic spread of silencing signals; and even modulating endogenous small RNAs and microRNAs to perturb host gene regulation. These suppressors are structurally and evolutionarily diverse, often unrelated in sequence, and sometimes multifunctional, including serving roles in pathogenicity, movement, or symptom development in addition to silencing suppression. Understanding the mechanisms and diversity of VSRs in vegetable viruses has both theoretical and practical implications: it illuminates the arms-race dynamics between host and pathogen, aids in predicting viral virulence, and suggests approaches for resistance breeding or silencing-based antiviral strategies. Future work should aim to

U. Kashyap · A. Saikia · D. Goswami · A. Bordoloi · N. D. Namsa (✉)
Department of Molecular Biology and Biotechnology, Tezpur University, Tezpur, Assam, India
e-mail: namsa@tezu.ernet.in



R. Kumar et al. (eds.), *Vegetable Crops and Viral Pathogens Volume I*, Plant Pathology for Sustainable Agriculture, https://doi.org/10.1007/978-981-95-7265-6_6

characterize lesser-known VSRs in emerging vegetable viruses, map precise host targets, and explore how VSRs affect plant development and immunity.

**Keywords**

Viral suppressors · RNA silencing · Vegetable viruses · Diversity · siRNAs

## 6.1 Introduction

Vegetable crops hold the position of being the most significant dietary component in global agriculture. Vegetables are widely recommended for their health-promoting properties, as they are essential for maintaining a balanced diet. They are rich in vitamins, minerals, dietary fiber, and bioactive substances and hence enhance the overall health of an individual. Vegetables are also considered to be important economic crops, particularly in Asia and Africa, that sustain and support smallholder farmers and contribute significantly to the rural livelihood and improve the quality of their lives. Among these cash crops, tomato, cauliflower, potato, cucumber, pepper, and cabbage are some of the most widely grown vegetables. They have high value both in domestic and international export markets. Let us take the tomato as an example; it is the most consumed vegetable in the world and can be grown in almost every country. In addition to supporting a sizable food processing business that produces purees, sauces, and canned goods, it provides vitamin C, anti-cancer components like phenolic constituent ferulic acid, and vital minerals like lycopene and folates (Sarkar et al. 2024; Mazumder et al. 2024). After rice, wheat, and maize, the potato ranked as the fourth most important vegetable crop. In addition to being a part of our day-to-day meals, potatoes are also used to make chips, starch, and animal feed, which results in complex economic value chains (Kumar et al. 2020a, b, 2022; Jeevalatha et al. 2021). The antioxidant qualities of brassica vegetables, like cauliflower and cabbage, are also crucial to their traditional diets in Asia and Europe. In addition to being important in fresh markets, cucumbers are also important in processed forms like pickles, which support both individual home nutrition and commercial exchange. Factors like urbanization, population growth, and growing knowledge of the health advantages of diets rich in vegetables have caused a significant increase in the value of vegetable crops globally over the past few decades (Collins et al. 2022). In view of their importance, vegetable crops still face major constraints from biotic stresses. The most severe destructive factor would be the viral diseases. These viruses, which affect the plants, have no direct curative measures. They can be efficiently transmitted by vectors such as flies, aphids, and leafhoppers, making their prevention and control highly challenging. In open-field vegetable systems, viruses can propagate quickly and frequently, thus destroying the whole harvest.

### 6.1.1 Plant Viruses

Plant viruses can be defined as sub-microscopic rod-shaped, spherical, or filamentous particles containing different kinds of genomes. The majority of plant-virus genomes consist of ssRNA (single-stranded RNA). The viral infection of plant cells is mainly mediated by vectors (e.g., aphids, nematodes, whiteflies, cytrids, or plasmodiophorids), but they can also be spread through mechanical wounds as well as vertical transmission via seeds. Multiplication of the virus inside the plant depends on the cellular mechanism of the host that supports the viral genome replication and cell-to-cell, as well as systemic transport of the virus via plasmodesmata and connected phloem. Once viruses have invaded once viruses have invaded the plant system, they infect a small region, causing, for example, the formation of small patterns of a greenish yellow mosaic, which can spread as the virus circulates through the plant vascular system to distant plant organs or to other host plants by vector-mediated transmission (Gaffar and Koch 2019). Depending on the crop, virus, and environmental factors, the yield loss of vegetables caused by viral infections can vary in a wide range. For example, one of the begomoviruses, namely tomato yellow leaf curl virus (TYLCV), transmitted by whiteflies, causes severe yellowing, stunting, and curling of tomato leaves, which often reduces the yields by more than half in heavily infested fields. Other viruses like cucumber mosaic virus (CMV) cause distortion of fruit, mosaic symptoms, and low market value of cucurbits. Potato viruses like potato leafroll virus (PLRV) and potato virus Y (PVY) result in lower quality and tuber yield and compromise seed potato certification systems and global transportation (Kumar et al. 2021, 2023). Another example could be the turnip mosaic virus (TuMV), which possesses severe effects on *Brassica* crops, causing production penalties, vein clearing, and mottling (Gaffar and Koch 2019). In addition to a reduction in yield, viral infections have some hidden economic costs, too. In order to manage vectors, farmers have to spend more money on pesticides, which in turn drives up production costs and increases environmental concerns. The expense of clean planting materials is increased by the stringent viral indexing required by the seed and the certification system of seedlings. Besides, sometimes virus-infected plants (even if they are edible) are destined to lose their market value due to the appearance of symptoms like curling of leaves, mosaic pattern, and abnormalities in fruit development. It has been known to the global trade business that these viral outbreaks create tension in the setting of the worldwide supply of food, as diseased crops are strictly prohibited for transportation by the laws and government to control the spread of infection. It is, therefore, can be said that controlling viral infections continues to be an agro-economic and scientific issue requiring innovative and ground-breaking strategies to meet the demands of an expanding world population. The ability of viruses to bypass host defense systems, their rapid mutation, and the emergence of new strains make them serious obstacles to stable and sustainable vegetable production. The foundation for making efficient strategies to protect these valuable crops is to understand the main antiviral response of plants, how viruses interact with host resistance, and how they actually execute RNA silencing, so that it can assure the safeguard of these high-value crops (Gaffar and Koch 2019; Sundaresha et al. 2022, 2024).

## 6.2 Antiviral Plant Immunity

Plants have developed a multilayered antiviral defense mechanism. The pattern receptors located on the plant cell membrane present in the plants are able to detect the presence of pathogen- and microbe-associated molecular patterns (PAMPs/MAMPs) and lead to the activation of PAMP-triggered immunity (PTI). Though PTI is often sufficient to prevent the colonization of pathogens, some pathogens possess evolved effector proteins or small RNA effectors that suppress PTI. In turn, plants have developed R-proteins that trigger effector-triggered immunity (ETI; also known as R-gene-mediated resistance) after direct or indirect interaction with their cognate pathogen-encoded effector, also called avirulence (Avr) proteins. Both PTI and ETI are linked with the activation of immune responses in the inoculated tissue, such as the production of anti-microbial compounds, the generation of ROS (reactive oxygen species), defense-associated gene expression, including the pathogenesis-related (PR)-1, and the accumulation of the defense signaling hormone salicylic acid. However, RNA silencing (commonly referred to as RNAi (RNA interference) is the central and crucial defense mechanism for plant viruses (Gaffar and Koch 2019).

### 6.2.1 RNA Silencing-Based Plant Antiviral Immunity

#### 6.2.1.1 Biogenesis of Small RNAs

RNA silencing (also called RNA interference, RNAi) can be termed as a conserved regulatory mechanism of gene expression in eukaryotic species, which is initiated by dsRNA-provoking gene silencing by inhibition of transcription (transcriptional gene silencing, TGS) or by sequence-specific complementary mRNA transcripts degradation (post-transcriptional gene silencing, PTGS). RNA silencing plays an important role in diverse cellular, physiological, and developmental processes, which regulate gene expression through sRNAs (small non-coding RNAs) and are associated with the development of protection against viral infection, the control of epigenetic modification, genome stability regulation, curbing of transposon movement, and regulation of heterochromatin synthesis.

#### 6.2.1.2 Assembly of the RISC

The ultimate triggering/eliciting factor of RNA silencing is dsRNA, which is present in the plant-infecting RNA viruses, i.e., an intermediate of replication generated by viral RNA-dependent RNA polymerases (RdRPs). The mechanism of RNA silencing starts with the initial processing or cleavage of a dsRNA precursor into short 21–24 nt (nucleotide) miRNA (microRNA) or siRNA (small interfering) duplexes by an RNase III-like enzyme known as Dicer-like proteins (DCL).

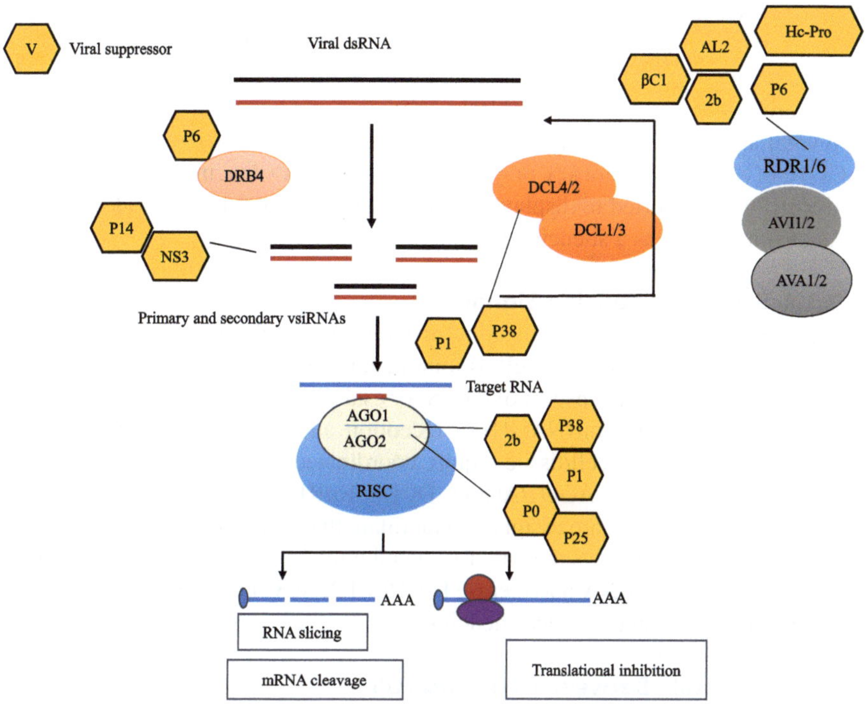


**Fig. 6.1** Antiviral RNA silencing starts with the initial processing or cleavage of a precursor viral dsRNA into short 21–24 nucleotide vsiRNA duplexes by RNaseIII-like enzymes called DCLs. Double-stranded siRNAs are incorporated into an RNA-induced silencing complex (RISC), and the activated RISC subsequently unwinds the vsiRNA, thereby generating an antisense (or guide) strand that targets complementary mRNA transcripts via base-pairing interactions. Subsequent degradation of the targeted viral mRNA or inhibition of translation can interfere with protein biosynthesis

Double-stranded siRNAs are integrated into double-stranded siRNAs are integrated into an RNA-induced silencing complex (RISC), which contains an Argonaute (AGO) protein that possesses a small RNA-binding domain and an endonucleolytic activity to cleave the target RNAs. The RISC activation subsequently unwinds siRNAs, thereby producing a sense (passenger) and an antisense (guide) strand in an ATP-dependent reaction. When the degradation of the sense strand occurs, the RISC containing the antisense strand subsequently targets a complementary transcript of mRNA through base-pairing interaction, degrades the mRNA, and in turn inhibits protein biosynthesis (Fig. 6.1).

#### 6.2.1.3 Amplification of Silencing

In plants, RNA silencing is stated as the best-studied antiviral defense mechanism, and notably similar antiviral functions in mammals are also reported. However, its diversity depends mainly on the existence of numerous copies of the core RNA-silencing pathway components, which presumably are the result of gene duplication followed by specialization. As an example, we can take the model organism *Arabidopsis thaliana,* which has four DCLs, six RdRPs, and ten AGOs, and they are involved in multiple silencing-related pathways. Here, the DCL2, DCL3, and DCL4 play a crucial role in virus-induced RNA silencing, and among all the DCLs, DCL3 is pivotal against DNA viruses. AGO1, AGO2, AGO3, AGO4, AGO5, AGO7, and AGO10 of *Arabidopsis* are key players in antiviral RNA silencing. RdRP1, RdRP2, and RdRP6 are reported to display antiviral activity, amplifying virus-derived siRNAs or functioning as a silencing signal.

#### 6.2.1.4 Systemic Silencing Signal

Besides RDR mediation, amplification of vsiRNA (small interfering viral RNA), short distance (cell-to-cell) as well as long-distance (phloem) expansion of silencing signals is an important aspect of PTGS for establishing systemic antiviral immunity. The long-distance mRNA PTGS reception and the synthesis of secondary siRNAs in the recipient cell/tissue require coordinated activity of a set of proteins. These proteins comprise AGO1/AGO2, DCL2, RdRP6, and SGS3 (suppressor of gene silencing 3), which are assigned to facilitate the silencing signal amplification process. Moreover, proteins that are associated with TGS pathways also contribute to systemic PTGS, as RNA polymerase Iva, RdRP2, AGO4, and DCL3 are involved in the reception of long-distance silencing.

#### 6.2.1.5 Crosstalk Between PTGS and TGS

RNA-directed DNA methylation (RdDM) is considered one of the prominent antiviral mechanisms in plants. 24-nucleotide siRNAs, which are primarily, direct the AGO4/AGO6 complexes to nascent transcripts linked to viral DNA. This induces modification in histone and cytosine methylation, which results in the formation of heterochromatin and the viral mini-chromosomes' transcriptional suppression. PTGS and TGS (transcriptional gene silencing) are the first line of defense against RNA and DNA viruses, respectively. Both PTGS and TGS are often triggered by viral genomes; for instance, a geminivirus infection produces 24-nt siRNAs (TGS) and 21/22-nt vsiRNAs (PTGS). Virus frequently encodes suppressors that block both arms through disrupting processes like methylation, destroying AGOs, or sequestering siRNAs.

## 6.3 Concept and Functional Roles of Viral Suppressors of RNA Silencing

### 6.3.1 What Is VSR?

Antiviral silencing is surely no exception, as was primarily established when the potyviral helper-component proteinase (Hc-Pro) and cucumoviral 2b protein, which had both previously been identified as virulence factors, were noted to suppress transgene silencing (Incarbone and Dunoyer 2013). To escape the antiviral silencing-based immune response, viruses have developed various countermechanisms (e.g., Phytoviruses have evolved ways to counteract, attenuate, or escape this response), and the most widely used strategy among these is the generation of viral suppressors of RNA silencing (VSRs) that target several stages of the silencing machinery. Now, VSRs have been isolated from approximately all the phytovirus families and have independently arisen several times to target RNA silencing at different points and via diverse mechanisms. This is an example of convergent evolution. VSRs are largely diversified in sequence, structure, and activity within and across the virus families, indicating that they are acquired rapidly through evolutionary convergence as a necessary adaptation to the RNA-silencing-dependent immune response. In accordance with this, expression of VSR is frequently a prerequisite for virus replication as well as the systemic infection of the host in both plants and insects. While it has been reported that VSRs may target many stages of the antiviral silencing pathway, their mechanism of action is typically categorized into:

- Binding to long dsRNA can result in inhibition of Dicer processing
- The binding and sequestration of sRNA duplexes, which prevents RISC assembly
- Effectors or processing factors are directly targeted, which leads to their inhibition or destabilization

The study of VSRs not only enhances our understanding of plant-virus interaction but also sheds light on fundamental aspects of the host silencing mechanism itself. The recent findings from *in planta* have given us new insights into the mode of action of VSR and unravelled the fact that plants react to the perturbation of the pathways of silencing caused by viral infection by employing a battery of counter-defense systems (Incarbone and Dunoyer 2013).

## 6.4 Mechanism of Action

### 6.4.1 VSRs Targeting RNA-Silencing Effector Protein

The first class of viral suppressors acts by direct inactivation, destabilization, or competition with the core silencing effectors of the host, such as AGO proteins, DCLs, RdRPs (RNA-dependent RNA polymerase), and cofactors like SGS3, DRB4, and HEN1. Targeting these proteins, viruses can effectively invade the antiviral silencing machinery (Incarbone and Dunoyer 2013).

#### 6.4.1.1 Argonaute-Interacting VSRs

- TCV (turnip crinkle virus) P38 forms homodimers and directly binds to AGO1 (and possibly AGO2). This type of interaction requires GW motifs (AGO hook), which imitate host AGO-binding proteins, and are essential for this interaction. The binding of P38 spares miRNA loading and inhibits effective loading of siRNA into AGO1. This results in antiviral silencing being compromised, but developmental regulation through miRNA is maintained (Pumplin and Voinnet 2013).

- An F-box protein named Polerovirus P0 recruits the host SCF E3 ubiquitin ligase complex (SKP1-CULLIN1) to induce the degradation of AGO1. Notably, this degradation relies on a pathway that is normally employed by the plant to recycle unloaded AGO proteins, called autophagy, and is proteosome independent.
- The PVX (potato virus X P25) virus promotes degradation of both AGO1 and AGO2 proteins, but this process is sensitive to proteosome inhibitors such as MG132, suggesting a different pathway for degradation.
- Multifunctionality of VSR can be seen in CMV (cucumber mosaic virus 2b), where the virus binds to RNA and also to the AGO proteins.

#### 6.4.1.2 DCL-Targeting VSRs

- Direct interaction occurs between the CaMV (cauliflower mosaic virus) P6/TAV and a DCL4 dsRNA-binding cofactor, DRB4. This inhibits the initiation of PTGS by preventing the efficient processing of viral dsRNA into 21-nt siRNAs.

#### 6.4.1.3 RDR/SGS3-Targeting VSRs

- By specifically targeting the RdRP6, it suppresses systemic (long-distance) silencing instead of silencing local intracellular and also blocks amplification of secondary siRNA.
- Another virus, namely TYLCV (tomato yellow leaf curl virus V2), interacts with the cofactor of RdRP6, SGS3, and prevents the formation of the precursors of dsRNA required for secondary siRNA production.

#### 6.4.1.4 Stability Interference of vsiRNA

- The replicase subunit P126 of TMV (tobacco mosaic virus) inhibits HEN1–mediated vsiRNA methylation, which in turn stabilizes siRNA duplexes. Loss of methylation leads to the degradation of vsiRNAs and weakened silencing.
- A viral endoribonuclease, i.e., RNase III of Sweet potato chlorotic stunt virus, directly cleaves siRNA duplexes into ~14 nt fragments, which no longer possess the ability to load RISC and hence neutralize antiviral small RNAs.

### 6.4.2 Sequestration of Small RNAs

Protein-based sequestration: As a "molecular caliper," the tombusvirus P19 protein generates homodimers that tightly bind to 21-nt siRNA duplexes. This results in the prevention of siRNA and miRNA incorporation into AGO1, silencing RISC activity. Here, P19 protein selectively spares miR168, which in turn increases its loading into AGO1/AGO10. As miR168 causes the downregulation of AGO1 mRNA, this generates a feedback loop that reduces the levels of AGO1 and further weakens the antiviral defense mechanism. Moreover, tombusvirus infection induces MIR168 transcription and amplifies this mechanism of suppression. Similar strategies are taken by unrelated viruses, demonstrating convergent evolution in pursuit of endogenous silencing pathways.

#### 6.4.2.1 RNA-Based Structural Evasion

Some of the viruses or the subviral pathogens depend on structured RNAs. For example, PSTVd (Potato Spindle Tuber Viroid) contains a circular rod-shaped RNA genome that can be cleaved by DCL, but cannot be efficiently targeted by AGO protein because of its rigid secondary structure. Another example is the CaMV (cauliflower mosaic virus) 8S RNA leader, which produces an abundant amount of vsiRNAs that serve as a decoy, redirecting the silencing away from the vital viral RNAs. In order to contribute to the symptom development, some vsiRNAs may even be led to target the host mRNAs.

## 6.5 Diversity of VSRs

Virus suppressors involved in RNA silencing are found in almost all plant-virus families. The vegetable-infecting viruses encode a broad range of VSRs. In addition to securing viral multiplication and movement, these VSR proteins lead to symptom development and determine the range of the host. The major VSRs arranged by virus families on the basis of their effects on economically important vegetable crops will be discussed in this chapter.

### 6.5.1 *Potyviridae* Family

The Potyviridae family represents the broadest group of known plant RNA viruses and accounts for more than half of the viral crop damage. The leading proteases of viruses within the family vary considerably in size and arrangement and play important roles during the infection. The fusion protein, namely the potyviral P1/HC-Pro, is the first identified VSR (viral suppressor of RNA silencing). Multiple studies have shown that P1 enhances the ability to suppress HC-Pro, and hence the fusion protein P1/HC-Pro has been utilized to investigate the mechanism of silencing suppression (Hong et al. 2023). The HC-Pro (helper-component proteinase) inhibits the regulatory pathway of siRNA (short-interfering RNA) and miRNA (microRNA) by the process called methylation and prevents the incorporation into AGO protein. HC-Pro is also responsible for the disruption of microRNA pathways, leading to some kind of developmental abnormalities. It also might perturb different steps of the RNA-silencing pathway by disturbing the RdRP6-mediated amplification of viral dsRNA. Another component of the above-discussed fusion protein is "P1." P1 contributes to the suppression of silencing by modulating the accumulation of HC-Pro and its function. It promotes as well as enhances the suppression activity and balances the replication procedure with host tolerance (Qin et al. 2020).

## 6.5.2 *Bromoviridae* Family

Includes 2b protein encoded by CMV (cucumber mosaic virus) was among the first suppressors identified that might inhibit PTGS (post-transcriptional gene silencing), but with little or no effect on the function of miRNA. 2b directly interacts with AGO1 both in vivo and in vitro, and this interaction occurs essentially on one surface of the PAZ-containing module and part of the PIWI-box of AGO1. Along with this interaction, 2b specifically inhibits the cleavage activity of AGO1 in the RISC reconstitution assay. It also interferes with RdDM, which in turn weakens the transcriptional host defense gene silencing. 2b blocks the cleavage activity of AGO1 to inhibit miRNA pathways, RNA-silencing attenuation, and counter host defense (Zhang et al. 2006).

## 6.5.3 *Tombusviridae* Family

The tombusvirus encodes p19 protein, which is involved in several activities such as pathogenesis and virus transportation. Recently, studies have demonstrated that p19 is a plant suppressor of RNA silencing, which can bind to siRNA with great affinity (Xia et al. 2009). P19 is known to form a "molecular caliper homodimer" that in turn binds to duplexes of 21-nt siRNA with high affinity. This sequestration blocks siRNA loading into AGO as well as alters the pathways of endogenous silencing by specifically ignoring miR168. In the end, tomato disease symptoms become prominent, and the suppression of antiviral silencing occurs.

## 6.5.4 *Geminiviridae* Family

### 6.5.4.1 V2 Protein

The V2 ORF encodes the pre-coat protein. V2 protein, along with CP, is needed for efficient accumulation of viral DNA forms and, therefore, facilitates successful infection in the host. It is also essential for efficient viral movement. V2 is known to suppress both PTGS and TGS. V2 protein of TYLCV (tomato yellow leaf curl virus), which interacts with the host SGS3 (suppressor of gene silencing 3) that assists RdRP6 in the dsRNA production. The V2-SGS3 interaction affects the silencing amplification step (Rishishwar and Dasgupta 2019). Tomato yellow leaf curl China virus-coded V2 protein possibly sequesters the siRNA produced against the virus and hinders the silencing pathway. Similarly, it has been assumed that CLCuMV (cotton leaf curl Multan virus) V2 also sequesters long double-stranded RNA and prevents its Dicer-mediated cleavage. The V2 of TYLCV also facilitates its function as a suppressor by interfering with the innate immunity of the plant by binding to CYP1 protein, a member of the family of papin-like cystein proteases (Fig. 6.2).

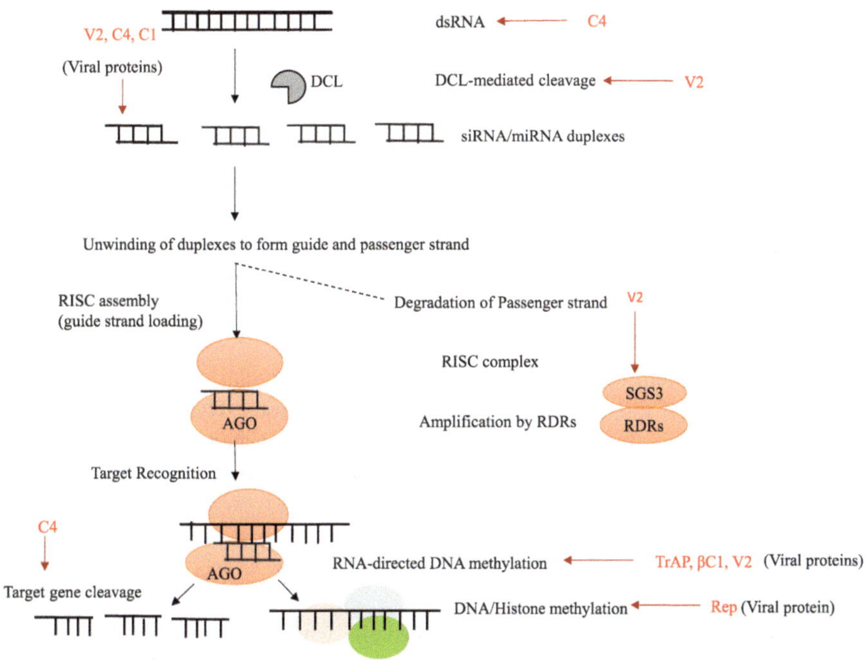


**Fig. 6.2** The major steps in the RNA-silencing pathway are mediated by the geminiviral suppressor proteins. The dsRNA generated from several sources is cleaved into ~21–30 nucleotide small RNA duplexes (siRNA/mRNA). One of the strands belongs to this duplex, called the guide strand, which is integrated in the RISC complex containing an AGO protein. Then, this sRNA guides the AGO protein to the target sequence and mediates the silencing of the gene by various machinery. The RDRs are known to amplify/maintain the gene silencing. (Rishishwar and Dasgupta 2019)

#### 6.5.4.2 Replication-Associated Protein (AC1, C1)

The C1 or AC1 ORF encoded replication-associated protein (Rep) plays an indispensable role in the replication of the virus. Rep has been reported to be a part of all the main steps of RCR (rolling-circle replication). The C2 has also been known to act both as a TGS and PTGS suppressor. Moreover, C2 inhibits the components of the methylation mechanism, such as SNF1 (Sucrose Nonfermenting1) kinase, adenosine kinase, and certain histone methyltransferases. Thereby, AC2 acts as a TGS suppressor.

#### 6.5.4.3 AC4/C4 Protein

The C4 ORF exists completely within the C1 ORF, which is a different reading frame. AC4/C4 acts as a suppressor of RNA silencing in both bipartite and monopartite begomoviruses. The ACMV-coded AC4 specifically binds to the single-stranded sRNA. The MYMV-AC4 binds to sRNA and causes the inhibition of the cleavage of target mRNA. On the other hand, CLCuMV-C4 binds to both long and short RNAs, with a preference particularly for the dsRNA form.

#### 6.5.4.4 βC1 Protein

The βC1 proteins from various betasatellites have been shown to act as RNA-silencing suppressors. The βC1gene belongs to the tomato yellow leaf curl China virus (TYLCCNB), which can interact with S-adenosyl homocysteine hydrolase, a methyl cycle enzyme that is required in methylation, and in turn, suppresses. The βC1 encoded by TYLCCNB has been reported to improve the expression of *N. benthamiana* calmodulin-like protein Nbrgs-CaM. The enhanced Nbrgs-CaM expression hindered the synthesis of secondary siRNAs, likely through inhibiting the RNA-dependent RNA polymerase 6 (RdRP6) expression (Ammara et al. 2015).

#### 6.5.4.5 Alphasatellites

Similar to betasatellites, alphasatellites comprise a circular, ssDNA genome of approximately half the size of their helper begomoviruses. They possess a single gene that encodes the Alpha-Rep, a replication-associated protein from two alphasatellites, which are reported to act as PTGS suppressors.

### 6.5.5 *Tospoviridae* Family

The Tospoviridae family encodes the NS protein, which acts as an RNA-silencing suppressor. In TSWV (tomato spotted wilt virus), the multifunctional protein NSs is necessary for systemic movement in plants, as well as establishment of persistent infection and transmission by flower thrips (*Frankliniella occidentalis*), and is also a hypersensitive response activator in resistant plants. It has been confirmed that NSs are responsible for the suppression of RNA silencing by various mechanisms, including interaction with AGO proteins and binding to long dsRNAs and small RNAs. The NS of different orthotospoviruses can suppress silencing of a GFP transgene and interfere with the spreading of systemic RNA-silencing signal (Garcia-Ruiz et al. 2018).

### 6.5.6 *Luteoviridae* Family

Polerovirus, a member of the Luteoviridae family, encodes the P0 protein that carries a motif, namely the F-box motif, required for SCF-like complex formation, which is crucial for P0's silencing suppressor function. F-box protein (P0) then targets the PAZ motif and its adjacent upstream in AGO1 and mediates its degradation. Without affecting signal movement, P0 protein is able to block silencing and may contribute to the phloem restriction of viruses in Polerovirus group (Baumberger et al. 2007).

### 6.5.7 *Tobamoviridae* Family

The silencing-suppressing activity of tobamovirus resides in the viral 126-kDa small replicase subunit. Its infection leads to interference with HEN1-mediated methylation of miRNA and siRNA. Interference with methylation, mediated by HEN1, of small RNAs and that this interference and the formation of virus-induced symptoms of disease are associated with the silencing suppression activity of the 126 kDa protein (Vogler et al. 2007).

### 6.5.8 Other Viral Suppressors

*Closteroviridae* family member CCYV (cucurbit chlorotic yellows virus) RNA 1-encoded protein p22 has previously been reported to have the property of RNA-silencing suppressor locally, for which an interaction with cucumber SKP1LB via an F-box-like motif was found to be crucial (Salavert et al. 2020). CCYV p22 is capable of binding both single- and double-stranded long RNAs, as well as single- and double-stranded small interfering RNA (siRNA) molecules.

## 6.6 Identification and Characterization of VSR: Experimental Approaches

The identification and characterization of VSR (viral suppressors of RNA silencing) depend on a broad range of experimental methods. These methods include the strategies of molecular biology, biochemistry, and genetics in order to determine whether a viral protein can cause RNA silencing and also identify the mechanism and host variables that are involved. A detailed discussion is carried out below.

### 6.6.1 Transient Expression Assay

The agroinfiltration-based GFP silencing suppression test is one of the ancient and most effective strategies for determining silencing suppression. This test can be well defined in the *Nicotiana benthamiana* plant. The leaves of *N. benthamiana* are used to transiently express the reporter transgene of GFP (green fluorescent protein). Generally, the RNA-silencing machinery of the host rapidly suppresses the GFP expression, which in turn leads to loss of fluorescence. Silencing of this suppression, due to coexpression of viral protein, can cause prolonged expression of GFP fluorescence (Beritza et al. 2024; Li et al. 2017).

### 6.6.2 Biochemical Assay

Biochemical assays such as Co-IP (co-immunoprecipitation) and EMSA (Electrophoretic Mobility Shift Assay) are among the useful approaches that are taken to identify and characterize VSRs. The Co-IP is useful in identifying viral protein and host silencing protein interaction. EMSA is broadly used to detect the interaction between VSR and small RNAs (Sanobar et al. 2021).

### 6.6.3 Mutational Analysis

Engineering viral genomes to make them devoid of the VSR gene, known as VSR knockout mutants, can be a powerful tool for VSR identification. The reduced infection rate and attenuated symptoms can confirm the gene suppression. Site-directed mutagenesis of functional motifs is an efficient way to understand VSR function, such as mutating the GW/WG motif (AGO hook), which can be helpful to test AGO-binding dependency (Murray et al. 2013).

### 6.6.4 Small RNA Profiling

Functional evidence can be obtained from profiling strategies by exploiting the RNA population alteration ability of VSR. Northern blotting is used traditionally to detect vsiRNA and host miRNA in infected tissues. Presence of VSRs can be confirmed by the absence or reduced amount of these molecules.

### 6.6.5 RdDM and Methylation Assay

RdDM (RNA-directed DNA methylation) probing assays are crucial for VSR detection and characterization (Wu et al. 2010). It includes bisulfite sequencing, which is used to ascertain the methylation status of cytosine in viral or host DNA. Another popular method, the ChIP assay (Chromatin Immunoprecipitation), is also found to be successful in terms of characterization of VSR. It can detect the recruitment of silencing proteins to viral DNA by measuring the changes in histone modification. Failure in recruitment indicates VSR presence. Co-ChIP is also used to detect chromatin-associated AGO proteins.

## 6.7 Evolutionary Perspectives and Ecological Implications

### 6.7.1 Co-evolution of VSR and Plant Silencing Pathways

The interests of viruses and hosts frequently conflict: for a virus, host infection is essential for replication, whereas for a host, infection can result in disease. The virus and host face the same selective pressure from this relationship, which leads to reciprocal adaptation and counter-adaptation in the form of an evolutionary "arms race." The interaction between the antiviral immune system of RNAi of many plants and viral suppressors of RNAi appears to have the potential to trigger such an arms race. Plant hosts have to face strong selection pressure from diverse viral VSRs, which drives the rapid evolution of genes involved in the RNAi pathway (Murray et al. 2013). In turn, viruses adapt intermittently, though they lack sequence homology across VSR families, showing episodic co-evolutionary dynamics.

### 6.7.2 Horizontal Gene Transfer and Convergent Evolution of Suppression Mechanism

Amidst the diversity of VSRs, similar functions are also performed by different VSRs. As discussed above, siRNA sequestration by tombusvirus P19, NS of tospovirus, and p22 of closteovirus illustrates convergent evolution by displaying independent viral suppressor of RNA-silencing strategies facilitated by the development of different RNA blocking proteins. In other instances, the presence of domains in a virus with host-like motifs is nothing but an indication of horizontal transfer or host protein mimicry. This suggests that few chemical configurations are available to switch off RNA silencing (de la Casa-Esperón 2012).

### 6.7.3 Role of VSR in Mixed Viral Infection

#### 6.7.3.1 Viral Synergy

Interaction between two independent viruses in the same host can cause synergistic viral diseases in higher plants, and they can be characterized by dramatic increases in symptoms and in the accumulation of one of the co-infecting viruses. In PVX (Potato Virus X)/potyviral synergism, the increase in pathogenicity and accumulation of PVX are facilitated by the expression of potyviral 5′ proximal sequences encoding P1, the helper-component proteinase (HC-Pro), and a fraction of P3 (Pruss et al. 1997). The same sequence of potyvirus (called P1/HC-Pro) enhances the pathogenicity and accumulation of two other heterologous viruses, such as cucumber mosaic virus and tobacco mosaic virus.

#### 6.7.3.2 Host Range Expansion

With the help of vectors like insects (aphids, whiteflies), viruses can expand their host range. The promiscuity of vectors exposes viruses to new hosts that were previously uninfected. Sustainable transmission can be done through spill-over infection, i.e., viral infection of a secondary or alternative host. Genetic variation and adaptation: Events like recombination and mutation can enable viruses to invade the host barriers (for example, binding to the receptor, host defense mechanisms), allowing them to infect a new host.

## 6.8 Application: Crop Protection and Breeding

VSRs have direct implications for designing strategies of long-term disease management in the vegetable crops. These VSRs are important proteins for pathogenicity, and targeting these proteins and their functions can give new possibilities in traditional breeding as well as biotechnology.

### 6.8.1 Breeding for Resistance Targeting VSR Activity

Host genes that VSRs take advantage of can be altered to reduce their susceptibility. As an example, Polerovirus P0 hijacks the SCF family of ubiquitin E3 ligase complex of the host for the degradation of AGO1. Similarly, RDRs (RNA-dependent RNA polymerase) mediated antiviral immune defense has a crucial role in resistance to many plant viruses. In RdRP6-mediated defense against viral infection, SGS3, a plant-specific RNA-binding protein, was found to fight against some viruses in *Arabidopsis*. Identifying these types of host factors, less susceptible to VSR, can be useful for the selection of tomato and cucurbits (Pazhouhandeh et al. 2006).

### 6.8.2 Transgenic RNAi Approaches

VSR encoding viral genes can be directly suppressed by several biotechnological treatments. Artificially designed siRNA and miRNAs target the VSR's conserved regions, such as 2b of CMV betasatellites, HC-Pro of Potyviridae. These molecules are known to suppress the viral accumulation and enhance the symptoms in the host (transgenic plant) when expressed. Tomato lines, as an example, produce hairpin RNAi constructs against TYLCV betasatellite have improved tolerance (Bortolamiol et al. 2007). But the challenge is here to design broad-spectrum constructs that remain effective against rapid viral evolution.

### 6.8.3 Genome Editing

CRISPR (Clustered Regularly Interspersed Short Palindromic Repeats) has greatly expanded the ability to probe virus–host interaction genetically. CRISPR systems enable systemic or focused, genomewide studies of almost all aspects of a virus lifecycle (Gebre et al. 2018). Combined with its relative ease of use and high reproducibility, CRISPR is becoming a pivotal tool in the studies of the host factors essential for viral pathogenesis. CRISPR can be useful to edit the AGO1 regulatory sequences to inhibit excessive miR168-mediated downregulation by P19 protein of tombusvirus (Várallyay et al. 2014). The interaction of V2 protein with the host SGS3 can lead to a reduction in VSR binding by tomato yellow leaf curl virus (Glick et al. 2008).

### 6.8.4 Field-Level Management Considerations

VSR has real-time applications in field-level disease management. As discussed above, VSR-driven synergism is often the product of mixed infection. In PVX (Potato Virus X)/potyviral synergism, the increase in pathogenicity and accumulation of PVX are facilitated by the expression of potyviral 5′ proximal sequences encoding P1, the helper-component proteinase (HC-Pro), and a fraction of P3 (Pruss et al. 1997). Therefore, high VSR-generated primary infection control should be the topmost priority of integrated management methods so that the ill effects of infection can be reduced. Gaining the knowledge of what synergic combination could be the deadliest can be helpful to improve vector control, resistant rootstocks, and crop rotation.

## 6.9 Future Prospects and Challenges

Major advancements have been made in VSR characterization, but there still exist significant gaps in linking the VSRs to the outcomes of their respective disease at the field level. In order to close these gaps, critical disease resilient strategies should be made.

### 6.9.1 Gaps in Mechanistic Understanding

A huge number of VSRs have been identified, but it is yet to unravel their way of interaction with the host precisely (Burgyán and Havelda 2011). For instance, though it has been known to us already that proteins like P0 and P25 can degrade AGO protein, little is known about their regulatory circuits and host cofactors. Likewise, VSRs that bind to long dsRNA (e.g., P38 and P14) can comprise the DCLs' activity, and AGO1-targeting VSRs (e.g., P0, P1, P38, and 2b) inhibit the RISCs, which in turn can lead to altering the expression of an unpredictable number of genes involved in the development of plants. Indeed, the VSR expression in transgenic plants leads to a phenotype that mimics the viral symptoms. However, the transgenic expression of VSRs does not necessarily reflect the effects of viral infection on pathways of endogenous silencing, since in natural viral infections, the VSR expression is restricted to virus-infected tissue and compartments, and it is also limited by time. It requires a deeper mechanistic framework to predict the durability of resistance.

### 6.9.2 Predicting Emergent VSR in New Viral Strains

Recombination or host shift sometimes leads to the production of emerging viral variants that may encode novel suppressors or modified versions of preexisting ones. The current methodology we have at present is not suitable to detect VSR

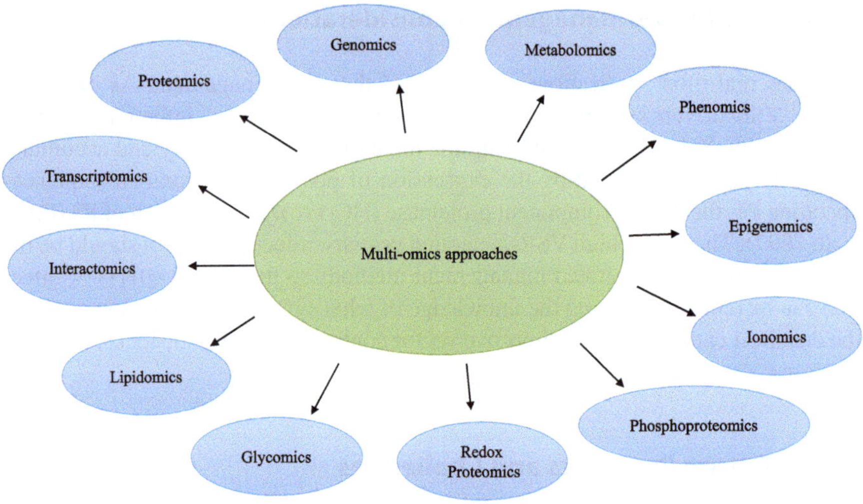


**Fig. 6.3** The role of diverse omics technologies, which include genomics, proteomics, transcriptomics, metabolomics, epigenomics, phenomics, and multi-omics, in understanding the plant-virus interactions and machinery underlying virus resistance. The high-throughput omics technologies could be exploited for more effective management and monitoring of plant disease

diversity specifically, making it challenging to interpret the severity of one lineage among many. Using a predictive model, the scale of risk factors, such as the suppressors, can be determined.

### 6.9.3 Integrating Multi-omics Approaches

The past decade has witnessed the development and deployment of sophisticated high-throughput omics technologies, including different omics such as genomic, proteomics, transcriptomics, and metabolomics, in order to understand the machineries underlying plant and virus interactions and implement strategies to prevent crop losses (Fig. 6.3). Integrating several fields of omics, such as genomics, proteomics, transcriptomics, and metabolomics, allows researchers to elucidate complex machinery underlying the interaction of plant and virus, which allows the formulation of critical insights into how specific expression of genes correlates with metabolic reprogramming in virus-infected plants (Samarskaya et al. 2025).

### 6.9.4 Translating Molecular Knowledge into Field Resilience

Linking molecular knowledge with sustainable crop protection is the ultimate challenge. Despite being the most promising strategies, RNAi and gene editing show several drawbacks, such as RNAi-mediated resistance facing durability, small RNAs might affect the off-target areas in the genome, viral variants having rapid viral mutations can sometimes escape the CRISPR-based targeting, and engineered plants face ethical obligations. Therefore, an integrated approach is much needed, i.e., RNAi, gene editing, and field surveillance to improve the quality of plants (Halder et al. 2022).

## 6.10 Conclusion

VSRs play a crucial role in the outcome of plant-viral interaction, especially in terms of vegetable crops, where the quality and yield are significantly controlled by viral diseases. To better understand the molecular silencing suppression mechanisms of VSRs, we need more detailed information about the replication, subcellular localization, and viral gene expression regulation, including VSRs, which are still not known for many of the plant viruses. Many VSRs have multiple functions in the viral life cycle; hence, the separate analysis of their suppression on silencing activities can lead to misinterpretation. Here, in this chapter, we try to highlight the basic principles of RNA silencing as a primary antiviral defense, the diversity of VSRs and the strategies they employ, and the dynamics of host-viral interaction. The complicated molecular arms race between viruses and plants is also demonstrated through the different strategies VSRs have taken to evade plant defense mechanisms, such as siRNA sequestration and AGO degradation. New advancements can be made to design sustainable resistance by doing further research on VSR diversity and their mechanism of action, host cofactors. Multi-omics analysis is also helpful to translate molecular insights into useful field-level resilience. As a whole, VSRs are a crucial target for viral pathogenicity analysis as well as innovation. New secure possibilities in the agricultural sector can be witnessed by studying these viral suppressors of RNA silencing.

## References

Ammara Ue, Mansoor S, Saeed M, Amin I, Briddon RW, Al-Sadi AM (2015) RNA interference-based resistance in transgenic tomato plants against tomato yellow leaf curl virus-Oman (TYLCV-OM) and its associated betasatellite. Virol J 12:38. https://doi.org/10.1186/s12985-015-0263-y

Baumberger N, Tsai CH, Lie M, Havecker E, Baulcombe DC (2007) The Polerovirus silencing suppressor P0 targets ARGONAUTE proteins for degradation. Curr Biol 17(18):1609–1614. https://doi.org/10.1016/j.cub.2007.08.039

Beritza K, Watts EC, van der Hoorn RAL (2024) Improving transient protein expression in agroinfiltrated Nicotiana benthamiana. New Phytol 243(3):846–850. https://doi.org/10.1111/nph.19894

Bortolamiol D, Pazhouhandeh M, Marrocco K, Genschik P, Ziegler-Graff V (2007) The Polerovirus F box protein P0 targets ARGONAUTE1 to suppress RNA silencing. Curr Biol 17(18):1615–1621. https://doi.org/10.1016/j.cub.2007.07.061

Burgyán J, Havelda Z (2011) Viral suppressors of RNA silencing. Trends Plant Sci 16(5):265–272. https://doi.org/10.1016/j.tplants.2011.02.010

Collins EJ, Bowyer C, Tsouza A, Chopra M (2022) Tomatoes: an extensive review of the associated health impacts of tomatoes and factors that can affect their cultivation. Biology 11(2):239. https://doi.org/10.3390/biology11020239

de la Casa-Esperón E (2012) Horizontal transfer and the evolution of host-pathogen interactions. Int J Evol Biol 2012:679045. https://doi.org/10.1155/2012/679045

Gaffar FY, Koch A (2019) Catch me if you can! RNA silencing-based improvement of antiviral plant immunity. Viruses 11(7):673. https://doi.org/10.3390/v11070673

Garcia-Ruiz H, Gabriel Peralta SM, Harte-Maxwell PA (2018) Tomato spotted wilt virus NSs protein supports infection and systemic movement of a Potyvirus and is a symptom determinant. Viruses 10(3):129. https://doi.org/10.3390/v10030129

Gebre M, Nomburg JL, Gewurz BE (2018) CRISPR-Cas9 genetic analysis of virus-host interactions. Viruses 10(2):55. https://doi.org/10.3390/v10020055

Glick E, Zrachya A, Levy Y, Mett A, Gidoni D, Belausov E, Citovsky V, Gafni Y (2008) Interaction with host SGS3 is required for suppression of RNA silencing by tomato yellow leaf curl virus V2 protein. Proc Natl Acad Sci USA 105(1):157–161. https://doi.org/10.1073/pnas.0709036105

Halder K, Chaudhuri A, Abdin MZ, Majee M, Datta A (2022) RNA interference for improving disease resistance in plants and its relevance in this clustered regularly interspaced short palindromic repeats-dominated era in terms of dsRNA-based biopesticides. Front Plant Sci 13:885128. https://doi.org/10.3389/fpls.2022.885128

Hong SF, Fang RY, Wei WL, Jirawitchalert S, Pan ZJ, Hung YL, Pham TH, Chiu YH, Shen TL, Huang CK, Lin SS (2023) Development of an assay system for the analysis of host RISC activity in the presence of a potyvirus RNA silencing suppressor, HC-Pro. Virol J 20(1):10. https://doi.org/10.1186/s12985-022-01956-2

Incarbone M, Dunoyer P (2013) RNA silencing and its suppression: novel insights from in planta analyses. Trends Plant Sci 18(7):382–392. https://doi.org/10.1016/j.tplants.2013.04.001

Jeevalatha A, Vanishree G, Siddappa S, Kumar R, Kaundal P, Kumar A, Chakrabarti SK (2021) Molecular characterization and infectivity analysis of tomato leaf curl New Delhi virus isolates infecting potato. 3 Biotech 11:203. https://doi.org/10.1007/s13205-021-02752-5

Kumar R, Jeevalatha A, Baswaraj R, Tiwari RK (2020a) Viral and viroid diseases of potato and their management. In: Singh AK, Chakrabarti SK, Singh B, Sharma J, Dua VK (eds) Potato science & technology for sub tropics. New India Publishing Agency, New Delhi, pp 267–292. ISBN: 978-93-89571-93-6

Kumar R, Kaundal P, Jeevalatha A, Sharma S, Chakrabarti SK (2020b) Development of a visual detection of Potato virus S by reverse transcription loop-mediated isothermal amplification. 3 Biotech 10:213. https://doi.org/10.1007/s13205-020-02214-4

Kumar R, Tiwari RK, Jeevalatha A, Sundaresha S, Shah MA, Sharma S, Sagar V, Kumar M, Chakrabarti SK (2021) Potato apical leaf curl disease: current status and perspectives on a disease caused by tomato leaf curl New Delhi virus. J Plant Dis Prot 128:897–911. https://doi.org/10.1007/s41348-021-00463-w

Kumar R, Tiwari RK, Sundaresha S, Kaundal P, Raigond B (2022) Potato viruses and their management. In: Chakrabarti SK, Sharma S, Shah MA (eds) Sustainable management of potato pests and diseases. Springer, Singapore, pp 309–335. https://doi.org/10.1007/978-981-16-7695-6_12

Kumar R, Lal MK, Tiwari RK, Chourasia KN, Kumar A, Kumar R, Sharma S, Singh B (2023) Investigating the interplay between tomato leaf curl New Delhi virus infection, starch metabolism and antioxidant Defence system in potato (Solanum tuberosum L.). Antioxidants 12:1447. https://doi.org/10.3390/antiox12071447

Li F, Wang Y, Zhou X (2017) SGS3 cooperates with RDR6 in triggering Geminivirus-induced gene silencing and in suppressing Geminivirus infection in Nicotiana Benthamiana. Viruses 9(9):247. https://doi.org/10.3390/v9090247

Mazumder P, Mondal F, Sarkar M, Majumdar A, Biswas KK, Sharma SK, Lal MK, Tiwari RK, Kumar R, Roy A (2024) Biological and molecular characterization of a new isolate of tomato mottle mosaic virus causing severe shoestring and fruit deformities in tomato plants in India. Plants 13:2811. https://doi.org/10.3390/plants13192811

Murray GG, Kosakovsky Pond SL, Obbard DJ (2013) Suppressors of RNAi from plant viruses are subject to episodic positive selection. Proc Biol Sci 280(1765):20130965. https://doi.org/10.1098/rspb.2013.0965

Pazhouhandeh M, Dieterle M, Marrocco K, Lechner E, Berry B, Brault V, Hemmer O, Kretsch T, Richards KE, Genschik P, Ziegler-Graff V (2006) F-box-like domain in the polerovirus protein P0 is required for silencing suppressor function. Proc Natl Acad Sci USA 103(6):1994–1999. https://doi.org/10.1073/pnas.0510784103

Pruss G, Ge X, Shi XM, Carrington JC, Bowman Vance V (1997) Plant viral synergism: the potyviral genome encodes a broad-range pathogenicity enhancer that transactivates replication of heterologous viruses. Plant Cell 9(6):859–868. https://doi.org/10.1105/tpc.9.6.859

Pumplin N, Voinnet O (2013) RNA silencing suppression by plant pathogens: defence, counter-defence and counter-counter-defence. Nat Rev Microbiol 11(11):745–760. https://doi.org/10.1038/nrmicro3120

Qin L, Shen W, Tang Z, Hu W, Shangguan L, Wang Y, Tuo D, Li Z, Miao W, Valli AA, Wang A, Cui H (2020) A newly identified virus in the family *Potyviridae* encodes two leader cysteine proteases in tandem that evolved contrasting RNA silencing suppression functions. J Virol 95(1):e01414–e01420. https://doi.org/10.1128/JVI.01414-20

Rishishwar R, Dasgupta I (2019) Suppressors of RNA silencing encoded by geminiviruses and associated DNA satellites. Virus 30(1):58–65. https://doi.org/10.1007/s13337-018-0418-8

Salavert F, Navarro JA, Owen CA, Khechmar S, Pallás V, Livieratos IC (2020) Cucurbit chlorotic yellows virus p22 suppressor of RNA silencing binds single-, double-stranded long and short interfering RNA molecules in vitro. Virus Res 279:197887. https://doi.org/10.1016/j.virusres.2020.197887

Samarskaya V, Spechenkova N, Kalinina NO, Love AJ, Taliansky M (2025) The emerging role of omics-based approaches in plant virology. Viruses 17(7):986. https://doi.org/10.3390/v17070986

Sanobar N, Lin PC, Pan ZJ, Fang RY, Tjita V, Chen FF, Wang HC, Tsai HL, Wu SH, Shen TL, Chen YH, Lin SS (2021) Investigating the viral suppressor HC-pro inhibiting small RNA methylation through functional comparison of HEN1 in angiosperm and bryophyte. Viruses 13(9):1837. https://doi.org/10.3390/v13091837

Sarkar M, Gupta D, Singh OW, Paul S, Kumar R, Mandal B, Roy A (2024) Identification of pathogenicity determinants in ToLCNDV and their RNAi-based knockdown for disease management in Nicotiana benthamiana and tomato plants. Front Microbiol 15:1481523. https://doi.org/10.3389/fmicb.2024.1481523

Sundaresha S, Jeevalatha A, Kumar R, Sood S, Sharma S, Bhardwaj V, Singh BP, Chakrabarti SK (2022) RNA interference: a versatile tool to augment plant protection strategies in potato. In: Chakrabarti SK, Sharma S, Shah MA (eds) Sustainable management of potato pests and diseases. Springer, Singapore, pp 369–388. (Hardcover ISBN 978-981-16-7694-9; Softcover ISBN978-981-16-7697-0; eBook ISBN978-981-16-7695-6). https://doi.org/10.1007/978-981-16-7695-6_15

Sundaresha S, Tiwari JK, Bhardwaj V, Sharma S, Jeevalatha A, Kumar R, Sagar V, Singh BP, Chakrabarti SK, Kumar M (2024) Genetic engineering in potato: status and future perspectives. In: Kalia P (ed) Genetic engineering of vegetable crops. CABI, Wallingford, pp 133–140. https://doi.org/10.1079/9781800627123.0008

Várallyay É, Oláh E, Havelda Z (2014) Independent parallel functions of p19 plant viral suppressor of RNA silencing required for effective suppressor activity. Nucleic Acids Res 42(1):599–608. https://doi.org/10.1093/nar/gkt846

Vogler H, Akbergenov R, Shivaprasad PV, Dang V, Fasler M, Kwon MO, Zhanybekova S, Hohn T, Heinlein M (2007) Modification of small RNAs associated with suppression of RNA silencing by tobamovirus replicase protein. J Virol 81(19):10379–10388. https://doi.org/10.1128/JVI.00727-07

Wu Q, Wang X, Ding SW (2010) Viral suppressors of RNA-based viral immunity: host targets. Cell Host Microbe 8(1):12–15. https://doi.org/10.1016/j.chom.2010.06.009

Xia Z, Zhu Z, Zhu J, Zhou R (2009) Recognition mechanism of siRNA by viral p19 suppressor of RNA silencing: a molecular dynamics study. Biophys J 96(5):1761–1769. https://doi.org/10.1016/j.bpj.2008.11.047

Zhang X, Yuan YR, Pei Y, Lin SS, Tuschl T, Patel DJ, Chua NH (2006) Cucumber mosaic virus-encoded 2b suppressor inhibits Arabidopsis Argonaute1 cleavage activity to counter plant defense. Genes Dev 20(23):3255–3268. https://doi.org/10.1101/gad.1495506

# Viral Infection and Altered Nutrient Dynamics in Vegetable Crops

7

Munmi Borah, Devayani Sarmah, Swapnil Borah, Greeshma Varghese, Puja Dey, Preetam Baruah, Sanjukta Chakraborty, Priyam Kafle, and Pronoy Kalita

**Abstract**

Vegetable crops are vital for global food and nutritional security; however, their productivity and quality are severely constrained by viral diseases. Beyond visible symptoms and yield losses, viral infections profoundly alter plant physiological, biochemical, and molecular processes, particularly nutrient uptake, translocation, and assimilation. Plant viruses disrupt source–sink relationships, vascular transport, photosynthesis, and redox balance, leading to imbalances in both macro- and micronutrients such as nitrogen, phosphorus, potassium, calcium, iron, zinc, and sulfur. These nutrient perturbations not only compromise plant growth and defense capacity but also deteriorate the nutritional and functional quality of harvested produce. The magnitude of these effects depends on host–virus interactions, nutrient availability, infection stage, and environmental conditions. This chapter provides a comprehensive overview of major viral pathogens of vegetable crops, their transmission mechanisms, and their influence on nutrient dynamics. It critically examines virus-induced changes in nutrient uptake, metabolism, and allocation, along with associated physiological and biochemical consequences. Management strategies integrating balanced nutrition, resistant cultivars, biotechnological interventions, and integrated disease

M. Borah (✉) · S. Borah · G. Varghese · P. Dey · P. Baruah · S. Chakraborty · P. Kafle · P. Kalita
Department of Plant Pathology, Assam Agricultural University, Jorhat, Assam, India
e-mail: munmi.borah@aau.ac.in

D. Sarmah
Department of Plant Pathology, Assam Agricultural University, Jorhat, Assam, India

Division of Plant Pathology, ICAR-Central Rice Research Institute, Cuttack, Odisha, India



R. Kumar et al. (eds.), *Vegetable Crops and Viral Pathogens Volume I*, Plant Pathology for Sustainable Agriculture, https://doi.org/10.1007/978-981-95-7265-6_7

management are discussed to mitigate viral impacts. Understanding virus–nutrient interactions is essential for developing sustainable approaches to safeguard vegetable productivity, quality, and food security under changing climatic conditions.

**Keywords**

Plant viruses · Nutrient dynamics · Macronutrients · Micronutrients · Virus–host interactions · Physiological stress · Biochemical changes · Photosynthesis · Oxidative stress · Crop productivity · Food quality · Integrated disease management

## 7.1 Introduction

Vegetables contribute to food security due to their high nutritional value and economic returns, as they provide essential vitamins, minerals, dietary fiber, and phytochemicals that support human health, while also serving as a sustainable component of diverse farming systems (Parajuli et al. 2019). Global vegetable production exceeds 1.15 billion metric tons annually, as they are an integral part of the diet worldwide and cultivated across different climates. The rising global population projects demand for food, which intensifies the pressure on vegetable production, which already faces challenges from climate change, land degradation, and environmental pollution. Plant diseases are a major threat, causing substantial yield and quality losses. Bacterial, fungal, and viral diseases undermine both agricultural productivity and food security, highlighting the need for effective detection and management strategies. Viral infections in vegetable crops can alter the nutrient dynamics and impact crop health and productivity. Viral infections exhibit symptoms such as chlorosis, leaf deformation, stunted growth, and yield loss, which may be attributed to disrupted host's physiological and biochemical processes, including nutrient uptake, assimilation, and allocation (Kumar et al. 2023, 2025a, b). The nutrient dynamics affect photosynthesis, carbohydrate metabolism, and mineral balance, ultimately influencing both plant growth and the nutritional quality of harvested produce. As the demand for vegetables continues to rise, understanding virus–nutrient interactions is crucial for management strategies that protect yields, ensure food quality, and support agricultural sustainability under changing environmental conditions. This chapter will discuss the effect of plant viruses in vegetable crops with respect to the nutrient uptake and distribution, and the resulting physiological and biochemical changes. It also examines the impact on crop productivity and food quality and discusses strategies to manage viral infections.

## 7.2 Major Viral Pathogens and Host Range

Vegetable crops are affected by several viral diseases that severely impact their yield and quality, with an average yield loss estimated to be about 70–80%, which under severe conditions can reach 90–100%. The severity of viral diseases can be determined by the symptoms exhibited, such as stunted growth, yellowing of leaves, mottling or streaking, necrosis, distorted leaves or flowers, wilting, and reduced yield. Globally, around 1000 plant viruses have been described, of which 200 are known to infect vegetable crops (Nagendran et al. 2017). The major virus groups infecting vegetables include *Geminiviridae, Potyviridae, Cucumoviridae*, tospoviruses, and tobamoviruses.

### 7.2.1 Transmission Mechanisms

Viruses can be spread in plants via two mechanisms: horizontal transmission, which is by insects, humans, or mechanical means, or vertical transmission by seed, pollen, or ovule.

- *Insect Vectors*: Almost 55% of plant viruses are transmitted by insect vectors, and the major ones include aphids, leafhoppers, thrips, whiteflies, flea beetles, and scale insects (Lei et al. 2025). Aphids are the most important vectors, transmitting viruses such as cucumber mosaic virus (CMV) and potato virus Y (PVY) in crops like tomato, pepper, and cucurbits (Varghese et al. 2024; Karasev and Gray 2013). Whiteflies (*Bemisia tabaci*) are key vectors of begomoviruses, and thrips transmit orthotospoviruses, which infect a wide range of vegetable crops (Pappu et al. 2009). Leafhoppers spread viruses such as Beet curly top virus and maize streak virus in certain solanaceous and cucurbit hosts (Alkhatib et al. 2024).
- *Seed-Mediated*: 20% of plant viruses are spread through seeds (Mink 1993). Two possible pathways have been suggested for the invasion of plant seed embryos by viruses: direct infection of the embryo following fertilization or indirect infection of the zygote during fertilization through transmission from virus-infected gametes (Wang and Maule 1997).
- *Mechanical Transmission*: Mechanical transmission is a significant mode of virus spread in vegetable crops, particularly in greenhouses and nurseries. This method involves the transfer of plant viruses through direct physical contact, such as handling infected plants, using contaminated tools, or through abrasions on plant surfaces. Notably, viruses like TMV, tomato mosaic virus (ToMV), and cucumber green mottle mosaic virus (CGMMV) are efficiently transmitted via this route (Chanda et al. 2021).

## 7.3 Nutrient Dynamics in Healthy Plants

Plant growth and productivity depend on the acquisition and efficient utilization of mineral nutrients. Nutrients are classified into macronutrients, required in larger amounts, and micronutrients, which are needed in trace quantities but equally essential for plant metabolism (Mengel and Kirkby 2001). The processes of nutrient uptake, assimilation, and translocation are tightly regulated to ensure metabolic balance and adaptability to environmental conditions (Hawkesford et al. 2012).

### 7.3.1 Nutrient Uptake

Nutrient uptake occurs primarily facilitated by plant roots, mediated by transport proteins in the plasma membrane. Both active and passive transport systems are required for the uptake of macronutrients like nitrogen, phosphorus, potassium, calcium, magnesium, and sulfur. For example, nitrate is absorbed via high- and low-affinity nitrate transporters (Crawford and Glass 1995), and potassium uptake involves voltage-gated channels and transporters responsive to cellular energy status (Maathuis 2009). Micronutrient uptake also requires specific transport mechanisms, as ferric reductases aid in the iron uptake in dicots by the reduction of $Fe^{3+}$ to $Fe^{2+}$ before the transport across the root epidermis (Kobayashi and Nishizawa 2012). In contrast, grasses utilize phytosiderophores to chelate and transport iron. The ZIP family transporters absorb zinc, manganese, and copper, keeping a check on the intracellular concentrations despite their low availability in soils (Grotz et al. 1998).

### 7.3.2 Nutrient Assimilation

Nutrients are assimilated into organic molecules for metabolic processes after they are absorbed. Nitrate reductase converts nitrate to nitrite during nitrogen assimilation, and nitrite reductase then reduces nitrite to ammonium (Campbell 1999). The glutamine synthetase/glutamate synthase cycle then incorporates the ammonium into amino acids (Lea and Miflin 2006). Similar processes underlie sulfur assimilation, wherein sulfate is reduced and activated prior to being incorporated into cysteine and methionine (Takahashi et al. 2011). Nucleotides, phospholipids, and energy molecules like ATP, which are all essential for cellular functions, incorporate phosphorus, which is absorbed as orthophosphate (Raghothama 1999). Magnesium is assimilated into chlorophyll, facilitating light capture during photosynthesis. Calcium plays a role in structural stabilization of cell walls and acts as a secondary messenger in signal transduction (Hepler et al. 2005). Micronutrients serve as cofactors in enzymatic reactions during assimilation. For example, molybdenum is required for nitrate reductase activity (Mendel and Hänsch 2002), while copper is essential for plastocyanin in photosynthetic electron transport (Burkhead et al. 2009). Manganese participates in the water-splitting reaction of photosystem II (Dau and Zaharieva 2008), and zinc is indispensable for the structural stability of transcription factors and enzymes.

### 7.3.3 Nutrient Translocation

After assimilation, nutrients must be distributed to different plant organs to support growth and development. Xylem transport is primarily responsible for the upward movement of minerals from roots to shoots, driven by transpiration and root pressure. Mobile nutrients such as nitrate, potassium, and magnesium are readily translocated to actively growing tissues (Mengel and Kirkby 2001). In contrast, calcium and boron exhibit poor phloem mobility, often leading to localized deficiencies in young tissues (Brown et al. 2002). Phloem transport redistributes photoassimilates and certain nutrients, such as nitrogen in the form of amino acids and sulfur in the form of glutathione. Micronutrients also undergo phloem-mediated redistribution depending on their chemical form and binding to organic ligands. For instance, iron can be mobilized in the phloem as complexes with nicotianamine (von Wirén et al. 1999), while zinc is translocated as metal-nicotianamine complexes to sink tissues (Curie et al. 2009).

### 7.3.4 Role of Nutrients in Plant Physiology and Defense

Nutrients are fundamental drivers of plant growth, development, and adaptation to environmental challenges. They not only support primary physiological processes but also contribute to structural and biochemical defense mechanisms against pathogens and herbivores (Table 7.1).

## 7.4 Impact of Viral Infection on Nutrient Uptake of Plants

Viral infections are one of the biggest biotic stresses that affect crop production around the world. Their effects go far beyond visible disease symptoms and can change the way plants work in ways that are not visible. One of the most important but least studied effects of viral infection is the disruption of nutrient uptake and transport systems, which is very important for plant health, productivity, and resilience. Viruses invade plant tissues, initiating a series of physiological and metabolic alterations that hinder the plant's capacity to acquire, transport, and utilize vital macronutrients and micronutrients (Chen et al. 2025). Viral pathogens systematically reprogram plant metabolism by hijacking cellular machinery and redirecting resources toward viral replication rather than normal physiological functions. This metabolic reprogramming affects both macronutrient and micronutrient dynamics, with nitrogen and phosphorus metabolism being particularly vulnerable due to their central roles in protein and nucleic acid synthesis required for viral reproduction (Seabloom et al. 2017). Furthermore, viral infections significantly impact the plant's vascular transport system, with viruses utilizing both phloem and xylem vessels for systemic movement while simultaneously disrupting normal nutrient flow. Research on citrus yellow vein clearing virus, for example, has revealed that infection leads to substantial accumulation of starch granules in vascular tissues, with vessel blockage severely obstructing iron transport and causing characteristic chlorosis

**Table 7.1** Macronutrients in plant physiology and defense

| Macronutrient | Key physiological function | Role in defense | Reference |
|---|---|---|---|
| Nitrogen (N) | Building block of amino acids, proteins, nucleotides, and chlorophyll; central to growth and photosynthesis | N status modulates immunity: N availability changes allocation between growth and defense, and N-related peptides/signals (e.g., CEPs) directly cross-talk with immune receptors to tune defense responses | Rzemieniewski et al. (2024) |
| Phosphorus (P) | A constituent of ATP, nucleic acids, and phospholipids; essential for energy transfer and signaling | P limitation alters immune signaling (e.g., ATP/eATP signaling and $Ca^{2+}$ signatures) and shifts trade-offs between growth and defense, so phosphate status strongly influences resistance phenotypes | Khan et al. (2023) |
| Potassium (K) | Major osmotic regulator and enzyme cofactor; controls stomata, turgor, and phloem transport | Adequate K strengthens physical and biochemical defenses (cell wall integrity, secondary metabolites, ROS-scavenging), reducing susceptibility to some pathogens and herbivores; K deficiency often increases disease/herbivore damage | Kommineni et al. (2024) |
| Calcium (Ca) | Structural role in cell walls/membrane stability and a universal intracellular second messenger | Cytosolic $Ca^{2+}$ transients are central to immune signaling (PTI/ETI); $Ca^{2}$-permeable channels and $Ca^{2+}$ sensors decode pathogen-triggered signals to activate downstream defense gene expression | Köster et al. (2022) |
| Magnesium (Mg) | Central atom in chlorophyll, cofactor for many enzymes (ATP-related processes), important for ribosome stability | Mg supports defense indirectly by sustaining photosynthesis/energy for defense, stabilizing cell walls, and promoting the synthesis of defense metabolites and antioxidant systems that limit pathogen damage | Ahmed et al. (2023) |
| Sulfur (S) | Components of cysteine, methionine, glutathione, and numerous defense compounds (e.g., glucosinolates, S-containing peptides) | Sulfur metabolism supplies thiol-based antioxidants (glutathione) and S-rich defense compounds; S availability strongly affects synthesis of antimicrobial metabolites and pathogen vs. host S-competition | Wang et al. (2022) |

symptoms (Liu et al. 2024). The disruption extends to specific nutrient transporters, with viral infection altering the expression of genes encoding phosphorus, potassium, and micronutrient transporters, ultimately affecting nutrient uptake and distribution (Sato et al. 2024). The severity of these effects often correlates with the extent of vascular system disruption and the specific nutrient pathways targeted by different viral pathogens. More concerning is the increased susceptibility of virus-infected plants to secondary abiotic stresses such as drought and salinity (Yang et al. 2024). This enhanced vulnerability occurs because viral infection compromises the plant's ability to maintain proper osmotic balance and ion homeostasis, processes that are crucial for stress tolerance and require adequate nutrient availability (Paudel and Sanfaçon 2022).

## 7.5 Viral Infection and Altered Nutrient Dynamics in Vegetable Crops

### 7.5.1 Virus–Nutrient Interactions

Viral infections disturb normal plant functions such as photosynthesis, hormone regulation, phloem flow, and source–sink balance, which in turn modify nutrient absorption, movement, and accumulation in plant tissues. The outcomes vary according to host plant species, the nature of the virus (whether phloem-restricted or systemic), the level of nutrient availability, and interactions with soil microorganisms and vectors, leading to either an increase, reduction, or altered distribution of specific elements.

#### 7.5.1.1 Viral Effects on Macronutrients (N, P, K, Mg, S)

##### 7.5.1.1.1 Effect on Nitrogen (N)

Viral infections often alter leaf nitrogen levels along with total plant nitrogen, where the extent and nature of these changes rely on the specific virus plus the plant's initial nitrogen condition (Scharf et al. 2002). Sometimes, infection hampers growth together with chlorophyll production, resulting in lower shoot nitrogen because of poor uptake or dilution; however, in different situations, nitrogen builds up in foliage owing to limited leaf size or altered distribution between sources and sinks (Rahoutei et al. 2000).

##### 7.5.1.1.2 Effect on Phosphorus (P)

Changes in plant phosphorus levels and corresponding shifts in N:P ratios are common outcomes of viral infection. The impact of reduced phosphorus differs across host–virus systems, sometimes enhancing and at other times constraining viral replication (Shalitin and Wolf 2000). Experimental nutrient additions show nitrogen-phosphorus ratios influence how infections develop as well as viral loads within hosts (Hessen et al. 2013).

##### 7.5.1.1.3 Effect on Potassium (K)

The role of potassium in plant defense and symptom expression is well established for a wide range of pathogens, and viral infections tend to follow similar patterns through their effects on host physiology. Given its involvement in stomatal functioning, water relations, and carbohydrate transport, imbalances in potassium are closely linked with characteristic viral symptoms, including leaf yellowing and tissue collapse (Amtmann et al. 2008).

##### 7.5.1.1.4 Effect on Magnesium (Mg)

As an essential element for chlorophyll synthesis and enzymatic activity, magnesium often shows variable concentrations in plants challenged by viral pathogens. Certain studies report declines linked to leaf yellowing and impaired photosynthesis, while others note tissue-specific increases that vary with the progression of infection (Rahoute et al. 2000).

##### 7.5.1.1.5 Effect on Sulfur (S)

The metabolism of sulfur and the presence of sulfur-rich compounds, including glutathione and thiols, are integral to redox regulation and antiviral defense mechanisms in plants. Alterations in foliar sulfur levels have been observed, and sulfur nutrition can modulate the intensity of disease by impacting antioxidant capacity and defense compound synthesis (Falk et al. 2007).

#### 7.5.1.2 Viral Effects on Micronutrients (Fe, Zn, Mn, Cu, B, Mo)

##### 7.5.1.2.1 Effect on Iron (Fe)

Iron (Fe) plays an essential role in electron transport, redox homeostasis, and serves as a cofactor in numerous enzymatic processes. Disruption of Fe availability disturbs reactive oxygen species (ROS) regulation, deters chlorophyll formation, and interferes with multiple metabolic activities (Briat et al. 2015).

##### 7.5.1.2.2 Effect on Zinc (Zn)

Zinc (Zn) contributes to the structural stability of proteins and enzymes associated with antiviral defense, potentially hindering viral replication or disrupting the activity of viral replication complexes. It also enhances the plant's antioxidant capacity. So, disruption of Zn availability will lead to a weak defense system as well as rapid viral replication. Foliar application of $ZnSO_4$ has been shown to reduce symptoms and lower the concentration of potato viruses PVY and PLRV by activating phenolic compound production. Zinc-based nanoparticles (ZnO-NPs) effectively suppressed TMV by reducing oxidative damage through ROS generation, activation of antioxidant defenses, and upregulation of SA-mediated resistance genes PR1 and PR2 (Bastakoti 2023).

##### 7.5.1.2.3 Effect on Manganese (Mn)

It acts as a vital cofactor for enzymes involved in managing oxidative stress and in the phenylpropanoid and lignin biosynthetic pathways. A lack of this element can weaken structural defense barriers, and it also plays a role in regulating photosynthetic activity and reactive oxygen species (ROS) detoxification (Millaleo et al. 2010).

#### 7.5.1.2.4 Effect on Copper (Cu)

Copper (Cu) functions as an important component of antiviral defense, as infected plants enhance the expression of copper transporter genes such as *HMA5* and members of the *COPT* family to raise Cu accumulation in leaves. This response contributes to the suppression of specific regulatory RNAs and activates protective defense mechanisms (Bernal et al. 2012). However, excessive Cu levels can lead to toxicity and damage in certain plant tissues.

#### 7.5.1.2.5 Effect on Boron (B)

Boron (B) plays a crucial role in maintaining cell wall structure, facilitating carbohydrate and sugar transport, and possibly participating in signaling processes. During viral infection, adequate B supply aids in nutrient redistribution, reinforces cell wall and membrane stability, and may influence reactive oxygen species (ROS) signaling (Brown et al. 2002). When boron is lacking or its activity is impaired during viral infection, the structural integrity of cell walls is compromised, making them more permeable. This facilitates easier viral movement through plasmodesmata, potentially resulting in more rapid systemic infection and intensified symptoms such as wilting, leaf yellowing, and tissue necrosis (Goldbach and Wimmer 2007).

#### 7.5.1.2.6 Effect on Molybdenum (Mo)

Molybdenum (Mo) is a key component of several molybdo enzymes, such as nitrate reductase and aldehyde oxidase, which are involved in redox regulation and defense-associated metabolic processes. Adequate Mo availability supports reactive oxygen species (ROS) detoxification and contributes to the activation of plant defense signaling pathways. A decrease in molybdenum availability during viral infection can diminish the activity of key enzymes, impairing nitrogen metabolism and overall metabolic stability. This often results in slower plant growth and decreased protein production, thereby reducing the plant's capacity to resist viral stress (Bittner 2014).

## 7.6 Physiological and Biochemical Consequences

Viral infections can lead to changes in physiological and biological processes. In vegetable crops, these changes can negatively impact nutritional quality and ultimately cause huge yield loss. Some of the physiological and biochemical consequences of viral infection in vegetable crops are discussed below.

### 7.6.1 Changes in Photosynthesis and Carbon Metabolism

One of the earliest and most prominent effects of a viral attack is an alteration in the function of the photosynthetic apparatus. Viruses can change the structure of the chloroplast and can divert carbon flux to support viral replication. Viral infection often causes a decline in chlorophyll content (chlorosis). It is also responsible for the damage to the thylakoid membranes, ultimately reducing the efficiency of light capture (Bahar et al. 2020). Viruses generally target the proteins that are associated with the chloroplast, especially those in photosystems I and II and in carbon fixation (e.g., Rubisco, Calvin cycle enzymes). In many cases, it was seen that viral infection can reduce the net photosynthetic rate. For example, when cucumber mosaic virus (CMV) infects tomato, it can lower the rate of electron transport and decrease the photochemical efficiency in the leaves. In addition to light reactions, viruses alter carbon partitioning; generally assimilated carbon is used for growth and storage reserves, but in this case, it is relocated for the synthesis of viral nucleic acids and proteins (Jiang et al. 2025). It is also common to observe starch or soluble sugar accumulation in infected leaves (Bahar et al. 2020). The carbon balance inside the host plant may change due to increased metabolic "sink" strength in the infected tissues, which pulls photo assimilates from healthy areas (Jiang et al. 2025).

### 7.6.2 Oxidative Stress and Nutrient-Related Defense Pathways

Viral infection triggers oxidative stress in the host plant by producing reactive oxygen species (ROS), defense signaling, and antioxidant responses—events that are closely related to nutrient status and homeostasis. The generation of ROS, such as superoxide and hydrogen peroxide, is one of the indicators of pathogen identification and stress. Also, viruses can disturb the redox balance in two ways: directly (by interfering with electron transport) or indirectly (by eliciting host defense) (Jiang and Zhou 2023). Upon a viral attack, host plants respond by upregulating antioxidative enzymes such as superoxide dismutase (SOD), catalase (CAT), guaiacol peroxidase (GPX), ascorbate peroxidase (APX), and polyphenol oxidase (PPO) (Bahar et al. 2020). Oxidative damage is responsible for the alteration of uptake and relocation of nutrients, especially micronutrients like Fe, Mn, Cu, and Zn, which are cofactors for antioxidant enzymes. Again, deficiency of Fe and Mn can lead to difficulties in the synthesis or activity of SOD and peroxidases. Moreover, virus infection may alter root–shoot signaling and nutrient transport. For example, when the alfalfa mosaic virus infects the alfalfa plant, it has been documented that it causes reductions in Fe, Mn, Cu, P, and Zn in foliar tissues and an increase in the nitrogen level as part of the viral load. The redox disruption also influences defense hormone signaling (salicylic acid, jasmonic acid, ethylene) and the activation of nutrient-based defense pathways (e.g., increased need for N and S for defense protein synthesis or phenolic biosynthesis).

### 7.6.3 Influence on Secondary Metabolites and Nutritional Quality

Viral infections affect the secondary metabolic network in the host, which is responsible for plant defense, nutritional quality, and flavor of the produce. Viral infection affects the upregulation and downregulation of compounds like phenolics, flavonoids, alkaloids, terpenes, and other specialized metabolites (Bahar et al. 2020). Phenolic compounds increase in infected tissues, causing enhanced phenolic polymerization and lignification as a part of the defense response. Changes in these metabolites can alter the antioxidant capacity, flavor profiles, and micronutrient-binding properties of vegetable organs. Virus can also reduce the amount of vitamins, carotenoids, or other phytonutrients in the host plant by directly interfering with biosynthesis or by depleting precursors (such as sugars or amino acids).

## 7.7 Implications for Crop Productivity and Food Quality

Viral infections are significant contributors to global crop losses. Unlike bacterial and fungal diseases, they cannot be managed by simply adopting chemical measures; their management depends on vector control, genetic resistance, and an integrative approach. Viruses affect not only crop yield but also quality at multiple levels: nutritional composition, organoleptic traits, and postharvest shelf life. With rising demand for nutritious food and growing global populations, understanding how viruses degrade both yield and quality is important for food security, trade, and public health. Even though the impact of viral infections on crop yield is understudied, the disastrous effects of viral infections have a long history. In the 1920s, the curly top virus in sugar beets ruined the entire sugar beet industry in the United States of America, and the swollen shoot of cacao caused the destruction of cacao plantations in West Africa (Thresh 1980). For viral infections, the time of infection is a factor modulating the yield loss. But the effect of viral disease on crop yield does not always increase linearly with the infection duration. The older plants show better resistance than the young plants. Also, plants infected at an early stage of development have a better capacity to recover from the infection as compared to those infected at later stages (Rao and Reddy 2020). But this is not the case always, as observed by Spence et al. (2007), wherein early infection of tulip mosaic virus (TuMV) in cabbages led to production of crops that were 50% lighter than healthy plants, whereas TuMV infection at a later stage was found to be less damaging. Also, mixed infection of tulip mosaic virus and cauliflower mosaic virus (CaMV) significantly reduced the marketability of cabbage heads by 25%, causing a 20 times greater crop loss as compared to non-infected plants. In the same study, *beet mosaic virus* (BtMV) was found to cause about 50% reduction in marketable leaf production in sweet chard (Spence et al. 2007).

## 7.8 Management Strategies

### 7.8.1 Nutrient Supplementation for Mitigating Viral Effects

The global population is growing quickly and is expected to exceed 7.6 billion, reaching about 10.4 billion by 2050, according to the Census Bureau (www.census.gov). To meet the food needs of future generations, we must improve agricultural productivity. Two main factors that help achieve higher crop yields are the availability of nutrients and effective protection against diseases caused by pathogens (Amtmann et al. 2008). Complete and balanced nutrition has always been the main defense for plants because mineral elements play a direct role in protection. Plant health can be impacted by these elements in two ways: either directly by activating enzymes that aid in the production of defense compounds such as callose, glucosinolates, lignin, phenols, and phytoalexins, or indirectly by altering microbial activity, the composition of root exudates, and the pH of the rhizosphere (Datnoff et al. 2007). Phosphorus (P) is listed as the second component in fertilizer analysis, and when phosphorus is lacking, plants may become more susceptible to diseases. Phosphorus significantly reduces yellow and blast disease, as well as bacterial leaf blight in rice, and tobacco leaf curl virus, mold, and pod and stem blight in soybean (Dordas 2008). Potassium (K) promotes root growth, increases cellulose and protein content, and reduces lodging by activating over 60 different plant growth enzymes (Datnoff et al. 2007). Perrenoud (1977) conducted a review of 2449 references and found that applying potassium resulted in a 41% reduction in viral disease incidence. K is required for the proper working of stomata during various pathogen infections, especially from bacteria and viruses, as they aid in the quick closure to prevent pathogen invasion (Pervez et al. 2007). Additionally, it has been found that higher potassium levels can reduce the multiplication of tobacco mosaic virus (TMV) in tobacco plants. In a study with groundnut bud necrosis virus (GBNV), adding phosphorus and potassium during infection can reduce the reactive oxygen species (ROS) production and cell death at the infection site and systemically, which led to reduced viral multiplication and spread within cowpea plants. The reduction in disease severity from micronutrients is because they actively participate in plant physiology and biochemistry. Essential micronutrients affect various processes that can change how plants respond to pathogens. Zinc (Zn) is one of these crucial micronutrients. It helps plants respond to phytopathogens mainly by activating or stabilizing metalloenzymes. However, using protective Zn concentrations against certain pathogens can increase susceptibility to others within the same host.

### 7.8.2 Breeding and Biotechnological Interventions

Traditional methods for plant breeding are slow, costly, and labor-intensive; various tools and methods have been created and used in genetics and genomics to improve plant breeding programs. For many crop species, including several vegetable types, whole-genome sequences, genetic linkage maps, marker assays, and molecular markers have been developed and published (Singh and Lebeda 2006). MAS integrates virus resistance genes with nutrient efficiency QTLs to sustain uptake under infection stress. Like in the case of tomatoes, MAS pyramids *Ty-1/Ty-3* (RNA-dependent RNA polymerase for TYLCV resistance) and *Ty-2* (NLR protein on chromosome 11) from wild species like *S. chilense* and *S. habrochaites*, maintaining photosynthesis and nutrient allocation despite begomovirus-induced stunting. Potato breeders use RYSC3 markers for PVY resistance alongside QTLs for nitrogen/phosphorus efficiency, reducing virus-mediated respiration spikes that divert carbohydrates from roots (Bhardwaj et al. 2019; Shahriari et al. 2023; Gautam et al. 2024). CRISPR/Cas can be applied to generate virus resistance in crops by targeting either plant genes that confer susceptibility to viruses or the viral genomes themselves (Kalinina et al. 2020). For example, Baltes et al. (2015) demonstrated the application of CRISPR/Cas9 for protecting plants against geminiviruses. CRISPR/Cas9 edits host susceptibility genes like SlPelo (Pelota homolog) or eIF4E1 in tomatoes, conferring TYLCV/PVY resistance without foreign DNA, preventing amino acid accumulation and membrane permeability changes from infection (Shahriari et al. 2023). Although gene-editing approaches provide durable genetic resistance, nutrient-based antiviral approaches act through a different but complementary mechanism, creating a multidimensional defense system that reduces viral pressure.

### 7.8.3 Integrated Disease and Nutrient Management Practices

Key nutrients supporting antiviral defense include phosphorus (P), potassium (K), zinc (Zn), calcium (Ca), and manganese (Mn). Phosphorus and potassium help restrict viral replication and movement by sustaining cellular energy metabolism and regulating osmotic balance. Zinc contributes by maintaining cell wall integrity and activating resistance genes, particularly those containing zinc-finger domains. Calcium and manganese further enhance defense by stabilizing cellular membranes and triggering hypersensitive responses, thereby limiting virus spread through signaling pathways such as MAPK activation and reactive oxygen species (ROS) production (Zvereva et al. 2023; Cabot et al. 2019; Dhaka et al. 2025). Hence, the application of balanced fertilizers targeting P, K, Zn, Ca, and Mn during vulnerable growth stages, integrating soil tests to prevent imbalances. Combine with cultural practices like vector control, as nutrition supports but does not replace integrated pest management for viruses. Climate change is expected to intensify viral epidemics in crops due to increased vector activity and accelerated viral replication under higher temperatures. Therefore, effective management strategies must integrate the influence of climate change when planning for future disease outbreaks.

## 7.9 Knowledge Gaps and Future Perspectives

The literature review suggests that the studies on viral diseases are mainly concentrated on yield loss and subsequent management practices, but are fewer on the detailed nutrient changes (micronutrients, vitamins, phytochemicals) and postharvest quality parameters across different virus–crop combinations. A knowledge gap exists in understanding how viral infections influence postharvest physiology in realistic storage and transport conditions, including cold chain and modified atmosphere. Also, information on the cumulative impact of mixed virus infections on yield, quality, and postharvest behavior is complex and lacking. Field validation of the models predicting the impact of virus-induced changes in nutrient uptake is a missing link. Key insights on the molecular cross-talk between viral pathogens and nutrient cycles are lagging. Partial information is available specifically on the disruption of metabolic pathways such as nitrogen, potassium, and calcium during disease progression. There is limited availability of knowledge on the temporal shifts prevalent in nutrient allocation during viral infection. The availability of data on the impact of crop productivity and quality is mainly based on laboratory or greenhouse level; data under real field changing dynamics are available only to a limited extent. The future holds integration of various omics approaches with physiological profiling, precision phenotyping, and field-based omics platforms to unravel how vegetable crops and their viral pathogens interact to ultimately impact nutrient allocation and, subsequently, the yield and quality parameters. A multistress resilience approach: An approach to understand crop behavior under biotic (viral) and abiotic stress (e.g., salinity, drought, etc.) is needed to protect crops against compounded stress impacts.

## 7.10 Conclusion

Viral infection is one of the major constraints in vegetable production. It causes disruption in vital processes in host tissue, such as photosynthesis, carbon metabolism, and the redox balance, by making changes in the process of uptake, translocation, and partitioning of essential macro- and micronutrients, inducing oxidative stress. Virus infections in vegetable crops reduce both the quality and quantity of harvested produce and diminish marketable value, thus reducing global food security. Research in the future should be focused on predictive models under changing climates, integrating nutrient management with disease control through balanced fertilization, developing resistant genotypes, and adopting new technologies like RNAi for vector management. A multidisciplinary approach is necessary for efficient management of viral diseases in vegetable crops.

## References


Ahmed N, Zhang B, Bozdar B, Chachar S, Rai M, Li J, Li Y, Hayat F, Chachar Z, Tu P (2023) The power of magnesium: unlocking the potential for increased yield, quality, and stress tolerance of horticultural crops. Front Plant Sci 14:1285512. https://doi.org/10.3389/fpls.2023.1285512

Alkhatib BM, Belteton S, Creamer R (2024) Immunolocalization of beet curly top virus (BCTV) and GroEL chaperon protein of endosymbionts in beet leafhopper (*Circulifer tenellus*) vector tissue. Viruses 16(10):1571. https://doi.org/10.3390/v16101571

Amtmann A, Troufflard S, Armengaud P (2008) The effect of potassium nutrition on pest and disease resistance in plants. Physiol Plant 133(4):682–691

Bahar MH, Singh A, Tiwari RK (2020) Changes in phyto-chemical status upon viral infections in plant: a critical review. J Pharmacogn Phytochem 9(1):275–283. Retrieved from https://www.researchgate.net/publication/348018525

Baltes NJ, Hummel AW, Konecna E, Cegan R, Bruns AN, Bisaro DM, Voytas DF (2015) Conferring resistance to geminiviruses with the CRISPR–Cas prokaryotic immune system. Nat Plants 1(10):1–4

Bastakoti S (2023) Role of zinc in management of plant diseases: a review. Cogent Food Agric 9(1). https://doi.org/10.1080/23311932.2023.2194483

Bernal M, Casero D, Singh V, Wilson GT, Grande A, Yang H et al (2012) Transcriptome sequencing identifies SPL7-regulated copper acquisition genes FRO4/FRO5 and the copper dependence of iron homeostasis in Arabidopsis. Plant Cell 24(2):738–761

Bhardwaj V, Sood S, Kumar A, Vanishree G, Sharma S, Sundaresha S, Raigond B, Kumar R, Bairwa A, Lal M, Chakrabarti SK (2019) Efficiency and reliability of marker assisted selection for resistance to major biotic stresses in potato. Potato J 46:56–66

Bittner F (2014) Molybdenum metabolism in plants and crosstalk to iron. Front Plant Sci 5:28. https://doi.org/10.3389/fpls.2014.00028

Briat JF, Dubos C, Gaymard F (2015) Iron nutrition, biomass production, and plant product quality. Trends Plant Sci 20(1):33–40

Brown PH, Bellaloui N, Wimmer MA, Bassil ES, Ruiz J, Hu H, Pfeffer H, Dannel F, Römheld V (2002) Boron in plant biology. Plant Biol 4(2):205–223

Burkhead JL, Reynolds KAG, Abdel-Ghany SE, Cohu CM, Pilon M (2009) Copper homeostasis. New Phytol 182(4):799–816

Cabot C, Martos S, Llugany M, Gallego B, Tolrà R, Poschenrieder C (2019) A role for zinc in plant defense against pathogens and herbivores. Front Plant Sci 10:1171. https://doi.org/10.3389/fpls.2019.01171

Campbell WH (1999) Nitrate reductase structure, function and regulation: bridging the gap between biochemistry and physiology. Annu Rev Plant Physiol Plant Mol Biol 50:277–303

Chanda B, Shamimuzzaman M, Gilliard A, Ling K (2021) Effectiveness of disinfectants against the spread of tobamoviruses: tomato brown rugose fruit virus and cucumber green mottle mosaic virus. Virol J 18(1). https://doi.org/10.1186/s12985-020-01479-8

Chen L, Zhang Y, Wang M (2025) Reprogrammed plant metabolism during viral infections: mechanisms, pathways and implications. Mol Plant Pathol 70(1):45–62

Crawford NM, Glass ADM (1995) Molecular and physiological aspects of nitrate uptake in plants. Trends Plant Sci 1(10):389–395

Curie C, Cassin G, Couch D, Divol F, Higuchi K, Le Jean M, Misson J, Schikora A, Czernic P, Mari S (2009) Metal movement within the plant: contribution of nicotianamine and yellow stripe 1-like transporters. Ann Bot 103(1):1–11

Datnoff LE, Elmer WH, Huber DM (2007) Mineral nutrition and plant disease. The American Phytopathological Society, St. Paul, pp vi+–278

Dau H, Zaharieva I (2008) Principles, efficiency, and blueprint character of solar-energy conversion in photosynthetic water oxidation. Acc Chem Res 42(12):1861–1870

Dhaka AS, Narwal S, Kumar RR, Mishra GP, Meena NL, Maheshwari C, Praveen S (2025) Mineral nutrient content restricts viral RNA accumulation and the systemic spread of

groundnut bud necrosis virus in Vigna unguiculata. Australas Plant Pathol 54(3):357–369. https://doi.org/10.1007/s13313-025-01053-2
Dordas C (2008) Role of nutrients in controlling plant diseases in sustainable agriculture. A review. Agron Sustain Dev 28(1):33–46
Falk KL, Tokuhisa JG, Gershenzon J (2007) The effect of sulfur nutrition on plant glucosinolate content: physiology and molecular mechanisms. Plant Biol 9(05):573–581
Gautam R, Dhital S, Abhishek, Pahal P, Reetika, Himanshu, Beniwal V (2024) Breeding strategies of virus resistance in solanaceous crops. Int J Adv Biochem Res 8(3):848–853
Goldbach HE, Wimmer MA (2007) Boron in plants and animals: is there a role beyond cell-wall structure? J Plant Nutr Soil Sci 170(1):39–48
Grotz N, Fox T, Connolly E, Park W, Guerinot ML, Eide D (1998) Identification of a family of zinc transporter genes from Arabidopsis that respond to zinc deficiency. Proc Natl Acad Sci USA 95(12):7220–7224
Hawkesford M, Horst W, Kichey T, Lambers H, Schjoerring J, Møller IS, White P (2012) Functions of macronutrients. In: Marschner P (ed) Marschner's mineral nutrition of higher plants. Elsevier, pp 135–189
Hepler PK (2005) Calcium: a central regulator of plant growth and development. Plant Cell 17(8):2142–2155
Hessen DO, Elser JJ, Sterner RW, Urabe J (2013) Ecological stoichiometry: an elementary approach using basic principles. Limnol Oceanogr 58(6):2219–2236
Jiang T, Zhou T (2023) Unraveling the mechanisms of virus-induced symptom development in plants. Plants 12(15):2830. https://doi.org/10.3390/plants12152830
Jiang S, Chen Q, Zhang L, Wang X (2025) Reprogramming of plant metabolism by viruses: from symptom development to defense modulation. Mol Plant Pathol 26(3):421–439. https://doi.org/10.1111/mpp.70066
Kalinina NO, Khromov A, Love AJ, Taliansky ME (2020) CRISPR applications in plant virology: virus resistance and beyond. Phytopathology 110(1):18–28
Karasev AV, Gray SM (2013) Continuous and emerging challenges of potato virus Y in potato. Annu Rev Phytopathol 51:571–586
Khan F, Siddique AB, Shabala S, Zhou M, Zhao C (2023) Phosphorus plays key roles in regulating plants' physiological responses to abiotic stresses. Plants 12(15):2861. https://doi.org/10.3390/plants12152861
Kobayashi T, Nishizawa NK (2012) Iron uptake, translocation, and regulation in higher plants. Annu Rev Plant Biol 63:131–152. https://doi.org/10.1146/annurev-arplant-042811-105522
Kommineni V, Bhandari AB, Schuster G, Nelson SD (2024) Cotton response to foliar potassium application in South Texas dryland. Agronomy 14(10):2422
Köster P, DeFalco TA, Zipfel C (2022) $Ca^{2+}$ signals in plant immunity. EMBO J 41(12):e110741
Kumar R, Lal MK, Tiwari RK, Chourasia KN, Kumar A, Kumar R, Sharma S, Singh B (2023) Investigating the interplay between tomato leaf curl New Delhi virus infection, starch metabolism and antioxidant Defence system in potato (Solanum tuberosum L.). Antioxidants 12:1447. https://doi.org/10.3390/antiox12071447
Kumar R, Lal MK, Tiwari RK, Kumar D, Sagar V, Jena R, Kumar R, Singh G, Singh B (2025a) Antioxidant defence mechanism of potato (Solanum tuberosum) by regulation of physiological and biochemical response under potato virus Y necrotic strain (PVYNTN) infection. J Plant Growth Regul. https://doi.org/10.1007/s00344-025-11850-0
Kumar R, Lal MK, Tiwari RK, Kumar D, Sagar V, Kumar R, Roy A, Singh VK, Singh G, Singh B (2025b) Alteration of source-sink dynamics and antioxidant responses in potato (Solanum tuberosum L.) under potato Leafroll virus (PLRV) infection. J Plant Growth Regul 44:4727–4743. https://doi.org/10.1007/s00344-025-11719-2
Lea PJ, Miflin BJ (2006) Nitrogen assimilation and its relevance to crop improvement. Annu Plant Rev 42:1–40
Lei J, Yuan J, Chen M, Mao Q (2025) Insect-specific viruses and their emerging role in plant disease mitigation. Viruses 17(9):1269. https://doi.org/10.3390/v17091269

Liu X, Wang L, Chen H, Zhang K, Wang Y, Li S (2024) Citrus yellow vein clearing virus infection triggers phloem remobilization of iron- and zinc-nicotianamine in citrus. Plant Physiol 186(4):2234–2251

Maathuis FJM (2009) Physiological functions of mineral macronutrients. Curr Opin Plant Biol 12(3):250–258

Mendel RR, Hänsch R (2002) Molybdoenzymes and molybdenum cofactor in plants. J Exp Bot 53(375):1689–1698. https://doi.org/10.1093/jxb/erf038

Mengel K, Kirkby EA (2001) Principles of plant nutrition, 5th edn. Kluwer Academic Publishers, Dordrecht

Millaleo R, Reyes-Díaz M, Ivanov AG, Mora ML, Alberdi M (2010) Manganese as essential and toxic element for plants: transport, accumulation and resistance mechanisms. J Soil Sci Plant Nutr 10(4):470–481

Mink GI (1993) Pollen and seed-transmitted viruses and viroids. Annu Rev Phytopathol 31(1):375–402. https://doi.org/10.1146/annurev.py.31.090193.002111

Nagendran K, Pandey KK, Rai AB, Singh B (2017) Viruses of vegetable crops: symptomatology, diagnostics and management, Technical Bulletin No. 75. ICAR—Indian Institute of Vegetable Research

Pappu HR, Jones RAC, Jain RK (2009) Global status of tospovirus epidemics in diverse cropping systems: successes achieved and challenges ahead. Virus Res 141:219–236

Parajuli R, Thoma G, Matlock MD (2019) Environmental sustainability of fruit and vegetable production supply chains in the face of climate change: a review. Sci Total Environ 650:2863–2879. https://doi.org/10.1016/j.scitotenv.2018.10.019

Paudel DB, Sanfaçon H (2022) Plant mineral nutrition and disease resistance: a significant linkage for sustainable crop protection. Front Plant Sci 13:1003721

Perrenoud S (1977) Potassium and plant health, No. 3, pp. 218-pp. International Potash Institute, Berne

Pervez H, Ashraf M, Makhdum MI, Mahmood T (2007) Potassium nutrition of cotton (Gossypium hirsutum L.) in relation to cotton leaf curl virus disease in aridisols. Pak J Bot 39(2):529

Raghothama KG (1999) Phosphate acquisition. Annu Rev Plant Physiol Plant Mol Biol 50:665–693

Rahoutei J, García-Luque I, Barón M (2000) Inhibition of photosynthesis by viral infection: effect on PSII structure and function. Physiol Plant 110(2):286–292

Rao G, Reddy M (2020) Overview of yield losses due to plant viruses. https://doi.org/10.1016/B978-0-12-818654-1.00038-4

Rzemieniewski J, Leicher H, Lee HK, Broyart C, Nayem S, Wiese C, Maroschek J, Camgöz Z, Olsson Lalun V, Djordjevic MA, Vlot AC, Hückelhoven R, Santiago J, Stegmann M (2024) CEP signaling coordinates plant immunity with nitrogen status. Nat Commun 15:Article 10686

Sato M, Nakahara KS, Yoshii M, Ishikawa M, Uyeda I (2024) Integrated biological experiments and proteomic analyses of Nicotiana tabacum xylem sap revealed the host response to tomato spotted wilt Orthotospovirus infection. Int J Mol Sci 25(20):10907

Scharf PC, Wiebold WJ, Lory JA (2002) Corn yield response to nitrogen fertilizer timing and deficiency level. Agron J 94(3):435–441

Seabloom EW, Malmstrom CM, Mitchell CE, Power AG (2017) Environmental nutrient supply directly alters plant traits but indirectly determines virus growth rate. Front Microbiol 8:2116. https://doi.org/10.3389/fmicb.2017.02116

Shahriari Z, Su X, Zheng K, Zhang Z (2023) Advances and prospects of virus-resistant breeding in tomatoes. Int J Mol Sci 24(20):15448. https://doi.org/10.3390/ijms242015448

Shalitin D, Wolf S (2000) Cucumber mosaic virus infection affects sugar transport in melon plants. Plant Physiol 123(2):597–604

Singh RJ, Lebeda A (2006) Landmark research in vegetable crops. In: Genetic resources, chromosome engineering, and crop improvement: vegetable crops, volume 3. CRC Press, Boca Raton

Spence NJ, Phiri NA, Hughes SL, Mwaniki A, Simons S, Oduor G, Chacha D, Kuria A, Ndirangu S, Kibata GN, Marris GC (2007) Economic impact of *Turnip mosaic virus*, *Cauliflower mosaic virus* and *Beet mosaic virus* in three Kenyan vegetables. Plant Pathol 56:317–323. https://doi.org/10.1111/j.1365-3059.2006.01498.x

Takahashi H, Kopriva S, Giordano M, Saito K, Hell R (2011) Sulfur assimilation in photosynthetic organisms: molecular functions and regulations of transporters and assimilatory enzymes. Annu Rev Plant Biol 62:157–184. https://doi.org/10.1146/annurev-arplant-042110-103921
Thresh JM (1980) The origin and epidemiology of some important plant virus diseases. Appl Biol 5:1–65
Varghese G, Sarmah D, Borah M (2024) Molecular detection of ChiVMV infecting Bhut Jolokia crop in Assam [conference presentation]. National conference on plant health for food security: threats and promises. https://doi.org/10.13140/RG.2.2.26998.46408
von Wirén N, Klair S, Bansal S, Briat JF, Khodr H, Shioiri T, Leigh RA, Hider RC (1999) Nicotianamine chelates both Fe(III) and Fe(II): implications for metal transport in plants. Plant Physiol 119(3):1107–1114
Wang D, Maule AJ (1997) Contrasting patterns in the spread of two seed-borne viruses in pea embryos. Plant J 11(6):1333–1340. https://doi.org/10.1046/j.1365-313x.1997.11061333.x
Wang W, Liu J, Mishra B, Mukhtar MS, McDowell JM (2022) Sparking a sulfur war between plants and pathogens. Trends Plant Sci 27(12):1253–1265
Yang H, Zhang M, Kent ML, Berger PH, Gray SM (2024) A transition from enemies to allies: how viruses improve drought resilience in plants. Plant Physiol 195(3):1678–1692
Zvereva AS, Klingenbrunner M, Teige M (2023) Calcium signaling: an emerging player in plant antiviral defense. J Exp Bot 75(5):1265–1273. https://doi.org/10.1093/jxb/erad442

# Seed-Borne Viruses in Vegetable Crops: Challenges and Solutions

# 8

Kamal A. M. Abo-Elyousr, Muhammad Imran, Munirah F. Aldayel, and Nashwa Sallam

**Abstract**

Plant viruses propagated from seeds compromise vegetable crop development and world food security. Field viral infections go unreported and cost seed companies and producers' losses since infected seeds are invisible. The current research looks at vegetable crop seed-borne viruses. More precisely, it investigates the intricate interaction of environmental factors, plant biology, and virus characteristics that controls vertical transmission. This study investigates the many viruses carried by seeds in key vegetable crops. We study the contraction of several illnesses in seeds and seedlings. Still another topic is how early embryonic infection influences plant form and genes. The study also exposes the limits of approaches to seed-transmitted virus detection. Emphasizing the requirement of more sensitive, fast, reasonably priced testing tools helps to ensure virus-free

K. A. M. Abo-Elyousr (✉)
Department of Plant Pathology, Faculty of Agriculture, University of Assiut, Assiut, Egypt

Department of Agriculture, Faculty of Environmental Sciences, King Abdulaziz University, Jeddah, Saudi Arabia
e-mail: kaaboelyousr@agr.au.edu.eg; elyousr@aun.edu.eg

M. Imran
Department of Biological Engineering, Henan University of Technology, Zhengzhou, Henan, China
e-mail: imranpathologist@cau.edu.cn

M. F. Aldayel (✉)
Department of Biological Sciences, College of Science, King Faisal University, Al-Ahsa, Saudi Arabia
e-mail: maldayel@kfu.edu.sa

N. Sallam
Department of Plant Pathology, Faculty of Agriculture, University of Assiut, Assiut, Egypt
e-mail: nashwasallam@aun.edu.eg



R. Kumar et al. (eds.), *Vegetable Crops and Viral Pathogens Volume I*, Plant Pathology for Sustainable Agriculture, https://doi.org/10.1007/978-981-95-7265-6_8

seed production and dissemination. With an eye on their shortcomings, we go over chemical treatments, thermoscopy, and seed health testing. The research examines these issues in concert with novel concepts. This covers molecular diagnostics for precise and early identification of the viruses of seeds, biological control agents, breeding methods to raise host tolerance, and this chapter seeks to increase world vegetable crop productivity and seed-borne virus management. This will be accomplished by means of an exhaustive existing and future environment study.

**Keywords**

Seed-borne viruses · Vegetable · Seed transmission · Detection · Management · Plant pathology

## 8.1 Introduction

As parasites/pathogens, viruses reproduce through the host's cellular machinery. They use different replication mechanisms to allow latency in an appropriate host for infection, replication, and the virus particle to move inside the host (Escalante et al. 2024). Viruses can be naturally transmitted from plant to plant by horizontal transmission, which can involve assistance from insects, humans, or mechanical spread through agricultural tools or equipment (Pagán 2022). All viruses in nature must replicate and disseminate vertically. Seed transmission of plant viruses depends on their structures, biological and ecological characteristics, as well as the seed–virus–plant interaction. Vertically transmissible viruses, genome structures, and movement strategies assist in the dependability and success of this transmission pathway (Jayasinghe et al. 2022). Plant virus transmission in seeds is a major issue for seed companies. Unnoticed contaminated seeds can cause viral disease outbreaks, threatening food security. Plant viruses in early embryo development may produce morphological or genetic changes that reduce yield and economic value (Vishunavat et al. 2023). Viruses can infect seeds depending on their replication mechanism, host age, and resistance. Vyse, an RNA virus, can be detected neither in the mother plant's tissues nor in the adult progeny plants; however, it can be detected in the immature seed, thus representing a pinhole for the virus to pass on to the next generation. Six of the ten viruses analyzed (a multiple infection) have since been detected in F1 seeds (Dong et al. 2024). In general, due to their complex structure and amplification strategies, viruses are considered bad candidates for long-distance seed transmission, with seed transmission rates in transmission studies lower than for other virus types. Nevertheless, some of the recent successes in seed transmission follow-up mechanisms of infection, in combination with tinkering with plant's genetic machinery during the seed filling processes, make this topic increasingly important (Fortes et al. 2023). Estimating yield loss and damage from plant viruses is difficult due to crop yield and damage variability; however, experimental methods have shown reductions in physiological indicators such as plant growth and vigor. In a basic yield loss experiment, plants are inoculated with a virus or virus strain and compared to healthy plants (Tatineni and Hein 2023).

## 8.2 Overview of Seed-Borne Viruses

Seed-borne viruses, transmitted on seeds in the seed coat or, in a few rare cases, within seed cells, are an increasing concern to the global vegetable industry. This chapter presents challenges and management strategies for commonly studied seed-borne viruses in vegetable crops, including Avran, broomrape, and Gemini viruses (Tripathi et al. 2024a). Strategies include developing virus-free production systems at the seed, seedling, and mother plant levels, along with beneficial microorganisms that can inhibit virulence and virulence-associated viruses affecting crop plants (Escalante et al. 2024). The occurrence of seed-borne viruses is worldwide, affecting many agricultural commodities and natural ecosystems, including cereal grains, pulses, and oilseeds. Among them, vegetable crops such as tomato, onion, carrot, chili, capsicum, cucumber, and cabbage are of considerable economic importance, producing vegetables for fresh consumption or growing and selling in seedling form for juvenile vegetative growth before harvest (Sandra and Mandal 2024). There was limited publication of research related to seed-borne viruses in these crops, even though they are considered the ultimate challenge to biosecurity and food safety with respect to virus and virus-vector management. Greater epidemiological, molecular, genomic, biometrical, and diagnostic understanding of seed-borne viruses, such as cucurbit yellow stunt virus and tomato seed-borne mottle virus, among others, has increased the complexity of the responsible agents (Dell'Olmo et al. 2023; Fortes et al. 2023). The challenges associated with vegetable seed-borne and seed-transmitted viruses (SVBTs) are discussed, along with the biology, epidemiology, risk assessment, diagnostic detection, and management strategies of viruses of vegetables-borne viruses (VBTs) and their vectors. (Sandra and Mandal 2024) Additionally, the status of SVBTs in vegetables, their challenges and threats to the global economy and food security, virus transmission through seeds, methods and techniques for detection and diagnosis, and management strategies for disease prevention, control, or mitigation are summarized. Concerns and gaps in knowledge are also identified regarding the current and future development of vegetative transmission of viruses affecting vegetable crops (Wopereis and Kuo 2022).

Viruses can be naturally transmitted from plant to plant in two ways: by horizontal and vertical transmission. Horizontal transmission can involve succession of events which may include overcoming of physical and biological barriers, exposure of uninfected host, mechanical damage to the tissue, infection of cell(s) and replication spread via the host cell's machinery, and propagation and movement through a compatible species in which genetic defense mechanisms are not effective (Wu et al. 2022; García-Ordóñez and Pagán 2024). Viral acquisition in insects can only occur when virions are present in their mouthparts and then interact with Vm, which accepts, uncoats, translocates, and imports them into the cell. Seed transmission of plant viruses represents a huge problem for producers and seed companies (Tatineni and Hein 2023; Jones et al. 2021). One main problem is that the presence of infected or infested seeds might go unnoticed, resulting in outbreaks of viral diseases, representing a risk for food security. Testing seeds for virus presence and acting on those that go to planting is difficult (Tatineni and Hein 2023; Gomathi Devi et al. 2023). Virus-infected seeds might be decontaminated by heat treatment or via loss of viability by fungicide treatment. In addition, early embryo infection by plant viruses may produce morphological or genetic changes that reduce yield and earnings (Metwally et al. 2024). Producers and seed transmission face a major issue with plant virus transmission. Infected seeds may go unreported, causing viral disease epidemics that threaten food security (Tatineni and Hein 2023; Zhang et al. 2022). Early plant virus infection in embryos may produce morphological or genetic changes that reduce yield and economic value. Some seed-transmitted viruses are economically important for crop yield, seed multiplication, and seed and plantation material export (Sandra and Mandal 2024; Dell'Olmo et al. 2023). Seed-transmissible viruses are mostly ssRNA viruses, with a small number of dsRNA and ssDNA viruses. Plant viruses can infect seeds depending on their replication mechanism, host age, and resistance. Even at modest rates, seed transmission may help viruses proliferate, overwinter, and transmit vast distances (Sandra and Mandal 2024).

## 8.3 Impact on Vegetable Crops

Vegetables are rich in vitamins and micronutrients that are vital for maintaining good health, and certain vegetables are seen as protectors against diseases such as atherosclerosis, ulcers, cancer, and cataracts. Of the total vegetable production in India, 33% is lost due to virus diseases or other diseases. Parsley, coriander, chervil, and celery have become the most important spices and culinary herbs globally (Kaushik et al. 2021). European countries and the USA are the major producers, exporters, and consumers. However, the export volume of herbs and spices has been slowly declining as herbs have become a target for pests and diseases. The effects include seed germination, attempted control, impaired growth, malformation, crop loss, or in extreme cases, complete loss (Escalante et al. 2024) (Fig. 8.1).

### 8.3.1 Economic Consequences

Agricultural crops are the basis for food security and human wellbeing. During the last decades, increases in yield have been achieved thanks to genotypes and cultivation methods. However, the continuing rise of the world population poses challenges to food production, and crop losses are inevitable due to several biotic and abiotic stress factors (Altaf et al. 2025). Of the numerous biotic stress factors, viruses have a major impact on crops such as vegetables, fruits, and cereals. They produce a plethora of symptoms resulting in the reduction in yields, among others

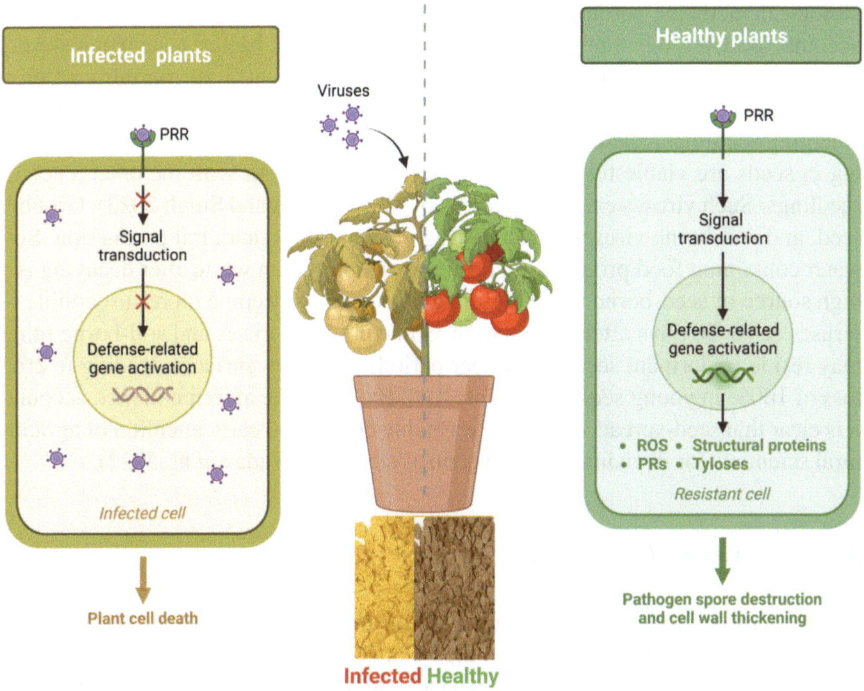


**Fig. 8.1** Comparison of using virus-infected and healthy seeds

and represent a huge economic loss if not properly controlled. However, seed health is crucial in agriculture, and the seed health standard provides better production and crop yields (Sokhandan-Bashir and Ighani 2024).

## 8.4 Major Seed-Borne Viruses in Vegetable Crops

Seed-spread viruses (SBV) trigger growing problems in agricultural cash crops. When compared to aphid or other insect-spread viruses, seed-spread viruses are still less well documented and fewer than 50 seed-borne viruses on vegetable crops have been reported. All viruses and their native plants have a unique relationship; some viruses induce malformations, stunting of growth, and leaf curling, leading to wilting of plants. Some viruses remain dormant and may emerge later to kill the crop (Jones and Congdon 2024). Vegetable production may be throttled by seed-borne transmission of viruses. There is no doubt that the soil-borne viruses will emerge in the near future, prompting calamities in crop production (Escalante et al. 2024). Field directions and transmission procedures of seed viruses have become a prime need of agricultural scientists to avert calamities. Very few studies have been undertaken on the morphological observation of viruses on seeds, seedlings, and other plant parts. Seed-spread viruses have been categorized based on their detection by biological methods, serologically, and nucleic acid detection. Seed viruses have become food for the transmission of soil-borne fungi, bacteria, nematodes, etc. (Suhendar et al. 2023). If the seeds are not treated, 30–80% of mortality of crop plants could be incurred and they would become heavy sources of infection to emerging seedlings, leading to devastating crop losses. Seed-tinged viruses remaining in seeds are viable for 14–91 years and may come out with the emergence of seedlings. Such viruses cause crop losses of 89% (Swami and Singh 2023). Ground, seed, and soil-borne viruses are a growing problem in agricultural production. Soil water containing food products, domestic waste, or human waste after decaying is a high source of seed-borne viruses. Emerging seedlings become more susceptible to viruses at 10–20 days after germination. Some poultry viruses and soil-borne fungi may reside in dormant seeds for longer periods or on dry surfaces, leading to crop loss of 100% in young seedlings (Table 8.1). Based on the aforementioned account, it is clear that seed-spread viruses on vegetable crops need early attention of agricultural scientists for devising effective control measures (Yadav et al. 2022).

### 8.4.1 Tomato Mosaic Virus

Tomato mosaic virus (ToMV) is a member of the *Tobamovirus* genus and has been known since 1927 to infect plants in the Solanaceae. Tomato mosaic virus can also infect *Capsicum annuum* (sweet pepper) and *Solanum melongena* (eggplant), causing disease symptoms and reducing crop yield. Infected seeds are a major means of spreading ToMV and can survive up to 14 years in the soil in an infective state. For

**Table 8.1** Examples of seed-borne virus diseases in vegetable crops and their control methods

| Name of disease | Method of control | References |
|---|---|---|
| Tomato Mosaic Virus (ToMV) | Using virus-free certified seed, sanitation practices, and crop rotation | Jones et al. (2021), Panno et al. (2021) |
| Cucumber Mosaic Virus (CMV) | Breeding for resistant varieties, vector control, and removal of infected plants | Lim et al. (2022), Yosef et al. (2022) |
| Pepper Mild Mottle Virus (PMMV) | Seed disinfection, strict hygiene, and use of resistant cultivars | Ochar et al. (2023) |
| Cucumber Green Mottle Mosaic Virus (CGMMV) | Hot water treatment of seeds, certification, and sanitation measures | Miao et al. (2021), Lovelock et al. (2023) |
| Bean Common Mosaic Virus (BCMV) | Virus-free seed use, roguing infected plants, vector management | Sokhandan-Bashir and Ighani (2024) |
| Lettuce Mosaic Virus (LMV) | Use of tested, clean seed, resistant varieties, and exclusion of infected seed lots | Dell'Olmo et al. (2023), Sandra and Mandal (2024) |
| Squash Vein Yellowing Virus (SqVYV) | Removal of infected plants, vector (whitefly) control, and sanitation | Akhtar and Kumar (2024) |
| Soybean Mosaic Virus (SMV) | Certified seed, roguing, resistant varieties | Gao et al. (2022), Tomitaka et al. (2024) |

this reason, ToMV has been extensively studied worldwide. There are no effective treatments for ToMV infection in plants (Jones 2021; Panno et al. 2021).

### 8.4.2 Cucumber Mosaic Virus

Cucumber mosaic virus (CMV) is an economically important virus, infecting more than 1200 plant species, which is more than other viruses. It has been reported that CMV infections lead to a significant decline in the yield of cucurbit crops in most parts of the world (Lim et al. 2022; Yosef et al. 2022). Gherkin, a commercial vegetable crop in India, is affected by CMV, but resistant germplasm is not available. In recent years, attempts have been made to transfer one of the sources of resistance, Gherkin (a wild species, *Cucumis anguria*), to Cucumber (*Cucumis sativus*), the premier gourd crop with several genetic markers. Screening of a set of true-breeding lines of gherkin accessions, including five lines from the International Plant Genetic Resources Institute, against local strains of CMV (Petrov et al. 2024).

### 8.4.3 Pepper Mild Mottle Virus

Pepper mild mottle virus (PMMV), the first virus to be isolated from a pepper plant, is responsible for economic losses in pepper plantations around the world. This virus, which is distributed worldwide, can infect many members of the family *Solanaceae*, including capsicum and Nicotiana species (Ochar et al. 2023; Kumari et al. 2023).

## 8.5 Detection Methods

Few detection techniques for seed-borne viruses are available, and most of them focus on a limited set of target viruses and crop hosts. Serological methods are usually able to detect a broad spectrum of viruses and have become the standard laboratory method and/or routine protocol of many national and international laboratories. The most used serological techniques combine the acquisition of data in a microplate format and an enzymatic or fluorescent detection response (Xiao et al. 2022; Andrew et al. 2022). Particularly, ELISA is widely used, especially for humic viruses and ruboviruses, but most of the larger sequences and different rod infections have not, until now, produced satisfactory results. It is necessary that before the first analysis, an immunoaffinity column (IC) minimizes the potential interferences associated with current extraction protocols (Gomathi Devi et al. 2023; Srivastava et al. 2024). Additionally, denaturing and native gel electrophoresis are mentioned for confirmatory detection of seed-borne viruses in case of serological doubts, and new molecular techniques are increasingly being assessed for developing detection methods for non-detectable viruses. However, there is little information regarding the availability of protocols adapted to DNA viruses (Song et al.

2022). Additional protocols are available for assaying infectivity with plant bioassays. However, this is a time-consuming process, and in situ infectivity tests are not as easily performed as the automatic olfactometry and freshly prepared extracts. Furthermore, infectivity assays can generally evaluate only a set of a limited number of hosts (Burton 2023). Recent trends have focused on novel and rapid detection techniques (Rubio et al. 2020), including several auspicious approaches based on diverse parameters such as electrochemistry, fluorescence, LCR, mass spectrometry, low-cost tiling array technology, gold nanoparticles, and a combination of tools such as microarrays (Zhang et al. 2023).

### 8.5.1 Molecular Techniques

Virus detection, the determination of the virus species and strain present in a sample, their concentration, and their distribution within the host and on seed surfaces, is fundamental for the prevention and control of crop infections by seed-borne viruses. The methods most frequently used for detection until recently were serological assays, particularly ELISA (Zhang et al. 2023; Tatineni and Hein 2023). These methods are rather specific as they previously required the production of a suitable polyclonal or monoclonal antibody. Other methods with a much wider range of application are nucleic acid hybridization methods, where a labeled nucleic acid probe, gene-specific PCR primers, or primers that amplify conserved regions of viral genomes are used for the detection of viruses (Eksin and Erdem 2023). However, these methods also have some limitations, such as sequences matching different viruses or non-transmissible viroids. Another drawback is that the target viruses and viroids may have a high genetic variability, making diagnosis difficult (Fontdevila Pareta et al. 2023).

### 8.5.2 Serological Methods

Serological methods, which utilize antigens derived from viruses, viruses' coats, and their polyclonal or monoclonal antibodies, are widely used to detect seed-borne viruses (SBVs) in vegetable seeds either in their native or recombinant forms (Srivastava et al. 2024). Direct antigen-coated plate ELISA is used with plant virus modules. Indirect and triple-antibody sandwich ELISA, developed with healing peptide technology, could be used for di- and tri-assembled icosackets. An immunochromatography assay, which is faster, more sensitive, and simpler (Charlermroj et al. 2013), could also be adopted. Very recently, the antibody microarray method, utilizing either chemiluminescent or fluorescence-immunofluorescence detection, could be employed (Noel Culver 1987). Since non-virus antigen-derived recombinant protein may cross-react with other plant viruses' antibodies, delicate screening against cross-reactivity, either by dot-blotting or by ELISA methods, in the initial production stage of antibodies is always necessitated before mass production (Guo et al. 2021). Virus plants, even at the cloning stage, should be carefully grown in a greenhouse without any other possible contamination. The indirect ELISA may be a rapid, routine use for mass detection in vegetable seeds. Still, however, sandwich-type or IC-NFELISA might bring clean data in the amplification stage of SBV testing; for example, in a third party on certification of seed stocks (Kwon et al. 2023). A sandwich-type ELISA and an immunofluorescence technique have been recently developed against cucurbit BCs (CCs), Cucumber tristeza virus (CTV), and Watermelon mosaic virus (WMV). A twin-antibody sandwich-type ELISA and Fluorescence enzyme assay (FEA) method detection procedures were automated to save time and effort in the routine testing of these viruses. The immunofluorescence method offered a fast and visual detection protocol, which could also be carried out at various settings, from a laboratory to a field (Kang 2023). A simultaneous quantification of WSMoV and CMLV was developed on the SEA micro-platform applying hp-ELISA with signal amplification by reduction of Ag-Fe nano-disc. The method achieved a limit of detection of 10(-5) mg mL (-1) for each virus and was a flexible alternative to molecular methods in seed-borne viruses testing (Zhang et al. 2022).

### 8.5.3 Visual Inspection

Visual inspection of seed lots is one of the simplest and oldest methods to detect and identify seedborne diseases (Davino et al. 2020). It should be performed on all seed lots before further testing. Although it has limitations, it may yield the best results when disease symptoms are expressed on the seeds and the environment in which the inspected seed was grown, harvested, and processed is known (Constable et al. 2019). Although many seedborne bacteria are non-culturable, the detection of seedborne diseases based on the symptoms exhibited by diseased seeds during visual inspection is possible. Dry seed lots exhibiting symptoms of decay, rot, spots, or softening should be discarded (Srivastava et al. 2024). All seed lots that are incubated wet should be inspected and, at the same time, examined for rot. The edge of the napkin or moistened paper holds the seeds, which may accumulate water nearby. The seeds and the napkin should be inspected within 4 h for soft decay, cloudy water, and spots in the napkin. Infected seeds can be disinfected or discarded in commercial seed lots, depending on the pathology laboratory level (Gilbert et al. 2023; Xhemali et al. 2024).

## 8.6 Challenges in Management

In India, being the fifth largest vegetable producer in the world, however, adequate mechanization has not been given for critical interventions like seed production, because of the labor-intensive characteristic of production and harvesting processes, which has a clearly visible reflection in the post richness of biodiversity of vegetables (Solankey and Kumari 2023). The very basis of seed-borne viruses as a major constraint in crop production was the availability of a large and diverse germplasm of vegetables in India. Seed-borne viruses caused 100% losses of vegetables at times. The plants were not only devastated by viruses but also aided in the spread of viruses in the neighboring areas (Escalante et al. 2024). Seed-borne viruses infect only seeds and young seedlings and express their symptoms after transmitting to new plants. Seed-borne viruses are having ill effects on agriculture and horticulture such as (Jones et al. 2021): (i) seeds become unhealthy and deteriorated; (ii) plants become unhealthy, less vigor and poor yields; (iii) infected seeds are being used for growing seedlings and thus perpetuation and very wide spread of the diseases; (iv) market demand decreases because of poor quality of produce; (v) wastage of chemicals, planting material and labor due to consequent diseases; (vi) viral diseases spread from innocuous to severe form, and (vii) loss in genetic purity of plant varieties; hence, control of seed-borne viruses is a challenging task (Xhemali et al. 2024).

### 8.6.1 Lack of Awareness

Viruses can be transmitted through the seed in several ways, according to virus type, whether they replicate within the seed or not. Five types of seed transmission routes have been proposed based on the type of virus involved and tissue in which the virus resides after plant infection, i.e., as a systemic and non-seedborne component. Seed-borne viruses are a major threat to the horticultural plant industry (plant propagation and sale) (Akhtar and Kumar 2024). Unlike other industries, the economic cost of seed-borne infections is often severe. There is growing concern that awareness and knowledge about the symptoms, signs, epidemiology, and management of seed-borne viruses may be lacking. Therefore, many growers are unknowledgeable about the risks associated with seed-borne viruses. Newly emerging seed-borne viruses may be detected on low-value crops such as weeds and ornamentals, where management is difficult (Jones et al. 2021). To assess the impacts on horticulture, a policy should be in place. In addition, the chance of discovery and elimination of weed or ornamental transmission reservoirs is low, if not impossible. Subsequently, a rapid infection or aggressive strain could result in significant economic losses. Finally, infected seeds are likely to introduce an infection into a new region, which may threaten non-target crops such as natives and organics (Jones 2021). The lettuce mosaic virus was first found in the United States of America in 1914 and in Japan in 1916 and was once a devastating disease in many green-shaded vegetables such as lettuce, celery, and spinach. Seed-borne viruses were eradicated from those vegetables in Japan during the latter country's first national seed-borne virus eradication program initiated in 1960. So far, no seed-borne virus disease has been reported in Japan (Sandra and Mandal 2024). Nevertheless, outbreaks of soybean mosaic virus, a seed-borne virus, have been on the increase in Japan. The widespread usage of soybean seeds rotated and exported from the USA, where the introduction of new strains of mosaic viruses has remained possible, is a major concern (Gao et al. 2022; Tomitaka et al. 2024).

Small seed-borne viruses such as those belonging to the genus *Masavirus* should be a concern. Those viruses were toxicogenic and accumulated in the seed and plant pollen. It has recently been reported that new potential seed-borne viruses of the family *Tospoviridae* were emergent soilborne plant viruses in Japan (Escalante et al. 2024). The impacts of new and/or hitherto unnoticed seed-borne viruses on special crops should be assessed.

### 8.6.2 Limited Resources

Many virus-resistant Living Modified Organisms (LMOs) underperform in the field due to limited resources and incomplete characterization of resistance traits (Kwibuka Bisimwa et al. 2022). Field tests are often in high-risk areas and short-term, targeting early-season Recombinant Inbred Lines (RILs), while large-scale insect-proof testing is impractical (Migueres et al. 2021). Political and regulatory challenges have prevented LMO releases in recent years (Snuzik 2024). Sustainable management requires risk assessment of new varieties, timely control measures, surveillance for new strains, and strong networks linking biodiversity centers, Non-Governmental Organization (NGO), and ICT tools to reduce delays and errors (Leke et al. 2016). Seed-borne viruses remain a global threat to the highly diversified vegetable sector, affecting both domestic and export markets (Jones et al. 2021; Jones and Congdon 2024). Implementing the suggested strategies may mitigate virus impacts or prevent their entry into production fields, the most reliable solution (Jones et al. 2021).

### 8.6.3 Resistance Development

Vegetable crops are major sources of seed-borne virus epidemiology. Although viruses can remain latent in seeds, they pose serious threats to crops such as tomato, green pea, cucumber, gherkin, and capsicum. Their bio-geo-diversity and ethnobotanical significance attract research, yet studies on seed-borne viruses remain limited (Pagán 2022; Jones and Congdon 2024). Accurate identification and quantification of viruses in seeds are crucial but challenging. Non-selective isolation can capture single or multiple infections (Jones et al. 2021). Research has focused on virus identification, hosts, and vectors (Li et al. 2025). Seed-borne viruses threaten productivity by infecting seedlings and driving endemic epidemics (Sandra and Mandal 2024). Simple ELISAs are sensitive, reliable, and cost-effective for diagnosis, while studies on banana streak virus clarify early infection timing (Venbrux et al. 2023). Knowledge gaps persist, partly due to the lower economic importance of some vegetables, leaving them vulnerable despite innate defense mechanisms (Sandra and Mandal 2024; Jones et al. 2021). Transmission mechanisms of emerging seed-borne viruses remain poorly understood.

## 8.7 Cultural Control Strategies

Strategies to manage seed-borne viruses and their transmitted pathogens should include cultural practices that remove infection sources, reduce seed transmission, and lower the spread from infected hosts. These practices are inexpensive and practical, particularly in developing countries (Wang et al. 2024; Jones and Congdon 2024). Effective control requires raising farmer awareness of infected seeds and applying correct measures at pre- and post-harvest stages (Kumar et al. 2021a, b). Seed certification agencies and border monitoring can reduce transmission (Kumar et al. 2022), while larger countries can enforce quarantine more easily than smaller ones. Seed treatments, such as high temperature, are effective on a large scale but harder to enforce in smaller nations (Anum et al. 2024). Cultural practices are more effective when promoted to farmers, and large-scale management, including crop rotation, can drastically reduce soil-borne pathogen transmission (Dou et al. 2025). Site-specific recommendations are easier to implement than individual farm exemptions, and cost-effective seed treatments are appropriate (Kumar et al. 2024a, b). Inspections take time to enforce, so low-cost practices requiring minimal compliance or monitoring are prioritized (Ohene et al. 2022).

### 8.7.1 Crop Rotation

Plant viruses have survived for years in the seeds and propagative parts of the vegetable. In addition, seed-borne viruses pose a major threat to vegetable crops grown in gardens and farms worldwide. It also causes global production loss as well as economic loss. To control viruses, several measures should be followed by vegetable seed manufacturers to supply virus-free seeds for vegetable crops (Kumar et al. 2020; Kumar and Singh 2023; Tripathi et al. 2024a). However, apart from the seed-based measures, there are some cultural practices that can be followed to control the seed-borne viruses of vegetable crops, but such practices are generally neglected by the growers or farmers. Thus, awareness of such measures is necessary (Jones and Congdon 2024; Jones 2021). Crop rotation controls diseases transmitted by soil sustainably. Diverse cropping strategies help manage soil-borne crop diseases in many agricultural settings. Indian mustard and wild rocket rotations reduced cucumber Fusarium wilt and associated pathogen, *Fusarium oxysporum* f.sp. *cucumerinum* (Jin et al. 2019; Gong et al. 2021).

### 8.7.2 Sanitation Practices

To develop proper seed health-management strategies based on sanitation practices, it is necessary to identify the source and pathways of viruses. In many cases, wild relatives and closely associated plant species provide primary sources of viruses (Kumar and Singh 2023; Yang et al. 2022). Because some viruses have a weed- or grass-host, chemical control has limited impact, while sanitation practices and roguing infected plants help considerably. Several virus sources are assumed for a new seed-borne virus, including foreign gene acquisition pathways through vegetable germplasm (Jacobsen et al. 2019). In many cases, the domestic reservoirs of seed-borne viruses are within the crop species itself. After viruses establish obvious symptoms, seed degeneration occurs most dramatically, thereby demonstrating how seed-borne transmission is completely lost (Khan et al. 2025). Managing healthy genetic resources through quarantine, roguing, and careful field management is the simplest and most effective solution for this type of seed-borne virus. Alternatively, sanitation practices such as thermotherapy, chemical treatment, and shoot-tip grafting are commonly used to eliminate viruses (Giovanardi et al. 2018). In vegetable crops, the produce that becomes infected from virus spread through seed sources is delivered immediately into the food chain, and the virus infection goes unnoticed by farmers. Virus-free seed production is the only method of prevention known to be effective and applicable. In many developing countries, it is expensive to produce virus-free seeds using tissue culture methods (Xu et al. 2024; Li et al. 2024; Fuji et al. 2022).

## 8.8 Biological Control Approaches

Eco-friendly crop disease management is crucial for the sustainability of farming practices in the present day, as they are based on a system of dependent interrelated factors of climate, soil, water, selection of optimum crops, and utilization of eco-friendly/benign measures, warranting input costs, perennial productivity, and health. Plant disease is a major constraint to agricultural production, causing a reduction in the quantity and quality of produce (Collinge et al. 2022). In plants, diseases are caused by fungi, bacteria, viruses, viroids, phytoplasmas, and nematodes. Pathogen infestation by various chromists, saprophytic fungi, pathogenic fungi, slime molds, higher fungi, bacteria, viruses, etc., causes a large amount of crop loss each year during both pre- and post-harvest stages, involving a wide variety of pathogens. These plant pathogens induce direct losses (40%) in crop yield by causing mottled or necrotic leaves or streaking or wilting of whole plants. In addition, indirect losses are also incurred (40 billion dollars/year worldwide) (Chai et al. 2022). Over 80% of plant viruses are composed of single-stranded RNA genomes. Plant viruses are major threats to the food supply and food security of the world, causing about 30% of the worldwide crop diseases. They contribute to nearly half the number of emerging plant disease epidemics (Manjunatha et al. 2025). Viruses are intracellular pathogens, thus rendering chemical control nascent. Use of insecticides has resulted in insect vector resistance. Nevertheless, viruses are mostly transmissible in an aphid non-persistent manner. As a result, prophylactic measures involve the removal of affected crop debris, and adoption of better crop rotation practices, and increased applications of insecticides (with a combination of systemic ones) in four to six spray schedules, depending upon local conditions (Sharma et al. 2024).

### 8.8.1 Use of Beneficial Microorganisms

Identifying the potential hosts and assessing the loss estimates associated with varietal quality are vital against seed-borne viruses. Such studies provide clues for horizon scouting sources which may pave the way for pre-seed checking the seeds microbially before preparations for the ensuing cropping season (Jones et al. 2021). The rapid computerized ELISA-based Research wing can form the frontline troops against the surveillance of these viruses in the farmer's condition before rabi sowing. Research must require thorough pursuits on acaricides and their safe application techniques vis-à-vis their impacts on all legs of the agricultural sustainability (Cheravitch 2025).

### 8.8.2 Plant Growth-Promoting Rhizobacteria

Plant growth-promoting rhizobacteria (PGPR) have attracted significant attention in recent decades for their potential ability to improve the growth of a variety of plants (Imran et al. 2022a; Imran et al. 2023; Abo-Elyousr et al. 2024). PGPRs are commonly free-living bacteria that colonize the plant rhizosphere and promote plant growth through biochemical environmental modification (Tripathi et al. 2024b; Imran et al. 2025a). The use of PGPRs in agriculture would reduce the dependency on chemical fertilizers, which not only add nutrients but also cause environmental problems, including contamination of water bodies and soil deterioration. Relevant to the history of plant microbiology, traits of PGPRs include biological control of phytopathogens, phytohormone production, mineral solubilization, hydrogen cyanide production, and antifungal properties (Islam et al. 2016). Among various PGPRs, plant growth-promoting bacteria (PGPB) that colonize the root surface are a major focus of research and products in the context of environmentally friendly alternatives to chemical fertilizers (Gupta et al. 2022; Imran et al. 2022b). PGPB may prevent crop damage due to biotic stresses caused by pathogens or phytophagous nematodes and enhance crop productivity by alleviating abiotic stresses. Alternatively, PGPB may be inoculated into the seed or soil to colonize the root microenvironment, where they may benefit the plant (Natsagdorj et al. 2019). Interaction with a fungal pathogen in the absence of abiotic stress might be considered a holistic process, and such a situation needs to be clarified. Hence, it is important to examine PGPB that works effectively under natural conditions (Consentino et al. 2022).

## 8.9 Chemical Control Options

Historically, viral diseases have caused devastating losses in major vegetable crops in developed and developing countries. In many instances, plant virus infections lead to a complete loss of yield and quality of horticultural produce (Jones 2021; Zhang et al. 2022; Kumar et al. 2022). Most plant viruses are obligate pathogens, relying on their living host plants for multiplication and transmission. Plant viruses are generally transmitted through viruliferous insects, nematodes, infected seeds, and infected planting materials. Once in contact with a suitable host, viruses enter plant cells, where they multiply and subsequently spread locally and systemically through the vascular system (Tatineni and Hein 2023; Xhemali et al. 2024).

### 8.9.1 Fungicides and Insecticides

In general, seeds can be considered as sensitive targets for the use of products during the seed treatment and seed production process (Escalante et al. 2024). This phase is the only one that allowed the use of a systematic approach due to the very stringent I and X conditions inherent to the seed quality control and production processes. Economically, seed as carriers of diseases is one of the major issues for growers and seed companies (Westengen et al. 2023; Zhang et al. 2022). As a responsible problem holder, growth chamber tests by seeding infected seeds, and biological, biochemical, and physical methods have been established for disinfection of seeds, which allowed the reduction in vegetative-based transmission issues in seedling production. Also, based on severity ranking, HPR seeds have been sent to producers as safe plant material and other new vegetable species with similar technologies as a new business strategy with added value for growers (Bošnjak Mihovilović et al. 2024). Insects have a key role as the primary source for the virus, but the other factors—vegetative propagation, pest management, soil hygiene, and crop rotation—are soft targets for this type of virus issue. The processing of pesticides for beneficial insects has a crucial role in any pest control protocol; e.g., the oil of the China tree has been used (Gomathi Devi et al. 2023; Denancé and Grimault 2022). Cool storage of infected plants is also one important soft target against aphids, as this technique has also been used successfully. Monitoring of pests along with biotechnological control has to consider certain climate conditions. Monitoring of peach mosaic on seedlings has been done by visual inspection as a simple tool, but, as different races used for monitoring potato or hypersensitivity reaction, special micro-chips were developed for washable plates (De Schutter et al. 2022; Čirjak et al. 2022). In seeds, virus disinfestation is generally not resistant to oils, ethanol, and fiery methods; however, they are time-consuming, labor-intensive, and can also cause seed damage. But this is not a high-tech topic, but is practical (Kundu et al. 2022; Liu et al. 2021) (Fig. 8.2).

### 8.9.2 Seed Treatments

Seed transmission of plant viruses represents a huge problem for producers and seed companies. There is a range of possible problems brought about by seed transmission of viruses, which include the inability to detect infected seed, outbreaks of viral diseases and ensuing high losses of yield and/or quality, or the presence of infected seed but the inability to visually detect disease. All of these represent an

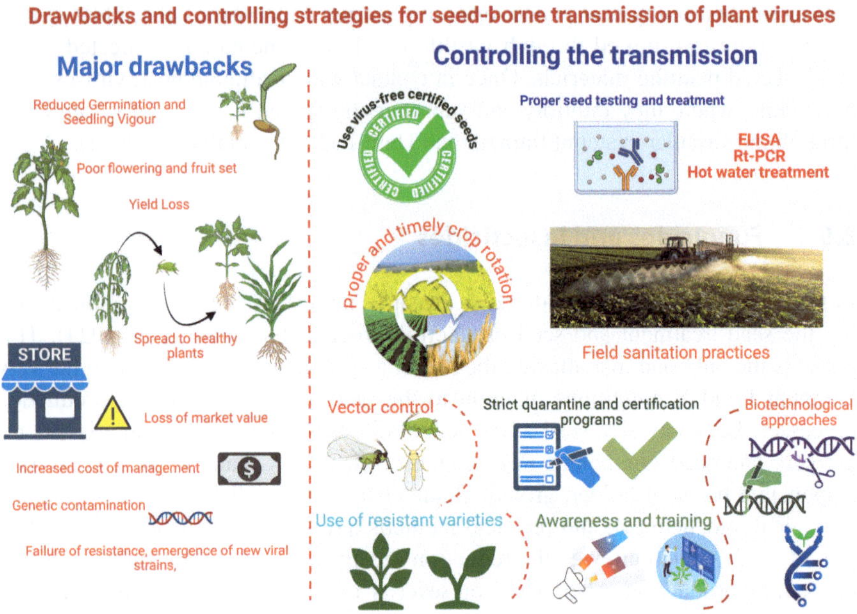


**Fig. 8.2** Seed-borne transmission of plant viruses

economic risk for producers, seed companies, and ultimately, for food security (Escalante et al. 2024). Virus-induced changes during seed development may go undetected, resulting in later infection and subsequently cannot predict the occurrence of an epidemic within the new variety. Some viruses infect seeds, but their presence do not elicit subsequent disease development in the seedlings. These viruses can remain dormant in the seed, but once environmental conditions are appropriate, disease may occur that was not apparent earlier (Devi et al. 2024; Beyaz 2023). These viruses present a danger in newly established fields. Seed companies monitoring viruses in the seeds may be unable to detect all viruses present, especially with the use of the single antibody technique frequently used for testing seeds for viruses. It is estimated that laboratory testing of 2–5000 seeds per lot is usually insufficient to detect seed-borne virus when the incidence is less than 1%. Most lab procedures do not deal with the presence of ascitic viruses as they require living tissue for symptom development (Saif et al. 2022). These treatments should not directly kill the virus in the seed, but reduce seed viability and/or prevent the emergence of seedlings before a virus-free period. Growers should realize that some of the chemicals commonly used to treat seed for insect or fungal diseases can be phytotoxic. Treatments with hot water, fluorohydrin, or DA 09867 could all offer solutions to this problem (Shelar et al. 2023; Chen et al. 2025a, b).

## 8.10 Integrated Pest Management (IPM)

Further work is needed to ensure the sustainability of no-till and IPM technologies. Case studies highlight the need to evaluate no-till agriculture across social, economic, and environmental dimensions (Chawla et al. 2022; Vacchi et al. 2021). Plant-borne diseases are exacerbated by climate extremes, and IPM complements plant breeding for climate-resilient crops under salinity, cold, and drought stress. In Ethiopia, IPM with natural inputs in tomato reduced nematodes, *Tuta absoluta*, and leaf curl virus (Mohapatra et al. 2022; Zhou et al. 2024), while bio-augmentative control and soil health benefits were demonstrated (Ghosal and Sahu 2023; Shingote et al. 2022). Engineered water reservoirs supporting *G. margarita* improved biocontrol of *M. incognita*, and organogenesis studies inform drought and infection responses for high-value crops (Clavero-Camacho et al. 2024; Pervez et al. 2024). Climate change impacts insects, pests, and natural enemies, influencing pest dynamics and control strategies (Demirozer et al. 2012). Integrating these insights enables practical, cost-effective adaptation for food security and sustainable agriculture (Benitez-Alfonso et al. 2023; Wijerathna-Yapa and Pathirana 2022).

## 8.11 Role of Seed Certification Programs

Concerns regarding seed-borne virus transmission have driven interested parties to adopt numerous measures and management options to reduce or eradicate levels of seed-borne virus. Seed Health Programs or Seed Certification Programs are the most widely used and accepted mechanism to identify, categorize, and regulate seed-borne organisms that can have a detrimental impact on seed health and crop establishment (Dell'Olmo et al. 2023; Klaedtke et al. 2022). Standard Protocols developed by various National Seed Councils and Seed Certification Boards often focus efforts solely on seed health inspection, testing, tagging, and billing procedures. Such protocols generally have a limited scope that does not allow for taking a more holistic view of the potential seed-borne transmission threats, as well as those that may arise from other pathways (Powell 2024). As seed movements continue to globalize, the popularization and wider acceptance of additional seed health assurance measures, such as Integrated Seed Systems, can be beneficial to many regions. The expansion of the seed certification programs of many countries is limited by resources and available labor. Utilizing seed-seed synergy initiatives to pursue regulatory collaborations in the most accessible regions supplemented by follow-up and supportive arrangements may allow for expedited future entry into seed certification programs of countries where a critical mass of seed production and certification interest exists (Mulesa et al. 2021).

### 8.11.1 Standards and Regulations

Knowledge is required to be in a good estate as to the quality of the seeds, the way they are produced and how all these are checked before dispatching for trading (Constable et al. 2019). The MVF can help in this matter by using ISF seed methodologies and other relevant available methodologies. However, the MVF needs to be aware of the types of plant viruses that are seed-borne and that prevent disasters from happening (Fumero et al. 2024). The necessary procedures based on the seed and the location need to be established in terms of screening, trading, dispatching, storing, and distribution. (Shelar et al. 2023) For crop plants, screening of all propagation material is needed to sift out all strains of viruses that are seed-borne. Consideration has to be made of the fact that viruses may not be homogeneous and may differ at the same seed lots as to their titers, strains, cells, tissues, and other properties. Hence, sampling is to be established and this has to be done within the MVF by trained scientists (Srivastava et al. 2024). Seed shipment from Israel to Australia of tomato and capsicum seeds containing the viroid PSTVd has resulted in questions being raised about future shipments of any similar types of seeds from Israel. Since these viroids were discovered, in all probability, they have been proliferated and cultivated in Israel, so they justify the rejection of future shipment (Bozbuga et al. 2022). But under which terms can this viroid be exterminated once it is present in the greenhouses? There is not enough information about PSTVd infecting the whole tomato genome, whether the viroid sequence is incorporated into the goat genome, and for how long, and whether part of the viroid might be converted to RNA, which can again propagate the viroid, should be researched (Zhang et al. 2023).

### 8.11.2 Benefits of Certification

Seed-borne viruses (SBVs) are a worldwide danger to vegetable seed systems, which seriously reduce yields in susceptible vegetable crops. Seed certification action plans were discussed according to best practices from various organizations and countries, which led to technical assistance toward the quarantine authorities and exporting countries (Jones 2021; Dell'Olmo et al. 2023). Vegetables are increasingly traded over long distances, which results in a focus on breeding new resistant cultivars since no vaccines or chemical control methods exist (Warda et al. 2025). Modern virus diagnostics used for eradication and testing were discussed, as well as combined seed testing and evaluation methods to guarantee safe seeds free from SBVs. The setup of these quarantine systems is constrained by a low level of funded human resources on the government side, the plant health officials, as well as on the exporting side, the vegetable seed companies (Chen and Kong 2023). Subsequently, this again leads to serious economic losses for the vegetable crops fielding high-quality seeds. An overview of the given points and time steps was summarized and used as support documents for the countries involved in the project. Initial bioinformatics studies using miRNA profiling were used to detect tomato yellow leaf curl viruses from six continents (Dong et al. 2023; Srivastava et al. 2024).

## 8.12 Future Directions

Seed development is an ideal model to study virus transmission and fate, because the model plant species used are amenable to a variety of high-throughput, powerful, and efficient experimental and analytical techniques. It would be potentially more feasible to examine the impact of viruses on either seed structure and germination or seed-preserved viromes in a seed purifying study, because many horticulture plants produce quite different, relatively large and distinct seeds (Escalante et al. 2024). Because seeds are also germinated in agricultural practice, it would be worthwhile to determine an efficient and effective pre-germination treatment for the attenuation of viruses on seed integrity and viruses in cotyledons (Ekren et al. 2023). More research is needed on the role of seed-associated small RNAs in seed transmission to other tissues for DNA and RNA viruses. These investigations are expected to uncover new viruses, novel mechanisms for specific interactions, and exploitable viral vulnerabilities that can pave the way to virus-resistant vegetables (Ortega-Cuadros et al. 2024). Further, seed treatment with dsRNA-based formulations could also be considered as a promising tool (Imran et al. 2025b; Chen et al. 2025a, b) to control the emerging seed-borne viruses, as in emerging nanotechnology, dsRNA has also been used to control plant diseases. dsRNA controls seedborne viruses by triggering RNA interference, which specifically degrades viral RNA and prevents infection in seeds.

## 8.13 Conclusion

The occurrence of seed-borne viruses in crop plants remains a big challenge in the production of healthy seeds, seedlings, and crops. Many vegetable crops are known to be infected and distributed through seeds, and most of these viruses insert their genetic material into the germ cells of the seeds. Such viruses are transmitted even during the production of the seed, in the seedling/plant stage and in crop management. The external and internal seed structures act as effective barriers against the entry of viruses that infect seeds. However, there is a potential for viruses to enter the developing seed during its formation in the ovule. One of the well-studied examples is the Tomato yellow leaf curl virus in the *Solanaceae* family. The virus is taken up by the ovule and transmitted vertically to the offspring seedling. The efficient and systemic infection of seeds by this virus requires helper factors from the plant, such as long non-coding RNA, DNA type III-A methyl-transferase and histone methyltransferase. The rumination of this knowledge will help in studying other viruses in the family or those viruses infecting the seeds of other crops.

**Acknowledgments** The authors extend their appreciation for the support of the Deanship of Scientific Research, Vice Presidency for Graduate Studies and Scientific Research, King Faisal University, Saudi Arabia [Grant No. KFU254311].

## References

Abo-Elyousr KA, Imran M, Sallam NM, Abdel-Aal AM, Assiri ME, Abdel-Rahim IR (2024) Sustainable biocontrol of purple blotch disease in *Allium cepa* L. by biocontrol yeasts, *Pichia kluyveri* and *Filobasidium wieringae*. Egypt J Biol Pest Control 34(1):11. https://doi.org/10.1186/s41938-024-00776-6

Akhtar J, Kumar P (2024) Seed-borne fungi: challenges in seed health testing for biosecurity and agricultural sustainability. Indian Phytopathol 77:257–269. https://doi.org/10.1007/s42360-024-00723-3

Altaf MT, Liaqat W, Ali A, Jamil A, Fahad M, Rahman MAU, Mohamed HI (2025) Advancing chickpea breeding: omics insights for targeted abiotic stress mitigation and genetic enhancement. Biochem Genet 63:1063–1115

Andrew A, Navien TN, Yeoh TS, Citartan M, Mangantig E, Sum MS, Tang TH (2022) Diagnostic accuracy of serological tests for the diagnosis of chikungunya virus infection: a systematic review and meta-analysis. PLoS Negl Trop Dis 16:e0010152

Anum H, Tong Y, Cheng R (2024) Different Preharvest diseases in garlic and their eco-friendly management strategies. Plants (Basel) 13(2):267. https://doi.org/10.3390/plants13020267

Benitez-Alfonso Y, Soanes BK, Zimba S, Sinanaj B, German L, Sharma V, Foyer CH (2023) Enhancing climate change resilience in agricultural crops. Curr Biol 33:R1246–R1261. https://doi.org/10.1016/j.cub.2023.10.028

Beyaz R (2023) Functional gene silence technique in plants: virus-induced gene silencing (VIGS). In: Advance concepts on natural and agricultural sciences. Iksad Publications, Ankara, pp 161–196

Bošnjak Mihovilović A, Kereša S, Lazarević B, Topolovec Pintarić S, Martinko K, Marković Z, Habuš Jerčić I (2024) The use of sodium hypochlorite and plant preservative mixture significantly reduces seed-borne pathogen contamination when establishing in vitro cultures of wheat (Triticum aestivum L.) seeds. Agriculture 14:556. https://doi.org/10.3390/agriculture14040556

Bozbuga, R., Yalcin Ates, S., Gok Guler, P., Nilufer Yildiz, H., Aridici Kara, P., Bulent Arpaci, B., & Imren, M. (2022). Host-pathogen and pest interactions: virus, nematode, viroid, bacteria, and pests in tomato cultivation. IntechOpen https://doi.org/10.5772/intechopen.106064

Burton DR (2023) Antiviral neutralizing antibodies: from in vitro to in vivo activity. Nat Rev Immunol 23:720–734. https://doi.org/10.1038/s41577-023-00858-w

Chai Y, Senay S, Horvath D, Pardey P (2022) Multi-peril pathogen risks to global wheat production: a probabilistic loss and investment assessment. Front Plant Sci 13:1034600. https://doi.org/10.3389/fpls.2022.1034600

Charlermroj R, Himananto O, Seepiban C, Kumpoosiri M, Warin N, Oplatowska M, Gajanandana O, Grant IR, Karoonuthaisiri N, Elliott CT (2013) Multiplex detection of plant pathogens using a microsphere immunoassay technology. PLoS One 8(4):e62344. https://doi.org/10.1371/journal.pone.0062344

Chawla S, Varghese BS, Hussain CG, Keçili R, Hussain CM (2022) Environmental impacts of post-consumer plastic wastes: treatment technologies towards eco-sustainability and circular economy. Chemosphere 308:135867. https://doi.org/10.1016/j.chemosphere.2022.135867

Chen Z, Kong G (2023) Hospital admission, facility-based isolation, and social distancing: An SEIR model with constrained medical resources. Prod Oper Manag 32(5):1397–1414. https://doi.org/10.1111/poms.13702

Chen C, Imran M, Feng X, Shen X, Sun Z (2025a) Spray-induced gene silencing for crop protection: recent advances and emerging trends. Front Plant Sci 16:1527944. https://doi.org/10.3389/fpls.2025.1527944

Chen Y, Liu B, Bian Y, Gao Y, Liu Y, Cheng M, Hang L (2025b) Effects of soybean urease induced carbonate precipitation on the seed emergence and seedling growth of Caragana korshinskii Kom and its application in wind erosion control. Plant Soil 507:629–653. https://doi.org/10.1007/s11104-024-06758-9

Cheravitch J (2025) Commissars with keyboards: the lingering relevance of the military-political origins of Chinese and Russian psychological warfare. Intell Natl Secur 40(1):58–78. https://doi.org/10.1080/02684527.2024.2405256

Čirjak D, Miklečić I, Lemić D, Kos T, Pajač Živković I (2022) Automatic pest monitoring systems in apple production under changing climatic conditions. Horticulturae 8(6):520. https://doi.org/10.3390/horticulturae8060520

Clavero-Camacho I, Ruiz-Cuenca AN, Cantalapiedra-Navarrete C, Castillo P, Palomares-Rius JE (2024) Diversity of microbial, biocontrol agents and nematode abundance on a susceptible Prunus rootstock under a Meloidogyne root gradient infection. Front Plant Sci 15:1386535

Collinge DB, Jensen DF, Rabiey M, Sarrocco S, Shaw MW, Shaw RH (2022) Biological control of plant diseases–what has been achieved and what is the direction? Plant Pathol 71:1024–1047

Consentino BB, Aprile S, Rouphael Y, Ntatsi G, De Pasquale C, Iapichino G, Sabatino L (2022) Application of PGPB combined with variable N doses affects growth, yield-related traits, N-fertilizer efficiency and nutritional status of lettuce grown under controlled condition. Agronomy 12:236

Constable F, Chambers G, Penrose L, Daly A, Mackie J, Davis K, Rodoni B, Gibbs M (2019) Viroid-infected tomato and capsicum seed shipments to Australia. Viruses 11(2):98. https://doi.org/10.3390/v11020098

Davino S, Caruso AG, Bertacca S, Barone S, Panno S (2020) *Tomato Brown rugose fruit virus*: seed transmission rate and efficacy of different seed disinfection treatments. Plants (Basel) 9(11):1615. https://doi.org/10.3390/plants9111615

De Schutter K, Taning CNT, Van Daele L, Van Damme EJ, Dubruel P, Smagghe G (2022) RNAi-based biocontrol products: market status, regulatory aspects, and risk assessment. Frontiers in Insect Science 1:818037

Dell'Olmo E, Tiberini A, Sigillo L (2023) Leguminous seedborne pathogens: seed health and sustainable crop management. Plants 12:2040. https://doi.org/10.3390/plants12102040

Demirozer O, Tyler-Julian K, Funderburk J, Leppla N, Reitz S (2012) Frankliniella occidentalis (Pergande) integrated pest management programs for fruiting vegetables in Florida. Pest Manag Sci 68(12):1537–1545. https://doi.org/10.1002/ps.3389

Denancé N, Grimault V (2022) Seed pathway for pest dissemination: the ISTA reference pest list, a bibliographic resource in non-vegetable crops. EPPO Bull 52(2):434–445. https://doi.org/10.1111/epp.12834

Devi BM, Guruprasath S, Balu P, Chattopadhyay A, Thilagar SS, Dhanabalan KV, Jailani AAK (2024) Dissecting diagnostic and management strategies for plant viral diseases: what next? Agriculture 14:284

Dong Y, Zhu H, Qiu D (2023) Hrip1 enhances tomato resistance to yellow leaf curl virus by manipulating the phenylpropanoid biosynthesis and plant hormone pathway. 3 Biotech 13:11. https://doi.org/10.1007/s13205-022-03426-6

Dong J, Zhu T, Lv R, Dong K, Li Y, Zhang B, Wu K (2024) Occurrence and characterization of viruses infecting Amorphophallus in Yunnan, China. Sci Rep 14:12948

Dou Y, Yu S, Liu S, Cui T, Huang R, Wang Y, Li X (2025) Crop rotations reduce pathogenic fungi compared to continuous cropping. Rhizosphere 34:101074

Ekren S, Paylan IC, Gokcol A (2023) Seed quality improvement applications in black cumin seeds (Nigella sativa L.). Front Sustain Food Syst 7:1212958. https://doi.org/10.3389/fsufs.2023.1212958

Eksin E, Erdem A (2023) Recent progress on optical biosensors for nucleic acid detection related to infectious viral diseases. Micromachines 14:295. https://doi.org/10.3390/mi14020295

Escalante C, Sanz-Saez A, Jacobson A, Otulak-Kozieł K, Kozieł E, Balkcom KS, Zhao C, Conner K (2024) Plant virus transmission during seed development and implications to plant defense system. Front Plant Sci 15:1385456

Fontdevila Pareta N, Khalili M, Maachi A, Rivarez MPS, Rollin J, Salavert F, Massart S (2023) Managing the deluge of newly discovered plant viruses and viroids: an optimized scientific and regulatory framework for their characterization and risk analysis. Front Microbiol 14:1181562

Fortes IM, Pérez-Padilla V, Romero-Rodríguez B, Fernández-Muñoz R, Moyano C, Castillo AG, Moriones E (2023) Begomovirus tomato leaf curl New Delhi virus is seedborne but not seed transmitted in melon. Plant Dis 107:473–479

Fuji SI, Mochizuki T, Okuda M, Tsuda S, Kagiwada S, Sekine KT, Suzuki N (2022) Plant viruses and viroids in Japan. J Gen Plant Pathol 88:105–127

Fumero MV, Garis SB, Alberione EJ, Jofré E, Vanzetti LS (2024) Genomic analysis identifies five pathogenic bacterial species in Argentinian wheat. Trop Plant Pathol 49:1–12. https://doi.org/10.21203/rs.3.rs-4010493/v1

Gao X, Jia ZQ, Tao HZ, Xu Y, Li YZ, Liu YT (2022) Use of deep sequencing to profile small RNAs derived from tomato spotted wilt orthotospovirus and hippeastrum chlorotic ringspot orthotospovirus in infected Capsicum annuum. Virus Res 309:198648. https://doi.org/10.1016/j.virusres.2021.198648

García-Ordóñez L, Pagán I (2024) vertical and horizontal transmission of plant viruses: two extremes of a continuum? NPJ Viruses 2:18. https://doi.org/10.1038/s44298-024-00030-8

Ghosal A, Sahu NC (2023) Border netting technology with integrated pest management (IPM) strategies for sustainable chilli leaf curl management. In: Environmental management and sustainability in India: case studies from West Bengal, pp 479–487

Gilbert GS, Diaz A, Bregoff HA (2023) Seed disinfestation practices to control seed-borne fungi and bacteria in home production of sprouts. Foods 12:747. https://doi.org/10.3390/foods12040747

Giovanardi D, Biondi E, Ignjatov M, Jevtic R, Stefani E (2018) Impact of bacterial spot outbreaks on the phytosanitary quality of tomato and pepper seeds. Plant Pathol 67(5):1168–1176. https://doi.org/10.1111/ppa.12839

Gomathi Devi R, Jothika C, Sankari A, Lakshmi S, Malathi VG, Renukadevi P (2023) Seed transmission of begomoviruses: a potential threat for bitter gourd cultivation. Plants 12:1396

Gong X, Shi J, Zhou X, Yuan T, Gao D, Wu F (2021) Crop rotation with cress increases cucumber yields by regulating the composition of the rhizosphere soil microbial community. Front Microbiol 12:631882

Guo Y, Tong Y, Zhu H, Xiao Y, Guo H, Shang L, Bai Y (2021) Quercetin suppresses pancreatic ductal adenocarcinoma progression via inhibition of SHH and TGF-β/Smad signaling pathways. Cell Biol Toxicol 37:479–496

Gupta A, Mishra R, Rai S, Bano A, Pathak N, Fujita M, Hasanuzzaman M (2022) Mechanistic insights of plant growth promoting bacteria mediated drought and salt stress tolerance in plants for sustainable agriculture. Int J Mol Sci 23:3741

Imran M, Abo-Elyousr KA, Mousa MA, Saad MM (2022a) A study on the synergetic effect of Bacillus amyloliquefaciens and dipotassium phosphate on Alternaria solani causing early blight disease of tomato. Eur J Plant Pathol 162(1):63–77. https://doi.org/10.1007/s10658-021-02384-8

Imran M, Abo-Elyousr KA, Mousa MA, Saad MM (2022b) Screening and biocontrol evaluation of indigenous native Trichoderma spp. against early blight disease and their field assessment to alleviate natural infection. Egypt J Biol Pest Control 32(1):40. https://doi.org/10.1186/s41938-022-00544-4

Imran M, Abo-Elyousr KA, Mousa MA, Saad MM (2023) Use of Trichoderma culture filtrates as a sustainable approach to mitigate early blight disease of tomato and their influence on plant biomarkers and antioxidants production. Front Plant Sci 14:1192818. https://doi.org/10.3389/fpls.2023.1192818

Imran M, Alomran MM, Khalifa HA, Bilal MS, Ahmed AF, Ali EF, Alqurashi AS, Aldayel MF, Abo-Elyousr KA (2025a) Exploring smart solutions for stressed soils: dual-action of culture filtrates and nanoparticles for sustainable onion cultivation. Eur J Plant Pathol:1–22. https://doi.org/10.1007/s10658-025-03113-1

Imran M, Feng X, Sun Z, Al Omari H, Zhang G, Zhu J, Aldayel MF, Li C (2025b) Nanotechnology-driven gene silencing: advancements in SIGS-dsRNA technology for sustainable disease management. Chem Biol Technol Agric 12(1):31. https://doi.org/10.1186/s40538-025-00738-6

Islam S, Akanda AM, Prova A, Islam MT, Hossain MM (2016) Isolation and identification of plant growth promoting Rhizobacteria from cucumber rhizosphere and their effect on plant growth promotion and disease suppression. Front Microbiol 6:1360. https://doi.org/10.3389/fmicb.2015.01360

Jacobsen K, Omondi BA, Almekinders C, Alvarez E, Blomme G, Dita M, Iskra-Caruana MI, Ocimati W, Tinzaara W, Kumar PL, Staver C (2019) Seed degeneration of banana planting materials: strategies for improved farmer access to healthy seed. Plant Pathologyy 68(2):207–228. https://doi.org/10.1111/ppa.12958

Jayasinghe WH, Akhter MS, Nakahara K, Maruthi MN (2022) Effect of aphid biology and morphology on plant virus transmission. Pest Manag Sci 78:416–427

Jin X, Wang J, Li D, Wu F, Zhou X (2019) Rotations with Indian mustard and wild rocket suppressed cucumber Fusarium wilt disease and changed rhizosphere bacterial communities. Microorganisms 7(2):57. https://doi.org/10.3390/microorganisms7020057

Jones RAC (2021) Global plant virus disease pandemics and epidemics. Plants 10:233. https://doi.org/10.3390/plants10020233

Jones RAC, Congdon BS (2024) Australian cool-season pulse seed-borne virus research: 1. Alfalfa and cucumber mosaic viruses and less important viruses. *Viruses* 16(1):144. https://doi.org/10.3390/v16010144

Jones RAC, Sharman M, Trębicki P, Maina S, Congdon BS (2021) Virus diseases of cereal and oilseed crops in Australia: current position and future challenges. Viruses 13:2051

Kang S (2023) Nanobiotechnology enabled environmental sensing of water and wastewater. Doctor of Philosophy In Civil Engineering, p 310

Kaushik P, Sharma V, Saini I, Yadav VK, Jayaswal D, Singh G, Srivastava A (2021) Impact of COVID-19 pandemic on vegetable sector and its allies. Indian J Tradit Knowl 19:S-177

Khan R, He P, Chen X, He P, Ahmed A, Wu Y, He Y (2025) Bacillus endophytes for sustainable management of tomato spotted wilt virus and yield production. Pest Manag Sci 81:2073–2085

Klaedtke SM, Rey F, Groot SPC (2022) Designing a seed health strategy for organic cropping systems, based on a dynamic perspective on seed and plant health. Sustainability 14(17):10903. https://doi.org/10.3390/su141710903

Kumar R, Singh BP (2023) Diagnostics for production of disease-free planting material of potato. Int J Innov Hortic 12:31–49. https://doi.org/10.5958/2582-2527.2023.00003.9

Kumar R, Jeevalatha A, Baswaraj R, Tiwari RK (2020) Viral and viroid diseases of potato and their management. In: Singh AK, Chakrabarti SK, Singh B, Sharma J, Dua VK (eds) Potato science & technology for sub tropics. New India Publishing Agency, New Delhi, pp 267–292. (ISBN: 978-93-89571-93-6)

Kumar PL, Cuervo M, Kreuze JF, Muller G, Kulkarni G, Kumari SG, Negawo AT (2021a) Phytosanitary interventions for safe global germplasm exchange and the prevention of transboundary pest spread: the role of CGIAR germplasm health units. Plants 10:328

Kumar R, Tiwari RK, Jeevalatha A, Sundaresha S, Shah MA, Sharma S, Sagar V, Kumar M, Chakrabarti SK (2021b) Potato apical leaf curl disease: current status and perspectives on a disease caused by tomato leaf curl New Delhi virus. J Plant Dis Protect 128:897–911. https://doi.org/10.1007/s41348-021-00463-w

Kumar R, Tiwari RK, Sundaresha S, Kaundal P, Raigond B (2022) Potato viruses and their management. In: Chakrabarti SK, Sharma S, Shah MA (eds) Sustainable management of potato pests and diseases. Springer, Singapore, pp 309–335. https://doi.org/10.1007/978-981-16-7695-6_12. (Hardcover ISBN 978-981-16-7694-9; Softcover ISBN978-981-16-7697-0; eBook ISBN978-981-16-7695-6)

Kumar P, Kushwaha HL, Kumar A, Parray RA, Singh T, Meena MC, Rathod SK (2024a) Efficiency and cost-effectiveness analysis of developed embedded system-controlled seed and fertilizer applicator. J Exp Agric Int 46:626–635

Kumar V, Sharma KV, Kedam N, Patel A, Kate TR, Rathnayake U (2024b) A comprehensive review on smart and sustainable agriculture using IoT technologies. Smart Agricultural Technology 8:100487

Kumari N, Sharma V, Patel P, Sharma PN (2023) Pepper mild mottle virus: a formidable foe of capsicum production a review. Front Virol 3:1208853. https://doi.org/10.3389/fviro.2023.1208853

Kundu A, Ganguli S, Pal A (2022) Genomic designing towards biotic stress resistance in mungbean and urdbean. In: Genomic designing for biotic stress resistant pulse crops, pp 381–414

Kwibuka Bisimwa Y, Nyirakanani C, Pierre Bizimana J, Bisimwa E, Brostaux Y, Lassois L, Vanderschuren H, Massart S (2022) Risk factors associated with cassava brown streak disease dissemination through seed pathways in Eastern D.R. Congo. Front Plant Sci 13:803980. https://doi.org/10.3389/fpls.2022.803980

Kwon EY, Ruan X, Yu F, Lin Y, Du D, Van Wie BJ (2023) Simultaneous detection of two herbicides in fruits and vegetables with nanoparticle-linked immunosorbent and lateral flow immunoassays. Food Chem 399:133955. https://doi.org/10.1016/j.foodchem.2022.133955

Leke WN, Mignouna DB, Brown JK, Kvarnheden a (2016) Begomovirus disease complex: emerging threat to vegetable production systems of West and Central Africa. Agric Food Secur 4:1. https://doi.org/10.1186/s40066-014-0020-2

Li J, Shang Q, Luo Y, Wei S, Zhao C, Ban L (2024) Transmission from seed to seedling and elimination of alfalfa viruses. Front Plant Sci 15:1330219

Li H, Chen X, Wang JJ, Shen J, Abuduwufuer K, Zhang Z, Wang DW (2025) Spatiotemporal transcriptomics elucidates the pathogenesis of fulminant viral myocarditis. Signal Transduct Target Ther 10:5

Lim K, Whye L, Hung M, Chung HH (2022) Cucumber mosaic virus: Global genome comparison and beyond. Malays J Microbiol. https://doi.org/10.21161/mjm.211276

Liu Y, Zhang X, Qi W, Yang Y, Liu Z, An T, Chen J (2021) Prevention and control strategies of African swine fever and progress on pig farm repopulation in China. Viruses 13:2552

Lovelock DA, Mintoff SJ, Kurz N, Neilsen M, Patel S, Constable FE, Tran-Nguyen LT (2023) Ability of non-hosts and cucurbitaceous weeds to transmit cucumber green mottle mosaic virus. Viruses 15:683

Manjunatha L, Rajashekara H, Saisree Uppala L, Siddesh Ambika D, Patil B, Subbanna Shankarappa K, Sukumari Nath V, Rooplanaik Kavitha T, Kumar Mishra A (2025) Tomato spotted wilt virus. In: Amaresan N, Kumar K (eds) Compendium of phytopathogenic microbes in agro-ecology. Springer, Cham. https://doi.org/10.1007/978-3-031-81884-4_22

Metwally RA, Taha MA, El-Moaty NMA, Abdelhameed RE (2024) Attenuation of Zucchini mosaic virus disease in cucumber plants by mycorrhizal symbiosis. Plant Cell Rep 43:54

Miao S, Liang C, Li J, Baker B, Luo L (2021) Polycistronic artificial microRNA-mediated resistance to cucumber green mottle mosaic virus in cucumber. Int J Mol Sci 22(22):12237. https://doi.org/10.3390/ijms222212237

Migueres M, Lhomme S, Izopet J (2021) Hepatitis A: epidemiology, high-risk groups, prevention and research on antiviral treatment. Viruses 13:1900

Mohapatra SD, Zaidi NW, Mohapatra M, Nagothu US, Senapati R, Mohapatra M, Nayak AK (2022) Integrated pest management in rice and the potential to contribute to climate-neutral and resilient farming systems. In: Climate neutral and resilient farming systems, pp 69–86

Mulesa TH, Dalle SP, Makate C, Haug R, Westengen OT (2021) Pluralistic seed system development: a path to seed security? Agronomy 11:372

Natsagdorj O, Sakamoto H, Santiago DMO, Santiago CD, Orikasa Y, Okazaki K, Ikeda S, Ohwada T (2019) *Variovorax* sp. has an optimum cell density to fully function as a plant growth promoter. Microorganisms 7(3):82. https://doi.org/10.3390/microorganisms703008

Noel Culver J (1987) Serological detection of peanut stripe virus. PhD thesis http://hdl.handle.net/11244/16926

Ochar K, Ko HC, Woo HJ, Hahn BS, Hur O (2023) Pepper mild mottle virus: An infectious pathogen in pepper production and a potential indicator of domestic water quality. Viruses 15:282. https://doi.org/10.3390/v15020282

Ohene E, Chan APC, Darko A (2022) Prioritizing barriers to net-zero carbon building sector. Build Environ 223:109437. https://doi.org/10.1016/j.buildenv.2022.109437

Ortega-Cuadros M, Aligon S, Arias T, Vasco-Palacios AM, Rosier-Pennevert C, Guschinskaya N, Grappin P (2024) Fungal necrotrophic interaction: a case study of seed immune response to a seed-borne pathogen. Seeds 3:216–227

Pagán I (2022) Transmission ransmission throughseeds: the unknown life of plant viruses. PLoS Pathog 18(8):e1010707. https://doi.org/10.1371/journal.ppat.1010707

Panno S, Davino S, Caruso AG, Bertacca S, Crnogorac A, Mandić A, Matić S (2021) A review of the most common and economically important diseases that undermine the cultivation of tomato crop in the Mediterranean basin. Agronomy 11:2188

Pervez R, Danish M, Verma N (2024) Interaction of plant-parasitic nematode and filamentous fungi: an insight story of mechanism involved and tool for sustainable agriculture. In: Industrial applications of soil microbes, pp 240–254

Petrov NM, Stoyanova MI, Gaur R (2024) Viruses as stress factors and their management in vegetable crops. In: Molecular dynamics of plant stress and its management, pp 331–350

Powell AA (2024) Seed vigor and its assessment. In: Handbook of seed science and technology, p 46

Rubio L, Galipienso L, Ferriol I (2020) Detection of plant viruses and disease management: relevance of genetic diversity and evolution. Front Plant Sci 11:1092. https://doi.org/10.3389/fpls.2020.01092

Saif A, Cuccurullo A, Gallipoli D, Perlot C, Bruno AW (2022) Advances in enzyme induced carbonate precipitation and application to soil improvement: a review. Materials 15(3):950. https://doi.org/10.3390/ma15030950

Sandra N, Mandal B (2024) Emerging evidence of seed transmission of begomoviruses: implications in global circulation and disease outbreak. Front Plant Sci 15:1376284. https://doi.org/10.3389/fpls.2024.1376284

Sharma R, Singh S, Singh SK (2024) A mathematical model for prevention the population growth of aphids in plants. Unpublished. https://doi.org/10.13140/RG.2.2.21577.15206

Shelar A, Nile SH, Singh AV, Rothenstein D, Bill J, Xiao J, Patil R (2023) Recent advances in nano-enabled seed treatment strategies for sustainable agriculture: challenges, risk assessment, and future perspectives. Nano Micro Lett 15:54

Shingote PR, Wasule DL, Parma VS, Holkar SK, Karkute SG, Parlawar ND, Senanayake DMJB (2022) An overview of chili leaf curl disease: molecular mechanisms, impact, challenges, and disease management strategies in Indian subcontinent. Front Microbiol 13:899512

Snuzik A (2024) Assessing CRISPR/Cas9 potential in SDG3 attainment: malaria elimination regulatory and community engagement landscape. Snuzik Malar J 23:192. https://doi.org/10.1186/s12936-024-04996-x

Sokhandan-Bashir N, Ighani H (2024) Economic significance of viruses in horticultural crops. In: Viral diseases of field and horticultural crops, pp 331–343

Solankey SS, Kumari M (2023) In: Solankey SS, Kumari M (eds) Advances in research on vegetable production under a changing climate vol. 2. Springer

Song P, Wu LR, Yan YH, Zhang JX, Chu T, Kwong LN, Zhang DY (2022) Limitations and opportunities of technologies for the analysis of cell-free DNA in cancer diagnostics. Nat Biomed Eng 6:232–245

Srivastava A, Pandey V, Singh N, Marwal A, Shahid MS, Gaur RK (2024) In silico identification of papaya genome-encoded microRNAs to target begomovirus genes in papaya leaf curl disease. Front Microbiol 15:1340275

Suhendar MA, Koswanudin D, Hidayatun N, Diantina S, Manzila I, Zulchi PT (2023) Detection of pathogens to test seed health of some food crops in storage. IOP Conf Ser Earth Environ Sci 1160:012041

Swami V, Singh N (2023) Screening and synergistic action of medicinal plant powders against cowpea weevil Callosobruchus maculatus as adulticidal and fecundity and fertility inhibitor. Int J Entomol Res 9(10):20–24

Tatineni S, Hein GL (2023) Plant viruses of agricultural importance: current and future perspectives of virus disease management strategies. Phytopathology 113(2):117. https://doi.org/10.1094/PHYTO-05-22-0167-RVW

Tomitaka Y, Shimomoto Y, Ryang BS, Hayashi K, Oki T, Matsuyama M, Sekine KT (2024) Development and application of attenuated plant viruses as biological control agents in Japan. Viruses 16:517

Tripathi A, Pandey VK, Jain D, Singh G, Brar NS, Taufeeq A, Rustagi S (2024a) An updated review on significance of PGPR-induced plant signalling and stress management in advancing sustainable agriculture. J Agric Food Res 16:101169

Tripathi AN, Maurya S, Pandey KK, Behera TK (2024b) Global scenario of vegetable fungal diseases. Vegetable Sci 51:54–65. https://doi.org/10.61180/vegsci.2024.v51.spl.06

Vacchi M, Siligardi C, Demaria F, Cedillo-González EI, González-Sánchez R, Settembre-Blundo D (2021) Technological sustainability or sustainable technology? A multidimensional vision of sustainability in manufacturing. Sustainability 13:9942

Venbrux M, Crauwels S, Rediers H (2023) Current and emerging trends in techniques for plant pathogen detection. Front Plant Sci 14:1120968. https://doi.org/10.3389/fpls.2023.1120968

Vishunavat K, Prabakar K, Anand T (2023) Seed health: testing and management. Malavika Dadlani. https://doi.org/10.1007/978-981-19-5888-5

Wang R, Bai B, Li D, Wang J, Huang W, Wu Y, Zhao L (2024) Phytoplasma: a plant pathogen that cannot be ignored in agricultural production—research progress and outlook. Mol Plant Pathol 25:e13437. https://doi.org/10.1111/mpp.13437

Warda M, Tekin S, Khafaga N, Sengul E, Çelebi F, Laçin BB (2025) Comprehensive insights into genetically modified foods: technological advances, nutritional benefits, therapeutic applications, and health implications. Food Saf Health 3(3):334–355. https://doi.org/10.1002/fsh3.70011

Westengen OT, Dalle SP, Mulesa TH (2023) Navigating toward resilient and inclusive seed systems. Proc Natl Acad Sci USA 120:e2218777120. https://doi.org/10.1073/pnas.2218777120

Wijerathna-Yapa A, Pathirana R (2022) Sustainable agro-food systems for addressing climate change and food security. Agriculture 12(10):1554. https://doi.org/10.3390/agriculture12101554

Wopereis MCS, Kuo CG (2022) Vegetable research matters for Southeast Asia. Int Symp Southeast Asia Veg 2021:601–613. https://doi.org/10.2991/978-94-6463-028-2_62

Wu W, Shan HW, Li JM, Zhang CX, Chen JP, Mao Q (2022) Roles of bacterial symbionts in transmission of plant virus by hemipteran vectors. Front Microbiol 13:805352. https://doi.org/10.3389/fmicb.2022.805352

Xhemali B, Giovanardi D, Biondi E, Stefani E (2024) Tomato and pepper seeds as pathways for the dissemination of Phytopathogenic bacteria: a constant challenge for the seed industry and the sustainability of crop production. Sustainability 16:1808. https://doi.org/10.3390/su16051808

Xiao M, Tian F, Liu X, Zhou Q, Pan J, Luo Z, Yi C (2022) Virus detection: from state-of-the-art laboratories to smartphone-based point-of-care testing. Adv Sci 9:2105904. https://doi.org/10.1002/advs.202105904

Xu Y, Wang L-X, Chen C, Ma S-S, Zhou R, Xiong A-S (2024) Virus-free sweet potato industry: development status and production suggestions. Horticulturae 10(9):979. https://doi.org/10.3390/horticulturae10090979

Yadav GK, Dadhich SK, Bhateshwar MC (2022) In: Yadav GK, Dadhich SU, Bhateshwar MC (eds) Recent innovative approaches in agricultural science. Bhumi Publishing, Maharashtra, pp 1–218

Yang S, Mao Q, Wang Y, He J, Yang J, Chen X, Zhang W (2022) Expanding known viral diversity in plants: Virome of 161 species alongside an ancient canal. Environ Microbiome 17:58

Yosef MAR, El Adly AM, Abd El Rady AW, El-Shanawany AEA (2022) Some ornamental and weed reservoir for cucumber mosaic virus in Egypt. J Environ Stud 27:35–41

Zhang S, Griffiths JS, Marchand G, Bernards MA, Wang A (2022) Tomato brown rugose fruit virus: An emerging and rapidly spreading plant RNA virus that threatens tomato production worldwide. Mol Plant Pathol 23:1262–1277

Zhang M, Shi Z, Chen G, Cao A, Wang Q, Yan D, Li Y (2023) Detection and identification methods and control techniques for crop seed diseases. Agriculture 13:1786

Zhou W, Arcot Y, Medina RF, Bernal J, Cisneros-Zevallos L, Akbulut ME (2024) Integrated pest management: An update on the sustainability approach to crop protection. ACS Omega 9:41130–41147

# Role of Planting Material and Legislative Regulations in Virus Disease Management in Vegetables

9

Leny Galvez and Mark Angelo O. Balendres

**Abstract**

Viruses, which rely on their host to cause infection, pose a significant threat to global vegetable production, resulting in substantial yield losses, deteriorated vegetable quality, and economic losses for farmers. Managing viral diseases is a complex process due to the intricate biological mechanisms of the virus and its ability to cause latent infections, making early detection challenging. Virus spread is exacerbated by unrestricted global trade and global warming, which favors virus-vector interactions and the possible emergence of new virus strains. Several disease management options are already in place. This includes using clean and virus-free planting materials, as well as the application of disease detection technologies. These strategies, however, must be supported by robust legislative policies covering seed production and distribution, quarantine and certification, and stricter trade policies.

**Keywords**

Legislative regulation · Planting material · Quarantine and regulations · Vegetable viruses · Virus disease management

L. Galvez
Philippine Fiber Industry Development Authority, Department of Agriculture, Quezon City, Philippines

M. A. O. Balendres (✉)
Department of Biology, College of Science, De La Salle University, Manila, Philippines
e-mail: mark.angelo.balendres@dlsu.edu.ph



R. Kumar et al. (eds.), *Vegetable Crops and Viral Pathogens Volume I*, Plant Pathology for Sustainable Agriculture, https://doi.org/10.1007/978-981-95-7265-6_9

## 9.1 Introduction

Vegetable production plays a crucial role in global food security by providing essential nutrients, supporting diverse agricultural economies, and serving as a primary livelihood for millions of smallholder farmers worldwide (Schreinemachers et al. 2018). Despite its critical importance, this sector is increasingly threatened by viral diseases. These plant viruses depend entirely on their host for survival and replication (Zhang et al. 2025). However, their presence can cause devastating impacts, such as significant yield losses, severe deterioration in the quality of vegetables, and substantial economic hardship for farming communities (Medina-Puche and Lozano-Duran 2019). Crops such as tomato, pepper, and cucurbits are particularly vulnerable to various economically impactful viruses (FAO 2019; Rao and Reddy 2020). The economic repercussions extend beyond direct crop failure, encompassing increased production costs due to disease management efforts, reduced marketability of affected products, and even trade restrictions that can cripple national agricultural exports (Tatineni and Hein 2023). Managing viral diseases in vegetables is an inherently complex process, primarily due to the intricate biological mechanisms of viruses themselves. A significant challenge is their capacity to induce latent infections, where plants carry the virus without showing symptoms (Takahashi et al. 2019). This asymptomatic phase makes early detection extremely challenging, allowing infected material to serve as silent reservoirs for disease dissemination (Richert-Pöggeler and Minarovits 2014). Furthermore, the various modes of virus transmission, such as through insect vectors (e.g., whiteflies and aphids), mechanical contact, and infected seeds, collectively complicate virus disease management (Gray and Banerjee 1999; Medina-Puche and Lozano-Duran 2019). The global movement of plant viruses is intensified by two overarching scenarios: unrestricted global trading (Hamborg and Blystad 2022; R. A. C. Jones 2021) and the pervasive effects of climate change (Jeger et al. 2023; Montes and Pagan 2022). The rapid increase in volume of international trade in planting materials, including seeds and vegetative propagules, created major pathways for the rapid dissemination of viruses and their associated vectors across geographical and continental barriers (Jeger et al. 2023; Montes and Pagan 2022). Consequently, viruses that originate in one region can rapidly emerge as global threats. Concurrently, the changing climate is reshaping environmental conditions that influence vector dynamics, virulence, abundance, distribution, and evolution of viral strains. These changes in weather patterns intensify disease pressure and challenge the established management strategies (Lahlali et al. 2024; Ma et al. 2022). Additionally, the constant evolutionary pressure on the virus population resulting from host and vector shifts, and genetic recombination events leading to novel virus strains or species emergence, overcoming existing plant resistance mechanisms or available diagnostic tools, poses continuous new challenges to disease management (LaTourrette and Garcia-Ruiz 2022; H. Wang et al. 2022). The presence of latent infections, combined with the rapid increase in global trade, poses a substantial challenge to phytosanitary security (Ilyas et al. 2022; Sáez and Pagán 2023). The rate and volume of global trade present logistical challenges for consignment inspections, as they can evade detection of asymptomatic samples that pose significant global risks. This situation requires a significant shift in phytosanitary protocols, from symptom-based inspections to robust molecular testing, before the issuance of export certification. Implementing these measures necessitates advanced technical capabilities and established regulatory guidelines that mandate rigorous testing of every consignment. It should be mandatory for every country to have strong phytosanitary systems to prevent the global spread of plant pathogens (Kumar et al. 2021a). This chapter examines the foundational role of virus-free planting material and recent advances in disease detection technologies to effectively manage plant viral diseases, as well as the necessity of supporting these technical strategies with robust legislative frameworks that govern seed production, certification, quarantine, and trade. The interplay between clean planting materials and regulatory policies is explored as an indispensable pillar of comprehensive virus management in vegetables.

## 9.2 Understanding Vegetable Virus Diseases: Biology, Epidemiology, and Impact

Effective management of vegetable virus diseases requires a deeper understanding of virus biology, transmission, and the tangible losses they cause to agricultural systems. Plant viruses are obligate intracellular microorganisms that depend solely on the cellular machinery of their host plant for their survival and replication (Louten 2016; Robinson et al. 2018). Viruses typically enter plants through wounds caused by mechanical injury, insect feeding, or environmental stress. However, they are most commonly introduced directly into plant cells by vectors such as white flies or aphids (Gray and Banerjee 1999). Once inside the cell, viruses hijack the host's ribosomes and other cellular machinery to replicate and synthesize new viral proteins, which are then used to form new virions to infect neighboring cells, thereby affecting systemic infection throughout the plant. While the replication mechanisms vary between virus families (e.g., RNA viruses versus DNA viruses), the overarching goal is to produce new viral particles that can then move throughout the plant and be transmitted to new hosts (Hull 2014; Rampersad and Tennant 2018).

### 9.2.1 Modes of Virus Transmission

The spread of vegetable viruses is facilitated by several distinct modes of transmission, which are classified into horizontal and vertical transmission. Horizontal transmission by arthropods, particularly aphids, is the most extensively studied mode of plant virus transmission (Fereres and Raccah 2015). There are at least 25 virus genera transmitted via this pathway (Fereres and Raccah 2015). Aphids are recognized as the most successful vectors of plant viruses, including potyviruses such as potato virus Y (PVY) and bromoviruses like cucumber mosaic virus (CMV) (Gadhave et al. 2020; Palukaitis and García-Arenal 2019). On the other hand, whiteflies are primary vectors of devastating viruses, such as tomato yellow leaf curl virus (TYLCV) (Navot et al. 1991) and tomato leaf curl New Delhi virus (ToLCNDV) (Naga et al. 2021). Thrips are responsible for spreading viruses such as tomato spotted wilt virus (TSWV) (LaBonte et al. 2025). By feeding on an infected plant, these vectors acquire the virus particles. After acquiring the virus, they transmit it to other plants during subsequent feeding. The efficiency and specificity of transmission depend on the virus-vector relationship, whether it involves non-persistent transmission (brief acquisition and transmission) or persistent transmission (where the virus circulates and replicates within the vector) (Gray and Banerjee 1999). Soil-inhabiting microorganisms are also responsible for the horizontal transfer of viruses (Roberts 2014), such as beet necrotic yellow vein virus (BNYVV) (Dehnen-Schmutz et al. 2020), grapevine fanleaf virus (GFLV) (Roy et al. 2025), or barley yellow mosaic virus (BYMV) (Toyama and Kusaba 1970) (Zhu et al. 2024). Plant viruses can also spread through vegetative propagation by invading storage organs, such as tubers, rhizomes, stolons, corms, or suckers. This is a common route for PVY in potatoes, banana bunchy top virus in banana, and abaca (Da Silva et al. 2020). Additionally, virus transmission via grafting of infected tissue into healthy plants is widespread in grapevine, *Prunus* spp., and citrus orchards worldwide (Mandal et al. 2025). Finally, mechanical transmission, which occurs through direct contact with infected plant sap, is the primary mode of transmission during field epidemics of economically important viruses. Viruses like tobacco mosaic virus (TMV) and tomato mosaic virus (ToMV) are prime examples, readily spreading through contaminated tools, human handling, or even plant-to-plant contact (Sacristán et al. 2011). Another mechanism of virus dissemination is through vertical transmission, where the virus is transmitted across generations, such as through pollen or seeds. This mode of transmission is particularly critical for the long-distance dissemination of viruses. Viruses such as TMV (Ellis et al. 2020), tomato brown rugose fruit virus (ToBRFV) (Davino et al. 2020), and PVY can be carried within or on seeds from an infected parent plant, leading to the infection of seedlings upon germination. Seed-borne viruses underscore the critical importance of rigorous pathogen-free seed production, as infected seed can serve as a primary inoculum source, facilitating the introduction of viruses into uninfected fields and new geographic regions.

### 9.2.2 Economic and Quality Losses in Key Vegetable Crops

Viral diseases pose a significant economic constraint on vegetable production, leading to substantial yield and quality losses (Rao and Reddy 2020). Yield reductions can range from moderate to complete crop failure, with *Tomato yellow leaf curl virus* (TYLCV) alone capable of causing up to 100% loss in susceptible tomato varieties (Prasad et al. 2020). In addition to reducing yield, viral infections degrade marketable quality by inducing symptoms such as stunted growth, leaf mosaics, and distortions, as well as fruit malformation, discoloration, and reduced shelf life, resulting in unmarketable quality (Rao and Reddy 2020). These vector-mediated, mechanical, and seed-borne transmission modes of viruses function as interconnected networks rather than independent pathways, thereby enhancing viral spread. For example, a seed-borne virus can introduce a pathogen into an uninfected field, which may spread mechanically during cultivation or be acquired by insect vectors. These vectors can then disseminate the virus locally or over long distances. Global trade facilitates the transcontinental spread, while climate change amplifies vector-mediated transmission by increasing the abundance and activity of vectors (Ma et al. 2022). No single control strategy is sufficient due to the complex interplay of transmission pathways. Approaches that focus solely on clean seed systems are ineffective without concurrent control of vector and mechanical transmission. Similarly, vector control is undermined by the continual introduction of infected planting material. Therefore, effective disease management requires integrated, multi-pronged strategies that simultaneously address all transmission routes, supported by enabling legislative frameworks (Coalition for Operational Research on Neglected Tropical Diseases 2022; Krenz et al. 2024).

## 9.3 The Cornerstone of Management: Clean and Virus-Free Planting Material

Using clean and virus-free planting material is fundamental for effective plant virus management (Krishna et al. 2022). Introducing infected material can initiate and sustain epidemics across multiple growing seasons and regions, as many plant viruses are transmitted through seed or vegetative propagation. Ensuring the status of planting material prevents early infection and enhances the efficacy of other control strategies, such as vector management and host resistance (FAO 2020; Rizzo et al. 2021). In this context, a clean seed system is the foundation for integrated and sustainable virus management programs. Producing authentic virus-free planting material demands advanced scientific techniques and strict adherence to rigorous protocols such as meristem culture (Alavijeh et al. 2025; Zhang et al. 2019), thermotherapy (Wang et al. 2018; Zhang et al. 2024a, 2024b), chemotherapy (Hu et al. 2012; Zhang et al. 2024a, 2024b), cryotherapy (Wang et al. 2003; Wang and Valkonen 2008), indexing and diagnostic testing (Baranwal et al. 2020; Kumar 2013), and controlled environment production (Antignus et al. 2001). Accurate, sensitive, and reliable virus detection is crucial for ensuring that planting materials are free from viruses. Modern molecular and serological tools have transformed disease management, addressing the limitations of traditional methods. Emerging technologies, including precise molecular tools, imaging, remote sensing, and AI-assisted analysis, offer improved sensitivity, specificity, and scalability (Singh et al. 2025). Enzyme-Linked Immunosorbent Assay (ELISA) is a biochemical method that detects viral proteins (antigens) using specific antibodies. In this test, the antibodies are linked to enzymes that produce a visible color change upon reacting with a specific substrate. This color change enables the detection and quantification of the target substance in a sample (Aydin et al. 2025). Dot blot immunobinding assay (DBIA) enables the detection of specific nucleic acids and proteins. This is being done by spotting or immobilizing the antigen on a nitrocellulose membrane, followed by incubation with a primary antibody, a peroxidase-conjugated secondary antibody, and visualization using 4-chloro-1-naphthol. DBIA could detect much lower amounts of virus compared to ELISA (Powell 1987), but requires a higher volume of antiserum. Tissue blotting immunoassay (TBIA) is performed by gently pressing freshly cut tissue onto the nitrocellulose membrane. Viral antigens are subsequently identified using enzyme-labeled immunoprobes (Freeman et al. 2013). The Polymerase Chain Reaction (PCR) assay sensitive and specific molecular method detects viral nucleic acids (DNA) by producing or photocopying millions of copies of the target DNA fragments from a small sample (Iralu et al. 2025). Reverse Transcription PCR (RT-PCR) is another variation that uses reverse transcriptase to detect RNA viruses (Pallás et al. 2018). Quantitative PCR (qPCR), also known as real-time PCR, enables the detection and quantification of the target sequence in real-time as it is being synthesized, eliminating the need for agarose gel electrophoresis (Jalali et al. 2017). PCR technologies can detect very low concentrations of viruses, making them crucial for identifying latent infections where virus titers may be low or symptoms are absent (Boubourakas et al. 2009). Loop-mediated isothermal amplification (LAMP) is a nucleic acid amplification method performed under isothermal conditions (60–65 °C). (Notomi 2000). It utilizes a strand-displacing DNA polymerase and four to six primers (F3, B3, FIP, BIC, F1, B1) that target six distinct sequence regions. Its key advantage lies in its operational simplicity, requiring only basic equipment such as a water bath or heat block, yet obtaining high specificity, efficiency, and applicability for on-site detection. The reaction can be observed by the naked eye or through agarose gel electrophoresis (Fukuta et al. 2003; Panno et al. 2020). Next-Generation Sequencing (NGS) has emerged as a powerful high-throughput diagnostic tool for detecting multiple and novel viruses in plant samples, including those associated with asymptomatic or mixed infections. Its ability to sequence all nucleic acids in a sample makes it a powerful tool for virus profiling, diagnosis, and certification of crops, nuclear stocks, and imported germplasm (Adams et al. 2009; S. Jones et al. 2017). Despite significant progress in diagnostic technologies, achieving absolute certainty in virus detection remains challenging, particularly with latent infections where virus titers are very low or unevenly distributed within host tissues (Tromas et al. 2014; MacKenzie et al. 1997). Additionally, the high throughput required for testing large volumes of material in commercial seed and vegetative propagation systems imposes substantial logistical and economic constraints (Maree et al. 2018). Even the most analytically sensitive tests may fail if the virus is absent from the specific sampled tissue or present at a very low concentration below the limit of detection, leading to false negatives even under rigorous testing protocols (Cardwell et al. 2018; Massart et al. 2014). As a result, a "virus-free" certification must be interpreted as "not detected using current methods," rather than a definitive assertion of viral absence. Improving diagnostic confidence requires a multi-layered certification strategy. This could include longitudinal indexing of mother plants, repeated testing over time, and the use of complementary detection methods to maximize both sensitivity and specificity (Cardwell et al. 2018; Rithesh et al. 2025). Regulatory frameworks must highly consider the limitations of these assays and should mandate the most appropriate and sensitive detection protocols. Moreover, continuous post-certification surveillance is critical, as low-titer or unevenly distributed infections may emerge after planting, potentially subverting initial negative test results (Al Rwahnih et al. 2025).

### 9.3.1 Challenges in Ensuring Widespread Access to Clean Materials

Although technically feasible, producing certified virus-free planting material is often limited by its high demand for labor and infrastructure, such as specialized facilities, highly trained personnel, and sophisticated diagnostic equipment, which leads to higher production costs. It was demonstrated that through meristem culture, 92.5% virus-free banana plantlets can be produced; however, the same study also highlighted the labor-intensive nature of this approach (Thanuja et al. 2025). Several studies highlighted that virus eradication and germplasm sanitation protocols, such as meristem tip culture, thermotherapy, chemotherapy with repeated diagnostic assays, are technically rigorous and resource-intensive, underscoring the high operational demands required to ensure virus-free planting material (Abdelwahab Elansary et al. 2024; Bettoni et al. 2024; Kumar et al. 2021b). The high production costs associated with virus-free seeds and planting material often result in elevated market prices, a consequence that makes them less accessible to smallholder farmers, particularly those in developing regions. This creates a significant disconnect, where the most effective solutions are not readily available to target farmer beneficiaries. This access limitation to virus-free planting material can intensify inequities in agricultural productivity and compromise global food security efforts, as disease incidence persists in regions where farmers rely on traditional, often pathogen-laden, seed production practices (Kibiki et al. 2020). Beyond financial constraints, a major barrier for smallholder farmers in adopting virus-free and certified planting materials is the limited awareness of their long-term agronomic and economic advantages. This knowledge gap often results in a continued use of easily and readily available, but potentially virus-infected, uncertified sources (Quarshie et al. 2021). Without proper training and education, farmer groups struggle to distinguish between certified and uncertified sources, thereby lowering the credibility of certified planting materials (Dey et al. 2022). Bridging the accessibility gap calls for robust legislative and policy interventions. Both government and international organizations need to implement policies that will provide subsidies for clean planting material, invest in infrastructure for local production and distribution, and promote knowledge transfer and training programs to capacitate farmer beneficiaries.

## 9.4 Legislative and Regulatory Frameworks for Virus Disease Management

The effectiveness of various technical interventions in virus disease management, such as producing and using certified and clean planting material, is critically dependent on the robustness of the legislative and regulatory frameworks. These frameworks provide the essential legal and institutional infrastructure that mandates, standardizes, and enforces phytosanitary measures across national seed systems to global trade (Mabunda et al. 2025). National seed laws and regulatory frameworks are the backbone of ensuring the health and quality of planting materials. These laws govern the entire seed value chain, encompassing varietal registration, quality assessment, and market oversight. These legal instruments are particularly critical for virus management, as they order mandatory seed-health testing for known seed-borne viruses, safeguarding national and international seed systems (Kumar et al. 2022; Vishunavat et al. 2023). These programs ensure that the available planting material provided meets the established criteria for genetic purity, germination, and overall health. Certification involves a series of processes, including rigorous field inspections of seed crops, laboratory testing for the presence of pathogens (including viruses), and strict adherence to defined production protocols. Certified seeds are guaranteed to be of quality, often signifying that they are "virus-free" or contain virus levels below a specified detection threshold. This provides assurance to farmers regarding the health status of their planting materials (Bogdanović et al. 2015; Ojdowska et al. 2025). International frameworks help harmonize certification practices. The OECD Seed Schemes facilitate cross-border movement of high-quality seed through standardized identity and purity regulations across participating nations (OECD 2020). Complementing this, the International Seed Testing Association (ISTA) develops and validates global sampling and testing methodologies, ensuring uniformity and reliability via accredited laboratories (FAO and ISTA 2023). Despite these systems being in place, the centralization of seed governance can limit its impact and reduce effectiveness, particularly in remote agricultural areas. For example, the maize seed system in low-income regions of West Africa has observed that centralized seed governance often fails to adapt to local conditions and meet local demand. They emphasized that expanding certification without strategic regional targeting diminished yield benefits, underscoring the importance of a more decentralized seed distribution network (Adigoun et al. 2022; Quarshie et al. 2021; Takeshima et al. 2025). Varietal registration systems require a thorough evaluation of the new cultivars. Before a variety can be officially released and commercialized, it is carefully assessed for its agronomic performance, disease resistance, uniformity, stability, and value for cultivation and use. Embedding criteria for virus resistance or tolerance within this evaluative framework provides a strategic mechanism to promote the development and deployment of more resilient crop cultivars. A recent Strength, Weakness, Opportunities, and Threats (SWOT)-analysis study highlights how Value for Cultivation and Use (VCU) testing is instrumental in evaluating cultivar quality traits such as nutritional content and disease resistance under varied conditions, including organic trials in multiple European countries. This underscores how VCU assessments can incorporate resistance traits into registration processes (Mendler-Drienyovszki et al. 2024). Many countries require multi-location field trials and committee review processes before granting release status (Gergely and Poós 2015; Mekonen et al. 2025). In India, for instance, a two-tier evaluation, including Initial and Advanced Varietal Trials under the AICRP, precedes national release notification and subsequent eligibility for certification under the Seeds Act (Chand et al. 2021). Regional harmonization initiatives further simplify the process of releasing a new cultivar. In Eastern and Southern Africa, the DTMA program recommends that the release of improved maize germplasm in one country be recognized regionally. This approach facilitates the dissemination and adoption of improved maize germplasm (Bishaw and Gastel 2009). Innovative models are emerging. In Laos, an alternative, participatory varietal registration system enables farmer-selected "promotion varieties" to be registered regionally without full Distinctness, Uniformity and Stability(DUS) compliance, allowing locally adapted lines to enter production more efficiently (Amare and Kassahun 2021; De Jonge et al. 2025)

Private seed companies serve as critical conduits for certified seed production and distribution, leveraging their operational agility, established distribution networks, and capacity for innovation, particularly in high-multiplication or hybrid crops, to meet evolving market demands and enhance seed system efficiency (Donovan et al. 2021; Hambloch et al. 2021). Legislative frameworks must establish clear, enforceable guidelines, incentives, and oversight mechanisms to ensure private actors adhere to seed health and quality standards. This ensures the integrity, safety, and reliability of the seed supply chain. In many low- and middle-income countries, the integration of private enterprises with public breeding programs and regulatory bodies through contracts, cooperatives, and multi-stakeholder platforms has enhanced seed delivery to smallholder farmers, building both market reach and trust (Spielman et al. 2025; Rachkara et al. 2017). Furthermore, public-private platforms significantly improve the supply of high-quality certified seed by fostering investment, facilitating knowledge transfer, and creating predictable supply chains (Jaffee and Srivastava 1994; Loch and Boyce 2003). Without robust regulatory frameworks and incentives, the strategic potential of private seed companies may be undermined by fragmentation, inequitable access, and diminished confidence among farmers and stakeholders.

### 9.4.1 Quarantine and Phytosanitary Regulations

Quarantine and phytosanitary regulations are the frontline defense against the spread of plant pests and diseases. They provide a clear regulatory framework and empower National Plant Protection Organizations (NPPOs) to conduct key activities, such as inspection, surveillance, and pest risk analysis (PRA) (Carvajal-Yepes et al. 2019). These measures become vital for vegetatively propagated crops, such as potatoes, sweet potatoes, and cassava, which are prone to accumulating viruses over successive generations unless subjected to rigorous phytosanitary controls (Kumar et al. 2020; Spielman et al. 2025). International cooperation through frameworks like the International Plant Protection Convention (IPPC) and associated standards (ISPMs) underpins coordinated efforts to safeguard national and global agricultural resources from transboundary pest incursions (Kumar et al. 2022). The NPPOs play a critical role in protecting plant health. They manage border controls and issue phytosanitary certificates to prevent the introduction of exotic pests and diseases, such as Tomato brown rugose fruit virus (ToBRFV), through international trade. Operating under the International Plant Protection Convention (IPPC), NPPOs are authorized to conduct inspections, testing, and certification of regulated plant materials to ensure they meet the phytosanitary standards of importing countries. These efforts are especially vital given that over 70% of invasive pests in the U.S. and 57% in Europe have been introduced via live plant trade (Mabunda et al. 2025; Yoe et al. 2021), and 68% of national food supplies are derived from crops with a foreign origin (Khoury et al. 2016). By assessing the pest risk of various imports, and collaboratively working with global networks like the CGIAR Germplasm Health Units, NPPOs help prevent the spread of viruses and other pathogens, especially in countries with limited biosecurity infrastructure (Kumar et al. 2022). Pest Risk Analysis (PRA) is a systematic, evidence-based process used by National Plant Protection Organizations (NPPOs) to assess the probability and potential consequences of quarantine pests, including viruses such as tomato yellow leaf curl virus (TYLCV), entering, establishing, and spreading within a country (EPSA Panel on Plant Health 2014; Ramos et al. 2019). Moreover, the International Plant Protection Convention (IPPC) has established international standards for pest management (ISPMs). Among these standards is ISPM 2, a framework that describes the Pest Risk Assessment (PRA) processes, starting from initiation, through pest risk assessment, and concluding with pest risk management. Moreover, ISPM 38 guides NPPOs on identifying, assessing, and managing pest risks associated with the international movement of seeds. This framework also guides procedures for establishing phytosanitary import requirements to facilitate the international movement of seeds, as well as the inspection, sampling, and testing of seeds, and the phytosanitary certification of seeds for export and re-export. Additionally, ISPM 11 describes PRA for quarantine pests, including environmental risks and living modified organisms (LMOs) (International Plant Protection Convention 2023). For example, species distribution models have been applied to predict TYLCV invasion risks under future climate scenarios (Ramos et al. 2019), while global spread and evolution of TYLCV have been explored in depth (Lefeuvre et al. 2010; Mabvakure et al. 2016). This comprehensive approach ensures that phytosanitary measures, like import restrictions, inspection protocols, and quarantine treatments, are science-based, transparent, and aligned with the needs of international trade. Post-entry quarantine mandates the temporary isolation and monitoring of high-risk plant materials in a controlled environment (Food and Agriculture Organization of the United Nations 2016; Whattam et al. 2024). This enables the detection of latent infections that are not visible at the point of origin or during the initial border inspection, ensuring that plants are safe before being released into the country (Portales et al. 2023; Whattam et al. 2021). However, it is important to consider that the effectiveness of PEQ systems is heavily dependent on the quality of diagnostic infrastructure, personnel capability, and robust legislative frameworks governing plant importation and quarantine procedures (Gauthier et al. 2022; Iwamae and Maekawa 2025; Magarey et al. 2009).

### 9.4.2 International Trade Policies and Agreements

The current globalized agricultural system, which manages plant health risks across borders, necessitates coordinated international cooperation and harmonized standards. The World Trade Organization's Agreement on the Application of Sanitary and Phytosanitary Measures (SPS Agreement) provides a legal framework that allows countries to implement necessary steps to protect their plant health, provided they are scientifically justified and not unnecessarily restrictive to trade (World Trade Organization 2021). To complement this Agreement, the International Plant Protection Convention (IPPC) promotes international cooperation in plant protection by developing International Standards for Phytosanitary Measures (ISPMs). These standards guide countries to manage pest risk, seed health, and the issuance of phytosanitary certificates (International Plant Protection Convention 2024). These standards are designed to promote consistency and transparency in phytosanitary practices, promoting safe and fair trade. Additionally, regional trade blocs often adopt harmonized phytosanitary protocols within their memberships, supporting smoother intra-regional trade while maintaining shared external plant health safeguards (Carvajal-Yepes et al. 2019; Santeramo and Lamonaca 2022). Despite these frameworks, global trade in plant materials continues to face persistent challenges. Non-compliance with regulations, the continued presence of illegal trade, and the overwhelming volume of plant imports and exports make comprehensive inspection and enforcement difficult (EFSA Panel on Plant Health (PLH) et al. 2019; Sirohi 2024). These issues are further compounded by the ongoing tension between the need to promote efficient international trade and the necessity of preventing the spread of pests and diseases. Policymakers and phytosanitary authorities must continuously balance, ensuring that trade remains open while enforcing strict measures to safeguard agricultural and ecological systems. While international agreements, such as the SPS Agreement and IPPC, provide essential guidance and structure, implementing them remains a complex and evolving challenge in the face of increasing globalization and biological threats (Hinsley et al. 2025; Pollard and Marzano 2023).

## References


Abdelwahab Elansary DW, Gürcan K, Roumi V, Şimşek Ö (2024) Efficacy of plant tissue culture techniques for eliminating Black Mulberry Idaeovirus (BMIV) from infected Black Mulberry (Morus nigra). Plants 13(21):2959. https://doi.org/10.3390/plants13212959

Adams IP, Glover RH, Monger WA, Mumford R, Jackeviciene E, Navalinskiene M, Samuitiene M, Boonham N (2009) Next-generation sequencing and metagenomic analysis: a universal diagnostic tool in plant virology. Mol Plant Pathol 10(4):537–545. https://doi.org/10.1111/j.1364-3703.2009.00545.x

Adigoun RFR, Houdegbe AC, Fassinou Hotegni NV, Segnon AC, N'Danikou S, Adjé CAO, Adadja RPM, Achigan-Dako EG (2022) Enabling effective maize seed system in low-income countries of West Africa: insights from Benin. Front Sustain Food Syst 6:1045629. https://doi.org/10.3389/fsufs.2022.1045629

Al Rwahnih M, Klaassen V, Erickson T, Alabi OJ, Stevens K, Hwang MS, Port L (2025) A new era in federal quarantine and state certification diagnostics at clean plant centers in the United States. Plant Dis 109(7):1392–1403. https://doi.org/10.1094/PDIS-10-24-2104-FE

Alavijeh MK, Bayat H, Kianpour D, Kalantari S, Zarei A (2025) Optimization of in vitro propagation and virus eradication using meristem culture and thermotherapy in two geranium species Pelargonium X hortorum ('Zonal') and Pelargonium × domesticum ('Regal'). BMC Plant Biol 25(1):9. https://doi.org/10.1186/s12870-024-06027-y

Amare K, Kassahun A (2021) Participatory variety selection for released white common bean varieties in South Gondar Zone, Ethiopia. Heliyon 7(12):e08610. https://doi.org/10.1016/j.heliyon.2021.e08610

Antignus Y, Nestel D, Cohen S, Lapidot M (2001) Ultraviolet-deficient greenhouse environment affects whitefly attraction and flight-behavior. Environ Entomol 30(2):394–399. https://doi.org/10.1603/0046-225X-30.2.394

Aydin S, Emre E, Ugur K, Aydin MA, Sahin İ, Cinar V, Akbulut T (2025) An overview of ELISA: A review and update on best laboratory practices for quantifying peptides and proteins in biological fluids. J Int Med Res 53(2):03000605251315913. https://doi.org/10.1177/03000605251315913

Baranwal VK, Kapoor R, Kumar S, Srivastava N (2020) Recent advances of virus diagnostics in horticultural crops. In: Applied plant virology. Elsevier, pp 27–37. https://doi.org/10.1016/B978-0-12-818654-1.00002-5

Bettoni JC, Wang M-R, Li J-W, Fan X, Fazio G, Hurtado-Gonzales OP, Volk GM, Wang Q-C (2024) Application of biotechniques for in vitro virus and viroid elimination in pome fruit crops. Phytopathology 114(5):930–954. https://doi.org/10.1094/PHYTO-07-23-0232-KC

Bishaw Z, & van Gastel AJG (2009) Seed production and distribution systems in developing countries. In R. E. Bennet (Ed.), Plant breeding and farmer participation (pp. 35–52). Food Agric Organ U N (FAO)

Bogdanović S, Mladenov V, Balešević-Tubić S (2015) The importance of using certified seed. Selekcija i Semenarstvo 21(2):63–67. https://doi.org/10.5937/SelSem1502063B

Boubourakas IN, Fukuta S, Kyriakopoulou PE (2009) Sensitive and rapid detection of peach latent mosaic viroid by the reverse transcription loop-mediated isothermal amplification. J Virol Methods 160(1–2):63–68. https://doi.org/10.1016/j.jviromet.2009.04.021

Cardwell K, Dennis G, Flannery AR, Fletcher J, Luster D, Nakhla M, Rice A, Shiel P, Stack J, Walsh C, Levy L (2018) Principles of diagnostic assay validation for plant pathogens: a basic review of concepts. Plant Health Progress 19(4):272–278. https://doi.org/10.1094/PHP-06-18-0036-RV

Chand S, Chandra K, Indu & Lal Khatik C (2021) Varietal Release, Notification and Denotification System in India. IntechOpen. https://doi.org/10.5772/intechopen.94212

Carvajal-Yepes M, Cardwell K, Nelson A, Garrett KA, Giovani B, Saunders DGO, Kamoun S, Legg JP, Verdier V, Lessel J, Neher RA, Day R, Pardey P, Gullino ML, Records AR, Bextine B, Leach JE, Staiger S, Tohme J (2019) A global surveillance system for crop diseases. Science 364(6447):1237–1239. https://doi.org/10.1126/science.aaw1572

Coalition for Operational Research on Neglected Tropical Diseases (2022) Multi-pronged approach to NTD surveillance: breakout session report

Da Silva W, Kutnjak D, Xu Y, Xu Y, Giovannoni J, Elena SF, Gray S (2020) Transmission modes affect the population structure of potato virus Y in potato. PLoS Pathog 16(6):e1008608. https://doi.org/10.1371/journal.ppat.1008608

Davino S, Caruso AG, Bertacca S, Barone S, Panno S (2020) Tomato Brown rugose fruit virus: seed transmission rate and efficacy of different seed disinfection treatments. Plants 9(11):1615. https://doi.org/10.3390/plants9111615

De Jonge B, Dey B, Visser B (2025) Developing a registration system for farmers' varieties. Agric Syst 222:104183. https://doi.org/10.1016/j.agsy.2024.104183

Dehnen-Schmutz K, Di Serio F, Gonthier P, Jacques M, Jaques Miret JA, Fejer Justesen A, MacLeod A, Magnusson CS, Milonas P, Navas-Cortes JA, Parnell S, Potting R, Reignault PL, Thulke H, Van der Werf W, Civera AV, Yuen J, Zappalà L, Candresse T et al (2020) Pest categorisation of beet necrotic yellow vein virus. EFSA J 18(12). https://doi.org/10.2903/j.efsa.2020.6360

Dey B, Visser B, Tin H, Mahamadou Laouali A, Baba Toure Mahamadou N, Nkhoma C, Alonzo Recinos S, Opiyo C, Bragdon S (2022) Strengths and weaknesses of organized crop seed production by smallholder farmers: a five-country case study. Outlook Agric 51(3):359–371. https://doi.org/10.1177/00307270221115454

Donovan J, Rutsaert P, Spielman D, Shikuku KM, Demont M (2021) Seed value chain development in the global south: key issues and new directions for public breeding programs. Outlook Agric 50(4):366–377. https://doi.org/10.1177/00307270211059551

EFSA Panel on Plant Health (PLH), Bragard C, Dehnen-Schmutz K, Gonthier P, Jacques M, Jaques Miret JA, Justesen AF, MacLeod A, Magnusson CS, Milonas P, Navas-Cortes JA, Parnell S, Potting R, Reignault PL, Thulke H, Van der Werf W, Vicent Civera A, Yuen J, Zappalà L et al (2019) Pest categorisation of non-EU viruses and viroids of Cydonia Mill., Malus Mill And Pyrus L. EFSA J 17(9). https://doi.org/10.2903/j.efsa.2019.5590

Ellis MD, Hoak JM, Ellis BW, Brown JA, Sit TL, Wilkinson CA, Reed TD, Welbaum GE (2020) Quantitative real-time PCR analysis of individual flue-cured tobacco seeds and seedlings reveals seed transmission of tobacco mosaic virus. Phytopathology 110(1):194–205. https://doi.org/10.1094/PHYTO-06-19-0201-FI

EPSA Panel on Plant Health (2014) Scientific opinion on the pest categorisation of tomato yellow leaf curl virus and related viruses causing tomato yellow leaf curl disease in Europe. EFSA J 10(3850). https://doi.org/10.2903/j.efsa.2014.3850

FAO (2019) Averting risks to the food chain: a compendium of proven emergency prevention methods and tools. Food and Agriculture Organization of the United Nations Knowledge Reporsitory. https://doi.org/10.4060/ca2346en

FAO. (2020). Recognizing the importance of plant health. Food and Agriculture Organization of the United Nations

FAO, ISTA (2023, June 19) Guidelines for the establishment and management of seed testing laboratories. FAO; International Seed Testing Association (ISTA). https://doi.org/10.4060/cc6103en

Fereres A, Raccah B (2015) Plant virus transmission by insects. In: Encyclopedia of life sciences, 1st edn. Wiley, pp 1–12. https://doi.org/10.1002/9780470015902.a0000760.pub3

Food and Agriculture Organization of the United Nations (2016) ISPM 34: Design and operation of post-entry quarantine stations for plants (International Standards for Phytosanitary Measures No. 34. FAO. https://www.ippc.int/static/media/files/publication/en/2016/01/ISPM_34_2010_En_2015-12-22_PostCPM10_InkAmReformatted.pdf

Freeman AJ, Spackman ME, Aftab M, McQueen V, King S, Van Leur JAG, Loh MH, Rodoni B (2013) Comparison of tissue blot immunoassay and reverse transcription polymerase chain

reaction assay for virus-testing pulse crops from a South-Eastern Australia survey. Australas Plant Pathol 42(6):675–683. https://doi.org/10.1007/s13313-013-0252-9
Fukuta S, Kato S, Yoshida K, Mizukami Y, Ishida A, Ueda J, Kanbe M, Ishimoto Y (2003) Detection of tomato yellow leaf curl virus by loop-mediated isothermal amplification reaction. J Virol Methods 112(1–2):35–40. https://doi.org/10.1016/S0166-0934(03)00187-3
Gadhave KR, Gautam S, Rasmussen DA, Srinivasan R (2020) Aphid transmission of potyvirus: the largest plant-infecting RNA virus genus. Viruses 12(7):773. https://doi.org/10.3390/v12070773
Gauthier M-EA, Lelwala RV, Elliott CE, Windell C, Fiorito S, Dinsdale A, Whattam M, Pattemore J, Barrero RA (2022) Side-by-side comparison of post-entry quarantine and high throughput sequencing methods for virus and viroid diagnosis. Biology 11(2):263. https://doi.org/10.3390/biology11020263
Gergely L, Poós B (2015) The role of disease resistance in the registration of crop varieties in Hungary. Acta Agrar Debrecen 66:27–29. https://doi.org/10.34101/actaagrar/66/1886
Gray SM, Banerjee N (1999) Mechanisms of arthropod transmission of plant and animal viruses. Microbiol Mol Biol Rev 63(1):128–148. https://doi.org/10.1128/mmbr.63.1.128-148.1999
Hambloch C, Kahwai J, Mugonya J (2021) Contextualizing private sector-based seed system development: the case of sorghum in Eastern Africa. Outlook Agric 50(4):378–391. https://doi.org/10.1177/00307270211056357
Hamborg Z, Blystad D-R (2022) First report of tomato brown rugose fruit virus in tomato in Norway. Plant Dis 106(7):2004. https://doi.org/10.1094/pdis-10-21-2208-pdn
Hinsley A, Hughes AC, Van Valkenburg J, Stark T, Van Delft J, Sutherland W, Petrovan SO (2025) Understanding the environmental and social risks from the international trade in ornamental plants. Bioscience 75(3):222–239. https://doi.org/10.1093/biosci/biae124
Hu GJ, Hong N, Wang LP, Hu HJ, Wang GP (2012) Efficacy of virus elimination from in vitro-cultured sand pear (Pyrus pyrifolia) by chemotherapy combined with thermotherapy. Crop Prot 37:20–25. https://doi.org/10.1016/j.cropro.2012.02.017
Hull R (2014) Replication of plant viruses. In: Plant virology. Elsevier, pp 341–421. https://doi.org/10.1016/b978-0-12-384871-0.00007-8
Ilyas R, Rohde MJ, Richert-Pöggeler KR, Ziebell H (2022) To be seen or not to be seen: latent infection by Tobamoviruses. Plants 11(16):2166. https://doi.org/10.3390/plants11162166
International Plant Protection Convention (2023) Pest risk analysis. IPPC. https://www.ippc.int/en/centre-of-excellence/phytosanitary-system/pest-risk-analysis/pra-ispms/
International Plant Protection Convention (2024) IAdopted standards (ISPMs). Food and Agriculture Organization of the United Nations
Iralu N, Wani S, Meghanath D, Hamid A (2025) Polymerase chain reaction (PCR) for virus detection. In: Hamid A, Ali G, Shikari A, Saleem S, Wani SH, Wani S (eds) Detection of plant viruses. Springer US, pp 111–113. https://doi.org/10.1007/978-1-0716-4390-7_20
Iwamae T, Maekawa A (2025) Development of a two-step RT-multiplex PCR assay for the simultaneous detection of six viruses with a wide host range, including fruit-tree, for use in post-entry plant quarantine inspections in Japan. Virus Dis 36(1):97–103. https://doi.org/10.1007/s13337-025-00911-3
Jaffee S, Srivastava J (1994) The roles of the private and public sectors in enhancing the performance of seed systems. World Bank Res Obs 9(1):97–117. https://doi.org/10.1093/wbro/9.1.97
Jalali M, Zaborowska J, Jalali M (2017) The polymerase chain reaction. In: Basic science methods for clinical researchers. Elsevier, pp 1–18. https://doi.org/10.1016/B978-0-12-803077-6.00001-1
Jeger MJ, Fereres A, Malmstrom CE, Mauck KE, Wintermantel WM (2023) Epidemiology and management of plant viruses under a changing climate. Phytopathology 113(9):1620–1621. https://doi.org/10.1094/phyto-07-23-0262-v
Jones RAC (2021) Global plant virus disease pandemics and epidemics. Plants 10(2):233. https://doi.org/10.3390/plants10020233
Jones S, Baizan-Edge A, MacFarlane S, Torrance L (2017) Viral diagnostics in plants using next generation sequencing: computational analysis in practice. Front Plant Sci 8:1770. https://doi.org/10.3389/fpls.2017.01770

Khoury CK, Achicanoy HA, Bjorkman AD, Navarro-Racines C, Guarino L, Flores-Palacios X, Engels JMM, Wiersema JH, Dempewolf H, Sotelo S, Ramírez-Villegas J, Castañeda-Álvarez NP, Fowler C, Jarvis A, Rieseberg LH, Struik PC (2016) Origins of food crops connect countries worldwide. Proc R Soc B Biol Sci 283(1832):20160792. https://doi.org/10.1098/rspb.2016.0792

Kibiki C, Daud Mutabazi K, Tairo F (2020) Profit analysis of virus free sweet potato and vine multiplication by smallholder farmers in selected regions of Tanzania. Int J Agric Econ 5(4):99. https://doi.org/10.11648/j.ijae.20200504.12

Krenz B, Niehl A, Krczal G (2024) Emerging strategies in plant virus disease control: insights from the 56th meeting of the DPG working group "Viruskrankheiten der Pflanzen". J Plant Dis Prot 131(6):1761–1768. https://doi.org/10.1007/s41348-024-00992-0

Krishna R, Ansari WA, Khandagale K, Benke AP, Soumia PS, Manjunathagowda DC, Gawande SJ, Ade AB, Mokat DN, Singh M (2022) Meristem culture: a potential technique for in vitro virus-free plants production in vegetatively propagated crops. In: Advances in plant tissue culture. Elsevier, pp 325–343. https://doi.org/10.1016/B978-0-323-90795-8.00017-5

Kumar PL (2013) Virus indexing and certification in clonal crops: the case of banana. Unpublished. https://doi.org/10.13140/2.1.1970.2080

Kumar R, Jeevalatha A, Baswaraj R, Tiwari RK (2020) Viral and viroid diseases of potato and their management. In: Singh AK, Chakrabarti SK, Singh B, Sharma J, Dua VK (eds) Potato science & technology for sub tropics. New India Publishing Agency, New Delhi, pp 267–292. (ISBN: 978-93-89571-93-6)

Kumar PL, Cuervo M, Kreuze JF, Muller G, Kulkarni G, Kumari SG, Massart S, Mezzalama M, Alakonya A, Muchugi A, Graziosi I, Ndjiondjop M-N, Sharma R, Negawo AT (2021a) Phytosanitary interventions for safe global germplasm exchange and the prevention of transboundary Pest spread: the role of CGIAR germplasm health units. Plants 10(2):328. https://doi.org/10.3390/plants10020328

Kumar R, Tiwari RK, Jeevalatha A, Sundaresha S, Shah MA, Sharma S, Sagar V, Kumar M, Chakrabarti SK (2021b) Potato apical leaf curl disease: current status and perspectives on a disease caused by tomato leaf curl New Delhi virus. J Plant Dis Protect 128:897–911. https://doi.org/10.1007/s41348-021-00463-w

Kumar R, Tiwari RK, Sundaresha S, Kaundal P, Raigond B (2022) Potato viruses and their management. In: Chakrabarti SK, Sharma S, Shah MA (eds) Sustainable management of potato pests and diseases. Springer, Singapore, pp 309–335. (Hardcover ISBN 978-981-16-7694-9; Softcover ISBN978-981-16-7697-0; eBook ISBN978-981-16-7695-6). https://doi.org/10.1007/978-981-16-7695-6_12

LaBonte P, Packer R, McInnes H, Rotenberg D, Whitfield AE (2025) Methods for Thrips transmission and maintenance of tomato spotted wilt virus (TSWV). In: Methods in molecular biology. Springer US, pp 119–136. https://doi.org/10.1007/978-1-0716-4338-9_10

Lahlali R, Taoussi M, Laasli S-E, Gachara G, Ezzouggari R, Belabess Z, Aberkani K, Assouguem A, Meddich A, El Jarroudi M, Barka EA (2024) Effects of climate change on plant pathogens and host-pathogen interactions. Crop Environ 3(3):159–170. https://doi.org/10.1016/j.crope.2024.05.003

LaTourrette K, Garcia-Ruiz H (2022) Determinants of virus variation, evolution, and host adaptation. Pathogens 11(9):1039. https://doi.org/10.3390/pathogens11091039

Lefeuvre P, Martin DP, Harkins G, Lemey P, Gray AJA, Meredith S, Lakay F, Monjane A, Lett J-M, Varsani A, Heydarnejad J (2010) The spread of tomato yellow leaf curl virus from the mddle east to the world. PLoS Pathog 6(10):e1001164. https://doi.org/10.1371/journal.ppat.1001164

Loch DS, Boyce KG (2003) Balancing public and private sector roles in an effective seed supply system. Field Crop Res 84(1–2):105–122. https://doi.org/10.1016/S0378-4290(03)00144-8

Louten J (2016) Features of host cells. In: Essential human virology. Elsevier, pp 31–48. https://doi.org/10.1016/b978-0-12-800947-5.00003-x

Ma J, Guo Y, Gao J, Tang H, Xu K, Liu Q, Xu L (2022) Climate change drives the transmission and spread of vector-borne diseases: an ecological perspective. Biology 11(11):1628. https://doi.org/10.3390/biology11111628

Mabunda GP, Nemukondeni N, Selaledi L (2025) Sanitary and phytosanitary (SPS) measures and their implications for international agricultural trade: challenges and opportunities; comprehensive review. Discov Agric 3(1):117. https://doi.org/10.1007/s44279-025-00301-9

Mabvakure B, Martin DP, Kraberger S, Cloete L, Van Brunschot S, Geering ADW, Thomas JE, Bananej K, Lett J-M, Lefeuvre P, Varsani A, Harkins GW (2016) Ongoing geographical spread of Tomato yellow leaf curl virus. Virology 498:257–264. https://doi.org/10.1016/j.virol.2016.08.033

MacKenzie DJ, McLean MA, Mukerji S, Green M (1997) Improved RNA extraction from woody plants for the detection of viral pathogens by reverse transcription-polymerase chain reaction. Plant Dis 81(2):222–226. https://doi.org/10.1094/PDIS.1997.81.2.222

Mandal B, Iralu N, Wani S, Meghanath D, Hamid A (2025) Transmission of plant virus by grafting. In: Hamid A, Ali G, Shikari A, Saleem S, Wani SH, Wani S (eds) Detection of plant viruses. Springer US, pp 13–15. https://doi.org/10.1007/978-1-0716-4390-7_4

Magarey RD, Fowler GA, & Borchert DM (2009) Plant biosecurity in the global economy. Annu Rev Phytopathol 47:357–378

Maree HJ, Fox A, Al Rwahnih M, Boonham N, Candresse T (2018) Application of HTS for routine plant virus diagnostics: state of the art and challenges. Front Plant Sci 9:1082. https://doi.org/10.3389/fpls.2018.01082

Massart S, Olmos A, Jijakli H, Candresse T (2014) Current impact and future directions of high throughput sequencing in plant virus diagnostics. Virus Res 188:90–96. https://doi.org/10.1016/j.virusres.2014.03.029

Medina-Puche L, Lozano-Duran R (2019) Tailoring the cell: A glimpse of how plant viruses manipulate their hosts. Curr Opin Plant Biol 52:164–173. https://doi.org/10.1016/j.pbi.2019.09.007

Mekonen M, Chala A, Tesfaye K, Mengiste T, Nida H, Mekonnen T, Tirfessa A, Girma G, Mehary Z, Alemu H, Kuru B, Geleta M (2025) Multi-location and multi-year field trials revealed broad-spectrum resistance of Sorghum ( Sorghum bicolor (L.) Moench) to Anthracnose (Colletotrichum sublineola). J Phytopathol 173(2):e70065. https://doi.org/10.1111/jph.70065

Mendler-Drienyovszki N, Magyar-Tábori K, Mancinelli R, Black L, Brown H, Allam M, Udupa SM, Atait M, Novarina E, Bardelli T, Hansen PK, Cottney P, Giulini A (2024) Assessing the opportunities and risks of DUS and VCU variety testing for sustainable production through SWOT analysis results. Agriculture 14(10):1817. https://doi.org/10.3390/agriculture14101817

Montes N, Pagan I (2022) Challenges and opportunities for plant viruses under a climate change scenario. In: Advances in virus research. Elsevier, pp 1–66. https://doi.org/10.1016/bs.aivir.2022.08.001

Naga KC, Sundaresha S, Kumar R, Tiwari RK, Subhash S, Verma G, Buckseth T, Bairwa A, Sharma S, Katare S, Srivastava RM (2021) A new record of Asia II 5 genetic group of Bemisia tabaci (Gennadius) in the major potato growing areas of India and its relationship with tomato leaf curl New Delhi virus infecting potato. 3 Biotech 11:421. https://doi.org/10.1007/s13205-021-02966-7

Navot N, Pichersky E, Zeidan M, Zamir D, Czosnek H (1991) Tomato yellow leaf curl virus: A whitefly-transmitted geminivirus with a single genomic component. Virology 185(1):151–161. https://doi.org/10.1016/0042-6822(91)90763-2

Notomi T (2000) Loop-mediated isothermal amplification of DNA. Nucleic Acids Res 28(12):63e–663e. https://doi.org/10.1093/nar/28.12.e63

OECD (2020) Boosting productivity and trade through seed certification: OECD seed schemes statistics 2019. OECD Publishing. https://doi.org/10.1787/185cedb6-en

Ojdowska P, Oleksiak T, Studnicki M, Iwańska M (2025) Certified seed use enhances yield stability in cereal production under temperate climate conditions. Agronomy 15(8):1886. https://doi.org/10.3390/agronomy15081886

Pallás V, Sánchez-Navarro JA, James D (2018) Recent advances on the multiplex molecular detection of plant viruses and viroids. Front Microbiol 9:2087. https://doi.org/10.3389/fmicb.2018.02087

Palukaitis, P., & García-Arenal, F. (Eds.). (2019). Cucumber Mosaic virus. The American Phytopathological Society. https://doi.org/10.1094/9780890546109

Panno S, Matić S, Tiberini A, Caruso AG, Bella P, Torta L, Stassi R, Davino S (2020) Loop mediated isothermal amplification: principles and applications in plant virology. Plants 9(4):461. https://doi.org/10.3390/plants9040461

Pollard CRJ, Marzano M (2023) On a handshake: business-to-business trust in the biosecurity behaviours of the UK live plant trade. Biol Invasions 25(8):2531–2547. https://doi.org/10.1007/s10530-023-03054-y

Portales LA, Yago JI, Dimayacyac AC (2023) Fungal plant pathogens of quarantine importance in The Philippines. In: Mycology in the tropics. Elsevier, pp 325–342. https://doi.org/10.1016/B978-0-323-99489-7.00013-5

Powell CA (1987) Detection of three plant viruses by dot-immunobinding assay. Phytopathology 77(2):306. https://doi.org/10.1094/Phyto-77-306

Prasad A, Sharma N, Hari-Gowthem G, Muthamilarasan M, Prasad M (2020) Tomato yellow leaf curl virus: impact, challenges, and management. Trends Plant Sci 25(9):897–911. https://doi.org/10.1016/j.tplants.2020.03.015

Quarshie PT, Abdulai A-R, Fraser EDG (2021) Africa's "seed" revolution and value chain constraints to early generation seeds commercialization and adoption in Ghana. Front Sustain Food Syst 5:665297. https://doi.org/10.3389/fsufs.2021.665297

Ramos RS, Kumar L, Shabani F, Picanço MC (2019) Risk of spread of tomato yellow leaf curl virus (TYLCV) in tomato crops under various climate change scenarios. Agric Syst 173:524–535. https://doi.org/10.1016/j.agsy.2019.03.020

Rampersad S, Tennant P (2018) Replication and expression strategies of viruses. In: Viruses. Elsevier, pp 55–82. https://doi.org/10.1016/b978-0-12-811257-1.00003-6

Rao GP, Reddy MG (2020) Overview of yield losses due to plant viruses. In: Applied plant virology. Elsevier, pp 531–562. https://doi.org/10.1016/b978-0-12-818654-1.00038-4

Richert-Pöggeler KR, Minarovits J (2014) Diversity of latent plant–virus interactions and their impact on the virosphere. In: Plant virus–host interaction. Elsevier, pp 263–275. https://doi.org/10.1016/b978-0-12-411584-2.00014-7

Rithesh L, Paul A, Amanthra Keloth MR, Jose S, Kumar A (2025) Detection and identification of plant viruses, viroids, and phytoplasma based on high-throughput molecular approaches. In: Chen J-T, Khan M, Parveen A, Patra JK (eds) Molecular and biotechnological tools for plant disease management. Springer Nature Singapore, pp 235–256. https://doi.org/10.1007/978-981-97-7510-1_8

Rizzo DM, Lichtveld M, Mazet JAK, Togami E, Miller SA (2021) Plant health and its effects on food safety and security in a One Health framework: four case studies. One Health Outlook 3(1):6. https://doi.org/10.1186/s42522-021-00038-7

Roberts AG (2014) Plant viruses: soil-borne. In: Encyclopedia of life sciences, 1st edn. Wiley. https://doi.org/10.1002/9780470015902.a0000761.pub3

Robinson M, Schor S, Barouch-Bentov R, Einav S (2018) Viral journeys on the intracellular highways. Cell Mol Life Sci 75(20):3693–3714. https://doi.org/10.1007/s00018-018-2882-0

Roy BG, Demangeat G, Meyer S, Gertz C, Komar V, Martin IR, Fuchs MF (2025) A soil-borne virus modifies the root system architecture of a plant host via a single amino acid and influences nematode transmission. Phytopathology®. https://doi.org/10.1094/phyto-03-25-0105-r

Sacristán S, Díaz M, Fraile A, García-Arenal F (2011) Contact transmission of tobacco mosaic virus: a quantitative analysis of parameters relevant for virus evolution. J Virol 85(10):4974–4981. https://doi.org/10.1128/JVI.00057-11

Sáez C, Pagán I (2023) When plants are Trojan horses for viruses. New Phytol 237(4):1071–1073. https://doi.org/10.1111/nph.18571

Santeramo FG, Lamonaca E (2022) Standards and regulatory cooperation in regional trade agreements: what the effects on trade? Appl Econ Perspect Policy 44(4):1682–1701. https://doi.org/10.1002/aepp.13276

Schreinemachers P, Simmons EB, Wopereis MCS (2018) Tapping the economic and nutritional power of vegetables. Glob Food Sec 16:36–45. https://doi.org/10.1016/j.gfs.2017.09.005

Singh A, Yasheshwar, Kaushik NK, Kala D, Nagraik R, Gupta S, Kaushal A, Walia Y, Dhir S, Noorani MS (2025) Conventional and cutting-edge advances in plant virus detection: emerging trends and techniques. 3 Biotech 15(4):100. https://doi.org/10.1007/s13205-025-04253-1

Sirohi R (2024) Regulatory issues: challenges and implications. Pharma Bioprocess 12(6):236–237

Spielman DJ, Gatto M, Wossen T, McEwan M, Abdoulaye T, Maredia MK, Hareau G (2025) Policy and regulation in seed sector development for vegetatively propagated crops: insights from Kenya, Nigeria, and Vietnam. Agric Syst 229:104441. https://doi.org/10.1016/j.agsy.2025.104441

Takahashi H, Fukuhara T, Kitazawa H, Kormelink R (2019) Virus latency and the impact on plants. Front Microbiol 10. https://doi.org/10.3389/fmicb.2019.02764

Takeshima H, Ragasa C, Bamiwuye T, Andam KS, Spielman DJ, Edeh HO, Fasoranti A, Haile B, Nwagboso C, Omoigui LO, Abdoulaye T, Kumar PL, Wossen T (2025) Seed certification, certified seeds use and yield outcomes in Nigeria: insights from nationally-representative farm panel data and seed company location data. Agric Syst 224:104268. https://doi.org/10.1016/j.agsy.2025.104268

Tatineni S, Hein GL (2023) Plant viruses of agricultural importance: current and future perspectives of virus disease management strategies. Phytopathology 113(2):117–141. https://doi.org/10.1094/phyto-05-22-0167-rvw

Thanuja K, Arulmozhiyan R, Saraswathi MS, Selvarajan R, Jegadeeswari V, Babu VR (2025) Production of virus-free quality planting material in banana (Musa spp.) variety CO 3 (ABB) through meristem culture. Plant Sci Today. https://doi.org/10.14719/pst.6136

Toyama A, Kusaba T (1970) Transmission of soil-borne barley yellow mosaic virus. Jpn J Phytopathol 36(4):223–229. https://doi.org/10.3186/jjphytopath.36.223

Tromas N, Zwart MP, Lafforgue G, Elena SF (2014) Within-host spatiotemporal dynamics of plant virus infection at the cellular level. PLoS Genet 10(2):e1004186. https://doi.org/10.1371/journal.pgen.1004186

Vishunavat K, Prabakar K, Anand T (2023) Seed health: testing and management. In: Dadlani M, Yadava DK (eds) Seed science and technology. Springer Nature Singapore, pp 335–364. https://doi.org/10.1007/978-981-19-5888-5_14

Wang QC, Valkonen JPT (2008) Elimination of two viruses which interact synergistically from sweetpotato by shoot tip culture and cryotherapy. J Virol Methods 154(1–2):135–145. https://doi.org/10.1016/j.jviromet.2008.08.006

Wang Q, Mawassi M, Li P, Gafny R, Sela I, Tanne E (2003) Elimination of grapevine virus A (GVA) by cryopreservation of in vitro-grown shoot tips of Vitis vinifera L. Plant Sci 165(2):321–327. https://doi.org/10.1016/S0168-9452(03)00091-8

Wang MR, Li J-W, Hao X-Y, Zhao L, Wang Q-C (2018) In vitro thermotherapy-based methods for plant virus eradication. Plant Methods 14(1):87. https://doi.org/10.1186/s13007-018-0355-y

Wang H, Cui X, Cai X, An T (2022) Recombination in positive-strand RNA viruses. Front Microbiol 13. https://doi.org/10.3389/fmicb.2022.870759

Whattam M, Dinsdale A, Elliott CE (2021) Evolution of plant virus diagnostics used in Australian post entry quarantine. Plants 10(7):1430. https://doi.org/10.3390/plants10071430

Whattam M, Azzopardi S, Nehl D, Maxwell A, Davis K (2024) Protecting Australia's plant health: plant quarantine in an evolving biosecurity system†. Hist Rec Aust Sci 36(1). https://doi.org/10.1071/HR24012

World Trade Organization (2021) WTO's SPS Agreement: Sanitary and phytosanitary measures. In: World Trade Organization, Easing Trade Bottlenecks of Landlocked Developing Countries. WTO, pp 54–67. https://doi.org/10.30875/9789287054227c006

Yoe C, Griffin R, Bloem S (2021) A new direction for phytosanitary risk management. Crop Prot 149:105781. https://doi.org/10.1016/j.cropro.2021.105781

Zhang Z, Wang Q-C, Spetz C, Blystad D-R (2019) In vitro therapies for virus elimination of potato-valuable germplasm in Norway. Sci Hortic 249:7–14. https://doi.org/10.1016/j.scienta.2019.01.027

Zhang AL, Bettoni JC, Shi X, Liu Y, Yang B, Liu Z (2024a) In vitro chemotherapy-based methods for virus elimination from Actinidia macrosperma. Sci Hortic 337:113543. https://doi.org/10.1016/j.scienta.2024.113543

Zhang AL, Hao M, Shi Z, Gao Z, Xu Y, Liu Z (2024b) Combining thermotherapy with shoot tip culture or cryotherapy for improved virus eradication from in vitro Actinidia macrosperma. Plant Dis 108(10):3072–3077. https://doi.org/10.1094/PDIS-03-24-0546-RE

Zhang Q, Zhang K, Li D, Wang X, Zhang Y (2025) Replication organelles of plant positive-strand RNA viruses: a boost in knowledge following new imaging approaches. Annu Rev Phytopathol. https://doi.org/10.1146/annurev-phyto-121823-032017

Zhu H, Okiyama T, Mishina K, Kikuchi S, Sassa H, Komatsuda T, Kato T, Oono Y (2024) Identification of *Barley yellow mosaic virus* isolates breaking rym3 resistance in Japan. Genes 15(6):697. https://doi.org/10.3390/genes15060697

# Revolutionizing Vegetable Virus Detection: A Systematic Review of Cas9, Cas12, and Cas13 Diagnostic Platforms

# 10

Nilim Kumar Saikia, Arpita Talukdar, Debojit Debnath, Channakeshavaiah Chikkaputtaiah, and Shridhar Shivakumar Hiremath

**Abstract**

Vegetables are an important part of human diets but are frequently threatened by a broad range of viral pathogens that cause severe losses in yield, quality, and ultimately, food security. With increasing global trade and climate variability, the incidence of mixed and emerging infections such as begomoviruses, tobamoviruses, and potyviruses in tomato, pepper, cucumber, and other vegetables has surged, necessitating rapid and on-site diagnostic capabilities. While current diagnostic methods, such as qPCR and HTS, are reliable, they remain impractical for routine surveillance in both seed health testing and disease management at the farm level due to high cost, time-intensive workflow, and infrastructure requirements. The recent advances in CRISPR-based diagnostics have provided a revolutionary alternative in the detection of vegetable viruses. Cas9, Cas12, and Cas13 are used on platforms that can detect a range of DNA and RNA viruses with great sensitivity and specificity. Also, the development of one-pot assays that combine amplification and detection in a single reaction, as well as amplifi-

N. K. Saikia · A. Talukdar · S. S. Hiremath (✉)
Center for Infectious Diseases, Biological Sciences and Technology Division, CSIR-North East Institute of Science and Technology (CSIR-NEIST), Jorhat, Assam, India

Academy of Scientific and Innovative Research (AcSIR), Ghaziabad, India
e-mail: nilimsaikia1999@gmail.com

D. Debnath
Center for Infectious Diseases, Biological Sciences and Technology Division, CSIR-North East Institute of Science and Technology (CSIR-NEIST), Jorhat, Assam, India

C. Chikkaputtaiah
Academy of Scientific and Innovative Research (AcSIR), Ghaziabad, India

Center for Biotechnology, Biological Sciences and Technology Division, CSIR-North East Institute of Science and Technology (CSIR-NEIST), Jorhat, Assam, India



R. Kumar et al. (eds.), *Vegetable Crops and Viral Pathogens Volume I*, Plant Pathology for Sustainable Agriculture, https://doi.org/10.1007/978-981-95-7265-6_10

cation-free technologies that offer visual outputs with fluorescence or lateral-flow assays, is making point-of-care diagnostics more available. Also, the combination of artificial intelligence (AI)/machine learning (ML), microfluidic-based equipment, and portable fluorescence sensors is improving speed, accuracy, and scalability. This chapter provides a systematic review of recent developments in CRISPR diagnostics of vegetable viruses. It puts focus on major technical advances in the Cas9, Cas12, and Cas13 platforms, along with case studies. Also, it addresses the issues and outlook of the translation of these technologies into agricultural biosecurity that could be practised and scaled into tangible tools.

**Keywords**

CRISPR/Cas diagnostics · Biosensing · Nucleic acid detection · Pathogen diagnosis · Point-of-care testing (POCT)

## 10.1 Introduction

Vegetables are one of the most diverse and economically significant groups of crops grown worldwide, having excellent sources of nutrients, proteins, dietary fibres, phytochemicals, and vitamins (Gopalakrishnan 2007). However, the production of vegetables is severely affected by viral diseases, which substantially affect agricultural production, impacting the yield and quality (Hiremath et al. 2025). It is estimated that plant viruses are responsible for global crop losses of more than US$30 billion annually, and account for nearly half of all emerging and re-emerging plant pathogens (Sastry and Zitter 2014). Unlike bacterial or fungal pathogens, viruses lack direct chemical control measures such as antibacterial or antifungal agents. Hence, their management largely depends on early, rapid, and accurate detection followed by prophylactic measures such as the use of virus-free planting material, vector management, removal of infected plants, and following strict field sanitation practices to prevent further spread (Rubio et al. 2020). This challenge is particularly pronounced in vegetable crops, where multiple viruses such as cucumoviruses, orthotospoviruses, poleroviruses, tospoviruses, and begomoviruses are widespread and frequently occur as mixed infections (Netla et al. 2025). The ability to diagnose these infections rapidly and reliably is, therefore, critical to maintain both the yield and quality of the produce. Currently, diagnostic methods, such as polymerase chain reaction (PCR), quantitative real-time PCR (qPCR), and high-throughput sequencing (HTS), are regarded as gold standards due to their high sensitivity and specificity. However, these methods are limited by the high costs of thermal cyclers, the need for skilled personnel, and the need for high-quality nucleic acid templates, which often restrict their use in field applications (Kumar et al. 2017, 2023a, b; Jeevalatha et al. 2018). Furthermore, the rate of high mutation and variability of the sequence that are characteristic of viral genomes in PCR-based testing may nullify the effects of binding of a primer, thus reducing sensitivity and sometimes leading to false positives or false negatives (Malinowski 2005; Marais et al. 2024). Recently, HTS-based methods have revolutionised virus discovery and surveillance by enabling unbiased detection of both known and unknown viruses without the knowledge of prior sequencing data (Netla et al. 2025). However, this method remains exorbitantly expensive, requires advanced bioinformatics expertise, and involves long processing times, making it impractical for routine crop health surveillance and on-site virus diagnostics, particularly in low-resource settings.

Given these limitations, it is important to improve the existing nucleic acid detection methods that combine speed, convenience, and accessibility. Ideally, such approaches would eliminate the requirements of sophisticated laboratory facilities while reducing the use of both manpower and costs. In this context, the limitations of conventional methods are addressed by isothermal nucleic-acid amplification (INA) methods, such as loop-mediated isothermal amplification (LAMP) (Kumar et al. 2020) and recombinase polymerase amplification (RPA) assays (Kumar et al. 2021; Paul et al. 2025). These techniques are fast, portable, and require minimal equipment, and work at a constant temperature; however, achieving specificity remains more challenging compared to PCR. Since the LAMP requires using more than one primer set, chances of primer-dimer formation and non-specific amplification are high, and this often results in false positives. Likewise, RPA exhibits a high level of tolerance to mismatches between primers and target sequences, which can potentially affect its diagnostic specificity (Bhat et al. 2022). Although the enzymes involved in INAs are resistant to inhibitors present in complex sample matrices, they are still unable to discriminate closely related viral strains. Recent advances in CRISPR (Clustered regularly interspaced short palindromic repeats)/Cas-based technologies have begun to address these diagnostic bottlenecks by providing a layer of programmable, sequence-specific recognition that enhances both sensitivity and specificity (Li et al. 2022). The CRISPR/Cas system, originally discovered as an adaptive immune mechanism in bacteria (Barrangou et al. 2007), has been repurposed into powerful tools for nucleic acid detection (Li et al. 2022). This was possible due to the collateral cleavage activity shown by Cas12 and Cas13 enzymes, in which the enzyme, upon binding to the target sequence, activates indiscriminate cleavage of single-stranded DNA or RNA reporters, respectively. This feature has been exploited to produce visible signals, such as fluorescence or colour changes on lateral-flow strips, thereby allowing rapid and instrument-free detection. In contrast to conventional molecular assays, CRISPR/Cas systems possess the ability to distinguish single-nucleotide mismatches, offering unparalleled diagnostic specificity for the detection of viral variants (Chen et al. 2018; Gootenberg et al. 2017; Kellner et al. 2019).

Several CRISPR-based platforms developed in recent times have shown strong diagnostic potential by harnessing the collateral cleavage activity and high target specificity. DNA Endonuclease Targeted CRISPR Trans Reporter (DETECTR) is among them and involves the use of the Cas12a enzyme to detect DNA viruses with fluorescent reporters, whereas Specific High-sensitivity Enzyme Reporter UnLOCKing (SHERLOCK) is an approach that involves the use of the Cas13a enzyme to detect RNA viruses through lateral-flow or fluorescent detection

(Chen et al. 2018; Kellner et al. 2019). Likewise, the One-Hour Low-cost Multipurpose Highly Efficient System (HOLMES) (Li et al. 2018) has shown the versatility of the Cas12a enzyme, demonstrating the utility for the detection of both RNA and DNA targets. These assays, combined with INA, can typically be completed within an hour or so. Notably, they can be used with complex sample matrices, including crude plant sap or paper-based nucleic acid extraction systems, eliminating a long sample preparation process and significantly lowering the time and cost (Bernabé-Orts et al. 2022; Zhang et al. 2020). These attributes make CRISPR-based diagnostics much more appropriate as a point-of-care tool in the agricultural fields, where rapid decision-making can be of significant importance in the management of crops. Moreover, CRISPR-Dx (CRISPR-based diagnostics) assays also offer excellent sensitivity, and studies have shown that they can detect nucleic acids at attomolar concentrations, surpassing PCR-based methods (Li et al. 2018). This level of sensitivity is especially beneficial in vegetable crops, where the initial infection is often accompanied by low viral titres and produces non-specific symptoms that can be difficult to distinguish visually from nutrient deficiencies, fungal or bacterial diseases, or other abiotic stresses (Wang et al. 2022). By enabling detection at early stages, CRISPR diagnostics provide an opportunity for timely intervention, preventing widespread virus transmission. In this chapter, we explored how CRISPR-based platforms are being adapted for rapid detection of major vegetable viruses and why these systems are attracting so much attention. The flexibility of CRISPR diagnostics, along with their ability to deliver highly specific readouts, has already been demonstrated in several practical examples, some of which are beginning to bridge the gap between laboratory development and field use. At the same time, we have also described the major challenges that need to be addressed to enable the practical development and deployment of these CRISPR-Dx as a transformative approach for strengthening disease management and supporting sustainable vegetable production.

## 10.2 Overview of CRISPR/Cas Systems in Diagnostics

CRISPRs are the short DNA repeats that are separated by unique spacers and were first discovered in *Escherichia coli* in 1987 as unusual repetitive sequences (Ishino et al. 2018). Along with the Cas proteins, these systems use RNA as a guiding molecule to recognise and cleave foreign nucleic acids, serving as an adaptive defence/immunity mechanism in bacteria and archaea (Barrangou et al. 2007). These CRISPR/Cas systems are broadly divided into two classes, which are further categorised into several types and subtypes based on signature effector proteins, Cas gene composition, and locus organisation. Class I systems (types I, III, and IV) are characterised by a multi-subunit CRISPR RNA (crRNA)-effector complex, whereas class II systems (types II, V, and VI) are characterised by a single-subunit crRNA-effector complex (Makarova et al. 2020). Among these, Class II effectors such as Cas9 (type II), Cas12a and Cas14/Cas12f (type V), and Cas13 (type VI) are the most widely used in gene editing, and are now being further adapted for diagnostic applications (Das et al. 2022; Li et al. 2022).

CRISPR/Cas9 was the first Cas variant that was used for disease diagnosis. This class II type II effector protein complex has the ability to recognise double-stranded DNA (dsDNA) and exploit its sequence-specific DNA binding and cleavage properties for targeted detection (Makarova et al. 2020; Pardee et al. 2016). Cas9 was initially thought to lack collateral cleavage activity, so in the early diagnostic experiments, an additional auxiliary signal amplification strategy, such as nucleic acid sequence-based amplification (NASBA), was used to achieve detectable signals (Pardee et al. 2016). This limitation was overcome by using the Cas12a, Cas14/Cas12f, and Cas13 systems in disease diagnostics, which undergo a conformational change upon target recognition that unleashes non-specific collateral cleavage activity. Cas12a and Cas14/Cas12f indiscriminately cleave single-stranded DNA (ssDNA) substrates, whereas Cas13 degrades single-stranded RNA (ssRNA) substrates (Li et al. 2022; Makarova et al. 2020). However, recent studies have revealed a paradigm shift in which CRISPR/Cas9 systems guided by crRNA and *trans*-activating crRNA can also mediate RuvC domain-dependent *trans*-cleavage of both ssDNA and ssRNA substrates (Chen et al. 2025a). In CRISPR-Dx, these collateral activities are harnessed by incorporating a reporter molecule. During collateral cleavage, the Cas nucleases cut the reporter molecule. This generates a detectable signal that can be measured through different platforms, such as fluorescence-based readouts or paper-based assays (Li et al. 2022).

CRISPR/Cas12 (also known as Cpf1) enzymes are class II, type V effector proteins guided by a single crRNA (Makarova et al. 2015). The Cas12-crRNA complex first binds specifically to a 5′-TTTV-3′ protospacer adjacent motif (PAM) on the dsDNA and, upon PAM recognition and crRNA-DNA hybridisation, it undergoes a conformational shift that activates its RuvC nuclease domain. This activation mediates two distinct activities. First, it involves precise cleavage of the target (*cis*-cleavage), and then it unleashes collateral, non-specific degradation of nearby ssDNA (*trans*-cleavage) (Swarts and Jinek 2019). This dsDNA target-mediated *trans*-cleavage activity is used for molecular diagnostics.

Utilising this *trans*-cleavage activity of Cas12a, Chen et al. (2018) developed the DETECTR (DNA Endonuclease Targeted CRISPR Trans Reporter) platform for molecular diagnostics. In this approach, Cas12a is programmed with a specific crRNA to recognise a target dsDNA sequence, and a synthetic fluorophore quencher (FQ)-labelled ssDNA reporter probe is added to the reaction. Upon target recognition and cleavage, Cas12a's collateral *trans*-cleavage activity non-specifically degrades the reporter, separating the fluorophore from the quencher and producing a detectable fluorescent signal. DETECTR was demonstrated only for DNA virus detection, although in principle, the platform can also be adapted for RNA viruses through reverse transcription. Simultaneously, Li et al. (2018) developed the HOLMES (One-Hour Low-cost Multipurpose highly Efficient System) platform. It also exploits the *trans*-cleavage activity of Cas12a for nucleic acid detection. Unlike DETECTR, HOLMES explicitly showed broader versatility. It was applied not only to DNA viruses but also to RNA viruses and to single-nucleotide polymorphism

(SNP) genotyping (by using shorter guide sequences). HOLMES further enhanced flexibility by designing PCR primers that introduced PAM sequences adjacent to SNP loci, thereby enabling sequence-independent SNP genotyping. In subsequent refinements, HOLMESv2 (Li et al. 2019) was developed, replacing Cas12a with Cas12b to enhance thermal stability and detection efficiency. Teng et al. (2019) have also developed cDETECTION (Cas12b-mediated DNA Detection) based on Cas12b. CRISPR/Cas12f (originally denoted as Cas14) enzymes are class II, subtype V-F effector proteins, which are half the size of other class II enzymes that primarily cleave ssDNA, while also unleashing dsDNA cleavage activity (Karvelis et al. 2020; Makarova et al. 2020). Unlike Cas12a, Cas12f functions in a PAM-independent manner, although it also has an RuvC nuclease domain for its activity (Harrington et al. 2018). Based on this characteristic, Harrington et al. (2018) developed the Cas14-DETECTR platform, which exploits Cas14's ability to mediate targeted ssDNA cleavage without restrictive sequence requirements. Upon target recognition, Cas12f also induces a ssDNA *trans*-cleavage activity. Later, it was disclosed that Cas12f also possesses dsDNA target cleavage activity in a PAM-dependent manner (Karvelis et al. 2020), which has since enabled the development of Cas12f-based dsDNA detection platforms such as PDTCTR (PAM-dependent dsDNA Target-activated Cas12f1 Trans Reporter) (He et al. 2024).

CRISPR/Cas13a (originally denoted as C2c2) enzymes are class II type VI effector proteins guided by a single crRNA, but unlike Cas12, they specifically recognise and cleave ssRNA targets without the requirement for a PAM sequence/site (Makarova et al. 2015). Upon binding to the complementary RNA sequence, Cas13 undergoes a conformational shift that activates its ribonuclease activity, mediated by two conserved HEPN (Higher Eukaryotes and Prokaryotes Nucleotide-binding) domains. This activation mediates two distinct activities. First, precise cleavage of the target RNA (*cis*-cleavage) and then, a collateral *trans*-cleavage activity that results in nonspecific degradation of nearby ssRNAs (East-Seletsky et al. 2016). Based on this *trans*-cleavage RNase activity, Gootenberg et al. (2017) developed SHERLOCK (Specific High-Sensitivity Enzymatic Reporter UnLOCKing) and subsequently SHERLOCKv2 (Gootenberg et al. 2018; Kellner et al. 2019). SHERLOCK also uses an FQ-labelled reporter probe, but in contrast to DETECTR's ssDNA reporter, it uses an ssRNA reporter. It is degraded upon *trans*-cleavage to release a fluorescent signal (Gootenberg et al. 2017). Molecular detection of plant viruses is often challenged by the presence of low titres in plant matrices. To overcome this limitation, a nucleic acid preamplification step is usually required to achieve relevant limits of detection (LoD). Although PCR remains the gold standard, isothermal amplification methods such as RPA and LAMP are more commonly used in combination with CRISPR-Dx, as they are faster, simpler, and more suitable for field applications compared to PCR (Li et al. 2022). Depending on the nucleic acid type and the Cas effectors used in the platform, CRISPR-Dx is combined with RT-RPA or RT-LAMP for RNA viruses, and with RPA or LAMP for DNA viruses.

## 10.3 CRISPR/Cas Systems in Vegetable Virus Diagnostics

### 10.3.1 CRISPR/Cas9 in Vegetable Virus Diagnostics

Although the CRISPR/Cas9 system has been extensively harnessed to confer plant resistance against viruses, bacteria, and fungi (Saikia et al. 2024), its role in crop disease diagnostics and pathogen detection is still limited. Before 2024, only two main Cas9-based diagnostic techniques were reported for the identification of phytopathogens and plant nucleic acids. The first is a colourimetric assay in which the Cas9/sgRNA complex specifically recognises and cleaves target pathogen DNA, followed by strand displacement amplification (SDA) and rolling circle amplification (RCA) to generate numerous single-stranded DNA products. These amplified products then hybridise with oligonucleotide-functionalised gold nanoparticles (AuNPs), causing particle aggregation and a visible colourimetric shift, enabling naked-eye detection of plant pathogens with high specificity for single-base mismatches (Cao et al. 2021). The second approach utilises the binding precision of Cas9, using a biotinylated catalytically inactive variant (bio-dCas9) that binds to FAM-labelled target DNA. The resulting dual-labelled complexes are subsequently visualised on lateral flow assays (LFA) through functionalized AuNPs. BIO-SCAN is one such example of this approach (Ali et al. 2022). Based on this, Sánchez et al. (2022) used Bio-SCAN for the detection of major vegetable-infecting viruses, including tomato yellow leaf curl virus (TYLCV), tobacco mosaic virus (TMV), and potato virus Y (PVY), along with other plant pathogens, in *Nicotiana benthamiana* as a model plant. Initially, these viruses were pre-amplified through RPA (for TYLCV) or RT-RPA (for TMV and PVY) using FAM-labelled primers. The resulting amplicons were subsequently used for detection through the CRISPR/Cas9 system, where the sgRNAs were designed for coat protein (CP) regions of TYLCV, TMV, and PVY, directing a biotinylated dCas9 complex to bind the viral DNA. The resulting amplicons were then recognised by the CRISPR/Cas9 system, in which sgRNAs targeted the coat protein genes. This resulted in dual-labelled complexes (Bio-dCas9–sgRNA-DNA), which were applied to streptavidin-biotin-based lateral flow strips, where streptavidin-coated AuNPs interacted with the biotin tag and anti-FAM antibodies captured the FAM label, which produced visible test and control bands. The presence of both bands confirmed successful detection, while a single control band indicated a negative result. Moreover, Bio-SCAN has also been successfully applied for the detection of diverse phytopathogens, including viruses, bacteria, fungi, as well as resistance alleles. In recent developments, a study by J. Chen et al. (2025a) disclosed that Cas9 also unleashes *trans*-cleavage activity, which is expected to enable the development of Cas9-based diagnostic methods comparable to those based on Cas12 or Cas13. Based on this, Besati et al. (2025) developed a CRISPR/Cas9 assay for the detection of RNA virus tomato brown rugose fruit virus (ToBRFV) using *Streptococcus pyogenes* Cas9 (spCas9). First, viral RNA was reverse-transcribed into cDNA, and RPA primers targeting the CP gene were designed. Then, a crRNA containing a spacer sequence (AUCUCAAGAUGCAGGUUGCAG) and a *trans*-activating crRNA (tracrRNA) were assembled into a ribonucleoprotein complex, which recognised NGG PAM sites within RPA-amplified viral DNA. Upon target recognition, Cas9's collateral nuclease activity was activated, leading to non-specific cleavage of an 18-nucleotide poly(T) reporter probe labelled with fluorescein and biotin. Upon cleavage of the reporter, it was visualised on lateral flow strips as test bands, enabling rapid detection within 30 min.

## 10.3.2 CRISPR/Cas12 in Vegetable Virus Diagnostics

Most CRISPR/Cas12a-based diagnostic systems for phytopathogens incorporate a nucleic acid amplification step before enzymatic cleavage to enhance target specificity and are coupled with various signal transduction processes (Fig. 10.1). For instance, one of the earliest CRISPR/Cas12-based assays for vegetable virus detection was developed by Alon et al. (2021) for the differential detection of two closely related tobamoviruses, tomato mosaic virus (ToMV), and ToBRFV, by using Cas12a from *Lachnospiraceae bacterium* (LbCas12a). RT-PCR was used as a preamplification step, with species-specific crRNAs designed to target a region within the

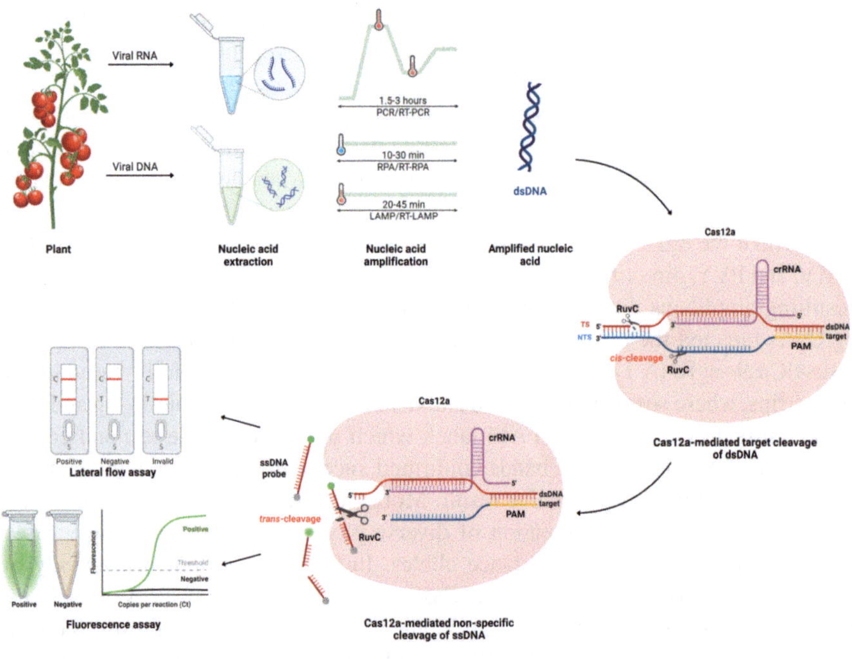


**Fig. 10.1** Schematic of basic CRISPR/Cas12a-mediated plant virus detection: Viral nucleic acids are extracted and converted into dsDNA (with reverse transcription for RNA viruses). Cas12a-crRNA complexes recognise the target strand (TS) of dsDNA adjacent to a PAM site on the non-target strand (NTS). Upon binding, the RuvC domain mediates *cis*-cleavage of both strands, which allosterically activates Cas12a to indiscriminately cleave ssDNA reporters (*trans*-cleavage), generating detectable signals via fluorescence or lateral flow assay

tobamovirus ORF1. Detection was carried out using an FQ ssDNA reporter probe measured on a fluorescence plate reader. The sensitivity assay of the CRISPR/Cas12 system revealed the detection limit of 15 ng RT-PCR product for ToMV and 15–30 ng for ToBRFV. Further, the viruses were successfully detected in the greenhouse-collected samples. In another study, the detection of two geminiviruses, tomato yellow leaf curl virus (TYLCV), and tomato leaf curl New Delhi virus (ToLCNDV), was achieved using a CRISPR/Cas12a assay combined with a LAMP reaction by Mahas et al. (2021) within 25–30 min. As a pre-amplification step, the viral CP regions were amplified at 65 °C for 40–45 min using LAMP primers. Further, the amplified products were incubated with LbCas12a and virus-specific crRNAs along with a HEX-labelled ssDNA reporter. This produced a fluorescent signal upon collateral cleavage, resulting in visualisation. This assay showed no cross-reactivity with other common geminiviruses and also showed higher sensitivity than conventional PCR for TYCLV. It successfully detected both TYLCV and ToLCNDV in *N. benthamiana* and tomato plants within 1 h, and the assay can be readily combined with a simple, portable, low-cost P51 Molecular Fluorescence Viewer instrument for easy visualisation of fluorescence. Using this LAMP-coupled CRISPR/Cas12a diagnostic assay, Bernabé-Orts et al. (2022) further adapted the LbCas12a system for the detection of ToBRFV at the field level by incorporating a rapid, 30-s RNA extraction method that eliminated the need for conventional laboratory equipment, along with column-mediated RNA extraction. Two gRNAs (MP1 and MP2) were designed targeting the movement protein (MP) ORF together with specific RT-LAMP primer sets. Detection was achieved either through fluorescence using an FQ ssDNA reporter probe or by naked-eye visualisation on lateral flow strips with a biotinylated reporter. The assay showed high specificity, with no cross-reactivity with other tobamoviruses such as pepper mild mottle virus (PMMoV), tobacco mild green mosaic virus (TMGMV), ToMV, and TMV. Sensitivity tests showed a detection limit of ~100 copies per reaction for column-mediated RNA extraction, comparable to RT-qPCR. It was also observed that using column-based extraction, infection could be detected as early as 1 day post-inoculation (dpi), while the dipstick extraction enabled consistent detection only from 4 dpi, highlighting the trade-off between sensitivity and field applicability. Based on these advancements of isothermal, field-deployable systems, Paul et al. (2025) developed an RPA-coupled CRISPR/Cas12a DETECTR assay for chilli leaf curl virus (ChiLCV) by targeting the AC1 region with a highly specific LbCas12a gRNA. By optimising FAM-biotin reporter concentrations using a logistic model, the assay showed consistent fluorescence and lateral-flow results with ~7 femtograms sensitivity with no cross-reactivity to related begomoviruses and accurately detected ChiLCV directly from crude extracts across different geographical locations. Recently, Besati et al. (2025) also developed another assay for ToBRFV using LbCas12a, along with CRISPR/Cas9. Viral RNA was first reverse-transcribed and amplified by RT-PCR, followed by RPA for secondary amplification. The *trans*-cleavage activity of LbCas12a was then carried out, and the results were visualised through a lateral flow assay. Taken together, the above studies relied on a two-step workflow, in which nucleic acid amplification, CRISPR cleavage assay, and detection were performed separately. While effective, such approaches require liquid transfer between reactions, which not only makes the process cumbersome but also increases the risk of aerosol contamination and false-positive results (Chen et al. 2025b). To overcome these limitations, Aman et al. (2020) developed a true one-pot RT-RPA-CRISPR/Cas12a assay named iSCAN-OP (in vitro Specific CRISPR-based Assay for Nucleic acids detection—One Pot) for plant RNA viruses—potato virus X (PVX), PVY, and TMV, along with a two-pot assay (iSCAN-TP). In the one-pot assay, all reagents, reverse transcriptase, RPA primers, LbCas12a-crRNA RNPs, and an ssDNA FQ reporter, were combined in a single tube and incubated isothermally at 42 °C (endpoint readout in ~20 min; compatible with a low-cost P51 fluorescence viewer or real-time monitoring on a qPCR instrument). This single-tube format reduces carryover contamination and false positives while maintaining high performance: the assay showed virus-specific detection (including in mixed PVX/TMV infections) and sensitivity to the picomolar range, improving to femtomolar with extended runtime. An alternative to one-pot amplified workflows is to completely omit the nucleic acid amplification step, thereby reducing both contamination risks and resource usage. In this direction, Shashikala et al. (2025) developed an amplification-free CRISPR/Cas12a assay for tomato leaf curl Karnataka virus (ToLCKV), in which pre-assembled LbCas12a-crRNA RNPs were applied directly to extracted genomic DNA. This assay showed that Cas12a collateral cleavage on a fluorescent ssDNA reporter probe could detect PCR-generated templates down to 0.1 ng, and can also identify the virus directly from plant genomic DNA (detection limit ≈ 10 ng), i.e., without any thermal or isothermal pre-amplification. The authors reported a clear fluorescence signal within 60 min for higher-concentration templates, but also noted that low-titre samples required extended incubation (up to ~3 h) to reach detectable signal levels, so time-to-result depends on input concentration. Guide selection was critical: CPgRNA1 outperformed other guides in both PCR and genomic DNA templates, and CPgRNA1/CPgRNA2 also detected field samples with high accuracy, showing the approach can work on real material. The amplification-free format removes the amplification step (and the tube-to-tube transfers it often entails), simplifying workflow and reducing a major source of carryover contamination, but it comes with trade-offs—notably a higher practical detection limit on crude genomic DNA and, for low viral loads, longer incubation times compared with amplified CRISPR assays. Similarly, Navyatejaswini et al. (2026) developed an amplification-free Cas12a assay for ChiLCV that also applied pre-assembled LbCas12a-gRNA complexes directly to CTAB-extracted genomic DNA. Another important advance in the Cas12a-based diagnostics for plants was the demonstration of multiplexed virus detection. Marqués et al. (2022) used LbCas12a to detect three RNA viruses—tobacco etch virus (TEV), TMV, and PVX in *N. benthamiana*. First viral CP regions were amplified using RT-PCR or RT-RPA, which was followed by Cas12a cleavage assays with virus-specific crRNAs. Multiplexing was achieved by amplifying multiple viral targets simultaneously and then running parallel Cas12a detection reactions. The system reliably distinguished between single, dual, and triple infections, with fluorescence kinetics comparable to single-virus infections. Interestingly, the

PVX signal was enhanced in co-infections with TMV or TEV, reflecting known biological synergism between PVX and potyviruses. Compared to RT-PCR, RT-RPA reduced assay time to ~35 min, but it increased background signals. This indicates a sensitivity-specificity trade-off. This system was also modified for field-deployability by using alkaline lysis for rapid RNA extraction and lateral flow strips for naked-eye readout. Together, these results highlight Cas12a's capacity not only for sensitive diagnostics but also for multiplexed detection in mixed infections, a critical feature for real-world crop pathology.

### 10.3.3 CRISPR/Cas13 in Vegetable Virus Diagnostics

As with Cas12a diagnostics, earlier Cas13-based assays for plant viruses also incorporated a preamplification step to enhance sensitivity. Zhang et al. (2021) developed a CRISPR/Cas13a assay for the detection of tomato spotted wilt virus (TSWV), where they targeted the conserved *N* gene. RPA was used as a preamplification step at 39 °C to amplify the viral cDNA, which was reverse-transcribed from extracted total RNA. RPA products were then transcribed by T7 in vitro transcription to generate ssRNA, which was then detected using Cas13a from *Leptotrichia wadei* (LwCas13a) and virus-specific crRNAs. The RNA probe produced a markedly enhanced fluorescence signal through Cas13a-mediated collateral cleavage activity (Fig. 10.2). The assay showed a sensitivity of ~$2.26 \times 10^2$ copies/μL, about ten-fold higher than RT-PCR, with no cross-reactivity with other common vegetable viruses such as tomato zonate spot virus (TZSV), TMV, cucumber mosaic virus (CMV), and PVY. Following the early amplification-dependent Cas13 assays, researchers also tried to develop amplification-free CRISPR/Cas13 diagnostics for vegetable viruses to simplify diagnostics and reduce contamination risks. Marqués et al. (2022) provided one of the first studies of direct Cas13-based detection in plants, using Cas13a and Cas13d to detect TEV, TMV, and PVX from *N. benthamiana*. In this system, viral RNA from crude or purified extracts was directly targeted by crRNAs without preamplification. Reporter cleavage produced measurable fluorescence within 30 min, with signal intensity scaling with viral load, thereby enabling semi-quantitative readouts of infection levels. While the method confirmed virus detection even in crude alkaline-lysis extracts, sensitivity was lower compared with Cas12a-based amplified workflows, and incomplete reporter processing was observed (particularly for TEV). Moreover, attempts to adapt Cas13 detection to lateral flow strips were unsuccessful, limiting field visualisation. Nonetheless, this study was pivotal in establishing amplification-free Cas13 diagnostics for plant viruses (Table 10.1). In parallel, Ueda et al. (2023) developed direct-SATORI (CRISPR-based amplification-free digital RNA detection), which integrates microchamber arrays with *Leptotrichia buccalis* Cas13a (LbuCas13a) detection for rapid RNA virus diagnostics. Viral RNA is directly introduced into femtoliter-scale chambers on a CD (compact disc)-based microchamber device preloaded with Cas13a-crRNA complexes, enabling single-molecule resolution without the need for nucleic acid amplification. Using this system, eight plant RNA viruses, namely CMV, PVX,

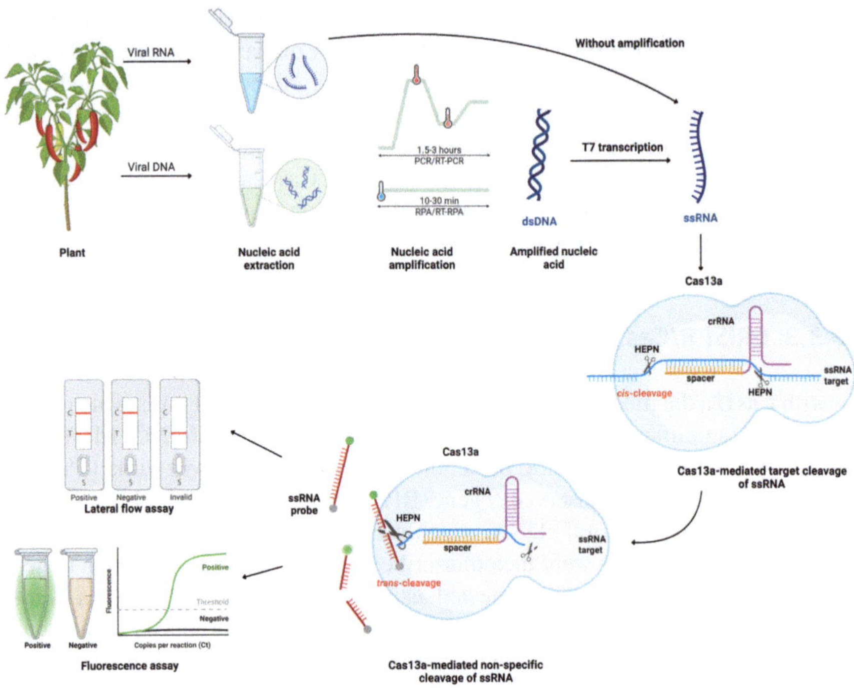


**Fig. 10.2** Schematic of basic CRISPR/Cas13a-mediated plant virus detection: For RNA viruses, viral RNA is either detected directly or reverse transcribed into cDNA, which can then be transcribed into RNA using T7 RNA polymerase if required. For DNA viruses, T7 transcription is essential to generate RNA substrates. Cas13a-crRNA complexes specifically recognise the resulting ssRNA. Upon binding, the HEPN domain mediates *cis*-cleavage of the target RNA and simultaneously triggers indiscriminate *trans*-cleavage of nearby ssRNA reporters, producing detectable signals via fluorescence or lateral flow assay

PVY, ToMV, TSWV, tomato aspermy virus (TAV), TMV, and tobacco necrosis virus (TNV) were detected within 15 min, with high specificity, with LoD values ranging from 98 to 7038 copies/μL. Although direct-SATORI demonstrates the potential of ultra-rapid, amplification-free Cas13 diagnostics, its dependence on specialised microfluidic devices currently limits its broader applicability. In contrast, a study by Hak et al. (2024) focused on developing a more field-applicable amplification-free Cas13 assay for ToBRFV. For field deployment, a simple 5-min protocol was followed by using a lysis buffer (15% PEG-4000, 20 mM NaOH) through vortexing and incubating at room temperature. This method yielded crude RNA extracts suitable for direct use in the Cas13 cleavage assay. The system used LbuCas13a with crRNAs targeting the ORF1/ORF2 overlap region. The assay was able to detect synthetic ToBRFV RNA down to ~20 pM and virus in plant RNA diluted up to 1:100 in laboratory conditions, while the field-deployable workflow was able to detect up to 1:5 dilutions. Then, the assay was coupled with a low-cost (~$35), portable, battery-operated fluorescence viewer, and the reaction tubes were imaged

**Table 10.1** Application of CRISPR-based tools for rapid detection of vegetable viruses

| CRISPR enzyme | Target virus | Virus genome | Plant | Nucleic acid extraction process | Preamplification | Target region | Quantification strategies | Limit of detection | Assay name | References |
|---|---|---|---|---|---|---|---|---|---|---|
| dCas9 | Tomato yellow leaf curl virus (TYLCV), | DNA | *Nicotiana benthamiana* | Tris-HCl based | RPA | Coat protein (CP) gene | Lateral-flow assay | – | Bio-SCAN | Sánchez et al. (2022) |
| | Tobacco mosaic virus (TMV), and potato virus Y (PVY) | RNA | | TRIzol-based / Direct-zol kit | RT-RPA | | | | | |
| SpCas9<br>LbCas12a | Tomato brown rugose fruit virus (ToBRFV) | RNA | Tomato | Spin-column | RT-PCR followed by RPA (39 °C, 30 min) | CP gene | Lateral-flow assay | 0.1 ng RPA product | – | Besati et al. (2025) |
| LbCas12a | Potato virus X (PVX), PVY, TMV | RNA | *N. Benthamiana* | TRIzol-based | RT-RPA | CP gene | Fluorescence readout | ~$10^3$–$10^4$ copies (picomolar range) for PVX, PVY, and ~ $10^2$–$10^3$ copies (femtomolar range) for TMV | iSCAN-OP and iSCAN-TP | Aman et al. (2020) |
| LbCas12a | Tomato mosaic virus (ToMV), (ToBRFV) | RNA | Tomato | Spin-column | RT-PCR | ORF1 | Fluorescence readout | 15–30 ng of PCR products | – | Alon et al. (2021) |

(continued)

**Table 10.1** (continued)

| CRISPR enzyme | Target virus | Virus genome | Plant | Nucleic acid extraction process | Preamplification | Target region | Quantification strategies | Limit of detection | Assay name | References |
|---|---|---|---|---|---|---|---|---|---|---|
| LbCas12a | TYLCV, tomato leaf curl New Delhi virus (ToLCNDV) | DNA | *N. benthamiana*, tomato | Spin-column | LAMP (65 °C, 40–45 min) | CP gene of DNA-A | Fluorescence readout | ~100 aM synthetic dsDNA (≈60 copies/reaction) | – | Mahas et al. (2021) |
| LbCas12a | ToBRFV | RNA | Tomato | Spin-column | RT-LAMP (25–45 min at 62 °C) | Movement protein ORF (MP1 and MP2 sites) | Fluorescence readout | 10–100 copies RNA/reaction (equivalent to RT-qPCR) | – | Bernabé-Orts et al. (2022) |
| | | | | Fast dipstick protocol (cellulose strip in lysis buffer) | | | Lateral-flow assay | Lower than the spin-column RNA extraction method | | |
| LbCas12a | Tomato leaf curl Karnataka virus (ToLCKV) | DNA | Tomato | CTAB-based | None (direct detection) | CP and Rep (AC1) genes | Fluorescence readout | 0.1 ng PCR product; 10 ng genomic DNA | – | Shashikala et al. (2025) |
| LbCas12a | Chilli leaf curl virus (ChLCV) | DNA | Chili | CTAB-based | None (direct detection) | AV1 and AC1 genes | Fluorescence readout | 0.1 ng PCR product; 1 ng genomic DNA | – | Navyatejaswini et al. (2026) |

| | | | | | | | | | | |
|---|---|---|---|---|---|---|---|---|---|---|
| LbCas12a | ChiLCV | DNA | Chili | CTAB-based | RPA | AC1 gene | Lateral-flow assay | 7 fg cloned viral genome | DETECTR | Paul et al. (2025) |
| | | | | Crude sap (TE buffer) | | | | | | |
| LbCas12a | Tobacco etch virus (TEV), TMV, PVX | RNA | *N. benthamiana* | Spin-column | RT-PCR (90 min) or RT-RPA (30 min) | CP gene | Fluorescence readout, Lateral-flow assay | ~$10^3$ copies/reaction with PCR amplicon | – | Marqués et al. (2022) |
| LbuCas13a/d | | | | Extraction-free crude RNA sampling (15% PEG/NaOH, 5 min) | None (direct detection) | CP gene | Fluorescence readout only | – | | |
| LwCas13a | Tomato spotted wilt virus (TSWV) | RNA | Tomato | TRIzol-based | RT-qPCR followed by RPA (39 °C, 30 min) | Nucleo-capsid (N) gene | Fluorescence readout | 2.26 × $10^2$ copies/μL | – | Zhang et al. (2021) |
| LtrCas13a | ToBRFV, ToMV, Tomato aspermy virus (TAV), TSWVCucumber mosaic virus (CMV), PVX, PVY, TMV, Tobacco necrosis virus (TNV) | RNA | Tomato | Extraction-free: 1 mg leaf +700 μM SDS, 80 °C, 5 min | None (direct detection) | CP gene | Fluorescence readout | 98–7038 copies/μL | Direct-SATORI | Ueda et al. (2023) |

(continued)

**Table 10.1** (continued)

| CRISPR enzyme | Target virus | Virus genome | Plant | Nucleic acid extraction process | Preamplification | Target region | Quantification strategies | Limit of detection | Assay name | References |
|---|---|---|---|---|---|---|---|---|---|---|
| LbuCas13a | ToBRFV | RNA | Tomato | Extraction-free crude RNA sampling | None (direct detection) | ORF1/2 region for ToBFRV and CGMMV; HC-Pro and CP region for TuMV | Portable fluorescence readout (P51 + phone) | Detected in 1:100 dilution; early detection (3 dpi) in 15 min | – | Hak et al. (2024) |
| | Cucumber green mottle mosaic virus (CGMMV) | | Cucumber | | | | | | | |
| | Turnip mosaic virus (TuMV) | | *Brassica* sp. | | | | | | | |
| Cas14a | Tomato yellow leaf curl China virus (TYLCCNV), TYLCV, ToLCNDV, and Tobacco curly shoot virus (TbCSV) | DNA | Tomato | Spin-column | Asymmetric multienzyme isothermal rapid amplification (aMIRA) | V1 gene of TYLCV, AV1 gene of TYL-CCNV, ToL-CNDV, and TbCSV | Centrifugal microfluidic chip, portable optical detection, and machine learning-based signal interpretation | 12-fold improvement over conventional fluorescence methods | MaC14a | Huang et al. (2025) |

using a mobile phone camera. Hak et al. (2024) further developed amplification-free CRISPR/Cas13 diagnostics assays with the same approach for cucumber green mottle mosaic virus (CGMMV) in cucumber and turnip mosaic virus (TuMV) in *Brassica* sp. by targeting the ORF1/ORF2, and HC-Pro, CP region, respectively, which showed its cross-pathosystem applicability.

## 10.4 Challenges in CRISPR Diagnostics for Plant Viruses

In recent times, CRISPR/Cas-based diagnostics have emerged as powerful and promising tools for detecting plant viruses, including those affecting vegetables. However, their broad practical implementation remains constrained by several critical challenges. A major challenge in CRISPR/Cas-based plant virus detection is nucleic acid extraction from complex plant matrices. At present, this step is usually carried out in laboratories, where the conditions and equipment needed are available. However, effective field application requires nucleic acid extraction and purification methods that are not only robust but also inexpensive, rapid, and user-friendly. Hence, the development of point-of-care diagnostics requires optimisation of nucleic acid extraction through simplified steps, minimal portable tools, better sample stability, and low-cost reagents. Studies are being conducted in this direction. For example, Zou et al. (2017) developed an affordable cellulose-based filter paper method for the purification of nucleic acids from crude plant sap in less than 30 s. This method has been successfully applied to extract nucleic acids from tomato and pepper leaves for the detection of CMV (Zou et al. 2017). More recently, this method has also been applied to CRISPR/Cas-based vegetable virus detection (Table 10.1). Bernabé-Orts et al. (2022) used it to extract RNA from tomato leaves for successful detection of ToBRFV using CRISPR/Cas12a system at the field level. However, challenges remain since contaminants such as polysaccharides, proteins, and polyphenols can interfere with amplification processes (Rezadoost et al. 2016). Therefore, these crude extraction methods must be extensively validated and optimised to ensure reliable application across different plant samples and under field conditions. However, using CRISPR-Dx methods with the crude extracts has a major challenge in achieving sensitivity and specificity comparable to established techniques, such as qPCR. Standard CRISPR-Dx methods, which do not include a pre-amplification step, typically exhibit a limit of detection (LoD) in the picomolar (pM) to femtomolar (fM) range (Shi et al. 2021). Although amplification-free approaches, such as Direct-SATORI, have reported attomolar sensitivity of approximately 98 copies/μL (~163 aM) using tomato viruses (Ueda et al. 2023). Still, this level of sensitivity may not be sufficient for detecting viruses present at low titres. Consequently, most CRISPR-Cas platforms are integrated with amplification methods such as PCR, RT-PCR, qPCR, or isothermal techniques like RPA, RT-RPA, or LAMP, RT-LAMP. Methods incorporating isothermal techniques like DETECTR and SHERLOCK can detect up to ~2–20 aM (~1–10 copies/μL), which are comparable to RT-qPCR; however, they still require incubation times of more than 30 min (Gootenberg et al. 2017; Kellner et al. 2019). However, integration with amplification processes also adds the complexity of designing and the cost of primers and probes. Additionally, the RPA technology, which has been incorporated with the use of CRISPR-Dx methods, is a patented technology developed by Piepenburg et al. (2006) and commercialised by TwistDx. This limits its potential for wide usage. Besides amplification, the patent issues of the CRISPR enzymes, including Cas9, Cas12, and Cas13, remain a major barrier to agricultural disease diagnostics. The ongoing legal battles and the divided nature of CRISPR patent rights among the large institutions create a sense of doubt and limitation to wider access. These patent bottlenecks would hinder the transfer of technology and would make it harder to develop low-cost, large-scale CRISPR assays of crops, especially in less developed and developing countries. Another major challenge facing CRISPR-Dx is multiplexing. Although combining isothermal amplification methods with CRISPR-Dx increases viral titres and makes them more sensitive, it comes with trade-offs that complicate multiplex detection. As an example, the SHERLOCKv2 system has already shown the ability to detect up to four targets simultaneously; however, its multiplexing capacity is constrained by the limitations of the RPA pre-amplification step (Gootenberg et al. 2018). It is even more cumbersome to achieve higher levels of multiplexing using LAMP-based CRISPR assays, because LAMP requires 4–6 primers per target (Mahas et al. 2021). This setup also increases the risk of primer-dimer formation, cross-reactivity, and unspecific amplification. Stability and storage of CRISPR components are another critical issue for CRISPR-Dx. Cold storage of most CRISPR-associated proteins and guide RNAs is needed to maintain the activity and stability of these proteins and guide RNAs. Such a dependence on refrigeration not only increases the cost of operation, but it also limits the field deployment, which is a specific challenge in rural or resource-limited farming environments.

## 10.5 Future Perspectives

CRISPR-Dx has rapidly advanced from proof-of-concept studies to field-oriented applications for vegetable virus detection. Yet, several important frontiers remain to be addressed before such technologies become mainstream in agricultural practice.

I. *Developing robust one-pot assays, simplified extraction methods and engineering new Cas orthologs:* Development of one-pot assays by combining with rapid nucleic acid extraction methods, such as the paper-based dipstick (Zou et al. 2017), alkaline PEG lysis buffer (Hwang et al. 2013), can offer a promising path for CRISPR-Dx. Also, advances in protein engineering to design Cas enzymes tolerant to crude extracts, and the discovery of novel variants (Tong et al. 2025) and thermostable Cas orthologs (Pan et al. 2025) can enable one-pot assay to work with isothermal amplification techniques operating above 55 °C (e.g., LAMP) for true point-of-care diagnostics.

   For detailed strategies on one-pot CRISPR assays, readers may refer to a recent review by Chen et al. (2025b) (DOI: 10.1016/j.tibtech.2024.07.007).

II. *Portability and accessible field-deployable readouts/visualisations:* For real-world adoption, CRISPR-DX must be portable, inexpensive, and operable by minimally trained personnel. Advances such as inexpensive fluorescence readers, such as the P51 molecular fluorescence viewer, paper-based microfluidics, and lateral flow strips are promising. The next generation of systems should combine crude extract compatibility with disposable, electricity-free readouts, ensuring ease of use in farms and low-resource conditions.

III. *Minimise off-target effects:* Designing highly specific gRNAs requires detailed knowledge of viral and host genomes, as well as access to reliable databases. While numerous computational tools and libraries exist for gRNA design in genome editing and diagnostics, they are mostly focused on human pathogens, lacking for plant pathogens. Developing dedicated databases of conserved viral regions in vegetable crops will be crucial for enhancing gRNA specificity and minimising false positives in CRISPR-Dx.

IV. *Stable reagent formulations:* Cas proteins, gRNAs, and reporter molecules, which generally require an ultra-low temperature for storage, can be stored at ambient temperatures in a stable form by lyophilising. A study by Lee et al. (2020) reported that lyophilised components can also improve assay sensitivity by allowing larger sample input volumes, a feature that could be particularly useful for detecting low-titre viral infections in plants.

V. *Advancing multiplexing strategies:* Multiplexing is important for vegetable virus diagnostics, as it lowers costs, improves efficiency, and enables simultaneous detection of multiple pathogens or variants. Since mixed infections are common in crops, this capability is particularly valuable. Recently, Lin et al. (2025) developed a Cas12a *cis*-cleavage mediated lateral flow assay (cc-LFA) that enables ultra-specific, multiplex detection with single-base resolution, in contrast to traditional *trans*-cleavage-based CRISPR/Cas12a assays for human pathogens. Such *cis*-cleavage strategies could be adapted for highly specific, multiplexed detection of plant viruses directly in the field.

   Readers interested in advancement strategies for CRISPR multiplexing may refer to two recent reviews by Zhang et al. (2024) (DOI: 10.1016/j.trac.2023.117431) and Wei et al. (2025) (DOI: 10.1021/acs.analchem.4c04428).

VI. *Expanding access through indigenous Cas variants and IP-free platforms:* Much of the CRISPR diagnostic landscape is tied to intellectual property restrictions, which can limit accessibility in regions such as India. Developing and characterising indigenous Cas variants from local microbial genomes could provide not only novel biochemical properties (e.g., enhanced thermostability or activity in crude plant extracts) but also reduce dependency on patented Cas enzymes, fostering local innovation and accessibility.

VII. *CRISPR-based seed testing for quarantine centres*: Seed testing is an important part of quarantine programs to prevent the spread of seed-borne viruses in vegetables and other crops. While HTS provides comprehensive low-cost detection for bulk samples, its exorbitantly huge initial investment and time-exhaustive workflows limit accessibility, especially in developing countries.

PCR remains a common alternative but is also relatively costly and time-intensive. CRISPR-Dx can offer high sensitivity and rapid turnaround, which can be used as a promising solution for routine seed testing in vegetables and other crops.

VIII. *Integration with artificial intelligence (AI) and machine learning (ML):* AI can be used not only to enhance gRNA design by predicting off-target interactions more accurately (Chuai et al. 2018; Huang et al. 2024; Metsky et al. 2022), but also to optimise assay sensitivity, specificity, and signal interpretation. By machine learning from large-scale genomic and diagnostic datasets, AI algorithms can identify conserved viral targets across diverse strains, predict cross-reactivity risks, and dynamically adjust detection thresholds to improve the analytical sensitivity of CRISPR-Dx in field conditions. A recent paradigm shift in crop diagnostics is the MaC14a platform, which uses this potential (Huang et al. 2025). In this platform, asymmetric multienzyme isothermal rapid amplification (aMIRA) with CRISPR-Cas14a on a centrifugal microfluidic chip was coupled with an AI-driven Long Short-Term Memory (LSTM) neural network for real-time signal interpretation. It reduced the time to detect geminiviruses in tomato leaves and whiteflies from an hour to just 5 min, while still giving results that were comparable to qPCR.

## 10.6 Conclusions

The CRISPR diagnostics have emerged as one of the powerful tools for the rapid/quick detection of plant pathogens in vegetable ecosystems, including viruses. This system is successful in addressing the key limitations of conventional molecular methods, especially their use in point-of-care detection. Even though the first usage of CRISPR diagnostics started in 2017 with the SHERLOCK platform, recent studies have improved the system with exceptional sensitivity, programmability, and versatility, with the ability to integrate into point-of-care platforms such as lateral flow assays and portable fluorescence readers. The capacity to deliver quick, accurate, and field-deployable results provides more value in agricultural fields where timely management decisions are critical to protect the produce. CRISPR-Dx system represents a paradigm shift in plant disease diagnostics with the potential to enhance sustainable farming, mitigate economic losses, and strengthen global food security.

## References

Ali Z, Sánchez E, Tehseen M, Mahas A, Marsic T, Aman R, Sivakrishna Rao G, Alhamlan FS, Alsanea MS, Al-Qahtani AA, Hamdan S, Mahfouz M (2022) Bio-SCAN: a CRISPR/dCas9-based lateral flow assay for rapid, specific, and sensitive detection of SARS-CoV-2. ACS Synth Biol 11(1):406–419. https://doi.org/10.1021/acssynbio.1c00499

Alon DM, Hak H, Bornstein M, Pines G, Spiegelman Z (2021) Differential detection of the Tobamoviruses tomato mosaic virus (ToMV) and tomato brown rugose fruit virus (ToBRFV) using CRISPR-Cas12a. Plants 10(6):6. https://doi.org/10.3390/plants10061256

Aman R, Mahas A, Marsic T, Hassan N, Mahfouz MM (2020) Efficient, rapid, and sensitive detection of plant RNA viruses with one-pot RT-RPA–CRISPR/Cas12a assay. Front Microbiol 11:610872. https://doi.org/10.3389/fmicb.2020.610872

Barrangou R, Fremaux C, Deveau H, Richards M, Boyaval P, Moineau S, Romero DA, Horvath P (2007) CRISPR provides acquired resistance against viruses in prokaryotes. Science 315(5819):1709–1712. https://doi.org/10.1126/science.1138140

Bernabé-Orts JM, Hernando Y, Aranda MA (2022) Toward a CRISPR-based point-of-care test for tomato brown rugose fruit virus detection. PhytoFrontiers™ 2(2):92–100. https://doi.org/10.1094/PHYTOFR-08-21-0053-TA

Besati M, Safarnejad MR, Aliahmadi A, Farzaneh M, Ruiz R, Montagud-Martínez R, Rodrigo G, Rafati H (2025) Detection of tomato brown rugose fruit virus through CRISPR-Cas12a and CRISPR-Cas9 systems. Sci Rep 15(1):25638. https://doi.org/10.1038/s41598-025-11825-x

Bhat AI, Aman R, Mahfouz M (2022) Onsite detection of plant viruses using isothermal amplification assays. Plant Biotechnol J 20(10):1859–1873. https://doi.org/10.1111/pbi.13871

Cao Y, Wu J, Pang B, Zhang H, Le XC (2021) CRISPR/Cas12a-mediated gold nanoparticle aggregation for colorimetric detection of SARS-CoV-2. Chem Commun 57(56):6871–6874. https://doi.org/10.1039/d1cc02546e

Chen JS, Ma E, Harrington LB, Da Costa M, Tian X, Palefsky JM, Doudna JA (2018) CRISPR-Cas12a target binding unleashes indiscriminate single-stranded DNase activity. Science 360(6387):436–439. https://doi.org/10.1126/science.aar6245

Chen J, Chen Y, Huang L, Lin X, Chen H, Xiang W, Liu L (2025a) Trans-nuclease activity of Cas9 activated by DNA or RNA target binding. Nat Biotechnol 43(4):558–568. https://doi.org/10.1038/s41587-024-02255-7

Chen L, Hu M, Zhou X (2025b) Trends in developing one-pot CRISPR diagnostics strategies. Trends Biotechnol 43(1):98–110. https://doi.org/10.1016/j.tibtech.2024.07.007

Chuai G, Ma H, Yan J, Chen M, Hong N, Xue D, Zhou C, Zhu C, Chen K, Duan B, Gu F, Qu S, Huang D, Wei J, Liu Q (2018) DeepCRISPR: optimized CRISPR guide RNA design by deep learning. Genome Biol 19(1):80. https://doi.org/10.1186/s13059-018-1459-4

Das D, Singha DL, Paswan RR, Chowdhury N, Sharma M, Reddy PS, Chikkaputtaiah C (2022) Recent advancements in CRISPR/Cas technology for accelerated crop improvement. Planta 255(5):109. https://doi.org/10.1007/s00425-022-03894-3

East-Seletsky A, O'Connell MR, Knight SC, Burstein D, Cate JHD, Tjian R, Doudna JA (2016) Two distinct RNase activities of CRISPR-C2c2 enable guide-RNA processing and RNA detection. Nature 538(7624):270–273. https://doi.org/10.1038/nature19802

Gootenberg JS, Abudayyeh OO, Lee JW, Essletzbichler P, Dy AJ, Joung J, Verdine V, Donghia N, Daringer NM, Freije CA, Myhrvold C, Bhattacharyya RP, Livny J, Regev A, Koonin EV, Hung DT, Sabeti PC, Collins JJ, Zhang F (2017) Nucleic acid detection with CRISPR-Cas13a/C2c2. Science 356(6336):438–442. https://doi.org/10.1126/science.aam9321

Gootenberg JS, Abudayyeh OO, Kellner MJ, Joung J, Collins JJ, Zhang F (2018) Multiplexed and portable nucleic acid detection platform with Cas13, Cas12a, and Csm6. Science 360(6387):439–444. https://doi.org/10.1126/science.aaq0179

Gopalakrishnan TR (2007) Vegetable crops (No. 4). New India Publishing

Hak H, Ostendorp S, Reza A, Ishgur Greenberg S, Pines G, Kehr J, Spiegelman Z (2024) Rapid on-site detection of crop RNA viruses using CRISPR/Cas13a. J Exp Bot:erae495. https://doi.org/10.1093/jxb/erae495

Harrington LB, Burstein D, Chen JS, Paez-Espino D, Ma E, Witte IP, Cofsky JC, Kyrpides NC, Banfield JF, Doudna JA (2018) Programmed DNA destruction by miniature CRISPR-Cas14 enzymes. Science 362(6416):839–842. https://doi.org/10.1126/science.aav4294

He J, Hu X, Weng X, Wang H, Yu J, Jiang T, Zou L, Zhou X, Lyu Z, Liu J, Zhou P, Xiao X, Zhen D, Deng Z (2024) Efficient, specific and direct detection of double-stranded DNA targets using

Cas12f1 nucleases and engineered guide RNAs. Biosens Bioelectron 260:116428. https://doi.org/10.1016/j.bios.2024.116428

Hiremath S, Bharath M, Muttappagol M, Kumar HDV, Kumar TLM, Shankarappa KS, Venkataravanappa V, Basha CRJ, Reddy CNL (2025) Identification of potential plant secondary metabolites targeting begomovirus-associated betasatellite virulence factor βC1 protein through molecular docking, simulation and MM-PBSA studies. Discov Plants 2(1):52. https://doi.org/10.1007/s44372-025-00128-0

Huang B, Guo L, Yin H, Wu Y, Zeng Z, Xu S, Lou Y, Ai Z, Zhang W, Kan X, Yu Q, Du S, Li C, Wu L, Huang X, Wang S, Wang X (2024) Deep learning enhancing guide RNA design for CRISPR/Cas12a-based diagnostics. iMeta 3(4):e214. https://doi.org/10.1002/imt2.214

Huang Z, Ma G, Hu R, Xu J, Fu Y, Tang J, Li D, Chen Z, Shentu X, Ye Z, Sun K, Yu X (2025) MaC14a: an AI-enhanced CRISPR-Cas14a microfluidic platform for accurate on-site detection of geminiviruses in tomato plants and whiteflies. Chem Eng J 520:165904. https://doi.org/10.1016/j.cej.2025.165904

Hwang H, Bae S-C, Lee S, Lee Y-H, Chang A (2013) A rapid and simple genotyping method for various plants by direct-PCR. Plant Breed Biotechnol 1(3):290–297. https://doi.org/10.9787/PBB.2013.1.3.290

Ishino Y, Krupovic M, Forterre P (2018) History of CRISPR-Cas from encounter with a mysterious repeated sequence to genome editing technology. J Bacteriol 200(7). https://doi.org/10.1128/jb.00580-17

Jeevalatha A, Kaundal P, Kumar R, Raigond B, Kumar R, Sharma S, Chakrabarti SK (2018) Optimized loop-mediated isothermal amplification assay for Tomato leaf curl New Delhi virus-[potato] detection in potato leaves and tubers. Eur J Plant Pathol 150:565–573. https://doi.org/10.1007/s10658-017-1300-z

Karvelis T, Bigelyte G, Young JK, Hou Z, Zedaveinyte R, Budre K, Paulraj S, Djukanovic V, Gasior S, Silanskas A, Venclovas Č, Siksnys V (2020) PAM recognition by miniature CRISPR–Cas12f nucleases triggers programmable double-stranded DNA target cleavage. Nucleic Acids Res 48(9):5016–5023. https://doi.org/10.1093/nar/gkaa208

Kellner MJ, Koob J, Gootenberg JS, Abudayyeh OO, Zhang F (2019) SHERLOCK: nucleic acid detection with CRISPR nucleases. Nat Protoc 14(10):2986–3012. https://doi.org/10.1038/s41596-019-0210-2

Kumar R, Jeevalatha A, Baswaraj R, Sharma S, Nagesh M (2017) A multiplex RT-PCR assay for simultaneous detection of five viruses in potato. J Plant Pathol 99:37–45. https://doi.org/10.4454/jpp.v99i1.3824

Kumar R, Kaundal P, Jeevalatha A, Sharma S, Chakrabarti SK (2020) Development of a visual detection of Potato virus S by reverse transcription loop-mediated isothermal amplification. 3 Biotech 10:213. https://doi.org/10.1007/s13205-020-02214-4

Kumar R, Kaundal P, Tiwari RK, Sundaresha S, Kumari H, Naga KC, Sharma S, Kumar M (2021) Rapid and sensitive detection of potato virus X by one-step reverse transcription-recombinase polymerase amplification method in potato leaves and dormant tubers. Mol Cell Probes 58:101743. https://doi.org/10.1016/j.mcp.2021.101743

Kumar R, Kaundal P, Tiwari RK, Lal MK, Kumari H, Kumar R, Naga KC, Kumar A, Singh B, Sagar V, Sharma S (2023a) Development of reverse transcription recombinase polymerase amplification (RT-RPA): a methodology for quick diagnosis of potato Leafroll viral disease in potato. Int J Mol Sci 24:2511. https://doi.org/10.3390/ijms24032511

Kumar R, Kaundal P, Tiwari RK, Lal MK, Kumari H, Kumar R, Sagar V, Singh B (2023b) Optimization of a simple, low-cost one-step reverse transcription recombinase polymerase amplification method for real-time detection of potato virus A in potato leaves and tubers. 3 Biotech 13:373. https://doi.org/10.1007/s13205-023-03791-w

Lee RA, Puig HD, Nguyen PQ, Angenent-Mari NM, Donghia NM, McGee JP, Dvorin JD, Klapperich CM, Pollock NR, Collins JJ (2020) Ultrasensitive CRISPR-based diagnostic for field-applicable detection of Plasmodium species in symptomatic and asymptomatic malaria. Proc Natl Acad Sci 117(41):25722–25731. https://doi.org/10.1073/pnas.2010196117

Li S-Y, Cheng Q-X, Wang J-M, Li X-Y, Zhang Z-L, Gao S, Cao R-B, Zhao G-P, Wang J (2018) CRISPR-Cas12a-assisted nucleic acid detection. Cell Discov 4(1):20. https://doi.org/10.1038/s41421-018-0028-z

Li L, Li S, Wu N, Wu J, Wang G, Zhao G, Wang J (2019) HOLMESv2: a CRISPR-Cas12b-assisted platform for nucleic acid detection and DNA methylation quantitation. ACS Synth Biol 8(10):2228–2237. https://doi.org/10.1021/acssynbio.9b00209

Li L, Shen G, Wu M, Jiang J, Xia Q, Lin P (2022) CRISPR-Cas-mediated diagnostics. Trends Biotechnol 40(11):1326–1345. https://doi.org/10.1016/j.tibtech.2022.04.006

Lin M, Qiu Z, Hao M, Qi W, Zhang T, Shen Y, Xiao H, Liang C, Xie L, Jiang Y, Cheng M, Tian T, Zhou X (2025) Cas12a Cis-cleavage mediated lateral flow assay enables multiplex and ultra-specific nucleic acid detection. Nat Commun 16(1):5597. https://doi.org/10.1038/s41467-025-60917-9

Mahas A, Hassan N, Aman R, Marsic T, Wang Q, Ali Z, Mahfouz MM (2021) LAMP-coupled CRISPR–Cas12a module for rapid and sensitive detection of plant DNA viruses. Viruses 13(3):466. https://doi.org/10.3390/v13030466

Makarova KS, Wolf YI, Alkhnbashi OS, Costa F, Shah SA, Saunders SJ, Barrangou R, Brouns SJJ, Charpentier E, Haft DH, Horvath P, Moineau S, Mojica FJM, Terns RM, Terns MP, White MF, Yakunin AF, Garrett RA, Van Der Oost J et al (2015) An updated evolutionary classification of CRISPR–Cas systems. Nat Rev Microbiol 13(11):722–736. https://doi.org/10.1038/nrmicro3569

Makarova KS, Wolf YI, Iranzo J, Shmakov SA, Alkhnbashi OS, Brouns SJJ, Charpentier E, Cheng D, Haft DH, Horvath P, Moineau S, Mojica FJM, Scott D, Shah SA, Siksnys V, Terns MP, Venclovas Č, White MF, Yakunin AF et al (2020) Evolutionary classification of CRISPR–Cas systems: a burst of class 2 and derived variants. Nat Rev Microbiol 18(2):67–83. https://doi.org/10.1038/s41579-019-0299-x

Malinowski T (2005) Potential problems with the reliability of PCR based diagnostic methods related to plant viruses sequence variation. https://www.semanticscholar.org/paper/POTENTIAL-PROBLEMS-WITH-THE-RELIABILITY-OF-PCR-TO-Malinowski/39bcb1da1883b75a58bf8889850ccd3eb6e68863

Marais A, Gentit P, Brans Y, Renvoisé JP, Faure C, Saison A, Cousseau P, Castaing J, Chambon F, Pion A, Calado G, Lefebvre M, Garnier S, Latour F, Bresson K, Grasseau N, Candresse T (2024) Comparative performance evaluation of double-stranded RNA high-throughput sequencing for the detection of viral infection in temperate fruit crops. Phytopathology 114(7):1701–1709. https://doi.org/10.1094/PHYTO-12-23-0480-R

Marqués M-C, Sánchez-Vicente J, Ruiz R, Montagud-Martínez R, Márquez-Costa R, Gómez G, Carbonell A, Daròs J-A, Rodrigo G (2022) Diagnostics of infections produced by the plant viruses TMV, TEV, and PVX with CRISPR-Cas12 and CRISPR-Cas13. ACS Synth Biol 11(7):2384–2393. https://doi.org/10.1021/acssynbio.2c00090

Metsky HC, Welch NL, Pillai PP, Haradhvala NJ, Rumker L, Mantena S, Zhang YB, Yang DK, Ackerman CM, Weller J, Blainey PC, Myhrvold C, Mitzenmacher M, Sabeti PC (2022) Designing sensitive viral diagnostics with machine learning. Nat Biotechnol 40(7):1123–1131. https://doi.org/10.1038/s41587-022-01213-5

Navyatejaswini B, Dhawane Y, Venkataravanappa V, Shashikala T, Manamohan M, Jha GK, Asokan R, Ashok K (2026) Fast-track detection of Chilli Leaf Curl Virus using CRISPR/Cas12a without amplification. Physiol Mol Plant Pathol 141:102973. https://doi.org/10.1016/j.pmpp.2025.102973

Netla VR, Hiremath S, Muttappagol M, Vinay Kumar HD, Koti Prasanna S, Kumar TLM, Basha CRJ, Venkataravanappa V, Shankarappa KS, Maruthi MN, Lakshminarayana Reddy CN (2025) Virome analysis of field-collected chilli samples reveals diverse viruses. Virol J 22(1):116. https://doi.org/10.1186/s12985-025-02713-x

Pan J, Mabuchi M, Kneller DW, Fuchs RT, Curcuru JL, Tamanaha E, Tanner NA, Robb GB, Corrêa IR Jr (2025) Kinetic analysis and engineering of thermostable Cas12a for nucleic acid detection. Nucleic Acids Res 53(11):gkaf509. https://doi.org/10.1093/nar/gkaf509

Pardee K, Green AA, Takahashi MK, Braff D, Lambert G, Lee JW, Ferrante T, Ma D, Donghia N, Fan M, Daringer NM, Bosch I, Dudley DM, O'Connor DH, Gehrke L, Collins JJ (2016) Rapid, low-cost detection of Zika virus using programmable biomolecular components. Cell 165(5):1255–1266. https://doi.org/10.1016/j.cell.2016.04.059

Paul S, Emmadi V, Saxena S, Sarkar M, Mandal B, Kumar R, Sinha P, Roy A (2025) RPA-assisted CRISPR-Cas12a-enabled point-of-care diagnostic platform for chili leaf curl virus with fluorescent and colorimetric readouts. Front Microbiol 16:1644322. https://doi.org/10.3389/fmicb.2025.1644322

Piepenburg O, Williams CH, Stemple DL, Armes NA (2006) DNA detection using recombination proteins. PLoS Biol 4(7):e204. https://doi.org/10.1371/journal.pbio.0040204

Rezadoost MH, Kordrostami M, Kumleh HH (2016) An efficient protocol for isolation of inhibitor-free nucleic acids even from recalcitrant plants. 3 Biotech 6(1):61. https://doi.org/10.1007/s13205-016-0375-0

Rubio L, Galipienso L, Ferriol I (2020) Detection of plant viruses and disease management: relevance of genetic diversity and evolution. Front Plant Sci 11:1092. https://doi.org/10.3389/fpls.2020.01092

Saikia B, Remya S, Debbarma J, Maharana J, Sastry GN, Chikkaputtaiah C (2024) CRISPR/Cas9-based genome editing and functional analysis of SlHyPRP1 and SlDEA1 genes of Solanum lycopersicum L. in imparting genetic tolerance to multiple stress factors. Front Plant Sci 15. https://doi.org/10.3389/fpls.2024.1304381

Sánchez E, Ali Z, Islam T, Mahfouz M (2022) A CRISPR -based lateral flow assay for plant genotyping and pathogen diagnostics. Plant Biotechnol J 20(12):2418–2429. https://doi.org/10.1111/pbi.13924

Sastry KS, Zitter TA (2014) Management of virus and viroid diseases of crops in the tropics. In: Sastry KS, Zitter TA (eds) Plant virus and viroid diseases in the tropics: volume 2: epidemiology and management. Springer Netherlands, pp 149–480. https://doi.org/10.1007/978-94-007-7820-7_2

Shashikala T, Yogi D, Akshay K, Ashok K, Nagesha SN, Manamohan M, Jha GK, Asokan R (2025) CRISPR/Cas12a mediated rapid and efficient detection of tomato leaf curl Karnataka virus without amplification. Biocatal Agric Biotechnol 64:103528. https://doi.org/10.1016/j.bcab.2025.103528

Shi K, Xie S, Tian R, Wang S, Lu Q, Gao D, Lei C, Zhu H, Nie Z (2021) A CRISPR-Cas autocatalysis-driven feedback amplification network for supersensitive DNA diagnostics. *Sci Adv* 7(5):eabc7802. https://doi.org/10.1126/sciadv.abc7802

Swarts DC, Jinek M (2019) Mechanistic insights into the cis- and trans-acting DNase activities of Cas12a. Mol Cell 73(3):589–600.e4. https://doi.org/10.1016/j.molcel.2018.11.021

Teng F, Guo L, Cui T, Wang X-G, Xu K, Gao Q, Zhou Q, Li W (2019) CDetection: CRISPR-Cas12b-based DNA detection with sub-attomolar sensitivity and single-base specificity. Genome Biol 20(1):132. https://doi.org/10.1186/s13059-019-1742-z

Tong X, Li T, Zhang K, Zhao D, Zhang Y, Yin H (2025) Structure-guided design of Cas12a variants improves detection of nucleic acids. Cell Insight 4(2):100228. https://doi.org/10.1016/j.cellin.2025.100228

Ueda T, Shinoda H, Makino A, Yoshimura M, Iida T, Watanabe R (2023) Purification/amplification-free RNA detection platform for rapid and multiplex diagnosis of plant viral infections. Anal Chem 95(25):9680–9686. https://doi.org/10.1021/acs.analchem.3c01691

Wang YM, Ostendorf B, Gautam D, Habili N, Pagay V (2022) Plant viral disease detection: from molecular diagnosis to optical sensing technology—a multidisciplinary review. Remote Sens 14(7):1542. https://doi.org/10.3390/rs14071542

Wei L, Wang Z, She Y, Fu H (2025) CRISPR/Cas multiplexed biosensing: advances, challenges, and perspectives. Anal Chem 97(23):11943–11958. https://doi.org/10.1021/acs.analchem.4c04428

Zhang Y, Zhang Y, Xie K (2020) Evaluation of CRISPR/Cas12a-based DNA detection for fast pathogen diagnosis and GMO test in rice. Mol Breed 40(1):11. https://doi.org/10.1007/s11032-019-1092-2

Zhang W, Jiao Y, Ding C, Shen L, Li Y, Yu Y, Huang K, Li B, Wang F, Yang J (2021) Rapid detection of tomato spotted wilt virus with Cas13a in tomato and Frankliniella occidentalis. Front Microbiol 12. https://doi.org/10.3389/fmicb.2021.745173

Zhang C, Huang Z, Li Z, Hu J, Liu R, Lv Y (2024) CRISPR-associated "genetic scissors" for multiplexing analysis. TrAC Trends Anal Chem 170:117431. https://doi.org/10.1016/j.trac.2023.117431

Zou Y, Mason MG, Wang Y, Wee E, Turni C, Blackall PJ, Trau M, Botella JR (2017) Nucleic acid purification from plants, animals and microbes in under 30 seconds. PLoS Biol 15(11):e2003916. https://doi.org/10.1371/journal.pbio.2003916

# Exploitation of CRISPR-Based Tools for Rapid Detection of Vegetable Viruses

# 11

Mark Gabriel S. Sagarbarria and Mark Angelo O. Balendres

**Abstract**

Plant viruses are a threat to global vegetable production, as they infect commercial crops. The risk of plant viruses spreading to different locations increases as vegetables are transported through the supply chain. National authorities have established integrated pest management approaches and plant quarantine control systems to limit the spread of plant pathogens, including viruses. Despite these countermeasures, plant viruses may still spread, as they are easily transmitted through infected planting materials and vectors, such as insects. Hence, there must also be tools to rapidly diagnose plant materials for any viral infections. Combined with other preventive measures, early detection of viral pathogens will enable growers and national authorities to devise more effective disease prevention strategies. However, the rapid and accurate detection of vegetable viruses poses a great challenge. Available techniques may be time-consuming, expensive, have inadequate sensitivity, and/or not apply to bulk plant samples. The CRISPR/Cas system, most known for genome editing, has been used as a technique for identifying viral pathogens. It is a bacterial defense system that facilitates resistance to and interference with viruses. Due to its ability to recognize and degrade specific target nucleic acids, it has also been harnessed as a molecular diagnostic tool. This chapter presents recent innovations in using CRISPR/Cas in molecular diagnostics of viral infections in vegetables. Challenges to rapid viral detection are also discussed despite the recent advances in molecular diagnostics of plant viruses enabled by CRISPR/Cas.

M. G. S. Sagarbarria
Department of Biology, College of Arts and Sciences, University of the Philippines Manila, Metro Manila, Philippines

M. A. O. Balendres (✉)
Department of Biology, College of Science, De La Salle University, Manila, Philippines
e-mail: mark.angelo.balendres@dlsu.edu.ph



R. Kumar et al. (eds.), *Vegetable Crops and Viral Pathogens Volume I*, Plant Pathology for Sustainable Agriculture, https://doi.org/10.1007/978-981-95-7265-6_11

**Keywords**

CRISPR/Cas system · Genome editing · Pathogen diagnostics · Plant viruses

## 11.1 Introduction

Plant viruses are responsible for many devastating plant diseases in vegetables. They typically cause considerable yield losses in many vegetable crops by causing physiological disorders, resulting in diseases that reduce plant vigor and growth. Plant viruses are widespread, accounting for almost half of the observed emerging plant diseases caused by them (Anderson et al. 2004). Yield losses from viral diseases vary but can reach up to 100% depending on the crop (Rao and Reddy 2020). Effective control of plant viruses remains challenging due to their high transmissibility. Common transmission pathways include insect vectors, asexual propagation, mechanical transmission, and seed transmission from parent to offspring (García-Ordóñez and Pagán 2024). Furthermore, there is no chemical treatment for plant viral diseases besides treatments directed at vectors. A practical strategy for improving crops involves incorporating host resistance to the virus, which can be achieved through traditional plant breeding or genetic engineering. This approach is combined with prophylactic measures, such as plant quarantine legislation and vector control, to prevent the spread of viral diseases (Nicaise 2014). These preventive measures rely on early and accurate identification of infected plants, followed by their prompt removal and destruction to prevent further spread (Kalimuthu et al. 2022). Growers often rely on visual assessment to identify symptoms of viral diseases in plants. This method is the easiest and most reliable when conducted with the aid of expert plant pathologists. However, this approach can be challenging, as viral symptoms often resemble those caused by other pathogens or abiotic stresses, such as poor soil conditions, nutrient deficiencies, and adverse weather conditions. Identifying viral diseases can become even more challenging, as different virus species or virus variants can infect a single plant, a phenomenon known as mixed infections (Rubio et al. 2020; Sánchez-Tovar et al. 2025). In vegetables, an example of this is the tomato streak disease, which results from the co-infection of two viruses, potato virus X (PVX) and tobacco mosaic virus (TMV). Because of these problems, quick, accurate, and precise diagnostic tools are crucial for verifying the presence of viral diseases in the field. CRISPR/Cas genome editing technology has recently been applied for molecular diagnostics of viral diseases. Hence, this chapter presents: (i) an overview of available detection strategies used for plant viruses in vegetables, (ii) an explanation on how CRISPR/Cas technology can be used to detect plant viruses and how it can improve upon current molecular diagnostic techniques, and (iii) the remaining challenges in plant virus diagnostics even with the advancements made possible by CRISPR/Cas.

## 11.2 Detection Strategies for Plant Viruses in Vegetables

Due to the limitations of visual assessment in the field, scientists have developed molecular diagnostic tools that can detect viral particles. The different strategies are discussed here to provide the context of these techniques, leading to the advent of CRISPR/Cas. Serological techniques utilize enzyme-linked immunosorbent assays (ELISA) to detect viral proteins (antigens) in plant samples or antibodies produced by infected plants. Specifically, ELISA utilizes specific antibody-antigen reactions, resulting in high specificity and sensitivity in screening. ELISA has been used to detect numerous viral diseases in vegetables, including tomato (Jacquemond et al. 2009; Xie et al. 2013), potato (Kumar et al. 2014), cucumber (Shang et al. 2011), and cauliflower (Hekmat Kashtiban and Ertunç 2020). Most ELISA protocols utilize a 96-well microtiter plate coated with virus-specific antibodies (also known as capture antibodies). Afterward, homogenized plant tissue, typically from leaves, is added so that any possible viral antigens can bind to the capture antibodies. An enzyme-linked detection antibody is then added that binds specifically to the viral antigen. To visualize this process, the detection antibody is conjugated to an enzyme that catalyzes a colorimetric reaction upon the addition of a specific substrate. This results in a color change when a virus is present. ELISA can also detect multiple viruses from the same sample by using separate wells with different virus-specific antibodies. ELISA is a robust tool, but it has several downsides. Among these are its dependence on capture antibodies. These can be expensive to procure and require refrigerated storage, posing logistical challenges. Furthermore, it is prone to false positives due to the non-specific binding of the capture antibodies (Sakamoto et al. 2018). Before adding homogenized plant tissue, the plate must be incubated with a blocking buffer to improve assay accuracy. Furthermore, ELISA may also fail to capture all infections if more than one virus is present, as the assay is specific for each viral antigen. If viral antigens also change due to mutation, new capture antibodies must be developed. Nevertheless, ELISA has been commonly used for field detection of plant viruses, especially when diagnostic testing is done routinely and on a large scale (Xie et al. 2013). Polymerase chain reaction (PCR), and its variants, reverse transcription PCR (RT-PCR) and quantitative PCR (qPCR) are another set of commonly used molecular diagnostic tools for plant viruses. These methods enable the amplification of viral DNA copies in the plant sample, utilizing the principle of DNA replication. PCR involves multiple temperature steps and typically takes about 2–3 h, including the time it takes for detection, which is usually done using gel electrophoresis. PCR and its variants are more sensitive than ELISA for detecting plant viral infections (Mekuria et al. 2003; Peiman and Xie 2006). This makes them more effective at detecting low viral loads or latent infections. However, the process requires more specialized equipment, such as thermal cyclers and gel documentation systems, limiting the use of these techniques for on-field assays. Quantitative PCR (qPCR or qRT-PCR) enables the real-time detection of viral nucleic acids during amplification, thereby eliminating the need for post-PCR gel electrophoresis. However, these techniques require more advanced thermal cyclers that use fluorescent dyes or probes to track amplification in real-time, further restricting their use in field conditions.

PCR and its variations can be used to detect multiple viral infections using different primer sets, though this step-by-step approach can be time-consuming. To overcome this, the simultaneous detection of multiple infections can be achieved through multiplex PCR. However, multiplexing presents challenges, such as the complex design of primers and the optimization of amplification conditions (James et al. 2006). PCR is considered the gold standard for identifying viruses. However, the equipment requirements, long reaction times, and need for trained technical personnel limit its application as a rapid technique (Park 2022). An alternative nucleic acid amplification method, loop-mediated isothermal amplification (LAMP), can also detect nucleic acids without the need for thermal cyclers. LAMP is an isothermal technique, meaning DNA is amplified at a constant temperature, unlike PCR. This method utilizes Bst DNA polymerase, which exhibits strong strand displacement activity, enabling it to synthesize a new DNA strand while displacing downstream strands, thereby eliminating the need for high-temperature denaturation (Notomi et al. 2000). This enables all LAMP reactions at a constant temperature (around 60–65 °C), removing the need for cycling and allowing reactions in simple block heaters or water baths. LAMP also leverages clever primer design to eliminate the need for temperature cycling, utilizing 4–6 specially designed primers instead of the typical two in PCR (Park 2022). Inner forward and backward primers contain sequences complementary to the target's internal regions, enabling the formation of a self-hybridizing loop structure after strand displacement. After initial amplification, the copied DNA folds into a dumbbell-shaped structure due to the self-annealing of the complementary sequences, forming loop regions that facilitate further amplification. These loops serve as new priming sites, enabling rapid, exponential DNA synthesis and the generation of large amounts of product, which can be detected through various methods (e.g., fluorescence or colorimetric changes). Many of these detection reagents can be added to the reaction mix before amplification begins, eliminating the need for repeated opening of reaction tubes and reducing the risk of contamination. A typical LAMP workflow for virus detection involves the crude extraction of nucleic acids from plant tissues, followed by the addition of the extract to a reaction mix containing Bst DNA polymerase, primers, and detection dyes. This reaction is heated to 60–65 °C and can be completed within 60 min. LAMP has been developed as a field-deployable diagnostic tool for vegetable viruses due to its simplicity compared to ELISA and PCR. By using a portable fluorometer, LAMP has been successfully used to detect tomato leaf curl New Delhi virus (ToLCNDV), pepper yellow leaf curl Indonesia virus (PepYLCIV), and tomato yellow leaf curl Kanchanaburi virus (TYLCKaV) infecting *Cucurbitaceae* and *Solanaceae* plants (Wilisiani et al. 2019). Other studies have utilized a portable battery-powered thermal cycler to conduct reverse transcription loop-mediated isothermal amplification (RT-LAMP) in field conditions to detect ToLCNDV (Caruso et al. 2023) and tomato spotted wilt virus (TSWV) (Caruso et al. 2024). In field conditions, RT-LAMP was also used to detect tomato chlorotic spot virus (TCSV) using dye-based colorimetric methods (Yilmaz et al. 2023). Another advantage of LAMP in field conditions is that even crude extracts of nucleic acids may be used. Some methods can even amplify nucleic acids obtained through the toothpick method (Wilisiani et al. 2019), wherein a toothpick is used to pierce plant tissue to extract DNA, and then dipped into the LAMP mixture. Bst polymerase is even more robust than other polymerases as it can tolerate potential inhibitors found in crude extracts (Francois et al. 2011). While robust and straightforward, the main challenge with LAMP is primer design. Due to the number of primers used in LAMP (4–6) and the numerous factors to consider when designing primers, there is less flexibility in choosing a target site than with PCR. This is alleviated by software and online tools to help with LAMP primer design (Soroka et al. 2021). Furthermore, due to the number of primers used in LAMP, primer design for multiplexing poses an even greater challenge, considering the numerous primer–primer interactions.

Recombinase Polymerase Amplification (RPA) is another isothermal amplification technique. RPA is quite like LAMP in that both are robust and straightforward techniques. This method amplifies specific DNA fragments without temperature cycling, typically at a temperature range of 37–42 °C, often using two primers similar to those used in PCR. However, instead of relying on Taq polymerase, RPA combines three key enzymes: a recombinase, a strand-displacing DNA polymerase, and single-stranded binding proteins. This setup enables efficient amplification of DNA and RNA templates through strand extension, which takes about 20 min (Kumar et al. 2021; Kumar et al. 2022; Kumar et al. 2023a, b). Like LAMP, field-deployable tools for detecting vegetable viruses have been developed using RPA. Notably, the nucleic acids of tomato chlorotic dwarf viroid (TCDVd) (Hammond and Zhang 2016) and tomato yellow leaf curl virus (TYLCV) (Zhou et al. 2022) were amplified using RPA and subsequently detected using a lateral flow dipstick (LFD) assay. These studies used primers and an internal probe designed to generate an amplicon labeled with biotin and fluorophores. After amplification, the amplicon was transferred into an Amplicon Detection Chamber (AmplifyRP®) containing a lateral flow strip. The strip has antibodies that capture the labeled amplicons: One antibody binds the fluorophore label, while another binds the biotin. If the target is present, a test line and a control line appear on the strip, and if only the control line appears, the test is negative (Rigano et al. 2014). Results from RPA-LFD were observable after just 20 min, making the method rapid and field-friendly, with no need for complex instruments or technical expertise. An advantage of RPA over LAMP is its simpler primer design, which is more similar to PCR, allowing multiplex RPA assays to be done, which was successfully used to simultaneously detect three viruses in cucurbits (Jailani and Paret 2023). Previously described methods require knowledge of the viral genome. There are cases where testing for known viruses does not produce results. In instances where unaccounted symptoms are observed in the field, high-throughput sequencing methods can be employed to determine if new virus species or variants are emerging. Rubio et al. (2020) have extensively reviewed sequencing-based methods for detecting plant viruses. However, high-throughput sequencing is not used as a routine diagnostic method due to the high costs and technical expertise required. Instead, sequencing-based methods can be combined with diagnostic techniques and the expertise of plant pathologists. When more rapid diagnostic methods fail to detect plant viruses, yet symptoms are not linked to other pathogens or abiotic stresses, as verified by expert plant pathologists, sequencing-based methods may be used to pinpoint and determine the possible emergence of genetic diversity and evolution of plant viruses. High-throughput sequencing of new viruses also paves the way for the development of rapid detection tools, such as ELISA, PCR, LAMP, and RPA (Kumar et al. 2023b; Paul et al. 2025).

## 11.3 CRISPR/Cas and Its Application in Vegetable Virus Detection

### 11.3.1 Biology of CRISPR/Cas and Application in Molecular Diagnostics

The CRISPR/Cas system started as an observed biological phenomenon that acted as an adaptive immune system in bacteria (Bolotin et al. 2005). Clustered, regularly interspaced short palindromic repeats (CRISPRs) are segments in the bacterial genome that are transcribed into CRISPR RNAs (crRNAs). These crRNAs are complementary to sequences from previously encountered phages. The crRNA activates and guides Cas (CRISPR-associated) proteins, typically endonucleases, forming a ribonucleoprotein. This complex scans for complementary target sequences in invading viruses, and upon binding, the Cas protein cleaves the foreign nucleic acids, thereby neutralizing the threat. This system evolved into a gene-editing tool when it was discovered that crRNA sequences could be reprogrammed to direct Cas proteins to virtually any DNA target (Jinek et al. 2012). Numerous Cas proteins are used for genome engineering (reviewed by Hillary and Ceasar (2023)). The Cas9 is the most commonly used for gene editing. At the same time, other Cas variants, such as Cas12a, offer greater target flexibility, and Cas12b/Cas13 enable the targeting of RNA. These diverse capabilities have expanded CRISPR/Cas from a gene-editing tool into a powerful platform for molecular diagnostics, particularly in viral detection (Srivastava et al. 2020). Notably, platforms such as HOLMES (Li et al. 2019), SHERLOCK (Kellner et al. 2019), and DETECTR (Chen et al. 2018), which were initially developed for detecting human viral pathogens, have been successfully adapted for plant virus diagnostics (Aman et al. 2020; Paul et al. 2025). These tools harness the *trans* cleavage activity of Cas12 and Cas13. These Cas proteins become activated when they bind to their specific target nucleic acid (DNA for Cas12a, DNA/RNA for Cas12b, RNA for Cas13). Once activated, they begin to cleave nearby single-stranded DNA or RNA molecules non-specifically. Fluorescent reporter probes are included in the detection assay, which are cleaved by Cas proteins after target recognition, resulting in a signal that indicates the presence of the virus.

## 11.3.2 CRISPR/Cas Technology Improves Diagnostics

### 11.3.2.1 Specificity

CRISPR/Cas technology is highly specific. Aside from needing the crRNA to have sufficient complementation to the target region, some Cas proteins (Cas9 and Cas12) must also recognize a sequence known as the protospacer adjacent motif (PAM) for endonuclease activity to occur. The PAM immediately follows the target region, and each Cas protein has a different PAM requirement. Sufficient complementation of the crRNA to the target and the PAM requirement improves the specificity of isothermal techniques, such as the LAMP and RPA assays. LAMP uses 4–6 primers, which increases the chances of mispriming or primer–primer interactions and has been previously noted to exhibit non-specific background amplification at later times, even with careful assay design (Jang and Kim 2022). RPA, on the other hand, reacts at low temperatures (37–42 °C), which may cause primers to partially bind with partial complementarity, potentially amplifying off-targets (Daher et al. 2015; Lobato and O'Sullivan 2018). Furthermore, both LAMP and RPA may broadly detect a virus species. However, they may not be able to distinguish between variants within a species or different pathotypes. For example, an RT-LAMP protocol to detect tomato spotted wilt virus (TSWV) (Caruso et al. 2024; Paul et al. 2021) may generally detect TSWV but may not differentiate between resistance-breaking (RB) and non-RB isolates. Differentiating between virus variants can be important, as in this case, the RB isolates of TSWV are known to overcome host resistance in tomato due to only two single-nucleotide mutations (Batuman et al. 2017). Typically, techniques such as RT-PCR and sequencing are required to detect differences of only two nucleotides between samples. However, CRISPR/Cas12a has enabled specific detection of a TSWV-RB strain that first appeared in California (Shymanovich et al. 2024).

Their protocol first used RT-LAMP to amplify TSWV nucleic acids, followed by an RB-specific LAMP/Cas12a assay to detect the two nucleotide insertions. Furthermore, separate crRNAs were designed for the detection of RB-specific and wild-type-specific targets. Once the crRNA matched its target sequence, the Cas12a enzyme activated and non-specifically cleaved fluorescent ssDNA reporters present in the reaction mixture, producing a signal detected via a qPCR machine (Fig. 11.1). The use of Cas12a can be adapted to detect other plant viruses that have similar variants resulting from a few nucleotide changes, or perhaps closely related viruses that infect the same plants. Cas12 has also already been used to differentially detect tomato mosaic virus (ToMV) and tomato brown rugose fruit virus (ToBRFV) after amplification using RT-PCR (Alon et al. 2021).

### 11.3.2.2 Multiplex Capability

CRISPR/Cas technology is also capable of simultaneously detecting multiple viruses in one sample, which is particularly useful for identifying mixed infections. Specifically, Cas12a has been successfully used to identify Tobacco plants that had been infected with up to three viruses: tobacco mosaic virus, tobacco etch virus and potato virus X (Marqués et al. 2022). The study amplified each virus using

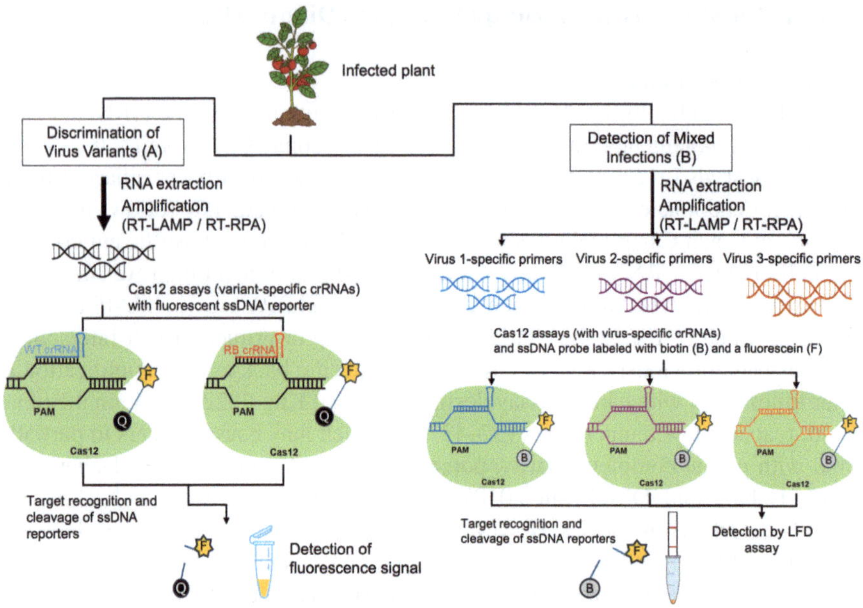


**Fig. 11.1** Applications of Cas12 in improving plant virus diagnostics in combination with LAMP or RPA. (**a**) Enhanced sensitivity of viral diagnostics using RT-LAMP or RT-RPA for amplifying viral nucleic acids, followed by a CRISPR/Cas12a assay, allowing differentiation of resistance-breaking (RB) versus wild-type (WT) variants of TSWV. Target recognition by Cas12a results in the cleavage of fluorescent ssDNA reporter probes, separating quencher molecules (Q) from the fluorophore (F), resulting in a detectable fluorescent signal. (**b**) Multiplex application of CRISPR/Cas technology by simultaneously detecting multiple viruses in mixed infections using a Cas12 assay after amplification via RT-LAMP or RT-RPA. Detection of collateral cleavage of ssDNA reporters labeled with biotin (B) and a fluorophore (F) can be done in the field using the LFD assay

virus-specific primers, generating up to three fragments. The amplified DNA was incubated with Cas12a and a fluorescent reporter probe. The assay was performed on a 96-well plate, with each well containing a different crRNA specific for each virus. The study observed fluorescence signals, measured in a qPCR machine, for each type of infection (single, double, or triple). Furthermore, their protocol was adapted to be field-deployable, utilizing detection via a lateral flow dipstick, which employed an ssDNA probe labeled with both a fluorophore and biotin (Fig. 11.1). This study has demonstrated that the concurrent detection of multiple viral pathogens in suspected mixed infections can be significantly accelerated by leveraging the programmability of crRNAs, enabling multiplex detection without the need for complex equipment or specialized technical personnel.

#### 11.3.2.3 Potential for Amplification-Free Detection

CRISPR/Cas may detect viruses by targeting the viral genome directly, rather than relying on amplified products. In the case of RNA viruses, this approach leverages the ability of Cas13 to target RNA directly. In the study of Marqués et al. (2022),

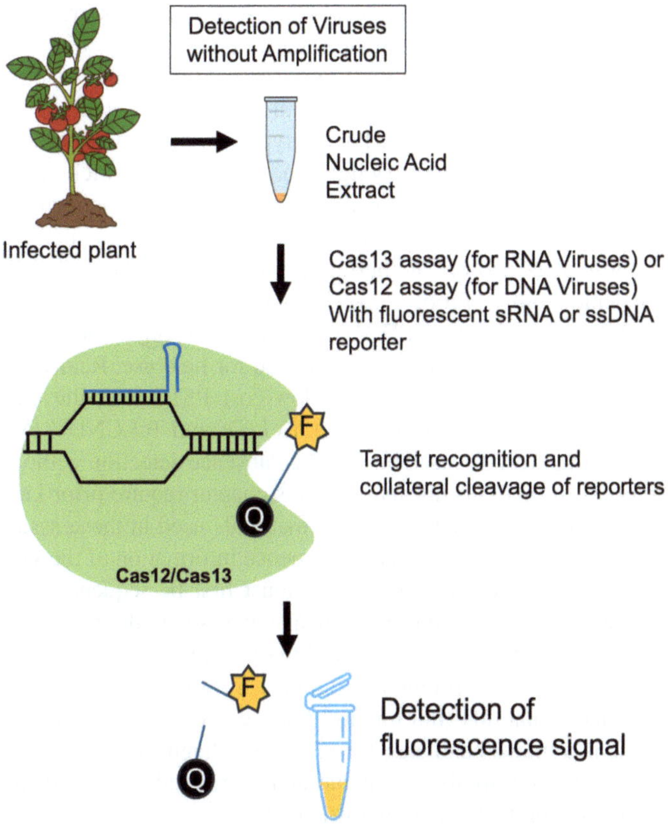


**Fig. 11.2** Detection of plant viruses, either RNA or DNA-based, using CRISPR/Cas13 or CRISPR/Cas12 assays without amplification

Cas13a/d was used to detect the tobacco mosaic virus, tobacco etch virus, and potato virus X in tobacco plants, even in crude extracts. Using fluorescent sRNAs, fluorescent signals were detected in crude extracts added with the crRNACas13 complex, signifying virus detection. It was observed that fluorescence signals were substantially different in infected plants compared to control plants, indicating that this method could detect plant viruses without the need for amplification. In the case of DNA viruses, detection of tomato leaf curl Karnataka virus (ToLCKV) through a CRISPR/Cas12a assay without an amplification step has also been reported (Shashikala et al. 2025). A general overview of this process is shown in Fig. 11.2.

#### 11.3.2.4 Remaining Challenges in Virus Diagnostics in Vegetables

The above examples show that CRISPR/Cas technology shows great promise as a tool for rapid and field-deployable vegetable virus diagnostics, especially when combined with LAMP and RPA. These technologies enable highly sensitive, specific, and fast assays. However, quantifying viral load in field conditions remains challenging when using these techniques. Quantitative assessment of viral titer is important for evaluating plant resistance, understanding the variability of virus accumulation in different organs, monitoring seasonal dynamics of infection, and identifying co-infections with multiple viral genotypes (Mehle et al. 2020). Detection methods for LAMP and RPA are typically qualitative, utilizing colorimetric changes, fluorescence under UV light, and lateral flow dipsticks. Modifications to LAMP and RPA have been developed to enable real-time monitoring of amplification (Mori et al. 2004; Piepenburg et al. 2006). However, these methods still rely on bulky turbidimeters and qPCR thermal cyclers (Papadakis et al. 2022), which hinder field deployability. Proof-of-concept studies have explored smartphone-based RT-LAMP monitoring for TSWV (Paul et al. 2021; Shymanovich et al. 2024). However, these approaches are not yet optimized for field use. Real-time detection of Cas12a activity was reported using a low-cost P51 Molecular Fluorescence Viewer in a LAMP-Cas12a assay to detect TYLCV and ToLCNDV (Mahas et al. 2021). However, this work focused on presence/absence detection without quantifying viral titers. Most methods discussed in this chapter require prior knowledge of the plant virus being diagnosed. To design materials used in these assays, such as crRNAs, primers, and probes, molecular sequence information of the virus must be available. As such, new viruses or variants must first be sequenced and studied, which may be expensive and require technical expertise. Aside from technical limitations, another challenge is the reproducibility of results from these assays. Ovesná et al. (2023) highlighted the problems with data from LAMP assays, wherein few studies report on repeatability (intra-assay variance) and reproducibility (inter-assay variance). This is a major challenge for molecular diagnostics in general, as published assays should be highly repeatable and reproducible, should they ever be adopted as routine, point-of-care diagnostic methods.

## 11.4 Conclusion

Molecular diagnostics for plant viruses can be performed rapidly with enhanced specificity and flexibility, thanks to the advent of CRISPR/Cas genome editing. CRISPR/Cas, in combination with isothermal amplification techniques and simple detection tools such as LFD, has simplified diagnostic protocols while offering high sensitivity, specificity, multiplex potential, and rapid turnaround times. Furthermore, qualitative assays that focus on determining the presence or absence of a viral pathogen are highly likely to be field-deployable and suited for use by non-technical personnel. However, these methods can currently be modified to be only semi-quantitative at best. Should precise quantification of viral be required for applications such as in resistance screening or monitoring seasonal changes, more sophisticated techniques such as qPCR should still be employed. Although CRISPR/Cas has improved diagnostic techniques used for plant virus detection, the greatest challenge remains that these technologies must be validated and standardized before they can be integrated into routine diagnostic workflows. Moreover, these tools should be compared with conventional gold standards, such as ELISA and qPCR. Another obstacle lies in scaling these technologies for widespread use. Currently, there is no commercially available kit using CRISPR/Cas available based on https://platform.smartprotect-h2020.eu/en (accessed June 17, 2025). Bridging this translational gap will require collaboration between plant pathologists, molecular biologists, engineers, and extension services to co-develop platforms tailored for field deployment in vegetable production systems.

## References


Alon DM, Hak H, Bornstein M, Pines G, Spiegelman Z (2021) Differential detection of the Tobamoviruses Tomato Mosaic Virus (ToMV) and Tomato Brown Rugose Fruit Virus (ToBRFV) using CRISPR-Cas12a. Plants 10(6):1256. https://doi.org/10.3390/plants10061256

Aman R, Mahas A, Marsic T, Hassan N, Mahfouz MM (2020) Efficient, rapid, and sensitive detection of plant RNA viruses with one-pot RT-RPA–CRISPR/Cas12a assay. Front Microbiol 11:610872. https://doi.org/10.3389/fmicb.2020.610872

Anderson PK, Cunningham AA, Patel NG, Morales FJ, Epstein PR, Daszak P (2004) Emerging infectious diseases of plants: pathogen pollution, climate change and agrotechnology drivers. Trends Ecol Evol 19(10):535–544. https://doi.org/10.1016/j.tree.2004.07.021

Batuman O, Turini TA, Oliveira PV, Rojas MR, Macedo M, Mellinger HC, Adkins S, Gilbertson RL (2017) First report of a resistance-breaking strain of tomato spotted wilt virus infecting tomatoes with the Sw-5 Tospovirus-resistance gene in California. Plant Dis 101(4):637–637. https://doi.org/10.1094/PDIS-09-16-1371-PDN

Bolotin A, Quinquis B, Sorokin A, Dusko Ehrlich S (2005) Clustered regularly interspaced short palindrome repeats (CRISPRs) have spacers of extrachromosomal origin. Microbiology 151(8):2551–2561. https://doi.org/10.1099/mic.0.28048-0

Caruso AG, Ragona A, Bertacca S, Montoya MAM, Panno S, Davino S (2023) Development of an in-field real-time LAMP assay for rapid detection of tomato leaf curl New Delhi virus. Plants 12(7):1487. https://doi.org/10.3390/plants12071487

Caruso AG, Ragona A, Agrò G, Bertacca S, Yahyaoui E, Galipienso L, Rubio L, Panno S, Davino S (2024) Rapid detection of tomato spotted wilt virus by real-time RT-LAMP and in-field application. J Plant Pathol 106(2):697–712. https://doi.org/10.1007/s42161-024-01613-3

Chen JS, Ma E, Harrington LB, Da Costa M, Tian X, Palefsky JM, Doudna JA (2018) CRISPR-Cas12a target binding unleashes indiscriminate single-stranded DNase activity. Science 360(6387):436–439. https://doi.org/10.1126/science.aar6245

Daher RK, Stewart G, Boissinot M, Boudreau DK, Bergeron MG (2015) Influence of sequence mismatches on the specificity of recombinase polymerase amplification technology. Mol Cell Probes 29(2):116–121. https://doi.org/10.1016/j.mcp.2014.11.005

Francois P, Tangomo M, Hibbs J, Bonetti E-J, Boehme CC, Notomi T, Perkins MD, Schrenzel J (2011) Robustness of a loop-mediated isothermal amplification reaction for diagnostic applications. FEMS Immunol Med Microbiol 62(1):41–48. https://doi.org/10.1111/j.1574-695X.2011.00785.x

García-Ordóñez L, Pagán I (2024) Vertical and horizontal transmission of plant viruses: two extremes of a continuum? Npj Viruses 2(1):18. https://doi.org/10.1038/s44298-024-00030-8

Hammond RW, Zhang S (2016) Development of a rapid diagnostic assay for the detection of tomato chlorotic dwarf viroid based on isothermal reverse-transcription-recombinase polymerase amplification. J Virol Methods 236:62–67. https://doi.org/10.1016/j.jviromet.2016.06.013

Hekmat Kashtiban A, Ertunç F (2020) Serological and molecular detection of Cauliflower mosaic virus and its vectors infecting cold-season vegetables in Ankara province. Bitki Koruma Bülteni 60(3):64–71. https://doi.org/10.16955/bitkorb.694019

Hillary VE, Ceasar SA (2023) A review on the mechanism and applications of CRISPR/Cas9/Cas12/Cas13/Cas14 proteins utilized for genome engineering. Mol Biotechnol 65(3):311–325. https://doi.org/10.1007/s12033-022-00567-0

Jacquemond M, Verdin E, Dalmon A, Guilbaud L, Gognalons P (2009) Serological and molecular detection of Tomato chlorosis virus and Tomato infectious chlorosis virus in tomato. Plant Pathol 58(2):210–220. https://doi.org/10.1111/j.1365-3059.2008.01959.x

Jailani AAK, Paret ML (2023) Development of a multiplex RT-RPA assay for simultaneous detection of three viruses in cucurbits. Mol Plant Pathol 24(11):1443–1450. https://doi.org/10.1111/mpp.13380

James D, Varga A, Pallas V, Candresse T (2006) Strategies for simultaneous detection of multiple plant viruses. Can J Plant Pathol 28(1):16–29. https://doi.org/10.1080/07060660609507267

Jang M, Kim S (2022) Inhibition of non-specific amplification in loop-mediated isothermal amplification via tetramethylammonium chloride. Biochip J 16(3):326–333. https://doi.org/10.1007/s13206-022-00070-3

Jinek M, Chylinski K, Fonfara I, Hauer M, Doudna JA, Charpentier E (2012) A programmable dual-RNA-guided DNA endonuclease in adaptive bacterial immunity. Science 337(6096):816–821. https://doi.org/10.1126/science.1225829

Kalimuthu K, Arivalagan J, Mohan M, Christyraj JRSS, Arockiaraj J, Muthusamy R, Ju H-J (2022) Point of care diagnosis of plant virus: current trends and prospects. Mol Cell Probes 61:101779. https://doi.org/10.1016/j.mcp.2021.101779

Kellner MJ, Koob JG, Gootenberg JS, Abudayyeh OO, Zhang F (2019) SHERLOCK: nucleic acid detection with CRISPR nucleases. Nat Protoc 14(10):2986–3012. https://doi.org/10.1038/s41596-019-0210-2

Kumar R, Jeevalatha A, Sharma NN, Sharma S, Chakrabarti SK, Singh BP (2014) Development of PCR based methods for detection of Potato aucuba mosaic virus in India. Potato J 41:166–174

Kumar R, Kaundal P, Tiwari RK, Sundaresha S, Kumari H, Naga KC, Sharma S, Kumar M (2021) Rapid and sensitive detection of potato virus X by one-step reverse transcription-recombinase polymerase amplification method in potato leaves and dormant tubers. Mol Cell Probes 58:101743. https://doi.org/10.1016/j.mcp.2021.101743

Kumar R, Kaundal P, Tiwari RK, Sundaresha S, Kumari H, Lal MK, Naga KC, Sharma S, Sagar V, Kumar M (2022) Establishment of a one-step reverse transcription recombinase polymerase amplification assay for the detection of potato virus S. J Virol Methods 307:114568. https://doi.org/10.1016/j.jviromet.2022.114568

Kumar R, Kaundal P, Tiwari RK, Lal MK, Kumari H, Kumar R, Naga KC, Kumar A, Singh B, Sagar V, Sharma S (2023a) Development of reverse transcription recombinase polymerase amplification (RT-RPA): a methodology for quick diagnosis of potato leafroll viral disease in potato. Int J Mol Sci 24:2511. https://doi.org/10.3390/ijms24032511

Kumar R, Kaundal P, Tiwari RK, Lal MK, Kumari H, Kumar R, Sagar V, Singh B (2023b) Optimization of a simple, low-cost one-step reverse transcription recombinase polymerase amplification method for real-time detection of potato virus a in potato leaves and tubers. 3 Biotech 13:373. https://doi.org/10.1007/s13205-023-03791-w

Li L, Li S, Wu N, Wu J, Wang G, Zhao G, Wang J (2019) HOLMESv2: a CRISPR-Cas12b-assisted platform for nucleic acid detection and DNA methylation quantitation. ACS Synth Biol 8(10):2228–2237. https://doi.org/10.1021/acssynbio.9b00209

Lobato IM, O'Sullivan CK (2018) Recombinase polymerase amplification: basics, applications and recent advances. TrAC Trends Anal Chem 98:19–35. https://doi.org/10.1016/j.trac.2017.10.015

Mahas A, Hassan N, Aman R, Marsic T, Wang Q, Ali Z, Mahfouz MM (2021) Lamp-coupled crispr–cas12a module for rapid and sensitive detection of plant dna viruses. Viruses 13(3):466. https://doi.org/10.3390/v13030466

Marqués M-C, Sánchez-Vicente J, Ruiz R, Montagud-Martínez R, Márquez-Costa R, Gómez G, Carbonell A, Daròs J-A, Rodrigo G (2022) Diagnostics of infections produced by the plant viruses TMV, TEV, and PVX with CRISPR-Cas12 and CRISPR-Cas13. ACS Synth Biol 11(7):2384–2393. https://doi.org/10.1021/acssynbio.2c00090

Mehle N, Gregur L, Bogožalec Košir A, Dobnik D (2020) One-step reverse-transcription digital PCR for reliable quantification of different Pepino mosaic virus genotypes. Plants 9(3):326. https://doi.org/10.3390/plants9030326

Mekuria G, Ramesh SA, Alberts E, Bertozzi T, Wirthensohn M, Collins G, Sedgley M (2003) Comparison of ELISA and RT-PCR for the detection of Prunus necrotic ring spot virus and prune dwarf virus in almond (Prunus dulcis). J Virol Methods 114(1):65–69. https://doi.org/10.1016/j.jviromet.2003.08.014

Mori Y, Kitao M, Tomita N, Notomi T (2004) Real-time turbidimetry of LAMP reaction for quantifying template DNA. J Biochem Biophys Methods 59(2):145–157. https://doi.org/10.1016/j.jbbm.2003.12.005

Nicaise V (2014) Crop immunity against viruses: outcomes and future challenges. Front Plant Sci 5(NOV):660. https://doi.org/10.3389/fpls.2014.00660

Notomi T, Okayama H, Masubuchi H, Yonekawa T, Watanabe K, Amino N, Hase T (2000) Loop-mediated isothermal amplification of DNA. Nucleic Acids Res 28(12):e63

Ovesná J, Kaminiaris MD, Tsiropoulos Z, Collier R, Kelly A, De Mey J, Pollet S (2023) Applicability of Smart Tools in Vegetable Disease Diagnostics. Agronomy, 13(5):1211

Papadakis G, Pantazis AK, Fikas N, Chatziioannidou S, Tsiakalou V, Michaelidou K, Pogka V, Megariti M, Vardaki M, Giarentis K, Heaney J, Nastouli E, Karamitros T, Mentis A, Zafiropoulos A, Sourvinos G, Agelaki S, Gizeli E (2022) Portable real-time colorimetric LAMP-device for rapid quantitative detection of nucleic acids in crude samples. Sci Rep 12(1):3775. https://doi.org/10.1038/s41598-022-06632-7

Park J-W (2022) Principles and applications of loop-mediated isothermal amplification to point-of-care tests. Biosensors 12(10):857. https://doi.org/10.3390/bios12100857

Paul R, Ostermann E, Chen Y, Saville AC, Yang Y, Gu Z, Whitfield AE, Ristaino JB, Wei Q (2021) Integrated microneedle-smartphone nucleic acid amplification platform for in-field diagnosis of plant diseases. Biosens Bioelectron 187:113312. https://doi.org/10.1016/j.bios.2021.113312

Paul S, Emmadi V, Saxena S, Sarkar M, Mandal B, Kumar R, Sinha P, Roy A (2025) RPA-assisted CRISPR-Cas12a-enabled point-of-care diagnostic platform for chili leaf curl virus with fluorescent and colorimetric readouts. Front Microbiol 16:1644322. https://doi.org/10.3389/fmicb.2025.1644322

Peiman M, Xie C (2006) Sensitive detection of potato viruses, PVX, PLRV and PVS, by RT-PCR in potato leaf and tuber. Aust Plant Dis Notes 1(1):41. https://doi.org/10.1071/DN06017

Piepenburg O, Williams CH, Stemple DL, Armes NA (2006) DNA detection using recombination proteins. PLoS Biol 4(7):e204. https://doi.org/10.1371/journal.pbio.0040204

Rao GP, Reddy MG (2020) Overview of yield losses due to plant viruses. In: Awasthi LP (ed) Applied plant virology. Elsevier, pp 531–562. https://doi.org/10.1016/B978-0-12-818654-1.00038-4

Rigano LA, Malamud F, Orce IG, Filippone MP, Marano MR, do Amaral AM, Castagnaro AP, Vojnov AA (2014) Rapid and sensitive detection of Candidatus Liberibacter asiaticus by loop mediated isothermal amplification combined with a lateral flow dipstick. BMC Microbiol 14(1):86. https://doi.org/10.1186/1471-2180-14-86

Rubio L, Galipienso L, Ferriol I (2020) Detection of plant viruses and disease management: relevance of genetic diversity and evolution. Front Plant Sci 11:1092. https://doi.org/10.3389/fpls.2020.01092

Sakamoto S, Putalun W, Vimolmangkang S, Phoolcharoen W, Shoyama Y, Tanaka H, Morimoto S (2018) Enzyme-linked immunosorbent assay for the quantitative/qualitative analysis of plant secondary metabolites. J Nat Med 72(1):32–42. https://doi.org/10.1007/s11418-017-1144-z

Sánchez-Tovar MR, Rivera-Bustamante RF, Saavedra-Trejo DL, Guevara-González RG, Torres-Pacheco I (2025) Mixed plant viral infections: complementation, interference and their effects, a review. Agronomy 15(3):620. https://doi.org/10.3390/agronomy15030620

Shang H, Xie Y, Zhou X, Qian Y, Wu J (2011) Monoclonal antibody-based serological methods for detection of Cucumber green mottle mosaic virus. Virol J 8(1):228. https://doi.org/10.1186/1743-422X-8-228

Shashikala T, Yogi D, Akshay K, Ashok K, Nagesha SN, Manamohan M, Jha GK, Asokan R (2025) CRISPR/Cas12a mediated rapid and efficient detection of Tomato Leaf Curl Karnataka Virus without amplification. Biocatal Agric Biotechnol 64:103528. https://doi.org/10.1016/j.bcab.2025.103528

Shymanovich T, Saville AC, Mohammad N, Wei Q, Rasmussen D, Lahre KA, Rotenberg D, Whitfield AE, Ristaino JB (2024) Disease progress and detection of a California resistance-breaking strain of tomato spotted wilt virus in tomato with LAMP and CRISPR-Cas12a assays. PhytoFrontiers™ 4(1):50–60. https://doi.org/10.1094/PHYTOFR-05-23-0058-FI

Soroka M, Wasowicz B, Rymaszewska A (2021) Loop-mediated isothermal amplification (LAMP): the better sibling of PCR? Cells 10(8):1931. https://doi.org/10.3390/cells10081931

Srivastava S, Upadhyay DJ, Srivastava A (2020) Next-generation molecular diagnostics development by CRISPR/Cas tool: rapid detection and surveillance of viral disease outbreaks. Front Mol Biosci 7:582499. https://doi.org/10.3389/fmolb.2020.582499

Wilisiani F, Tomiyama A, Katoh H, Hartono S, Neriya Y, Nishigawa H, Natsuaki T (2019) Development of a LAMP assay with a portable device for real-time detection of begomoviruses under field conditions. J Virol Methods 265:71–76. https://doi.org/10.1016/j.jviromet.2018.10.005

Xie Y, Jiao X, Zhou X, Liu H, Ni Y, Wu J (2013) Highly sensitive serological methods for detecting tomato yellow leaf curl virus in tomato plants and whiteflies. Virol J 10(1):142. https://doi.org/10.1186/1743-422X-10-142

Yilmaz S, Adkins S, Batuman O (2023) Field-portable, rapid, and low-cost RT-LAMP assay for the detection of tomato chlorotic spot virus. Phytopathology 113(3):567–576. https://doi.org/10.1094/PHYTO-08-22-0319-R

Zhou Y, Zheng HY, Jiang DM, Liu M, Zhang W, Yan JY (2022) A rapid detection of tomato yellow leaf curl virus using recombinase polymerase amplification-lateral flow dipstick assay. Lett Appl Microbiol 74(5):640–646. https://doi.org/10.1111/lam.13611

# The Modern Era of On-Site Diagnostic Kits for Virus Detection in Vegetable Crops

12

Varshini Pillai and A. Swapna Geetanjali

**Abstract**

Vegetable crops account for about 4–5% of global agricultural production volume and are indispensable for the agricultural economy. Like any other plant, these crops are also vulnerable to a variety of viral infections. Unfortunately, treating these viral diseases remains a major challenge with no specific cure. Once infected, plants typically show symptoms such as leaf distortion, stunted growth, lower yield, and eventually crop failure, resulting in significant financial losses to the farmers. Early and regular diagnosis is therefore essential for maintaining healthy crops. Recently, on-site diagnostic kits have emerged as a promising tool for quickly identifying viral diseases right in the field. Traditional laboratory-based techniques such as ELISA and PCR are highly accurate but require specialized equipment and time-consuming procedures. In contrast, on-site kits based on techniques like lateral flow assays, loop-mediated isothermal amplification (LAMP), recombinase polymerase amplification (RPA), and CRISPR offer faster results with similar sensitivity and specificity. These methods also minimize sample handling and can be used effectively even in lower-resource settings like agricultural farms. This chapter delves into how the latest on-site diagnostic tools, such as LAMP and CRISPR-based kits, work in detecting plant viruses, reviews real-world uses on farms, current limitations, and promising innovations pushing field detection forward.

V. Pillai · A. Swapna Geetanjali (✉)
Department of Genetic Engineering, SRM Institute of Science and Technology, Kattankulathur, Tamil Nadu, India
e-mail: swapnaga@srmist.edu.in


R. Kumar et al. (eds.), *Vegetable Crops and Viral Pathogens Volume I*, Plant Pathology for Sustainable Agriculture, https://doi.org/10.1007/978-981-95-7265-6_12

**Keywords**

On-site diagnostics · Lateral flow assay · Recombinase polymerase assay · CRISPR diagnostics · Rapid detection

## 12.1 Introduction

Vegetables are one of the important and nutritious sources of vitamins, minerals, and dietary fibers. Usually short-duration and cultivated in all seasons throughout the year, vegetable crops are highly profitable to farmers. Among the highest producers of vegetable crops in the world, Asia ranks at one with 40% of the total vegetable production. China and India are the leading producers of vegetables in Asia, with an annual production of 588.26 million metric tons and 132 million metric tons (Raveena et al. 2022). Like any crop, vegetable crops are also susceptible to diseases and biotic and abiotic stresses. One of the greatest challenges faced by growers today is the devastating impact of plant viruses. Unlike fungal or bacterial pathogens, viruses cannot be controlled with chemical treatments once a plant is infected. It starts to spread through seeds, insect vectors, or contaminated tools, often reducing yields and market quality by alarming proportions. This leads to detrimental losses to the agriculturalists and farmers who expect higher economic returns from vegetable crops. Traditionally, approaches to managing viral diseases include crop rotation, where different crops are cultivated in an alternating or circulate manner, using pest/virus-resistant varieties, and methods to control vectors, such as insecticides. As efficient as these methods are, the disease spreads, and infection remains largely uncontrollable (Pandit et al. 2022). Apart from the control of disease, the diagnosis of viral diseases also remains constrained due to the variable resources and manpower required (Tatineni and Hein 2023). In an era of modernization and rapidly evolving technology, there is a growing need for faster and simpler ways to detect plant viruses directly in the field. Traditional methods include laboratory-based techniques such as enzyme-linked immunosorbent assay (ELISA) and polymerase chain reaction (PCR), which are considered gold standards due to their sensitivity and specificity (Mosolygó et al. 2022). These assays also come with limitations such as cost, time-consuming, and not always accessible to farmers in rural or resource-limited areas. This is where the modern on-site diagnostic kits step in for easier diagnosis. These diagnostic kits are similar to pathogen diagnostic kits for human or zoonotic viral diseases that are currently in use, such as the rapid COVID detection kits. These diagnostic kits allow direct, on-field detection of the virus within minutes and use the latest molecular techniques such as recombinase polymerase amplification (RPA), lateral flow assays, loop-mediated isothermal amplification (LAMP), and CRISPR-based detection platforms (Bhat et al. 2022; Gong et al. 2025). The major advantage here is that these kits not only allow simple and efficient detection of viral diseases but also allow for continuous early detection even before visible symptoms appear, unlike traditional methods, where only upon symptoms, the crop is tested. Plant viruses spread invisibly during the early stages of infection, with symptoms either delayed or confused with nutrient deficiencies, abiotic stress, or other pathogens. By the time visual signs appear, not only is there a way to save the plant, but also to prevent significant secondary spread through vectors such as whiteflies, aphids, or thrips. Early and accurate detection enables farmers to remove infected plants, adjust vector control strategies, and prevent contaminated planting material from being disposed of or sterilized properly (Kanapiya et al. 2024). Looking at the bigger picture, tracking these viruses early helps track outbreaks across regions and even globally. Given the current climatic changes and how they have impacted farming and the agricultural economy, such data on early diagnosis and outbreak patterns would largely help the community. With cutting-edge science being brought straight to the field, it would help in the continuous flow of farming and give farmers the timely intel they need to stay ahead of infection.

## 12.2 Economic Impact of Viral Diseases in Vegetables

Worldwide, viral infections in plants total to tens of billions of dollars in loss per year (Tatineni and Hein 2023). For vegetables alone, the impacts can be detrimental through a reduction in yield and quality, which tends to make the produce unsellable. Several hundred plant viruses affect vegetative crops. Susceptible vegetable crops include Solanaceous plants such as tomato, pepper, eggplant, and Cucurbitaceous plants such as cucumber, squash, and melon (Hančinský et al. 2020). In contrast to other pathogens, plant viruses are not curable by chemical treatment once infection is established. They are widely spread via insect vectors like whiteflies, aphids, and thrips. Key examples for devastating plant viruses include tomato yellow leaf curl virus (TYLCV), which significantly lowers tomato yields (Li et al. 2022); cucumber mosaic virus (CMV), which attacks a broad host range of cucurbits, peppers, and leafy greens; and chilli leaf curl virus (ChiLCV) affecting chilli peppers largely (Nalla et al. 2023). Viral diseases can lower vegetable yields by 20–80%, depending on the crop and severity of the outbreak (Tatineni and Hein 2023). Apart from yield loss, viruses also induce deformation, discoloration, and stunting, resulting in produce that is unsuitable for consumer or export purposes (Jiang and Zhou 2023). Since prevention and early detection are the only viable control options, viruses in vegetables are not only a biological problem but also a socio-economic constraint to producers worldwide. It impacts not only the immediate front (monetary expenses of farmers) but also affects the supply chains, disrupts market prices, and endangers farmer livelihoods. To individual farmers, whose profit margins are thin, a single viral outbreak causes years of debt or financial ruin.

## 12.3 Previous Techniques Employed and Limitations of Conventional Laboratory-Based Diagnostics

Viral disease detection in plants has evolved together with the latest technological innovations of that time. Previously, researchers primarily relied on serological assays such as ELISA, first popularized in the late 1970s (Clark and Adams 1977).

Clark & Adams described how the microplate ELISA method, using enzyme-labeled antibodies, detected viruses in both purified samples and crude extracts from infected plants, thus establishing a foundational tool for routine virus screening. Over time, further precision was developed with techniques like DAS-ELISA and magnetic microsphere enzyme immunoassays, which provided specificity and throughput. Yet, these serology-based techniques have their limits with specificity, where the detection depends on the antibody quality and the diversity of the viruses. Simultaneously, nucleic acid detection methods emerged, beginning with nucleic acid spot hybridization (NASH) (Maule et al. 1983), which used DNA/RNA probes to bind viral genetic material. This offered improved sensitivity over serological tests. Though these techniques were more sensitive than serology and were able to identify viral genomes even at lower concentrations, they came with a downside of time consumption and intense labor. In the following decade, PCR revolutionized the entire field of biotechnology (National Laboratory of Enteric Pathogens 1991). Disease diagnosis leveraged the relatively simple, traditional PCR and RT-PCR techniques to detect DNA and RNA viruses with high specificity and sensitivity (Singh 1998). Further, multiplex RT-PCR and nested PCR improved the system by allowing detection of several viruses simultaneously (Bertolini et al. 2001). The shift to real-time qPCR with fluorescent probes such as TaqMan provided not only detection but viral load quantification with reproducibility and lower risk of contamination (Jeevalatha et al. 2015). Likewise, detection was made possible with the development of microarray technologies in the 1990s, where a diagnostic chip could test for several viral signatures at once (Wang et al. 2002). In the 2000s, high-throughput sequencing (HTS) or next-generation sequencing (NGS) became the gold standard of diagnosis. This revolutionized virus detection by making it possible to sequence all viral sequences in a sample regardless of whether they are novel or previously identified (Maree et al. 2018). HTS processes typically involve library preparation (fragmentation, adapter ligation), sequencing on platforms such as Illumina, Oxford Nanopore, or PacBio, and bioinformatics tools to sort through the sheer data for viral signatures (Kutnjak et al. 2021). HTS offers the highest sensitivity and unbiased detection, including for new viruses, but remains limited by high expense, long library preparation, and bioinformatics analysis time, and demands for sophisticated infrastructure and expertise, making it inaccessible for routine on-site crop surveillance (Massart et al. 2014). In total, viral plant diagnostics have evolved from observation of symptomatic appearance historically in the earlier periods, to serological (ELISA, immunoassays), to nucleic acid hybridization (NASH), to PCR-based assays (conventional, multiplex, nested, qPCR), to microarrays, and finally to HTS/NGS as laboratory-based detection techniques (Kumar et al. 2021, 2022, 2023b; Jeevalatha et al. 2018; Raigond et al. 2022). Although increasingly becoming more sensitive and specific, each had substantial limitations that hindered their use in practical applications, especially under field conditions.

## 12.4 Current Developments in On-Site Diagnostics Kits for Virus Detection

Recent advancements in molecular biology have led to the development of several on-site diagnostic platforms that are portable, user-friendly, and allow on-field diagnostics. These on-site diagnostic platforms represent a significant advancement in the field of plant virology, offering rapid, reliable, and cost-effective solutions for detecting viral infections in vegetable crops. By enabling early detection, these technologies help farmers to intervene in a timely and possibly reduce the spread of diseases, minimizing crop losses. As these technologies continue to evolve, they hold the promise of transforming plant disease management, making it more proactive and accessible to farmers worldwide.

The following are a few of the molecular-based techniques (Fig. 12.1) employed on-field for the detection of viral diseases.

(a) *Isothermal Amplification*—Isothermal amplification methods, such as Loop-Mediated Isothermal Amplification (LAMP), Recombinase Polymerase Amplification (RPA), and Helicase-Dependent Amplification (HDA), have gained popularity for their simplicity and speed. These techniques amplify DNA or RNA at a constant temperature, such as in the field, eliminating the need for complex thermal cycling equipment. For instance, a field-portable RT-LAMP assay has been developed to detect tomato chlorosis virus (TCSV) in tomato plants under 35 min, with a detection limit as low as 0.9 pg/μl of total RNA. Similarly, a portable LAMP device has been used for real-time detection of begomoviruses, demonstrating the feasibility of on-site diagnostics in vegetable crops (Yilmaz et al. 2023). Another study with potato leafroll virus (PLRV) aimed at developing a one-step reverse transcription recombinase polymerase

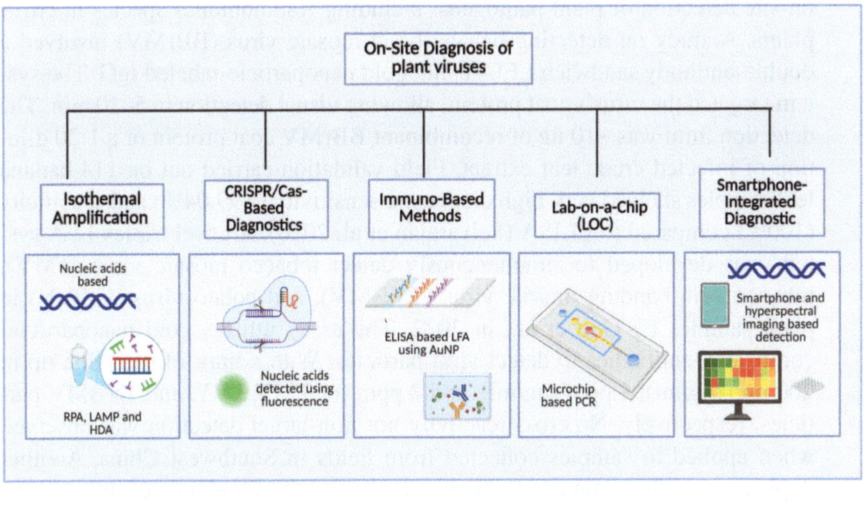


**Fig. 12.1** Current developments in on-site diagnostic kits for virus detection

amplification (RT-RPA) assay targeting the coat protein gene (Kumar et al. 2023a). Tomato chlorotic spot virus (TCSV), an emerging Orthotospovirus, threatens tomato and ornamental crops. Yilmaz and Batuman in 2023 developed an RT-RPA-LFA assay combined with lateral flow detection. Crude RNA extracts were used to detect the virus, and the amplification was carried out via body heat (palm ~36 °C), and gave visible results in approximately 15 min in the field. It was highly specific and sensitive down to ~6 pg/μl of total RNA.

(b) *CRISPR/Cas-Based Diagnostics*—The world of molecular genetics was revolutionized by the identification of the CRISPR/Cas system. Recently, this technique has been harnessed, particularly Cas13a, for plant virus detection due to its high specificity and sensitivity (Paul et al. 2025). A study demonstrated the use of CRISPR/Cas13a for the rapid detection of various crop viruses, including tomato brown rugose fruit virus (ToBRFV), directly from plant RNA. This method allows for early detection, even before visible symptoms appear, and can be integrated with portable devices for on-site diagnostics (Bernabé-Orts et al. 2022). Another study developed a field deployable Cas12a and Cas13a/d-based system to detect tobacco mosaic virus (TMV), tobacco etch virus (TEV), and potato virus X (PVX) in *Nicotiana benthamiana*. For the Cas12a-based system, viral RNA was first amplified by RT-PCR or RT-RPA, followed by the detection of DNA amplicon via fluorescence/lateral flow readouts (Liu et al. 2023; Paul et al. 2025). Using CRISPR-Cas13a/d, a similar study demonstrated amplification-free detection of viral RNA. Overall, diagnostics can be achieved within 30 min of leaf sampling, thus exhibiting speed, specificity, and multiplexing, with minimal equipment (Marqués et al. 2022).

(c) *Immuno-Based Methods*—Immunoassays, such as Lateral Flow Immunoassays (LFIA), have been adapted for field use due to their simplicity and rapid results since the 1990s. These assays detect viral proteins using antibodies and provide visual results within minutes. Portable LFIA devices have been developed for on-site detection of plant pathogens, including Xanthomonas species in citrus plants. A study on detecting banana bract mosaic virus (BBrMV) involved a double-antibody sandwich LFIA using gold nanoparticle-labeled IgG. The system targeted the virus's coat protein, allowing visual detection in 5–10 min. The detection limit was ~10 ng of recombinant BBrMV coat protein or a 1:20 dilution of infected crude leaf extract. Field validation carried out on 114 banana leaf samples showed very high diagnostic sensitivity (≈99.04%) and specificity (100%) compared to ELISA (Selvarajan et al. 2020). A novel triplex LFA system was developed to simultaneously detect tobacco mosaic virus (TMV), tobacco vein banding mosaic virus (TVBMV), and potato virus Y (PVY) in plant samples by Guo et al., in 2022. The assay utilizes gold nanoparticle-conjugated antibodies to detect viral particles. With a limit of detection up to 200 ppb (ng/ml), 1 ppm (μg/ml), and 2 ppm for TMV, PVY, and TVBMV particles, respectively. No cross-reactivity nor non-target detection was observed when applied to samples collected from fields in Southwest China. Another duplex lateral flow immunoassay was developed for simultaneous detection of sugarcane mosaic virus (SCMV) and sugarcane streak mosaic virus (SCSMV) using cysteamine-modified gold nanoparticles. The on-field diagnosis required 7–10 min to determine whether the samples were positive or negative with the ultrasensitive detection range of 10 ng to 1 pg, comparable to that of RT-PCR (Thangavelu et al. 2022).

(d) *Portable Lab-on-a-Chip (LOC) systems*—Miniaturized PCR devices have been developed to bring the power of traditional PCR to the field. These compact systems can amplify DNA and provide results on-site. A portable, standalone real-time microchip PCR system would operate without the need for external computers or auxiliary instruments. It utilizes a microchip with an integrated thin-film heater, a compact fluorescence detector, and a microcontroller for thermocycle control and data acquisition (Koo et al. 2013). Another study developed a portable and battery-operated PCR system for on-site detection of sweet potato viruses. This system utilized thermoelectric modules to convert electrical energy for heating and cooling, which allowed for rapid DNA amplification without the need for bulky equipment. Field testing of the system demonstrated its effectiveness in detecting sweet potato viruses directly from the crude leaf samples, providing results within a short timeframe in remote agricultural areas (Ssengo et al. 2020). A similar study developed a Lab-on-a-Chip kit using spectroscopy and a microfluidic design for the detection of grapevine leafroll-associated virus 3 (GLRaV-3) and grapevine fanleaf virus (GFLV). The LOC exhibited a higher level of detection than that of ELISA, about 15–20X greater sensitivity (1:50 to 1:100 dilutions). It supports simultaneous, multiplexed detection (both viruses in one chip), with replicate chambers to increase reliability, and easier readout via portable electronics. Overall, the LOC is proposed as a promising alternative to ELISA for large-scale (Buja et al. 2022).

(e) *Smartphone-Integrated Diagnostic Systems*—Smartphone-based diagnostic systems have been developed to leverage the widespread availability of smartphones. Recently, a smartphone-integrated RPA-CRISPR-Cas12a detection system has been developed for point-of-care diagnosis of potato late blight. It integrates a polyvinyl alcohol (PVA) microneedle (MN) patch with an RPA-CRISPR-Cas12a assay and smartphone-based fluorescence detection. The microneedle patch extracts DNA from plant leaves within 1 min, achieving a DNA extraction efficiency of 56 μg/mg, approximately three times higher than traditional CTAB methods (Zhao et al. 2025). Another study worked with MN patches as well, where the patches extract sufficient RNA from plant samples within a minute, with extracted RNA remaining stable on the MN surface for up to 3 days, and achieving detection limits as low as 1 pg (Paul et al. 2021). Another smartphone/tech-based module is the remote sensing technologies, including hyperspectral and multispectral imaging, which have been explored for early detection of plant viral infections. These methods capture detailed images of crops and can detect subtle changes in plant health, potentially indicating viral infections. They are non-invasive and can cover large areas quickly, making them suitable for large-scale monitoring of vegetable crops. One study used hyperspectral imaging of wheat leaves in laboratory and field settings to capture fine spectral differences among healthy, symptomatic, and non-symptomatic portions of disease-affected leaves (Peng et al. 2022).

## 12.5 Design and Workflow of On-Site Diagnostic Kits

The focus of on-site plant virus diagnostic kits has been developed with the requirements of simplicity, portability, and accuracy in mind, so that they can be utilized efficiently by inexperienced farmers or field technicians without laboratory-level resources. Most kits are constructed around a few essential components such as sample collection, nucleic acid or antigen preparation, amplification or detection, and visualization of the result (Liu and Tsutsui 2023). The process normally starts with crude extraction and sample collection, whereby a portion of infected stem, leaf, or fruit tissue is crushed manually within an extraction buffer in the kit; the buffer not only disrupts viral particles but also stabilizes nucleic acids and proteins while reducing inhibitors that might hinder downstream reactions. Other kits employ a basic pestle-and-bag system, so special laboratory equipment is unnecessary (Ivanov et al. 2021). Once extracted, the destination varies according to the detection mechanism of the kit. For kits based on serology, e.g., lateral flow immunoassays, the extract is added directly to the sample pad of a strip, where capillary action transfers the liquid along a nitrocellulose membrane with antibodies to the target virus. If the virus is present, antigen–antibody complexes form and accumulate at the test line, producing a visible colored band within 10–20 min, much like a pregnancy test (Mirica et al. 2022). This design makes LFIAs extremely user-friendly and rapid, though less sensitive than molecular approaches. In comparison, molecular-based platforms like RPA, LAMP (Kumar et al. 2021, 2022), or CRISPR-Cas detection have a bit more complex workflow. Following crude extraction, an aliquot of the sample is added to a lyophilized reaction pellet or pre-mixed tube with enzymes, primers, nucleotides, and buffers (Song et al. 2022). Since the processes are run at constant low temperatures (37–65 °C), the kits can be combined with portable heating blocks or even battery-powered units and are thus suited for field applications (Lei et al. 2023). Amplification is typically 20–40 min, producing a huge amount of target DNA or RNA if the virus is present. The amplified product is then either sensed through simple colorimetric probes or lateral flow strips (Zhou et al. 2022) that bind biotin- or fluorescein-labeled amplicons, or CRISPR-mediated fluorescence readouts (Lei et al. 2022), which exploit Cas enzymes cutting reporter molecules upon identifying viral sequences. Most kits combine amplification and detection in a one-pot reaction, minimizing handling steps and associated contamination risks. Lastly, the visibility of results is based on a color change to the naked eye, a band on a strip, or a fluorescent tag with a handheld UV or blue-light flashlight (Bhat et al. 2022). To increase the ease of use for farmers, commercial kits may include pictorial instructions, pamphlets, pre-aliquoted reagents, disposable plastic droppers, and reaction tubes. Certain newer platforms even include smartphone readers based on strip intensity or fluorescence that provide semi-quantitative analysis and cloud-based surveillance of disease. Overall, the design philosophy of the on-site kits includes ease, speed, and affordability, bringing top-tier molecular technology to the needs of real-world farming (Fig. 12.2).

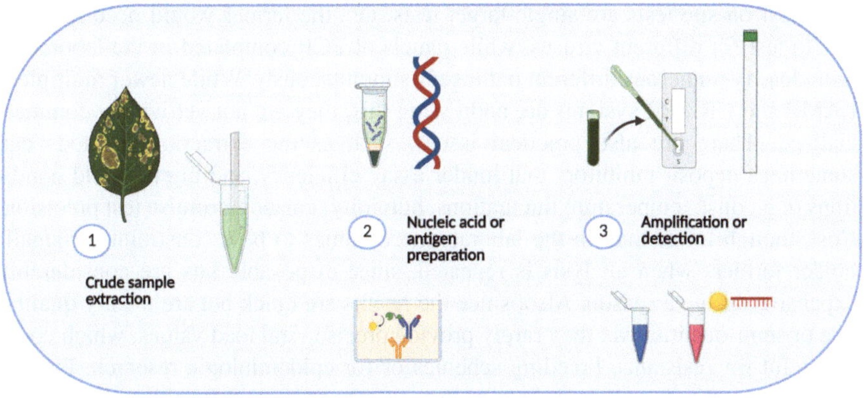


**Fig. 12.2** Design and workflow of on-site diagnostic kits

## 12.6 Advantages and Disadvantages of On-Site Diagnostic Kits

Plant virus diagnostic kits used on-site have advantages that make them invaluable in their use on farms, but there are equally important limitations that must be considered before widespread use. Their primary advantage is speed and accessibility. All the kits, including LFIAs, LAMP, RPA, and CRISPR-based technology, can produce results in 20–60 min without the need for sophisticated laboratory facilities (Raigond et al. 2019; Kumar et al. 2020; Kumar et al. 2023b). This rapid turnaround enables farmers, extension workers, and field pathologists to make quicker decisions about controlling disease, e.g., removing infected plants or adjusting crop protection strategies, thus reducing economic loss. Another advantage is that they are simple to use since the format is often a few steps, like tissue extraction into a buffer and then onto a test strip or amplification tube, that can be performed by nonspecialists. These kits could be easily made portable with heating blocks, which is an added advantage for use in remote agricultural regions with little laboratory infrastructure. Recently, some of the platforms are smartphone-app compatible or portable reader-compatible, such that they can be used to detect and update in real time on the regional or community level for disease mapping. Significantly, most of the kits are cheaper than centralized laboratory testing, particularly with regard to reduced demand for transport, cold storage, and professional services. Despite the positives, some drawbacks impede their widespread application. Sensitivity and specificity remain topics of concern since laboratory PCR and ELISA can recover very low viral particle concentrations; field kits have poor performance with samples where the viral titers are low, either early in infection or in asymptomatic plants (Rubio et al. 2020). This may result in false negatives, causing undue delay in applying proper disease control (Lane et al. 2007). Cross-reactivity with closely related viruses or background plant material likewise generates false positives and destroys confidence in the results (Mrkvová et al. 2022). Another disadvantage is that most on-site tests are single-target tests, i.e., the farmer would need multiple kits to test for different viruses, while panels of PCR completed in the laboratory will identify numerous different pathogens simultaneously. While newer multiplex-LAMP and CRISPR systems are addressing this, they are not yet widely commercialized. There are also practical issues, such as the extraction methods can sometimes deposit inhibitors that hinder assay efficiency, and uneven field conditions (e.g., dust, temperature fluctuations, humidity) can compromise test precision. Cost, though lower than in the laboratory, continues to be a constraint for smallholder farmers when analysis is repeated, since disposable kits are considerably expensive during a season. Also, since the results are quick but are mostly qualitative or semi-quantitative, they rarely provide precise viral load values, which could be useful for resistance breeding schemes or for epidemiologic research. Finally, though the technology has drastically changed, the farmers may still prefer sending samples to laboratory-qualified testing centers for confirmation, particularly where major economic decisions are made. So on-site diagnostic kits are a revolutionary step toward farmer-friendly, real-time plant virus identification, but the sensitivity, multiplexing, and dependability limitations render them most valuable.

## 12.7 Future Directions and Innovations

The future for on-site diagnostic kits to identify plant viruses in vegetable crops is toward increasing precision, portability, and connectivity, with innovations aiming to improve beyond the limitations of existing platforms. One of the key avenues is the construction of multiplexed assays, which will enable the simultaneous detection of multiple viruses in a single reaction, responding to the problem of mixed infections, which are prevalent in tomato, pepper, and cucurbit crops. Methods such as multiplex-LAMP, RPA, and CRISPR-Cas systems are being developed to read multiple targets with color-coded outputs or fluorescent signals using a single system (Wang et al. 2025). Current focus to improve the sensitivity and specificity is the integration of CRISPR-based diagnostics, which covers not only the former but also is designed to give visual readouts through lateral flow strips or even naked-eye detection through colorimetric shifts (Karimi et al. 2025). These systems are being developed rapidly, with scientists developing lyophilized CRISPR reagents that can be transported without cold chains and activated with basic sample inputs, thus making it very practical for field conditions. Apart from molecular techniques, nanotechnology has also played a hand in on-site diagnosis with the use of nanoparticles like gold, carbon, or quantum dots to increase signal intensity in lateral flow tests and thus enhance detection sensitivity even at low titers of the virus (Mousavi et al. 2023; Narware et al. 2025). Simultaneously, another tech development gaining momentum is diagnostics with digital systems. Smartphone-enabled readers and handheld fluorescence detectors are being engineered to read test results and report them immediately to centralized databases. As with AI and machine learning, it is becoming a subsidiary to molecular diagnosis, by symptom visual analysis through smartphone cameras (Dolatabadian et al. 2025). These "plug-and-play" platforms would significantly eliminate human error in sample preparation as well. Further, this technology could be more environmentally friendly and sustainable, with biodegradable test strips and reagents that reduce plastic waste in mass use in agriculture. Ultimately, the way forward is the development of integrated diagnostics rather than single-purpose kits. Through the integration of molecular specificity (CRISPR, LAMP), intelligent detection (nanoparticles, AI, IoT-enabled readers), and farmer-focused design (low cost, few steps, stability at room temperature), the next generation of on-site diagnostics for plant viruses will be more sensitive, scalable, and sustainable. Rather than replacing laboratory diagnostics altogether, these new technologies are likely to supplement them, offering a first line of defense. In this manner, the technologies of the coming decade may fundamentally change vegetable crop protection by making early, accurate, and extensive virus surveillance feasible in the field itself.

## 12.8 Conclusion

On-site diagnostic kits have revolutionized the ease and simplicity of diagnosing infection in both animal and plant pathology. The timely advancements in recent years serve the farmers, agriculturalists, etc., in not only timely diagnosing or preventing their crops from viral infections but also in understanding outbreak patterns and combating climate change effects. With advancements in molecular techniques such as CRISPR/Cas system and so on, the gap between diagnosis and intervention has been expedited. These tools focus on making farmers self-reliant instead of solely depending on centralized laboratory testing, which might delay timely and accurate diagnostics. By enabling quick and accessible diagnosis, such technologies also promote sustainable agricultural practices, improve plant health management, and strengthen food security. However, the promise of on-site diagnostics comes with its own set of challenges, such as reagent stability under field conditions, designing affordable consumables, and achieving multiplexing, which remain key areas requiring further innovation. Additionally, scaling these technologies for widespread field adoption will require improvements in supply chains, user training, inclusive policy frameworks, as well as the willingness of farmers to adapt to new-age technologies. Despite these challenges, the trajectory of on-site diagnostic kits seems positive. The tools available today already outperform earlier generations of diagnostic systems in sensitivity, portability, and making it user-friendly. Further research continues to push these boundaries through the integration of digital technologies, which has accelerated both accuracy and scalability. In summary, on-site diagnostic platforms represent one of the most transformative breakthroughs in early plant virus detection and crop management. They encapsulate laboratory-level accuracy while delivering field-level practicality, empowering stakeholders across the agricultural sector to safeguard crop health more effectively. From LAMP, RPA, to CRISPR, and lateral flow assay-based devices, the continued evolution will play a critical role in advancing sustainable farming practices, strengthening food security, and preparing agricultural systems to support the hardworking farming community.

## References


Bernabé-Orts JM, Hernando Y, Aranda MA (2022) Toward a CRISPR-based point-of-care test for tomato brown rugose fruit virus detection. PhytoFrontiers 2(2):92–100. https://doi.org/10.1094/PHYTOFR-08-21-0053-TA

Bertolini E, Olmos A, Martínez MC, Gorris MT, Cambra M (2001) Single-step multiplex RT-PCR for simultaneous and colourimetric detection of six RNA viruses in olive trees. J Virol Methods 96(1):33–41. https://doi.org/10.1016/S0166-0934(01)00313-5

Bhat AI, Aman R, Mahfouz M (2022) Onsite detection of plant viruses using isothermal amplification assays. Plant Biotechnol J 20(10):1859. https://doi.org/10.1111/PBI.13871

Buja I, Sabella E, Monteduro AG, Rizzato S, Bellis LD, Elicio V, Formica L, Luvisi A, Maruccio G (2022) Detection of Ampelovirus and Nepovirus by Lab-on-a-Chip: A Promising Alternative to ELISA Test for Large Scale Health Screening of Grapevine. BiosensS, 12(3):147. https://doi.org/10.3390/bios12030147

Clark MF, Adams AN (1977) Characteristics of the microplate method of enzyme linked immunosorbent assay for the detection of plant viruses. J Gen Virol 34(3):475–483. https://doi.org/10.1099/0022-1317-34-3-475/CITE/REFWORKS

Dolatabadian A, Neik TX, Danilevicz MF, Upadhyaya SR, Batley J, Edwards D (2025) Image-based crop disease detection using machine learning. Plant Pathol 74(1):18–38. https://doi.org/10.1111/PPA.14006

Gong XY, Wang ZH, Bashir M, Tang T, Gan X, Yang WC (2025) Recent application of CRISPR/Cas in plant disease detection. TrAC Trends Anal Chem 189:118251. https://doi.org/10.1016/J.TRAC.2025.118251

Guo Y, Wu W, Zhang X, Ding M, Yu J, Zhang J, Jia MA, Tian Y (2022) Triplex lateral flow immunoassay for rapid diagnosis of tobacco mosaic virus, Tobacco Vein Banding Mosaic Virus, and Potato Virus Y. Plant Dis 106(12):3033–3039. https://doi.org/10.1094/PDIS-08-21-1756-RE

Hančinský R, Mihálik D, Mrkvová M, Candresse T, Glasa M (2020) Plant viruses infecting Solanaceae family members in the cultivated and wild environments: a review. Plants 9(5):667. https://doi.org/10.3390/PLANTS9050667

Ivanov A v, Safenkova I v, Zherdev A v, Dzantiev BB (2021) The potential use of isothermal amplification assays for in-field diagnostics of plant pathogens. Plants 10(11):2424. https://doi.org/10.3390/PLANTS10112424/S1

Jeevalatha A, Kumar R, Raigond B, Sundaresha S, Sharma S, Singh BP (2015) Duplex realtime RT-PCR assay for the detection of potato spindle tuber viroid (PSTVd) along with ef 1-α gene of potato. Phytoparasitica 43:317–325. https://doi.org/10.1007/s12600-014-0452-z

Jeevalatha A, Kaundal P, Kumar R, Raigond B, Kumar R, Sharma S, Chakrabarti SK (2018) Optimized loop-mediated isothermal amplification assay for tomato leaf curl New Delhi virus-[potato] detection in potato leaves and tubers. Eur J Plant Pathol 150:565–573. https://doi.org/10.1007/s10658-017-1300-z

Jiang T, Zhou T (2023) Unraveling the mechanisms of virus-induced symptom development in plants. Plants 12(15):2830. https://doi.org/10.3390/PLANTS12152830

Kanapiya A, Amanbayeva U, Tulegenova Z, Abash A, Zhangazin S, Dyussembayev K, Mukiyanova G (2024) Recent advances and challenges in plant viral diagnostics. Front Plant Sci 15:1451790. https://doi.org/10.3389/FPLS.2024.1451790

Karimi M, Ghorbani A, Niazi A, Rostami M, Tahmasebi A (2025) CRISPR-Cas13a as a next-generation tool for rapid and precise plant RNA virus diagnostics. Plant Methods 21(1):1–16. https://doi.org/10.1186/S13007-025-01401-9/FIGURES/4

Koo C, Malapi-Wight M, Kim HS, Cifci OS, Vaughn-Diaz VL, Ma B, Kim S, Abdel-Raziq H, Ong K, Jo YK, Gross DC, Shim WB, Han A (2013) Development of a real-time microchip PCR system for portable plant disease diagnosis. PLoS One 8(12):e82704. https://doi.org/10.1371/JOURNAL.PONE.0082704

Kumar R, Kaundal P, Jeevalatha A, Sharma S, Chakrabarti SK (2020) Development of a visual detection of Potato virus S by reverse transcription loop-mediated isothermal amplification. 3 Biotech 10:213. https://doi.org/10.1007/s13205-020-02214-4

Kumar R, Kaundal P, Tiwari RK, Sundaresha S, Kumari H, Naga KC, Sharma S, Kumar M (2021) Rapid and sensitive detection of potato virus X by one-step reverse transcription-recombinase polymerase amplification method in potato leaves and dormant tubers. Mol Cell Probes 58:101743. https://doi.org/10.1016/j.mcp.2021.101743

Kumar R, Kaundal P, Tiwari RK, Sundaresha S, Kumari H, Lal MK, Naga KC, Sharma S, Sagar V, Kumar M (2022) Establishment of a one-step reverse transcription recombinase polymerase amplification assay for the detection of potato virus S. J Virol Methods 307:114568. https://doi.org/10.1016/j.jviromet.2022.114568

Kumar R, Kaundal P, Tiwari RK, Lal MK, Kumari H, Kumar R, Naga KC, Kumar A, Singh B, Sagar V, Sharma S (2023a) Development of reverse transcription recombinase polymerase amplification (RT-RPA): a methodology for quick diagnosis of potato Leafroll viral disease in potato. Int J Mol Sci 24(3). https://doi.org/10.3390/IJMS24032511

Kumar R, Kaundal P, Tiwari RK, Lal MK, Kumari H, Kumar R, Sagar V, Singh B (2023b) Optimization of a simple, low-cost one-step reverse transcription recombinase polymerase amplification method for real-time detection of Potato Virus A in potato leaves and tubers. 3 Biotech 13:373. https://doi.org/10.1007/s13205-023-03791-w

Kutnjak D, Tamisier L, Adams I, Boonham N, Candresse T, Chiumenti M, de Jonghe K, Kreuze JF, Lefebvre M, Silva G, Malapi-wight M, Margaria P, Pleško IM, McGreig S, Miozzi L, Remenant B, Reynard JS, Rollin J, Rott M et al (2021) A primer on the analysis of high-throughput sequencing data for detection of plant viruses. Microorganisms 9(4):841. https://doi.org/10.3390/MICROORGANISMS9040841

Lane CR, Hobden E, Walker L, Barton VC, Inman AJ, Hughes KJD, Swan H, Colyer A, Barker I (2007) Evaluation of a rapid diagnostic field test kit for identification of Phytophthora species, including P. ramorum and P. kernoviae at the point of inspection. Plant Pathol 56(5):828–835. https://doi.org/10.1111/J.1365-3059.2007.01615.X

Lei R, Li Y, Li L, Wang J, Cui Z, Ju R, Jiang L, Liao X, Wu P, Wang X (2022) A CRISPR/Cas12a-based portable platform for rapid detection of Leptosphaeria maculans in Brassica crops. Front Plant Sci 13:976510. https://doi.org/10.3389/FPLS.2022.976510/BIBTEX

Lei R, Kuang R, Peng X, Jiao Z, Zhao Z, Cong H, Fan Z, Zhang Y (2023) Portable rapid detection of maize chlorotic mottle virus using RT-RAA/CRISPR-Cas12a based lateral flow assay. Front Plant Sci 14:1088544. https://doi.org/10.3389/FPLS.2023.1088544/BIBTEX

Li F, Qiao R, Yang X, Gong P, Zhou X (2022) Occurrence, distribution, and management of tomato yellow leaf curl virus in China. Phytopathol Res 4(1):1–12. https://doi.org/10.1186/S42483-022-00133-1/FIGURES/3

Liu C-W, Tsutsui H (2023) Sample–to-answer sensing technologies for nucleic acid preparation and detection in the field. SLAS Technol 28(5):302–323. https://doi.org/10.1016/J.SLAST.2023.06.002

Liu H, Hu X, Zeng H, He C, Cheng F, Tang X, Wang J (2023) A rapid and high-throughput system for the detection of transgenic products based on LAMP-CRISPR-Cas12a. Curr Res Food Sci 7. https://doi.org/10.1016/j.crfs.2023.100605

Maree HJ, Fox A, al Rwahnih M, Boonham N, Candresse T (2018) Application of HTS for routine plant virus diagnostics: state of the art and challenges. Front Plant Sci 9:1082. https://doi.org/10.3389/FPLS.2018.01082

Marqués MC, Sánchez-Vicente J, Ruiz R, Montagud-Martínez R, Márquez-Costa R, Gómez G, Carbonell A, Daròs JA, Rodrigo G (2022) Diagnostics of infections produced by the plant viruses TMV, TEV, and PVX with CRISPR-Cas12 and CRISPR-Cas13. ACS Synth Biol 11(7):2384–2393. https://doi.org/10.1021/ACSSYNBIO.2C00090

Massart S, Olmos A, Jijakli H, Candresse T (2014) Current impact and future directions of high throughput sequencing in plant virus diagnostics. Virus Res 188:90–96. https://doi.org/10.1016/j.virusres.2014.03.029

Maule AJ, Hull R, Donson J (1983) The application of spot hybridization to the detection of DNA and RNA viruses in plant tissues. J Virol Methods 6(4):215–224. https://doi.org/10.1016/0166-0934(83)90048-4

Mirica AC, Stan D, Chelcea IC, Mihailescu CM, Ofiteru A, Bocancia-Mateescu LA (2022) Latest trends in lateral flow immunoassay (LFIA) detection labels and conjugation process. Front Bioeng Biotechnol 10:922772. https://doi.org/10.3389/FBIOE.2022.922772/FULL

Mosolygó T, Laczi K, Spengler G, Burián K (2022) A practical approach for quantitative polymerase chain reaction, the gold standard in microbiological diagnosis. Sci 4(1):4. https://doi.org/10.3390/SCI4010004

Mousavi SM, Kalashgrani MY, Gholami A, Omidifar N, Binazadeh M, Chiang WH (2023) Recent advances in quantum dot-based lateral flow immunoassays for the rapid, point-of-care diagnosis of COVID-19. Biosensors 13(8):786. https://doi.org/10.3390/BIOS13080786

Mrkvová M, Hančinský R, Grešíková S, Kaňuková Š, Barilla J, Glasa M, Hauptvogel P, Kraic J, Mihálik D (2022) Evaluation of new polyclonal antibody developed for serological diagnostics of tomato mosaic virus. Viruses 14(6):1331. https://doi.org/10.3390/V14061331

Nalla MK, Schafleitner R, Pappu HR, Barchenger DW (2023) Current status, breeding strategies and future prospects for managing chilli leaf curl virus disease and associated begomoviruses in Chilli (Capsicum spp.). Front Plant Sci 14:1223982. https://doi.org/10.3389/FPLS.2023.1223982/FULL

Narware J, Chakma J, Singh SP, Prasad DR, Meher J, Singh P, Bhargava P, Sawant SB, Pitambara, Singh JP, Manzar N, Kashyap AS (2025) Nanomaterial-based biosensors: a new frontier in plant pathogen detection and plant disease management. Front Bioeng Biotechnol 13:1570318. https://doi.org/10.3389/FBIOE.2025.1570318/BIBTEX

National Laboratory of Enteric Pathogens, B. of M. L. C. for D. C (1991) The polymerase chain reaction: an overview and development of diagnostic PCR protocols at the LCDC. Can J Infect Dis 2(2):89. https://doi.org/10.1155/1991/580478

Pandit MA, Kumar J, Gulati S, Bhandari N, Mehta P, Katyal R, Rawat CD, Mishra V, Kaur J (2022) Major biological control strategies for plant pathogens. Pathogens 11(2):273. https://doi.org/10.3390/PATHOGENS11020273

Paul R, Ostermann E, Chen Y, Saville AC, Yang Y, Gu Z, Whitfield AE, Ristaino JB, Wei Q (2021) Integrated microneedle-smartphone nucleic acid amplification platform for in-field diagnosis of plant diseases. Biosens Bioelectron 187:113312. https://doi.org/10.1016/J.BIOS.2021.113312

Paul S, Emmadi V, Saxena S, Sarkar M, Mandal B, Kumar R, Sinha P, Roy A (2025) RPA-assisted CRISPR-Cas12a-enabled point-of-care diagnostic platform for chili leaf curl virus with fluorescent and colorimetric readouts. Front Microbiol 16:1644322. https://doi.org/10.3389/fmicb.2025.1644322

Peng Y, Dallas MM, Ascencio-Ibáñez JT, Hoyer JS, Legg J, Hanley-Bowdoin L, Grieve B, Yin H (2022) Early detection of plant virus infection using multispectral imaging and spatial–spectral machine learning. Sci Rep 12(1):1–14. https://doi.org/10.1038/S41598-022-06372-8;SUBJMETA

Raigond B, Verma A, Roach S, Kochhar T, Shilpa, Jeevalatha A, Kumar R, Sharma S, Chakrabarti SK (2019) One-step reverse transcription loop-mediated isothermal amplification: a simple, sensitive and rapid assay for detection of potato virus X in potato leaves and tubers. Indian Phytopathol 72:321–328. https://doi.org/10.1007/s42360-019-00147-4

Raigond B, Verma G, Kumar R, Tiwari RK (2022) Serological and molecular diagnosis of potato viruses: an overview. In: Chakrabarti SK, Sharma S, Shah MA (eds) Sustainable management of potato pests and diseases. Springer, Singapore, pp 337–352. https://doi.org/10.1007/978-981-16-7695-6_13

Raveena K, Yadvika, Kumar A, Garg MK, Chanakaewsomboon I (2022) A review on global energy use patterns in major crop production systems. Environ Sci Adv 1(5):662–679. https://doi.org/10.1039/D2VA00126H

Rubio L, Galipienso L, Ferriol I (2020) Detection of plant viruses and disease management: relevance of genetic diversity and evolution. Front Plant Sci 11:539737. https://doi.org/10.3389/FPLS.2020.01092/FULL

Selvarajan R, Kanichelvam PS, Balasubramanian V, Sethurama Subramanian S (2020) A rapid and sensitive lateral flow immunoassay (LFIA) test for the on-site detection of banana bract mosaic virus in banana plants. J Virol Methods 284:113929. https://doi.org/10.1016/J.JVIROMET.2020.113929

Singh RP (1998) Reverse-transcription polymerase chain reaction for the detection of viruses from plants and aphids. J Virol Methods 74(2):125–138. https://doi.org/10.1016/S0166-0934(98)00074-3

Song X, Coulter FJ, Yang M, Smith JL, Tafesse FG, Messer WB, Reif JH (2022) A lyophilized colorimetric RT-LAMP test kit for rapid, low-cost, at-home molecular testing of SARS-CoV-2 and other pathogens. Sci Rep 12(1):7043. https://doi.org/10.1038/S41598-022-11144-5

Ssengo J, Wasswa P, Mukasa SB, Okiror A, Kyamanywa S (2020) Portable PCR field-based detection of sweetpotato viruses. Afr Crop Sci J 28(3):363–374. https://doi.org/10.4314/ACSJ.V28I3.3

Tatineni S, Hein GL (2023) Plant viruses of agricultural importance: current and future perspectives of virus disease management strategies. Phytopathology 113(2):117–141. https://doi.org/10.1094/PHYTO-05-22-0167-RVW

Thangavelu RM, Kadirvel N, Balasubramaniam P, Viswanathan R (2022) Ultrasensitive nano-gold labelled, duplex lateral flow immunochromatographic assay for early detection of sugarcane mosaic viruses. Sci Rep 12(1):1–14. https://doi.org/10.1038/S41598-022-07950-6;SUBJMETA

Wang D, Coscoy L, Zylberberg M, Avila PC, Boushey HA, Ganem D, DeRisi JL (2002) Microarray-based detection and genotyping of viral pathogens. Proc Natl Acad Sci 99(24):15687–15692. https://doi.org/10.1073/PNAS.242579699

Wang J, Yang W, Fang J, Xu H, Li J, Zhang Y, Liang Q, Xu B, Li H, Niu E (2025) Field-parallel six-sample microfluidic detection of plant viruses via raffinose-assisted one-pot LAMP-CRISPR/Cas12b. J Adv Res. https://doi.org/10.1016/J.JARE.2025.08.030

Yilmaz S, Batuman O (2023) Development of a reverse transcription recombinase polymerase amplification combined with lateral flow assay for equipment-free on-site field detection of tomato chlorotic spot virus. Virol J 20(1):1–11. https://doi.org/10.1186/S12985-023-02097-W/FIGURES/5

Yilmaz S, Adkins S, Batuman O (2023) Field-portable, rapid, and low-cost RT-LAMP assay for the detection of tomato chlorotic spot virus. Phytopathology 113(3):567–576. https://doi.org/10.1094/PHYTO-08-22-0319-R

Zhao J, Xu H, Xu C, Yin W, Luo L, Liu G, Wang Y (2025) Smartphone-integrated RPA-CRISPR-Cas12a detection system with microneedle sampling for point-of-care diagnosis of potato late blight in early stage. BioRxiv, 2025.06.17.660259. https://doi.org/10.1101/2025.06.17.660259

Zhou Y, Zheng HY, Jiang DM, Liu M, Zhang W, Yan JY (2022) A rapid detection of tomato yellow leaf curl virus using recombinase polymerase amplification-lateral flow dipstick assay. Lett Appl Microbiol 74(5):640–646. https://doi.org/10.1111/LAM.13611

# Nanotechnology-Enabled Diagnostics and Therapeutics in Virus Management

# 13

Bhumika Prasad Kumar, P. J. Krishnapriya, and B. S. Sandeep

**Abstract**

Plant viruses represent one of the significant menaces to the global food supply, as they diminish agricultural output and compromise the standard of food products. Conventional detection and control of plant viruses frequently lack the speed necessary to address sudden outbreaks, are susceptible to misinterpreting visual symptoms, and often fall short in managing disease-spreading vectors. The ongoing emergence of new variants resulting from genetic evolution also hinders successful detection and management. Metal and engineered nanomaterials silver, zinc, iron, gold, silicon, titanium, etc., are reported to hinder virus transmission and vector spread and boost innate plant defense, suggesting their multifarious response to combat viruses. Nanophytovirology is the cutting-edge, multidisciplinary approach wherein nanoparticles are used for diagnostics and therapeutics in the division of plant viral diseases. Nanobiosensors, namely DNA-based, quantum dot-based, antibody-based, optical, and magnetic, are employed for the identification of plant viruses. In the field of disease management, nanoparticles (NPs) can be used as a carrier of antiviral chemicals, therapeutic agents, plant defense inducers, and biostimulants, and for the development of transgenic crops. NPs move through nonprotoplasmic (apoplastic) and protoplasmic (symplastic) pathways of plants, and their effect varies with the nanoparticle's type, size, charge, and concentration, leading to elevated reactive oxygen species production. The greater prolonged impacts of nanoparticles on flora and fauna, optimal treatment kinetics, potential for retreatment, and the optimization of dose/concentration are factors that should be taken into account for their

B. P. Kumar · P. J. Krishnapriya (✉) · B. S. Sandeep
College of Agriculture, Kerala Agricultural University, Trivandrum, Kerala, India
e-mail: krishnapriya.pj@kau.in

 
R. Kumar et al. (eds.), *Vegetable Crops and Viral Pathogens Volume I*, Plant Pathology for Sustainable Agriculture, https://doi.org/10.1007/978-981-95-7265-6_13

acceptance. Emerging concepts such as nanofabrication, nano-enabled dsRNA delivery, nanocomposites, RNAi-aided nanoparticle delivery, and nanobased diagnostic kits can offer enhanced possibilities in this context.

**Keywords**

Nanoparticles · Synthesis · Virus · Biosensors · Antiviral · Biostimulant

## 13.1 Introduction

Plant diseases are a severe peril to food safety by decreasing farm productivity both quantitatively and qualitatively. Savary et al. (2019) reported that 10–40% of yield loss worldwide in various crops, involving maize, rice, wheat, soybean, and potato, was caused by pathogens and pests. Plant viruses contribute to 10–12% of emerging diseases, and virus variants could lead to losses of up to 47% of all plant diseases (Warghane et al. 2024). The degree of output declined due to viral infections relies on many factors, including mutations, high vector reproduction rates, and fluctuating climates. Many detection techniques have been improved to detect and determine plant viruses. Some conventional methods relied on symptoms such as mottling, enations, yellowing, interveinal chlorosis, and mosaic; however, this approach can be misleading because nutrient deficiency disorders can produce similar symptoms (Kumar et al. 2020, 2023; Mazumder et al. 2025; Sarkar et al. 2024). The transmission study is a traditional detection method commonly used in laboratories where the virus is mechanically transmitted into a healthy or indicator plant, and diagnosis is based on symptom expression in the host. However, a drawback of this procedure is that it takes an extended time for symptoms to appear, which allows the virus to spread quickly. Methods predicted from the physical properties of viruses, such as examination of morphological features via Electron Microscopy and immune-molecular analyses like Enzyme-Linked Immunosorbent Assay (ELISA) and Polymerase Chain Reaction (PCR), lack the sensitivity needed for the early detection of low-titer diseased samples (Jeevalatha et al. 2018; Kumar et al. 2022; Naidu and Hughes 2003). These approaches are not suitable for large-scale screening and are not effective in differentiating between two closely related virus groups. Such detection methods require skilled personnel and well-equipped labs, making them unsuitable for field-level use. Conventional methods of virus management include using virus-free seeds or planting material, resistant varieties, growing during periods of minimal vector activity, removing alternate hosts, controlling vectors with traps and insecticides, and integrated disease management involving cultural, chemical, mechanical, and biological practices (Tripathi and Verma 2017). However, these strategies take time to produce results, and they might not be effective against novel variants. Therefore, better methods are needed for accurate plant virus identification and management to ensure reliable results.

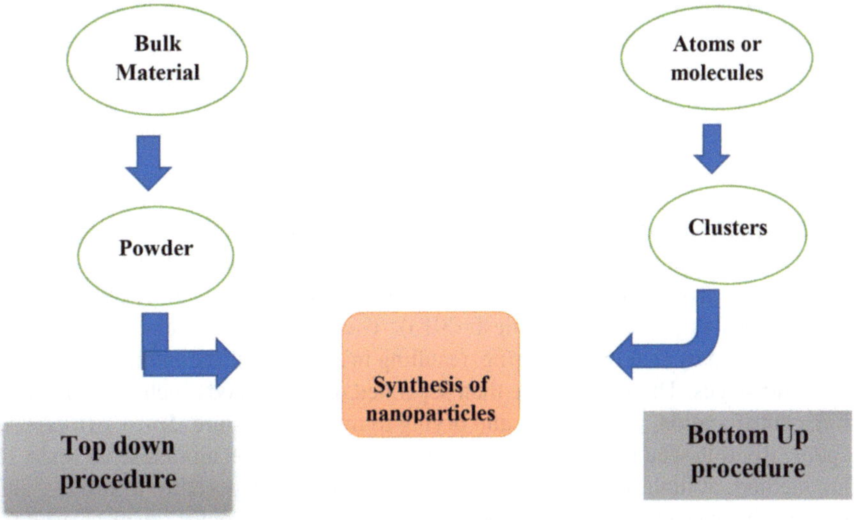


**Fig. 13.1** Synthesis of nanoparticles

## 13.2 Nanoparticles (NPs) Synthesis

Nanomaterials are tiny particles with sizes that vary from 1 to 100 nm. NPs show outlandish characteristics attributed to their larger surface area, which results in increased strength and chemical reactivity. NPs are arranged into three categories: nonmetal-based, metal-based, and carbon-based nanoparticles based on their constitution (Ealias and Saravanakumar 2017). Creation of NPs is categorized as the top-down and the bottom-up systems (Fig. 13.1).

**Top-Down Method** The top-down procedure, also known as the destructive technique, includes breaking down bulk or huge materials into powdered form and then transforming them again into nanoscale materials. Some popular top-down methods are given below: *Thermal decomposition* is an endothermic reaction in which the complex metal compounds, having chemical bonds, are broken by heat (Rahimi-Nasrabadi et al. 2013). At a specific decomposition temperature, the complex metal compounds are chemically broken down, resulting in the formation of small-scale nanomaterials. For instance, the fabrication of copper NPs using copper oxalate. *Sputtering* refers to the ejection of minute particles from a target surface when bombarded with energized ions, causing NPs to pile up on its surface (Shah and Gavrin 2006). *Laser ablation* synthesis in solution is a technique in which a laser beam condenses a plasma plume to synthesize NPs by irradiating various metals submerged in a liquid solution (Amendola and Meneghetti 2009). *Nanolithography* involves studying the patterning or fabrication of structures at the nanoscale. There are multiple methods of nanolithography, including scanning probe, nano-imprint, electron beam, and optical nanolithography, for nanoparticle synthesis (Sebastian et al. 2020). The capability to engrave the size and shape of NPs is an advantage of this approach. *The mechanical milling method* is one of the most accepted, cost-effective, and simplest top-down approaches. It involves synthesizing NPs through milling and postannealing of elemental or compound powders, followed by agitation involving centrifugal or centripetal forces on these activated materials (Yadav et al. 2012).

**Bottom-Up Method** This approach is also known as the constructive method. In this process, minute particles are used to create nanoparticles. Some bottom-up methods are given below: *Sol-gel* involves using chemicals such as chloride and metal oxides as precursors (Ramesh 2013). The predecessor is distributed in the liquid solution by shaking or stirring, resulting in a mixture that contains both liquid and solid stages. These stages are then separated using methods such as sedimentation, centrifugation, purification, and drying to remove moisture. *Spray pyrolysis* is a generally followed protocol for producing nanoparticles on an industrial scale. In this method, at high pressure, the chemical precursor is exposed to a furnace, and then the byproduct gases are classified to restore nanoparticles, such as Al2O3, ZnO, and ZrO2 (Kammler et al. 2001). Plasma and laser can be used instead of a furnace and flame, respectively, at high temperatures to make easy evaporation. The *spinning method* uses the Spinning Disc Reactor (SDR), which mixes inert gases, nitrogen, precursor, and water, resulting in the precipitation of NPs (Chianese et al. 2021). This process is controlled by factors such as disc rotation speed, ratio of the precursor to water, and the disc's surface area. *Chemical Vapor Deposition (CVD)* is a system where a gaseous reactant is introduced into a chamber where the reaction occurs to form a narrow or thin film on a substrate, which is then recovered after heating (Bhaviripudi et al. 2007).

## 13.3 Biosynthesis of Nanoparticles

### 13.3.1 Synthesis of Nanoparticles from Bacteria

Nanoparticle synthesis from bacteria is carried out in two ways: both exocellular or extracorporeal and endocellular or intracorporeal synthesis. In exocellular or extracorporeal synthesis, metal ions are taken up by bacteria in the culture medium when metal particles are added, followed by an interaction between the added metal particles and extracellular components of the bacteria, such as proteins and biomolecules. As a result, these metal particles are converted into their elemental or neutral form by cell wall proteins, extracellular metabolites, and enzymes. These unstable metal nanoparticles, which form clusters through homogeneous nucleation, were then stabilized by capping and recovered by centrifugation. Impurities were removed by resuspending the sample in water or ethanol (Table 13.1).

In endocellular or intracorporeal synthesis of nanoparticles, metal ions are transported into bacterial cells through proteins or porins present on the cell surface. Metal ions were converted into their elemental form by metabolites, enzymes, and

**Table 13.1** Nanoparticles prepared exocellularly from bacteria

| Nanoparticles | Bacteria | References |
| --- | --- | --- |
| Ag (silver) | *Bacillus subtilis* | Tariq et al. (2020) |
| CuO (copper oxide) | *Lactobacillus casei subsp. casei* | Koukhan et al. (2020) |

**Table 13.2** Nanoparticles prepared endocellularly from bacteria

| Nanoparticles | Bacteria | References |
| --- | --- | --- |
| Au (gold) | *Lactobacillus kimchicus* DCY51T | Markus et al. (2016) |
| Ag and cadmium sulfide NPs | *Staphylococcus epidermidis* (ATCC 12228) | Rezvani Amin et al. (2016) |

electron donors. These ions aggregate to form nanoparticle clusters, mainly in the periplasmic space or cell wall, serving as templates for nanoparticle assembly. Protein capping followed by recovery of nanoparticles using physical methods (ultrasonication, grinding, and bead-beating), thermal methods (autoclaving to burst the cells), or chemical methods (detergents, salts, or enzymes to lyse bacterial cell walls) is performed. Centrifugation followed by washing is used to remove cell debris from nanoparticles (Table 13.2).

### 13.3.2 Synthesis of Nanoparticles from Fungi

This approach to nanoparticle fabrication includes two methods: the exocellular method and the endocellular method. In the exocellular method, fungi secrete biomolecules, such as NADH-dependent nitrate reductase enzymes, which reduce metal ions outside the cell membrane, leading to the generation of stable nanoparticles (Moghaddam et al. 2015). In the intracellular production of NPs from fungi, cations (positively charged metal ions) bind to anionic (negatively charged) functional groups on the fungal cell wall surface. They are transported into the cytoplasm via metal-ion transporters or endocytosis. Inside the cytoplasm, such metal ions were reduced to their elemental or neutral form by reducing metabolites or intracellular enzymes. These reduced metal ions then aggregated to form nuclei, which were stabilized by protein capping. The nanoparticles were then recovered through centrifugation and washing to remove impurities (Table 13.3).

### 13.3.3 Virus-Mediated Nanoparticles Synthesis

Nowadays, virus particles are used as biotemplates for nanoparticle synthesis. This approach is emerging and precise, especially with plant viruses due to their surface morphology (size and shape), structural uniformity, and nonpathogenic nature to humans. The methodology involves steps such as metal ion binding to specific amino acids on the virus capsid. In this process, nucleation occurs as bound metals

**Table 13.3** NPs prepared from fungi

| Nanoparticles | Fungi | References |
|---|---|---|
| AgNPs | *Talaromyces purpureogenus* | Sharma et al. (2022) |
| Zinc oxide (ZnO) | *Phanerochaete chrysosporium* | Sharma et al. (2021) |

**Table 13.4** Nanoparticles prepared using virus templates

| Nanoparticles | Virus template | References |
|---|---|---|
| {HAuCl4 (gold chloride) and AgNO3 (silver nitrate)} metallic-hybrid nanomaterials | Squash leaf curl China virus (SLCCNV) | Thangavelu et al. (2020) |
| Au | Cowpea chlorotic mottle virus (CCMV) | Slocik et al. (2004) |

are reduced to their zero-oxidation state, aided by chemical, enzymatic, or physicochemical reduction that can occur externally or within the capsid. This is followed by the organization of nanoparticles at the nucleation sites, stabilization with protein capping, and finally, functionalization to prevent nanoparticle aggregation (Table 13.4).

### 13.3.4 Production of NPs Using Phytoextracts

"Phyto-nanotechnology" is the process of developing NPs from plants or plant-based sources in an eco-friendly manner. This method involves grinding plant extract, suspending it in a solvent under ambient conditions, and mixing it with a metal solution. The alteration in the mixture's color indicated the generation of NPs. Afterwards, the liquid (solution) was centrifuged, and impurities were removed by washing. The nanoparticles were then collected after drying and characterization. This approach is found to be cheaper and has low or no toxic effects for the fabrication of NPs (Table 13.5).

Commonly exploited methods for nanoparticle synthesis used in plant virology are green synthesis from plants, bacteria, fungi, and algae, as well as sol-gel, chemical reduction, laser ablation, and microemulsion techniques (Table 13.6).

## 13.4 Nanophytovirology

Nanoscience is an emerging technology with broader utilization in many disciplines, involving medicine, viral disease detection and management, and fertilizers. Similarly, nanoparticles can help in the efficient identification and control of plant viruses, which paved the way for the evolution of nanophytovirology.

**Table 13.5** Nanoparticles from plant extracts against plant viruses

| Nanoparticle produced | Plant extracts | Virus | References |
|---|---|---|---|
| CuO | Orange peel extract | Tobacco mosaic virus (TMV) | Liu et al. (2024) |
| Ag | *Punica granatum* (pomegranate) peel | TMV | Al-Askar et al. (2023) |

**Table 13.6** Nanoparticles used against plant viruses

| Synthesis method | NPs synthesized | Virus inhibition | References |
|---|---|---|---|
| Biological synthesis | Ag | Bean yellow mosaic virus inhibition | Elbeshehy et al. (2015) |
| Chemical reduction | Au | Barley yellow dwarf virus detection | Alkubaisi and Aref (2017) |

## 13.5 Nanoparticles in the Diagnostics of Plant Viral Diseases

### 13.5.1 Detection of Plant Viruses Using Biosensors

A biosensor is an instrument or device used to detect biological or chemical analytes in a given sample, with key components as a biodetector, a transducer, and an electronic circuit. The detector is the biological receptor where the interaction occurs with the biological analyte, acting as the first division in the biosensor. The transducer transforms the biochemical signal into an accessible signal. The electronic circuit, also called the signal processing unit, processes, records, amplifies, and displays the signal on the monitor (Kirsch et al. 2013). Generally used biosensors (Table 13.7) and types of biosensors are listed below (Fig. 13.2):

**Optical Biosensors** Optical biosensors offer an alternative testing method for detecting viruses, providing results that are more rapid and reliable (Maddali et al. 2021). This type of biosensor uses illumination to identify the reaction of a chemical in a biological system. It can sense intercellular, intracellular, biological, and physicochemical parameters within the microecosystem due to its small size. The interferometric reflectance imaging sensor technique is utilized in several instances, including the multiplexed detection of pathogens, which are functionalized at different sensing regions to distinguish them from other parts and targets due to its high-speed capability (Daaboul et al. 2010).

**Electrochemical Biosensors** Such biosensors work based on electron mobility in redox reactions, which are identified when a potential is applied over the two electrodes. This depends on the electron generation, where either the movement of electrons occurs over the electric double layer of the working electrode, generating current, or a deviation in the double layer potential leads to voltage generation. The

**Table 13.7** Biosensors used against plant viruses

| Type of biosensor | Nanoparticles | Virus | References |
|---|---|---|---|
| Optical biosensor (quantum dot (QD)-based biosensor) | CdTe, CdSe/ZnS QDs | Citrus tristeza virus | Shojaei et al. (2016) |
| Resonant biosensor (surface plasmon resonance biosensor) | Gold nanorods (AuNRs) | Cymbidium mosaic virus | Lin et al. (2014) |
| Optical biosensor (quantum dot (QD)-based biosensor) | Zinc oxide films | Grapevine virus type A | Tereshchenko et al. (2017) |

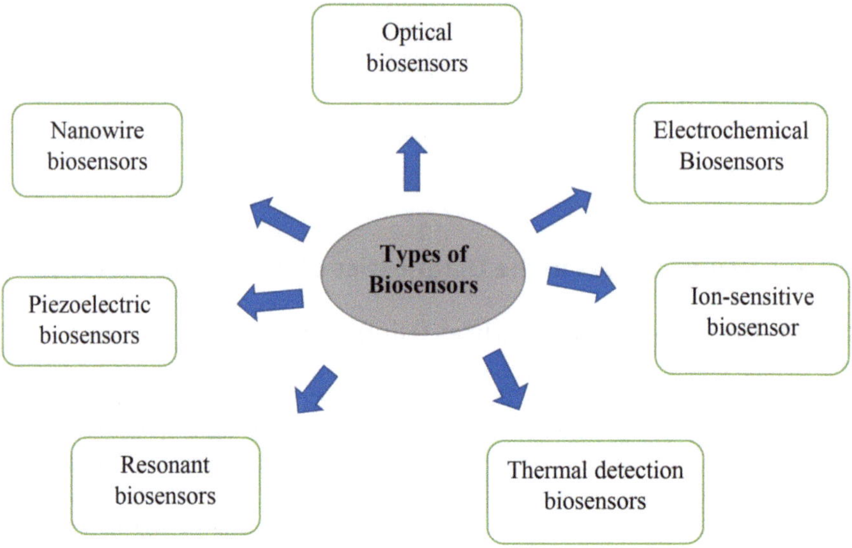


**Fig. 13.2** Types of biosensors

alteration in charge scattering is spotted by ion-selective electrodes such as pH meters, which are designed to sense these changes (Shanbhag et al. 2023).

**Ion-Sensitive Biosensor** These are ion-sensitive semiconductor field-effect transistors (FETs). Alteration in the surface electrical potential is due to interactions between the semiconductor and ions. This possible change in the potential enables measurement. The sensor electrode can be coated with a polymer layer to create an ion-sensitive field-effect transistor. The transfer of four analyte ions through this polymer layer is preferred. The FET surface potential changes as a result of diffusing through the polymer layer; these changes were measured. These biosensors are mainly used for pH measurement and are also known as enzyme field-effect transistors (Sellappan et al. 2022).

**Thermal Detection Biosensor** This type of biosensor focuses on heat production or absorption, resulting in a change in the medium temperature where the reaction takes place. These biosensors were created by combining temperature sensors with damaged/immobilized enzymes. After the analyte is added to the sensor, it comes into contact with the damaged/immobilized enzymes, resulting in a heat reaction that is measured and estimated based on the analyte concentration (Syam et al. 2012).

**Resonant Biosensors** This type of biosensor has an antibody serving as a biological component that combines with an acoustic wave transducer. Variation in the mass of the membrane, when the analyte molecule binds to the antibody. This biological interaction changes the mass of functional membrane layers, shifting the resonant frequency, which is the key parameter measured in resonant biosensors (Higson et al. 1994).

**Piezoelectric Biosensor** In such a biosensor, at a specific frequency, the quartz crystal oscillates when an alternating voltage is applied, and its mass and viscoelastic properties change if target analytes from the sample are bound to the functionalized gold electrode surface, producing a measurable frequency shift (Skládal 2024).

**Antibody-Based Biosensor** Antibody-based detection of virus particles using nanoparticles is a highly accurate and unique approach that has revolutionized viral diagnostics. Here, a specific antibody is bound to an electrode transducer having immobilized antigen, which changes the binding event into an electrical signal. Frietas et al. (2019) experimented with an antibody-based biosensor to detect citrus tristeza virus (CTV). In his experiment, the first step involved modifying the electrode surface by adding polydiallyl dimethyl ammonium chloride, along with the primary antibody and gold nanoparticles, using Bovine Serum Albumin (BSA). Later, functionalized magnetic beads were combined with a secondary antibody and Horse Radish Peroxidase, resulting in a complex formation. The complex was then added to the sample, having the CTV coat protein to form a complex unit. This bound unit was applied over the immobilized primary antibody on the electrode surface, where interaction occurs with the substrate, hydrogen peroxide. The conversion of benzoquinone into hydroquinone produces a measurable electrical signal, which is catalyzed by the enzyme Horse Radish Peroxidase.

**DNA/RNA-Based Biosensor** Millan and Mikkelson (1993) described the working principle of this type of biosensor, in which the first step is modifying the electrode by adding a DNA/RNA probe, which is then immobilized on the electrode surface. After adding DNA/RNA from a plant sample, it binds with the immobilized probe, resulting in a hybridized complex that generates a measurable electrochemical signal. Gold nanoparticle-modified DNA hybridization for CTV detection was performed by Khater et al. (2019). The process, similar to the one discussed earlier, began with the modification of the electrode surface by adding AuNPs and a DNA probe. When the sample containing viral DNA was added to the immobilized electrode surface with probe DNA, a DNA-DNA hybridization complex was formed. The resulting changes in electrical properties were visualized by the Nyquist plot and charge transfer resistance (Rct) values obtained from an equivalent surface, indicating the presence of the virus through DNA hybridization.

**Quantum Dot-Based Biosensor** Such a biosensor is fabricated using tiny semiconductor nanocrystals. Fluorescence Resonance Energy Transfer (FRET) is the unique optical phenomenon used here. Quantum dots, compared to organic dyes and fluorescent proteins, exhibit special highlights such as a wider absorption spectrum, size-tunable light emission, and photobleaching resistance (Wegner and Hildebrandt 2015). Safari (2023) described the work pattern of quantum dots, initiated with synthesis and functionalization, followed by the immobilization of the biorecognition element, signal transduction, and the assessment of the calculable signal. Identification of Sorghum mosaic virus was accomplished using cadmium telluride and carbon quantum dots (Han et al. 2023).

**Surface Plasmon Resonance (SPR)-Based Biosensor** Topor et al. (2023) explained the protocol of an SPR-based biosensor, where a beam of light passes through a prism containing a thin gold layer, generating an evanescent wave that interacts with surface plasmons. After the sample is added, the analyte present in the sample binds with the receptor over the biosensor, resulting in a change in the refractive index (RI). Changes in the RI, in succession, altered the resonance condition of the surface plasmons, leading to a shift in the angle of reflected light, which further modifies the resonance condition in a detectable way. Boltovets et al. (2004) used a SPR biosensor for the detection of TMV.

**Quartz Crystal Microbalances (QCM)-Based Biosensors** Lim et al. (2020) narrated the methodology of QCM biosensors, based on the biological interaction occurring between the target analyte and bioreceptor. The interaction causes changes in the masses of the selective analyte in the plant sample and the biological receptor on its surface, which can be measured. Dickert et al. (2004) used molecularly imprinted polymers (MIPs) created through surface imprinting technology for detecting TMV. This MIP layer was formed by imprinting acrylate-based polymers with a size of 30–100 microgram/ml, which created nanoscale virus-shaped pits that fit into these gaps. Virus recognition was based on the change in the mass of TMV particles after rebinding over the pits, which can be detected by shifts in the frequency in QCM measurements.

**Nano-PCR**

AuNPs are capable of enhancing the specificity of PCR owing to their properties, such as polymerase and primer adsorption, which alter polymerase enzyme availability and lower the melting temperature of the primer, allowing smooth dissociation of PCR products (Shamsian et al. 2025). An ultra-sensitive nano-PCR using magnesium oxide NPs formulations was improved for early identification of Tomato leaf curl virus (ToLCV). This method demonstrated a 13-fold increase in PCR efficiency, reducing the number of cycles from 35 to 25 compared to standard PCR (Devika et al. 2024).

**Immunoassays**

The nano-based immunoassay, specifically Lateral Flow Immunoassay (LFIA), is a rapid, responsive, portable, and economical method for the early identification of plant viruses. In this method, a plant sample containing viral particles is added to the sample well strip. The sample moves by capillary action across the conjugate pad, which consists of labeled gold nanoparticles along with antibodies. These labeled NPs bind to viral particles in the plant sample, forming an antigen-antibody complex. This complex is detected by the test line, resulting in color development that indicates a positive result. The occurrence of a control line establishes the accuracy of the flow and reagent function. Absence or lack of viral particles in the given sample indicates a negative result, where only the control line develops color (Mirica et al. 2022). Maize chlorotic mottle virus was detected using immunochromatographic strips through the LFIA technique, with a sensitivity of up to 1:25,600 dilution, which was 12.1 times more responsive than RT-PCR (Zhang et al. 2023).

## 13.6 Role of NPs in Therapeutics

### 13.6.1 Antiviral Activity of Nanoparticles

**Silver NPs** Al-Askar et al. (2023) substantiated the antiviral activity of Ag NPs against TMV in tomato plants. Ag + ions liberated from AgNPs attach to phosphorus and sulfur components of nucleic acids and proteins of virions, leading to inhibition of replication and transcription of viral particles along with synthesis of reactive oxygen species (ROS), which directly caused oxidative damage to viral constituents. The size, concentration, and time of exposure of NPs influence the antiviral activity of AgNPs. El Gamal et al. (2022a, b) noticed that the foliar application of AgNPs was practical in faba beans against bean yellow mosaic virus (BYMV). When the plants were treated with silver nanoparticles, there was a complete disruption of viral structural integrity and replication in both preventive and curative treatments, i.e., at 200 mg/L 24 hours before virus inoculation and 100 mg/L 48 h after virus inoculation. Additionally, AgNPs completely prevented the virus acquisition rate in the vector, *Aphis craccivora*, when plants were treated at a concentration of 250–300 mg/L. Shafie et al. (2018) narrated that treatment with 200 ppm silver NPs to potato plants 24 h after tomato spotted wilt virus inoculation completely inhibited virus multiplication. Mahfouze et al. (2020) used fabricated silver nanoparticles in banana plants to treat banana bunchy top virus (BBTV) at different concentrations, like 40 ppm, 50 ppm, and 60 ppm. Plants that were treated at 50 ppm relatedly showed a lower level of exterior symptom expression of less than 36% infection rate. Treated plants at 50 ppm showed improved dry weight, an increase in the phenol and proline amount, and leaf area, along with increased carotenoid content and high enzyme activities like PO and PPO when compared with untreated plants. Plants that are treated at 60 ppm showed phytotoxicity in treated banana plants.

**Table 13.8** Effective doses of nanoparticles against viruses in host plants

| Nanoparticles (NPs) | Dose of NPs | Virus | Host plant | References |
|---|---|---|---|---|
| Copper oxide | 100 mg/l | TMV | *Nicotiana benthamiana* | Liu et al. (2024) |
| Zinc oxide NPs | 100 mg/L | Tomato mosaic virus | *Solanum lycopersicum* | Sofy et al. (2021) |
| Iron oxide | 100 mg/L | TMV | *Nicotiana benthamiana* | Cai et al. (2020) |
| Chitosan NPs<br>Chitosan NPs+ Ag NPs | Chitosan NPs (400 ppm)<br>Chitosan NPs + silver nanoparticles (200 ppm) | Alfalfa mosaic virus | *Capsicum annum* | El-Ganainy et al. (2023) |
| ZnO NPs | 100 μ g/mL | TMV | Tomato plants | Abdelkhalek and Al-Askar (2020) |

**Chitosan NPs** Chitosan NPs are a biopolymer derived from chitin, showcasing broad antagonistic activities against pathogens and pests (Ke et al. 2021). Its antiviral action depended on the surface charge and size of the nanoparticle and the grade of acetylation. The positive surface charges in chitosan nanoparticles and the chitosan glucosamine polymer chain could attract viral RNA particles, since the primary chain of virus particles contains negatively charged phosphate groups, thereby suppressing viral activity in the host by blocking multiplication. Chitosan can also induce the expression of PR-1 protein, phenylalanine lyase, and polyphenol oxidase activity, promoting growth in treated plants. El Gamal et al. (2022a, b) evaluated the antiviral effectiveness of chitosan nanoparticles on BYMV-infected faba bean plants, showing that they outperform bulk chitosan particles. Foliar application of chitosan nanoparticles at a dosage of 300–400 mg/L, 48 h after virus inoculation, resulted in complete retardation of disease symptoms. Electron microscopy revealed defective and shortened viral particles along with the inhibition of virus replication. Increased activity of phenylalanine ammonia lyase by 3.14 times and polyphenol oxidase by 11% was also observed, along with a 16.22-fold increase in PR-1 gene upregulation. Different types of nanoparticles have been used for the management of viruses (Table 13.8).

### 13.6.2 Nanoparticles as Biostimulants

Biostimulants are products that enhance one or more traits of the host plant, such as nutrient utilization, tolerance, nutrient uptake in the root zone, and yield quality. NPs can improve the innate immunity of the host, promoting disease tolerance or suppression. Hao et al. (2018) observed a 40% increase in phytohormone, 40–68% increase in zeatin riboside, 24–45% reduction in abscisic acid, and 50% increase in shoot biomass of tobacco infected by turnip mosiac virus after treatment with 50 mg/L of ferric oxide and 200 mg/L of titanium dioxide NPs for 21 days. Cai et al. (2019) demonstrated that zinc oxide and silicon dioxide nanoparticles at a concentration of 100 mg/L were effective against TMV. These nanoparticles enhanced the pr1 and pr2 gene expression levels by 2.03 to 2.29 times and 1.24 to 3.96 times, respectively. These increases corresponded with fresh and dry weights of 0.05 and 0.87 mg for ZnO and 0.049 and 0.81 mg for $SiO_2$ NPs. Tobacco plants infected by TMV showed normal photosynthetic rates comparable to healthy plants when the plants were treated with Carbon Nanotubes (CNT) and C-60 fullerenes at 200 mg/L. Salicylic acid and abscisic acid levels were also enhanced by 33–52% (Adeel et al. 2021). When cerium dioxide nanoparticles at a concentration of 100 mg/L were applied on alfalfa plants infected with AMV, a 58.87% reduction in disease symptoms, along with increased plant height (15.17%), crude fat (10.09%), crude protein (71.19%), and plant dry biomass (37.12%) was observed (Guo et al. 2025).

### 13.6.3 Nanoparticles in Genetic Modification

RNAi technology is one of the antiviral protocols widely used for virus management. In this protocol, double-stranded RNA mimics the action of viral RNA in the host plant, preventing the translation and other necessary mechanisms required by the virus to survive and multiply in the host plants (Ambesajir et al. 2012). NPs can be used as a nucleic acid carrier for efficient delivery to specific targeted areas. The use of nanocarriers in RNA gene silencing aids in protecting delivered nucleic acids from degradation, extending the gene silencing effects, facilitating easy transport between cells in the host plant, and increasing the efficiency of RNA interference (Zhou et al. 2013). An experiment was conducted using chitosan-based quaternary ammonium salt nanoparticles (HACC)-dsRNA nanoparticles for effective nucleic acid delivery through various approaches, including root soaking and pollen internalization, which were found to be effective in TMV-infected plants. TMV inoculum in seeds was reduced from 95.1% to 61.1% when nanoparticles were introduced through pollen internalization, indicating a reduction in the vertical transmission of virus-infected seeds. HACC enhanced dsRNA stability and uptake across vascular tissue via the root soaking method, and delivered nanoparticles-initiated RNA interference by generating small interfering RNAs of size between 21 and 24 nucleotides that selectively silenced targeted TMV genes (Xu et al. 2023).

## 13.7 Uptake of Nanoparticles Via Foliar and Root Application

The treatment of nanoparticles to plants was made using two ways: root and foliar applications. Generally, higher plants are protected by a thick cuticular layer, which serves as one of the defensive barriers against the entry of nanoparticles into plant tissue (Wang et al. 2023). Particles move through the cuticular layer in two ways:

the lipophilic pathway and the hydrophilic pathway. Studies have shown that, in addition to the cuticular pathway for the movement of molecules, there is another route called the stomatal pathway, which occurs through stomatal pores for hydrophilic molecules. This is the only demonstrated pathway for the movement of nanoparticles from the leaf surface to internal plant tissues in foliar applications. Vascular transport is a prerequisite for the long-distance translocation of nanoparticles once they enter the apoplast of the leaf through the stomatal pathway. The surface charge of NPs influences their movement through xylem and phloem (Schreiber 2005). Wang et al. (2013) reported that foliar uptake of NPs, namely iron oxide, titanium dioxide, magnesium oxide, and zinc oxide, ranging in size from 24 to 47 nm, was observed, followed by their movement through the stomatal pathway detected in sieve tube elements of the root. The nanoparticles then moved through the phloem tissue of the root, reached the shoot of the watermelon plants, and the cycle repeated. Wang et al. (2012) observed the movement of copper oxide NPs, measuring 20–40 nm in diameter, in both the phloem and xylem of maize plants. The translocation of NPs from maize roots to shoots via xylem and back from shoots to roots via phloem was confirmed through split-root experiments and high-resolution Transmission Electron Microscopy (TEM). Similarly, Ma et al. (2017) observed the movement of cerium dioxide ($CeO_2$) nanoparticles through both xylem and phloem tissue of cucumber using a split-root experiment. They found that 15% of cerium particles were reduced to Ce3+ during movement through xylem, while Ce2+ particles were transported through the phloem.

NPs come in contact with the epidermis when applied to developing roots and secondary roots due to an underdeveloped cuticular layer in this region. Typically, NPs are positively charged and adhere to the negatively charged surface of the root hairs due to organic acids and mucilaginous particles secreted by the follicles. Additionally, NPs move into the internal tissues of the root by passing through barriers such as the cuticle, epidermis, cortex, endodermis, and casparian strips, then travel through the xylem to reach the shoot. Two pathways for nanoparticle translocation through roots: the nonprotoplasmic (apoplastic) pathway and the protoplasmic (symplastic) pathway. In the nonprotoplasmic tract, nanoparticles penetrate pores in the cell wall and then move to the space between the wall and the plasma membrane or pass through intercellular spaces without crossing the cell membrane. Numerous studies have demonstrated that nanoparticles can form aggregations in the apoplastic pathway, as observed using Transmission Electron Microscopy (TEM). The movement through apoplast pathways depends on pore size and nanoparticle diameter. For example, 20 nm ZnO NPs moved through the apoplast in rye grass roots (Lin and Xing 2008), 22 nm lanthanum trioxide NPs in cucumber roots (Ma et al. 2011), 20–80 nm Ag NPs in Arabidopsis thaliana roots (Geiser-Lee et al. 2014), 12 nm Au NPs in Arabidopsis roots (Avellan et al. 2017), and 43 nm CuO NPs in *Elsholtzia splendens* roots (Shi et al. 2014). A limitation of this pathway is that the cell wall pore size is generally between 5 and 20 nm. Nanoparticles smaller than 20 nm can traverse through these pores, while larger ones linger in intercellular spaces. The deterioration of cell walls and expansion of pore size are primary entryways for nanoparticles (Nair et al. 2010). In the symplastic pathway, nanoparticles are delivered through the cytoplasm from one cell to another. The leading barriers to motion in this context are the penetration of the cell membrane and entry through plasmodesmata (Wang et al. 2023). Numerous hypotheses have been forwarded to explain how host plant cells translocate nanoparticles across membranes, involving mechanisms such as endocytosis, porins, and interconnected ion channels. Nanoparticle uptake in the symplastic route is endocytosis. Recent studies using isolated plant cells have shown endocytic internalization of nanoparticles. Clathrin-dependent and clathrin-independent endocytosis of gold nanoparticles into tobacco protoplasts was noticed via Confocal Laser Scanning Microscopy (Onelli et al. 2008).

## 13.8 Challenges and Prospects in Nanophytovirology

Nanotechnology aids in disease diagnosis and treatment, but research also suggests that nanoparticles can be detrimental to plants, a phenomenon known as nanophytotoxicity. For instance, CeO nanoparticles have been found to decrease nitrogen fixation and reduce pod size in soybeans (López-Moreno et al. 2010). AgNPs penetrating rice cell walls led to cell wall and vacuole damage, as well as interference in plant metabolism (Mirzajani et al. 2013). Reports have shown that nanoparticles such as aluminum, carbon, zinc, and zinc oxide exhibit nonphytotoxicity when applied to various plants (Francis et al. 2024). Uneven distribution/deposition of NPs in various plant parts, variability in dosage levels, application timing, and absence of tailored characterization protocols are other supplementary limitations. Scientists are now improving on nanocomposites that combine diverse active agents, such as antiviral nanoparticles targeting various viruses, along with integrated pest management strategies. Nanobots are being designed for targeted applications, such as the early determination and treatment of viral infections (Zhou et al. 2021). A reduction in manual effort and costs is possible with the development of AI-driven nanobiosensors.

## 13.9 Conclusion

Nanophytovirology has facilitated an important way in plant virology, focusing on early detection, targeted treatment, enhanced plant immunity, and growth promotion by utilizing biosensors, nano-PCR, LFIA, and nanobiostimulants. These novel technologies enable higher sensitivity, specificity, and field applicability. Nanocarriers enable the precise delivery of nucleic acids, facilitating the initiation of RNA interference and contributing to eco-friendly control strategies. Nevertheless, NPs pose obstacles, such as phytotoxicity risks, unstable uptake, lack of standardized protocols, and nonspecific targeting effects. Such limitations rely on selecting the correct NPs and applying at an accurate dosage following specific protocols. The integration of nanotechnology with artificial intelligence represents an innovative delivery system for sustainable agricultural practices, holding immense ability to revolutionize plant virology by safeguarding agricultural output and enhancing global food security.

## References

Abdelkhalek A, Al-Askar AA (2020) Green-synthesized ZnO nanoparticles mediated by Mentha spicata extract induce plant systemic resistance against Tobacco mosaic virus. Appl Sci 10(15):5054. https://doi.org/10.3390/app10155054

Adeel M, Farooq T, White JC, Hao Y, He Z, Rui Y (2021) Carbon-based nanomaterials suppress tobacco mosaic virus (TMV) infection and induce resistance in Nicotiana benthamiana. J Hazard Mater 404(Pt A):124167. https://doi.org/10.1016/j.jhazmat.2020.124167

Al-Askar AA, Aseel DG, El-Gendi H, Sobhy S, Samy MA, Hamdy E, El-Messeiry S, Behiry SI, Elbeaino T, Abdelkhalek A (2023) Antiviral activity of biosynthesized silver nanoparticles from pomegranate (Punica granatum L.) peel extract against Tobacco Mosaic Virus. Plants (Basel) 12(11):2103. https://doi.org/10.3390/plants12112103

Alkubaisi NA, Aref NMA (2017) Dispersed gold nanoparticles potentially ruin the gold barley yellow dwarf virus and eliminate virus infectivity hazards. Appl Nanosci 7:31–40. https://doi.org/10.1007/s13204-016-0540-0

Ambesajir A, Kaushik A, Kaushik JJ, Petros ST (2012) RNA interference: a futuristic tool and its therapeutic applications. Saudi J Biol Sci 19(4):395–403. https://doi.org/10.1016/j.sjbs.2012.08.001

Amendola V, Meneghetti M (2009) Laser ablation synthesis in solution and size manipulation of noble metal nanoparticles. Phys Chem Chem Phys 11:3805–3821. https://doi.org/10.1039/b900654k

Avellan A, Schwab F, Masion A, Chaurand P, Borschneck D, Vidal V, Rose J, Santaella C, Levard C (2017) Nanoparticle uptake in plants: gold nanomaterial localized in roots of Arabidopsis thaliana by X-ray computed nanotomography and hyperspectral imaging. Environ Sci Technol 51(15):8682–8691. https://doi.org/10.1021/acs.est.7b01133

Bhaviripudi S, Mile E, Steiner SA 3rd, Zare AT, Dresselhaus MS, Belcher AM, Kong J (2007) CVD synthesis of single-walled carbon nanotubes from gold nanoparticle catalysts. J Am Chem Soc 129(6):1516–1517. https://doi.org/10.1021/ja0673332

Boltovets PM, Snopok BA, Boyko VR, Shevchenko TP, Dyachenko NS, Shirshov Yu M (2004) Detection of plant viruses using a surface plasmon resonance via complexing with specific antibodies. Biosens Bioelectron 20(3)

Boroumand Moghaddam A, Namvar F, Moniri M, Md Tahir P, Azizi S, Mohamad R (2015) Nanoparticles biosynthesized by fungi and yeast: a review of their preparation, properties, and medical applications. Molecules 20(9):16540–16565. https://doi.org/10.3390/molecules200916540

Cai L, Liu C, Fan G, Liu C, Sun X (2019) Preventing viral disease by ZnO-NPs through directly deactivating TMV and activating the plant immunity in Nicotiana benthamiana. Environ Sci Nano 6:3653–3669. https://doi.org/10.1039/C9EN00850K

Cai L, Cai L, Jia H, Liu C, Wang D, Sun X (2020) Foliar exposure of $Fe_3O_4$ nanoparticles on Nicotiana benthamiana: evidence for nanoparticles uptake, plant growth promoter, and defense response elicitor against plant virus. J Hazard Mater 393:122415. https://doi.org/10.1016/j.jhazmat.2020.122415

Chianese A, Picano A, Stoller M (2021) Spinning disc reactor for producing nanoparticles: applications and optimal operating variables. Chem Eng Trans 84

Daaboul GG, Yurt A, Zhang X, Hwang GM, Goldberg BB, Ünlü MS (2010) High-throughput detection and sizing of individual low-index nanoparticles and viruses for pathogen identification. Nano Lett 10(11):4727–4731. https://doi.org/10.1021/nl103210p

Devika PP, Alex S, Soni KB, Sindura KP, Ayisha R, Manju RV (2024) Nano-PCR for the early detection of tomato leaf curl virus. 3 Biotech 14(1):5. https://doi.org/10.1007/s13205-023-03842-2

Dickert FL, Hayden O, Bindeus R, Mann KJ, Blaas D, Waigmann E (2004) Bioimprinted QCM sensors for virus detection—screening of plant sap. Anal Bioanal Chem 378(8):1929–1934. https://doi.org/10.1007/s00216-004-2521-5

Ealias AM, Saravanakumar MP (2017) A review on the classification, characterisation, synthesis of nanoparticles, and their application. IOP Conf Ser: Mater Sci Eng 263:032019. https://doi.org/10.1088/1757-899X/263/3/032019

El Gamal AYS, Atia MM, El Sayed T, Abou-Zaid MI, Tohamy MR (2022a) Antiviral activity of chitosan nanoparticles for controlling plant-infecting viruses. S Afr J Sci 118(1/2):10693. https://doi.org/10.17159/sajs.2022/10693

El Gamal AY, Tohamy MR, Abou-Zaid MI, Atia MM, El Sayed T, Farroh KY (2022b) Silver nanoparticles as a viricidal agent to inhibit plant-infecting viruses and disrupt their acquisition and transmission by their aphid vector. Arch Virol 167(85):97. https://doi.org/10.1007/s00705-021-05280-y

Elbeshehy EKF, Elazzazy AM, Aggelis G (2015) Silver nanoparticles synthesis mediated by new isolates of Bacillus spp., nanoparticle characterization and their activity against Bean Yellow Mosaic Virus and human pathogens. Front Microbiol 6:453. https://doi.org/10.3389/fmicb.2015.00453

El-Ganainy SM, Soliman AM, Ismail AM, Sattar MN, Farroh KY, Shafie RM (2023) Antiviral activity of chitosan nanoparticles and chitosan silver nanocomposites against Alfalfa mosaic virus. Polymers 15(13):2961. https://doi.org/10.3390/polym15132961

Francis DV, Abdalla AK, Mahakham W, Sarmah AK, Ahmed ZFR (2024) Interaction of plants and metal nanoparticles: exploring its molecular mechanisms for sustainable agriculture and crop improvement. Environ Int 190:108859. https://doi.org/10.1016/j.envint.2024.108859

Freitas TA, Proença CA, Baldo TA, Materón EM, Wong A, Magnani RF, Faria RC (2019) Ultrasensitive immunoassay for detection of citrus tristeza virus in citrus samples using a disposable microfluidic electrochemical device. Talanta 205:120110. https://doi.org/10.1016/j.talanta.2019.07.005

Geisler-Lee J, Brooks M, Gerfen JR, Wang Q, Fotis C, Sparer A, Ma X, Berg RH, Geisler M (2014) Reproductive toxicity and life history study of silver nanoparticle effect, uptake, and transport in Arabidopsis thaliana. Nanomaterials (Basel) 4(2):301–318. https://doi.org/10.3390/nano4020301

Guo Z, Zhang T, Chen Z, Niu J, Rasel M, Luo E, Li Y, Xu N, Hassan MU, Kareem HA, Wang Q, Yan X, Guo Y (2025) Multiple pathways revealing the CeO2 nanoparticle-biostimulant-based "stress training" strategy for enhanced Medicago sativa L. Antiviral Capability. ACS Nano 19(8):7677–7689. https://doi.org/10.1021/acsnano.4c10637

Han B, Lee J, Kim YJ (2023) Quantum dot-based biosensor for rapid and sensitive detection of Sorghum mosaic virus. J Plant Pathol 105(1):101–107

Hao Y, Yuan W, Ma C, White JC, Zhang Z, Adeel M, Zhou T, Rui Y, Xing B (2018) Engineered nanomaterials suppress Turnip mosaic virus infection in tobacco (Nicotiana benthamiana). Environ Sci Nano 5(7):1685–1693. https://doi.org/10.1039/C8EN00014J

Higson SP, Reddy SM, Vadgama P (1994) Enzyme and other biosensors: evolution of a technology. Eng Sci Educ J 3(1):41–48

Jeevalatha A, Kaundal P, Kumar R, Raigond B, Kumar R, Sharma S, Chakrabarti SK (2018) Optimized loop-mediated isothermal amplification assay for tomato leaf curl New Delhi virus-[potato] detection in potato leaves and tubers. Eur J Plant Pathol 150:565–573. https://doi.org/10.1007/s10658-017-1300-z

Kammler HK, Mädler L, Pratsinis SE (2001) Flame synthesis of nanoparticles. Chem Eng Technol 24(6):583–596. https://doi.org/10.1002/1521-4125(200106)24:6<583::AID-CEAT583>3.0.CO;2-H

Ke C-L, Deng F-S, Chuang C-Y, Lin C-H (2021) Antimicrobial actions and applications of chitosan. Polymers 13(6):904. https://doi.org/10.3390/polym13060904

Khater M, de la Escosura-Muñiz A, Quesada-González D, Merkoçi A (2019) Electrochemical detection of plant virus using gold nanoparticle-modified electrodes. Anal Chim Acta 1046:123–131. https://doi.org/10.1016/j.aca.2018.09.031

Kirsch J, Siltanen C, Zhou Q, Revzin A, Simonian A (2013) Biosensor technology: recent advances in threat agent detection and medicine. Chem Soc Rev 42(22):8733–8768. https://doi.org/10.1039/c3cs60141b

Kouhkan M, Ahangar P, Babaganjeh L, Allahyari-Devin M (2020) Biosynthesis of copper oxide nanoparticles using lactobacillus casei Subsp. Casei and its anticancer and antibacterial activities. Curr Nano 15. https://doi.org/10.2174/1573413715666190318155801

Kumar R, Jeevalatha A, Baswaraj R, Tiwari RK (2020) Viral and viroid diseases of potato and their management. In: Singh AK, Chakrabarti SK, Singh B, Sharma J, Dua VK (eds) Potato science and technology for sub tropics. New India Publishing Agency, New Delhi, pp 267–292. (ISBN: 978-93-89571-93-6)

Kumar R, Tiwari RK, Sundaresha S, Kaundal P, Raigond B (2022) Potato viruses and their management. In: Chakrabarti SK, Sharma S, Shah MA (eds) Sustainable management of potato pests and diseases. Springer, Singapore, pp 309–335. https://doi.org/10.1007/978-981-16-7695-6_12

Kumar R, Lal MK, Prasad P, Tiwari RK (2023) Editorial: current advancements in real-time plant pathogen diagnostics: from lab assays to in-field detection. Front Plant Sci 14:1255654. https://doi.org/10.3389/fpls.2023.1255654

Lim HJ, Saha T, Tey BT, Tan WS, Ooi CW (2020) Quartz crystal microbalance-based biosensors as rapid diagnostic devices for infectious diseases. Biosens Bioelectron 168:112513. https://doi.org/10.1016/j.bios.2020.112513

Lin D, Xing B (2008) Root uptake and phytotoxicity of ZnO nanoparticles. Environ Sci Technol 42(15):5580–5585. https://doi.org/10.1021/es800422x

Lin H-Y, Huang C-H, Lu S-H, Kuo I-T, Chau L-K (2014) Direct detection of orchid viruses using nanorod-based fiber optic particle plasmon resonance immunosensor. Biosens Bioelectron 51(371):378. https://doi.org/10.1016/j.bios.2013.08.009

Liu S, Tian W, Liu Z, Wei X, Yuan K, Du W, Chen S, Chen S, Zhou D, Cai L (2024) Biosynthesis of cupric oxide nanoparticles: its antiviral activities against TMV by directly destroying virion and inducing plant resistance. Phytopathol Res 6:30. https://doi.org/10.1186/s42483-024-00253-3

López-Moreno ML, de la Rosa G, Hernández-Viezcas JA, Castillo-Michel H, Botez CE, Peralta-Videa JR, Gardea-Torresdey JL (2010) Evidence of the differential biotransformation and genotoxicity of ZnO and $CeO_2$ nanoparticles on soybean (Glycine max) plants. Environ Sci Technol 44(19):7315–7320. https://doi.org/10.1021/es903891g

Ma Y, He X, Zhang P, Zhang Z, Guo Z, Tai R, Xu Z, Zhang L, Ding Y, Zhao Y, Chai Z (2011) Phytotoxicity and biotransformation of $La_2O_3$ nanoparticles in a terrestrial plant, cucumber (Cucumis sativus). Nanotoxicology 5(4):743–753. https://doi.org/10.3109/17435390.2010.545487

Ma Y, He X, Zhang P, Zhang Z, Ding Y, Zhang J, Wang G, Xie C, Luo W, Zhang J, Zheng L, Chai Z, Yang K (2017) Xylem and phloem-based transport of $CeO_2$ nanoparticles in hydroponic cucumber plants. Environ Sci Technol 51(9):5215–5221. https://doi.org/10.1021/acs.est.6b05998

Maddali H, Miles CE, Kohn J, O'Carroll DM (2021) Optical biosensors for virus detection: prospects for SARS-CoV-2/COVID-19. Chembiochem 22(7):1176–1189. https://doi.org/10.1002/cbic.202000744

Mahfouze HA, El-Dougdoug NK, Mahfouze SA (2020) Virucidal activity of silver nanoparticles against Banana bunchy top virus (BBTV) in banana plants. Bull Natl Res Cent 44:199. https://doi.org/10.1186/s42269-020-00433-6

Markus J, Mathiyalagan R, Kim YJ, Abbai R, Singh P, Ahn S, Perez ZEJ, Hurh J, Yang DC (2016) Intracellular synthesis of gold nanoparticles with antioxidant activity by probiotic Lactobacillus kimchicus DCY51T isolated from Korean kimchi. Enzym Microb Technol 95:85–93. https://doi.org/10.1016/j.enzmictec.2016.08.018

Mazumder P, Majumdar A, Roy B, Mohanbabu Y, Valli NS, Sarkar M, Mondal F, Verma N, Kumar R, Roy A (2025) RNA vaccination in plants: a non-transgenic approach to combat plant viruses with special reference to tomato leaf curl New Delhi virus. In: Tomato leaf curl New Delhi virus (ToLCNDV) insights into Virome dynamics and management. Springer Nature Singapore, Singapore, pp 339–352. https://doi.org/10.1007/978-981-95-2852-3_17

Millan KM, Mikkelsen SR (1993) Sequence-selective biosensor for DNA based on electroactive hybridization indicators. Anal Chem 65:2317–2323. https://doi.org/10.1021/ac00065a025

Mirica AC, Stan D, Chelcea IC, Mihailescu CM, Ofiteru A, Bocancia-Mateescu LA (2022) Latest trends in lateral flow immunoassay (LFIA) detection labels and conjugation process. Front Bioeng Biotechnol 10:922772. https://doi.org/10.3389/fbioe.2022.922772

Mirzajani F, Askari H, Hamzelou S, Farzaneh M, Ghassempour A (2013) Effect of silver nanoparticles on Oryza sativa L. and its rhizosphere bacteria. Ecotoxicol Environ Saf 88:48–54. https://doi.org/10.1016/j.ecoenv.2012.10.018

Naidu RA, Hughes JDA (2003) Methods for the detection of plant virus diseases. In: Proceedings of a Conference on Plant Virology in Sub-Saharan Africa, 4–8 June 2001. Ibadan, pp 233–253

Nair R, Varghese S, Nair B, Maekawa T, Yoshida Y, Kumar S (2010) Nanoparticulate material delivery to plants. Plant Sci 179(154):163. https://doi.org/10.1016/j.plantsci.2010.04.012

Onelli E, Prescianotto-Baschong C, Caccianiga M, Moscatelli A (2008) Clathrin-dependent and independent endocytic pathways in tobacco protoplasts revealed by labelling with charged nanogold. J Exp Bot 59(11):3051–3068. https://doi.org/10.1093/jxb/ern154

Rahimi-Nasrabadi M, Pourmortazavi SM, Dehaghani A, Hajimirsadeghi S, Zahedi M (2013) Synthesis and characterization of copper oxalate and copper oxide nanoparticles by statistically optimized controlled precipitation and calcination of precursor. CrystEngComm 15:4077–4086. https://doi.org/10.1039/c3ce26930b

Ramesh S (2013) Sol-gel synthesis and characterization of metal oxide nanoparticles. Int J Sci Res 2(8):348–351. https://doi.org/10.1155/2013/929321

Rezvani Amin Z, Khashyarmanesh Z, Fazly Bazzaz BS (2016) Different behavior of Staphylococcus epidermidis in intracellular biosynthesis of silver and cadmium sulfide nanoparticles: more stability and lower toxicity of extracted nanoparticles. World J Microbiol Biotechnol 32(9):140. https://doi.org/10.1007/s11274-016-2110-8

Safari M (2023) Recent advances in quantum dots-based biosensors. In: Quantum Dots—Recent Advances, New Perspectives and Contemporary Applications. IntechOpen. https://doi.org/10.5772/intechopen.108205

Sarkar M, Gupta D, Singh OW, Paul S, Kumar R, Mandal B, Roy A (2024) Identification of pathogenicity determinants in ToLCNDV and their RNAi-based knockdown for disease management in Nicotiana benthamiana and tomato plants. Front Microbiol 15:1481523. https://doi.org/10.3389/fmicb.2024.1481523

Savary S, Willocquet L, Pethybridge SJ, Esker P, McRoberts N, Nelson A (2019) The global burden of pathogens and pests on major food crops. Nat Ecol Evol 3:430–439. https://doi.org/10.1038/s41559-018-0793-y

Schreiber L (2005) Polar paths of diffusion across plant cuticles: new evidence for an old hypothesis. Ann Bot 95(7):1069–1073. https://doi.org/10.1093/aob/mci122

Sebastian EM, Jain SK, Purohit R, Dhakad SK, Rana RS (2020) Nanolithography and its current advancements. Mater Today Proc 26(Part 2):2351–2356. https://doi.org/10.1016/j.matpr.2020.02.505

Sellappan L, Manoharan S, Sanmugam A, Anh NT (2022) Role of nanobiosensors and biosensors for plant virus detection. In: Denizli A, Nguyen TA, Rajendran S, Yasin G, Nadda AK (eds) Nanosensors for smart agriculture. Elsevier, pp 493–506. https://doi.org/10.1016/B978-0-12-824554-5.00004-5

Shafie RM, Salama AM, Farroh KY (2018) Silver nanoparticles activity against tomato spotted wilt virus. Middle East J Agric Res 7(4):1251–1267

Shah P, Gavrin A (2006) Synthesis of nanoparticles using high-pressure sputtering for magnetic domain imaging. J Magn Magn Mater 301:118–123. https://doi.org/10.1016/j.jmmm.2005.06.023

Shamsian S, Siddique AB, Kordzadeh-Kermani V, de la Vega TL, Falcone F, Ray M, Ashrafizadeh SN, Martínez Chapa SO, Madou MJ, Madadelahi M (2025) Nanomaterials in PCR: exploring light-to-heat conversion mechanisms and microfluidic integration. Microsyst Nanoeng 11:127. https://doi.org/10.1038/s41378-025-00674-6

Shanbhag MM, Manasa G, Mascarenhas RJ, Mondal K, Shetti NP (2023) Fundamentals of bio-electrochemical sensing. Chem Eng J Adv 16:100516. https://doi.org/10.1016/j.ceja.2023.100516

Sharma JL, Dhayal V, Sharma RK (2021) White-rot fungus-mediated green synthesis of zinc oxide nanoparticles and their impregnation on cellulose to develop environmentally friendly antimicrobial fibers. 3 Biotech 11(6):269. https://doi.org/10.1007/s13205-021-02840-6

Sharma A, Sagar A, Rana J, Rani R (2022) Green synthesis of silver nanoparticles and its antibacterial activity using fungus Talaromyces purpureogenus isolated from Taxus baccata Linn. Micro Nano Syst Lett 10(1). https://doi.org/10.1186/s40486-022-00144-9

Shi J, Peng C, Yang Y, Yang J, Zhang H, Yuan X, Chen Y, Hu T (2014) Phytotoxicity and accumulation of copper oxide nanoparticles to the Cu-tolerant plant Elsholtzia splendens. Nanotoxicology 8(2):179–188. https://doi.org/10.3109/17435390.2013.766768

Shojaei TR, Salleh MAM, Sijam K, Rahim RA, Mohsenifar A, Safarnejad R, Tabatabaei M (2016) Fluorometric immunoassay for detecting the plant virus Citrus tristeza using carbon nanoparticles acting as quenchers and antibodies labeled with CdTe quantum dots. Mikrochim Acta 183:2277–2287

Skládal P (2024) Piezoelectric biosensors: shedding light on principles and applications. Microchim Acta 191. https://doi.org/10.1007/s00604-024-06257-9

Slocik JM, Naik RR, Stone MO, Wright DW (2004) Viral templates for gold nanoparticle synthesis. Chem Commun 24:2552–2553. https://doi.org/10.1039/b413074j

Sofy AR, Sofy MR, Hmed AA, Dawoud RA, Alnaggar AEA-M, Soliman AM, El-Dougdoug NK (2021) Ameliorating the adverse effects of tomato mosaic tobamovirus infecting tomato plants in Egypt by boosting immunity in tomato plants using zinc oxide nanoparticles. Molecules 26(5):1337. https://doi.org/10.3390/molecules26051337

Syam R, Davis KJ, Pratheesh M, Anoopraj R, Joseph BS (2012) Biosensors: a novel approach for pathogen detection. VETSCAN 7(1):14–18

Tariq F, Ahmed N, Afzal M, Khan MAU, Zeshan B (2020) Synthesis, characterization, and antimicrobial activity of Bacillus subtilis-derived silver nanoparticles against multidrug-resistant bacteria. Jundishapur J Microbiol 13(5):e91934. https://doi.org/10.5812/jjm.91934

Tereshchenko A, Fedorenko V, Smyntyna V, Konup I, Konup A, Eriksson M, Yakimova R, Ramanavicius A, Balme S, Bechelany M (2017) ZnO films formed by atomic layer deposition as an optical biosensor platform for the detection of Grapevine virus A-type proteins. Biosens Bioelectron 92:763–769

Thangavelu RM, Ganapathy R, Ramasamy P, Krishnan K (2020) Fabrication of virus metal hybrid nanomaterials: an ideal reference for biosemiconductor. Arab J Chem 13:2750–2765. https://doi.org/10.1016/j.arabjc.2018.07.006

Topor C-V, Puiu M, Bala C (2023) Strategies for surface design in surface plasmon resonance (SPR) sensing. Biosensors 13(4):465. https://doi.org/10.3390/bios13040465

Tripathi S, Verma R (2017) Management of viral diseases through conventional approaches in India. In: A century of plant virology in India. Springer, pp 729–755. https://doi.org/10.1007/978-981-10-5672-7_29

Wang Z, Xie X, Zhao J, Liu X, Feng W, White JC, Xing B (2012) Xylem- and phloem-based transport of CuO nanoparticles in maize (Zea mays L.). Environ Sci Technol 46(8):4434–4441. https://doi.org/10.1021/es204212z

Wang WN, Tarafdar J, Biswas P (2013) Nanoparticle synthesis and delivery by an aerosol route for watermelon plant foliar uptake. J Nanopart Res 15(1):1417. https://doi.org/10.1007/s11051-013-1417-8

Wang X, Xie H, Wang P, Yin H (2023) Nanoparticles in plants: uptake, transport and physiological activity in leaf and root. Materials (Basel) 16(8):3097. https://doi.org/10.3390/ma16083097

Warghane A, Saini R, Shri M, Andankar I, Ghosh DK, Chopade BA (2024) Application of nanoparticles for management of plant viral pathogens: current status and prospects. Virology 592:109998. https://doi.org/10.1016/j.virol.2024.109998

Wegner KD, Hildebrandt N (2015) Quantum dots: bright and versatile in vitro and in vivo fluorescence imaging biosensors. Chem Soc Rev 44(14):4792–4834. https://doi.org/10.1039/C4CS00532E

Xu X, Jiao Y, Shen L, Li Y, Mei Y, Yang W, Li C, Cao Y, Chen F, Li B, Yang J (2023) Nanoparticle-dsRNA treatment of pollen and root systems of diseased plants effectively reduces the rate of tobacco mosaic virus in contemporary seeds. ACS Appl Mater Interfaces 15(24):29052–29063. https://doi.org/10.1021/acsami.3c02798

Yadav TP, Yadav RM, Singh DP (2012) Mechanical milling: a top-down approach for the synthesis of nanomaterials and nanocomposites. Nanosci Nanotechnol 2(3):22–48. https://doi.org/10.5923/j.nn.20120203.01

Zhang C, Guo M, Dong J, Liu L, Zhou X, Wu J (2023) Visual and super-sensitive detection of maize chlorotic mottle virus by dot-ELISA and au nanoparticle-based immunochromatographic test strip. Viruses 15(7):1607. https://doi.org/10.3390/v15071607

Zhou J, Shum KT, Burnett JC, Rossi JJ (2013) Nanoparticle-based delivery of RNAi therapeutics: progress and challenges. Pharmaceuticals (Basel) 6(1):85–107. https://doi.org/10.3390/ph6010085

Zhou J, Krishnan N, Jiang Y, Fang RH, Zhang L (2021) Nanotechnology for virus treatment. Nano Today 36:101031. https://doi.org/10.1016/j.nantod.2020.101031

# AI and Machine Learning Applications for Virus Detection and Disease Forecasting in Vegetable Crops

# 14

Sneha Choudhary, Malini Ray, Sebabrata Pramanik, and Sonali Bhagat

**Abstract**

Plant viruses are a global threat to vegetable production, especially for members of *Cucurbitaceae* & *Solanaceae*. For decades, the preliminary diagnosis of plant viruses has been based on visual surveys, which many times have often proven to be inaccurate. Whereas laboratory-based detection and diagnosis are time-consuming and labor-intensive. Efficient plant disease forecasting and precise, as well as early disease diagnosis, are inextricably linked. Artificial Intelligence (AI) and Machine Learning (ML) have blended smoothly into the agriculture sector, bringing revolution in Phytopathology. ML, a branch of AI, has enabled precise diagnosis and detection of plant viral diseases at an early stage. Integration of AI/ML has offered timely plant health monitoring, disease forecasting, and data-driven management in vegetable crops. Various ML models/ applications have been analyzed across various vegetables like cucumber, sweet potato, tomato, pepper, cassava, etc., with higher precision. Numerous plant viruses,

S. Choudhary
Department of Plant Pathology, Sher-e-Kashmir University of Agricultural Sciences and Technology, Jammu, India

Department of Plant Pathology, School of Agriculture, Lovely Professional University, Phagwara, Punjab, India

M. Ray
Department of Plant Pathology, School of Agriculture, Lovely Professional University, Phagwara, Punjab, India

S. Pramanik
Department of Computer Science and Engineering, School of Computing, Lovely Professional University, Phagwara, Punjab, India

S. Bhagat (✉)
Department of Plant Pathology, Sher-e-Kashmir University of Agricultural Sciences and Technology, Jammu, India



R. Kumar et al. (eds.), *Vegetable Crops and Viral Pathogens Volume I*, Plant Pathology for Sustainable Agriculture, https://doi.org/10.1007/978-981-95-7265-6_14

viz. *cucumber mosaic virus, cassava mosaic virus, tomato spotted wilt virus, tobacco mosaic virus, tomato leaf curl virus, groundnut bud necrosis virus,* etc. have been detected with >90% accuracy using various ML models like convolutional neural networks (CNNs), Support Vector Machine (SVM), Successive projections algorithm (SPA) with extreme learning machine (ELM), etc. ML models can be coupled with imaging sensors like hyperspectral sensors to increase accuracy and the scope of detection. ML models can also be used to predict disease outbreaks based on the trends of weather and crop data. Therefore, our chapter emphasizes how AI/ML-led applications are strengthening next-generation plant virology.

**Keywords**

Artificial intelligence · Machine learning · Plant virus · Virus detection · Disease forecasting · Vegetable crops · Precision agriculture

## 14.1 Introduction

Agriculture is considered to be the foundation of the Indian economy that ensures food security for the increasing population. Within the agricultural sector, vegetables play a major role as they contribute substantially to the national economy. India is leading globally in the second position as a producer of vegetables, growing a wide range of vegetable crops like tomato, brinjal, okra, cucurbits, and crucifers across diverse agroclimatic zones (APEDA 2024). In spite of its great significance, vegetable productivity is critically endangered by numerous viral diseases (Fig. 14.1 & Table 14.1). Viruses like cucumber mosaic virus (CMV), tomato yellow leaf curl virus (TYLCV), and other begomoviruses that infect okra and cucurbits lead to 20–100% yield losses, depending on the crop and the environment (Hanssen et al. 2010; Mishra et al. 2017). Recurrent viral epidemics not only reduce yield and quality but also escalate dependence on chemical control measures, leading to increased production costs and environmental concerns (Fargette et al. 2006). Hence, viral diseases pose a critical threat to the sustainability and profitability of vegetable

**Fig. 14.1** Some important viral diseases on vegetable crops. (**a–c**) Mosaics of French bean, cucurbits, and chili respectively; (**d**) leaf curl of tomato

**Table 14.1** Some important viral diseases in various vegetable crops at the global level

| Disease/ symptom | Virus | Family | Crop |
|---|---|---|---|
| Mosaics | Cucumber mosaic virus (CMV) | *Bromoviridae* | Cucurbits, brinjal |
| | Zucchini mosaic virus (ZYMV), papaya ringspot virus (PRSV), squash mosaic virus (SqMV), etc. | *Potyviridae* | Cucurbits |
| | Cucumber green mottle mosaic virus (CGMMV) | *Virgaviridae* | |
| | Potato virus Y (PVY) | *Potyviridae* | Potato |
| | Potato virus X (PVX) | *Alphaflexiviridae* | |
| | Bean common mosaic virus (BCMV) | *Potyviridae* | French bean |
| | Pepper mild mottle virus | *Virgaviridae* | Chili, pepper |
| Leaf curl | Tomato leaf curl virus (TLCV) | *Geminiviridae* | Tomato, chili |
| Leaf roll | Potato leaf roll virus (PLRV) | *Luteoviridae* | Potato |
| Yellow mosaic | Bhendi yellow vein mosaic virus (BYVMV) | *Geminiviridae* | Okra |
| Tomato spotted wilt | Tomato spotted wilt virus (TSWV) | *Bunyaviridae* | Tomato |

cultivation, underscoring the need for reliable diagnostic tools, disease forecasting systems, and integrated management approaches (Tatineni and Hein 2023). Traditionally, disease detection has been totally dependent on field observations and expert diagnosis, which are complicated, labor-intensive, time-consuming, and often restricted to subjectivity. Any sort of delay or misidentification can hasten the spread of the disease spread which can cause significant economic losses, as various viral infections can devastate crops within days (Luvisi et al. 2016; Ampatzidis et al. 2017; Cruz et al. 2019). Furthermore, multiple factors, including climatic changes, poor soil fertility, pest pressure, and inappropriate crop selection, can influence crop health; among these, pest and pathogen infestations alone can result in yield losses up to 30–33% annually in India (Kumar et al. 2019a, b). With recent advancements, technologies like Artificial Intelligence (AI) and Machine Learning (ML) have transformed the field of crop health improvement and monitoring. By using advanced image processing and pattern recognition, AI and ML techniques facilitate precise, fast, and computerized identification and detection of viral diseases at the initial stages. These technologies not only reduce reliance on subjective visual assessment but also enhance the precision of disease forecasting, enabling timely interventions and improved crop management (Wani et al. 2022; Al Bashish et al. 2010). Therefore, the incorporation of AI and ML into plant disease detection methods holds great potential in minimizing crop losses, enhancing resilience against emerging viral threats, and encouraging sustainable vegetable production. In this context, this chapter provides a deep understanding of the application of AI and ML for virus detection and disease forecasting in vegetable crops.

## 14.2 Literature Survey

The systemic literature survey using keywords “Artificial Intelligence,” “Machine Learning,” “Plant Virus,” “Virus Detection,” and “Vegetable Crops” was conducted using the Scopus database assessing the significance of research done in this field currently (Fig. 14.2). The output gave four findings, all four from the year 2024, comprising one article, one book chapter, one conference paper, and one review paper, two of which were from India origin. This emerging body of work highlights the growing importance of AI/ML in plant disease detection and forecasting. Jafar et al. (2024), the most cited article, demonstrate the potential of AI/ML in accurate and rapid detection of major vegetable crop diseases, including fungal, bacterial, and viral infections. Advanced deep learning models, viz., CNNs, ResNet, and EfficientNet, have been highly successful in classifying viral pathogens, such as cucumber green mottle mosaic virus, tomato mosaic virus, and pepper mild mottle virus, as well as nonviral diseases. Such models have used image acquisition, preprocessing, segmentation, and feature extraction to distinguish the subtle viral symptoms and abiotic stresses. Moreover, the combination of AI and IoT systems (e.g., drones, sensors) improves disease prediction through the incorporation of environmental and spatiotemporal information, which allows early warning of epidemics. Together, these methods not only enhance the diagnostics of viral diseases but also offer a holistic design of management of all the major diseases in vegetable crops, thus enhancing yield and sustainability.

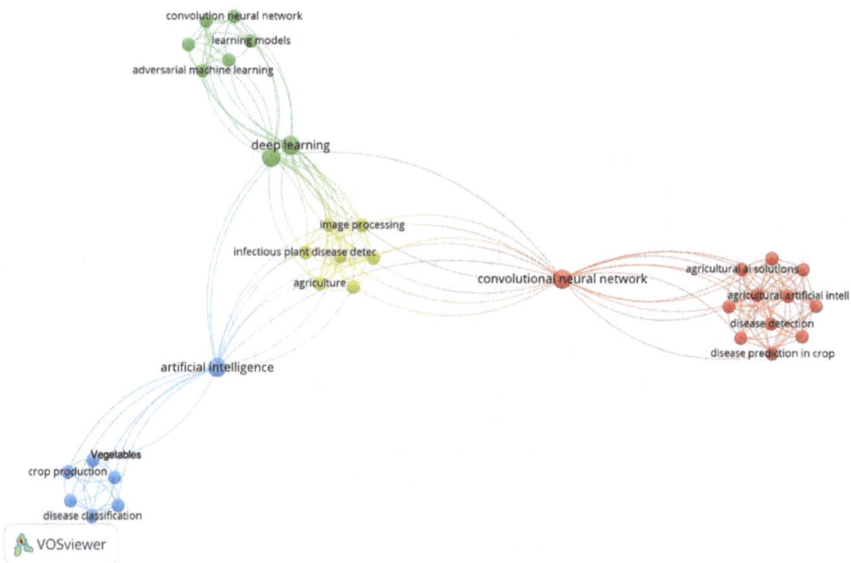


**Fig. 14.2** Bibliometric analysis of publications pertaining to artificial intelligence and plant viruses in vegetable crops, depicting trends in publication types, research volume, and regional contributions

## 14.3 Traditional/Conventional Detection & Forecasting Methods

Conventionally, identification begins in the field, i.e., based on the visual inspections & recognizing viral infections based on symptomatology. However, this method is not only less reliable but also time-consuming; moreover, in the world of plant virology, a single crop can exhibit similar symptoms for more than one virus. For example, cucurbits suffer from mosaic disease, which can be caused by >59 viruses belonging to various families like *Potyviridae* (ZYMV, SqMV, WMV, PRSV, etc.), *Virgaviridae* (CGMMV), *Bromoviridae* (CMV), etc. (Nagendran et al. 2017). The use of local assay hosts since 1929, helped in the confirmation of the presence of the viral titer; however, it could not accurately distinguish between two viruses. Later on, the invention of the electron microscope helped in the visualization of the viruses, but it is a costly method needing expertise & cannot be used to distinguish viral strains. The advent of molecular & serological techniques not only helped to resolve above above-stated issues, but also increased accuracy, specificity, and was also rapid. Polymerase Chain Reaction (PCR) is a widely utilized molecular technique and has enabled not only the detection of viruses at early infections but also to detection of specific viruses out of mixed infections. Today, various types of PCR are being used in plant virus detection, viz., reverse transcriptase PCR, multiplex PCR, quantitative/real-time PCR, etc. (Paul et al. 2025; Lal et al. 2023; Kumar et al. 2019a, b). Similarly, since the identification of the antigenic properties of viruses, serological tools have become an inseparable method, especially for the preliminary screening of large sample sizes. The serological methods include precipitation tests, agglutination tests, double diffusion methods, and the method that brought revolution to the viral diagnostics, i.e., Enzyme Linked Sorbent Assay (ELISA). ELISA can be direct, indirect, sandwich, or competitive. This technique is high-throughput, sensitive, and specific; however, when compared to the PCR technique, its main drawback is its high cost for producing antibodies (Buja et al. 2021). Forecasting methods have heavily relied on expert systems & statistical models based on (a) environmental factors like temperature, relative humidity, and rainfall; (b) disease incidence & disease index, and also on vector population. The forecasting system requires records of the outbreaks, weather conditions, and also crop stages for prediction. Few forecasting models for viral diseases are either vector-based or climate-based (Table 14.2).

Recent breakthroughs in the field of computer vision AI and ML have revolutionized crop disease forecasting, presenting groundbreaking solutions, especially against the drawbacks of the traditional methods (Table 14.3, Fig. 14.3). Internet of Things (IoT), sensing technology, machine learning techniques, Unmanned Aerial Vehicle (UAV), and Mobile Cloud Computing (MCC) have empowered innovative disease management strategies and smart farming. The following subsections focus on AI and ML and their utilization in the detection & forecasting of viral diseases.

**Table 14.2** Forecasting model used for viral diseases

| | Vector-based forecasting models | Climate-based forecasting models |
|---|---|---|
| Based on | Vector population & their epidemics | Temperature, humidity, rainfall, degree days |
| Example | Prediction of PLRV epidemics based on aphid dynamics | Forecasting TLCV & white fly abundance in tomato based on degree-day and climate-driven models |

**Table 14.3** Advantages and disadvantages of traditional detection and forecasting methods for viral diseases in vegetable crops

| Method | Advantages | Disadvantages |
|---|---|---|
| Visual inspection | Rapid & inexpensive | Low specificity, confounded by abiotic stress |
| Indicator host | Distinguishes similar viruses | Labor-intensive, time-consuming, limited sensitivity, & is not high-throughput |
| Electron microscopy | Direct visualization of virus particles | Requires costly equipment and expertise |
| ELISA | High specificity & sensitivity | Detects only known viruses, and also requires antibody production |
| Statistical forecasting | Useful for outbreak prediction, adaptable to local data | Relies on static datasets |
| Rule-based/ weather model | Easy to implement | Ignores emerging strains/environment |
| Vector surveillance | Early warning for vector-transmitted viruses | Needs extensive monitoring |

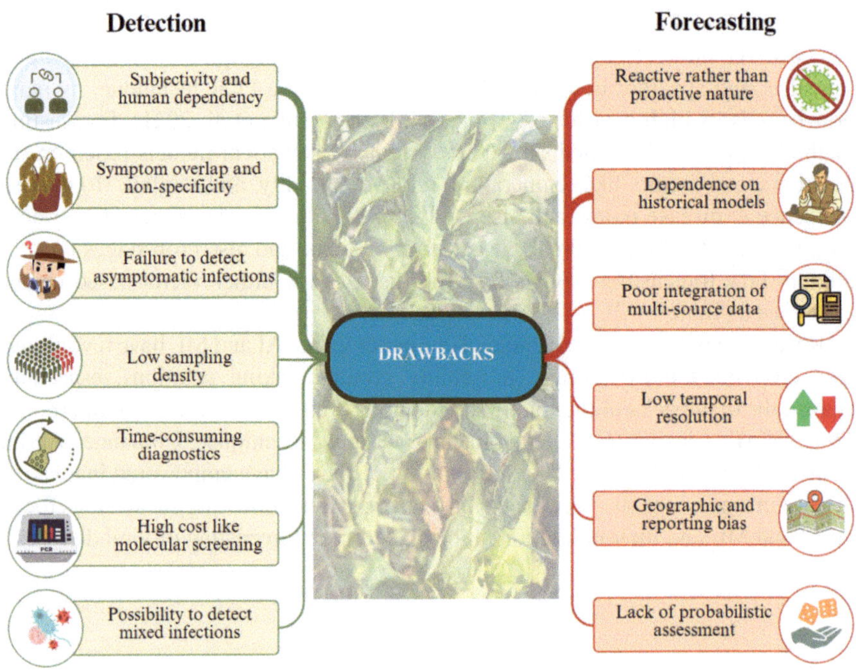


**Fig. 14.3** Various drawbacks of the conventional detection and forecasting techniques

## 14.4 Artificial Intelligence

The term artificial intelligence (AI) was first coined in 1956 by a workshop of Dartmouth under the leadership of McCarthy et al. (2006). Since then, it has adopted the contributions of various other research fields, including computer science and programming, neuroscience, robotics, linguistics, and futurology. Although the terminology is not well defined amongst different disciplines, Collins (2021) offers a holistic overview of definitions. AI is intelligence exhibited by machines in which systems are created to mimic human cognitive processes. In computer science, AI is the analysis of intelligent agents, wherein devices are capable of perceiving their surroundings and thereby acting in ways that are more likely to lead to the attainment of particular objectives. The term AI is colloquially applied when machines replicate human-like skills like learning and problem-solving. Current AI capabilities include speaking human language, playing strategic games (chess and Go), autonomous cars, network routing, simulations, and other complicated data interpretation tasks. The long-term goal is to achieve general intelligence in which machines will be able to work on different tasks to the same extent as humans do. The methods of AI are statistical, computational intelligence, and symbolic reasoning. The means used in AI include search and optimization tools as well as logic, probability, and economics. The study of AI cuts across several fields of study, viz., computer science, psychology, mathematics, philosophy, linguistics, neuroscience, agriculture, etc. Recent attempts also define AI as "the ability of a machine to engage in cognitive tasks similar to those of human minds, such as perceiving, learning and reasoning, interacting with its surroundings, problem-solving, decision-making, and also creativity" (Appio et al. 2023; Jha et al. 2019). Throughout the decades, AI research has been referred to as using two large dimensions, the "depth dimension," i.e., whether it is oriented to the human thought processes or to actual actions (thinking vs. acting), and the "objective dimension," i.e., replicating human decision-making or the purpose to find the most rational solution (human-like vs. rational). These dimensions resulted in various research lines, including cognitive modeling, in the search for machines that can think humanly, do reasoning steps similarly to humans, and come up with similar drawings (Liakos et al. 2018). With AI being stimulated, numerous new logics and methods are evolving and making problem-solving processes even simpler. The following are such methods as "Artificial neural networks (ANN)," "Fuzzy logic," "Expert systems," and "Neuro-fuzzy logic." Among all of these, ANN is the most popular and frequently utilized approach with regard to research purposes. ANNs are based on a biological model developed in 1959 by Nobel laureates David H. Hubel and Torsten Wiesel. Most ANNs are considered to be cascading models of cells (i.e., simple and complex cells in the primary visual cortex) (Weng et al. 1992). As the most complicated part of the body is our human brain, wherein the neural network is interrelated, and the synaptic signals pass via neurons using axons. The invention of the ANN method was done after remembering the same principle about how the human brain works. The algorithms of this method are of different types, based on various applications. ANN models can be trained using various algorithms like the Rprop algorithm,

Silva and Almeida, the Delta-bar-delta algorithm, Quickprop algorithm, Dynamic Adaptation algorithm, etc. ANN is a task-based approach, that is, it instructs the system to work according to a set of intrinsic tasks as opposed to a more traditional computational programming task. Furthermore, the architecture of an ANN consists of three layers, viz. the Input layer, the Hidden (middle) layer, Output layer. Artificial Intelligence (AI) is one of the key research topics in this highly technical field of computer science. Because of its rapid technological breakthroughs and robust application to problems, mostly those that are difficult for humans and standard computing structures to solve, this technology is rapidly becoming pervasive (Sharma 2021).

## 14.5 Machine Learning

Machine learning (ML), a subset of AI, is a technique for finding, comprehending, and analyzing patterns in data. It is a subset of artificial intelligence (Fig. 14.4). ML is often described as the "computer's ability to learn without being explicitly programmed" (Samuel 1959). ML is the study and development of algorithms (based on pattern recognition & computational learning theory) that can learn from and make predictions on data. Such algorithms overcome strictly static program instructions by making data-driven predictions or decisions through building a model from sample inputs (Jordan and Mitchell 2015). According to Tom Mitchell's more formal definition, a computer program is said to learn from experience (E) about a task (T) and a performance measure (P). If the program's performance on T, as

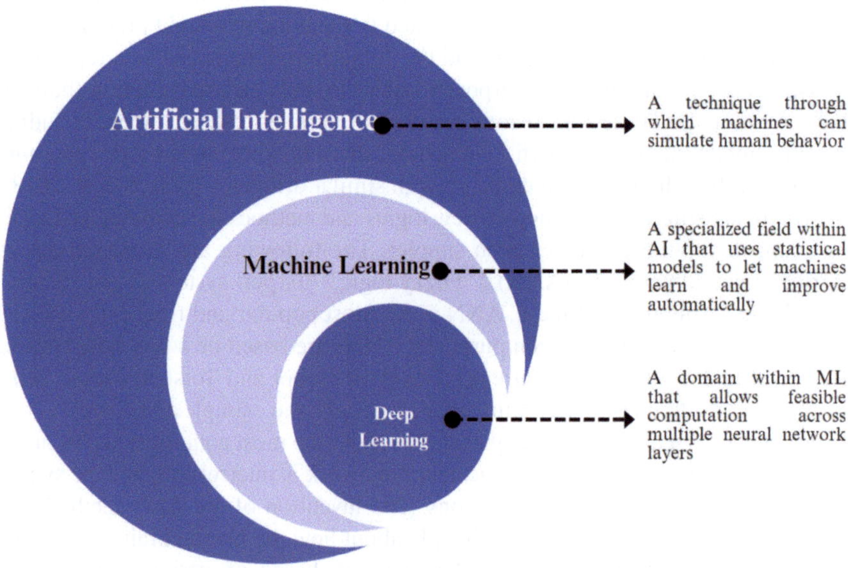


**Fig. 14.4** Subsets of AI

determined by P, increases with experience E, it is said to as a machine learning program (Samuel 1959; Das et al. 2015). ML is used when designing and programming explicit algorithms with good performance that are challenging or impractical, such as in the case of Email filtering, the detection of network intruders or malicious insiders working toward a data breach, optical character recognition (OCR), learning to rank, and computer vision, etc. It is closely related to mathematical optimization, which provides the field with theory, methods, and application domains. Data mining, a subfield that focuses more on exploratory data analysis and is called unsupervised learning, is sometimes confused with machine learning. Additionally, ML can be utilized in an unsupervised manner to identify significant anomalies by learning and establishing baseline behavioral profiles for different entities. Machine learning is a technique used in data analytics to create sophisticated models and algorithms that are predictive; this is referred to as predictive analytics in commercial contexts. By learning from past relationships and patterns in the data, these analytical models enable researchers, data scientists, engineers, and analysts to "produce reliable, repeatable decisions, and results" and unearth "hidden insights" (Ongsulee 2017).

### 14.5.1 Working Principle of ML

ML approaches usually comprise a learning process whose goal is to learn from "experience" (training data) to complete a given task. In ML, data is a set of examples, generally a set of attributes, also called variables or feature forms an example. This attribute can be nominal, i.e., enumeration, binary, i.e., 0 or 1, ordinal, like A+ or B−, or numeric (integer, number, etc.). To measure the efficiency of the ML model, a performance metric that is enhanced with experience over time is used. Various models (statistical & mathematical) have been employed to calculate the performance of ML. At the end, the trained model can be used to predict or group new testing data using the experience obtained during the training process. Figure 14.5 shows a typical ML approach. Depending on the learning model (classification, regression, clustering, and dimensionality reduction), learning type (supervised/unsupervised), and used to carry out the chosen task, machine learning tasks are usually divided into many broad groups (Samuel 1959; Das et al. 2015).

### 14.5.2 Types of Machine Learning Algorithms

ML algorithms are generally categorized into four major paradigms, viz., supervised learning, semi-supervised learning, unsupervised learning, and reinforcement learning (Fig. 14.6). Each category differs in the way training data is utilized, the nature of feedback available, and the kinds of tasks they are designed to solve (Xu and Wunsch 2005; Chapelle et al. 2009; Jordan and Mitchell 2015; Das et al. 2015; Silver et al. 2016; Ongsulee 2017).

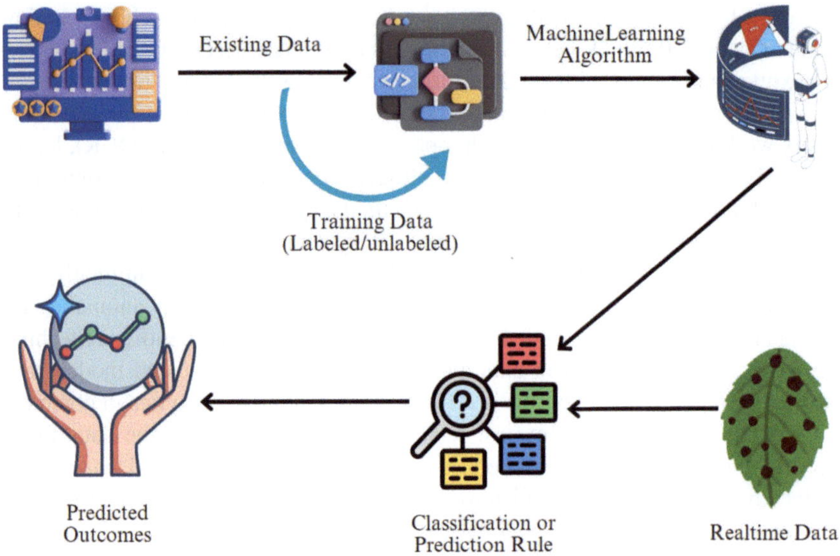


**Fig. 14.5** Graphical illustration of typical machine learning models

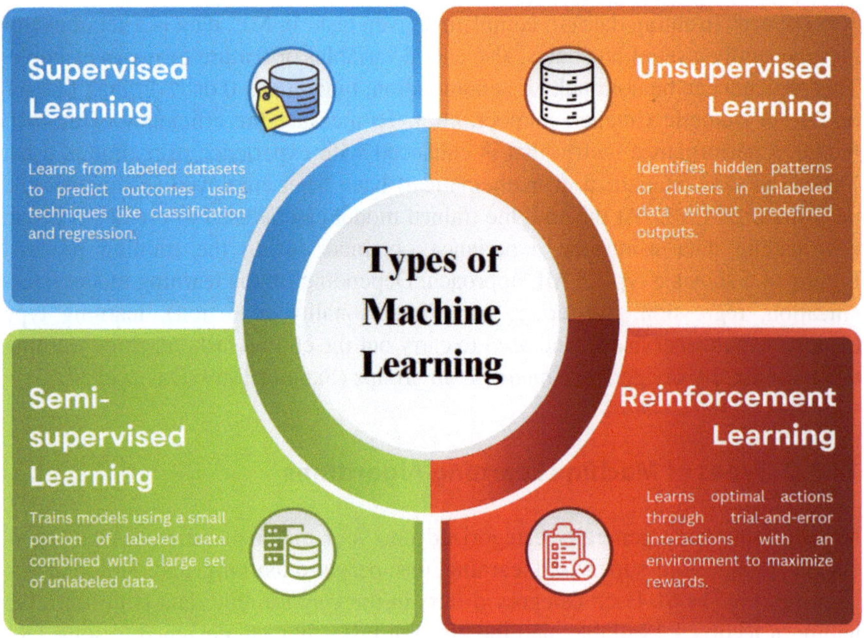


**Fig. 14.6** Types of machine learning algorithms

(i) *Supervised Learning Algorithms*: These are trained using labeled examples, such as an input where the desired output is known. Data points on a piece of equipment, for instance, might be marked "F" (failed) or "R" (runs). The algorithm is exposed to a set of inputs with their matching correct outputs. Such algorithms learn by comparing their actual output with the standard output entered in order to identify errors and modify the model accordingly. To understand Supervised, the best example is "facial recognition." This algorithm can be used for the detection of viruses.
(ii) *Unsupervised Learning*: No historical labels are required for the data to be utilized. The system needs to find out the "right answer" on its own based on the output. Such an algorithm explores the data set and finds the structure within using clustering and pattern discovery. It has been used in genomics and DNA microarray analysis.
(iii) *Semi-Supervised Learning*: Used for the same applications as supervised learning, but varies from it as it can utilize both labeled and unlabeled data for training. Methods like classification, regression, and prediction can be coupled with such learning.
(iv) *Reinforcement Learning*: Herein, the algorithm discovers through trial and error which actions yield the greatest rewards. It is often used for robotics, gaming, and navigation. This type of learning has been utilized for decision-making, skill-based learning simulations, forecasting, etc.

### 14.5.3 Deep Learning

Deep learning is a subset of ML algorithms that are trained to learn hierarchical representations of features in a multilayer neural network, and perform the task of perception, language, and decision-making without hand-authored features. It is placed in the domain of ML and AI, and its contemporary popularity indicates the progress in data and computing (with a special emphasis on GPUs), architectures, and optimization methods (Liakos et al. 2018).

## 14.6 Roles of AI/ML in Plant Disease Detection and Forecasting (Fig. 14.7)

(i) *Early and Precise Disease Identification*

AI and ML models are used to identify plant diseases quickly and accurately using visual, physiological, or spectral data. This reduces the errors in diagnosis that have been commonly linked with traditional manual evaluation and allows an early diagnosis at an initial stage when the signs of the condition are often not visible.

(ii) *Automated Image-Based Diagnosis*

Advanced deep learning models can be used to automatically classify plant diseases using digital images of leaves, stems, and fruits. These methods are

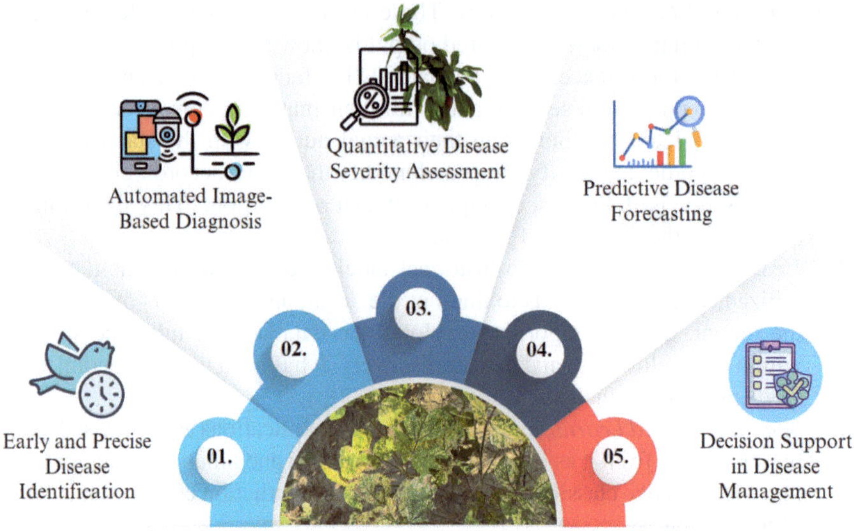


**Fig. 14.7** Role of AI/ML in plant disease detection & forecasting

scalable and can be deployed in the field in real-time with the use of a smartphone, drone, or remote sensing platform.

(iii) *Quantitative Disease Severity Assessment*

Beyond qualitative detection, ML algorithms support objective estimation of disease severity. These models can designate levels of infection (e.g., mild, moderate, severe) by segmenting diseased and healthy tissue regions, and can be used to more accurately improve practices of precision agriculture and direct site-specific interventions.

(iv) *Predictive Disease Forecasting*

AI/ML approaches integrate multiple data streams meteorological variables, soil health indicators, crop growth stage, and historical disease occurrence, to forecast the probability and temporal dynamics of disease outbreaks. This type of predictive analytics is necessary to implement preventive actions and minimize the loss of yields.

(v) *Decision Support in Disease Management*

By providing data-driven insights on disease progression and spatial distribution, AI/ML-based forecasting systems serve as decision support tools for farmers, extension workers, and researchers. These models suggest disease control measures that should be implemented in time and make the most of the use of fungicides or biocontrol agents to decrease the economic cost and environmental effects.

A variety of AI and ML models are utilized for plant virus detection in plants through image analysis. These models are either based on ML approaches or on more complex deep learning approaches.

### 14.6.1 Some Common Machine Learning Approaches in the Identification of Plant Viruses

These models often use features that are manually designed or extracted from images to classify or identify diseases.

- *Support Vector Machines (SVM)*: A supervised learning model that finds a hyperplane to classify data. SVMs use small to medium datasets and can be utilized to classify viral sequences or detect infected plants.
- *Random Forest (RF)*: This is an ensemble learning method that has high accuracy and robustness and can even detect virus-induced spectral changes in plants.
- *Decision Trees (DT)*: A supervised learning approach that creates a tree-like structure wherein each internal node represents a feature, each leaf node an outcome, and each branch a decision rule.
- *K-Nearest Neighbors (KNN)*: A nonparametric, basic image-based detection method that classifies a dataset based on the majority class of its "k" nearest neighbors in the feature space.
- *Naive Bayes (NB)*: A probabilistic classifier based on applying Bayes' theorem with strong (naive) independence assumptions between the features. It is used for the classification of genomic data.
- *XGBoost (XGB)*: An advanced implementation of gradient boosting that is known for its speed and performance.

### 14.6.2 Some Common Deep Learning Approaches in the Identification of Plant Viruses/Diseases

#### 14.6.2.1 Classification & Feature Extraction Networks

These networks are primarily used to assign a class label (e.g., a specific disease) to an entire image.

(a) *Artificial Neural Networks (ANN)*: Composed of interconnected nodes in layers, ANNs are used for various tasks, including pattern recognition and classification.

(b) *Convolutional Neural Networks (CNN or DCNN)*: CNNs are the most prominent deep learning models for image recognition. CNN from the input data automatically learns spatial hierarchies of features. Many specific, pretrained architectures (AlexNet, VGGNet, GoogleNet (Inception), ResNet (Residual Network), DenseNet, MobileNet) are commonly adapted for disease detection.

(c) *Object Detection Networks*

These models go a step further than classification by identifying *where* a disease is located within an image, typically by drawing a bounding box around it.

- *Faster R-CNN (Region-Based CNN)*: This DL approach is often used to directly locate diseases on plants utilizing 2 two-stage detection methods, i.e., stage I, proposing regions of interest; and stage II, classifying them.

- *YOLO (You Only Look Once)*: Unlike Faster R-CNN, this is a one-stage detector, wherein the entire image is processed in a single run, thereby making it suitable for rapid and real-time detection.
- *SSD (Single Shot MultiBox Detector)*: Another one-stage detector, which, in a single go, predicts and classifies directly from feature maps in a single run.

(d) *Image Segmentation Networks*

These models provide the most detailed analysis by classifying each pixel in an image. This allows for the precise outlining of the diseased area, which can be used to quantify the severity of the infection.

- *FCN (Fully Convolutional Network)*: A foundational model for semantic segmentation that replaces the fully connected layers in a CNN with convolutional layers to output a pixel-wise map.
- *U-Net*: An architecture known for its U-shaped design that includes skip connections between the down-sampling and up-sampling paths, making it very effective for segmenting biomedical and plant images.
- *Mask R-CNN*: Extends Faster R-CNN to perform instance segmentation, meaning it can detect and segment each instance of a disease lesion, even when they overlap.

Support Vector Machines (SVMs) and Convolutional Neural Networks (CNNs) are among the most widely used and consistently effective models in research and practice, warranting detailed coverage (Fig. 14.8, Table 14.5).

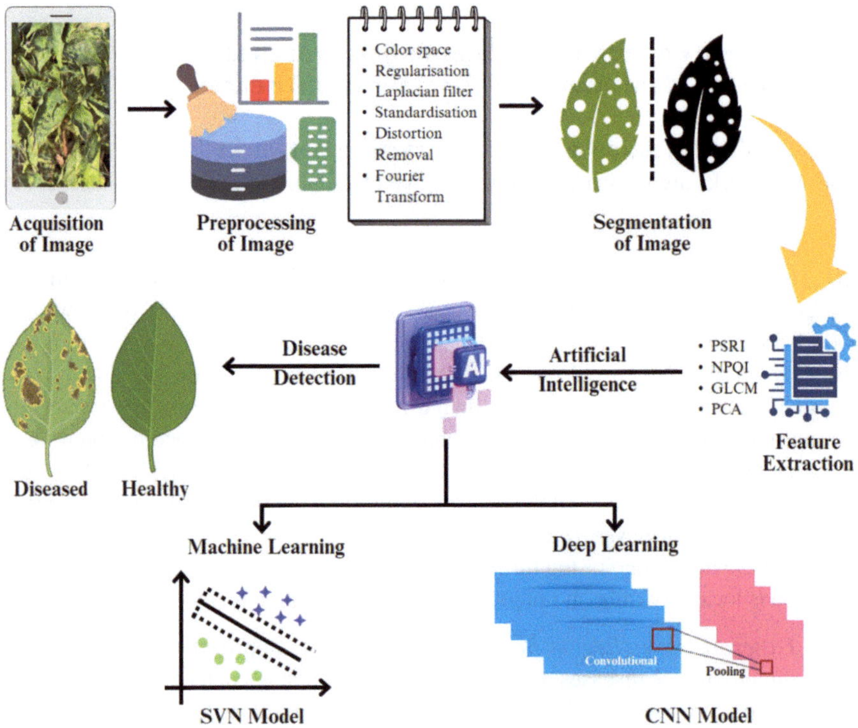


**Fig. 14.8** Working of various AI & ML/DL models for disease detection

### 14.6.3 Convolutional Neural Networks

The most common use of CNNs in detecting plant diseases is to classify leaf images of a variety of crops and datasets. To localize the lesions or symptomatic areas in the field of a patient, CNN-based object detectors are most popular, including YOLO and Faster R CNN. CNN is a DL architecture that is trained to directly learn visual patterns on top of images via convolutional filters, nonlinear activations, and pooling, and then to assign learned patterns to classes via fully connected networks. To classify plant disease, they learn the end-to-end discriminative patterns on pixels and match them to healthy and specific disease visual symptoms on leaves (spots, mosaics, chlorosis, lesions) (O'shea and Nash 2015; Ajit et al. 2020).

#### 14.6.3.1 Mechanism of CNN in Plant Disease Detection

The CNNs are used to process images of plant leaves by passing the image through a series of interconnected layers. Each layer is trained to extract the features of the image that are more complex and that differentiate between normal and diseased leaf images. The first part is the image acquisition, where the samples of the leaves are taken in the field or under laboratory conditions. Ahead of this, these images are preprocessed, i.e., they are scaled to standard size, pixel intensity is normalized, and often they are subjected to data augmentation methods, like color adjustment, flipping, and rotation to enhance the diversity of the dataset. A CNN has a convolutional step in its initial computational step. These layers slide local filters, or kernels, over the image to identify local features (Ajit et al. 2020). Under plant pathology, these characteristics can be outlined in the leaf edges, venation, and minor lesions or discoloration by the pathogen. The network starts to learn more complex and abstract features, e.g., the propagation of necrotic spots or the mosaic patterns characteristic of viral infections, as the data passes through more convolutional layers. To shrink the spatial dimension of the feature maps, convolution layers are separated by pooling layers. This step loses only the most crucial properties as the computational expenses are reduced, and the system becomes more resistant to minor deviations in the manifestation of the disease, like minor rotations or changes in the location of the leaf. Activation functions, typically the Rectified Linear Unit (ReLU), add nonlinearity and allow the CNN to learn how to represent complicated patterns between visual signs and disease types. Once enough features have been extracted, the CNN then flattens the multidimensional feature maps to vectors, which are then fed to fully connected layers. These layers are considered to be classifiers, which generate a probability of each possible class by uniting the acquired features. The final prediction of the disease (e.g., early blight, powdery mildew, or healthy leaf) is given by the output layer, which can be modeled as a Softmax function. The CNN is trained by showing large amounts of labeled images, comparing predicted values against actual labels, and correcting the filter weights via backpropagation to reduce the error as described by a loss function. Transfer learning has emerged as a major approach in plant disease detection. Instead of training a CNN on a blank system, which needs large datasets, researchers fine-tune existing models, including AlexNet, VGG16, ResNet, and Inception. Even models like ImageNet that were first trained on large-scale image datasets already have general feature extractors in them. With retraining of images only on plant-specific images, researchers can obtain high accuracy on smaller agricultural datasets, as well as save on computational costs and training time (Benos et al. 2021; Mohanty et al. 2016).

#### 14.6.3.2 CNN Versus PCR

CNN (in ML) and PCR (in molecular techniques) are the two most commonly utilized methods in their respective areas for the diagnosis of viruses in plants. In terms of accuracy, where CNN has shown an accuracy ranging from 90% to 99.7%, PCR can be 99% precise based on specific primers. However, the major drawback of CNN shall be its inability to detect early infections, therefore, limiting the early diagnosis of the disease, which is possible with quantitative PCR (Table 14.4).

**Table 14.4** Comparison between convolutional neural networks (CNNs) and PCR-based diagnostics for the detection of viral diseases in vegetable crops

| Feature | CNN-based detection | PCR-based detection |
|---|---|---|
| Accuracy | 99–99.7% accuracy for image-based detection | >99% sensitivity and specificity at the molecular detection level |
| Target | Visual symptoms (phenotypic) on plant tissues | Viral nucleic acid (DNA/RNA) sequences |
| Sensitivity | Dependent on symptom visibility, therefore, early or asymptomatic infection cannot be detected | Very high or low viral loads can be detected even before symptoms develop |
| Specificity | Limited by the similarity of symptoms among diseases and environmental effects | Very high |
| Infrastructure | Requires cameras or smartphones; deployable in the field or remotely | Requires laboratory equipment and trained personnel |
| Cost | Lower cost per test after model development; scalable to large datasets | Higher cost due to reagents, equipment, and labor |
| Speed | Rapid (real-time or near real-time imaging-based screening possible) | Slower (requires sample collection and lab processing) |

### 14.6.4 Support Vector Machines

Support Vector machines (SVMs) are based on the statistical theory of learning and are statistical learning models that construct an optimum separating hyperplane to maximize the distance between the classes, and can be extended to nonlinear decision surfaces through the use of kernel functions and to regression through epsilon-insensitive losses. It was initially proposed in the text of Cortes and Vapnik (1995) as part of the statistical learning theory. Practically, SVMs are also prized due to good generalization in high-dimensional space, resistance to over-fitting by using the max margin, and compatibility with several different tasks, including Least Squares SVM (LS SVM), Support Vector Regression (SVR), and pipelines that combine the Successive Projections Algorithm with SVM to provide better feature selection (Liakos et al. 2018). SVM is a strong and versatile supervised machine learning algorithm that is applicable in both classification as well as regression. The core idea behind SVM is that labeled data is used to train and determine an ideal boundary that would distinguish between the various types of objects. What is different about SVMs is that they not only find a separating boundary, but the optimum one, thus making them especially effective even in complex and high-dimensional data. This has made them effective in many applications, such as text categorization, image recognition, and many biological applications, such as classifying gene-expression profiles. The old method of visual inspection as the method of disease diagnosis is usually too late; when such symptoms as lesions, mosaic patterns, or yellowing appear, the pathogen has already been well-established and is difficult and costly to control. Effective intervention is the early detection—detection of stress in plants at an early presymptomatic phase. That is where machine learning could provide a revolutionary solution. Algorithms based on the analysis of complex data from sources such as hyperspectral cameras are able to identify subtle physiological changes in plants that cannot be observed with the human eye. The Support Vector Machine SVM is one of the strongest and most utilized algorithms for this classification challenge and is a good option with high-dimensional features and small labeled datasets typical of agricultural sensing (Benos et al. 2021).

#### 14.6.4.1 Mechanism of SVM in Plant Disease Detection

The conventional method of visual observation in the diagnosis of diseases is usually too late; before symptoms such as lesions, mosaics, or yellowing can be noticeable, the pathogen has already established itself and is hard and expensive to control. Timely intervention is the most important solution for the early detection of stress in plants before it manifests. This is the area where machine learning can provide a revolutionary solution. Algorithms treating complex data, such as hyperspectral camera data, can be used to observe minute changes in plants that are invisible to the human eye. The Support Vector Machine SVM is among the most powerful and popular algorithms to do this classification task, which is effective with high-dimensional features and moderate-sized labeled datasets typical of agricultural sensing (Benos et al. 2021). SVM usage in plant disease identification is a multi-step, systematic procedure that converts the raw sensor data into diagnoses that are actionable. It starts with the gathering of data on crops in the field or a controlled environment. This data should have indicators of plant health. This is normally done with advanced sensors:

- *Hyperspectral and Multispectral Imagers*: These gadgets are the key to early detection. They measure the reflectance of light of plant leaves in hundreds of small wavelength bands, including the visible to near-infrared range (e.g., 350–2500 nm). This is because plants that are healthy and stressed differ in spectral signature because of small alterations in pigments (such as chlorophyll), water content, and cell structure before visual symptoms emerge.
- *Digital and Thermal Cameras*: Both high-resolution RGB and thermal cameras have the potential to extract texture and color features as well as disease-induced changes in canopy temperature. Raw sensor data used is usually noisy and too complicated to be used directly. Hence, it needs to be narrowed down to a series of valuable attributes.
- *Preprocessing*: This is the initial step of the raw spectral data, which is cleaned to eliminate noise and irregularities due to atmospheric conditions or variability of equipment. Common techniques include applying mathematical corrections like the *Standard Normal Variate (SNV), Multivariate Scattering Correction (MSC)*, or smoothing filters like *Savitzky-Golay (SavGol)*.
- *Feature Extraction*: Based on the processed data, certain informative features are extracted. This helps cut down dimensions and concentrates the model on the most important information.

This is the learning phase. The features (a list of 20 vegetation indices) and the known labels (Healthy or Diseased) are put into the SVM algorithm. It is then fed into the SVM, which determines the best hyperplane that separates the classes in the high-dimensional feature space (Liakos et al. 2018). After the model has been trained, it is now able to make predictions. When a new unlabeled plant has its data gathered and run through the same feature extraction pipeline, the SVM model projects their feature vector into the existing dimensional space. Depending on which side of the hyperplace the new data point is on, the model will either define the plant as being Healthy or Diseased, giving a fast, nondestructive diagnosis (Table 14.5).

**Table 14.5** List of a few viral diseases in various vegetable crops and their detection or forecasting using ML/DL

| Crop | Virus | ML/DL model | Accuracy (%) | Detection/ forecasting | References |
|---|---|---|---|---|---|
| Tomato | Tomato leaf curl virus | Convolutional neural networks | – | Detection | Bupi et al. (2023) |
| | Tomato mosaic virus, tomato leaf curl virus | MobileNetV2 and ResNet50 | 99.91 | Detection | Sharma et al. (2025) |
| | | Convolutional neural networks | 99.18 | Detection | Brahimi et al. (2017) |
| Chilli | Chili leaf curl virus (ChiLCV) | EfficientNetB4 | 92 | Detection | Pratap and Kumar (2023) |
| Potato | PLRV | Convolutional neural networks | 99.23 | Forecasting | Tarik et al. (2021) and Kaur et al. (2024) |
| | Mosaic, leaf curl, tuber-like potato spindle tuber viroid (PSTVD), potato leafroll virus (PLRV), and potato virus Y (PVY) | Logistic regression (LR), support vector machine (SVM), decision tree (DT), and KNN | 81.05 | Detection | Raigond et al. (2022), Jeevalatha et al. (2015), Kumar et al. (2017), Kumar et al. (2023), and Kumar et al. (2025) |

## 14.7 Challenges and Limitations

Even though AI and ML are developing quickly, various obstacles prevent their widespread use in vegetable crop virus detection and forecasting. The datasets are frequently limited, dispersed, or regional, particularly for vegetable viruses. Model training and validation often face hurdles like poor image quality, insufficient data, lack of real-time ground truth validation, etc. The scalability and generalizability of AI models represent yet another significant constraint. Usually, the models perform well in controlled sets; however, in the operational field, the accuracy decreases. Many times, the viral symptoms are misclassified with abiotic stresses like nutritional deficiencies. Furthermore, farmers in many places lack the technical know-how to use these AI-driven models.

## 14.8 Conclusion

As we have entered the era of Artificial Intelligence, we have seen its major impact on how we detect, diagnose, and utilize the diagnosis for predicting viral diseases. AI and ML have the potential to change how we identify and predict diseases in vegetable crops. In crops like tomato, pepper, and cucumber, AI-driven models have proven their accuracy in prediction and detection, therefore solidifying their role in sustainable agriculture. However, in order to be successfully incorporated into more crops, issues like accessibility to farmers, cost, generation of models, and data availability should also be taken into consideration. AI coupled with sensors, IoT, etc., can provide early, precise, and scalable methods for managing viral outbreaks. Therefore, it is necessary to have AI tools that work in harmony with current farming methods in order to close the gap between recent innovations and their application.

## References

Ajit A, Acharya K, Samanta A (2020) A review of convolutional neural networks. In: 2020 international conference on emerging trends in information technology and engineering (ic-ETITE). IEEE, pp 1–5. https://doi.org/10.1109/ic-etite47903.2020.049

Al Bashish D, Braik M, Bani-Ahmad S (2010) A framework for detection and classification of plant leaf and stem diseases. In: 2010 international conference on signal and image processing. IEEE, pp 113–118. https://doi.org/10.1109/icsip.2010.5697452

Ampatzidis Y, De Bellis L, Luvisi A (2017) iPathology: robotic applications and management of plants and plant diseases. Sustainability 9:1010. https://doi.org/10.3390/su9061010

APEDA (2024) Agricultural and processed food products export development authority. https://apeda.gov.in/FreshFruitsAndVegetables

Appio FP, La Torre D, Lazzeri F, Masri H, Schiavone F (2023) Artificial intelligence in business: opportunities and challenges. In: Impact of artificial intelligence in business and society. Routledge, pp 83–167. https://doi.org/10.4324/9781003304616-6

Benos L, Tagarakis AC, Dolias G, Berruto R, Kateris D, Bochtis D (2021) Machine learning in agriculture: a comprehensive updated review. Sensors 21(11):3758. https://doi.org/10.3390/s21113758

Brahimi M, Boukhalfa K, Moussaoui A (2017) Deep learning for tomato diseases: classification and symptoms visualization. Appl Artif Intell 31(4):299–315. https://doi.org/10.1080/08839514.2017.1315516

Buja I, Sabella E, Monteduro AG, Chiriacò MS, De Bellis L, Luvisi A, Maruccio G (2021) Advances in plant disease detection and monitoring: from traditional assays to in-field diagnostics. Sensors 21(6):2129. https://doi.org/10.3390/s21062129

Bupi N, Sangaraju VK, Phan LT, Lal A, Vo TTB, Ho PT, Qureshi MA, Tabassum M, Lee S, Manavalan B (2023) An effective integrated machine learning framework for identifying severity of tomato yellow leaf curl virus and their experimental validation. Research 6:0016. https://doi.org/10.34133/research.0016

Chapelle O, Scholkopf B, Zien A (2009) Semi-supervised learning (introduction). IEEE Trans Neural Netw 20(3):542–542. https://doi.org/10.1109/tnn.2009.2015974

Collins H (2021) The science of artificial intelligence and its critics. Interdiscip Sci Rev 46(1–2):53–70. https://doi.org/10.1080/03080188.2020.1840821

Cortes C, Vapnik V (1995) Support-vector networks. Machine learning 20(3):273–297. https://doi.org/10.1023/a:1022627411411

Cruz A, Ampatzidis Y, Pierro R, Materazzi A, Panattoni A, De Bellis L et al (2019) Detection of grapevine yellows symptoms in Vitis vinifera L. with artificial intelligence. Comput Electron Agric 157:63–76. https://doi.org/10.1016/j.compag.2018.12.028

Das S, Dey A, Pal A, Roy N (2015) Applications of artificial intelligence in machine learning: review and prospect. Int J Comput Appl 115(9). https://doi.org/10.5120/20182-2402

Fargette D, Konaté G, Fauquet C, Muller E, Peterschmitt M, Thresh JM (2006) Molecular ecology and emergence of tropical plant viruses. Annu Rev Phytopathol 44:235–260. https://doi.org/10.1146/annurev.phyto.44.120705.104644

Hanssen IM, Lapidot M, Thomma BPHJ (2010) Emerging viral diseases of tomato crops. Mol Plant-Microbe Interact 23(5):539–548. https://doi.org/10.1094/mpmi-23-5-0539

Jafar A, Bibi N, Naqvi RA, Sadeghi-Niaraki A, Jeong D (2024) Revolutionizing agriculture with artificial intelligence: plant disease detection methods, applications, and their limitations. Front Plant Sci 15:1356260. https://doi.org/10.3389/fpls.2024.1356260

Jeevalatha A, Kumar R, Raigond B, Sundaresha S, Sharma S, Singh BP (2015) Duplex real-time RT-PCR assay for the detection of potato spindle tuber viroid (PSTVd) along with ef 1-α gene of potato. Phytoparasitica 43:317–325. https://doi.org/10.1007/s12600-014-0452-z

Jha K, Doshi A, Patel P, Shah M (2019) A comprehensive review on automation in agriculture using artificial intelligence. Artif Intell Agric 2:1–12. https://doi.org/10.1016/j.aiia.2019.05.004

Jordan MI, Mitchell TM (2015) Machine learning: trends, perspectives, and prospects. Science 349(6245):255–260. https://doi.org/10.1126/science.aaa8415

Kaur A, Randhawa GS, Abbas F, Ali M, Esau TJ, Farooque AA, Singh R (2024) Artificial intelligence-driven smart farming for accurate detection of potato diseases: a systematic review. IEEE Access https://doi.org/10.1109/access.2024.3510456

Kumar R, Jeevalatha A, Baswaraj R, Sharma S, Nagesh M (2017) A multiplex RT-PCR assay for simultaneous detection of five viruses in potato. J Plant Pathol 99:37–45. https://doi.org/10.4454/jpp.v99i1.3824

Kumar A, Sarkar S, Pradhan C (2019a) Recommendation system for crop identification and pest control technique in agriculture. In: 2019 international conference on communication and signal processing (ICCSP). IEEE, pp 185–189. https://doi.org/10.1109/iccsp.2019.8698099

Kumar R, Tiwari RK, Jeevalatha A, Kaundal P, Sharma S, Chakrabarti SK (2019b) Potato viruses and their diagnostic techniques-an overview. J Pharmacogn Phytochem 8:1932–1944

Kumar R, Kaundal P, Tiwari RK, Lal MK, Kumari H, Kumar R, Naga KC, Kumar A, Singh B, Sagar V, Sharma S (2023) Development of reverse transcription recombinase polymerase amplification (RT-RPA): a methodology for quick diagnosis of potato Leafroll viral disease in potato. Int J Mol Sci 24:2511. https://doi.org/10.3390/ijms24032511

Kumar R, Lal MK, Tiwari RK, Kumar D, Sagar V, Jena R, Kumar R, Singh G, Singh B (2025) Antioxidant defence mechanism of potato (Solanum tuberosum) by regulation of physiological and biochemical response under Potato Virus Y necrotic strain (PVYNTN) infection. J Plant Growth Regul. https://doi.org/10.1007/s00344-025-11850-0

Lal P, Tiwari RK, Kumar A, Altaf MA, Alsahli AA, Lal MK, Kumar R (2023) Bibliometric analysis of real-time PCR-based pathogen detection in plant protection research: a comprehensive study. Front Plant Sci 14:1129714. https://doi.org/10.3389/fpls.2023.1129714

Liakos KG, Busato P, Moshou D, Pearson S, Bochtis D (2018) Machine learning in agriculture: a review. Sensors 18(8):2674. https://doi.org/10.3390/s18082674

Luvisi A, Ampatzidis YG, De Bellis L (2016) Plant pathology and information technology: opportunity for management of disease outbreak and applications in regulation frameworks. Sustainability 8:831. https://doi.org/10.3390/su8080831

McCarthy J, Minsky ML, Rochester N, Shannon CE (2006) A proposal for the dartmouth summer research project on artificial intelligence, august 31, 1955. AI Mag 27(4):12

Mishra GP, Singh B, Seth T, Singh AK, Halder J, Krishnan N et al (2017) Biotechnological advancements and begomovirus management in okra (Abelmoschus esculentus L.): status and perspectives. Front Plant Sci 8:360. https://doi.org/10.3389/fpls.2017.00360

Mohanty SP, Hughes DP, Salathé M (2016) Using deep learning for image-based plant disease detection. Front Plant Sci 7:1419. https://doi.org/10.3389/fpls.2016.01419

Nagendran K, Mohankumar S, Aravintharaj R, Balaji CG, Manoranjitham SK, Singh AK et al (2017) The occurrence and distribution of major viruses infecting cucurbits in Tamil Nadu state, India. Crop Prot 99:10–16. https://doi.org/10.1016/j.cropro.2017.05.006

O'shea K, Nash R (2015) An introduction to convolutional neural networks. arXiv preprint arXiv:1511.08458

Ongsulee P (2017) Artificial intelligence, machine learning and deep learning. In: 2017 15th international conference on ICT and knowledge engineering (ICT&KE). IEEE, pp 1–6. https://doi.org/10.1109/ictke.2017.8259629

Paul S, Emmadi V, Saxena S, Sarkar M, Mandal B, Kumar R, Sinha P, Roy A (2025) RPA-assisted CRISPR-Cas12a-enabled point-of-care diagnostic platform for chili leaf curl virus with fluorescent and colorimetric readouts. Front Microbiol 16:1644322. https://doi.org/10.3389/fmicb.2025.1644322

Pratap VK, Kumar NS (2023) High-precision multiclass classification of chili leaf disease through customized EffecientNetB4 from chili leaf images. Smart Agric Technol 5:100295. https://doi.org/10.1016/j.atech.2023.100295

Raigond B, Verma G, Kumar R, Tiwari RK (2022) Serological and molecular diagnosis of potato viruses: an overview. In: Chakrabarti SK, Sharma S, Shah MA (eds) Sustainable management of potato pests and diseases. Springer, Singapore, pp 337–352. (Hardcover ISBN 978-981-16-7694-9; Softcover ISBN978-981-16-7697-0; eBook ISBN978-981-16-7695-6). https://doi.org/10.1007/978-981-16-7695-6_13

Samuel AL (1959) Some studies in machine learning using the game of checkers. IBM J Res Dev 3(3):210–229. https://doi.org/10.1147/rd.33.0210

Sharma R (2021) Artificial intelligence in agriculture: a review. In: 2021 5th international conference on intelligent computing and control systems (ICICCS). IEEE, pp 937–942. https://doi.org/10.1109/iciccs51141.2021.9432187

Sharma J, Al-Huqail AA, Almogren A, Doshi H, Jayaprakash B, Bharathi B et al (2025) Deep learning based ensemble model for accurate tomato leaf disease classification by leveraging ResNet50 and MobileNetV2 architectures. Sci Rep 15(1):13904. https://doi.org/10.1038/s41598-025-98015-x

Silver D, Huang A, Maddison CJ, Guez A, Sifre L, Van Den Driessche G et al (2016) Mastering the game of Go with deep neural networks and tree search. Nature 529(7587):484–489. https://doi.org/10.1038/nature16961

Tarik MI, Akter S, Al Mamun A, Sattar A (2021) Potato disease detection using machine learning. In Proceedings of the 3rd international conference on intelligent communication technologies and virtual mobile networks, pp 800–803. https://doi.org/10.1109/ICICV50876.2021.9388606

Tatineni S, Hein GL (2023) Plant viruses of agricultural importance: current and future perspectives of virus disease management strategies. Phytopathology 113(2):117–141. https://doi.org/10.1094/phyto-05-22-0167-rvw

Wani JA, Sharma S, Muzamil M, Ahmed S, Sharma S, Singh S (2022) Machine learning and deep learning based computational techniques in automatic agricultural diseases detection: methodologies, applications, and challenges. Arch Comput Methods Eng 29(1):641–677. https://doi.org/10.1007/s11831-021-09588-5

Weng J, Ahuja N, Huang TS (1992) Cresceptron: a self-organizing neural network which grows adaptively. In: [Proceedings 1992] IJCNN international joint conference on neural networks, vol 1. IEEE, pp 576–581. https://doi.org/10.1109/ijcnn.1992.287150

Xu R, Wunsch D (2005) Survey of clustering algorithms. IEEE Trans Neural Netw 16(3):645–678. https://doi.org/10.1109/tnn.2005.845141

[illegible]

# Participatory Plant Health Surveillance for Early Detection of Viral Outbreaks in Vegetables

15

Ubilfrido Vásquez-Gutierrez, Agustín Hernández-Juárez, and Juan Mayo-Hernandez

**Abstract**

Early detection of viral diseases in vegetable crops has been based on principles that include participatory surveillance (PS). This chapter integrates local knowledge, the role of farmers, and scientific technologies to contribute to and improve detection strategies. PS has been affected by genetic recombination of viruses, rapid spread, and a wide range of hosts, similar to what happened with ToBRFV and TYLCV. Late detection has led to the introduction of exotic plant viruses and yield losses of up to 100% in crop production. The participatory surveillance pyramid integrates farmers and extension workers who are responsible for inspecting the symptoms at the field level, while researchers, diagnostic laboratories, and the federal government verify the results, implement eradication actions, and establish new policies. The PS relies on information technology to develop sampling procedures, training programs for producers, and improve the efficiency of virus diagnosis. Cutting-edge molecular techniques and next-generation platforms for mass sequencing link two different scenarios, laboratory and field, to detect the presence of unknown viruses. However, participatory strategies face obstacles in terms of logistics, data validation, and sustainability, which affect the management of viral threats in vegetable production systems.

U. Vásquez-Gutierrez (✉)
Department of Parasitology, Universidad Autónoma Agraria Antonio Narro, Saltillo Coahuila, CP, Mexico

Department of Parasitology, Work Area of Plant Virology. Universidad Autónoma Agraria Antonio Narro, Saltillo Coahuila, CP, Mexico
e-mail: d_ubilfrido.vazquezG@uaaan.edu.mx

A. Hernández-Juárez · J. Mayo-Hernandez
Department of Parasitology, Universidad Autónoma Agraria Antonio Narro, Saltillo Coahuila, CP, Mexico



R. Kumar et al. (eds.), *Vegetable Crops and Viral Pathogens Volume I*, Plant Pathology for Sustainable Agriculture, https://doi.org/10.1007/978-981-95-7265-6_15

**Keywords**

Dissemination · Monitoring · Phytosanitary systems · Rapid recombination · Viral pathogens

## 15.1 Introduction to Participatory Surveillance in Plant Health

The participation of communities in identifying and monitoring events that affect stability, health, and well-being is called participatory pest surveillance (PS). This movement promotes the value of local knowledge and normalizes collaboration between the social population and phytosanitary authorities to streamline an effective response. PS considers five fundamental principles: (i) community autonomy; (ii) transparency and trust; (iii) equity and social justice; (iv) privacy protection; and (v) collective action and empowerment (McNeil et al. 2025). The context of participatory surveillance has gained relevance in recent years. Participatory surveillance provides descriptive information on the system approach, geography, users, technology, information sharing, and the perceived impact of participatory surveillance systems across the health spectrum (Soubeyrand et al. 2023; McNeil et al. 2022). The understanding of plant disease epidemiology, surveillance, and prevention arose from principles applied to zoonotic diseases, hence the similar concepts adopted in agroecology. Morris et al. (2022) described four knowledge challenges for plant health management: (i) discovering reservoirs and reviewing pathogen life cycles; (ii) elucidating virulence factors beyond the context of direct interactions between the host and pathogen; (iii) considering natural long-distance dissemination pathways; and (iv) updating disease forecasts in the face of environmental and cultural changes. The participation of stakeholders in disease surveillance and prevention is a key point for collaboration in biodiversity management and, consequently, human, animal, and plant health (Morris et al. 2022). McNeil et al. (2022) identified around 60 participatory surveillance systems in operation worldwide, of which 43% include data on environmental health, but not objectively for plants, although they mention the applicability for collecting data from associated and trained members for the early detection of invasive pests in plants.

### 15.1.1 Historical Background and Developments in Plant Pathology

Since 2012, the International Workshop on Participatory Surveillance (IWOPS) has been a pioneer in developing practices and methods for the direct participation of populations in the monitoring of endemic and epidemic diseases in humans (Smolinski et al. 2017). Grant et al. (2019) studied citizen participation in a socio-technical biosecurity system (STS) in New Zealand, based on surveillance technologies for forest protection. They concluded that the success of early detection will depend on the biosecurity authority of the central government and the general public, supported by a general surveillance network (GSN). Others have linked it to participatory epidemiology focused on disease prevention in animals (Alders et al. 2020). García-Figueroa et al. (2021) used the huanglongbing (HLB) epidemic as a study model and examined the institutions that guarantee citrus health in different countries around the world. They concluded that it is necessary to understand the collective action problem associated with phytosanitary provision and to combine institutional approaches over time to ensure food production. Sixty-five scientists participated in a research agenda to improve and modernize phytosanitary surveillance in order to anticipate and mitigate the emergence of diseases and insect pests. The approach integrated a wide range of scientific fields and identified key knowledge gaps in prevention, risk assessment, early detection, and multi-stakeholder collaboration. These research lines were organized into four thematic areas: i) anticipating the emergence of pests; ii) using versatile, broad-spectrum surveillance tools; iii) pest surveillance using novel approaches; and iv) plant health management through the establishment of multi-stakeholder and cooperative surveillance systems for early warning systems (Sawadogo et al. 2025). Plant crops provide more than 80% of food to consumers and are the main source of nutrition for meat-producing animals. However, pests threaten plant security, causing global losses of up to 30% and billions of dollars in food production. Most case studies focus on phytopathogenic fungi, food safety and mycotoxins, human pathogens associated with plants, and the use of pesticides in agriculture (Rizzo et al. 2021), but diseases caused by phytopathogenic viruses are overlooked (Savary et al. 2019). Virus-caused diseases reduce crop quality and cause yield losses, leading to ecosystem instability and affecting the ecological diversity of plant species (Tatineni et al. 2023). Global economic losses amount to US$30 billion annually, of which 50% are caused by emerging viruses (Salaudeen et al. 2016). The spread and dispersion of viruses has been facilitated by international trade in plant and propagative material, as well as environmental factors that promote the emergence of viral variants and new strains (Takahashi et al. 2019). After introduction into the country, viruses can easily spread through contact between plants, equipment, tools, weed reservoirs, raw materials, and insect vectors.

### 15.1.2 Relevance of Viral Diseases in Vegetables

The instability of climate change has resulted in the spread of plant viruses, with the disruption of homeostasis creating optimal conditions for the spread and dissemination of viruses (Tatineni and Hein 2023). Rising temperatures and changes in precipitation and humidity levels have led to the mobilization and adaptation of potential virus vectors, affecting crops such as rice, wheat, corn, and legumes. These factors are undermining the foundations of food security (Davino et al. 2020).

## 15.2 Importance of Early Detection in the Management of Viral Diseases in Plants

Epidemiological surveillance works in coordination with the early detection of plant viruses. Viral diseases could drastically reduce crop production if early and effective detection is not carried out, which would result in high production losses. The spread of emerging viruses affecting economically important crops could destabilize each country's established epidemiological surveillance program and, as a result, lead to the rapid spread of plant viruses, as was the case in 2018 with the tomato brown rugose fruit virus (ToBRFV) in México (Camacho-Beltran et al. 2019; Cambrón-Crisantos et al. 2019) and the tomato fruit blotch virus (ToBFV) in Brazil (Ciuffo et al. 2020). Therefore, phytosanitary agencies in each country must commit to strengthening surveillance against the introduction of emerging exotic viruses. Oxford Nanopore (ONT) sequencing is considered a third-generation technology that comes in the form of a portable MinION sequencer that requires minimal technical skills and bioinformatics knowledge, making it easy to apply and scale up in field conditions (Ranabhat et al. 2025). The approximate cost of ONT ranges from $4000 to $28,000, depending on model and version, while the annual investment cost per country for epidemiological surveillance of plant viruses ranges from $18,000 to $36 million per year. However, this cost increases exponentially when a new plant virus is introduced that poses a global threat, which can amount to an estimated cost of US$2.6 billion. Currently, sequencing technologies such as next-generation sequencing (NGS) and high-throughput sequencing (HTS) platforms offer advantages for the rapid detection of viral diseases through high-performance analysis (Lefterova et al. 2015). For many low-income and developing countries, it is difficult to cover the cost of these virological analyses, even though there is a pressing need for these revolutionary technologies to quickly and accurately diagnose known and unknown viral pathogens (Nezhad 2014). It has been mentioned that the primary problem in identifying new viral species lies in the assembly and mapping criteria. ONT is a promising technology for viral transcriptomic analysis, structural variation, and genomic resequencing, and the most innovative aspect is that it can be performed in the field, although the error rate remains an obstacle (5–10%) (Rang et al. 2018). Coupling with MinION nanopores improves the accuracy of rapid results in the field. ONT technology is not new, having already been demonstrated during the detection of an Ebola outbreak in Liberia in 2014 (Hoenen et al. 2016), which directly provides a promising way for the diagnosis of viral diseases in plants (Mehetre et al. 2021). Dennis et al. (2018) proposed a global virome project, whose objective was to sequence the complete genomes of viral species affecting humans and animals, given the emergence of outbreaks of new deadly viruses that spread rapidly and cause socioeconomic imbalance in humanity. However, this meta-project did not focus on plant viruses. Why? "The specific reason for this approach is unknown, although there are multiple arguments that could justify it." The use of a global virome meta-analysis focused on viral species could improve the application of tools based on new vaccines and virotherapies, which are ineffective due to the rapid spread and genetic mutation of new virus strains.

Knowing the identity of a plant virus would benefit the creation of infectious cDNA clones, which would support the detection of emerging virus mutations through nucleotide changes and promote the development of genetic materials resistant to viral variants (Claverie et al. 2023; Jones et al. 2021). Food security should be equal to human security. Climate change is causing temperatures to rise and, in turn, the emergence of new strains of plant viruses. Even if we invest in cutting-edge technology for the human sector, this does not guarantee the stability of the human community. Investment in equipment and infrastructure should focus on the surveillance and control of plant viruses that pose a daily threat to crops in different countries. As Norman Borlaug, Nobel Peace Prize winner (1970), said: "Civilization as we know it would not exist without agriculture."

### 15.2.1 Characteristics of Viral Pathogens that Require a Rapid Response

Since the first discovery of tobacco mosaic virus (TMV) by Ivanovsky (1892), it was thought that the causative agent of mosaic disease in tobacco was toxins. Subsequently, descriptors such as thermal inactivation point, isoelectric point, dilution endpoint, viral inclusions, differential plants, persistence, etc., began to be reported (Hamilton et al. 1981; Steere 1959). These were believed to be novel advances at the time, although, of course, many of these descriptors continue to be important for the management of plant viruses in field conditions. Vasquez-Gutierrez et al. (2024) studied the thermal inactivation of infectious sap from ToBRFV and reported alternate hosts, latency period, persistence, incubation period, and viral inclusions produced by infection of tobacco and tomato plants. Molad et al. (2024) studied the stability of ToBRFV virions under different pH conditions, demonstrating that above pH 10.0, the virus infectivity is reduced. Due to advances in bionanotechnology, descriptions of plant viruses have progressed over the years. The use of high-throughput sequencing (HTS) and next-generation sequencing (NGS) tools improves our understanding of new species and quasi-species that emerge through viro-genic recombinations (Naito et al. 2019). In this regard, it could be said that there must be specific characteristics to categorize a viral species with the ability to cause pandemics in susceptible crops, and then make important decisions to deny entry into pest-free territories. The following are specific characteristics of a viral species that require a rapid response to prevent introduction, implement management and eradication measures (Fig. 15.1): (a) molecular and physicochemical characterization; (b) origin and genetic variability; (c) rapid recombination; (d) symptoms; (e) transmission; and (f) host range.

I. *Molecular-physicochemical characterization.* Physicochemical and molecular characteristics, such as release and persistence of virions in solid and aqueous environments, have been described for many years. The Plant Virus Database (DPV) contains around 400 individual descriptions of plant viruses discovered between 1970 and January 2023 (Adams et al. 2005). These descriptors are

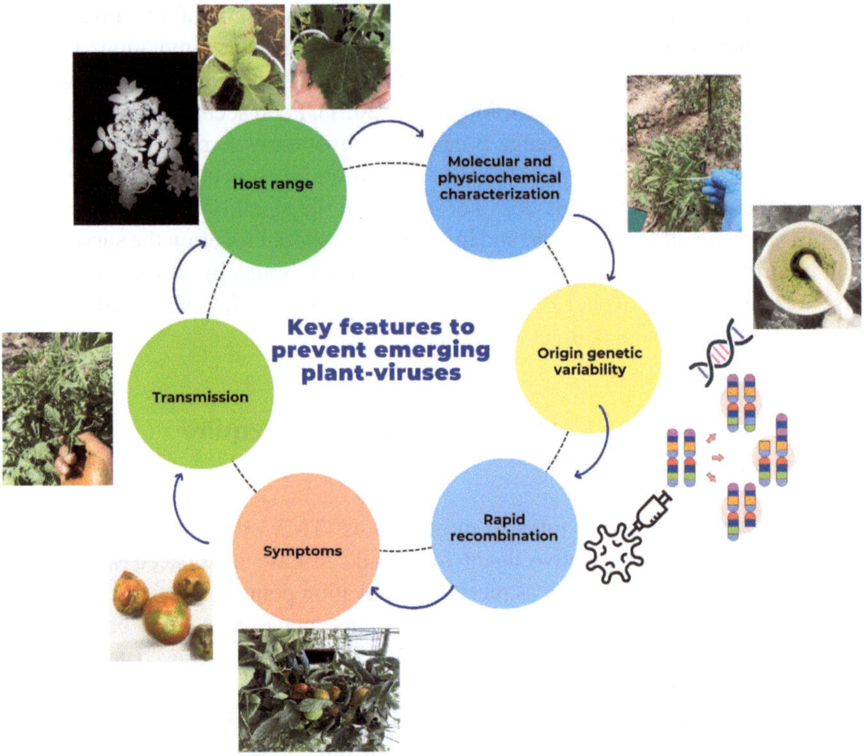


**Fig. 15.1** Key characteristics that confer emerging viruses

used to classify virus types according to the genetic material, stability or persistence, pH, dilution endpoint, latency period, incubation, etc. However, at present, the study of these characteristics is not taken into account as much (Mehle and Ravnikar 2012). Through a global analysis, it is proposed to include "descriptors" within specific characteristics to categorize potential emerging and re-emerging viruses that could represent serious pandemics and epidemics in the world.

II. *Genetic origin and variability (mutation rate).* Genetic variation involves two important terms: mutation and recombination. The former is the process that results in differences between nucleotides incorporated into a daughter strand during nucleic acid replication and those in the template. Mutation is the initial source of variation in populations, hence the interest in estimating the rate at which it occurs (Drake and Holland 1999; García-Arenal et al. 2001). Malpica et al. (2002) reported that most mutations in tobacco mosaic virus (TMV) and other RNA viruses do not adapt naturally, but rather due to the need for rapid replication of a chemically unstable genome, rather than being an evolutionary strategy among variants (Drake and Holland 1999). While recombination is the process in which various segments of genetic information are exchanged between the chains of different genetic variants, this produces a response to evolutionary variation (García-Arenal et al. 2001). Recombination is the result of random changes in the biological properties of viruses, causing the emergence of strains that break host resistance (Legg et al. 2000), which, in turn, is produced by the gene flow of transgenic plants resistant to virus derivatives and infectious virus populations (García-Arenal et al. 2001).

III. *Rapid recombination.* This occurs through the exchange of genetic information via small, unsegmented RNA fragments. Lai (1992) defined three types of RNA recombination: homologous recombination between viruses and cellular genes, which occurs during natural evolution; aberrant homologous recombination, which results from the duplication, insertion, or loss of genetic materials; and non-homologous recombination, which does not involve a homologous sequence. RNA recombination causes the emergence of viral variants, which leads to adaptation to new hosts and the emergence of pathogenic variants that break resistance. Sometimes RNA recombination increases the fitness of the virus, repairs the viral genome, and eliminates mutations that arise from erroneous replications (Cheng et al. 2005). Plant genes can affect RNA recombination, hindering the adaptation and evolution of the virus in the host (Nagy et al. 2008). Recombination is a relatively normal process in some plant viruses, especially positive-sense RNA viruses, such as potyviruses (Chare and Holmes 2006). Some reports suggest recombination between unrelated and distant viruses, suggesting that plant viruses undergo recombination with different genetic materials in the host (Sztuba et al. 2011). Geminivirus recombination adapts to changes in hosts as a result of transmission by polyphagous vectors (*Bemisia tabaci*) that facilitate the movement of native viruses between plants. These changes drive the emergence of new geminivirus variants, as most are descendants of interspecies or intraspecies recombination (Qin and Petty 2001; Lefeuvre and Moriones 2015). Despite the research that has been conducted, many questions remain unanswered. Some suggest that RNA recombination does not appear to function as a sexual form between viruses and that the frequency of crossover varies between viral species and is favored by natural selection (Bujarski 2013). Even so, next-generation and massive sequencing (NGS-HTS) will allow these paradigms to be resolved with precision (Beerenwinkel et al. 2012). Recombination occurs mainly in RNA viruses and is a key function during speciation and the emergence of viral variants, on which the potential for pathogenicity and virulence in the host depends (Domingo 2010).

IV. *Expressed symptoms:* Many centuries ago, plant viruses were identified by observed symptoms. For a long time, this was accepted as fact, but as technology advanced, several of the symptoms attributed to viruses were refuted, leading them to be called "phantom diseases" (Tzanetakis et al. 2025). Cutting-edge technologies such as high-throughput sequencing (HTS) through ONT platforms allow for the rapid detection of viruses in vegetables, leaving behind preventive visual diagnosis based on symptoms. Now, artificial intelligence (AI) enhances the preventive detection of viruses in field conditions, long before viral symptoms are observed in cultivated plants, as is the case with the recent integration of AI and CRISPR Cas13a for the rapid detection of ToBRFV (Karimi et al. 2025). The diagnosis of epidemic viruses through symptoms is currently going unnoticed. Otherwise, detection should focus on integrating mass detection technologies to identify viral species associated with seeds, seedlings, and plant materials that pose a phytosanitary risk to importing countries.

V. *Transmission/dissemination:* These two concepts are sometimes confused. Transmission refers to the mechanism of propagation between plants or from parent plants to their offspring, while dissemination refers to the spatial spread of the virus from an initial point of infection to other plants or areas within the same environment. Most epidemic viruses have persistent, non-persistent, and transovarial transmission mechanisms. Such is the case with rice stripe virus (RSV), a *Tenuivirus*, which is transmitted transovarially to grasshopper (*Laodelphax striatellus*) eggs (Huo et al. 2014). Plant viruses tend to spread rapidly. Sometimes, when a virus emerges, the mechanism of transmission and spread is unknown, which is why they tend to spread quickly. These characteristics are crucial for identifying viruses that pose a risk to horticultural crops.

VI. Range of hosts: Mass sequencing has made it possible to identify possible variations of mixed infections in horticultural crops (Hernández-Pérez et al. 2025; Mora-Ugalde et al. 2025). Identifying susceptible species through mechanical inoculation techniques and molecular characterization to determine viral species that coexist in crops, long before symptoms appear, promotes preventive surveillance of plant viruses (Vasquez-Gutierrez et al. 2025).

## 15.3 Parties Interested in Participatory Monitoring

In México, work programs are carried out for fruits and vegetables susceptible to quarantine pests or regulated imports. This is known as "conformity assessment." These procedures, carried out by third-party phytosanitary specialists (TEF) coordinated by México General Directorate of Plant Health (DGSV), must meet regulatory criteria to ensure the health of fruits/vegetables entering México. For example, when potatoes for consumption are moved from a region categorized as pest-free to a low-prevalence area, and the land transport passes through an area under phytosanitary control, the goods must meet specific requirements to guarantee their health. In other words, the goods must have a national transport certificate that certifies the absence in potatoes for consumption. Otherwise, the goods will be retained by a TEF until the DGSV orders their release or destruction if pests of quarantine interest are detected. Traceability measures need to be implemented for horticultural products that are moved via export and import. A conformity assessment scheme focused on plant products would reduce the spread of viral species. Most viruses are distributed via seeds and propagative materials, which is a good strategy for tracing the origin of the inoculum source and its subsequent distribution. This would enable early detection and ensure effective viro-epidemiological surveillance. The functional role of federal agencies, researchers, diagnostic laboratories, farmers, and extension workers ensures the surveillance of plant viruses. The development of high-throughput screening (HTS) platforms provides the opportunity to identify viral species co-infecting a single host, yet the importance of support from farmers, scientists, and the social community in detecting early viral outbreaks should not be overlooked. Following this principle, a "pyramid of participatory surveillance" (FARDEF) is proposed, where each step within the pyramid represents a level of responsibility with respect to the functional role they play in participatory surveillance (Fig. 15.2).

I. *Farmers:* When plant viruses first appear in greenhouses and commercial fields, farmers are the first to recognize the presence of viruses through the symptoms and damage observed. With the support of sales extension agents, management alternatives are applied to prevent the spread in the work area. Large and medium-sized producers collect samples of plant tissue with symptoms and send them to diagnostic laboratories for analysis.
II. Extension workers: Extension workers in the federal sector are responsible for notifying the federal government of the presence of non-concurrent symptoms in horticultural plants. When the symptoms observed match those described in foreign countries, they are sent to federal phytosanitary diagnostic laboratories

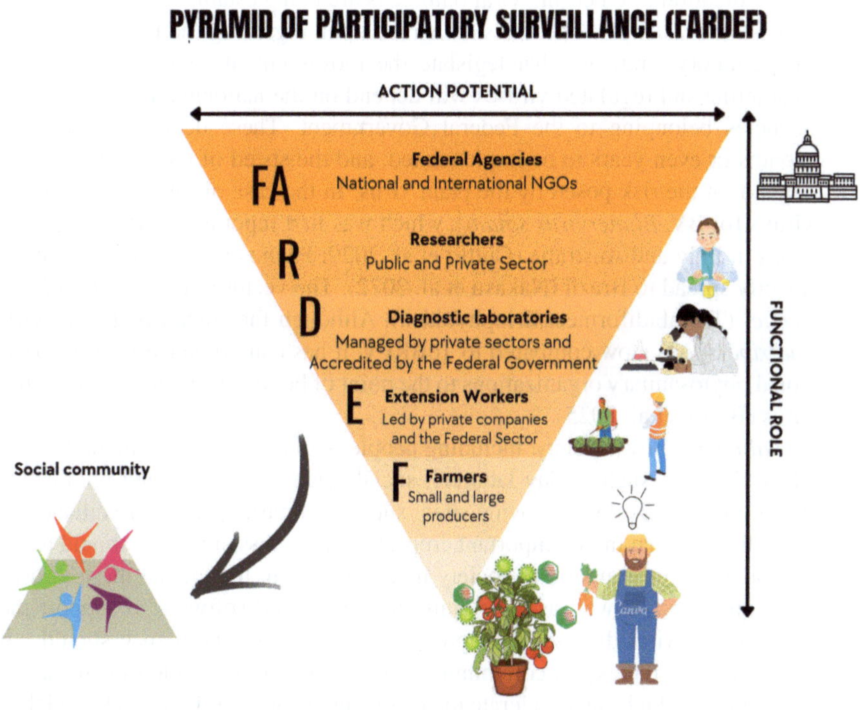


**Fig. 15.2** Pyramid of participatory surveillance (FARDEF)

for further analysis. In 2018, symptoms were observed in tomato and chili plants from commercial gardens (Cambrón-Crisantos et al. 2019).

III. *Diagnostic laboratories:* Extension workers from the public and private sectors, led by companies, are responsible for confirming the presence of plant viruses that pose a risk to horticultural crop production. Through private and public laboratories accredited by the federal government, the presence of a virus not previously reported in the country is notified for the first time, or, where applicable, its phytosanitary status within the national territory is reported. There are quarantine viruses, regulated viruses, non-quarantine regulated viruses, viruses in the process of eradication, and viruses of no economic importance. Regulatory decisions for the country are made according to the status of certain viruses.

IV. *Researchers:* With the presence of viral diseases that destabilize vegetable production, researchers in private and public sectors have a responsibility to seek alternatives to mitigate the phytosanitary risk posed by emerging viruses. The development of virus-resistant materials, evaluation and synthesis of molecules with antiviral characteristics (Gutierrez et al. 2024; Zohoursoleimani et al. 2025), integration of highly specific early detection systems (Nguyen et al. 2021), biogenic nanosystems carrying dsRNA for RNAi induction (Worral et al. 2019), and the use of AI to automate virus management strategies (Karimi et al. 2025). With technological advances in virus management, principles for the eradication of emerging viral outbreaks are being established.

V. *Federal Government*: The decision agreed upon regarding the implementation of regulatory strategies that legislate the movement of vegetables that host quarantine and regulated viruses will depend on the national and international agencies belonging to the Federal Government. These decisions may take months or even years to be implemented, and the speed of the agreement will depend on the risk posed by the plant virus. In the case of tomato fruit blotch virus (ToFBV, *Blunervirus solani*), which was first reported in 2020 in tomato crops in Italy and Australia (Ciuffo et al. 2020; Panno et al. 2024), and subsequently spread to Brazil (Nakasu et al. 2022). The vector mite is *Aculops lycopersici* (Trombidiformes: Eriophyoidea). Although the spread of *Blunervirus cuatripartita* is slow compared to ToBRFV, it has caused concern for international phytosanitary organizations to the point of being considered an emerging virus (Bertin et al. 2025).

VI. *Social community:* Society, including people with no experience in the field of plant virology, such as day laborers, social science students, and politicians. These individuals could provide silent support in participatory surveillance of viral diseases that infect important crops. Scientific dissemination should focus on spreading information to society in general about the impact these viruses have on crops, as well as developing platforms and software to support the detection of viral diseases through symptoms. Virus surveillance should not only focus on the expert community, but also on the integration of the social community, which can accelerate the reporting of early viral outbreaks and thus the eradication of epidemic outbreaks in horticultural crops.

## 15.4 Surveillance Frameworks and Methodologies

I. *Formal and informal surveillance systems.* Formal surveillance is carried out by phytosanitary authorities and accredited laboratories following official regulations (e.g., sample design, method validation, and standardized reporting). Informal (passive) surveillance integrates reports from farmers, extension workers, plant analysis, software, and other non-institutional sources (Mastin et al. 2020). The approach of this new system should be to integrate the formal (representative, traceability, diagnostic confirmation through HTS) with the informal (spatiotemporal coverage, early warning, reduced costs), and to distribute resources according to risk (e.g., routes of introduction, susceptible crops, phenological windows) to maximize early detection (Soubeyrand et al. 2023).

II. *Use of mobile applications and ICT in participatory data collection.* Mobile apps enable georeferenced reporting of symptoms, photo sharing, in-app guides, and user feedback. In tropical crops, Nuru (PlantVillage) showed that the combination of AI + farmers + extension workers improves the detection of plant viruses in cassava with higher performance compared to trained personnel when integrated into local workflows (Mrisho et al. 2020). In addition, remote sensing and multispectral imaging, combined with deep learning, can identify virus-related symptoms at the plot or crop level, confirming the sampling applied. Finally, digital participatory surveillance tactics in plant health represent a scalable alternative in the field (McNeil et al. 2022).

III. *Community monitoring networks and local knowledge***.** Training community networks, rural promoters, and local phytosanitary committees quickly captures ecological knowledge to identify symptoms, vectors, and alternate hosts (Ogunmodede et al. 2022). Programs such as Plantwise show that adapted diagnostic centers and standardized records can generate actionable data for extension and surveillance, provided there is feedback to the community and escalation to laboratories and the National Plant Protection Organization (NPPO) for confirmation and response (Wang et al. 2022).

IV. *Sampling protocols and symptom recognition guidelines***.** For participatory data to be useful, it must be organized within sampling protocols and symptom guidelines that are compatible with diagnostic standards: i) it is necessary to establish a sample size (batch/area) (Rubio et al. 2020), the optimal phenological stage, the choice of plant tissue (Zhao et al. 2025a, b), virus entry points, and sources of transmission, which may include propagative material, weeds, and insect vectors (Villamor et al. 2022). The international standard for phytosanitary measures, ISPM 31 (EPPO, 2021), proposes methodologies for sampling propagative materials and plants for early virus detection (Mastin et al. 2020).

V. *Traceability process.* Traceability guarantees the authenticity of detection. Tracking of the sample from the field to the final result in the laboratory and final report. The sequence includes: (a) initial registration: geographical location, crop and variety, phenology, and symptoms observed. (b) Type of sampling applied: code assigned per sample collected, brief description of plant tissue, and transport conditions. (c) Custody: prevents loss of information for samples. (d) Laboratory diagnosis: official results and methodology used during viral detection. (e) Feedback: sending results to local communities and federal phytosanitary authorities. (f) Traceability before and after: identification of areas of origin, possible routes of dissemination and mechanisms, including seeds, nurseries, and commercial greenhouses. This process ensures that participatory surveillance data collection is accurately integrated into the official system, enabling quick and transparent decision-making (EPPO, 2021; Rubio et al. 2020; Mastin et al. 2020). The effectiveness of the traceability process will depend on the involvement of key figures in early virus detection (Fig. 15.3). Farmers will observe viral symptoms in plants, which will be corroborated by extension workers, who will determine the economic importance of the plant virus and its impact on the crop. After this, plant tissue samples will be sent to phytosanitary diagnostic laboratories. If the analysis shows the presence of unknown viruses or viruses not reported within the national territory, the infected plant material will be sent to the federal phytosanitary laboratory to confirm the presence of an unknown virus. The federal laboratory decision is final. When the presence of unknown viruses is reported for the first time, the federal government eliminates the virus-infected plants from the area where the pathogen was detected, stops the importation of suspicious propagative materials, establishes preventive strategies to prevent dissemination, and renames the suspected virus with the status "actionable pest in process of eradication." It is the responsibility of the Plant Health Department to characterize the virus found using molecular biology and next-generation sequencing (NGS) to obtain the complete viral genome, which will confirm the difference between viral species of the same genus. As this is the first virus reported worldwide, the country is obliged to notify the North American Plant Protection Organization (NAPPO) of the phytosanitary alert and subsequently the International Committee on Taxonomy of Viruses (ICTV) to rename the viral species and have it reviewed by specialists in the field. The participation of researchers affiliated with the public sector promotes the study of the plant virus found, the possible risks that could arise, and preventive and corrective action strategies. When the virus is categorized with the status "Quarantine Pest," it means that it is only found in some main states of the country. Private sector researchers develop strategies that benefit farmers with the aim of reducing crop yield losses, including resistant materials, new products to reduce viral load, and compounds that inhibit viral transmission. As it is a mechanically and seed-transmitted virus, total eradication is difficult, so it may quickly advance to the status of "Regulated Non-Quarantine Pest," which means that the virus is distributed nationwide but still poses a risk to global trade. For plant viruses that have an impact on the production, work plans, communiques, circulars, and international standards are described to prevent global spread, and mandatory measures are established to prevent the risk of entry into countries with susceptible crops. The Commission on Phytosanitary Measures (CPM), the governing

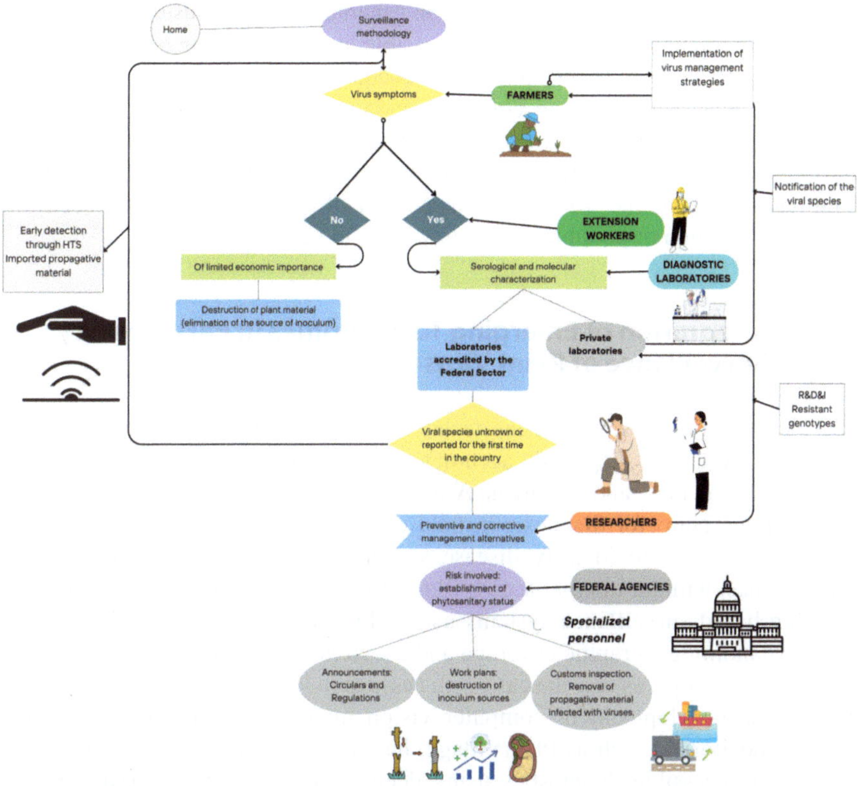


**Fig. 15.3** Participatory surveillance model for early detection and management of plant viruses

body of the International Plant Protection Convention (IPPC), is responsible for publishing the International Standards for Phytosanitary Measures (ISPM) focused on emerging epidemic viruses.

## 15.5 Capacity Building and Training for Farmers

Developing diagnostic and observational skills among farmers strengthens their ability to recognize early symptoms of viral diseases in vegetables (mosaics, deformities, or spots), which is essential for early detection and rapid response in participatory surveillance. The development of diagnostic and observation skills among farmers strengthens their ability to recognize the initial symptoms of viral diseases in vegetables (mosaics, deformities, or spots), which is essential for early detection and rapid response in participatory surveillance. The use and application of digital courses, workshops, and training that integrate practical and theoretical aspects: visual identification of viral symptoms in plant tissues, handling of tissue samples before sending them to the laboratory, protocols used for surveillance and communication with farmers and society in general. Gender inclusion is a key issue in designing training systems that are accessible to women, young people, children, and vulnerable groups. The adaptation of schedules, languages spoken in rural villages, and materials to ensure equal participation. In rural villages in southern México, children participate in the early detection of pine trees with yellowing leaves (*Pinus* sp.). The creation of brochures published in indigenous languages for rural regions around the world could support small producers in making decisions about the detection of possible emerging viruses in vegetables.

## 15.6 Integration of Emerging Technologies in Participatory Surveillance of Plant Viruses

The incorporation of cutting-edge technologies has benefited participatory surveillance, increasing diagnostic accuracy and strengthening interaction between farmers, technicians, and scientists. A novel case is the use of platforms such as Plant Village Nuro for diagnosis in field conditions, which allows farmers and extension workers to identify disease symptoms using images captured with smartphones, without internet access (Sawadogo et al. 2025). Some mass crowdsourcing-based surveillance systems use mobile devices to collect data in real time, expanding geographic coverage and accelerating the detection of viral outbreaks in key crops, as was the case with cassava in East Africa (Mutembesa et al. 2018). The development of computer vision models trained with agricultural image repositories, such as PlantVillage, has opened up new horizons for automatic and accessible diagnostics using deep learning algorithms (Hughes and Salathé et al. 2015). Next-generation sequencing, portable biosensors, CRISPR assays, and the integration of AI into predictive modeling offer new horizons for early detection and management of viruses in field and greenhouse production systems (Narsing et al. 2024). These technologies, integrated with drones and remote sensors, extend effective real-time monitoring to the open field scale and also aim to improve plant health with social, technological, and environmental dimensions. In line with this principle, multi-stakeholder digital platforms generate early warning systems based on open and cooperative data, increasing the capacity to respond to emerging diseases (Soubeyrand et al. 2023). State-of-the-art technologies such as CRISPR-Cas12a-based portable systems combined with mobile devices are advancing toward ultra-early detection, even before the presence of visible symptoms. The aggregation of biosensors connected to smartphones and invasive sampling techniques, such as microneedle patches, are recent examples with potential applications in crops such as potatoes (Zhao et al. 2025b). Finally, these innovations suggest that participatory surveillance of plant viruses is moving toward a more digital, inclusive, and predictive model. The combination of portable molecular diagnostics and artificial intelligence could support the early detection of emerging viruses in vegetables.

## 15.7 Case Studies of Successful Participatory Surveillance

I. *Monitoring TYLCV in small production systems.* Early identification of tomato yellow leaf curl virus (TYLCV) by small-scale producers through field observation and confirmation using molecular techniques reduced production losses and facilitated decision-making (Yan et al. 2021). Rapid field detection, complemented by laboratory analysis, enabled georeferenced mapping of outbreaks and the establishment of protocols for integrated virus management (Xie et al. 2013). Ongoing training for farmers in symptomatic diagnosis and standardized sampling was highlighted as essential for ensuring data quality and system effectiveness (McNeil et al. 2022).

II. *Reports and containment of ToLCNDV by farmers in Asia.* The rapid spread of tomato leaf curl New Delhi virus (ToLCNDV) in solanaceous and cucurbitaceous crops in Asia led to the improvement of participatory systems involving the coordination of teams by farmers (Jeevalatha et al. 2021; Kumar et al. 2021; Venkataravanappa et al. 2025). The integration of portable diagnostic tools and the use of agricultural extension teams facilitated immediate responses and prevented the spread to new production areas (Juárez et al. 2019). The aggregation of community data on digital platforms improved the traceability of outbreaks and allowed real-time feedback to farmers on risks and management measures (Sawadogo et al. 2025).

The study of these cases shows that farmer participation increases the spatial and temporal resolution of surveillance, strengthening early detection and outbreak containment (McNeil et al. 2022). The most transferable practices include ongoing training, standardization of sampling protocols, and secure data platform sharing, which enable surveillance models to be scaled to different regions and crops (Ramos et al. 2019).

## 15.8 Results of Late Detection of Plant Viruses

Late detection of plant viruses has had serious consequences for plant production systems, as evidenced by the cases of ToBRFV, TYLCV, and ToBFV (Table 15.1).

## 15.9 Strategic Criteria for Effective Participatory Monitoring

The AI integration and real-time data analysis make it possible to automatically classify symptom images, process massive information flows, and optimize the prioritization of areas for sampling (Jafar et al. 2024). The development of climate-smart surveillance approaches incorporates epidemiological models that consider climate and vector variables, anticipating outbreaks under climate-change scenarios (Ramos et al. 2019). Scaling models in different regions and crops depend on

**Table 15.1** Studies analyzed the late detection of emerging viruses

| Plant virus and disease caused | Consequences | Authors |
|---|---|---|
| *Tobamovirus fructirugosum* tomato brown rugose fruit diseases (ToBRFV) | ToBRFV causes roughness, necrosis, and brown spots on fruits, affecting commercial quality. It is transmitted mainly by mechanical contact and seeds, which facilitates its international dispersion. It has a CMRna+ genome belonging to the *Tobamovirus* genus. Its recent genetic origin is associated with point mutations that allowed it to overcome previously effective resistances in tomato. It exhibits limited variability and recombination, but with high environmental stability that favors its persistence. Its host range mainly includes tomato and chili, which has generated severe impacts in intensive agriculture. | Farizano et al. (2025); Rashid et al. (2019); Vasquez et al. (2025) |
| *Begomovirus coheni* Tomato yellow leaf curl disease (TYLCV) | TYLCV causes yellowing, leaf curling, and dwarfism in plants. It is transmitted by *Bemisia tabaci* in a persistent-circulative manner. It contains a circular ssDNA viral genome with high genetic variability and a rapid mutation rate, which hinders the development of resistant varieties. Recombination between isolates occurs frequently, which has given rise to virulent variants. It has a wide range of hosts, including solanaceous species and multiple weeds, which act as reservoirs. | Yan et al. (2021); Kumar et al. (2025); Xie et al. (2013); Zaidi et al. (2017) |
| *Blunervirus solani* (tomato fruit blotch virus) | ToFBV is considered an emerging pathogen for tomato crops. Late detection has caused significant losses in production. ToFBV produces irregular chlorotic spots and necrosis on leaves and fruits, drastically reducing the value of the crop. ToFBV is transmitted by *Aculops lycopersici* (Trombidiformes: Eriophyoidea) and belongs to the genus *Blunervirus* (family *Kitaviridae*), with a genome composed of four segments of mcRNA+. ToFBV originated from the diversification of blunerviruses associated with alternative hosts, showing a moderate mutation rate that gives it flexible adaptability. Blunerviruses show the ability to recombine between genomic segments, which increases the population diversity of ToFBV and favors the emergence of new variants. | Nakasu et al. (2022); Bertin et al. (2025); Ciuffo et al. (2020); Luigi et al. (2024) |

adaptation of protocols, platform interoperability, and community governance, factors that allow successful experiences to be replicated (Soubeyrand et al. 2023). Finally, federal government support is necessary to manage the various strategies within participatory surveillance and ensure efficiency in integrating research, extension services, and farmers in general (Smolinski et al. 2017).

## 15.10 Conclusions

Participatory monitoring presents challenges and limitations related to the integration of key actors collaborating in the FARDEF pyramid. Logistical and technical barriers such as limited connectivity, inputs, infrastructure for transporting plant tissue, and restrictions on support programs are influenced by socioeconomic factors, gender, and access to training, which must be considered in implementation designs. Stakeholder participation is largely due to concerns about data validation and reliability, which makes it necessary to establish quality metrics and continuous feedback. Even with these limitations, the importance of participatory surveillance is significant, as it promotes spatial-temporal expansion, reduces late detection, and promotes the resilience of horticultural systems. "FARDEF stakeholders are called upon to consolidate inclusive and proactive systems that integrate digital technologies, climate-smart approaches, and multi-stakeholder governance as pillars of modern plant health."

## References

Adams MJ, Antoniw JF (2005) DPVweb: an open access internet resource on plant viruses and virus diseases. Outlooks Pest Manag 16:268–270. https://doi.org/10.1564/16dec08

Alders RG, Ali SN, Ameri AA, Bagnol B, Cooper TL, Gozali A, Catley A (2020) Participatory epidemiology: principles, practice, utility, and lessons learnt. Front Vet Sci 7:532763. https://doi.org/10.3389/fvets.2020.532763

Beerenwinkel N, Günthard HF, Volker Roth Metzner KJ (2012) Challenges and opportunities in estimating viral genetic diversity from next-generation sequencing data. Front Microbiol 3:329. https://doi.org/10.3389/fmicb.2012.00329

Bertin S, Sybilska A, Luigi M et al (2025) Transmission of tomato fruit blotch virus by the tomato russet mite: epidemiological implications for an emerging/re-emerging tomato disease. Sci Rep 15:12079. https://doi.org/10.1038/s41598-025-97142-9

Bujarski JJ (2013) Genetic recombination in plant-infecting messenger-sense RNA viruses: overview and research perspectives. Front Plant Sci 4:68. https://doi.org/10.3389/fpls.2013.00068

Camacho-Beltrán E, Pérez-Villarreal A, Pérez-Leyva NE, Rodríguez-Negrete EA, Ceniceros-Ojeda EA, Méndez-Lozano J (2019) Occurrence of tomato brown rugose fruit virus infecting tomato crops in Mexico. Plant Dis 103(6). https://doi.org/10.1094/PDIS-11-18-1974-PDN

Cambrón-Crisantos JM, Rodríguez-Mendoza J, Valencia-Luna JB, Alcasio Rangel S, García-Ávila CDJ, López-Buenfil JA, Ochoa-Martínez DL (2019) First report of tomato brown rugose fruit virus (ToBRFV) in Michoacan, Mexico. Rev Mex Fitopatol 37(1):185–192. https://doi.org/10.18781/r.mex.fit.1810-5

Chare E, Holmes E (2006) A phylogenetic survey of recombination frequency in plant RNA viruses. Arch Virol 151:933–946. https://doi.org/10.1007/s00705-005-0675-x

Cheng CP, Panavas T, Luo G, Nagy PD (2005) Heterologous RNA replication enhancer stimulates in vitro RNA synthesis and template-switching by the carmovirus, but not by the tombusvirus, RNA-dependent RNA polymerase: implication for modular evolution of RNA viruses. Virology 341(1):107–121. https://doi.org/10.1016/j.virol.2005.06.042

Ciuffo M, Kinoti WM, Tiberini A, Forgia M, Tomassoli L, Constable FE, Turina M (2020) A new blunervirus infects tomato crops in Italy and Australia. Arch Virol 165(10):2379–2384. https://doi.org/10.1007/s00705-020-04760-x

Claverie S, Hoareau M, Chéhida SB, Filloux D, Varsani A, Roumagnac P, Martin DP, Lett JM, Lefeuvre P (2023) Metagenomics reveals the structure of *Mastrevirus*-host interaction network within an agro-ecosystem. Virus evolution 9(2):vead043. https://doi.org/10.1093/ve/vead043

Davino S, Caruso AG, Bertacca S, Barone S, Panno S (2020) Tomato Brown rugose fruit virus: seed transmission rate and efficacy of different seed disinfection treatments. Plants 9(11):1615. https://doi.org/10.3390/plants9111615

Dennis C, Daszak P, Wolfe ND, Gao GF et al (2018) The global Virome project. Science 359:872–874. https://doi.org/10.1126/science.aap7463

Domingo E (2010) Mechanisms of viral emergence. Vet Res 41(6):38. https://doi.org/10.1051/vetres/2010010

Drake JW, Holland JJ (1999) Mutation rates among RNA viruses. Proc Natl Acad Sci USA 96(24):13910–13913. https://doi.org/10.1073/pnas.96.24.13910

Farizano JV, Castagnaro E, Arroyo-Egea JT, Aparicio JD, Vallejos AC, Hebert EM, Saavedra L, Rapisarda VA, Villegas JM, Grillo-Puertas M (2025) Virulence traits and bacterial interactions within the complex microbial population in urinary double-J catheters. Front Microbiol 16:1624743. https://doi.org/10.3389/fmicb.2025.1624743

García-Arenal F, Fraile A, Malpica JM (2001) Variability and genetic structure of plant virus populations. Annu Rev Phytopathol 39:157–186. https://doi.org/10.1146/annurev.phyto.39.1.157

Garcia-Figuera S, Grafton-Cardwell EE, Babcock BA et al (2021) Institutional approaches for plant health provision as a collective action problem. Food Secur 13:273–290. https://doi.org/10.1007/s12571-020-01133-9

Grant A, Pawson SM, Marzano M (2019) Emerging stakeholder relations in participatory ICT design: renegotiating the boundaries of sociotechnical innovation in Forest biosecurity surveillance. Forests 10:836. https://doi.org/10.3390/f10100836

Gutierrez UV, Treviño GAF, Ortiz JCD, Uribe LAA, Olivas AF, Beache MB, Castillo FDH (2024) Chlorine dioxide: antiviral that reduces the spread of ToBRFV in tomato (Solanum lycopersicum L.) plants. Viruses 16(10):1510. https://doi.org/10.3390/v16101510

Hamilton RI, Edwardson JR, Francki RIB, Hsu HT, Hull R, Koenig R, Milne RG (1981) Guidelines for the identification and characterization of plant viruses. J Gen Virol 54(2):223–241. https://doi.org/10.1099/0022-1317-54-2-223

Hernández-Pérez DMJ, Obregón-Flores V, Diaz-Lara A, López-Cordova JP, Tovar-Pedraza JM (2025) Occurrence of mixed infections of cucurbit yellow stunting disorder virus (Crinivirus cucurbitae) and watermelon chlorotic stunt virus (Begomovirus citrulli) in watermelon (Citrullus lanatus) in Jalisco, Mexico. Plant Health Progress 26(3). https://doi.org/10.1094/PHP-10-24-0096-BR

Hoenen T, Groseth A, Rosenke K, Fischer RJ, Hoenen A, Judson SD, Martellaro C, Falzarano D, Marzi A, Squires RB et al (2016) Nanopore sequencing as a rapidly deployable Ebola outbreak tool. Emerg Infect Dis 22:331. https://doi.org/10.1094/PHP-10-24-0096-BR

Hughes D, Salathé M (2015) An open access repository of images on plant health to enable the development of mobile disease diagnostics. arXiv preprint arXiv:1511.08060

Huo Y, Liu W, Zhang F, Chen X, Li L, Liu Q, Zhou Y, Wei T, Fang R, Wang X (2014) Transovarial transmission of a plant virus is mediated by vitellogenin of its insect vector. PLoS Pathog 10(3):e1003949. https://doi.org/10.1371/journal.ppat.1003949

Ivanovsky D (1892) Ueber die Mosaikkrankheit der Tabakspflanze. St Petersb Acad Imp Sci Bull 35:67–70

Jafar A, Bibi N, Naqvi RA, Sadeghi-Niaraki A, Jeong D (2024) Revolutionizing agriculture with artificial intelligence: plant disease detection methods, applications, and their limitations. Front Plant Sci 15:1356260. https://doi.org/10.3389/fpls.2024.1356260

Jeevalatha A, Vanishree G, Siddappa S, Kumar R, Kaundal P, Kumar A, Chakrabarti SK (2021) Molecular characterization and infectivity analysis of tomato leaf curl New Delhi virus isolates infecting potato. 3 Biotech 11:203. https://doi.org/10.1007/s13205-021-02752-5

Jones RAC (2021) Global plant virus disease pandemics and epidemics. Plants 10:233. https://doi.org/10.3390/plants10020233

Juárez M, Rabadán MP, Martínez LD, Tayahi M, Grande-Pérez A, Gómez P (2019) Natural hosts and genetic diversity of the emerging tomato leaf curl New Delhi virus in Spain. Front Microbiol 10:140. https://doi.org/10.3389/fmicb.2019.00140

Karimi M, Ghorbani A, Niazi A, Rostami M, Tahmasebi A (2025) Integrating AI and CRISPR Cas13a for rapid detection of tomato brown rugose fruit virus. Sci Rep 15:25422. (2025). https://doi.org/10.1038/s41598-025-11405-z

Kumar R, Tiwari RK, Jeevalatha A, Sundaresha S, Shah MA, Sharma S, Sagar V, Kumar M, Chakrabarti SK (2021) Potato apical leaf curl disease: current status and perspectives on a disease caused by tomato leaf curl New Delhi virus. J Plant Dis Protect 128:897–911. https://doi.org/10.1007/s41348-021-00463-w

Kumar M, Bag S, McAvoy T, Torrance T, Cloud C, Simmons AM (2025) A shift in begomovirus Coheni populations associated with tomato yellow leaf curl disease infecting tomato cultivars in the Southeastern United States. Plant Pathol 74(5):1277–1289. https://doi.org/10.1111/ppa.14091

Lai MM (1992) RNA recombination in animal and plant viruses. Microbiol Rev 56(1):61–79. https://doi.org/10.1128/mr.56.1.61-79.1992

Lefeuvre P, Moriones E (2015) Recombination as a motor of host switches and virus emergence: geminiviruses as case studies. Curr Opin Virol 10:14–19. https://doi.org/10.1016/j.coviro.2014.12.005

Lefterova MI, Suarez CJ, Banaei N, Pinsky BA (2015) Next-generation sequencing for infectious disease diagnosis and management: a report of the Association for Molecular Pathology. J Mol Diagn 17(6):623–634. https://doi.org/10.1016/j.jmoldx.2015.07.004

Legg JP, Thresh JM (2000) Cassava mosaic virus disease in East Africa: a dynamic disease in a changing environment. Virus Res 71(1–2):135–149. https://doi.org/10.1016/s0168-1702(00)00194-5

Luigi M, Tiberini A, Taglienti A, Bertin S, Dragone I, Sybilska A, Tarchi F, Goggioli D, Lewandowski M, Simoni S, Faggioli F (2024) Molecular Methods for the Simultaneous Detection of Tomato Fruit Blotch Virus and Identification of Tomato Russet Mite, a New Potential Virus–Vector System Threatening Solanaceous Crops Worldwide. Viruses 16(5):806. https://doi.org/10.3390/v16050806

Malpica JM, Fraile A, Moreno I, Obies CI, Drake JW, García-Arenal F (2002) The rate and character of spontaneous mutation in an RNA virus. Genetics 162(4):1505–1511. https://doi.org/10.1093/genetics/162.4.1505

Mastin AJ, Gottwald TR, van den Bosch F, Cunniffe NJ, Parnell S (2020) Optimising risk-based surveillance for early detection of invasive plant pathogens. PLoS Biol 18(10):e3000863. https://doi.org/10.1371/journal.pbio.3000863

McNeil C, Verlander S, Divi N, Smolinski M (2022) The landscape of participatory surveillance systems across the one health Spectrum: systematic review. JMIR Public Health Surveill 8(8):e38551. https://doi.org/10.2196/38551

McNeil C, Divi N, Bargeron IVCT, Capobianco Dondona A, Ernst KC, Gupta AS, Fasominu O, Keatts L, Kelly T, Leal Neto OB, Lwin MO, Makhasi M, Mutagahywa EB, Montecino-Latorre D, Olson S, Pandit PS, Paolotti D, Parker MC, Samad MH, Sewalk K et al (2025) Data parameters from participatory surveillance systems in human, animal, and environmental health from around the globe: descriptive analysis. JMIR Public Health Surveill 11:e55356. https://preprints.jmir.org/preprint/55356

Mehetre GT, Leo VV, Singh G, Sorokan A, Maksimov I, Yadav MK, Upadhyaya K, Hashem A, Alsaleh AN, Dawoud TM, Almaary KS, Singh BP (2021) Current developments and challenges in plant viral Diagnostics: A systematic review. Viruses 13(3):412. https://doi.org/10.3390/v13030412

Mehle N, Ravnikar M (2012) Plant viruses in aqueous environment–survival, water mediated transmission and detection. Water Res 46(16):4902–4917. https://doi.org/10.1016/j.watres.2012.07.027

Molad O, Smith E, Luria N, Bakelman E, Lachman O, Reches M, Dombrovsky (2024) Studying tomato brown rugose fruit virus longevity in soil and virion susceptibility to pH treatments helped improve virus control by soil disinfection. Plant Soil 505:543–558. https://doi.org/10.1007/s11104-024-06690-y

Mora-Ugalde MDLÁ, Miranda-Campaña OA, Ramírez-Soto M, Cruz-Jaramillo JL, Espinosa de los Reyes Bolaños IJ, Tovar-Pedraza JM et al (2025) High throughput sequencing reveals mixed infections of impatiens necrotic spot virus and tomato brown rugose fruit virus in bell pepper crops in Mexico. J Plant Pathol 107(1):711–714. https://doi.org/10.1007/s42161-024-01806-w

Morris CE, Géniaux G, Nédellec C, Sauvion N, Soubeyrand S (2022) One health concepts and challenges for surveillance, forecasting, and mitigation of plant disease beyond the traditional scope of crop production. Plant Pathol 71(1):86–97. https://doi.org/10.1111/ppa.13446

Mrisho LM, Mbilinyi NA, Ndalahwa M, Ramcharan AM, Kehs AK, McCloskey PC, Murithi H, Hughes DP, Legg JP (2020) Accuracy of a smartphone-based object detection model, Plant Village Nuru, in identifying the foliar symptoms of the viral diseases of cassava-CMD and CBSD. Front Plant Sci 11:590889. https://doi.org/10.3389/fpls.2020.590889

Mutembesa D, Omongo C, Mwebaze E (2018) Crowdsourcing real-time viral disease and Pest information: A case of nation-wide cassava disease surveillance in a developing country. Proceedings of the AAAI Conference on Human Computation and Crowdsourcing 6(1):117–125. https://doi.org/10.1609/hcomp.v6i1.13322

Nagy PD (2008) Recombination in plant RNA viruses. In: Roossinck MJ (ed) Plant virus evolution. Springer, Berlin, Heidelberg. https://doi.org/10.1007/978-3-540-75763-4_8

Naito FYB, Melo FL, Fonseca MEN, Santos CAF, Chanes CR, Ribeiro BM, Gilbertson RL, Boiteux LS, de Cássia Pereira-Carvalho R (2019) Nanopore sequencing of a novel bipartite New World begomovirus infecting cowpea. Arch Virol 164(7):1907–1910. https://doi.org/10.1007/s00705-019-04254-5

Nakasu EY, Nagata T, Inoue-Nagata AK (2022) First report of tomato fruit blotch virus infecting tomatoes in Brazil. Plant Dis 106(8):2271. https://doi.org/10.1094/PDIS-07-21-1392-PDN

Narsing LP, Sushila C, Sanjay K, Mahima C, Priyanka KM, Samreen, Saumya, Saloni, Rajan G (2024) Advances in plant disease diagnostics and surveillance- A review. Plant Cell Biotechnol Mol Biol 25(11–12):137–150. https://doi.org/10.56557/pcbmb/2024/v25i11-128918

Nezhad AS (2014) Future of portable devices for plant pathogen diagnosis. Lab Chip 14(16):2887–2904. https://doi.org/10.1039/c4lc00487f

Nguyen C, Sagan V, Maimaitiyiming M, Maimaitijiang M, Bhadra S, Kwasniewski MT (2021) Early detection of plant viral disease using hyperspectral imaging and deep learning. Sensors 21(3):742. https://doi.org/10.3390/s21030742

Ogunmodede AM, Tambo JA, Adeleke AT, Gulak DM, Ogunsanwo MO (2022) Farmers' willingness to pay towards the sustainability of plant clinics: evidence from Bangladesh, Rwanda and Zambia. Int J Agric Sustain 20(7):1360–1372. https://doi.org/10.1080/14735903.2022.2082018

Panno, S., Ragona, A., Bertacca, S., Agrò, G., Yahyaoui, E., Dimauro, B.,. & Davino, S. (2024). Outbreak of tomato fruit blotch virus in the most relevant tomato greenhouse production area of Sicily. J Plant Pathol, 106(2), 789–792. doi:https://doi.org/10.1007/s42161-024-01623-1

Qin Y, Petty IT (2001) Genetic analysis of bipartite geminivirus tissue tropism. Virology 291(2):311–323. https://doi.org/10.1006/viro.2001.1205

Ramos RS, Kumar L, Shabani F, Picanço MC (2019) Risk of spread of tomato yellow leaf curl virus (TYLCV) in tomato crops under various climate change scenarios. Agric Syst 173:524–535. https://doi.org/10.1016/j.agsy.2019.03.020

Ranabhat NB, Fellers JP, Bruce MA, Rupp JLS (2025) High-throughput Oxford nanopore sequencing unveils complex viral population in Kansas wheat: implications for sustainable virus management. Viruses 17(1):126. 126. https://doi.org/10.3390/v17010126

Rang FJ, Kloosterman WP, de Ridder J (2018) From squiggle to basepair: computational approaches for improving nanopore sequencing read accuracy. Genome Biol 19(1):90. https://doi.org/10.1186/s13059-018-1462-9

Rashid M-O, Zhang X-Y, Wang Y, Li D-W, Yu J-L, Han C-G (2019) The three essential motifs in P0 for suppression of RNA silencing activity of potato leafroll virus are required for virus systemic infection. Viruses 11(2):170. https://doi.org/10.3390/v11020170

Rizzo DM, Lichtveld M, Mazet JAK, Togami E, Sally AM (2021) Plant health and its effects on food safety and security in a one health framework: four case studies. One Health Outlook 3:6. https://doi.org/10.1186/s42522-021-00038-7

Rubio L, Galipienso L, Ferriol I (2020) Detection of plant viruses and disease management: relevance of genetic diversity and evolution. Front Plant Sci 11:1092. https://doi.org/10.3389/fpls.2020.01092

Salaudeen MT, Adama CJ, Abdullahi AA, Ayeleke DA, Ibrahim AD (2016). Climate change and viral diseases in relation to crop productivity and food security: A review. http://irepo.futminna.edu.ng:8080/jspui/handle/123456789/1929

Savary S, Willocquet L, Pethybridge SJ, Esker P, McRoberts N, Nelson A (2019) The global burden of pathogens and pests on major food crops. Nature Ecology Evoltion 2013(3):430–439. https://doi.org/10.1038/s41559-018-0793-y

Sawadogo S, Tiendrebeogo F, Tibiri EB, Djigma F, Traoré L, Pita JS, Eni AO (2025) Artificial intelligence tool for cassava viral diseases diagnosis using participatory surveillance in Burkina Faso. Front Sustain Food Syst 9:1597039. https://doi.org/10.3389/fsufs.2025.1597039

Smolinski MS, Crawley AW, Olsen JM, Jayaraman T, Libel M (2017) Participatory disease surveillance: engaging communities directly in reporting, monitoring, and responding to health threats. JMIR Public Health Surveill 3(4):e62. https://doi.org/10.2196/publichealth.7540

Soubeyrand S, Claverie S, Hoareau M, Chéhida SB, Filloux D, Varsani A, Roumagnac P, Martin DP, Lett JM, Lefeuvre P (2023) Metagenomics reveals the structure of Mastrevirus-host interaction network within an agro-ecosystem. Virus evolution, 9(2), vead043. https://doi.org/10.1093/ve/vead043

Standard-Diagnostics, E. P. P. O (2021) PM 7/98 (5). Specific requirements for laboratories preparing accreditation for a plant pest diagnostic activity. EPPO Bull 51:468–498. https://doi.org/10.1111/epp.12780

Steere RL (1959) The purification of plant viruses. In: Advances in virus research, vol 6. Academic Press, pp 1–73. https://doi.org/10.1016/S0065-3527(08)60487-X

Sztuba-Solińska J, Urbanowicz A, Figlerowicz M, Bujarski JJ (2011) RNA-RNA recombination in plant virus replication and evolution. Annu Rev Phytopathol 49:415–443. https://doi.org/10.1146/annurev-phyto-072910-095351

Takahashi H, Fukuhara T, Kitazawa H, Kormelink R (2019) Virus latency and the impact on plants. Front Microbiol 10:2764. https://doi.org/10.3389/fmicb.2019.02764

Tatineni S, Hein GL (2023) Plant viruses of agricultural importance: current and future perspectives of virus disease management strategies. Phytopathology 113(2):117–141. https://doi.org/10.1094/PHYTO-05-22-0167-RVW

Tzanetakis IE, Aknadibossian V, Špak J, Constable F, Harper SJ, Hammond J et al (2025) Streamlining global germplasm exchange: integrating scientific rigor and common sense to exclude phantom agents from regulation. Plant Dis 109(4):736–755. https://doi.org/10.1094/PDIS-04-24-0745-FE

Vasquez-Gutierrez U, López-López H, Frías-Treviño GA, Delgado Ortiz JC, Flores Olivas A, Aguirre Uribe LA, Hernández Juarez A (2024) Biological exploration and physicochemical characteristics of tomato Brown rugose fruit virus in several host crops. Agronomy 14(2):388. https://doi.org/10.3390/agronomy14020388

Vásquez-Gutierrez U, Delgado-Ortiz JC, Frías-Treviño GA, Aguirre-Uribe LA, Flores-Olivas A (2025) Tobamovirus fructirugosum an emerging disease: review and current situation in Mexico. Mexican J Phytopathol 43(1):34. https://doi.org/10.18781/R.MEX.FIT.2401-7

Venkataravanappa V, Madhu GS, Muralidhara BM, Reddy MK, Reddy CN (2025) Tomato leaf curl New Delhi virus (Begomovirus solanumdelhiense) causes yellow mosaic disease in white pattypan squash (Cucurbita pepo L. ssp. pepo cv. Patissonina) in India. Indian Phytopathol:1–7. https://doi.org/10.1007/s42360-025-00887-6

Villamor DEV, Keller KE, Martin RR, Tzanetakis IE (2022) Comparison of high throughput sequencing to standard protocols for virus detection in berry crops. Plant Dis 106(2):518–525. https://doi.org/10.1094/PDIS-05-21-0949-RE

Wang YM, Ostendorf B, Gautam D, Habili N, Pagay V (2022) Plant viral disease detection: from molecular diagnosis to optical sensing technology—a multidisciplinary review. Remote Sens 14(7):1542. https://doi.org/10.3390/rs14071542

Worrall EA, Bravo-Cazar A, Nilon AT, Fletcher SJ, Robinson KE, Carr JP, Mitter N (2019) Exogenous application of RNAi-inducing double-stranded RNA inhibits aphid-mediated transmission of a plant virus. Front Plant Sci 10:265. https://doi.org/10.3389/fpls.2019.00265

Xie Y, Jiao X, Zhou X et al (2013) Highly sensitive serological methods for detecting tomato yellow leaf curl virus in tomato plants and whiteflies. Virology J 10:142. https://doi.org/10.1186/1743-422X-10-142

Yan Z, Wolters AA, Navas-Castillo J, Bai Y (2021) The global dimension of tomato yellow leaf curl disease: current status and breeding perspectives. Microorganisms 9(4):740. https://doi.org/10.3390/microorganisms9040740

Zaidi SS, Martin DP, Amin I, Farooq M, Mansoor S (2017) Tomato leaf curl New Delhi virus: a widespread bipartite begomovirus in the territory of monopartite begomoviruses. Mol Plant Pathol 18(7):901–911. https://doi.org/10.1111/mpp.12481

Zhao J, Xu H, Xu C, Yin W, Luo L, Liu G, Wang Y (2025a) Smartphone-integrated RPA-CRISPR-Cas12a Detection System with Microneedle Sampling for Point-of-Care Diagnosis of Potato Late Blight in Early Stage. bioRxiv 6. https://doi.org/10.1101/2025.06.17.660259

Zhao X, Xu Y, Xu X, Zhou H, Shi J, Yang C, Zhou X, Yang X (2025b) Comprehensive sampling and detection strategies for the field surveillance of tomato brown rugose fruit virus. Agronomy 15(2):318. https://doi.org/10.3390/agronomy15020318

Zohoursoleimani R, Aeini M, Ghodoum PMH (2025) Effects of abiotic and biotic elicitors on tomato resistance to tomato brown rugose fruit virus. Sci Rep 15:27216. https://doi.org/10.1038/s41598-025-13578-z

# Proteomic Insights into Host–Virus Interactions in Vegetable Crops

# 16

F. Neha Angelin, P. Jayanthi, and A. Swapna Geetanjali

**Abstract**

Viruses that infect crops are one of the major problems faced by farmers in recent times. Among all the abiotic and biotic stresses, including bacteria, fungi, and others, plant viruses are tagged as the most severe and unpredictable threat. The future of food security is at risk from plant viruses most chaotically. Due to their high recombination rate, viruses that infect crops have a very wide host range and have affected most crops, including vegetable crops. Vegetable crops, which are important for feeding people, need to be protected from viral infections right away. As a result, extensive research is in progress to understand the mechanisms regulating viral behavior inside hosts, any plant factors that are linked to viral diseases, the processes of viral recombination, and the strategies for their management. All these functions require viral and host proteins that actively interact with one another. Proteomics, in a way, provides greater insight into all these aspects of viral diseases, particularly the structure and function of both viral and host proteins. Studies have shown that proteomics has contributed to understanding the viral proteins involved in symptom development, as well as the host proteins that counteract these effects by triggering plant defense mechanisms. In this chapter, we will explore the vegetable crops and their associated viral diseases from a proteomics perspective. We attempt to provide a thorough comprehension of plant viral disease in crops such as cucumber, tomato, chilli, cassava, and potato, focusing on proteins involved in these infections. Each case study will include the host proteins activated during the viral infection, including upregulation of the pathogen-related proteins that increase secondary metabolite production (triggering the salicylic acid/jasmonic acid pathway) as well as the

F. Neha Angelin · P. Jayanthi · A. Swapna Geetanjali (✉)
Department of Genetic Engineering, SRM Institute of Science and Technology, Kattankulattur, Tamil Nadu, India
e-mail: swapnaga@srmist.edu.in



R. Kumar et al. (eds.), *Vegetable Crops and Viral Pathogens Volume I*, Plant Pathology for Sustainable Agriculture, https://doi.org/10.1007/978-981-95-7265-6_16

downregulation of essential proteins such as those involved in photosynthesis, alongside viral proteins that contribute to disease severity.

**Keywords**

Host–virus interactions · Proteomics · Cucurbits · Solanaceous crops · Leguminous crops

## 16.1 Introduction

According to FAO's *The State of Food and Agriculture 2023* report, all countries' economies and social well-being are closely tied to their agriculture and food systems. The prosperity of a nation is reflected in its agricultural strength and its ability to ensure food security. Food crops not only include grains and cereals, but also major cash crops like vegetables and fruits (Gunter et al. 2025). Vegetables are a powerhouse of essential nutrients, including vitamins, minerals, and antioxidants. Horticulture plays a major role within the broader fields of agriculture, providing nutritional values that make it high in demand. Vegetable crops also provide farmers with economic returns, making them highly preferred crops among many other varieties (Agregán et al. 2021). The rise of globalization and the impacts of climate change have significantly accelerated the transmission of plant viruses, allowing them to penetrate new areas and agricultural landscapes. This developing trend highlights the necessity for ongoing surveillance and sophisticated molecular techniques like proteomics to swiftly detect pathogen signatures. By understanding shifts in host responses over time, it is possible to build predictive models to support sustainable crop health management. However, in recent decades, vegetable farmers have been facing major difficulties in farming due to a lot of biotic and abiotic constraints (Verma et al. 2025). Among such biotic factors, viral pathogens present a serious risk to food security, since they considerably reduce production levels. Plant viruses are highly adaptable, rapidly evolving with a frequent recombination and mutation rate, which makes them resilient to infect a wide range of hosts and overcome resistance mechanisms (Pérez-Losada et al. 2015; Wang et al. 2022).

The virus often infects the major groups of vegetables as its host range expands:

- Solanaceous crops such as tomato (Ruiz et al. 2017) (Tomato leaf curl virus, tomato yellow mosaic virus) (Pratap et al. 2011), Peppers-*Capsicum annuum* (Chilli leaf curl virus, pepper mottle virus), and potato (*Potyvirus*) (Raigond et al. 2022),
- Cucurbits like cucumber (Cucumber mosaic virus), Melons (Cucumber green mottle virus),
- Legumes such as Mung bean (Mung bean yellow mosaic virus), Cowpea (Cowpea mosaic virus),
- Cruciferous crops, which are Cauliflower (Cauliflower mosaic virus), Cabbage, and Broccoli,
- Other vegetable crops like Beets, Okra, and Cassava.

Effective management of these viral pathogens requires deep knowledge and understanding of their genetic composition, molecular complexities such as the DNA/RNA, the proteins they encode, and recombination within the viral genome (Wu et al. 2024). Recent developments in plant pathology and the integration of OMICS have provided an excellent method for comprehending their genetic structure and associated functions. Omics is a holistic approach towards evaluating biological molecules, including DNA, RNA, proteins, and metabolites (Di Carli et al. 2012). It conducts high-throughput screening of numerous proteins or related molecules simultaneously, rather than analyzing them individually. Out of these, proteomics is very much reliable in understanding the functional part of the infection by actively investigating the proteins that are produced by the host or the viruses throughout the diseased phase (Samarskaya et al. 2025). Proteomics is concerned with proteins, particularly focusing on their structures, functions, changes, and interactions. In the vegetable crop viral infections, proteomics facilitates a complete analysis of host and pathogen proteins, elucidating their relationships, distribution, and alterations in host protein profiles during infection (Das et al. 2019a, b). Proteomics has been utilized in virus-infected vegetable crops to confirm viral coat protein structures, assess differential protein expression, and clarify host–viral protein interaction networks at the cellular functional level (Singh et al. 2018).

## 16.2 Experimental Proteomics in Understanding the Infection

A typical proteomics workflow starts with protein extraction from the infected plant tissues and the healthy plants in case of comparative studies. Isolation often involves TCA/acetone precipitation or phenol extraction to remove unwanted phenolics (Niu et al. 2018). If the proteomics analysis is to be done with subcellular components, it takes another fractionation step. This helps in tracking the localization of viral proteins. The isolated protein is then separated through 2D gel electrophoresis or Liquid chromatography based on its isoelectric point, molecular weight, and affinity. This can also be done with differential gel electrophoresis using multiple fluorescent dyes to differentiate proteins of infected from healthy. To identify proteins, the earlier developments were to use MALDI-TOF/TOF, LC-MS/MS, which are mass spectrometric techniques that can use the excised gel as samples (Cejnar et al. 2020; Vadivel 2015). Recently developed high-throughput proteomics technologies provide richer peptide coverage and more precise quantification. These technologies reduce sample loss and detect low-abundance viral proteins that would otherwise go undetected. In a study, the zucchini yellow mosaic virus (ZYMV) in *Cucurbita pepo* was used in 2D gel-assisted proteomics for understanding differential expression of stress-related proteins (Nováková et al. 2015). Label-free LC-MS/MS might require trypsin digestion followed by separation through Liquid chromatography (LC) based on their hydrophobicity. The detection tool is directly connected to the LC unit; the mass spectrometry usually measures the peptide based on the mass-to-charge ratio. This gives raw data, a fragmentation value that is then compared with the existing library to identify and infer the protein. There are methods like iTRAQ (Isobaric Tags for Relative and Absolute Quantification), which use isotopic tags to differentially quantify proteins, generally for multiplexed quantification. iTRAQ is used to label the peptides that may have the same mass but have different fragmentation values in MS/MS (Macharia et al. 2020). This technique is also followed by LC-MS/MS; later, based on the intensity of the tags, it gives the relative abundance of proteins. An iTRAQ-based proteomics study on *Solanum tuberosum* leaves infected by tobacco rattle virus (TRV) was done to identify stress-related proteins and metabolic proteins affected during the infection (Liu et al. 2021). The fragmentation values are then analyzed for the protein IDs based on the amino acid contents, thus identifying the protein and used for further bioinformatics-based studies.

## 16.3 Validation of Proteomics (Host–Virus Protein–Protein Interactions)

Proteomics does not end with identifying the proteins; the validation is carried out with the interaction studies or expression studies in the very own study group plants. This includes Co-immunoprecipitation to pull down the proteins that are produced against the virus using the antibodies (DeBlasio et al. 2017). This is also supported by MS to properly identify the proteins. Interaction studies also include the Yeast two-hybrid system or BiFC (Bimolecular fluorescence complementation) for the screening of potential related partners (Zilian and Maiss 2011). These validation tools let researchers map functional protein modules that regulate susceptibility or tolerance by distinguishing direct and indirect interactions. This insight is needed to build targeted host cell viral disruption techniques. Proteins that are tagged with fluorescent proteins are one other technique to study localization, which might ultimately lead the way in understanding the relative interaction group (Roshan et al. 2017; Fig. 16.1).

## 16.4 Molecular Mechanism of Plant–Virus Interactions

Regardless of their structurally precise genome, the viruses display complex evolutionary strategies like high mutation and recombination rates, making them challenging to control. Their compact genomes, made of either DNA or RNA, with genes that usually encode a limited number of multifunctional proteins, perform an extensive range of functions that are critical in the establishment of the diseases. These proteins carry out crucial roles as:

- Replication proteins: facilitate viral replication within the host cells (Maio et al. 2020).
- Coat protein: genome encapsidation to protect them from degradation and allow transmission as such.

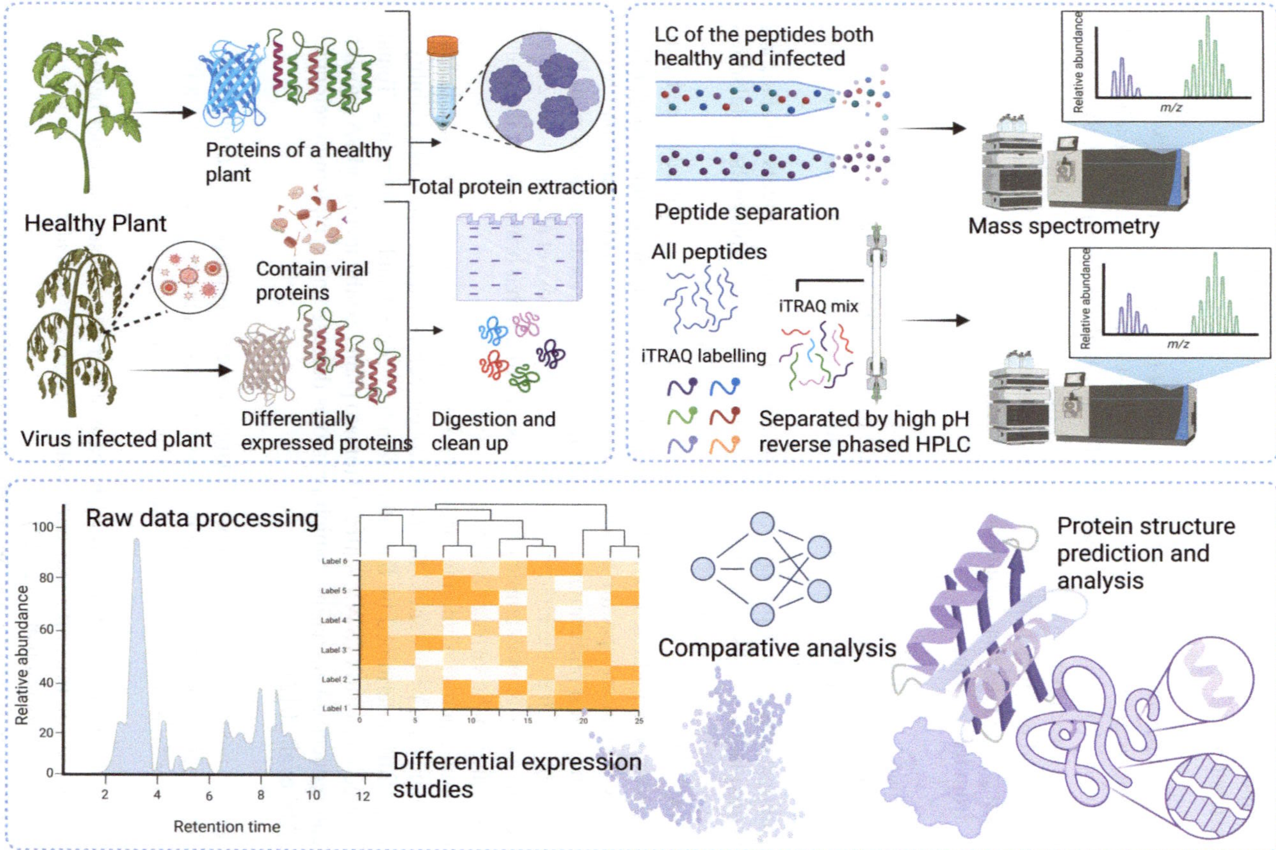


**Fig. 16.1** Steps involved in the proteomics of virus-infected crops

- Movement proteins: cell-to-cell movement through plasmodesmata and intracellular movement.
- Suppressor proteins: interfere with the host defense mechanism for survival.
- Symptom determinants: alter host physiology for symptom development and sustenance.

For example, the C4 protein of tomato yellow leaf curl virus (TYLCV) interacts with plant receptor-like kinases to decrease host defense signaling in addition to aiding in the development of symptoms (Roshan et al. 2017). Similarly, the 2b protein in the cucumber mosaic virus (CMV) has two functions: it suppresses RNA silencing and modifies hormone signaling pathways (Crawshaw et al. 2024).

During viral infection, plants deploy multiple layers of defense to eliminate the virus from the system:

- Secondary metabolites like Salicylic acid (SA) or Jasmonic acid (JA) make the internal system unsuitable for the pathogens to thrive and are a first line of defense. This works by keeping away the vectors that carry the virus or by acting against the virus as such.
- Pathogenesis-related proteins (PR proteins), heat shock proteins (HSPs), and antioxidant enzymes induced during the viral infection, stabilize the cellular process of mitigating oxidative stress (Stare et al. 2017).
- RNA silencing is the ultimate one-stop solution and a more advanced plant response to a pathogen. The host starts manufacturing small interfering RNAs (siRNAs) that are specific to the viral genome, causing its degradation, thereby restricting viral replication. However, many viruses encode suppressors of RNA silencing, such as the HC-Pro of potyviruses (Atabekova et al. 2023). AV2 of the begomovirus is an example of a suppressor protein that brings down the plant's resistance to the viruses (Sarkar et al. 2024).
- Cross-talk across signaling pathways involves a more sophisticated defense mechanism. For instance, the balance between resistance and susceptibility is frequently determined by the interaction between SA and JA signaling. Tomato spotted wilt virus (TSWV)-infected tomato plants exhibit increased SA signaling and reduced JA signaling, which promotes defense and symptom development (Abe et al. 2012).

Vectors are yet another problem to address since the pathogens are actively transmitted to gain access to enter the host system. Viruses are mostly spread via insect vectors, seeds, and mechanical contact. For example:

- White fly (*Bemisia tabaci*), as in the case of leaf curl viruses and other begomoviruses, including chilli leaf curl virus, tomato yellow leaf curl virus (Bragard et al. 2020; Janssen et al. 2022).
- Aphids transmit potyvirus and CMV in cucurbits (Jangra et al. 2024).

- Pepper mild mottle virus (PMMoV), cowpea severe mosaic virus, and others take seed transmission as a route for disease establishment (Patel et al. 2023).
- Mechanical transmission through physical touch is also reported in certain RNA viruses like TMV and CMV.

Once the virus finds a way to enter the system, it sheds the external structural proteins to release the internal genetic material. The genetic material, either DNA/ or RNA, as it is introduced in the host atmosphere, begins to translate into proteins. Proteins help them to invade the plant system by compromising the host protein networks and simultaneously reshaping the host metabolism.

## 16.5 Proteomics of Vegetable Crops with Viral Infections

Proteomics has emerged as a highly effective tool for studying proteins, the functional elements responsible for both infections and playing a pivotal role in defining the disease and defense. Unlike the other omics approaches, such as genomics and transcriptomics, which focus on DNA and RNA, proteomics focuses exclusively on proteins that directly carry out biological functions. This approach can selectively provide distinguishing insights on both the viral and host proteins, confirming their roles in viral infection and host defense mechanisms (Di Carli et al. 2012). During a viral attack, host plants often undergo significant changes in protein expression. These alterations reshape the plant's inner environment, either supporting viral invasion or activating defense mechanisms aimed at eliminating the pathogen from the system.

The pathogenic condition often induces large-scale reprogramming of host proteomes. This includes:

- Upregulation of defense-related proteins, such as proteins associated with the oxidative stress response, including superoxide dismutase and peroxidases.
- Downregulation of photosynthetic proteins, leading to reduced energy generation and visible symptoms such as chlorosis and stunted growth (Das et al. 2019a, b). For instance, proteomic studies in tomato plants infected with tomato yellow leaf curl virus have shown suppression of photosystem II, symptoms to the extent of yellowing of leaves, and reduced photosynthetic capacity (Farooq et al. 2019).
- Altered expression of metabolic enzymes allows the host to switch to defense mode rather than focusing on growth (Abe et al. 2012).

Proteomics assists in understanding such interaction networks, possibly producing a framework to identify potential targets for resistance breeding. Let us learn about vegetable viruses and proteome research on these plants to better comprehend infections in the sections that follow.

### 16.5.1 Viruses of Solanaceous Crops and Their Proteomics

Due to their highest susceptibility to viral infections, members of the *Solanaceae* family have been actively studied in research for viral pathogenesis. Viruses like the Begomovirus, Tobamovirus, and Potyvirus, being the cautiously significant viruses for infecting the members of this crop, have the worst effect in bringing down the yield of the crop, which affects the farmers' economy. This is where proteomics has its ultimate role in identifying the fundamental proteins involved in disease development, or even the proteins that can deploy virus defense in managing the disease. Speaking of the proteins involved in the establishment of viral diseases, there are certain proteins of the *Solanaceae* that have been extensively studied. In severe cases like leaf curl and mosaic conditions caused by begomoviruses, including tomato leaf curl virus, which infects crops like tomatoes and chilli, viral proteins play an active role in disease development. Involvement of viral proteins like Coat protein V1 of the DNA A complex and other proteins like the Regulatory protein C4 or the suppressor protein/precoat protein V2 intensively triggers plant response (Dai et al. 2022; Mazumder et al. 2025; Jeevalatha et al. 2021; Sarkar et al. 2025). Plants like tomato and chilli in response downregulate proteins like photosynthetic proteins, photosystem II oxygen-evolving enhancer protein (OEE1), Chlorophyll a-b binding proteins (LHCCII complex), and RUBISCO activase, to bring out impaired photosynthesis and thus develop mosaic or leaf curling symptoms (He et al. 2024a, b). As a response to leaf curl viruses or mosaic viruses, the plants of this family are known to respond with the upregulation of certain pathogen-related proteins like PR-1, chitinase, or Beta-1,3-glucanase as frontline defense signaling proteins (He et al. 2024a, b). The plant's attempt to rewire its defense mechanisms is also demonstrated by the high expression of pathogenesis-related (PR) and heat shock proteins (HSPs) in infected leaves caused by potato virus Y (PVY), a member of the Potyvirus group. Potyviruses also have a huge effect on the *Solanaceae* family, causing a greater impact, in which proteomics has shown a major role in addressing the proteins involved in the disease mechanism and control (Stare et al. 2017; Kumar et al. 2025a). In potato, the PVY infection triggers an immune response by upregulating PR-2 proteins such as Beta-1,3-glucanase and PR-5 thaumatin-like protein to prepare the plants for defense. Additionally, HSPs do not miss a chance in being consistently upregulated like the HSP70 and HSP90 to assist protein folding during stressful conditions but also being used by viral facilitators to assist the viral assembly (Makarova et al. 2018). Proteomic evidence also shows the expression of antioxidants like superoxide dismutase (SOD), catalase, and peroxidases to withstand the oxidative stress induced during viral infections (Jin et al. 2007; Kumar et al. 2025a). Tobamoviruses, like tobacco mosaic virus, on the other hand, also involve themselves in the infection of the *Solanaceae* family, interfering with the photosynthetic efficiency by inhibiting oxygen-evolving enhancer proteins and Rubisco activase in chloroplasts, which interferes with photosynthesis and drastically lowers plant vigor. Proteomic studies also put a spotlight on proteins like PsaA and PsbB, the components of Photosystem I and II, being downregulated, which resulted in photo inhibition and visible chlorosis in infected leaves (Das et al. 2019a, b). The upregulation of the thioredoxin H-type 1, which is evidence of the plant's active fight against oxidative stress by handling the reactive oxygen species (ROS). There are proofs that the tobamovirus actively infects other members of the family other than tobacco. Tomato, in such a case, affected by TMV has shown that the proteins are even differentially regulated in their fruits (Casado-Vela et al. 2006). The infected fruits showed upregulation of chitinase and endoglucanase to withstand the infection.

### 16.5.2 Proteomics in Cucurbits for Viral Infections

Cucurbit crops, such as cucumber, melon, pumpkin, and squash, are vulnerable to viral diseases, which result in severe yield losses worldwide. Reports identify viruses like cucumber mosaic virus (CMV), zucchini yellow mosaic virus (ZYMV), and squash yellow mosaic virus (SqYMV) as more common in the family of cucurbits. Detailed studies have been made in CMV, cucumber green mottle mosaic virus (CGMMV), and cucurbit chlorotic yellow virus (CCYV). The study of proteins in cucurbits, especially concerning cucumber mosaic virus (CMV), which is a fundamental virus affecting crops within the same family, encompasses various proteins involved in the progression of the disease. Recent findings indicate the differential expression of proteins in both susceptible and resistant varieties of melons (*Cucumis melo*), highlighting instances of both upregulation and downregulation of each protein. In this case, the plant's susceptibility was gained from the increased abundance of Glycosyltransferase, which likely functioned as a pro-viral factor. The proteins involved in the ER-associated degradation (ERAD) pathway and the ubiquitin-proteasome degradation system, which are responsible for managing protein quality, were likely manipulated by the virus. Thaumatin-like protein and carboxypeptidase, which serve as indicators of stress and defense, were found to interact with CMV proteins (Di Carli et al. 2010). Also, downregulation of ferredoxin, glyceraldehyde-3-phosphate dehydrogenase, and such strongly leads to understanding the symptoms developed during the time of infection. Resistant varieties, in contrast, had the upregulation of more defense proteins such as lipoxygenases and phloem-associated proteins. In another case of cucumber green mottle mosaic virus, a tobamovirus with two ORFs that are involved in virulence ultimately altered the protein pattern in plants like cucumber and *N. benthamiana* (Liu et al. 2015). The CGMMV infection significantly upregulates chitinase and thaumatin-like protein to enhance the plant immune response and peroxidases that alter the cell wall by strengthening it through phenylalanine metabolism to create a barrier to restrict the entry and spread of the viral pathogens. Localization of these proteins was confined to peroxisomes for viral replication, thus altering the mechanism of the organelle to promote infection (Gong et al. 2025). Cucurbit chlorotic yellow virus has four proteins involved in coding for methyltransferase and RNA helicase, an ORF that encodes RNA-dependent RNA polymerase and one coding p6 protein. There is also the p22 protein in another region of the bipartite RNA genome. The p22 protein is found to be an RNA silencing suppressor that counter-defends the plant's defense mechanism. p22 protein of the virus interacts with the ubiquitin-proteasome system to hijack the host's protein degradation mechanism to escape the antiviral defense (Kumar et al. 2025b). Its essential involvement in suppressor activity was understood on deletion of the F-box motif that resulted in the loss of p22's suppressing ability (Chen et al. 2019). In zucchini yellow mosaic virus (ZYMV) infections observed in the zucchini cultivars Zelena and Jaguar, both susceptible and resistant, respectively, showed proteomic changes that either disrupt the defense mechanism or those that improve them for their resistance. Key proteins like flavoproteins and catalase isoenzymes were extensively found in resistant lines to upregulate, thus managing and deploying protection against the virus infection (Nováková et al. 2015). In susceptible lines, unlike some plants, the upregulation of RuBisCO activase reflects the plant's repairing mechanism by preserving photosynthesis to withstand the viral infection and actin-7 upregulation that could be the virus's exploitation to facilitate movement across the cells and tissue to establish the symptoms and infection.

### 16.5.3 Leguminous Crops and Their Associated Viral Proteomics

Crops belonging to the leguminous family include many commercial crops like the soy bean, common bean, cowpea, chickpea, groundnut, mung bean, and others, which are significant for their nutritional value, especially as a vital source of protein. Beyond the nutritional value, these crops are more susceptible to viral diseases. Diseases showing chlorotic/mosaic and stunt symptoms are more commonly observed among these members. The most common ones are soybean mosaic virus (SMV), cowpea mosaic virus (CPMV), bean common mosaic virus (BCMV), chickpea chlorotic stunt virus (CpCSV), and groundnut necrotic virus. Proteomic insights into these viral diseases have provided evidence of proteins like metabolic proteins, the photosynthetic and defense-related proteins of the host.

Soybean mosaic virus is one of the viruses that infects the soybean crop and causes a destructive yield loss. Apart from this, there are also viruses causing other yellow mosaic viruses, like mung bean yellow mosaic viruses, that are also studied in soybean plants (Zeng et al. 2017). The research overview provides information on the differentially expressed proteins and their mechanism in establishing defense in the host system. Any viral disease is known to elicit the plant's defense to keep it living in such a stressful condition. As a direct response, it is known to produce Glutathione S-transferase upon the viral infection that is involved in detoxification and to protect the cells from any oxidative stress release during the infection (Pavan Kumar et al. 2017). Another such protein is Superoxide Dismutase (SOD), which scavenges the reactive oxygen species. Likewise, annexin, phenylalanine ammonia lyase (PAL), lipoxygenase (LOX), and cytochrome p450 are upregulated to guard the plants by working against the oxidative stress and eliciting the signaling pathway that ultimately synthesizes molecules such as phytoalexins, jasmonic acid, to keep up the plant defense (Yang et al. 2011). Downregulation in soybean, possibly like any other viral disease, brings down the essential protein members like

RuBisCO, Photosystem II OEEs, Glyceraldehyde-3-phosphate dehydrogenase (GAPDH) and such proteins of metabolic functions that enhance the symptom development (Huang et al. 2025). Cowpea is another major crop of this group facing viral pathogens like cowpea mosaic virus (CpMV) and cowpea severe mosaic virus (CpSMV). In cowpea, the early stages of CpSMV are known to trigger oxidative stress, resulting in the production of $H_2O_2$ and other antioxidant enzymes in order to resist the infection, and also known to have reduced Beta-1,3-glucanase in resistant varieties (Ribeiro et al. 2023). Mungbean yellow mosaic virus, on the other hand, in mung bean is known to release antioxidant enzymes like superoxide dismutase (SOD) and ascorbate peroxidase (APX) and reduced catalase activity, for which it might indirectly activate defense through a rapid surge of $H_2O_2$ (Kundu et al. 2013).

### 16.5.4 Viruses of Cruciferous Crops and Proteomics

Cruciferous crops are members of the *Brassicaceae* family, including key crops like cabbage, cauliflower, broccoli, and others. Cauliflower has a major threatening disease affecting it, caused by the cauliflower mosaic virus (CaMV). During the CaMV infection, host factors like those responsible for the abscisic acid biosynthesis are identified as important for the susceptibility of the host. Likewise, the NADPH oxidase production proves the accumulation of the reactive oxygen species (ROS), and Argonaut proteins are part of the host's defense mechanism to cleave the viral mRNAs with the RNA-induced silencing complex (RISC) (Bak and Emerson 2020; Pouresmaeil et al. 2023). Studies like the effect of turnip mosaic virus in Chinese cabbage have shown relative proteomic evidence of upregulation in Translational elongation factor EF1B, Heat shock proteins (HSPs) like HSP70 that are recruited by the viruses to assist assembly of viral particles, replication, and movement. Downregulation happens with the proteins involved in metabolism, like Beta-1,3-glucanase that might hydrolyze cellulose and thus facilitate movement, MYC2 to regulate jasmonate-responsive genes, Thaumatin super family, and Chitinase. In the same crops, cabbage leaf curl virus infections are known to induce the host proteome by enhancing the ubiquitin-proteosome system proteins that are exploited by the virus itself (Lyu et al. 2020).

### 16.5.5 Viruses of Beet and Cassava and the Proteomics

Apart from the viral diseases that are discussed in the above fields, there are a few that have been concerning crops such as beets, okra, and even cassava. These crops are common throughout the world and have a huge economic impact as well. As a crop, beets are facing wide loss due to viruses such as beet curly top virus (BCTV), beet necrotic yellow vein virus (BNYVV), beet chlorosis virus (BChV), and others. These viruses are mostly transmitted through insect vectors like aphids, as that of BChV, and there are even reports of diseases in beet that are transmitted via a soil-borne protist, *Polymyxa betae*, which can transmit necrotic yellow vein viruses in beet (Withycombe et al. 2024). This causes a special condition called rhizomania that affects the growth of the beets, with symptoms of hairy roots and stunted growth. Proteomics of viruses in beet has revealed several differentially expressed proteins in BNYVV-infected samples. Any virus-infected crops could primarily focus on the survival mechanisms to shut down the regulatory pathway and activate the defense pathway to survive the disease. In beets, proteins like superoxide dismutase, peroxidases, chitinases, and certain pathogenesis-related proteins were activated mostly in the susceptible variety to activate systemic acquired resistance during the period of infection. However, some proteins are expressed, which are involved in metabolism and energy production, just to keep the plants in their survival mode. That includes Photosystem II CP43 chlorophyll apoprotein, Glyceraldehyde-3-phosphate dehydrogenase B, Fructose-bisphosphate aldolase, etc., which are active participants of the plants surviving an active viral infection (Webb et al. 2014).

Cassava, being another vulnerable crop, is infected by viruses like cassava brown streak viruses and the mosaic virus, causing wide loss among the farmers by reducing the quality of the tuber. Among this group, plants infected mainly with the cassava mosaic virus create interest among the scientific group to know their mechanism for surviving the infection. Like most crops, cassava also responds to disease conditions with stress-related proteins and other growth-associated proteins. The proteomic analysis of cultivars of cassava exposed key proteins that are involved in the survival of the infection. In a case, the scientist could rule out proteins that are differentially expressed during infection, including metabolic proteins such as NADP-dependent sorbitol-6-phosphate dehydrogenase, oxygen-evolving enhancer protein, cytochrome c oxidase subunit II PS17, ATP synthase beta subunit, which are down-regulated mostly to survive the crisis by shutting down the growth metabolism pathway (Duraisamy et al. 2019). There is certainly protein upregulation that is involved in the event of disease progression, which is reflected in proteins like the plant defense mechanism. Such proteins are Ascorbate peroxidase APX2, a chaperone protein, and in tolerance varieties, such as KU50 proteins involved in plant hormone signal transduction, MAPK pathways, and plant pathogen interaction are found in abundance (Siriwan et al. 2022, 2023).

## 16.6 Conclusion

Plant viruses pose a major threat to food security because of their rapid evolution and adaptability, which allows them to infect diverse plants and overcome their defenses. Proteomics is essential for understanding plant pathology by analyzing the functions and interactions of both viral and host proteins. The typical proteomics workflow starts with protein extraction and then separation, followed by mass spectrometry for peptide identification and quantification. Finally, proteomics validation focuses on mapping host–virus, protein–protein interactions (PPIs) with fluorescent tagging aiding protein localization. Plant viruses use their compact genomes to encode multifunctional proteins for replication, movement, and defense suppression. Plants primarily defend with RNA silencing, supported by PR proteins and antioxidants, but these viruses employ suppressor proteins to counter the plant's defense. The outcome is often dictated by complex signaling pathway cross-talk, with vectors being the main route of entry.

Proteomic studies in vegetable crops have yielded crucial insights:

- Solanaceae crops: Viruses like Begomovirus and Tobamovirus downregulate photosynthetic proteins, while the host upregulates defense proteins.
- Cucurbits crops: Viruses like CMV exploit pro-viral factors and host degradation systems, while resistance relies on proteins like lipoxygenases and flavoproteins.
- Legumes: Infections trigger robust antioxidant defenses but suppress metabolic proteins.
- Crucifer crops: Viruses exploit host proteins for movement and assembly while overcoming RNA silencing.
- In tubers: Even these plants suffer severe outbreaks of viral infections, but have some quality mechanisms to fight or survive the infection using defense proteins against them.

By understanding these key host–virus interactions, proteomics drives the discovery of biomarkers for early detection and pinpoints genetic targets, providing the basis for innovative strategies to protect vegetable crops from devastating viral diseases. Also, the proteomic comparisons between the varieties of susceptible and resistant lines provide highlights on many defense-associated proteins that can be accommodated in bringing out disease-resistant cultivars. Such insights provide effective molecular targets for crop improvement strategies, including marker-assisted selection, RNAi-based silencing for crop line improvement, and gene editing of the susceptible varieties using CRISPR, which are notably beneficial. In summary, proteomics serves as a vital method for modern plant virology, allowing a deeper understanding of viral infections and enabling the creation of efficient, long-lasting disease-management approaches. With the ongoing emergence and resurgence of viral threats posing challenges to global agriculture, research aimed at proteomics will be essential for improving crop resilience, ensuring vegetable production, and maintaining future food security.

## References

Abe H, Tomitaka Y, Shimoda T, Seo S, Sakurai T, Kugimiya S, Tsuda S, Kobayashi M (2012) Antagonistic plant defense system regulated by phytohormones assists interactions among vector insect, Thrips and a Tospovirus. Plant Cell Physiol 53(1):204–212. https://doi.org/10.1093/pcp/pcr173

Agregán R, Echegaray N, López-Pedrouso M, Aadil RM, Hano C, Franco D, Lorenzo JM (2021) Proteomic advances in cereal and vegetable crops. Molecules 26(16):MDPI AG. https://doi.org/10.3390/molecules26164924

Atabekova AK, Solovieva AD, Chergintsev DA, Solovyev AG, Morozov SY (2023) Role of plant virus movement proteins in suppression of host RNAi defense. Int J Mol Sci 24(10) Multidisciplinary Digital Publishing Institute (MDPI). https://doi.org/10.3390/ijms24109049

Bak A, Emerson JB (2020) Cauliflower mosaic virus (CaMV) Biology, management, and relevance to GM Plant detection for sustainable organic agriculture. Front Sustain Food Syst 4). Frontiers Media S.A. https://doi.org/10.3389/fsufs.2020.00021

Bragard C, Dehnen-Schmutz K, Di Serio F, Gonthier P, Jacques MA, Jaques Miret JA, Justesen AF, MacLeod A, Magnusson CS, Milonas P, Navas-Cortes JA, Parnell S, Potting R, Reignault PL, Thulke HH, Van der Werf W, Vicent Civera A, Yuen J, Zappalà L et al (2020) Pest categorisation of tomato leaf curl New Delhi virus. EFSA J 18(7). https://doi.org/10.2903/j.efsa.2020.6179

Casado-Vela J, Sellés S, Martínez RB (2006) Proteomic analysis of tobacco mosaic virus-infected tomato (Lycopersicon esculentum M.) fruits and detection of viral coat protein. Proteomics 6 Suppl 1. https://doi.org/10.1002/pmic.200500317

Cejnar P, Kučková Š, Šantrůček J, Glasa M, Komínek P, Mihálik D, Slavíková L, Leišová-Svobodová L, Smirnova T, Hynek R, Kundu JK, Ryšánek P (2020) Efficient confirmation of plant viral proteins and identification of specific viral strains by Nanolc-Esi-Q-Tof using single-leaf-tissue samples. Pathogens 9(11):1–20. https://doi.org/10.3390/pathogens9110966

Chen S, Sun X, Shi Y, Wei Y, Han X, Li H, Chen L, Sun B, Sun H, Shi Y (2019) Cucurbit chlorotic yellows virus p22 protein interacts with cucumber SKP1LB1 and its F-box-like motif is crucial for silencing suppressor activity. Viruses 11(9). https://doi.org/10.3390/V11090818

Crawshaw S, Murphy AM, Rowling PJE, Nietlispach D, Itzhaki LS, Carr JP (2024) Investigating the interactions of the cucumber mosaic virus 2b protein with the viral 1a replicase component and the cellular RNA silencing factor Argonaute 1. Viruses 16(5). https://doi.org/10.3390/v16050676

Dai KW, Tsai YT, Wu CY, Lai YC, Lin NS, Hu CC (2022) Identification of crucial amino acids in Begomovirus C4 proteins involved in the modulation of the severity of leaf curling symptoms. Viruses 14(3). https://doi.org/10.3390/v14030499

Das PP, Chua GM, Lin Q, Wong S-M (2019a) iTRAQ-based analysis of leaf proteome identifies important proteins in secondary metabolite biosynthesis and defence pathways crucial to cross-protection against TMV. J Proteome 196:42–56. https://doi.org/10.1016/j.jprot.2019.02.002

Das PP, Lin Q, Wong S-M (2019b) Comparative proteomics of tobacco mosaic virus-infected Nicotiana tabacum plants identified major host proteins involved in photosystems and plant defence. J Proteome 194:191–199. https://doi.org/10.1016/j.jprot.2018.11.018

DeBlasio SL, Bereman MS, Mahoney J, Thannhauser TW, Gray SM, MacCoss MJ, Heck MC (2017) Evaluation of a Bead-Free coimmunoprecipitation technique for identification of virus-host protein interactions using High-Resolution mass spectrometry. J Biomol Tech 28(3):111–121. https://doi.org/10.7171/jbt.17-2803-002

Di Carli M, Villani ME, Bianco L, Lombardi R, Perrotta G, Benvenuto E, Donini M (2010) Proteomic analysis of the plant-virus interaction in cucumber mosaic virus (CMV) resistant transgenic tomato. J Proteome Res 9(11):5684–5697. https://doi.org/10.1021/pr100487x

Di Carli M, Benvenuto E, Donini M (2012) Recent insights into plant-virus interactions through proteomic analysis. J Proteome Res 11(10):4765–4780. https://doi.org/10.1021/pr300494e

Duraisamy R, Natesan S, Muthurajan R, Gandhi K, Lakshmanan P, Janaky JG, Karuppusamy N, Chokkappan M (2019) Proteomic analysis of cassava mosaic virus (CMV) responsive proteins in cassava leaf. Int J Curr Microbiol App Sci 8(04):2988–3005. https://doi.org/10.20546/ijcmas.2019.804.345

Farooq T, Liu D, Zhou X, Yang Q (2019) Tomato yellow leaf curl China virus impairs photosynthesis in the infected Nicotiana benthamiana with βC1 as an aggravating factor. Plant Pathol J 35(5):521–529. https://doi.org/10.5423/PPJ.OA.04.2019.0120

Gong P, Gao M, Chen Y, Zhang M, Huang Y, Hu X, Zhao S, Zhang H, Pan M, Cao B, Shen Q, Liu Y, Lozano-Durán R, Wang A, Zhou X, Li F (2025) Cucumber green mottle mosaic virus encodes additional small proteins with specific subcellular localizations and virulence function. Sci China Life Sci 68(6):1815–1827. https://doi.org/10.1007/s11427-024-2892-1

Gunter CC, Treadwell DD, Resende MFR, Whitaker VM, Balmant KM, Shelnutt KP, Borum PR, Andrade JE, Goodrich-Schneider R, Bernier AV, Donahoo WT, Jaffee MS, Angle JS, Hanson AD (2025) Food Rx: integrating horticulture research to improve nutrition and health. Proc Natl Acad Sci USA 122(21). https://doi.org/10.1073/pnas.2501723122

He WZ, Rong T, Liu XY, Rao Q (2024a) Transcriptomic profiling unravels the disruption of photosynthesis apparatuses and induction of immune responses by a bipartite Begomovirus in tomato plants. Plants 13(22). https://doi.org/10.3390/plants13223198

He WZ, Zhao L, Sun K, Feng Z, Zhou G, Rao Q (2024b) Transcriptomic profiling reveals the complex interaction between a bipartite begomovirus and a cucurbitaceous host plant. BMC Genomics 25(1). https://doi.org/10.1186/s12864-024-10781-6

Huang K, Sekine KT, Li S, Xu H, Song D, Kachroo P, Kachroo A, Luan H (2025) Comprehensive genomic characterization of the soybean G3PDH gene family and its role in virus resistance. BMC Plant Biol 25(1):1171. https://doi.org/10.1186/s12870-025-06579-7

Jangra S, Devendran R, Ghosh A (2024) Perspicacious insights into plant-virus-vector interactions applying omics. BMC Genomics 25(1) BioMed Central Ltd. https://doi.org/10.1186/s12864-024-10758-5

Janssen D, Simón A, Boulares M, Ruiz L (2022) Host species-dependent transmission of tomato leaf curl New Delhi virus-ES by Bemisia Tabaci. Plants 11(3). https://doi.org/10.3390/plants11030390

Jeevalatha A, Vanishree G, Siddappa S, Kumar R, Kaundal P, Kumar A, Chakrabarti SK (2021) Molecular characterization and infectivity analysis of tomato leaf curl New Delhi virus isolates infecting potato. 3 Biotech 11:203. https://doi.org/10.1007/s13205-021-02752-5

Jin Y, Ma D, Dong J, Jin J, Li D, Deng C, Wang T (2007) HC-pro protein of potato virus Y can interact with three Arabidopsis 20S proteasome subunits in planta. J Virol 81(23):12881–12888. https://doi.org/10.1128/jvi.00913-07

Kumar R, Lal MK, Tiwari RK, Kumar D, Sagar V, Jena R, Kumar R, Singh G, Singh B (2025a) Antioxidant defence mechanism of potato (Solanum tuberosum) by regulation of physiological and biochemical response under potato virus Y necrotic strain (PVYNTN) infection. J Plant Growth Regul. https://doi.org/10.1007/s00344-025-11850-0

Kumar R, Tiwari RK, Lal MK, Singh A, Jha J, Madival VA, Roy A, Singh G (2025b) Omics-based approaches in deciphering molecular mechanisms of tomato leaf curl New Delhi virus infection. In: Tomato leaf curl New Delhi virus (ToLCNDV) insights into Virome dynamics and management. Springer Nature Singapore, Singapore, pp 27–47. https://doi.org/10.1007/978-981-95-2852-3_3

Kundu S, Chakraborty D, Kundu A, Pal A (2013) Proteomics approach combined with biochemical attributes to elucidate compatible and incompatible plant-virus interactions between Vigna mungo and Mungbean Yellow Mosaic India Virus. Proteome Sci. http://www.proteomesci.com/content/11/1/15

Liu HW, Liang CQ, Liu PF, Luo LX, Li JQ (2015) Quantitative proteomics identifies 38 proteins that are differentially expressed in cucumber in response to cucumber green mottle mosaic virus infection. Virol J 12(1). https://doi.org/10.1186/s12985-015-0442-x

Liu L, Hao L, Liu N, Zhao Y, Zhong N, Zhao P (2021) Itraq-based proteomics analysis of response to solanum tuberosum leaves treated with the plant phytotoxin thaxtomin a. Int J Mol Sci 22(21). https://doi.org/10.3390/ijms222112036

Lyu S, Gao L, Zhang R, Zhang C, Hou X (2020) Correlation analysis of expression profile and quantitative iTRAQ-LC-MS/MS proteomics reveals resistance mechanism against TuMV in Chinese cabbage (Brassica rapa ssp. pekinensis). Front Genet 11. https://doi.org/10.3389/fgene.2020.00963

Macharia M, Das PP, Naqvi NI, Wong SM (2020) iTRAQ-based quantitative proteomics reveals a ferroptosis-like programmed cell death in plants infected by a highly virulent tobacco mosaic virus mutant 24A+UPD. Phytopathol Res 2(1). https://doi.org/10.1186/s42483-019-0043-5

Maio F, Helderman TA, Arroyo-Mateos M, van der Wolf M, Boeren S, Prins M, van den Burg HA (2020) Identification of tomato proteins that interact with replication initiator protein (rep) of the Geminivirus TYLCV. Front Plant Sci 11. https://doi.org/10.3389/fpls.2020.01069

Makarova S, Makhotenko A, Spechenkova N, Love AJ, Kalinina NO, Taliansky M (2018) Interactive responses of potato (Solanum tuberosum L.) plants to heat stress and infection with potato virus Y. Front Microbiol 9. https://doi.org/10.3389/fmicb.2018.02582

Mazumder P, Majumdar A, Roy B, Mohanbabu Y, Valli NS, Sarkar M, Mondal F, Verma N, Kumar R, Roy A (2025) RNA vaccination in plants: a non-transgenic approach to combat plant viruses with special reference to tomato leaf curl New Delhi virus. In: Tomato leaf curl New Delhi virus (ToLCNDV) insights into Virome dynamics and management. Springer Nature Singapore, Singapore, pp 339–352. https://doi.org/10.1007/978-981-95-2852-3_17

Niu L, Zhang H, Wu Z, Wang Y, Liu H, Wu X, Wang W (2018) Modified TCA/acetone precipitation of plant proteins for proteomic analysis. PLoS One 13(12). https://doi.org/10.1371/journal.pone.0202238

Nováková S, Flores-Ramírez G, Glasa M, Danchenko M, Fiala R, Skultety L (2015) Partially resistant Cucurbita pepo showed late onset of the Zucchini yellow mosaic virus infection due to rapid activation of defense mechanisms as compared to susceptible cultivar. Front Plant Sci 6(APR). https://doi.org/10.3389/fpls.2015.00263

Patel P, Kumari N, Sharma PN (2023) RT-PCR based detection of pepper mild mottle virus from capsicum seeds and seed transmission assay. VirusDisease 34(1):50–55. https://doi.org/10.1007/s13337-023-00807-0

Pavan Kumar BK, Kanakala S, Malathi VG, Gopal P, Usha R (2017) Transcriptomic and proteomic analysis of yellow mosaic diseased soybean. J Plant Biochem Biotechnol 26(2):224–234. https://doi.org/10.1007/s13562-016-0385-3

Pérez-Losada M, Arenas M, Galán JC, Palero F, González-Candelas F (2015) Recombination in viruses: mechanisms, methods of study, and evolutionary consequences. Infect Genet Evol 30:296–307. Elsevier. https://doi.org/10.1016/j.meegid.2014.12.022

Pouresmaeil M, Dall'Ara M, Salvato M, Turri V, Ratti C (2023) Cauliflower mosaic virus: virus-host interactions and its uses in biotechnology and medicine. In: Virology, vol 580. Academic Press Inc, pp 112–119. https://doi.org/10.1016/j.virol.2023.02.008

Pratap D, Kashikar AR, Mukherjee SK (2011) Molecular characterization and infectivity of a tomato leaf curl New Delhi virus variant associated with newly emerging yellow mosaic disease of eggplant in India. Virol J. http://www.virologyj.com/content/8/1/305

Raigond B, Verma G, Kumar R, Tiwari RK (2022) Serological and molecular diagnosis of potato viruses: an overview. In: Chakrabarti SK, Sharma S, Shah MA (eds) Sustainable management of potato pests and diseases. Springer, Singapore, pp 337–352. https://doi.org/10.1007/978-981-16-7695-6_13

Ribeiro DG, Bezerra ACM, Santos IR, Grynberg P, Fontes W, de Souza Castro M, de Sousa MV, Lisei-de-Sá ME, Grossi-de-Sá MF, Franco OL, Mehta A (2023) Proteomic insights of cowpea response to combined biotic and abiotic stresses. Plants 12(9). https://doi.org/10.3390/plants12091900

Roshan P, Kulshreshtha A, Hallan V (2017) Identification of host cellular targets of AC4 and AV2 proteins of tomato leaf curl palampur virus and their sub-cellular localization studies. VirusDisease 28(4):390–400. https://doi.org/10.1007/s13337-017-0405-5

Ruiz L, Simon A, Velasco L, Janssen D (2017) Biological characterization of tomato leaf curl New Delhi virus from Spain. Plant Pathol 66(3):376–382. https://doi.org/10.1111/ppa.12587

Samarskaya V, Spechenkova N, Kalinina NO, Love AJ, Taliansky M (2025) The emerging role of omics-based approaches in plant virology. Viruses 17(7). https://doi.org/10.3390/v17070986

Sarkar M, Gupta D, Singh OW, Paul S, Kumar R, Mandal B, Roy A (2024) Identification of pathogenicity determinants in ToLCNDV and their RNAi-based knockdown for disease management in Nicotiana benthamiana and tomato plants. Front Microbiol 15:1481523. https://doi.org/10.3389/fmicb.2024.1481523

Sarkar M, Mondal F, Majumdar A, Valli NS, Roy B, Verma N, Mohanbabu Y, Kumar R, Roy A (2025) Molecular arsenal of tomato leaf curl New Delhi virus: viral suppressors of RNA silencing and their roles in pathogenicity. In: Tomato leaf curl New Delhi virus (ToLCNDV) insights into Virome dynamics and management. Springer Nature Singapore, Singapore, pp 325–337. https://doi.org/10.1007/978-981-95-2852-3_16

Singh P, Singh Rajput R, Dev D, Maharshi A, Prachi Singh C (2018) Proteomics approaches to study host pathogen interaction. J Pharm Phytochem 7(4):1649–1654

Siriwan W, Hemniam N, Vannatim N, Malichan S, Chaowongdee S, Roytrakul S, Charoenlappanit S, Sawwa A (2022) Analysis of proteomic changes in cassava cv. Kasetsart 50 caused by Sri Lankan cassava mosaic virus infection. BMC Plant Biol 22(1). https://doi.org/10.1186/s12870-022-03967-1

Siriwan W, Vannatim N, Chaowongdee S, Roytrakul S, Charoenlappanit S, Pongpamorn P, Paemanee A, Malichan S (2023) Integrated proteomic and metabolomic analysis of cassava cv. Kasetsart 50 infected with Sri Lankan cassava mosaic virus. Agronomy 13(3). https://doi.org/10.3390/agronomy13030945

Stare T, Stare K, Weckwerth W, Wienkoop S, Gruden K (2017) Comparison between proteome and transcriptome response in potato (Solanum tuberosum L.) leaves following potato virus Y (PVY) infection. Proteomes 5(3). https://doi.org/10.3390/proteomes5030014

Vadivel AKA (2015) Gel-based proteomics in plants: time to move on from the tradition. Front Plant Sci 6(MAY):1–4. https://doi.org/10.3389/fpls.2015.00369

Verma A, Sindhu SR, Boora N, Rani R, Aarti, Devi R, Yashveer S (2025) Stress Biology in agriculture: a comprehensive review of abiotic and biotic challenges. Plant Cell Biotechnol Molecular Biol 26(9–10):128–147. https://doi.org/10.56557/pcbmb/2025/v26i9-109731

Wang H, Cui X, Cai X, An T (2022) Recombination in positive-Strand RNA viruses. Front Microbiol 13. https://doi.org/10.3389/fmicb.2022.870759

Webb KM, Broccardo CJ, Prenni JE, Wintermantel WM (2014) Proteomic profiling of sugar beet (Beta vulgaris) leaves during rhizomania compatible interactions. Proteomes 2(2):208–223. https://doi.org/10.3390/proteomes2020208

Withycombe J, Han J, MacWilliams J, Dorn KM, Nalam VJ, Nachappa P (2024) Transcriptomic profiling reveals distinct responses to beet curly top virus (BCTV) infection in resistant and susceptible sugar beet genotypes. BMC Genomics 25(1):1237. https://doi.org/10.1186/s12864-024-11143-y

Wu J, Zhang Y, Li F, Zhang X, Ye J, Wei T, Li Z, Tao X, Cui F, Wang X, Zhang L, Yan F, Li S, Liu Y, Li D, Zhou X, Li Y (2024) Plant virology in the 21st century in China: recent advances and future directions. J Integr Plant Biol 66(3):579–622). John Wiley and Sons Inc. https://doi.org/10.1111/jipb.13580

Yang H, Huang Y, Zhi H, Yu D (2011) Proteomics-based analysis of novel genes involved in response toward soybean mosaic virus infection. Mol Biol Rep 38(1):511–521. https://doi.org/10.1007/s11033-010-0135-x

Zeng W, Sun Z, Cai Z, Chen H, Lai Z, Yang S, Tang X (2017) Proteomic analysis by iTRAQ-MRM of soybean resistance to Lamprosema indicate. BMC Genomics 18(1). https://doi.org/10.1186/s12864-017-3825-0

Zilian E, Maiss E (2011) An optimized mRFP-based bimolecular fluorescence complementation system for the detection of protein-protein interactions in planta. J Virol Methods 174(1–2):158–165. https://doi.org/10.1016/j.jviromet.2011.03.032

# Metabolomic Approach to Identify Biomarkers in Managing Vegetable Viruses

17

Saswat Swarup Padhi, Laxmipriya Behera,
Bandana Mohapatra, S. Kruthika, Ravinder Kumar,
Rahul Kumar Tiwari, Awadhesh Kumar,
and Milan Kumar Lal

**Abstract**

Vegetable viruses pose a serious threat to global food security, as they significantly affect quality and production, which is quite challenging to manage using traditional diagnostic and control strategies. A potential approach for identifying biomarkers that can represent plant responses to viral infections is metabolomics. Metabolic reprogramming driven by viral invasion, includes significant changes in the levels of sugars, amino acids, organic acids, and secondary metabolites, including flavonoids and phenolics, which can serve as diagnostic, prognostic, and predictive biomarkers. Cutting-edge analytical tools, such as GC-MS, LC-MS, and NMR, combined with advanced statistical and machine learning techniques, help in the accurate identification of metabolic markers that can differentiate between resistant and susceptible cultivars and detect early stages of infections. Squash, tomato, cucumber, and other vegetable case studies depict unique metabolic changes connected to viral invasion and host defence. The combination of metabolomics with transcriptomics and proteomics improves the certainty of biomarker identification and creates new opportunities for early diagnosis and precision disease management. With the help of biosensors, artificial intelligence, and portable diagnostic platforms, the application of metabolomics presents revolutionary potential for sustainable management of vegetable

S. S. Padhi · L. Behera · B. Mohapatra · S. Kruthika · A. Kumar · M. K. Lal (✉)
ICAR-Central Rice Research Institute, Cuttack, Odisha, India

R. Kumar
Division of Plant Pathology, ICAR-Indian Agricultural Research Institute, New Delhi, Delhi, India

R. K. Tiwari
Division of Plant Protection, ICAR-Indian Sugarcane Research Institute, Lucknow, Uttar Pradesh, India



R. Kumar et al. (eds.), *Vegetable Crops and Viral Pathogens Volume I*, Plant Pathology for Sustainable Agriculture, https://doi.org/10.1007/978-981-95-7265-6_17

viruses in the face of climate change and evolving pathogen diversity, despite many challenges in validation and reproducibility.

**Keywords**

Pathogen · Horticultural crops · Biomarkers · Metabolites · Omics approach

## 17.1 Introduction

Vegetables have great significance in human nutrition and agricultural sustainability, contributing as an essential source of vitamins, minerals, dietary fibre, antioxidants, and bioactive compounds that play a crucial role in preventing various diseases (Ebert 2020). In middle-income or emerging economies, vegetables like cabbage, pepper, potato, tomato, and leafy greens are one of the major factors of food security, sustenance, and nutritional well-being (Kumar et al. 2023; Roy et al. 2025; Raigond et al. 2020). However, viruses pose a significant threat, resulting in the disruption of physiological processes and leading to substantial reductions in yield quantity and quality (Tomlinson 1987). Viruses like tomato yellow leaf curl virus (TYLCV), cucumber mosaic virus (CMV), turnip mosaic virus (TuMV), and beet mosaic virus (BtMV) are prevalent across global vegetable production systems. They are the major contributors to fatal epidemics and acute financial damage. Limited resources and traditional disease management strategies, such as the use of insecticides and crop rotation, extend limited and nonpermanent relief due to the adaptability of insect vectors, environmental fluctuations, and evolving viral strains (Rotbi et al. 2014). The significant effects of viral infections impact the vegetable crops, extending beyond visible symptoms such as mosaic patterns, stunted growth, and curling to hinder yield quality and quantity (Sastry and Zitter 2014). They sabotage physiological, biochemical processes and economic escalation. In most cases, viral infections contribute to the severe degradation of crops, making them highly nonmarketable and exacerbating food insecurity, especially in middle-income or low-income regions that chiefly rely on vegetable crops for nutrition and livelihood (Varma and Malathi 2003). The development of more precise and timely diagnostic tools that discover viral infections at early or presymptomatic stages is indeed a revolutionary move in the field of biosciences. The advancements in biomarkers extend beyond visible consequences (Marqués et al. 2022). Biomarkers are important indicators of physiological states that are highly probed as apparatuses for stress monitoring in plants, early virus detection, and breeding of genetically modified or trait-enhanced crops (Aina et al. 2024). Biomarkers also contribute to molecular changes such as protein profusion, metabolite concentrations, modification of gene activities, and enzymatic activities that mirror precise responses to viral pathogens. Selection of resistant lines, timely interventions, and evaluating treatment outcomes are essential to revolutionise plant disease management, although the recognition of robust biomarkers plays a significant role (Zhang et al. 2024).

Metabolomics is the comprehensive and high-throughput analysis of metabolites, which are micromolecules that serve as substrates, intermediates, and end products of cellular processes (Matthews et al. 2016). Metabolites are the final expression of products of the genome, transcriptome, and proteome, making them a close and dynamic representation of an organism's phenotype under specific environmental or pathological conditions. Metabolomics play a crucial role in studying the intricate biochemical alterations that occur in vegetable crops in response to viral infections (Salem et al. 2020). When a plant is affected by a virus, multiple defence mechanisms are triggered at the cellular level, including upregulation of defensive secondary metabolites such as flavonoids and alkaloids, changes in metabolic energy, oxidative bursts, and activation of signalling pathways such as salicylic acid and jasmonic acid (Sett et al. 2022). Metabolomic profiling helps to apprehend those changes and differentiate patterns and certain signs that are associated with symptom development, resistance, obstruction, and liability (Wishart 2019). The instances are tomato plants infected with TYLCV (tomato yellow leaf curl virus), which exhibit changes in the accumulation of specific and sugars that show host response and viral load and cucumber plants infected with CMV (cucumber mosaic virus), which reflect distinct accumulation of amino acids and antioxidant compounds during early infection stages (Hemmati et al. 2025). In vegetable virology, metabolomics-based methods are becoming potential platforms for the development of biomarkers, as it can provide a brief idea about the biochemical changes that occur in an infected plant. With the help of metabolomics, minute metabolic traces left by viruses can be identified, allowing early, sensitive, and accurate diagnosis in contrast to traditional techniques that mostly rely on visual symptom assessment or molecular identification of viral genomes (Tugizimana et al. 2018). Through the integration of advanced methods of analysis like nuclear magnetic resonance (NMR) spectroscopy, liquid chromatography and mass spectrometry (LC–MS), and gas chromatography and mass spectrometry (GC–MS), metabolomics enables high-resolution profiling of small molecules and their quantitative changes over various stages of disease (Castro-moretti et al. 2020). In order to distinguish between resistant and susceptible cultivars, as well as between viral species and strains, metabolomic platforms identify distinct markers that function as diagnostic signatures (Fig. 17.1).

The ability of metabolomics to correlate genotypic and phenotypic characteristics under viral stress conditions is one of its most remarkable benefits. For example, metabolite profiling has identified multiple pathways that are consistently altered in vegetable crops attacked by viruses, including phenylpropanoid metabolism, polyamine biosynthesis, and antioxidant defence networks (Villate et al. 2021). Identifying these pathway-specific metabolites enhances early diagnostic capabilities and provides molecular targets for genetic engineering or conventional breeding to enhance crop plants. Crucially, metabolomics is used for purposes other than diagnosis and resistance breeding, such as tracking the efficacy of biocontrol techniques (Razzaq et al. 2022). The responses of plants to elicitors, resistance inducers, or helpful microorganisms that lessen viral damage can all be assessed using metabolomic markers (Sade et al. 2015). Therefore, the incorporation of

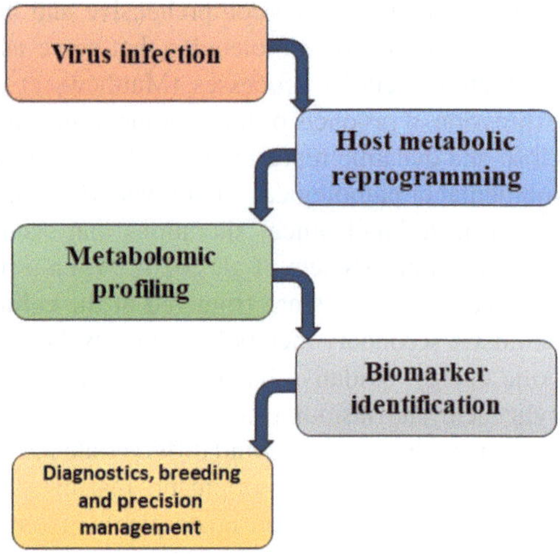


**Fig. 17.1** Conceptual framework of metabolomics in vegetable virus management

metabolomic markers into plant pathology pipelines has the potential to transform the management of vegetable viruses by guaranteeing early identification, enhancing host resistance, and expanding sustainable farming methods.

## 17.2 Plant-Virus Interactions

Plant viruses are obligatory intracellular parasites that severely threaten vegetable output all over the world. They use host cellular machinery to multiply, move systemically, and escape defence responses, leading to a wide spectrum of adverse consequences on crop health, production, and quality. It is essential to understand the basic mechanics of viral infection and the physiological reactions of the host to create accurate breeding, control, and diagnostic methods (Jones and Naidu 2019).

### 17.2.1 General Mechanisms of Viral Infection in Vegetable Crops

Mechanical damage or vector-mediated entrance, mainly through insects like aphids, whiteflies, nematodes, and thrips, is usually how plant virus infections start. When the virus tries to enter the host cell, it sheds its outer layer and releases its genetic material (which is often DNA or RNA) into the cytoplasm of the host cell. The viral genome commands the host's translational machinery to replicate and synthesise proteins, frequently creating replication complexes linked to organelle membranes (Bol et al. 1990; Culver and Padmanabhan 2008; Laliberté and Sanfaçon 2010). Coat proteins and helper components support long-distance movement across the phloem, while viral proteins like movement proteins (MPs) help in cell-to-cell transmission via plasmodesmata (Scott and Leisner 1993). Moreover, viruses use viral suppressors of RNA silencing (VSRs) to block host RNA interference pathways, which magnifies the effectiveness of infection (Díaz-Pendón and Ding 2008).

### 17.2.2 Host Responses at Cellular and Molecular Levels

In response to viral invasion, plants utilise a variety of defence mechanisms. Pattern recognition receptors (PRRs) mediate initial identification by identifying chemical patterns linked to viruses, thereby triggering pattern-triggered immunity (PTI). Effector-triggered immunity (ETI), which generally results in hypersensitivity response (HR) and localised cell death, is triggered when some plant resistance (R) genes recognise viral effectors.

RNA silencing is a fundamental antiviral defence of plants at the cellular level (DeFalco and Zipfel 2021). In order to direct argonaut proteins to break down complementary viral RNAs, dicer-like proteins convert viral dsRNA into small interfering RNAs (siRNAs). Additionally, viral infection also causes the expression of genes linked to defence and the modification of phytohormones such as ethylene, salicylic acid, and jasmonic acid (Katiyar-Agarwal et al. 2006; Liu et al. 2021). Metabolomic changes, such as modifications in sugars, amino acids, and secondary metabolites, are also noticeable during infection and represent the plant's reprogrammed biochemistry to prevent the spread of the virus or to deal with stress (Jiang et al. 2025).

### 17.2.3 Symptoms and Physiological Changes Induced by Viruses

Symptoms caused by viruses in vegetables differ based on the type of virus and the host plant. Stunted growth, decreased reproductive output, mosaic and mottling of leaves, leaf curling, vein clearing, and chlorosis are typical symptoms (Kostova et al. 2003). For instance, cabbage yield is severely reduced by the turnip mosaic virus (TuMV), particularly when infection occurs at the seedling stage. Early infection by beet mosaic virus (BtMV) can cut marketable leaf yield in Swiss chard by more than half (Wua et al. 2024). Photosynthesis, hormone signalling, and nutrient transport are among the vital physiological processes that viruses interfere with. These disruptions frequently result in energy imbalances, oxidative stress, and the build-up of reactive oxygen species (ROS), which affect the general health and productivity of plants (Jameson and Clarke 2002).

### 17.2.4 Limitations of Traditional Virus Detection and Management Approaches

Traditional methods for detecting and managing plant viruses include visual diagnosis, ELISA, PCR/RT-PCR, and vector control. However, these approaches have some critical limitations. Visual examination is not a reliable method for detecting

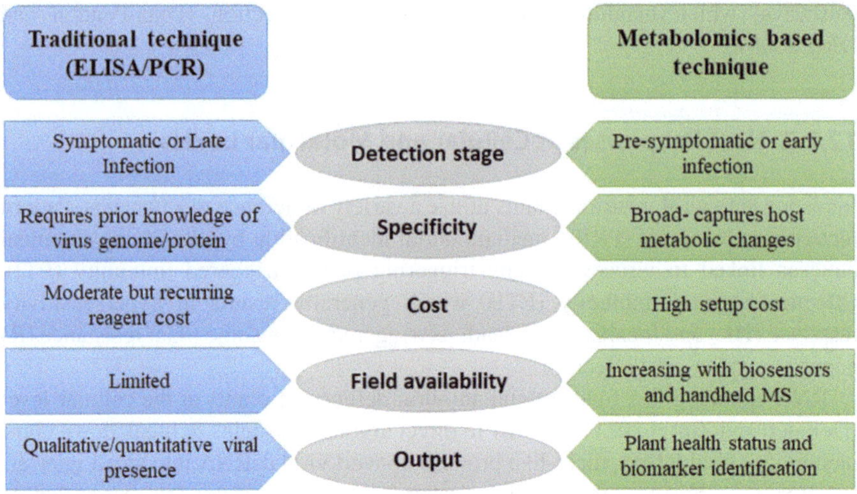


**Fig. 17.2** Comparison of traditional and metabolomics-based virus detection

undiscovered or asymptomatic infections, and it can mistakenly attribute virus symptoms to abiotic stress or nutritional deficits (Loebenstein and Katis 2014). The accuracy of nucleic acid-based and serological methods like ELISA is limited in the field due to the need for expensive reagents, skilled workers, and prior virus knowledge (Boonham et al. 2014). Pesticide-based vector management offers temporary relief, but it also aggravates environmental problems and multiplies vector resistance (Fig. 17.2). Although resistant cultivars are still a viable choice, crops with limited genetic variety may experience resistance breakdown due to viral mutation and evolution of strains (Van Den Berg et al. 2021). Cultural interventions like crop rotation and rogueing are often insufficient when disease pressure is high, particularly in environments with limited resources. Owing to these difficulties, omics-based methods, particularly metabolomics, are becoming increasingly effective in identifying early-stage biomarkers for disease prediction and management as well as detecting metabolic changes brought by viruses (Ullah et al. 2022).

## 17.3 Metabolomics: Concepts and Techniques

### 17.3.1 Definition and Scope of Metabolomics

Metabolomics is the methodical study of small metabolites, usually weighing less than 1.5–3 kDa, in biological systems, acting as a direct “functional readout” of physiological state and cellular activities (Resurreccion and Fong 2022). Both basic metabolites (such as carbohydrates, amino acids, organic acids, and lipids) and secondary metabolites (such as phenolics, flavonoids, alkaloids, and terpenoids) are necessary for growth and development in plants. They are also involved in signalling, defence, and ecological interactions (Zhang et al. 2022). When it comes to plant-virus interactions, metabolomics provides information about the metabolic reprogramming brought on by pathogen attack. Secondary metabolites frequently support antiviral defences or stress signalling, whereas primary metabolites may indicate energy imbalance or carbon/nitrogen redistribution (Okoń et al. 2024).

### 17.3.2 Types of Metabolites Involved in Plant-Virus Interactions

There are mainly two types of metabolites that are involved in plant-virus interaction, namely, primary metabolites and secondary metabolites.

*Primary metabolites* consist of lipids, carbohydrates, organic acids, and amino acids. Changes in carbon allocation and amino acid pools are frequently brought on by viral infection as a means of manipulating the pathogen and defending the host (Tomé 2021).

*Secondary metabolites* include phenolic acids, flavonoids, alkaloids, and terpenoids and are involved in signalling to herbivores or helpful microorganisms, oxidative stress reduction, and plant defence. As a result of the host response or metabolic rewiring, their levels often fluctuate during any viral infection (Jiang et al. 2025).

### 17.3.3 Analytical Platforms

#### 17.3.3.1 Gas Chromatography-Mass Spectrometry (GC-MS)

For metabolites that are volatile or derivatised (such as sugars, amino acids, organic acids, and volatile organic compounds), GC-MS is the best analysis method (Table 17.1). It provides good separation, better identification using large databases (Golm Metabolome Database) and high sensitivity (LOD ~$10^{-12}$ M). Despite the destructive nature of the analysis and the need for derivatisation, GC-MS is nevertheless widely used for both targeted and untargeted metabolomics in plants (Patel et al. 2021).

#### 17.3.3.2 Liquid Chromatography-Mass Spectrometry (LC-MS)

LC-MS is particularly good in profiling polar and thermo-labile metabolites without derivatisation, including flavonoids, carotenoids, glucosinolates, and some lipids (Razzaq et al. 2019). Electrospray (ESI) and APCI are examples of ionisation techniques that improve detection. LC-MS has extensive metabolite coverage and high sensitivity (LOD ~$10^{-15}$ M), but quantification can be less accurate (Table 17.1).

#### 17.3.3.3 Nuclear Magnetic Resonance (NMR) Spectroscopy

NMR is a highly reproducible, nondestructive analytical technique that offers absolute quantification and structural identification, frequently without the need for chromatographic separation or derivatisation (Table 17.1). In complicated mixes, 2D NMR aids in resolving overlap, while 1H NMR is frequently employed for high-throughput screening. Lower NMR sensitivity (LOD ~$10^{-6}$ M) requires higher-priced instrumentation and bigger sample quantities (Deborde et al. 2019).

**Table 17.1** List of analytical platforms used for plant-virus metabolomic studies, detailing their strength, limitations, sensitivity, and applications

| Platforms | Strengths | Limitations | LOD/ sensitivity | Applications | References |
|---|---|---|---|---|---|
| GC-MS | High sensitivity with extensive database | Requires derivatisation: destructive | $\sim 10^{-12}$ M | Used for volatile and primary metabolites like sugars, amino acids, and organic acids | Deng et al. (2004) |
| LC-MS | Broad metabolite coverage and high sensitivity | Quantification can be less accurate | $\sim 10^{-15}$ M | Secondary metabolite and lipid profiling | Cejnar et al. (2020) |
| NMR spectroscopy | Nondestructive and reproducible | Low sensitivity and expensive at cost | $\sim 10^{-6}$ M | Identifying and quantifying metabolites in biological mixtures | Maravi et al. (2022) |
| CE-MS (capillary electrophoresis mass spectrometry) | Appropriate for polar metabolites and ionic compounds | Limited libraries | $\sim 10^{-13}$ M | Mainly used for amino acid profiling | Ramautar et al. (2009) |
| Volatilomics (proton transfer reaction–mass spectrometry (PTR-MS), eNose) | Noninvasive, real-time measurements | Limited metabolite ID | – | Used in field detection of volatile biomarkers | Majchrzak et al. (2021) |

### 17.3.4 Targeted and Untargeted Metabolomics

Targeted metabolomics is concerned with predetermined groups of metabolites and frequently uses high precision and accuracy to quantify certain chemicals of interest (Roberts et al. 2012). To profile as many metabolites as possible, untargeted metabolomics employs broad-spectrum detection, which enables the identification of unknown metabolic alterations and the discovery of new biomarkers (Yang et al. 2024) 2018. The choice between targeted and untargeted procedures is dependent upon the goals of the study. Targeted methods validate particular known chemicals, whereas untargeted profiling is best suited for hypothesis development and biomarker discovery (Sobsey et al. 2020).

### 17.3.5 Data Preprocessing and Bioinformatics Tools

A standard workflow for metabolomics includes sample preparation, data acquisition, pre-processing, biomarker discovery, statistical analysis, identification, annotation, and biological interpretation. Utilising suitable solvents and techniques adapted to metabolite chemistry, samples are collected, preserved (freezing and drying), homogenised, and extracted. Data acquisition is done by producing chromatographic and spectral data using one or more analytical platforms (GC MS, LC MS, NMR) (Vinaixa et al. 2016). Preprocessing is done to lower technical variability and facilitate statistical comparison. It generally includes baseline correction, peak detection, alignment across samples, simplification, handling missing values, normalisation, and scaling (Antoniadis et al. 2007). During biomarker discovery and statistical analysis, multivariate methods such as machine learning, Principal Component Analysis (PCA) Partial Least Squares Discriminant Analysis (PLS-DA) Orthogonal Partial Least Squares Discriminant Analysis (OPLS-DA) enhance the ability to distinguish between healthy and infected states and search possible biomarkers (Wolfender et al. 2013). Using mass and retention indices or spectral libraries, tools like XCMS Online (in conjunction with METLIN) facilitate putative metabolite annotation, peak alignment, and feature discovery. Additional programmes such as iMet and ProbMetab provide computer annotation of unknown metabolites using Bayesian models or MS/MS spectra (Alaribe et al. 2025). Metabolic changes are linked to defence mechanisms and biological processes by pathway enrichment and network analysis (MetaboAnalyst). Reproducibility and workflow sharing in metabolomics can be improved by online systems like Workflow4Metabolomics (W4M) built on Galaxy, which offer unified pipelines for Liquid chromatography mass spectrometry (LC-MS), gas chromatography mass spectrometry (GC-MS), flow injection analysis mass spectrometry (FIA-MS), nuclear magnetic resonance (NMR) data from preprocessing to statistical analysis (Delporte et al. 2025).

## 17.4 Metabolic Reprogramming During Viral Infections

### 17.4.1 Changes in Primary Metabolism (Sugars, Amino Acids, Organic Acids)

Viral infection often triggers notable changes in primary metabolism, especially involving essential biomolecules such as carbohydrates, amino acids, and organic acids (Sumbria et al. 2021). For example, tomato mosaic virus (ToMV)-infected tomato leaves showed increased levels of sugar in young or top leaves and a general decrease in the level of organic acids and amino acids (except for asparagine) in older leaves (López-Gresa et al. 2012). This pattern was subsequently influenced by leaf position, developmental stage, and diurnal timing. Similarly, research on potato virus Y (PVY) and TMV in tobacco or potato revealed that TCA (Krebs cycle) intermediates, alongside activation of the gamma-aminobutyric acid (GABA) shunt, suggest enhanced respiratory activity to fight oxidative stress and restore energy (Xu et al. 2024). These modifications support defence responses and preserve NADH. Early cytomegalovirus (CMV) infection in squash cotyledons led to a decrease in starch storage and increased soluble sugars and amino acids, implying that the virus was responsible for the carbon allocation shifts from structural carbohydrates to energy-rich compounds that were available for viral reproduction (Ertunç 2020).

### 17.4.2 Alterations in Secondary Metabolites (Phenolics, Alkaloids, Terpenoids, Etc.)

Secondary metabolites, including terpenoids, alkaloids, flavonoids, and phenolics, are often reprogrammed during infection and are essential for antiviral defence (Dhamija and Mangla 2022). Chlorogenic acid, epicatechin, catechin, and gallic acid levels increased in CMV-infected squash at 6 dpi, whereas substances such as caffeic and ferulic acids were inhibited. This profile changed once more by 12 dpi, indicating dynamic temporal control (Abdelkhalek et al. 2022). Furthermore, phenylpropanoid pathway gene expression (PAL, CHS) and antioxidant enzyme levels (polyphenol oxidase, catalase) were dynamically altered during the course of infection, linking these metabolites to early defence signalling and oxidative stress mitigation. In tomato infected with ToMV, flavonoids accumulated in inoculated leaves; however, noninoculated systemic leaves displayed higher phenylpropanoids, indicating both local and systemic metabolic defence responses (Aseel et al. 2024).

### 17.4.3 Case Studies in Vegetables

#### 17.4.3.1 Tomato and Tomato Mosaic Virus (ToMV)

A time-course NMR study examined 32 metabolites, including sugars, amino acids, organic acids, flavonoids, and phenylpropanoids (Table 17.2). Organic acids and amino acids (except asparagine) decreased over time. Flavonoid accumulation was observed in inoculated leaves, whereas phenylpropanoids increased in systemic leaves. Diurnal patterns altered metabolite levels: midday amino/organic acids rose, whereas at night, sugar and secondary metabolite levels increased (Chen et al. 2002).

#### 17.4.3.2 Squash (*Cucurbita pepo*) and Cucumber Mosaic Virus (CMV)

Dynamic variations in phenolic and flavonoid components were discovered by HPLC and GC-MS profiling. At 6 dpi, chlorogenic acid and catechin were stimulated, while caffeic acid and rutin were suppressed. By 12 dpi, a new profile appeared (Table 17.2). Concurrent gene expression adjustments (PAL, CHS, PR genes) validated these biochemical changes, demonstrating a coordinated defence response (Abdelkhalek et al. 2022).

#### 17.4.3.3 Cucumber and Cucurbit Chlorotic Yellows Virus (CCYV)

Cucumber metabolomics revealed an early-stage reduction in flavonoids and terpenoids, as well as later-stage increases in lipids and amino acids (Table 17.2), supporting viral replication and vector attraction pathways versus defence-related secondary metabolism (Chen et al. 2021).

### 17.4.4 Metabolic Signatures Associated with Susceptibility or Resistance

Plant susceptibility is linked to specific metabolic signatures. In squash, early-stage transitory elevation of gallic acid and catechin was associated with defence signalling, but inhibited accumulation of important phenolics, such as caffeic acid and chlorogenic acid, was associated with rapid systemic infection (Abdelkhalek et al. 2022). Metabolite profiles varied by stage and leaf position in tomatoes. It increased phenylpropanoids in systemic leaves function as more general systemic defence markers, whereas higher flavonoids in inoculated leaves might indicate local resistance responses (Ramaroson et al. 2022). Genotypes with stronger resistance phenotypes frequently show a higher accumulation of antioxidants, GABA pathway intermediates, amino acids, or defensive secondary metabolites across systems (e.g. tobacco TMV, potato PVY), whereas susceptible plants are unable to maintain these responses efficiently (Khizar et al. 2020). Additionally, using PCA clustering and VIP score features, comparative metabolomics between resistant and susceptible cultivars, such as in cassava infected with mosaic virus and pepper infected with pepper yellow leaf curl virus (PYLCV), showed unique biomarker patterns that correlate with resistance phenotype (Agustika 2024).

**Table 17.2** Major vegetable viruses and associated metabolic biomarkers

| Host | Virus | Detection stage | Changes in primary metabolite levels | Changes in secondary metabolite levels | Associated biomarkers | References |
|---|---|---|---|---|---|---|
| Tomato | Tomato yellow leaf curl virus (TYLCV) | Early-mid | Rise in sucrose and proline levels | Quercetin and kaempferol levels increase | Sucrose and quercetin | Hemmati et al. (2025) |
| Cucumber | Cucumber mosaic virus (CMV) | Early | Gallic acid and catechin levels increase. | Transient flavonoids decrease. | Gallic acid and catechin | Abdelkhalek et al. (2022) |
| Squash | CMV | 6–12 days after infection | Reallocation of carbon to soluble sugars | Chlorogenic acid and ferulic acid levels decline. | Chlorogenic acid and ferulic acid | Abdelkhalek et al. (2022) |
| Chilly | Pepper yellow leaf curl virus (PYLCV) | Mid | Increased GABA (gamma aminobutyric acid) intermediates | Phenylpropanoid levels increased. | Ferulic acid and GABA | Agustika (2024) |
| Cabbage | Turnip mosaic virus (TuMV) | Early | The amino acid balance got disrupted. | Rise in kaempferol levels | Kaempferol and arginine | Mi et al. (2024) |
| Lettuce | Lettuce mosaic virus (LMV) | Early | Rise in the levels of sugars and organic acids | Phenolic levels increased. | Rutin | Aliyu et al. (2022) |
| Beans | Bean common mosaic virus (BCMV) | Early | Increased amino acid levels | Isoflavonoids got increased. | Isoflavonoids | Makalesi et al. (2020) |

## 17.5 Biomarkers: Concepts and Criteria

A biomarker is a measurable molecular attribute (such as a metabolite) that accurately represents a particular biological state, such as viral infection, disease progression, or host tolerance, in the context of plant virus metabolomics (Castro-moretti et al. 2020). Biomarkers can represent the complete range of disease, from the first signs to its advanced stages (Aronson and Ferner 2017). Characterised as modifications in the components of tissues, these markers provide a method for the uniform classification of diseases and their risk factors, hence enhancing the basic understanding of the biological basis of diseases. Based on their function, biomarkers are classified into three major types, that is, diagnostic biomarkers, prognostic biomarkers, and predictive biomarkers (Aina et al. 2024) (Bodaghi et al. 2023). Diagnostic biomarkers are the indicators that usually differentiate infected plants from healthy ones at a particular time. For instance, elevation in phenolic acids marks the early stage of infection (Espinoza et al. 2021). Predictive biomarkers predict how a plant will probably react to treatments (like elicitors) or genetic resistance mechanisms (such as identifying resistant cultivars based on baseline metabolic profiles) (Maia et al. 2020). When the difference between the effectiveness of two or more treatment options depends on the value of that biomarker, then the biomarker is referred to as predictive. These ideas, which originated in medical research, are immediately applicable to plant pathology. Profiles linked to resistance genotypes can be utilised predictively in breeding, and metabolites that regularly rise early during infection can be employed as diagnostic markers (Rahman and Schellhorn 2023).

### 17.5.1 Statistical Approaches for Biomarker Identification

In order to find reliable metabolic biomarkers, high-dimensional data must be statistically analysed. Diverse statistical methodologies have been suggested to evaluate the interaction between a continuous covariate and treatment (Ternès et al. 2017). The most popular chemometric techniques are as follows:

#### 17.5.1.1 Principal Component Analysis (PCA)

PCA is a technique for visualising grouping and identifying outliers in metabolomic data that reduces dimensionality without supervision. PCA suggests potential distinguishing characteristics by revealing the fundamental grouping of healthy versus infected samples. PCA components are frequently used to adjust for unmeasured confounder effects that are hidden in univariate testing (T'Kindt et al. 2010).

#### 17.5.1.2 Partial Least Squares Discriminant Analysis (PLS-DA)

PLS-DA is a supervised method that simulates the connections between sample class labels (Y vector; infected vs. control) and metabolite characteristics (X matrix). PLS-DA identifies the metabolites most responsible for class separation by combining predictive models with feature ranking using variable importance in projection (VIP) scores. To effectively explain the response variable Y, PLS-DA looks for the greatest possible breakdown of the predictor variable X (Blasco et al. 2015).

#### 17.5.1.3 Orthogonal PLS-DA (OPLS-DA)

OPLS-DA, or "orthogonal" PLS-DA, is an improved form of PLS-DA that has also become more popular in recent years. This model recapitulates the variance into "orthogonal" (or predictive) parts for the experimental groups (Khymenets et al. 2015). Therefore, OPLS-DA generates recognizable models more easily than its prior version.

#### 17.5.1.4 Support Vector Machines (SVM)

This supervised machine learning approach can be applied for classification and regression of nonlinear data sets (Ghosh et al. 2019). The versatility of SVM is demonstrated by its ability to identify nonlinear relationships in the data that defy the presumptions of PCA and OPLS. For classification with overlapping groups, multiclass classification, regression, and specialisations, SVM offers a variety of extensions. SVM's restriction to binary classification is one of its drawbacks (Nwosu 2021). Since SVM models are based on splitting the dataset into units of binary groups, they are oversimplified and may produce uninformative models. Other approaches have been put forth to expand the application of SVM to multiclass situations.

#### 17.5.1.5 Receiver Operating Characteristic (ROC) Curves

The receiver operating characteristics (ROC) curve plots the true positive rate (y axis) against the false-positive rate (x axis) to evaluate the specificity and sensitivity of a possible biomarker (Yang and Berdine 2017). It yields the area under the curve (AUC) metric, which shows how well a biomarker may differentiate between two research groups. These methods are usually used sequentially. PCA is used to investigate data structures, PLS-DA/OPLS-DA is used to find metabolite signatures that distinguish groups, and ROC analysis (frequently univariate or multivariate) is used to assess diagnostic performance and reliability (Brandolini-Bunlon et al. 2019).

### 17.5.2 Integration with Other Omics Layers

Metabolomic data alone might not adequately convey the complex nature of biological reactions. Strong biomarkers should ideally incorporate several molecular layers like transcriptomics and proteomics (Bansal et al. 2023). Transcriptomics identifies alterations in gene expression that underlie metabolic disturbances, such as the overexpression of genes involved in phenylpropanoid production after viral infection (Jiang et al. 2025). Proteomics helps in the identification of enzyme-level responses or signalling proteins linked to metabolite production, which transcript levels alone cannot resolve due to posttranslational regulation (Kukurugya et al. 2019). Causal and mechanistic connections can be established between virus-induced gene activation, protein function, and metabolite accumulation through the integration of transcriptome, proteome, and metabolome through multiomics (Wu et al. 2023). Due to the increased likelihood of replicable and biologically meaningful biomarkers backed by numerous omics layers, this layered approach improves confidence in biomarker selection. For multiomic integration and cross-lab reproducibility, recent systems biology frameworks highlight the significance of quantitative, high-quality datasets with established protocols (Zheng et al. 2025).

### 17.5.3 Challenges in Validation and Reproducibility

Although metabolomics provides a potent biomarker discovery tool, many challenges need to be overcome to get accurate and useful results.

#### 17.5.3.1 Technical and Biological Variability

Due to sample processing, extraction techniques, instrumentation, and batch effects, metabolic datasets are naturally variable. Due to small variations, even labs with similar protocols can produce different results (Han and Li 2022). Therefore, accurate quality control, standard operating procedures (SOPs), and reference standards are necessary for robust biomarker discovery.

#### 17.5.3.2 Model Overfitting and Statistical Pitfalls

If supervised approaches such as PLS-DA and OPLS-DA are not well verified, they are susceptible to overfitting. External validation datasets, permutation tests, and cross-validation are necessary to guarantee that the models' predictive ability is real and unaffected by noise (Worley and Powers 2016).

#### 17.5.3.3 Lack of Standardised Metadata

It is essential to carefully document metadata on plant age, cultivar, growth circumstances, sample time, environmental parameters, and infection timing (Ferreira et al. 2017). It is difficult to compare experiments and replicate results across circumstances when there is a lack of appropriate metadata.

#### 17.5.3.4 Validation in Independent Cohorts

Validation of candidate biomarkers found in discovery sets needs to be done in separate sample sets, preferably from other cultivars, regions, or seasons (Srikanth et al. 2025). Verifying that biomarkers are resilient in the face of changing real-world circumstances is crucial.

#### 17.5.3.5 Cross-Laboratory Reproducibility

Interlaboratory comparisons frequently show contradictory results because of minor variations in procedure or instrument quality. Standardising data and developing consensus biomarkers can be facilitated by taking part in ring trials and using interlab reference materials (Ntzifa and Lianidou 2023).

#### 17.5.3.6 Integration and Interpretation of Multiomic Data

There are conceptual and computational difficulties when combining metabolomic, transcriptomic, and proteomic data because the data formats vary in terms of scale, complexity, and quality (Sanches et al. 2024). The creation of integration pipelines and the maintenance of uniform sample metadata are significant obstacles. However, strong frameworks continue to appear to facilitate the development of mechanistic biomarkers in plant systems.

## 17.6 Future Perspectives

Metabolomics is playing an increasingly important role in understanding and regulating plant-virus interactions, as the threat of viral infections in vegetable crops is increasing due to global climatic stress and growing pathogen diversity. In order to properly utilise metabolomics, future research must shift from analysing individual metabolites to developing scalable, integrative, and field-ready solutions. Stimulating and influencing metabolic pathways include gene expression (transcriptomics), protein abundance and activity (proteomics), and epigenetic regulation. Thus, to fully capture the complexity of host-virus interactions, multiomics integration is crucial. System-level analyses that integrate metabolomics, transcriptomics, and proteomics will enable future plant virology research to identify causal biomarkers that are mechanistically involved in infection or resistance. For example, associating the accumulation of flavonoids with the upregulation of stress hormone or CHS (chalcone synthase) pathways may help in directing specific biocontrol tactics or improving resistance breeding. It is anticipated that network-based methods and integrative platforms like OmicsNet, MetaboAnalyst, or Cytoscape will be crucial for the visualisation and interpretation of these multilayered datasets. Artificial intelligence (AI) and machine learning (ML) are turning to be crucial tools for finding biomarkers, identifying patterns in the high-dimensional and nonlinear datasets produced by metabolomics. Random forests, support vector machines (SVMs), neural networks, and deep learning models are a few examples of algorithms that can disclose hidden metabolic patterns linked to resistance traits or viral infection. AI can reduce false positives in metabolite classification, enhance predictive modelling, and automate feature selection. Even before symptoms show up, ML-trained models can differentiate between healthy and diseased tomato plants using LC-MS fingerprints. In the future, these models can be implemented in field devices for real-time decision-making or linked into breeding pipelines to screen vast populations depending on metabolic potential.

Virus control in vegetables is about to undergo a revolution due to the use of metabolomics in precision agriculture. Designing biosensors that identify particular metabolic alterations, like increased phenolics or stress-related amino acids, can enable in-field real-time plant health monitoring. The development of biosensors based on electrochemical, optical, or microfluidic platforms is advancing to identify metabolic indicators such as flavonoids, glutathione, or $H_2O_2$. Combining these gadgets with cloud-based analytics, GPS information, and wireless technologies allows for precision intervention, like selectively harvesting infected plants or applying pesticides where they are needed. By lowering input consumption and environmental impact while increasing crop protection, this corresponds with the larger objective of sustainable agriculture. Field-deployable, portable devices act as the next frontier in applied plant metabolomics. In-field diagnosis may become feasible with the use of technologies like leaf-sap analysers, handheld mass spectrometers, and volatolomic sensors (which can identify volatile organic compounds released during infection). Early viral outbreaks can be detected through on-site metabolic fingerprinting provided by mobile diagnostic labs outfitted with miniature MS or GC systems. When paired with AI-driven disease mapping or drone-based sampling, this could change disease surveillance from a reactive to a proactive process. In areas with limited resources, cheap mobile metabolomic tests could give farmers and extension agents real-time information on crop virus health, enabling quick and localised management decisions. Despite such potential of metabolomics, the lack of consistency in sample collection, extraction procedures, data preparation, and validation standards is now obstructing its widespread use. The development of organised libraries of plant virus-specific biomarkers, worldwide metabolomics standards, and uniform reporting protocols is all necessary for reproducibility. Furthermore, to guarantee their strength, discovered biomarkers need to be cross-validated across several cultivars, environments, and geographical areas. To verify the functionality of metabolomic-based techniques in real-world situations, extensive field testing and interlaboratory cooperation will be essential.

## 17.7 Conclusion

Vegetable viruses continue to affect the global agricultural productivity by altering host physiology, reducing yield quality, and risking farmers' livelihood. The traditional detection techniques and management strategies are useful, but they get hampered by their dependence on symptomatic diagnosis, prior pathogen knowledge, and limited field applicability. Through extensive study of primary and secondary metabolites, metabolomics offers a greater understanding of host-virus interactions. Case studies of tomato, cucumber, squash, and other vegetable crops show that viral infections invariably cause changes in the levels of organic acids, flavonoids, phenolics, amino acids, and carbohydrates. These metabolic markers distinguish resistant from susceptible genotypes and reflect the intensity of the infection at its various stages. Most importantly, categorising biomarkers into diagnostic, prognostic, and predictive categories improves their use in early identification and predicting the progression of the disease. The mechanistic understanding of viral pathogenesis and defence pathways is further strengthened by the integration of metabolomic data with transcriptomics and proteomics, recognising particular metabolites and networks as targets for biocontrol or genetic improvement. Despite these advances, there are still several issues, most notably technical variability, reproducibility across laboratories, lack of standardised metadata, and lack of field validation. Developing reliable biomarker libraries requires addressing these problems through cross-environmental validation, interlaboratory trials, and standardised methods. In the future, metabolomics has the potential to revolutionise the management of vegetable viruses by facilitating portable field diagnostics, promoting resistance breeding, and directing precision agricultural strategies.

## References


Abdelkhalek A, Király L, Al-Mansori ANA, Younes HA, Zeid A, Elsharkawy MM, Behiry SI (2022) Defense responses and metabolic changes involving phenylpropanoid pathway and PR genes in squash (*Cucurbita pepo* L.) following Cucumber mosaic virus infection. Plants 11(15). https://doi.org/10.3390/plants11151908

Agustika D (2024) Early detection of pepper yellow leaf curl virus-infected chilli plants. PhD thesis, University of Warwick. http://webcat.warwick.ac.uk/record=b4074848

Aina O, Bakare OO, Fadaka AO, Keyster M, Klein A (2024) Plant biomarkers as early detection tools in stress management in food crops: a review. Planta 259, 60 https://doi.org/10.1007/s00425-024-04333-1

Alaribe SC, Titilayo BE, Ikpatt UN, Oladipupo AR, Alabi FA, Ikwugbado RI (2025) Computational methods for advanced mass spectrometry—a review. Phys Sci Rev 10(5–6):589–620. https://doi.org/10.1515/psr-2024-0006

Aliyu TH, Arogundade O, Essien EI, Balogun OS, Adeoti OM (2022) Detection of viruses infecting lettuce (*Lactuca sativa* L.) in the derived savanna of Nigeria and influence on proximate composition. J Food Agric 15(2):10. https://doi.org/10.4038/JFA.V15I2.5266

Antoniadis A, Bigot J, Lambert-Lacroix S, Letue F (2007) Nonparametric pre-processing methods and inference tools for analyzing time-of-flight mass spectrometry data. Curr Anal Chem 3(2):127–147. https://doi.org/10.2174/157341107780361718

Aronson JK, Ferner RE (2017) Biomarkers—a general review. Curr Protoc Pharmacol 2017(1):9.23.1–9.23.17. https://doi.org/10.1002/CPPH.19;CTYPE:STRING:JOURNAL

Aseel DG, Alaa A, Elsilk SE, Gaafar RM (2024) Protective and curative applications of some *Trichoderma* species to tomato plants infected by tomato mosaic virus. Egypt J Bot 64(3):109–129. https://doi.org/10.21608/ejbo.2024.265650.2679

Bansal H, Luthra H, Raghuram SR (2023) A review on machine learning aided multi-omics data integration techniques for healthcare. Stud Big Data 132:211–239. https://doi.org/10.1007/978-3-031-38325-0_10

Blasco H, Błaszczyński J, Billaut JC, Nadal-Desbarats L, Pradat PF, Devos D, Moreau C, Andres CR, Emond P, Corcia P, Słowiński R (2015) Comparative analysis of targeted metabolomics: dominance-based rough set approach versus orthogonal partial least square-discriminant analysis. J Biomed Inform 53:291–299. https://doi.org/10.1016/J.JBI.2014.12.001

Bol JF, Linthorst HJM, Cornelissen BJC (1990) Plant pathogenesis-related proteins induced by virus infection. 28: 113-138. https://doi.org/10.1146/annurev.py.28.090190.000553

Boonham N, Kreuze J, Winter S, van der Vlugt R, Bergervoet J, Tomlinson J, Mumford R (2014) Methods in virus diagnostics: from ELISA to next generation sequencing. Virus Res 186:20–31. https://doi.org/10.1016/J.VIRUSRES.2013.12.007

Brandolini-Bunlon M, Pétéra M, Gaudreau P, Comte B, Bougeard S, Pujos-Guillot E (2019) A new tool for multi-block PLS discriminant analysis of metabolomic data: application to systems epidemiology. np. Journées Scientifiques du Réseau Francophone de Métabolomique et Fluxomique RFMF, May 2019, Clermont-Ferrand, France. hal-02154660. https://hal.science/hal-02154660v1

Castro-moretti FR, Gentzel IN, Mackey D, Alonso AP (2020) Metabolomics as an emerging tool for the study of plant–pathogen interactions. Metabolites 10(2):52. https://doi.org/10.3390/metabo10020052

Cejnar P, Kučková Š, Šantrůček J, Glasa M, Komínek P, Mihálik D, Slavíková L, Leišová-Svobodová L, Smirnova T, Hynek R, Kundu JK, Ryšánek P (2020) Efficient confirmation of plant viral proteins and identification of specific viral strains by Nanolc-Esi-Q-Tof using single-leaf-tissue samples. Pathogens 9(11):1–20. https://doi.org/10.3390/pathogens9110966

Chen C, Zhou G, Chen J, Liu X, Lu X, Chen H, Tian Y (2021) Integrated metabolome and transcriptome analysis unveils novel pathway involved in the formation of yellow peel in cucumber. Int J Mol Sci 22(3):1494. https://doi.org/10.3390/IJMS22031494

Chen L-S, Lin Q, Nose A (2002) A comparative study on diurnal changes in metabolite levels in the leaves of three crassulacean acid metabolism (CAM) species, *Ananas comosus*, *Kalanchoë daigremontiana* and *K. pinnata*. J Exp Bot 53(367):341–350. https://doi.org/10.1093/jexbot/53.367.341

Coene KLM, Kluijtmans LAJ, van der Heeft E, Engelke UFH, de Boer S, Hoegen B, Kwast HJT, van de Vorst M, Huigen MCDG, Keularts IMLW, Schreuder MF, van Karnebeek CDM, Wortmann SB, de Vries MC, Janssen MCH, Gilissen C, Engel J, Wevers RA (2018) Next-generation metabolic screening: targeted and untargeted metabolomics for the diagnosis of inborn errors of metabolism in individual patients. J Inherit Metab Dis 41(3):337–353. https://doi.org/10.1007/S10545-017-0131-6/FIGURES/4

Culver JN, Padmanabhan MS (2008) Virus-induced disease: altering host physiology one interaction at a time. Annu Rev Phytopathol 45:221–243. https://doi.org/10.1146/annurev.phyto.45.062806.094422

DeFalco TA, Zipfel C (2021) Molecular mechanisms of early plant pattern-triggered immune signaling. Mol Cell 81(17):3449–3467. https://doi.org/10.1016/j.molcel.2021.07.029

Deborde C, Fontaine J, Jacob D, Botana A, Nicaise V, Richard-Forget F, Lecomte S, Decourtil C, Hamade K, Mesnard F, Moing A, Molinié R (2019) Optimizing 1D 1H-NMR profiling of plant samples for high throughput analysis: extract preparation, standardization, automation and spectra processing. Metabolomics 15. https://doi.org/10.1007/s11306-019-1488-3.

Delporte C, Tremblay-Franco M, Guitton Y, Canlet C, Weber RJM, Hecht H, Price EJ, Klánová J, Joly C, Dalle C, Saint-Vanne J, Thévenot E, Schmitz I, Chéreau S, Dechaumet S, Diémé B, Giacomoni F, Le Corguillé G, Pétéra M, Souard F (2025) Workflow4Metabolomics (W4M): a user-friendly metabolomics platform for analysis of mass spectrometry and nuclear magnetic resonance data. Curr Protoc 5(2):e70095. https://doi.org/10.1002/CPZ1.70095

Deng C, Zhang X, Zhu W, Qian J (2004) Investigation of tomato plant defence response to tobacco mosaic virus by determination of methyl salicylate with SPME-capillary GC-MS. Chromatographia 59(3–4):263–268. https://doi.org/10.1365/S10337-003-0144-1/METRICS

Dhamija N, Mangla AG (2022) Plant secondary metabolites in antiviral applications. In: Plant secondary metabolites: physico-chemical properties and therapeutic applications, pp 459–482. https://doi.org/10.1007/978-981-16-4779-6_15

Díaz-Pendón JA, Ding SW (2008) Direct and indirect roles of viral suppressors of RNA silencing in pathogenesis. In: Annual review of phytopathology, vol 46. Annual Reviews Inc, pp 303–326. https://doi.org/10.1146/annurev.phyto.46.081407.104746

Ebert AW (2020) The role of vegetable genetic resources in nutrition security and vegetable breeding. Plants 9(6):1–20. MDPI AG. https://doi.org/10.3390/plants9060736

Ertunç F (2020) Physiology of virus-infected plants. In: Applied plant virology: advances, detection, and antiviral strategies, pp 199–205. https://doi.org/10.1016/B978-0-12-818654-1.00015-3

Espinoza C, Bascou B, Calvayrac C, Bertrand C (2021) Deciphering prunus responses to PPV infection: a way toward the use of metabolomics approach for the diagnostic of Sharka disease. Metabolites 11(7):465. https://doi.org/10.3390/METABO11070465

Ferreira JD, Inácio B, Salek RM, Couto FM (2017) Assessing public metabolomics metadata, towards improving quality. J Integr Bioinform 14(4). 20170054. https://doi.org/10.1515/jib-2017-0054

Ghosh S, Dasgupta A, Swetapadma A (2019) A study on support vector machine based linear and non-linear pattern classification. In: Proceedings of the International Conference on Intelligent Sustainable Systems, ICISS, pp 24–28. https://doi.org/10.1109/ISS1.2019.8908018

Han W, Li L (2022) Evaluating and minimizing batch effects in metabolomics. Mass Spectrom Rev 41(3):421–442. https://doi.org/10.1002/MAS.21672;WGROUP:STRING:PUBLICATION

Hemmati F, Behjatnia SAA, Moghadam A, Afsharifar A (2025) Induction of systemic resistance against cucumber mosaic virus (CMV) and tomato yellow leaf curl virus (TYLCV) in tomato. Int J Pest Manag 71(3):307–320. https://doi.org/10.1080/09670874.2023.2202150

Jameson PE, Clarke SF (2002) Hormone-virus interactions in plants. Crit Rev Plant Sci 21(3):205–228. CRC Press LLC. https://doi.org/10.1080/0735-260291044241

Jiang T, Hao T, Chen W, Li C, Pang S, Fu C, Cheng J, Zhang C, Ghorbanpour M, Miao S (2025) Reprogrammed plant metabolism during viral infections: mechanisms, pathways and implications. Mol Plant Pathol 26(2) Wiley. https://doi.org/10.1111/mpp.70066

Jones RAC, Naidu RA (2019) Global dimensions of plant virus diseases: current tatus and future perspectives. Annu Rev Virol. 6(1), 387-409. https://doi.org/10.1146/annurev-virology-092818-015606

Katiyar-Agarwal S, Morgan R, Dahlbeck D, Borsani O, Villegas A, Zhu J-K, Staskawicz BJ, Jin H (2006) A pathogen-inducible endogenous siRNA in plant immunity. Proc Natl Acad Sci U S A 103:18002. https://doi.org/10.1073/pnas.0608258103

Khizar M, Shi J, Saleem S, Liaquat F, Ashraf M, Latif S, Haroon U, Waqas Hassan S, Ur Rehman S, Javed Chaudhary H, Masood Quraishi U, Farooq Hussain Munis M (2020) Resistance associated metabolite profiling of Aspergillus leaf spot in cotton through non-targeted metabolomics. PLoS One 15(2):e0228675. https://doi.org/10.1371/journal.pone.0228675

Khymenets O, Andres-Lacueva C, Urpi-Sarda M, Vazquez-Fresno R, Mart MM, Reglero G, Torres M, Llorach R (2015) Metabolic fingerprint after acute and under sustained consumption of a functional beverage based on grape skin extract in healthy human subjects. Food Funct 6(4):1288–1298. https://doi.org/10.1039/C4FO00684D

Kostova D, Lisa V, Milne RG, Vaira AM (2003) Virus diseases of vegetable crops in southern Bulgaria. Phytopathol Mediterr 42(1). 1-6. https://doi.org/10.1400/14524

Kukurugya MA, Mendonca CM, Solhtalab M, Wilkes RA, Thannhauser TW, Aristilde L (2019) Multi-omics analysis unravels a segregated metabolic flux network that tunes co-utilization of sugar and aromatic carbons in *Pseudomonas putida*. J Biol Chem 294(21):8464–8479. https://doi.org/10.1074/JBC.RA119.007885/ATTACHMENT/CCF21648-13CF-47CE-95AB-CF44E8343FD6/MMC1.PDF

Kumar R, Lal MK, Tiwari RK, Kumar A, Behera B, Nayak L, Dash GK, Sahoo SK, Jena J, Lal P, Altaf MA (2023) Traditional and emerging climate-resilient agricultural practices for enhancing food production and nutritional quality. In: Aftab T (ed) New Frontiers in plant-environment interactions: innovative technologies and developments, Environmental science and engineering. Springer, Cham, pp 551–570. https://doi.org/10.1007/978-3-031-43729-8_21

Laliberté JF, Sanfaçon H (2010) Cellular remodeling during plant virus infection. Annu Rev Phytopathol 48:69–91. https://doi.org/10.1146/annurev-phyto-073009-114239

Liu P, Zhang X, Zhang F, Xu M, Ye Z, Wang K, Liu S, Han X, Cheng Y, Zhong K, Zhang T, Li L, Ma Y, Chen M, Chen J, Yang J (2021) A virus-derived siRNA activates plant immunity by interfering with ROS scavenging. Mol Plant 14(7):1088–1103. https://doi.org/10.1016/j.molp.2021.03.022

Loebenstein G., Katis N. (Eds.). (2014). Control of plant virus diseases: seed-propagated crops. Academic Press. 90. https://doi.org/10.1016/B978-0-12-801246-8.09985-6

López-Gresa MP, Lisón P, Kim HK, Choi YH, Verpoorte R, Rodrigo I, Conejero V, Bellés JM (2012) Metabolic fingerprinting of tomato mosaic virus infected Solanum lycopersicum. J Plant Physiol 169(16):1586–1596. https://doi.org/10.1016/J.JPLPH.2012.05.021

Maia M, Ferreira AEN, Nascimento R, Monteiro F, Traquete F, Marques AP, Cunha J, Eiras-Dias JE, Cordeiro C, Figueiredo A, Sousa Silva M (2020) Integrating metabolomics and targeted gene expression to uncover potential biomarkers of fungal/oomycetes-associated disease susceptibility in grapevine. Sci Rep 10(1):1–15. https://doi.org/10.1038/s41598-020-72781-2

Majchrzak T, Wojnowski W, Wasik A (2021) Revealing dynamic changes of the volatile profile of food samples using PTR–MS. Food Chemistry 364:130404. https://doi.org/10.1016/j.foodchem.2021.130404

Makalesi A, Kılıç HÇ, Kök H, Yardımcı N (2020) Bean common mosaic virus and bean common mosaic necrosis virus infections in bean production areas in the lakes region of Turkey. Avrupa Bilim ve Teknoloji Dergisi 19:386–392. https://doi.org/10.31590/EJOSAT.705686

Maravi DK, Kumar S, Sahoo L (2022) NMR-based metabolomic profiling of mungbean infected with mungbean yellow mosaic India virus. Appl Biochem Biotechnol 194(12):5808–5826. https://doi.org/10.1007/s12010-022-04074-5

Marqués M, Sánchez-Vicente J, Ruiz R, Montagud-Martínez R, Márquez-Costa R, Gomez G, Carbonell A, Daròs J, Rodrigo G (2022) Diagnostics of infections produced by the plant viruses TMV, TEV, and PVX with CRISPR-Cas12 and CRISPR-Cas13. ACS Synthetic Biology 11:2384–2393. https://doi.org/10.1021/acssynbio.2c00090

Matthews H, Hanison J, Nirmalan N (2016) "Omics"-informed drug and biomarker discovery: opportunities, challenges and future perspectives. Proteomes. 4(3):28. https://doi.org/10.3390/proteomes4030028

Mi B, Xiao W, Tu N, Wu F (2024) Selection of pollution-safe head cabbage: interaction of multiple heavy metals in soil on bioaccumulation and transfer. Food Chem 452:139615. https://doi.org/10.1016/J.FOODCHEM.2024.139615

Ntzifa A, Lianidou E (2023) Pre-analytical conditions and implementation of quality control steps in liquid biopsy analysis. Crit Rev Clin Lab Sci 60(8):573–594. https://doi.org/10.1080/10408363.2023.2230290;REQUESTEDJOURNAL:JOURNAL:ILAB20;WGROUP:STRING:PUBLICATION

Nwosu K (2021) Multi-objective optimization in metabolomics/computational intelligence techniques (Doctoral dissertation, De Montfort University)

Okoń S, Zimowska B, Rai M (2024) Microbial genetics. In: Microbial genetics. CRC Press. https://doi.org/10.1201/9781003328933

Patel MK, Pandey S, Kumar M, Haque MI, Pal S, Yadav NS (2021) Plants metabolome study: emerging tools and techniques. Plants 10(11). https://doi.org/10.3390/PLANTS10112409

Rahman M, Schellhorn HE (2023) Metabolomics of infectious diseases in the era of personalized medicine. Front Mol Biosci 10:1120376. https://doi.org/10.3389/FMOLB.2023.1120376/BIBTEX

Ramaroson ML, Koutouan C, Helesbeux JJ, Le Clerc V, Hamama L, Geoffriau E, Briard M (2022) Role of phenylpropanoids and flavonoids in plant resistance to pests and diseases. Molecules 27(23):8371. https://doi.org/10.3390/MOLECULES27238371

Ramautar R, Somsen GW, de Jong GJ (2009) CE-MS in metabolomics. Electrophoresis 30(1):276–291. https://doi.org/10.1002/ELPS.200800512

Raigond B, Jeevalatha A, Kumar R, Verma G (2020) Virus detection and diagnosis. In: Chakrabarti SK, Bhardwaj V (eds) Potato in sub-tropics: a Saga of success. International Books & Periodical Supply Service (Publisher of Scientific Books) 38, Delhi, India, pp 83–96. ISBN: 978-93-88892-69-8; E-ISBN:978-93-88892-70-4

Razzaq A, Sadia B, Raza A, Hameed MK, Saleem F (2019) Metabolomics: a way forward for crop improvement. Metabolites 9(12):303. https://doi.org/10.3390/METABO9120303

Razzaq A, Wishart DS, Wani SH, Hameed MK, Mubin M, Saleem F (2022) Advances in Metabolomics-Driven Diagnostic Breeding and Crop Improvement. Metabolites. 12(6):511. https://doi.org/10.3390/metabo12060511

Resurreccion EP, Fong KW (2022) The integration of metabolomics with other omics: insights into understanding prostate cancer. Metabolites 12(6):488. https://doi.org/10.3390/METABO12060488

Roberts LD, Souza AL, Gerszten RE, Clish CB (2012) Targeted metabolomics. Curr Protoc Mol Biol 1(SUPPL.98):30.2.1–30.2.24. https://doi.org/10.1002/0471142727.MB3002S98;PAGE:STRING:ARTICLE/CHAPTER

Rotbi M, Ängskog P, Dez M J, Elmtili N (2014) Identification and distribution of Tomato yellow leaf curl virus TYLCV and Tomato yellow leaf curl Sardinia virus TYLCSV infecting vegetable crops in Morocco. Afr J Biotechnol 13(13):1476–1483. https://doi.org/10.5897/ajb2013.13407

Roy B, Verma N, Majumdar A, Mondal F, Valli NS, Sarkar M, Mazumder P, Chavan D, Kumar R, Roy A (2025) Advancements in diagnostic tools for rapid and sensitive detection of tomato leaf curl New Delhi virus in diverse crops. In: Tomato leaf curl New Delhi virus (ToLCNDV) insights into Virome dynamics and management. Springer, Singapore, pp 121–140. https://doi.org/10.1007/978-981-95-2852-3_8

Sade D, Shriki O, Cuadros-Inostroza A, Tohge T, Semel Y, Haviv Y, Willmitzer L, Fernie AR, Czosnek H, Brotman Y (2015) Comparative metabolomics and transcriptomics of plant response to Tomato yellow leaf curl virus infection in resistant and susceptible tomato cultivars. Metabolomics 11(1):81–97. https://doi.org/10.1007/S11306-014-0670-X/FIGURES/4

Salem MA, De Souza LP, Serag A, Fernie AR, Farag MA, Ezzat SM, Alseekh S (2020) Metabolomics in the context of plant natural products research: from sample preparation to metabolite analysis. Metabolites 10(1): 37. https://doi.org/10.3390/metabo10010037

Sanches PHG, de Melo NC, Porcari AM, de Carvalho LM (2024) Integrating molecular perspectives: strategies for comprehensive multi-omics integrative data analysis and machine learning applications in transcriptomics, proteomics, and metabolomics. Biology 13(11):848. https://doi.org/10.3390/BIOLOGY13110848

Sastry KS, Zitter TA (2014) Management of virus and viroid diseases of crops in the tropics. In: Plant virus and viroid diseases in the tropics. Springer, pp 149–480. https://doi.org/10.1007/978-94-007-7820-7_2

Scott M, Leisner RT. (1993) Movement of virus and photoassimilate in the phloem: a comparative analysis, 15(11), 741-748. https://doi.org/10.1002/bies.950151107.

Sett S, Prasad A, Prasad M (2022) Resistance genes on the verge of plant–virus interaction. Trends Plant Sci 27(12):1242–1252). https://doi.org/10.1016/j.tplants.2022.07.003

Sobsey CA, Ibrahim S, Richard VR, Gaspar V, Mitsa G, Lacasse V, Zahedi RP, Batist G, Borchers CH (2020) Targeted and untargeted proteomics approaches in biomarker development. Proteomics 20(9):1900029. https://doi.org/10.1002/PMIC.201900029

Srikanth P, Sivakumar D, Sharma A, Kaushik N (2025) Recent developments in omics techniques for improving plant abiotic stress using microbes. Int J Environ Sci Technol 22(5):3787–3810. https://doi.org/10.1007/S13762-024-05957-2/FIGURES/2

Sumbria D, Berber E, Mathayan M, Rouse BT (2021) Virus infections and host metabolism—can we manage the interactions? Front Immunol 11:594963. https://doi.org/10.3389/FIMMU.2020.594963/XML

Ternès N, Rotolo F, Heinze G, Michiels S (2017) Identification of biomarker-by-treatment interactions in randomized clinical trials with survival outcomes and high-dimensional spaces. Biom J 59(4):685–701. https://doi.org/10.1002/BIMJ.201500234;WGROUP:STRING:PUBLICATION

T'Kindt R, Scheltema RA, Jankevics A, Brunker K, Rijal S, Dujardin JC, Breitling R, Watson DG, Coombs GH, Decuypere S (2010) Metabolomics to unveil and understand phenotypic diversity between pathogen populations. PLoS Negl Trop Dis 4(11):e904. https://doi.org/10.1371/JOURNAL.PNTD.0000904

Tomé D (2021) Amino acid metabolism and signalling pathways: potential targets in the control of infection and immunity. Nutr Diabetes 11(1):1319–1327. https://doi.org/10.1038/S41387-021-00164-1

Tomlinson JA (1987) Epidemiology and control of virus diseases of vegetables. Ann Appl Biol 110: 661-681. https://doi.org/10.1111/j.1744-7348.1987.tb04187.x

Tugizimana F, Mhlongo MI, Piater LA, Dubery IA (2018) Metabolomics in plant priming research: the way forward? Int J Mol Sci 19(6) MDPI AG. https://doi.org/10.3390/ijms19061759

Ullah I, Yang L, Yin FT, Sun Y, Li XH, Li J, Wang XJ (2022) Multi-omics approaches in colorectal cancer screening and diagnosis, recent updates and future perspectives. Cancers 14(22):5545. https://doi.org/10.3390/CANCERS14225545

Van Den Berg H, Da Silva S, Bezerra H, Al-Eryani S, Chanda E, Nagpal BN, Knox TB, Velayudhan R, Yadav RS (2021) Recent trends in global insecticide use for disease vector control and potential implications for resistance management. Sci Rep 11:23867. https://doi.org/10.1038/s41598-021-03367-9

Varma A, Malathi VG (2003) Emerging geminivirus problems: a serious threat to crop production. Ann Appl Biol 142(2), 145-164. https://doi.org/10.1111/j.1744-7348.2003.tb00240.x

Villate A, San Nicolas M, Gallastegi M, Aulas PA, Olivares M, Usobiaga A, Etxebarria N, Aizpurua-Olaizola O (2021) Review: metabolomics as a prediction tool for plants performance under environmental stress. Plant Sci 303:110789. https://doi.org/10.1016/J.PLANTSCI.2020.110789

Vinaixa M, Schymanski EL, Neumann S, Navarro M, Salek RM, Yanes O (2016) Mass spectral databases for LC/MS- and GC/MS-based metabolomics: state of the field and future prospects. TrAC Trends Anal Chem 78:23–35. https://doi.org/10.1016/J.TRAC.2015.09.005

Wishart DS (2019) Metabolomics for investigating physiological and pathophysiological processes. Physiol Rev 99:1819–1875. https://doi.org/10.1152/physrev.00035.2018.-Metabolomics

Wolfender J, Rudaz S, Choi Y, Kim H (2013) Plant metabolomics: from holistic data to relevant biomarkers. Current medicinal chemistry 20(8):1056–90. https://doi.org/10.2174/0929867311320080009.

Worley B, Powers R (2016) PCA as a practical indicator of OPLS-DA model reliability. Curr Metabolomics 4(2):97–103. https://doi.org/10.2174/2213235X04666160613122429

Wu M, Northen TR, Ding Y (2023) Stressing the importance of plant specialized metabolites: omics-based approaches for discovering specialized metabolism in plant stress responses. Front Plant Sci 14:1272363. https://doi.org/10.3389/FPLS.2023.1272363/XML

Wua G, Fang X, Yu T, Chen J, Yana F (2024) Turnip mosaic virus pathogenesis and host resistance mechanisms in *Brassica*. Horticultural Plant Journal, 10(4), 947-960. https://doi.org/10.1016/j.hpj.2024.03.001

Xu Y, Zhang S, Zhang M, Jiao S, Guso Y, Jiang T (2024) The role of reactive oxygen species in plant-virus interactions. Plant Cell Rep 43(8):1–12. https://doi.org/10.1007/S00299-024-03280-1/METRICS

Yang S, Berdine G (2017) The receiver operating characteristic (ROC) curve. Southwest Respir Crit Care Chron 5(19):34. https://doi.org/10.12746/SWRCCC.V5I19.391

Yang J, Chen R, Wang C, Li C, Ye W, Zhang Z, Wang S (2024) A widely targeted metabolite modificomics strategy for modified metabolites identification in tomato. J integrative plant biology https://doi.org/10.1111/jipb.13629

Zhang T, Yu P, Li Y, Zhang Y, Deng H (2022) Metabolomics in the research of viral infectious disease mechanisms. Int J Front Med 5(7). 19-26. https://doi.org/10.25236/ijfm.2023.050704

Zhang ZY, Yang L, Yang WC (2024) Research progress in biomarker-targeted optical biosensors for plant disease detection. TrAC Trends Anal Chem. 178: 117841. https://doi.org/10.1016/j.trac.2024.117841

Zheng Y, Guo X, Chen Q, Zhang Y, Zhang Y, Liu Y, Duan S, Ma Y, Ni P, Wang J, Hou W, Yu Y, Fang X, Li J, Shi L, Zhang R, Dong L (2025) Quantitative methylation reference datasets of Quartet DNA reference materials for benchmarking genome-wide epigenome sequencing. Nat Commun 16:9202. https://doi.org/10.21203/RS.3.RS-6420165/V1

# Metabolomic-Driven Insights into Viral Impacts on Phytochemical Pathways in Tropical Vegetable Crops

18

Subham Dutta, Soumya Majumder, Moumita Panda, Mritunjoy Barman, and Jayanta Tarafdar

**Abstract**

The influence of viral infection on crop productivity or yield and quality has been widely studied by various researchers, but their effects on phytochemical profiles, more precisely metabolite biosynthesis, have been underexplored. Plant metabolomics is the subject of studying the phytomolecules present in the plant body, their enzymatic biosynthesis pathways, and classification of the molecules based on that, which helps to identify biomarkers associated with infections, nutrient deficiencies, and other diseases. Therefore, metabolomics can be an effective tool to understand plant-virus interactions. This chapter delivers the role of plant metabolomics and insights into alterations in metabolite profiles of tropical vegetables due to viral infection, which largely affect the yield and biochemical quality of the produce. It focuses on the changes in the biosynthesis pathways of compounds like alkaloids, phenolics, terpenoids, etc. and the role of

S. Dutta (✉)
School of Agriculture, Swami Vivekananda University, Barrackpore, West Bengal, India

Department of Plant Pathology, Bidhan Chandra Krishi Viswavidyalaya, Mohanpur, Nadia, West Bengal, India

S. Majumder
Department of Tea Science, University of North Bengal, Siliguri, Darjeeling, West Bengal, India

Division of Biochemistry, Darjeeling Tea Research & Development Centre, Tea Board of India, Kurseong, West Bengal, India

M. Panda · J. Tarafdar
Department of Plant Pathology, Bidhan Chandra Krishi Viswavidyalaya, Mohanpur, Nadia, West Bengal, India

M. Barman
Department of Entomology, University of Nebraska-Lincoln, Lincoln, NE, USA



R. Kumar et al. (eds.), *Vegetable Crops and Viral Pathogens Volume I*, Plant Pathology for Sustainable Agriculture, https://doi.org/10.1007/978-981-95-7265-6_18

different plant enzymes in these alterations. Contemporary chromatographic and spectroscopic tools such as GC-MS, LC-MS, HPLC, NMR, etc. have facilitated in-depth investigations on the effects of viruses on phytochemical profiles. Studies on vegetables like okra, tomato, chili, and cucurbits indicate that viral infections can alter metabolic pathways either by upregulating or by downregulating certain metabolite biosynthesis. Some specific metabolites showed increased or decreased levels, where the higher ones were associated with plant defense mechanisms and antimicrobial characteristics, serving as indicators of host-pathogenic interaction. By isolating these biomarker metabolites, we can reveal the chemical alterations of the host plant due to viral infections. It evidently establishes how hosts counter viruses and how the viral pathogens can modify host metabolisms to facilitate infection, rendering plants vulnerable to insect vectors. Therefore, a metabolomics study can aid in our comprehension of the processes underlying infection and viral transmission.

**Keywords**

Metabolic pathways · Phytochemistry · Plant metabolomics · Viral infection

## 18.1 Introduction

Vegetable crops supply nutrients, fiber, and phytochemicals to large tropical populations. Tomato, chili, okra, cucumber, and cucurbits support food security. These crops also supply raw materials for nutraceutical and pharmaceutical use because they contain defined secondary metabolites. Viral diseases reduce production in tropical and subtropical regions (Raigond et al. 2020). Whiteflies, aphids, and thrips spread viruses in warm and humid conditions (Naga et al. 2020, 2021). Viral infections reduce yield, lower fruit quality, deform plants, and cause crop failure during outbreaks (Umar et al. 2023). Traditional disease control fails to give timely or lasting results because viruses mutate fast and form new strains. Mixed infections increase pressure on producers (Varma and Malathi 2003). Metabolomics profiles small molecules in plant tissues. A clear view of biochemical shifts during infection can be obtained from metabolomics. Virus replication and spread are completely associated with the host metabolism. Viral not only alters plant metabolic pathways but also affects the redox balance and phytohormone signaling. These alterations induce metabolic signatures linked to infection or resistance (Cui et al. 2025). Novel biosynthesis of specific metabolites or switched off biosynthesis of certain metabolites or imbalance in metabolite ratios by upregulation or downregulation of production can help to separate undesired plant products from desired ones and resistant ones from the susceptible ones. These markers also help to identify different stages of infection, such as early and late stages (Allwood and Goodacre 2010; López-Gresa et al. 2017). According to past studies, in watermelon infected with cucumber green mottle mosaic virus, a resistant line showed an upsurge in flavonoid and phenylpropanoid biosynthesis, whereas salicylic acid, along with those flavonoids, was found to be involved in the defense mechanism against the infection (Liu et al. 2023). Okra infected with okra yellow vein mosaic virus showed downregulation of phenolic biosynthesis, while an increase in soluble sugars and antioxidant enzymes was noticed (Umar et al. 2023). Furthermore, NMR profiling revealed that mung bean yellow mosaic India virus resulted in altered profiles of amino acids, carbohydrates, and organic acids, including sucrose, GABA, and proline in the mung bean plant (Maravi et al. 2022). Metabolic alterations were also recorded in *Nicotiana tabacum* exposed to tobacco mosaic virus, which further helped to separate resistant from susceptible plants (Li et al. 2025). These studies show clear virus-driven metabolic shifts that differ by virus, host, tissue, and organ. Many shifts link primary metabolites with antioxidant responses or secondary metabolite pathways. This chapter explains how metabolomics supports research on tropical vegetable virus systems and describes changes in secondary metabolism, viral pressures, and enzyme functions. The chapter also explains how biomarkers support precision breeding, targeted metabolic engineering, and diagnostic development for virus-resistant vegetable lines. The chapter ends with an overview of current metabolomic tools, multiomic integration, and future directions for metabolomic-based virus management in tropical vegetable crops.

## 18.2 Basics of Plant Metabolomics

Metabolomics studies small molecules under 1500 Da in plant cells and tissues. These molecules reflect genetic, proteomic, and environmental inputs. In plant virus systems, metabolomics shows you which pathways shift during infection. These pathways include biosynthesis of sugar metabolites, peptides and amino acids, lipids and fatty acid derivatives, and redox compounds (Table 18.1). The approach also helps you find biomarkers or antiviral molecules and shows how viruses reprogram host metabolism (Jiang et al. 2025).

### 18.2.1 Primary and Secondary Metabolites

(a) Primary metabolites support growth, development, and basic metabolic processes. Examples include sugar metabolites, peptides and amino acids, lipids and fatty acid derivatives, and redox compounds, organic acids, nucleotides, etc.
(b) Secondary metabolites support defense, signaling, pest and pathogen interactions, color, and scent. Examples include phenolics, flavonoids, alkaloids, terpenoids, and glucosinolates. These compounds often provide the strongest biomarkers in plant pathogen studies (Salam et al. 2023) (Table 18.1).

**Table 18.1** Different classes of metabolites, their function, and few examples

| Class | Function | Examples | Sources |
|---|---|---|---|
| Primary metabolites | Essential for plant growth and development as well as for reproduction | Sugar metabolites, peptides and amino acids, lipids and fatty acid derivatives, redox compounds, organic acids, and nucleotides | Elshafie et al. (2023) |
| Terpenoids | Involved in defense, pigmentation, and signaling; largest class of secondary metabolites | Menthol, taxol, cannabinoids, carotenoids, and essential oils | Reshi et al. (2023) and Erb and Kliebenstein (2020) |
| Phenolic compounds | Provide antioxidant activity, UV protection, and defense against pathogens | Flavonoids, lignins, tannins, stilbenes, and anthocyanins | |
| Alkaloids | Serve as defense compounds; often toxic to herbivores | Caffeine, nicotine, morphine, quinine, and atropine | |
| Sulfur-containing | Involved in defense and signaling; include glucosinolates and related compounds | Glucosinolates, camalexin, and benzoxazinoids | |
| Glycosides | Serve in storage, transport, and defense; often release active aglycones upon hydrolysis | Saponins, cyanogenic glycosides, and cardiac glycosides | |
| Nitrogen-containing | Play roles in defense and signaling; include glucosinolates and related compounds | Glucosinolates and benzoxazinoids | |

## 18.2.2 Analytical Platforms

Analytical platforms help to study metabolomics by covering a wide range of metabolites belonging to varied classes. Each tool has its specific strengths and limits, whereas a multiplatform approach (e.g., MS attached to a chromatography machine) improves coverage. For example, liquid chromatography-mass spectrometry can identify polar, semipolar, and nonpolar metabolites with strong sensitivity, a wide dynamic range, and structural detail through MS or MSMS analysis, whereas gas chromatography-mass spectrometry only covers volatile and semivolatile metabolites and also supports a few derivatized sugars and amino acids (Serag et al. 2023). Nuclear magnetic resonance spectroscopy supports absolute quantification and structural analysis. It is nondestructive and reproducible, but it has lower sensitivity than mass spectrometry (Serag et al. 2023). High-performance liquid chromatography and UPLC support targeted analysis when reference standards are present. Other options include capillary electrophoresis MS, ion mobility, imaging metabolomics, and stable isotope flux analysis (Serag et al. 2023).

### 18.2.3 Data Processing and Bioinformatics

Data processing steps follow data collection. Preprocessing reduces noise and supports peak selection, alignment, normalization, and scaling. Multivariate analysis separates sample groups with PCA, PLS DA, and OPLS. Statistical tests and fold change analysis identify differential metabolites. Annotation links mass spectral features with HMDB, METLIN, KEGG, and PlantCyc. Pathway mapping links metabolites with biochemical routes. Integration with transcriptomics and proteomics defines regulatory links. Common tools include MetaboAnalyst, XCMS, and MZmine.

## 18.3 Virus-Induced Metabolic Reprogramming in Vegetables

Viruses reprogram host metabolism to support replication and spread while triggering defense responses.

### 18.3.1 Primary Metabolism Alterations

Carbohydrate and energy pathways shift in virus-infected plants. Source sink balance changes and soluble sugars, including glucose, fructose, and sucrose, rise. These sugars reflect viral replication and early symptoms such as vein clearing and leaf curling (Llave 2016). Glycolysis and TCA intermediates, including malic acid and citric acid, shift during begomovirus and CMV infections in okra and cucurbits (Jiang and Zhou 2023). Amino acid levels also rise. Proline supports osmotic balance and detoxifies reactive oxygen species. Aromatic amino acids support phenylpropanoid-based defense (Kogovšek et al. 2016). Nitrogen metabolism shifts include disruptions in glutamine and asparagine. These changes reduce protein synthesis but redirect nitrogen toward secondary metabolic defenses (Castro-Moretti et al. 2020; Jiang et al. 2025).

### 18.3.2 Secondary Metabolism Alterations

Viral infections increase phenolic acids, including ferulic, caffeic, and gallic acid. Flavonoids, including quercetin, kaempferol, and rutin, often rise. These compounds support antioxidant and antiviral responses (Kogovšek et al. 2016; Kumar et al. 2023). Alkaloid and terpenoid levels shift in solanaceous crops. Chili and tomato show changes in solanine and capsaicinoids after infection. Some terpenoid volatiles drop and weaken indirect defenses against insect vectors (Chen et al. 2022; Jiang et al. 2025). Cucurbits infected with CMV increase coumarins and stilbenes. These metabolites support antimicrobial and antiviral defense (Zhang et al. 2022).

### 18.3.3 Redox Homeostasis and Antioxidants

Virus infections increase reactive oxygen species. These compounds damage membranes and lead to chlorosis. Plants respond with increased ascorbic acid, glutathione, and tocopherols. Enzymes, including peroxidases and superoxide dismutases, also rise. Metabolomics often detects increased reduced glutathione and ascorbate (Kaur and Sharma 2022; Kogovšek et al. 2016).

### 18.3.4 Hormonal Crosstalk and Metabolite Changes

Salicylic acid rises in resistant plants and supports systemic acquired resistance and secondary metabolite pathways (Castro-Moretti et al. 2020). Jasmonic acid and ethylene often decline during viral infection and weaken defenses against vectors. Some tolerant genotypes show increased jasmonic acid (Jiang et al. 2025). Abscisic acid rises under stress and reduces salicylic acid-based defenses (Llave 2016).

### 18.3.5 Integration of Multipathway Responses

Virus-induced reprogramming reflects a balance between defense and viral resource use. Tomato infected with TYLCV shows increased flavonoids and phenolic acids, while polyamines, including putrescine and spermidine, rise and support viral replication (Sade et al. 2015). Cucurbits infected with CMV show sugar buildup that relates to symptoms and increased phenolic compounds that offer partial tolerance (Zhang et al. 2022). These shifts support markers that separate susceptible and resistant vegetable lines (Jiang et al. 2025; Castro-Moretti et al. 2020).

## 18.4 Biomarker Discovery Through Metabolomics

Biomarkers reflect biological states. They support diagnosis, prediction, and resistance screening.

(a) Diagnostic biomarkers show early infection. Shifts in sugars, amino acids, and organic acids appear early in virus-infected vegetables (Jiang et al. 2025).
(b) Prognostic biomarkers relate to symptom severity. In tomato infected with TYLCV, metabolic patterns relate to levels of leaf curling and chlorosis (Sade et al. 2015).
(c) Resistance biomarkers support breeding. Phenolic compounds and alkaloids rise in resistant genotypes after infection and support antioxidant responses and cell wall reinforcement (Ahmad et al. 2024; Dutta et al. 2025).

(d) Integration with Multiomic Approaches.

Integration with transcriptomics and proteomics supports biomarker discovery. Combined analysis identifies regulatory pathways linked with diagnosis, prediction, and resistance (Lu et al. 2022).

## 18.5 Case Studies

Some metabolomic alterations by various plant virus infections have been listed in Table 18.2.

**Table 18.2** Metabolomic alterations in vegetable viruses: host-specific responses and key findings

| Sl. No | Virus | Host | Metabolomic changes/key findings | Reference |
|---|---|---|---|---|
| 1 | TYLCV | Tomato | Flavonoid derivatives (quercetin, kaempferol) increased; polyamine levels altered; resistant cultivars showed increased phenolic acids | Alvarez (2014) and López-Gresa et al. (2017) |
| 2 | ChiLCV | Chili | Capsaicinoids decreased; phenolic acids (ferulic, caffeic) increased; gallic acid and rutin identified as tolerance biomarkers | Alghamdi et al. (2025) |
| 3 | OYVMV | Okra | Amino acids (proline) and organic acids (citric acid) increased; resistant varieties showed higher flavonoid glycosides | Karmakar et al. (2025) |
| 4 | CMV | Cucurbits | Sucrose and raffinose increased, correlating with severe symptoms; tolerant lines accumulated chlorogenic acid; distinct metabolic signatures confirmed by NMR | Abdelkhalek et al. (2022) |
| 5 | CCYV | Cucumber | Flavonoid and terpenoid metabolism repressed; lipid, amino acid, and nucleotide metabolism induced | Zhang et al. (2022) |
| 6 | ToMV | Tomato | Flavonoid and phenolic acid profiles altered; phenylpropanoid and flavonoid defense pathways activated | Li et al. (2023) |
| 7 | PepYLCV | Capsicum | Flavonoid and phenolic acid levels altered; genotype-specific resistance responses observed | Singh and Sharma (2023) |
| 8 | CGMMV | Cucumber | Amino acid and organic acid profiles altered; flavonoid levels increased | Li et al. (2023) |
| 9 | PVCV | Capsicum | Flavonoid and phenolic acid levels altered; metabolites linked to resistance identified | Singh and Sharma (2023) |
| 10 | TSWV | Tomato | Amino acid and organic acid levels altered; activation of flavonoid and phenolic defense pathways | Li et al. (2023) |

(continued)

**Table 18.2** (continued)

| Sl. No | Virus | Host | Metabolomic changes/key findings | Reference |
|---|---|---|---|---|
| 11 | PMMoV | Capsicum | Flavonoid and phenolic acid levels altered; genotype-specific resistance metabolites identified | Singh and Sharma (2023) |
| 12 | CGMMV | Cucurbits | Flavonoid levels increased; amino acid and organic acid profiles altered; defense-related metabolites detected | Li et al. (2023) |
| 13 | PYVCV | Capsicum | Flavonoid and phenolic acid levels altered; metabolites linked to resistance identified | Singh and Sharma (2023) |
| 14 | ToLCNDV | Tomato | Flavonoid and phenolic acid levels altered; phenylpropanoid pathways activated | Sánchez-Sánchez et al. (2024) |
| 15 | PepMoV | Capsicum | Flavonoid and phenolic acid levels altered; genotype-specific responses observed | Singh and Sharma (2023) |
| 16 | CVYV | Cucumber | Flavonoid levels increased; amino acid and organic acid profiles altered | Li et al. (2023) |
| 17 | Tomato chlorosis virus (ToCV) | Tomato | Phenolic acids increased; sugar metabolism altered; stress-related metabolites detected | López-Gresa et al. (2017) |
| 18 | Beet curly top virus (BCTV) | Tomato | Amino acids (proline, glutamine) increased; polyamine metabolism altered | Alvarez (2014) |
| 19 | Lettuce infectious yellows virus (LIYV) | Lettuce | Phenolic acids increased; carbohydrate metabolism altered | Zhang et al. (2022) |
| 20 | Papaya ringspot virus (PRSV) | Cucumber | Flavonoid glycosides increased; amino acid metabolism altered; resistant lines showed higher secondary metabolite levels | Kumari et al. (2021) |

*TYLCV* tomato yellow leaf curl virus, *ChiLCV* chili leaf curl virus, *OYVMV* okra yellow vein mosaic virus, *CMV* cucumber mosaic virus, *CCYV* cucurbit chlorotic yellows virus, *ToMV* tomato mosaic virus, *PepYLCV* papaya or pepper yellow leaf curl virus, *CGMMV* cucumber green mottle mosaic virus, *PVCV* papaya vein clearing virus, *TSWV* tomato spotted wilt virus, *PMMoV* pepper mild mottle virus, *PYVCV* papaya yellow vein clearing virus, *ToLCNDV* tomato leaf curl New Delhi virus, *PepMoV* pepper mottle virus, *CVYV* cucumber vein yellowing virus

## 18.6 Applications in Virus Management

Metabolomics enhances virus management strategies by improving detection, breeding, metabolic engineering, and biostimulant strategies.

### 18.6.1 Diagnostics

Metabolomics helps in the detection of infections early to identify metabolic signatures associated with viral infection before the onset of symptoms. Biosecurity practices are enhanced when metabolomics distinguishes viral infection from abiotic stress, which supports accurate risk assessment.

### 18.6.2 Breeding

Metabolomics-assisted selection links metabolic data with breeding lines. This supports the selection of varieties with stronger resistance. Predictive biomarkers help you forecast plant performance under viral pressure and support efficient breeding plans.

### 18.6.3 Metabolic Engineering

Targeted pathway modification uses metabolomic insights to guide changes in biosynthetic routes such as flavonoid and phenolic pathways to increase antiviral resistance. Stress response optimization uses metabolic knowledge to strengthen plant defenses through engineered pathways.

### 18.6.4 Biostimulants and Elicitors

Induced resistance involves biostimulants such as seaweed and alfalfa extracts that shift metabolism and raise tolerance to viral infection. Priming effects appear when metabolomics shows increased defense activity in plants pretreated with biostimulants.

## 18.7 Future Prospects and Challenges

Metabolomic intervention can identify a common metabolic route or pathways in several host ranges that are also necessary for virus replication. Therefore, any enzyme inhibitor or substrate analogue that stops this kind of virus-induced host metabolic reprogramming can be useful against a variety of viruses. The roles of different metabolites in plant growth, development, and responses to environmental signals will be clarified by connecting metabolite profiles with particular biological activities. Plant metabolomics is predicted to make an even greater contribution to our comprehension of plant systems and the creation of novel solutions for a sustainable future as technology and analytical techniques progress. Metabolite profiles might be altered or influenced by several environmental factors, including sunshine hours, weather, temperature, water levels, and soil conditions. Disparity occurs due to fluctuating stage and tissue type, demanding an accurate experimental design and sampling. Expenses associated with instrumentation limit access in resource-constrained environments. Processing, normalization, and interpretation are difficult due to the amount of data. Multiomic integration with transcriptomics, proteomics, and phenomics requires strong computational systems. Prospects remain strong. Multiomic integration supports regulatory network analysis for virus-host interactions. Machine learning and AI-based models support biomarker identification and the prediction of plant responses. Spatial metabolomics supports mapping of metabolic changes at tissue and cell levels. High-throughput screening and noninvasive sensors support real-time monitoring of viral infection in the field. Integration with gene editing, including CRISPR Cas9, supports targeted manipulation of metabolic pathways for improved resistance. Metabolite-informed biostimulants support primed defense pathways and link basic studies with applied outcomes in sustainable agriculture. Advances in analytical tools and computational systems support reliable diagnostic marker discovery and the development of virus-resistant crop varieties.

## 18.8 Conclusion

Metabolomics improved the understanding of virus-driven metabolic shifts in vegetable crops. Viral infection-induced alterations in primary and secondary metabolism involved metabolites like carbohydrates, amino acids, phenolics, flavonoids, alkaloids, terpenoids, fatty acids, etc. These metabolites not only reflect host defense responses and viral manipulation of cellular processes but also indicate early stages of infections by identifying diagnostic and prognostic markers linked with infection and resistance. The approach also strengthens breeding programs that target resistant cultivars. It guides metabolic engineering strategies that reinforce antiviral defenses. Several studies on tomato, chili, okra, and cucurbits showed how plant metabolomics has been utilized to separate susceptible and resistant genotypes based on their metabolite profiles, identifying metabolites associated with tolerance or resistance. Future research through integrated methods, including high-resolution analysis, single-cell tools, real-time monitoring, and AI-based prediction models, is needed to improve biomarker identification and strengthen disease surveillance for crop protection. Integration with other omic fields and computational tools, like molecular docking strengthened, might also support sustainable vegetable production.

## References

Abdelkhalek A, Király L, Al-Mansori ANA, Younes HA, Zeid A, Elsharkawy MM, Behiry SI (2022) Defense responses and metabolic changes involving phenylpropanoid pathway and PR genes in squash (Cucurbita pepo L.) following Cucumber mosaic virus infection. Plants 11(15):1908

Ahmad N, Xu Y, Zang F, Li D, Liu Z (2024) The evolutionary trajectories of specialized metabolites towards antiviral defense system in plants. Mol Hortic 4(1):2

Alghamdi M, Rathinasabapathy T, Komarnytsky S (2025) Capsaicinoid profiles, phenolic content, and antioxidant properties of chili peppers grown in urban settings. Int J Mol Sci 26(10):4916

Allwood JW, Goodacre R (2010) An introduction to liquid chromatography–mass spectrometry instrumentation applied in plant metabolomic analyses. Phytochem Anal 21(1):33–47

Alvarez MA (2014) Molecular farming in plants. In: Plant biotechnology for health: from secondary metabolites to molecular farming. Springer, Cham, pp 81–109

Castro-Moretti FR, Gentzel IN, Mackey D, Alonso AP (2020) Metabolomics as an emerging tool for the study of plant–pathogen interactions. Metabolites 10(2):52

Chen G, Su Q, Yuan G, Shi X, Turlings T, Zhang Y (2022) Suppression of terpenoid synthesis in tomato plants by a begomovirus enhances the attraction of its vector. Authorea Preprints

Cui R, Sun SS, Pei MN, Liu Y, Xiao YS, Feng YN et al (2025) Multiomics integration unveils rapeseed adaptative responses to potassium deprivation: a synergy of morpho-physiological, transcriptomic, and metabolic reprogramming. Plant Soil 517:325

Dutta S, Saha P, Barman M, Poorvasandhya R, Panda M, Ahmed T, Tarafdar J (2025) Exploring the physiological, biochemical, and enzymatic responses of Vigna mungo varieties to Mungbean Yellow Mosaic India Virus (MYMIV) infection. Scientific Reports, 15(1):1049

Elshafie HS, Camele I, Mohamed AA (2023) A comprehensive review on the biological, agricultural and pharmaceutical properties of secondary metabolites based-plant origin. Int J Mol Sci 24(4):3266

Erb M, Kliebenstein DJ (2020) Plant secondary metabolites as defenses, regulators, and primary metabolites: the blurred functional trichotomy. Plant Physiol 184(1):39–52

Jiang T, Zhou T (2023) Unraveling the mechanisms of virus-induced symptom development in plants. Plants 12(15):2830

Jiang T, Hao T, Chen W, Li C, Pang S, Fu C et al (2025) Reprogrammed plant metabolism during viral infections: mechanisms, pathways and implications. Mol Plant Pathol 26(2):e70066

Karmakar P, Maurya BK, Singh H, Sagar V, Majumder S, Kumar R et al (2025) Harnessing bioactive phytochemicals in okra: nutritional and therapeutic perspectives. In: Okra: status, challenges and opportunities. Springer, Singapore, pp 279–301

Kaur R, Sharma A (2022) Metabolomics of Brassica juncea. In: The Brassica juncea genome. Springer, Cham, pp 323–335

Kogovšek P, Pompe-Novak M, Petek M, Fragner L, Weckwerth W, Gruden K (2016) Primary metabolism, phenylpropanoids and antioxidant pathways are regulated in potato as a response to Potato virus Y infection. PLoS One 11(1):e0146135

Kumar AP, Bhasker K, Nikhil BSK, Srinivas P (2023) Role of phenylpropanoids and flavonoids in plant defense mechanism. Int J Environ Clim Change 13(9):2951–2960

Kumari S, Krishnan N, Dubey V, Das B, Pandey KK, Singh J (2021) Investigations on annual spreading of viruses infecting cucurbit crops in Uttar Pradesh State, India. Sci Rep, 11(1):17883

Li C, Zhang Q, Wang X, Liu Z, Zhang X, Jiang C et al (2025) Transcriptomic and metabolomic analyses reveal molecular mechanisms of tobacco mosaic virus (TMV) resistance in Nicotiana tabacum L. BMC Plant Biol 25(1):1029

Li Z, Tang Y, Lan G, Yu L, Ding S, She X, He Z (2023) Transcriptome and metabolome analyses reveal that jasmonic acids may facilitate the infection of cucumber green mottle mosaic virus in bottle gourd. Int J Mol Sci 24(23):16566

Liu M, Kang B, Wu H, Aranda MA, Peng B, Liu L et al (2023) Transcriptomic and metabolic profiling of watermelon uncovers the role of salicylic acid and flavonoids in the resistance to cucumber green mottle mosaic virus. J Exp Bot 74(17):5218–5235

Llave C (2016) Dynamic cross-talk between host primary metabolism and viruses during infections in plants. Curr Opin Virol 19:50–55

López-Gresa MP, Lisón P, Campos L, Rodrigo I, Rambla JL, Granell A et al (2017) A non-targeted metabolomics approach unravels the VOCs associated with the tomato immune response against Pseudomonas syringae. Front Plant Sci 8:1188

Lu X, Zhang L, Huang W, Zhang S, Zhang S, Li F et al (2022) Integrated volatile metabolomics and transcriptomics analyses reveal the influence of infection TuMV to volatile organic compounds in Brassica rapa. Horticulturae 8(1):57

Maravi DK, Kumar S, Sahoo L (2022) NMR-based metabolomic profiling of mungbean infected with mungbean yellow mosaic India virus. Appl Biochem Biotechnol 194(12):5808–5826

Naga KC, Buckseth T, Subhash S, Bairwa A, Verma G, Kumar R, Malik K, Sharma S, Chakrabarti SK (2020) Transmission efficiency of potato leafroll virus (PLRV) by potato aphid Aulacorthum solani and green peach aphid Myzus persicae. Indian J Entomol 71:68–71. https://doi.org/10.5958/0974-8172.2020.00017.6

Naga KC, Sundaresha S, Kumar R, Tiwari RK, Subhash S, Verma G, Buckseth T, Bairwa A, Sharma S, Katare S, Srivastava RM (2021) A new record of Asia II 5 genetic group of Bemisia tabaci (Gennadius) in the major potato growing areas of India and its relationship with tomato leaf curl New Delhi. 3 Biotech 11(9):421

Raigond B, Jeevalatha A, Kumar R, Verma G (2020) Virus detection and diagnosis. In: Chakrabarti SK, Bhardwaj V (eds) Potato in sub-tropics: a Saga of success. International Books & Periodical Supply Service (Publisher of Scientific Books) 38, Delhi, India, pp 83–96. ISBN: 978-93-88892-69-8; E-ISBN:978-93-88892-70-4

Reshi ZA, Ahmad W, Lukatkin AS, Javed SB (2023) From nature to lab: a review of secondary metabolite biosynthetic pathways, environmental influences, and in vitro approaches. Meta 13(8):895

Sade D, Shriki O, Cuadros-Inostroza A, Tohge T, Semel Y, Haviv Y et al (2015) Comparative metabolomics and transcriptomics of plant response to Tomato yellow leaf curl virus infection in resistant and susceptible tomato cultivars. Metabolomics 11(1):81–97

Salam U, Ullah S, Tang ZH, Elateeq AA, Khan Y, Khan J et al (2023) Plant metabolomics: an overview of the role of primary and secondary metabolites against different environmental stress factors. Life 13(3):706

Sánchez-Sánchez M, Vargas-Arispuro I, Tovar-Pedraza JM, González-Pérez CJ, Aispuro-Hernández E, Quintana-Obregón EA, Martínez-Téllez MÁ (2024) Mixed viral infections in vegetable crops: biochemical and molecular aspects. Rev Mex Fitopatol 42(3)

Serag A, Salem MA, Gong S, Wu JL, Farag MA (2023) Decoding metabolic reprogramming in plants under pathogen attacks, a comprehensive review of emerging metabolomics technologies to maximize their applications. Meta 13(3)

Singh S, Sharma N (2023) Indexing of various viruses infecting Capsicum and their impact on its Phytochemical attributes. Int J Biochem Res 8

Umar MA, Zeshan MA, Iftikhar Y, Umair M, Binyamin R, Ali M et al (2023) Assessment of biochemical contents in okra plants infected with okra yellow vein mosaic virus. Pak J Phytopathol 35(1)

Varma A, Malathi VG (2003) Emerging geminivirus problems: a serious threat to crop production. Ann Appl Biol 142(2):145–164

Zhang Y, Huang L, Liu L, Cao X, Sun C, Lin X (2022) Metabolic disturbance in lettuce (Lactuca sativa) plants triggered by imidacloprid and fenvalerate. Sci Total Environ 802:149764

# Bioinformatics Tools and Databases for Studying Vegetable-Infecting Viruses 19

Adrij Pawan Neog, Archita Bordoloi, Uchakankhi Kashyap, Darshana Goswami, and Nima D. Namsa

**Abstract**

Viruses infecting vegetables are one of the major menaces for global food security due to the associated reduction of crop yield, quality, and market price. With the rapid evolution and diversity of plant viruses, climate change, and global trade, the requirement for detection, identification, and management strategies has beenintensified. Bioinformatics has become a highly effective tool in plant virology, providing tools and databases to aid understanding of virus genomics, host-virusinteractions, and resistance. A summary of recent important bioinformatics tools and databases ispresented in this chapter that will aid in understanding the functional genomics associated with infection in crops. Several sequence analysis software tools, including BLAST and Clustal Omega, are available for the detection andphylogeny of virus genomes. GenBank, EMBL, and RefSeq storelarge collections of viral nucleotide and amino acid sequences, including specialized resources such as Virus-Host DB and ViPR (Virus Pathogen Resource). Furthermore, resources such as RNAcentral and sRNAtoolsfacilitate the analysis of viral small RNAs and their implication in host gene regulation. Recent progress in next-generation sequencing (NGS) and metagenomics, combined with pipelines like Galaxy, VirusDetect, and VirFind, is resourceful in detecting new and emerging viruses infecting vegetable crops. Genome editing and breeding programs, CRISPR-related prediction tools, and resistance gene databases can assist breeders in creating virus-resistant varieties. The ongoing development of bioinformatics applications, particularly in the field of plant virology, has made it more crucial for the long-term development of control and

A. P. Neog · A. Bordoloi · U. Kashyap · D. Goswami · N. D. Namsa (✉)
Department of Molecular Biology and Biotechnology, Tezpur University, Tezpur, Assam, India
e-mail: namsa@tezu.ernet.in



R. Kumar et al. (eds.), *Vegetable Crops and Viral Pathogens Volume I*, Plant Pathology for Sustainable Agriculture, https://doi.org/10.1007/978-981-95-7265-6_19

disease management strategies as well as the early detection of viruses in agricultural settings.

**Keywords**

Bioinformatics · Viral genomics · Genome annotation · Sequence databases · Plant virus taxonomy · High-throughput sequencing · Metagenomics · Resistance genes · Virus-resistant breeding

## 19.1 Introduction

Vegetable crops are of paramount importance both in the world as a whole and locally when considering their contribution to human nutrition, agricultural income, and economic development. Vegetables provide essential vitamins (such as A and C), minerals (including potassium), antioxidants, and fiber in the human diet and play an important role in preventing malnutrition and enhancing public health. Vegetables are an important component of horticultural crop production worldwide, which sustains millions of smallholder growers and contributes significantly to local and international markets. The researchers say there is a fast-growing demand for vegetables because of population expansion, urbanization, and increased consumer awareness of the importance of a healthy diet. However, vegetable crops are less well studied than staple cereals due to logistical reasons, and many people may not be aware of their importance and the continuous demand for more research, better breeding strategies, and enhanced disease evaluation tools leading toward sustainable cultivation (Duan and Lin 2024). Vegetables are an essential source of nutrition in global diets, contributing to the diversity and stability of the world's food systems, with a production volume of 11.5 billion tons from 27 major crops worldwide in 2020, resulting in a significant economic impact (Gong and Diao 2025). Breeding and crop management are now the major tools for advancing yield potential in vegetable crops with strong support from molecular biology and biofortification. Significant advancements in genome editing, genomic selection, and marker-assisted selection have led to a much faster pace for the production of new, better-quality vegetable cultivars. These new tools allow breeders to more accurately develop vegetables with nutrient-enhanced and functionally stable traits. Similarly, the use of molecular techniques has proved efficient for tagging and modifying genes that confer resistance to plant viruses as well as other pathogens and contributed to reduced crop losses and facilitated field performance (Yadav et al. 2023). Biofortification efforts to increase the levels of such vitamins, minerals, and health-promoting phytochemicals in crops will play an expanding role in ameliorating micronutrient deficiencies and allowing vegetable production to make substantial contributions toward public health objectives. Traditional breeding and transgenic methods are being employed to improve nutritional quality without loss of agronomic performance. In conjunction with protected cultivation, precision agriculture, and climate-smart practices, the market innovations stabilize yields and acceptable nutritional quality under unfavorable environmental conditions. These integrated approaches not only enhance the adaptive capacity of vegetable crops to climate stressors but also support their position at the core of sustainable food systems, with nutritious vegetables being available, productive, and health-promoting for a growing world population (Budhlakoti et al. 2022). Viral diseases are regarded as one of the most severe biotic limitations in vegetable production, as they lead to high losses in yield and quality, as well as losses after harvest. Compared to bacteria or fungal pathogens, plant viruses do not have curative remedies; therefore, infections remain systemic, causing stunted growth, chlorosis, deformed fruits, and overall loss of market value. Yield losses vary with host species, virus strain, and agro-ecological conditions but can range from 20% to 100% in susceptible cultivars during severe outbreaks. In solanaceous crops such as tomato and chilli, begomoviruses (e.g., tomato leaf curl New Delhi virus, ToLCNDV) have emerged as major threats, with outbreaks in Asia and the Mediterranean causing devastating production losses and threatening international trade (Sarkar et al. 2024; Kumar et al. 2025a). Similarly, tobamoviruses such as tomato brown rugose fruit virus (ToBRFV) compromise fruit quality and export potential due to strict quarantine regulations (Mehetre et al. 2021). Mixed infections, which are increasingly common under field conditions, exacerbate symptoms and accelerate disease spread through synergistic interactions (de Andrade et al. 2025; Cai et al. 2023). Historically, plant viruses have been characterized on the basis of host response as well as passage to indicator plants and microscopy. Nonetheless, the wide host range and arthropod-vectoring of these viruses improved the understanding of plant viruses through the detection of serological and molecular tools and eventually Sanger sequencing of plant viruses. The convergence of rapid innovations, for instance, fast, high-throughput sequencing (HTS), enables new opportunities for the risk-based identification and characterization of emerging harmful plant viruses. Bioinformatics has fundamentally changed the discipline of plant virology, as it represents the computational corner of bioinformatics to analyze the larger and more complex datasets produced by modern molecular biology. With the rise in the level of our understanding of plant viral diseases through one-gene investigations to genome-wide studies, bioinformatics has been a welcome companion in the further enhancement of our research and control of the diseases (Mehetre et al. 2021). The sequence repositories are used to support genome annotation and taxonomic classification by any of the curated resources offered by GenBank, RefSeq, or special databases such as ViPR and DPVweb. These databases can be used together with other analytical tools like BLAST, MAFFT, MEGA, and IQ-TREE to reconstruct phylogenetic relationships and study recombination and evolution of plant viruses. Such an analysis is important in tracking the spread of highly contagious viruses like tomato brown rugose fruit virus (ToBRFV) and begomoviruses that cause losses in world vegetable production. Bioinformatics also helps understand the host-virus interactions besides diagnostics and taxonomy. Design-based models include a weighted protein-protein interaction and resistance gene functions, which include if protein-protein interactions and resistance gene functions are identified, as a way to identify the candidate targets of breeding programs. New opportunities are also being taken to predict viral emergence and vector-virus interactions and mechanisms of host resistance using AI- and machine learning-based models, which are also being used to connect plant virology and genomics-assisted breeding to develop vegetables resistant to viruses. Since a viral genome has been sequenced, bioinformatics must be analyzed in the characterization and classification. Genome annotation programs also seek open reading frames (ORFs) and forecast protein activity and identify regulatory factors. Also, phylogenetic analysis represents one of the key aspects of modern virology, with the help of which a researcher can locate a new virus in the nomenclature of viruses created by the International Committee on Taxonomy of Viruses (ICTV). These studies can be used in research on viral evolution and recombination and in studies on the global spread of economically significant viruses (Finkel et al. 2018). By examining sequence data of numerous viral isolates collected at various times and locations, researchers can use that data to draw inferences about the dynamics of a virus in its population, draw inferences about the origin of an epidemic, and estimate which direction a virus is taking up an evolutionary course. On the host side, comparison of transcriptomic (RNA-seq) and proteomic data in infected versus mock plants also highlights the molecular battlefield between the virus and its host. Being entirely bioinformatic-based, these "omic" tools can pinpoint host genes implicated in resistance/susceptibility and represent precious targets for crop-breeding programs aimed at the development of resistant, virus-resistant vegetable cultivars. In reality, bioinformatics converts sequence data into biological meaning, leading to practical opportunities in disease and crop protection (Real et al. 2025).

## 19.2 Genomic Databases for Vegetable Viruses

### 19.2.1 General-Purpose Nucleotide and Protein Databases (NCBI GenBank, RefSeq, EMBL-EBI)

The fundamental tools for plant virology research remain the general-purpose nucleotide and protein databases, which provide the primary sequence repositories that underpin annotation, diagnostics, and evolutionary analyses of vegetable-infecting viruses. The most popular open-access nucleotide database is NCBI GenBank, which is constantly updated from submissions from all over the world and contains millions of viral genome entries, including complete genomes and partial sequences. Its integration with BLAST and taxonomy browsers allows researchers to quickly perform similarity searches and taxonomic assignments on novel viral isolates from vegetable crops. The list of databases commonly used for genome annotation of viruses are briefly summarized in Table 19.1.

**NCBI GenBank**

GenBank is also among the largest and most popular databases available to store DNA and RNA sequences of all organisms, including thousands of viral genomes of interest in the studies of plant diseases. NCBI virus is also regularly updated, and

**Table 19.1** Bioinformatic tools and databases are commonly used for understanding the genome and protein biology of viruses associated with diseases in plants, including the vegetable-infecting viruses

| Tool/database | Category | Description/function | References |
|---|---|---|---|
| NCBI GenBank | Sequence database | Central nucleotide repository with viral genome data; supports BLAST-based | Benson et al. (2013) |
| RefSeq | Sequence database | Curated collection of reference sequences for viral genomes and proteins | O'Leary et al. (2016) |
| EMBL-EBI | Sequence database | European archive for nucleotide sequences, supporting global virus submissions | Hingamp et al. (1999) |
| ICTV | Taxonomy | An authoritative virus classification system defines species, genera, and families. | Zerbini et al. (2025) |
| Virus Metadata Resource (VMR) | Metadata resource | Integrates virus metadata (host, geography, genome structure) with taxonomy | Ritsch et al. (2023) |
| VAPiD | Genome annotation | Automated pipeline for viral genome annotation and GenBank submission | Shean et al. (2019) |
| VIGA | Genome annotation | De novo viral genome annotator using homology and ab initio predictions | Fu et al. (2023) |
| Prokka | Genome annotation | Rapid genome annotation tool for prokaryotic and viral genomes | Seemann (2014) |
| BLAST | Sequence alignment | Detects sequence similarity for virus identification and variant discovery | Altschul et al. (1990) |
| MUSCLE | Sequence alignment | High-accuracy multiple sequence alignment for phylogenetic studies | Edgar (2004) |
| MAFFT | Sequence alignment | Fast and accurate alignment using FFT for large viral datasets | Katoh and Standley (2016) |
| Clustal Omega | Sequence alignment | Multiple sequence alignment tool revealing evolutionary and diagnostic markers | Sievers and Higgins (2018) |
| Geneious | Comparative genomics | Integrated platform for assembly, annotation, and phylogenetic analysis | Kearse et al. (2012) |
| UGENE | Comparative genomics | Open-source suite for sequence analysis, structure prediction, and visualization | Okonechnikov et al. (2012) |
| CLC Genomics Workbench | NGS analysis | Comprehensive tool for assembly, SNP calling, and comparative viral genomics | Schilbert et al. (2020) |

(continued)

**Table 19.1** (continued)

| Tool/database | Category | Description/function | References |
|---|---|---|---|
| VirFind | Virus detection | Identifies known and novel plant viruses from HTS data via host subtraction | Maree et al. (2018) |
| VirusDetect | Virus detection | Combines de novo assembly and mapping to detect low-abundance viruses | Zheng et al. (2017) |
| MetaViC | Virus detection | Metagenomic pipeline for discovery and classification of novel viral sequences. | Roux et al. (2021) |
| Kraken2 | Metagenomics | k-mer-based classifier for rapid taxonomic assignment in virome studies | Wood et al. (2019) |
| Centrifuge | Metagenomics | Lightweight classifier for virus identification from large metagenomic datasets | Kim et al. (2016) |
| RDP4 | Recombination detection | Detects recombination breakpoints in viral genomes using multiple algorithms | Martin et al. (2015) |
| SimPlot++ | Recombination analysis | Visualizes sequence similarity and identifies recombination regions | Samson et al. (2022) |
| MEGA | Phylogenetic analysis | Constructs evolutionary trees using distance and likelihood methods | Tamura et al. (2011) |
| IQ-TREE | Phylogenetics | Fast maximum likelihood phylogeny inference with model selection | Minh et al. (2020) |
| BEAST | Evolutionary analysis | Bayesian analysis for viral evolutionary rates, divergence, and spread | Baele et al. (2025) |
| STRING | Protein interaction | Predicts virus-host protein interaction networks for functional studies | Szklarczyk et al. (2023) |
| IntAct | Interaction database | Curated database of experimentally verified protein-protein interactions. | Kerrien et al. (2007) |
| InterProScan/ Pfam/GO | Functional annotation | Predicts domains, motifs, and biological roles in viral and host proteins | Blum et al. (2021) |
| SWISS-MODEL/ Phyre2/AlphaFold | Structural bioinformatics | Predicts three-dimensional structures of viral proteins for functional and therapeutic insights | Jumper et al. (2021) |

the main elements of this tool are the GenBank and RefSeq sequences and the information that is carefully chosen. Extra features, like the addition of new analytical or visualization features, are added. For example, in 2021, Glasa et al. examined the viromes of ten Slovakian tomato and sweet pepper samples and found that they were all infected with potato virus Y (PVY). The majority of datasets make it possible to determine almost the whole sequencing of a single-variant PVY genome from one of the recombinant strains (Glasa et al. 2021).

**RefSeq**

RefSeq is a publicly accessible database that contains feature and bibliographic annotations for nucleotide and protein sequences for major biological taxa, including key plant and vegetable viruses (Goldfarb et al. 2024). Research employing next-generation sequencing often uses RefSeq as a trusted baseline for genome mapping, identification of conserved motifs, and comparative genomics, ensuring high-quality standardization in analytical pipelines. Accession numbers from RefSeq are widely cited in epidemiological and genomic studies for identifying virus isolates, tracing new outbreaks, and understanding virus evolution.

**EMBL-EBI**

The European Molecular Biology Laboratory-European Bioinformatics Institute (EMBL-EBI) obtains sequence and sequence annotation data from patent offices, individual scientists, sequencing institutes, and genomic projects, which integrates sequence submissions from around the world and synchronizes data with GenBank and DDBJ daily (Hingamp et al. 1999). HTS's ability to accurately identify objects is entirely dependent on comparison with previously sequenced datasets. Global sequence databases like the National Center for Biotechnology Information (NCBI), the European Molecular Biology Laboratory (EMBL), and the DNA Data Bank of Japan (DDBJ) are frequently used by analysis standards that are referenced in both fundamental and applied virology research. The interpretation of HTS data and the establishment of baseline information for comprehending pathogens, biological systems, phenotypes, and genetic variation depend heavily on the availability of high-quality raw and reference sequences in these freely accessible worldwide databases (Maina et al. 2024).

## 19.2.2 Plant Virus–Specific Databases (DPVweb, CRPVDB)

While generalist databases like GenBank are essential archives, the complexity of virology research has spurred the development of specialized, curated databases. These resources provide value-added information, integrating raw sequence data with crucial biological context such as host range, symptomatology, and vector information. For researchers studying vegetable-infecting viruses, these platforms offer focused toolsets and expertly annotated data that streamline analysis and discovery.

**DPVweb (Descriptions of Plant Viruses)**

DPVweb is a comprehensive resource dedicated specifically to plant viruses, featuring annotated reference genomes, disease descriptions, host range, and distribution maps. DPVweb offers a comprehensive resource on viruses, viroids, and the satellites of fungi, plants, and protozoa, with authoritative descriptions of viruses affecting vegetable crops such as tomatoes, peppers, and cucurbits (Adams 2006).

**The Plant Virus Transmission Database**

The plant virus transmission database aims to provide access to information on the modes of transmission and vectors of over 1600 plant viruses. It was compiled from information as published in over 3500 papers, with the oldest publication dating back to 1924. The purpose of the plant viral transmission database, which presently includes approximately 1600 plant viruses, is to offer free online access to published information about the vectors and modes of transmission of these viruses. Despite its size, the database also shows that there are still a lot of unanswered questions regarding the potential vectors or modes of transmission of these plant viruses. The modes of transmission and potential vectors of agriculturally significant plant viruses have frequently been thoroughly investigated; however, this information is still lacking for over half of the viruses mentioned here.

### 19.2.3 Taxonomy and Classification Resources (ICTV, Virus Metadata Resource)

**ICTV (International Committee on Taxonomy of Viruses)**

All biological study is based on taxonomic classification, which also serves as the foundation for comparing biological creatures and, consequently, our comprehension of life. The official taxonomy for all viruses is created by the International Committee on Taxonomy of Viruses (ICTV). By providing an extensive online database containing taxonomic information, supporting metadata, and analytical tools to search and visualize the data, ICTV provides an essential resource for the scientific community and general public as well (Black et al. 2025). The ICTV is the principal reference for both naming and classifying the viruses that affect plants for plant virus researchers. It gives direction to a great number of groups of plant viruses, such as the *Geminiviridae*, *Potyviridae*, *Tombusviridae*, and many others. It has developed guidelines with the help of genomic criteria by which the researchers can better identify the species of virus and determine new viruses of the vegetable type (Wylie et al. 2017).

**Virus Metadata Resource (VMR)**

The Virus Metadata Resource (VMR) is an extension of ICTV work in which standard metadata (host, location, genome structure, sequence accession) of known virus species is being gathered. The Virus Metadata Resource (VMR) offers example virus isolates of each virus species, that is, examples of well-characterized isolates of that species. Current entries consist of the name of the virus, the isolate name, the recommended abbreviation, GenBank accession number, and the genome structure, as well as the source of the host (Black et al. 2025). To apply vegetable virology, VMR can use cross-referencing of new sequenced isolates with their taxonomic identity and representative genomes quickly.

## 19.3 Genome Annotation and Comparative Genomic Tools

### 19.3.1 Viral Genome Annotation Platforms (VAPiD, VIGA, Prokka)

Precise genome annotation is the prerequisite for viral genomics, by which coding sequences, regulatory elements, and functional motifs can be recognized for further applications such as comparative genomics, evolutionary biology, and diagnosis. For plant-infecting viruses with compact genomes containing multiple overlapping reading frames, dedicated annotation tools are necessary to allow for accurate genome annotations.

**VAPiD (Viral Annotation Pipeline and iDentification)**

Viral Annotation Pipeline and iDentification (VAPiD) is a lightweight, portable command-line program for annotating viral genomes and depositing them in GenBank. Almost all unsegmented viral genomes can be annotated using VAPiD. VAPiD has been successfully applied to annotate plant virus genomes, improving consistency in public repositories. With just one command, VAPiD enables users to convert any number of viral genome sequences into submission files (.sqn) that are ready for GenBank. When a user wants to submit a small number of sequences from various viral species, VAPiD makes it easier to enter sample metadata for viral genomic sequences. With the help of this software, public health, clinical virology, and research labs that lack the resources to create their own internal genome annotation and submission tools or modify existing ones will be able to share the sequence data they regularly produce with little effort. Because of this, it is a great all-in-one tool for the first characterization of a recently identified vegetable virus, offering a thorough report with functional annotations and its potential. This is an excellent tool for the initial characterization of new vegetable viruses because it produces a complete data set of functional annotations, chemical properties, and classification of the virus all at once (Shean et al. 2019).

**VIGA (VIral Genome Annotator)**

VIGA is a one-stop shop designed to identify eukaryotic viruses and assemble their genomes using NGS data. It combined the functions of genome assembly and viral detection using a variety of assembly techniques, including de novo and reference-based ones. VIGA could be utilized to assemble virus genomes and separate mixtures of virus strains from metagenomic and transcriptome data, according to evaluation on several simulated and actual virome datasets. For example, Fu et al. examined a sweet potato virome RNA-Seq dataset in 2023 that had ten samples (accession in the NCBI BioProject database: PRJNA517178). The sweet potato symptomless virus 1 (SPSMV), sweet potato latent virus (SPLV), sweet potato virus G (SPVG), sweet potato virus 2 (SPV2), sweet potato virus C (SPVC), sweet potato leaf curl virus (SPLCV), sweet potato feathery mottle virus (SPFMV), sweet potato virus E (SPVE), sweet potato virus F (SPVF), and sweet potato chlorotic fleck virus (SPCFV) were identified from the dataset using the homology-based method (Fu et al. 2023).

### 19.3.2 Sequence Alignment Tools (BLAST, MUSCLE, MAFFT)

Sequence alignment is a cornerstone of viral genomics, enabling accurate comparison of viral genomes, detection of conserved motifs, and inference of evolutionary relationships. Thus, alignment platforms form the backbone of plant virology pipelines, bridging raw sequence data with actionable insights for disease management and food security.

**BLAST: Identification of Novel Cucumber Mosaic Virus Variants**
High-throughput sequencing (HTS) was also employed in the study by Ismajli et al. (2025) as a first attempt to investigate plant viruses infecting solanaceous crops in Kosovo, including potatoes (*Solanum tuberosum*), tomatoes (*Solanum lycopersicum*), and peppers (*Capsicum annuum*). Cucumber mosaic virus, broad bean wilt virus 2 (BBWV2), potato virus Y, pepper cryptic virus 2 (PCV2), bell pepper endornavirus (BPEV), and ranunculus white mottle virus were the six viruses found in the sampled plants based on ribodepleted RNA sequencing and bioinformatics analysis, that is, a BLAST search of contigs to the reference genomes of the currently known plant-infecting viruses. For example, BLAST was used to identify tomato brown rugose fruit virus (ToBRFV) in symptomatic tomato plants in Italy by comparing its RNA-dependent RNA polymerase (RdRp) sequence with related tobamoviruses. Such applications highlight BLAST's role in outbreak detection and early diagnostics. A recent study by Asad et al. (2022) used BLAST to successfully identify genetic diversity among CGMMV isolates using sequencing and phylogenetics, revealing high variability and multiple divergent lineages (Asad et al. 2022).

**MUSCLE: Phylogenetic Analysis in Cucurbit Virus Diversity**
MUSCLE is a new computer application that is used to generate several protein sequence alignments. The components of the algorithm are the refinement with tree-dependent restricted partitioning, progressive alignment using a novel profile function we refer to as the log-expectation score, and fast distance calculation through k-mer counting (Edgar 2004). MUSCLE-based *Begomovirus* coat (CP) and replication-associated (Rep) protein alignments showed apparent trends of overall genetic diversity and evolutionary organization. CP and Rep sequences were found to have a high degree of variability, and the Rep proteins were found to be more divergent, reflecting higher adaptive pressures on the replication functions. Even with such diversity, MUSCLE found that there were some conserved functional motifs shared by both proteins, including the CP 2-barrel core and the Rep rolling-circle replication motifs, which indicate strong structural constraints. The alignments also showed clear regional clustering of sequences, indicating the evolution of *Begomovirus* is severely influenced by clustering of sequences by region and restricted gene flow. Besides this, there were also local patches of sudden differences in the sequences that were suggestive of recombination events, which is a known force of *Begomovirus* diversification. In general, the presented MUSCLE analysis revealed some evidence of various global lineages, recombination hotspots, and various forces of evolution on CP and Rep proteins (Crespo-Bellido et al. 2024).

**MAFFT: Large-Scale Population Genomics in Vegetable Virome Surveys**

MAFFT is a bioinformatic program that provides a fast and accurate way to do multiple sequence alignments for DNA, RNA, or protein sequence data. The basic functionality of MAFFT is that it identifies areas of overlap (similarities) between multiple sequences quickly using a fast Fourier transform (FFT). At the same time, it incorporates many different algorithms into its product, providing users with options based on speed versus accuracy (high-speed algorithms for large datasets versus accurate iterative refinement) (Katoh 2002). A MAFFT-based multiple sequence alignment study of the CMV coat protein (CP) gene sequences revealed that the chili isolate from Tamil Nadu, India, shares high sequence identity with other CMV subgroup IB isolates, confirming its subgroup placement with strong molecular support (Vinodhini et al. 2020).

### 19.3.3 Comparative Genomic Software (Geneious, UGENE, CLC Genomics Workbench)

Utilizing comparative genomics tools enables plant virus researchers to visualize and annotate viral genomes and conduct large-scale comparisons of multiple viral genomes (Liu and Di, 2020). For plant virus researchers, these comparative genomic platforms provide an opportunity to examine variability among plant viruses, how plant viruses adapt to different hosts, and how plant virus pathogens differ at the molecular level from nonpathogenic plant viruses.

**Geneious**

Geneious is a software platform based on an integrated approach, which is expected to enable the management of sequence data in a user-friendly manner. Because many of the necessary instruments needed to be implemented to analyze biological sequence data on a single desktop in a single platform, it is possible to perform complicated analyses without the need to install many software packages and command lines. Geneious has plug-ins to a broad range of popular algorithms, such as MAFFT, MUSCLE, BLAST, and BEAST, which can be as flexible and extensible as possible (Kearse et al. 2012). Geneious was applied in one of the studies to assemble, organize, and compare rice black-streaked dwarf virus (RBSDV) genomic and transcriptomic sequences of various host plants and insect vectors, which allowed the authors to find out unique genetic variations and host-related polymorphisms among viral isolates (Lagzian et al. 2024).

**UGENE**

The main goal of Unipro UGENE is to be a multiplatform open-source software to assist molecular biologists with no knowledge in bioinformatics to organize, analyze, and visualize their data. UGENE integrates well-known bioinformatic software into one user interface. The toolkit allows recovering data from remote data sources and allows for data type diversification. It provides visualization software of biological objects, such as phylogenetic trees, multiple sequence alignments, next-generation sequencing (NGS) assembly data, annotated genome sequences, and three-dimensional structures (Okonechnikov et al. 2012).

**CLC Genomic Workbench**

The researchers conducted high-quality read trimming, de novo assembly, and reference-based mapping (Liu et al. 2024) using CLC Genomics Workbench, which facilitated the correct reconstruction of several viral genomes and a viroid using a single sample of pokeweed. The assembly tools used by CLC demonstrated that the plant had a mixed infection with clear contigs representing different viruses and a viroid and, therefore, a complex community of pathogens. The analysis of coverage in CLC revealed that there was a high sequencing depth of various viral elements, which demonstrated active replication, and the analysis tools rated the coverage and annotation in the platform, which were used to classify the assembled sequence with high confidence. The hidden diversity revealed by HTS was also highlighted by the Workbench, which made it possible to detect coinfecting agents that could not be detected with only symptoms. Comprehensively, CLC Genomics Workbench played a fundamental role in uncovering the virus-viroid complex, fixing genomic structures, and confirming the existence of various pathogens in a single host plant (Kwak et al. 2024).

## 19.4 Phylogenetics and Evolutionary Analysis

### 19.4.1 Phylogenetic Tree Construction Tools (MEGA, IQ-TREE)

**MEGA (Molecular Evolutionary Genetics Analysis)**

The Molecular Evolutionary Genetics Analysis (MEGA) software was designed to provide a collection of biologist-friendly tools that are capable of statistically analyzing a set of DNA and protein sequence data in an evolutionary way. It has overgrown to provide tools to sequence alignment, phylogenetic tree building and visualization, testing diverse hypotheses of evolution, estimating sequence divergences, web-based sequence data shopping, and specialist automated systems generating natural language descriptions of the analysis techniques and data chosen by the user (Tamura et al. 2011). In the phylogenetic analysis with MEGA, the research revealed that the Hungarian isolates of garlic virus exhibit evident genetic segregation within known virus species, including GarV-A, GarV-B, GarV-C, GarV-D, OYDV, and LYSV, and this suggests their taxonomic position. Phylogenetic trees constructed using MEGA showed that there was great genetic diversity in the isolates, with some of them forming autonomous evolutionary branches, which showed that there were numerous instances of introduction and the subsequent splitting over an extended period of time in Hungary. The pairwise distance and nucleotide identity studies in MEGA further demonstrated the existence of a high degree of sequence variation among the same species of the virus isolates that can be attributed to the vegetative propagation and cross-border transmission of the garlic bulbs, which leads to the spread and diversification of the virus (Koczor et al. 2024).

**IQ-TREE**

An open-source program called IQ-TREE is a popular phylogenetic inference program based on the maximum likelihood (ML) criterion. The ability of IQ-TREE to perform well is due to the successful implementation of the new phylogenetic methods that improve the three fundamental steps of phylogenetic analysis: a special fast bootstrap approximation, a fast tree search algorithm, and rapidly selecting a model using ModelFinder (Minh et al. 2020). This phylogeny has enhanced phylogenetic resolution of several genera of vegetable-infecting geminiviruses, for which the researchers were able to propose new and more valid taxonomies (Minh et al. 2020).

### 19.4.2 Recombination and Mutation Detection Tools (RDP4, SimPlot)

The recombination and mutation are the significant factors that contribute to the evolution and the appearance of new viruses that infect vegetables. These events are important to detect and analyze the viral diversity, the ability of hosts to adapt, and the potential for new epidemics. Several bioinformatics applications, namely, RDP4 and SimPlot, have become commonplace in the study of plant virology.

**RDP4 (Recombination Detection Program v4)**

The latest recombination detecting software (RDP) is referred to as RDP4. RDP is a Windows program that relies on a large number of methods to detect and visualize recombination in virus genome sequence alignments and eliminates evidence of recombination. In comparison to previous versions of the program, RDP4 is able to analyze twice as many sequences (as many as 2500), approximately three times longer (up to 10 Mb) (Martin et al. 2015). It uses a variety of detection algorithms, that is, RDP, BootScan, MaxChi, and Chimaera, to provide powerful support for recombination signals. RDP4 comparison demonstrated that recombination is a significant evolutionary process that has influenced the diversity of tomato yellow leaf curl virus (TYLCV) throughout the world, as multiple statistically significant recombination events have been observed between whole-genome sequences. The Kuwaiti isolates displayed evidence of either being recombinant descendants or they may be parental contributors in observed recombinations, meaning that they are not isolated introductions, but they undergo regional evolution of viruses. In the study, RDP4 found that there were several recombination break sites in the coding and intergenic sequences, which indicated that TYLCV may often exchange genetic materials as it undergoes dissemination in various geographical locations (Akbar et al. 2024).

**SimPlot**

SimPlot++ is a sophisticated Python application that is used to visualize sequence similarities and recombination events with better speed, flexibility, and repeatability than the original SimPlot tool. The software provides similarity analysis by sliding window and bootscanning, whereby query sequences can be compared with a series of references, and recombination breakpoints can be identified easily (Samson et al. 2022). The SimPlot analysis has shown a convincing graphic image of discontinuity in the PVA genome at the region nearest to the 3′ end (3′-proximal) that suggests that multiple copies of PVA may exist in the Chinese isolate and not just one parent lineage. The similarity plots of the sliding window showed that the nucleotide similarity suddenly diverged at around the halfway position of the 3′ end of the isolate PVA genome, which coincided with the point at which the isolate was no longer clustering with the PVA variant that had high nucleotide similarity to it but began to cluster with a different variant of the PVA with high nucleotide similarity. This abrupt change in the similarity of the nucleotides is typical of a recombination event and is the reason why the other recombination detection methods supported the results and further evidence of the isolate being a true recombinant and not a consequence of a sequencing error or mutation (He et al. 2014).

## 19.5 Virus Detection from Sequencing Data

### 19.5.1 High-Throughput Sequencing Approaches for Plant Virus Discovery

The technologies of high-throughput sequencing (HTS) have fundamentally transformed the discipline of plant virology by allowing the detection of known and unknown viruses in large amounts in direct proportion to infected tissues of plants without any bias. As an alternative to more traditional serological or PCR-based methods, HTS can identify multiple viruses simultaneously, detect mixed infections, and define totally novel viral taxa, so it should not be avoided in the study of vegetable-infecting viruses. The most important HTS methods are small RNA sequencing, total RNA sequencing, and circular viral genome DNA sequencing (rolling circle amplification, RCA).

**Total RNA Sequencing (RNA-Seq)**

RNA-Seq, also known as total RNA sequencing, is an easy and effective technique for discovering viruses (Melnick et al. 2021). The course of total RNA-seq identified chlorotic dwarf virus of chickpea (CpCDV) as the vastly dominant viral agent in the watermelon plants with hard fruit syndrome, with a great percentage of total mapped reads comprising of viral reads. The deep-sequencing data allowed complete reconstruction of CpCDV genomes and ensured their high titers as well as eliminating other viruses or viroids as the stable causes of the disease. The mixed CpCDV variants in the same plant were also revealed by RNA-seq, which suggested the active diversity of the viral population and potential quasispecies interactions. The homogenous correlation of CpCDV reads with symptomatic plants, but not healthy controls, presented good molecular evidence that CpCDV causes the disease. On the whole, total RNA-seq served as an obligatory measure of unbiased pathogen detection, which helped confirm that CpCDV is the leading etiological organism and its intra-host RNA variability in watermelon (Zaagueri et al. 2017).

**Small RNA Sequencing (sRNA-Seq)**

The small RNA-seq can lead to the highly sensitive and reference-independent detection of plant viruses through capturing virus-derived small RNAs (vsRNAs) produced in the host process of RNA-silencing, and the vsRNA signatures can be utilized to identify known viruses, identify new ones, and detect mixed infections at a high level of accuracy. The sRNA-seq can be used to reconstruct the entire or nearly complete genome of the virus, even with a low abundance of the viral RNA or in the absence of an existing sequence, by mapping and assembling vsRNAs. The patterns of vsRNA size distribution and strand polarity can be used to understand the mechanisms of host antiviral defense and to understand the way in which the Dicer-like proteins process viral dsRNA intermediates. Also, the small RNA-seq reveals vsRNA hotspots, which show the dynamics of viral replication and structural characteristics of the viral genome. On the whole, the results indicate that sRNA-seq is an effective instrument in the study of the plant virome, assembling the viral genome, and defining RNA-silencing-mediated immunity (Pooggin 2018).

### 19.5.2 Bioinformatic Pipelines for Virus Detection (VirFind, VirusDetect)

**VirFind**

The diagnostic pipeline NGS involved fast diagnosis and virus detection using VirFind through the use of automation. With the help of this tool, the users of the pipeline could search the virus sequences in a relatively short period since it offers host-read subtraction, de novo assembly, and annotation tools based on BLAST. The authors were able to use VirFind to identify known and unknown viruses of various types of plant samples, with a very low percentage of viral sequence detection. VirFind was also used to verify the assembled contigs to confirm them to be correct and the actual viral reads of the background or nonviral sequences to clean up the NIEs. Furthermore, VirFind was instrumental in identifying previously unreported virus species from samples collected in the field, indicating its importance for the retrieval and discovery of virus species and new methods of surveillance. VirFind, therefore, represents an innovative approach to the use of NGS technology for the detection of plant viruses.

**VirusDetect**

VirusDetect has allowed us to verify that sRNA deep-sequencing can be used for extremely sensitive, high-throughput detection of plant viruses (even when the titer is very low or in the presence of other viroids), at least for the species tested here. The pipeline's use of host-read subtraction, de novo assembly, virus database mapping, and genome reconstruction in an integrated workflow effectively enabled the reliable detection of both known and new virus isolates from a variety of plant samples. VirusDetect could reconstruct nearly complete viral genomes through assembly of virus-derived short RNAs and enabled subsequent phylogenetic and diversity analysis. Benchmarking experiments demonstrated high sensitivity, specificity, and speed of VIP, and it was capable of detecting viruses that were either missed or not completely assembled by the standard pipelines (Zheng et al. 2017).

### 19.5.3 Classification and Metagenomic Tools (Kraken2, Centrifuge)

Metagenomic classification tools are essential in analyzing high-throughput sequencing datasets, where millions of reads may originate from diverse sources, including host, bacteria, fungi, and viruses. For plant virology, especially in vegetables, tools like Kraken2 and Centrifuge provide rapid and accurate classification of viral reads against large reference databases. These approaches are invaluable for virus discovery, epidemiological surveillance, and virome profiling.

**Kraken2**

Kraken2 can be described as a k-mer-based taxonomic classification system that attributes sequencing reads to organisms by matching k-mer matches with a reference database used to make high-throughput and rapid profiling of microbial communities. Kraken2 has three categories based on the indexed k-mers stored in the database that determine its accuracy. Although it is shown that Kraken2 is a rapid and very efficient code, the accuracy and sensitivity of this tool depend on the reference databases used and on the threshold used on the scores of the confidence in assigning the sequence reads (Liu et al. 2024). It was discovered that Kraken2 was among the fastest classifiers of taxonomic rankings in viruses and was effective when the virus genomes were well represented in the reference database but was less accurate on novel and incomplete viruses or viruses with high divergence. Since Kraken2 is based on k-mer matching, the tool tended to generate false positives or low-resolution assignments in case of low database coverage or when the viral sequences were highly taxon-specific. The authors demonstrated Kraken2 to be more competent at detecting highly abundant, well-characterized viruses, but it was not able to work as well with complex samples with the virome that had rare or unknown viruses (Cadenas-Castrejón et al. 2022).

**Centrifuge**

New microbial classification engine, called centrifuge, enables the possibility to quantify species in desktop computers and label readings faster, more accurately and sensibly. The system is specifically created to address the metagenomic classification problem and is indexed using the Ferragina-Manzini (FM) index and Burrows-Wheeler transform (BWT). The Centrifuge study shows that the fast, high-memory-efficient classification methodology is highly applicable to the detection of plant viruses in metagenomic data because the method can detect viral reads even in extremely low densities in mixed plant samples (Kim et al. 2016).

## 19.6 Host-Pathogen Interaction Resources

### 19.6.1 Databases on Pathogen–Host Interactions (PHI-base, Virus-Host Database)

**PHI-base (Pathogen-Host Interaction Database)**

PHI-base is offered as a curated open-access database that gathers information experimentally demonstrated to be involved in pathogen-host interaction in fungi, oomycetes, and bacteria and their plant or animal hosts. The PHI-base is annotated with extensive information on pathogenicity, virulence, effector capabilities, and host resistance mechanisms with peer-reviewed experimental data as backup. PHI-base is a useful source of information to determine virulence factors, predict members of the host, and inform functional studies in plant and animal pathology by combining genomic, phenotypic, and molecular interaction data. As an example, PHI-base is demonstrated to be a useful curated database of plant-pathogen interactions (including interactions with plant viruses), providing experimentally validated data on viral genes, host reactions, and molecular processes that influence the disease outcomes. Even though PHI-base is mainly fungal, bacterial, and oomycete, there are also entries of plant hosts becoming responsive to viral effectors or where viral pathogenicity factors have been experimentally defined (Urban et al. 2019).

**Virus-Host DB**

The VirusHostDatabase was used as the central source that connected viral genomes with experimentally verified host organisms to enable the authors to assess and test computational tools used to predict viral hosts. Viruses can have host-specific genomic signatures, including oligonucleotide composition and codon usage patterns, by combining virus-host DB-curated virus-host pairs with large-scale genome analysis, and such genomic signatures can confidently predict the host of an unknown viral sequence. It was also the virus-host DB that provided the possibility to identify hitherto unknown host-virus interactions to broaden the knowledge on viral ecology and coevolution. With the VirusHost Database, the research showed that plant viruses, such as bacteriophages and animal viruses, have host-specific genomic signatures that can be inferred in computational reconstruction to plant hosts (Mihara et al. 2016).

### 19.6.2 Protein-Protein Interaction Prediction Tools (STRING, IntAct)

**VirusSTRING**

The VirusesSTRING project presents a database that contains experimentally verified and computationally suggested protein-protein interactions between viruses and their hosts comprehensively. The authors demonstrate that this resource assists in understanding the systematic targeting of the viral proteins on major host pathways like immune regulation, replication machinery, and intracellular transportation to promote infection. VirusSTRING offers high-confidence interaction networks in many different viral families by amalgamating literature curation data, high-throughput experiment data, and state-of-the-art text mining. Comparison of conserved virus-host interaction strategies and functional prediction of understudied or newly discovered viral proteins can also be done in the database. VirusSTRING serves as an essential platform for exploring the molecular mechanisms of viral infection and identifying potential antiviral targets. Viruses.STRING also enables the identification of putative interactions for poorly characterized plant viral proteins by transferring interaction patterns from related viruses, thus supporting functional annotation and hypothesis generation. By linking plant viruses to specific host proteins, the database helps clarify how viruses manipulate plant physiology and identifies candidate host targets that could be explored for resistance breeding or antiviral strategies (Cook et al. 2018).

**IntAct**

An open-source database and software package called IntAct is used to model, store, and analyze data about molecular interactions. The database heavily utilizes restricted vocabularies and contains more than 126,000 binary interactions that were taken from more than 2100 scholarly papers. In addition to curating protein-protein interactions, IntAct now records an increasing amount of important research that provides information on interactions between DNA, RNA, and tiny molecules. In the case of IntAct, the Virus-Host Database is an extensive, trustworthy, and catalogued experimental source for studying the molecular interactions between viruses and their hosts. The IntAct Database not only contains the interactions that have been seen between virus effectors and their targets (i.e., viral proteins that interact with cellular proteins) but also enables scientists to determine the pathways within host cells (e.g., immune signaling, the process of synthesizing gene products (also called transcripts) and transport proteins to various parts of a cell) (Kerrien et al. 2007).

## 19.7 Structural Bioinformatics for Plant Viruses

### 19.7.1 Three-Dimensional Structure Prediction Tools

**AlphaFold2 and AlphaFold-Colab**

AlphaFold2, developed by DeepMind, is a revolutionary deep learning-based tool that can predict protein structures with astonishing, near-experimental accuracy (Jumper et al. 2021; Yang et al. 2023). It uses an AI model to analyze a protein's amino acid sequence and its coevolutionary relationships to directly infer the three-dimensional coordinates of its atoms. AlphaFold 3 is a next-generation biomolecular structure prediction system that extends beyond protein folding to accurately model complete biomolecular interactions, including protein-protein, protein-DNA, protein-RNA, RNA-RNA, and protein-ligand complexes (Abramson et al. 2024). Within the framework of plant viruses, AlphaFold can make structural predictions of viral coat proteins, viral movement proteins, viral polymerases, and viral transcriptional silencing suppressors with a high degree of reliability because these components are typically poorly studied due to experimental constraints. These inferred structures are used to reveal functional domains and interaction patterns that plant viruses employ to invade cells and move in and out of plasmodesmata and avoid being detected by plant immunity. AlphaFold models can also be used to predict the sites of interaction of the virus with the plant proteins, which gives insight into how viral effectors can control the process of the host and which proteins can be targeted to create resistance (Gutnik et al. 2023).

**RoseTTAFold**

RoseTTAFold is a potent predictor of protein structure, and its growing application to protein design is a foundation model. It is unbelievable that these networks can not only predict native protein structures accurately but also give a profound understanding of the sequence-structure relationship, which allows the researcher to design new proteins to have the desired shape and perform the desired functions. The scientists will be able to measure the foldability and the unstable regions and optimize the sequences by applying the structural constraints and the confidence metrics provided by AlphaFold and RoseTTAFold. Such models have been used to rapidly design new enzymes, binders, and therapeutic proteins by shrinking the experimental search space and allowing in silico screening to be done quickly. RoseTTAFold offer convenient structural prediction power that can significantly speed up studies of viral proteins, including coat proteins, movement proteins, replicases, and RNA-silencing suppressors, many of which do not have known structures. Through these models, researchers can predict interfaces of host interactions and determine structurally relevant residues by using which plant viruses traverse plasmodesmata or bind host factors or avoid antiviral immunity. (Wang et al. 2024).

**SWISS-MODEL**

SWISS-MODEL is an automated homology modeling web server, which is among the most reliable and proven structure prediction approaches (Waterhouse et al. 2018). It operates by identifying an experimentally determined protein structure in a related organism (a "template") with which the query protein is highly similar in terms of sequence. SWISS-MODEL then uses this template as a scaffold to build a model of the query protein. This method is extremely fast and accurate, provided that a good template (typically >30% sequence identity) is available. A researcher discovered a new strain of tobacco mosaic virus (TMV) infecting tomatoes and wanted to know if mutations in its coat protein would affect the virion's stability. Instead of using a computationally intensive deep learning method, they used SWISS-MODEL. Because thousands of structures of TMV and related viral coat proteins already exist in the Protein Data Bank (PDB), SWISS-MODEL quickly found an identical template with >99% sequence identity. In minutes, it generated a high-quality homology model of the new strain's coat protein. By mapping the mutations onto this model, the researcher could immediately see that they were located on the protein's surface, unlikely to disrupt its core fold, and thus would not compromise the structural integrity of the virus particle.

### 19.7.2 Structure-Guided Antiviral and Resistance Strategies

**Structure-Guided Antiviral Design**

Bioinformatics has allowed antivirals to be designed based on structures of virus proteins and RNA in order to identify the most highly conserved, but vulnerable, regions of viral genomes and proteins. By integrating structural modeling efforts that use platforms such as AlphaFold and RoseTTAFold and employing molecular docking and sequence conservation analysis as well as artificial intelligence models of modeling proteins, researchers can visualize the structures of viral proteins and identify conserved functional domains and predict the interaction between Cas proteins and viral RNA (Manavi, 2021). This allows researchers to rationally design CRISPR/Cas constructs that will allow for efficient destruction of viral replication without causing off-target effects in the host and predict virus escape mutations, which helps researchers determine the most durable antiviral target and create multispectrum resistance. In summary, the structure-based approach is an extremely powerful computational tool that will enable the development of highly effective and highly targeted antiviral strategies for plant viruses (Iksat et al. 2025).

**Engineering Resistance via Structural Knowledge**

By using RNA interference and CRISPR-based technologies to inhibit expression of these conserved regions, resistant pepper and tomato cultivars were developed that block viral replication and cell-to-cell movement (Sundaresha et al. 2022). This demonstrates how structural bioinformatics can be applied to guide the development of crop protection strategies. Many potyviruses, including turnip mosaic virus (TuMV), require the host protein eIF4E to initiate replication. The viral protein VPg binds directly to a pocket on the surface of eIF4E. A 2024 study used the crystal structure of the VPg-eIF4E complex to guide the engineering of resistance in *Arabidopsis thaliana* (a model for vegetable crops). Using PyMOL to visualize the interaction interface, they identified specific amino acids on eIF4E that made direct contact with the viral VPg protein. Using CRISPR-Cas9 gene editing, they mutated just a few of these contact residues in the plant's eIF4E gene. The resulting plants were perfectly healthy, but the modified eIF4E could no longer be recognized by the viral VPg. As a result, the engineered plants were completely resistant to TuMV infection, showcasing a highly precise, structure-guided resistance strategy (de Andrade et al. 2025).

## 19.8 Resistance Gene Databases and Breeding Applications

### 19.8.1 Plant Resistance Gene Databases (PRGdb, RGAugury)

**PRGdb (Plant Resistance Gene Database)**

PRGdb is a web-accessible open-source database that represents the first bioinformatic resource providing a comprehensive overview of resistance genes (R-genes) in plants (Calle García et al. 2021). More than 16,000 known and probable R-genes from 192 plant species that are threatened by 115 distinct diseases are stored in PRGdb and are connected to valuable biological data. The entire database consists of 10,463 computationally predicted putative R-genes, 6308 putative R-genes gathered from NCBI, and a collection of 73 hand-curated reference R-genes. PRGdb catalogs numerous NBS-LRR, receptor-like kinases (RLKs), and receptor-like proteins (RLPs) known to participate in viral resistance, enabling researchers to explore their conserved domains, evolutionary relationships, and species-wide distribution. The platform's comparative tools help pinpoint candidate antiviral R genes in crop genomes by matching structural motifs and domains associated with virus recognition. PRGdb supports the discovery of novel genes involved in resistance to economically important plant viruses, thereby facilitating breeding programs and molecular strategies aimed at enhancing viral immunity (Sanseverino et al. 2009).

**RGAugury**

To automate RGA prediction, an integrative pipeline called RGAugury was created. Nucleotide binding site (NB-ARC), leucine-rich repeat (LRR), transmembrane (TM), serine/threonine and tyrosine kinase (STTK), lysine motif (LysM), coiled-coil (CC), and Toll/Interleukin-1 receptor (TIR) are among the protein domains and motifs associated with RGA that were first identified by the pipeline. Based on the existence of combinations of these RGA domains and motifs, RGA candidates are discovered and categorized into four primary families: membrane-related RLP and RLK, TM-CC, and NBS-encoding. To enhance performance, all laborious pipeline analyses are parallelized. The workflow was evaluated against the well-annotated *Arabidopsis* genome, with 98.5%, 85.2%, and 100% of the RLPs, RLKs, and known NBS-encoding genes in the membrane being validated, respectively. The classification system provided by RGAugury enables the researcher to examine the domain architectures that are related to virus recognition and to study the different ways in which the RGAugury has evolved in various plant species in response to viral selection pressure. The scale of the pipeline to quickly find and compare RGAs across genomes is a potent means of learning more about natural variation in antiviral immunity and how to operate to achieve better immunity against plant viruses. The mode of bioinformatics has brought a revolution in the discovery of plant resistance (R) genes and the creation of molecular markers to breed virus-resistant vegetable crops. The discovery of new R-genes is made possible by high-throughput sequencing and comparative genomics and domain-based analysis, whereas the design of gene-targeted markers to be used in marker-assisted selection (MAS) can be achieved in silico (Li et al. 2016).

### 19.8.2 Bioinformatics in Resistance Gene Discovery and Marker Development

Bioinformatic approaches have revolutionized the discovery of plant resistance (R) genes and the development of molecular markers for breeding virus-resistant vegetable crops. High-throughput sequencing, comparative genomics, and domain-based analysis enable the identification of novel R-genes, while in silico prediction tools facilitate the design of gene-targeted markers for marker-assisted selection (MAS).

**Resistance Gene Discovery**

Genome-wide mining of disease resistance gene analogs (RGAs) strategy based on conserved domains is a potent system for identifying plant genes that can provide resistance against viral pathogens. Through the screening of entire genomes of characteristic immune-related motifs like NBS-LRRs, receptor-like kinases, and receptor-like proteins, this technique can be used to identify candidate genes that are implicated in identifying viral effectors and initiating antiviral defense signals. The conservation domain analysis is also used to show how these resistance genes have evolved under viral pressure, thereby indicating the evolutionary arms race between the plants and the virus. All in all, the given strategy enhances the work aimed at the identification of new antiviral resistance loci and contributes to breeding and biotechnological solutions that are going to help create virus-resistant crop varieties (Tirnaz et al. 2020).

**Marker Development**

One of the main aspects of breeding in the present day is the use of DNA molecular markers in an attempt to increase resistance to plant viruses. Markers have been highly developed to enable breeders to specifically detect and monitor resistance loci that protect against the significant viral pathogens, including begomoviruses, potyviruses, and tobamoviruses, including single-nucleotide polymorphisms (SNPs), simple sequence repeats (SSRs), and insertion-deletion polymorphisms (InDels). Besides genome-wide association studies, which are being utilized to discover and map new antiviral genes by quantitative trait locus (QTL) mapping, the fact that breeders can use marker-assisted selection (MAS) allows one to quickly and correctly introgress new genes to elite breeding lines. The application of these markers in the breeding of pipelines is resourceful in the breeding of long-lasting and wide-spectrum resistance to viruses and less reliance on the phenotype-based strategies to produce resistant types of crops (Kumar et al. 2025b).

## 19.9 Challenges, Limitations, and Future Directions

**Current Limitations in Databases and Tools for Vegetable Viruses**

The gains being achieved using data and bioinformatics on vegetable viruses have helped to an extent in aiding research and management, but there are limitations that limit the complete use of useful information. Examples of such challenges include a lack of complete representation of viral sequences in current databases (including bias) and problems achieving annotating accuracy, as well as difficulty integrating heterogeneous data at all levels/platforms.

**Incomplete Viral Genome Coverage**

Many vegetable viruses are not well represented in the public databases; they have been published or included in the public databases as fragmented, low-quality genomes. This prevents accurate phylogenetic analysis, primer development, and polymerase chain reaction (PCR) for diagnosed viruses, as well as performing comparative genomic analyses of all vegetable viruses, which limits our understanding of the evolutionary history of the vegetable crops that are infected by these viruses (Kumar et al. 2025b).

**Outdated or Minimal Genome Annotations**

Studies on host organism resistance to the virus and viral evolution will be facilitated by access to more features and annotations of viral genomes. Access to complete and current annotations of viral sequences will allow researchers to properly identify the areas within a genome that are responsible for pathogenicity, host adaptation, and/or resistance (Kumar et al. 2025b).

**Limited Host-Virus Interaction Information**

The databases that record interactions at the molecular level between vegetable viruses and plant hosts are not abundant. Lack of curated interaction data also makes it difficult to discern how viruses rewire host pathways or for plants to respond to antiviral defense and ultimately the synthesis of resistance (Kumar et al. 2025b).

**Few Computational Tools Are Tailored to Vegetable Virology**

Unfortunately, all of the available pathogen-analysis pipelines are developed primarily for fungal and bacterial pathogens and cannot effectively handle plant viral genomes with high mutation rates, small genome sizes, or mixed infections. Therefore, the vegetable virologists do not always have integral setups for their demanding work on sensitive detection, classification, and variant analysis (Kumar et al. 2025b).

**Poor Interoperability Among Databases**

Each virus, host, and omics database adopts its own data format and standards, impacting the integration of information between platforms. Such incompatible situations in multidatabase analysis are an obstacle when trying to perform a full-scale analysis on vegetable viruses.

**Lack of Standardized Metadata**

A large part of viral sequence records do not contain key metadata, including host species, cultivar, symptoms, sampling location, and the read platform from which they originated. Without standardized metadata collection across labs, it is more difficult to trace the spread of the virus or compare isolates or look at wider epidemiology in vegetable crops (Kumar et al. 2025b).

**Insufficient Multiomic Integration**
While genomics, transcriptomics, proteomics, and metabolomics are increasingly used, tools that integrate these datasets specifically for vegetable virus research are still limited. This gap restricts the ability to build complete models of virus infection and host responses (Kumar et al. 2025b).

**Limited Datasets for Machine Learning Applications**
Machine learning has great potential for virus detection, resistance prediction, and host-virus interaction modeling but the review notes that current datasets for vegetable viruses are too small, uneven, or poorly annotated to train high-performance models. This limits the advancement of AI-based approaches in vegetable virology (Kumar et al. 2025b).

**Challenges and Future Directions**
Although there has been great progress made with AI for vegetable virology, there are still several barriers, including a lack of dataset diversity, lack of model generalization for various crops/regions, limited computational capacity, and farmers adopting technology. In order for AI to fully maximize its capabilities within vegetable virology, there must be continuous development/improvement in algorithms, transfer learning, mobile deployment, and ethical data management.

## 19.10 Conclusion

Bioinformatics has gained the status of a fundamental tool in enhancing vegetable virus research and management. It enables extensive genome (re)analysis, functional annotation, and comparative studies, which help to gain insight into the diversity, evolution, and pathogenicity of viruses. Interconnected databases and tailored tools facilitate the high-throughput mining of viral sequences, host-pathogen interplays, and resistance genes for precision breeding as well as for targeted ATVs. Here, structural bioinformatics has revealed essential viral protein functions to shape structure-based antiviral design and resistance engineering. Running real-time surveillance using AI and machine learning technologies and methodologies has further revolutionary potential for real-time surveillance, early detection of viruses, and predicting outbreak locations to increase response time and lessen crop damage.. AI and ML methodologies have further revolutionary potential for real-time surveillance, early detection of viruses, and predicting outbreak locations to increase response time and lessen crop damage. Together, this bioinformatics framework facilitates the transfer of scientific understanding at the molecular level into long-term sustainable host resistance strategies for control of the vegetable viruses that protect food supply and support the agriculture sector. The future of vegetable virus management lies in integrated, multidimensional bioinformatic platforms that combine genomic, epidemiological, structural, and environmental data specific to vegetable crops. Advances in AI/ML will foster predictive diagnostics, enabling preemptive interventions and precision agriculture. Crop-specific virus resources with standardized, interoperable data will accelerate discovery and deployment of durable resistance traits. Coupled with gene editing and novel antiviral technologies informed by structural bioinformatics, these approaches promise adaptive, resilient crop protection amid evolving viral threats and climate change. Collaborative global data sharing and real-time surveillance systems integrated with big data and IoT sensors will enhance early warning and outbreak control. Continued investment in computational biology expertise and tool refinement is essential to unlock the full potential of bioinformatics-driven innovations, ensuring sustainable vegetable production and global food security in the coming decades.

## References


Abramson J, Adler J, Dunger J, Evans R, Green T, Pritzel A, Ronneberger O, Willmore L, Ballard AJ, Bambrick J, Bodenstein SW, Evans DA, Hung C-C, O'Neill M, Reiman D, Tunyasuvunakool K, Wu Z, Žemgulytė A, Arvaniti E et al (2024) Accurate structure prediction of biomolecular interactions with AlphaFold 3. Nature 630(8016):493–500. https://doi.org/10.1038/s41586-024-07487-w

Adams MJ (2006) DPVweb: a comprehensive database of plant and fungal virus genes and genomes. Nucleic Acids Res 34(90001):D382–D385. https://doi.org/10.1093/nar/gkj023

Akbar A, Al Hashash H, Al-Ali E (2024) Tomato yellow leaf curl virus (TYLCV) in Kuwait and global analysis of the population structure and evolutionary pattern of TYLCV. Virol J 21(1). https://doi.org/10.1186/s12985-024-02540-6

Altschul SF, Gish W, Miller W, Myers EW, Lipman DJ (1990) Basic local alignment search tool. J Mol Biol, 215(3):403–10. https://doi.org/10.1016/S0022-2836(05)80360-2

de Andrade FA, de Luna-Aragão MA, Ferreira JDC, Souza FF, da Rocha Oliveira AC, da Costa AF, Aragão FJL, dos Santos-Silva CA, Benko-Iseppon AM, Pandolfi V (2025) Deciphering cowpea resistance to potyvirus: assessment of eIF4E gene mutations and their impact on the eIF4E-VPg protein interaction. Viruses 17(8):1050. https://doi.org/10.3390/v17081050

Asad Z, Ashfaq M, Iqbal N, Parvaiz F, Mehmood MA, Hameed A, Malik AH, Kayani SB, Al-Kahtani MA, Ahmad Z (2022) Genetic diversity of cucumber green mottle mosaic virus (CGMMV) infecting cucurbits. Saudi J Biol Sci 29(5):3577–3585. https://doi.org/10.1016/j.sjbs.2022.02.027

Baele G, Ji X, Hassler GW, McCrone JT, Shao Y, Zhang Z, Holbrook AJ, Lemey P, Drummond AJ, Rambaut A, Suchard MA (2025) BEAST X for Bayesian phylogenetic, phylogeographic and phylodynamic inference. Nat Methods, 22(8):1653–1656. https://doi.org/10.1038/s41592-025-02751-x

Benson DA, Cavanaugh M, Clark K, Karsch-Mizrachi I, Lipman DJ, Ostell J, Sayers EW (2013) GenBank, NAR 41(D1):D36–D42. https://doi.org/10.1093/nar/gks1195

Black EJ, Powell CS, Dempsey DM, Hendrickson RC, Mims LR, Lefkowitz EJ (2025) Virus taxonomy: the database of the international committee on taxonomy of viruses. Nucleic Acids Res. https://doi.org/10.1093/nar/gkaf1159

Blum M, Chang HY, Chuguransky S, Grego T, Kandasaamy S, Mitchell A, Nuka G, Paysan-Lafosse T, Qureshi M, Raj S, Richardson L, Salazar GA, Williams L, Bork P, Bridge A, Gough J, Haft DH, Letunic I, Marchler-Bauer A, Mi H, Natale DA, Necci M, Orengo CA, Pandurangan AP, Rivoire C, Sigrist CJA, Sillitoe I, Thanki N, Thomas PD, Tosatto SCE, Wu CH, Bateman A, Finn RD (2021) The InterPro protein families and domains database: 20 years on. Nucleic Acids Res, 49(D1):D344–D354. https://doi.org/10.1093/nar/gkaa977

Budhlakoti N, Kushwaha AK, Rai A, Chaturvedi KK, Kumar A, Pradhan AK, Kumar U, Kumar RR, Juliana P, Mishra DC, Kumar S (2022) Genomic selection: a tool for accelerating the efficiency of molecular breeding for development of climate-resilient crops. Front Genet 13. https://doi.org/10.3389/fgene.2022.832153

Cadenas-Castrejón E, Verleyen J, Boukadida C, Díaz-González L, Taboada B (2022) Evaluation of tools for taxonomic classification of viruses. Brief Funct Genomics 22(1):31–41. https://doi.org/10.1093/bfgp/elac036

Cai L, Mei Y, Ye R, Deng Y, Zhang X, Hu Z, Zhou X, Zhang M, Yang J (2023) Tomato leaf curl New Delhi virus: an emerging plant begomovirus threatening cucurbit production. aBIOTECH 4(3):257–266. https://doi.org/10.1007/s42994-023-00118-4

Calle García J, Guadagno A, Paytuvi-Gallart A, Saera-Vila A, Amoroso CG, D'Esposito D, Andolfo G, Aiese Cigliano R, Sanseverino W, Ercolano MR (2021) PRGdb 4.0: an updated database dedicated to genes involved in plant disease resistance process. Nucleic Acids Res 50(D1):D1483–D1490. https://doi.org/10.1093/nar/gkab1087

Cook HV, Doncheva NT, Szklarczyk D, Von Mering C, Jensen LJ (2018) Viruses. STRING: a virus-host protein-protein interaction database. Viruses 10(10):519. https://doi.org/10.3390/v10100519

Crespo-Bellido A, Hoyer JS, Burgos-Amengual Y, Duffy S (2024) Phylogeographic analysis of Begomovirus coat and replication-associated proteins. J Gen Virol 105(10). https://doi.org/10.1099/jgv.0.002037

Duan Q, Lin Y (2024) Focus on vegetable crops. Plant Physiol 195(2):901–905. https://doi.org/10.1093/plphys/kiae246

Edgar RC (2004) MUSCLE: multiple sequence alignment with high accuracy and high throughput. Nucleic Acids Res 32(5):1792–1797. https://doi.org/10.1093/nar/gkh340

Finkel Y, Stern-Ginossar N, Schwartz M (2018) Viral short orfs and their possible functions. Proteomics 18(10). https://doi.org/10.1002/pmic.201700255

Fu P, Wu Y, Zhang Z, Qiu Y, Wang Y, Peng Y (2023) VIGA: a one-stop tool for eukaryotic virus identification and genome assembly from next-generation-sequencing data. Brief Bioinform 25(1). https://doi.org/10.1093/bib/bbad444

Glasa M, Hančinský R, Šoltys K, Predajňa L, Tomašechová J, Hauptvogel P, Mrkvová M, Mihálik D, Candresse T (2021) Molecular characterization of potato virus Y (PVY) using high-throughput sequencing: constraints on full genome reconstructions imposed by mixed infection involving recombinant PVY strains. Plants 10(4):753. https://doi.org/10.3390/plants10040753

Goldfarb T, Kodali VK, Pujar S, Brover V, Robbertse B, Farrell CM, Oh D-H, Astashyn A, Ermolaeva O, Haddad D, Hlavina W, Hoffman J, Jackson JD, Joardar VS, Kristensen D, Masterson P, McGarvey KM, McVeigh R, Mozes E et al (2024) NCBI RefSeq: reference sequence standards through 25 years of curation and annotation. Nucleic Acids Res 53(D1):D243–D257. https://doi.org/10.1093/nar/gkae1038

Gong C, Diao W (2025) Artificial intelligence in vegetable crops: recent advances and prospects. Vegetable Res 5(1):0–0. https://doi.org/10.48130/vegres-0025-0012

Gutnik D, Evseev P, Miroshnikov K, Shneider M (2023) Using alphafold predictions in viral research. Curr Issues Mol Biol 45(4):3705–3732. https://doi.org/10.3390/cimb45040240

He C, Zhang W, Hu X, Singh M, Xiong X, Nie X (2014) Molecular characterization of a Chinese isolate of potato virus A (PVA) and evidence of a genome recombination event between PVA variants at the 3′-proximal end of the genome. Arch Virol 159(9):2457–2462. https://doi.org/10.1007/s00705-014-2053-z

Hingamp P, van den Broek AE, Stoesser G, Baker W (1999) The EMBL nucleotide sequence database: contributing and accessing data. Mol Biotechnol 12(3):255–268. https://doi.org/10.1385/mb:12:3:255

Iksat N, Madirov A, Zhanassova K, Masalimov Z (2025) Artificial intelligence-assisted crispr/cas systems for targeting plant viruses. Genes 16(11):1258. https://doi.org/10.3390/genes16111258

Ismajli B, Galbács ZN, Takács AP, Várallyay É (2025) The first high-throughput sequencing-based study of viruses infecting solanaceous crops in Kosovo reveals multiple infections in peppers by six plant viruses. Plants 14(9):1273. https://doi.org/10.3390/plants14091273

Jumper J, Evans R, Pritzel A, Green T, Figurnov M, Ronneberger O, Tunyasuvunakool K, Bates R, Žídek A, Potapenko A, Bridgland A, Meyer C, Kohl SAA, Ballard AJ, Cowie A, Romera-Paredes B, Nikolov S, Jain R, Adler J, Back T, Petersen S, Reiman D, Clancy E, Zielinski M, Steinegger M, Pacholska M, Berghammer T, Bodenstein S, Silver D, Vinyals O, Senior AW,

Kavukcuoglu K, Kohli P, Hassabis D (2021) Highly accurate protein structure prediction with AlphaFold. Nat, 596(7873):583–589. https://doi.org/10.1038/s41586-021-03819-2

Katoh K (2002) MAFFT: a novel method for rapid multiple sequence alignment based on fast Fourier transform. Nucleic Acids Res 30(14):3059–3066. https://doi.org/10.1093/nar/gkf436

Katoh K, Standley DM (2016) A simple method to control over-alignment in the MAFFT multiple sequence alignment program. Bioinform, 32(13):1933–1942. https://doi.org/10.1093/bioinformatics/btw108

Kearse M, Moir R, Wilson A, Stones-Havas S, Cheung M, Sturrock S, Buxton S, Cooper A, Markowitz S, Duran C, Thierer T, Ashton B, Meintjes P, Drummond A (2012) Geneious basic: an integrated and extendable desktop software platform for the organization and analysis of sequence data. Bioinformatics 28(12):1647–1649. https://doi.org/10.1093/bioinformatics/bts199

Kerrien S, Alam-Faruque Y, Aranda B, Bancarz I, Bridge A, Derow C, Dimmer E, Feuermann M, Friedrichsen A, Huntley R, Kohler C, Khadake J, Leroy C, Liban A, Lieftink C, Montecchi-Palazzi L, Orchard S, Risse J, Robbe K et al (2007) IntAct--open source resource for molecular interaction data. Nucleic Acids Res 35(Database):D561–D565. https://doi.org/10.1093/nar/gkl958

Kim D, Song L, Breitwieser FP, Salzberg SL (2016) Centrifuge: rapid and sensitive classification of metagenomic sequences. Genome Res 26(12):1721–1729. https://doi.org/10.1101/gr.210641.116

Koczor Á, Ádám J, Ágoston J, Salánki K, Palkovics L (2024) Investigation of viral diseases of garlic (Allium sativum L.), new primers for RT-PCR detection, and diversity of garlic viruses in Hungary. Physiol Mol Plant Pathol 134:102394. https://doi.org/10.1016/j.pmpp.2024.102394

Kumar R, Tiwari RK, Lal MK, Majumdar SG, Swaraj K, Mall AK, Tripathi MK, Singh D, Singh G (2025a) Contemporary status of occurrence, spread, and yield loss due to tomato leaf curl New Delhi virus in global crop cultivation. In: Tomato leaf curl New Delhi virus (ToLCNDV) insights into virome dynamics and management. Springer, Singapore, pp 1–12. https://doi.org/10.1007/978-981-95-2852-3_1

Kumar R, Tiwari RK, Lal MK, Singh A, Jha J, Madival VA, Roy A, Singh G (2025b) Omics-based approaches in deciphering molecular mechanisms of tomato leaf curl New Delhi virus infection. In: Tomato leaf curl New Delhi virus (ToLCNDV) insights into virome dynamics and management. Springer, Singapore, pp 27–47. https://doi.org/10.1007/978-981-95-2852-3_3

Kwak M, Troiano E, Kil E-J, Parrella G (2024) High-throughput sequencing detected a virus–viroid complex in a single pokeweed plant. Front Plant Sci 15. https://doi.org/10.3389/fpls.2024.1435611

Lagzian A, Ghorbani A, Tabein S, Riseh RS (2024) Genetic variations and gene expression profiles of Rice Black-streaked dwarf virus (RBSDV) in different host plants and insect vectors: insights from RNA-Seq analysis. BMC Genomics 25(1). https://doi.org/10.1186/s12864-024-10649-9

Li P, Quan X, Jia G, Xiao J, Cloutier S, You FM (2016) RGAugury: a pipeline for genome-wide prediction of resistance gene analogs (RGAs) in plants. BMC Genomics 17(1). https://doi.org/10.1186/s12864-016-3197-x

Liu C-H, Di YP (2020) Analysis of RNA sequencing data using CLC genomics workbench. In: Methods in molecular biology.Springer US, pp 61–113. https://doi.org/10.1007/978-1-0716-0223-2_4

Liu Y, Ghaffari MH, Ma T, Tu Y (2024) Impact of database choice and confidence score on the performance of taxonomic classification using Kraken2. aBIOTECH 5(4):465–475. https://doi.org/10.1007/s42994-024-00178-0

Maina S, Donovan NJ, Plett K, Bogema D, Rodoni BC (2024) High-throughput sequencing for plant virology diagnostics and its potential in plant health certification. Front Hortic 3. https://doi.org/10.3389/fhort.2024.1388028

Manavi MA (2021) In silico study and virtual Drug Screening of the FDA-approved approach with Molecular Docking. Springer. https://doi.org/10.21203/rs.3.rs-433785/v1

Maree HJ, Fox A, Al Rwahnih M, Boonham N, Candresse T (2018) Application of HTS for routine plant virus diagnostics: state of the art and challenges. Front Plant Sci 9. https://doi.org/10.3389/fpls.2018.01082

Martin DP, Murrell B, Golden M, Khoosal A, Muhire B (2015) RDP4: detection and analysis of recombination patterns in virus genomes. Virus Evol 1(1). https://doi.org/10.1093/ve/vev003

Mehetre GT, Leo VV, Singh G, Sorokan A, Maksimov I, Yadav MK, Upadhyaya K, Hashem A, Alsaleh AN, Dawoud TM, Almaary KS, Singh BP (2021) Current developments and challenges in plant viral diagnostics: a systematic review. Viruses 13(3):412. https://doi.org/10.3390/v13030412

Melnick M, Gonzales P, LaRocca TJ, Song Y, Wuu J, Benatar M, Oskarsson B, Petrucelli L, Dowell RD, Link CD, Prudencio M (2021) Application of a bioinformatic pipeline to RNA-seq data identifies novel virus-like sequence in human blood. G3 Genes|Genomes|Genetics 11(9). https://doi.org/10.1093/g3journal/jkab141

Mihara T, Nishimura Y, Shimizu Y, Nishiyama H, Yoshikawa G, Uehara H, Hingamp P, Goto S, Ogata H (2016) Linking virus genomes with host taxonomy. Viruses 8(3):66. https://doi.org/10.3390/v8030066

Minh BQ, Schmidt HA, Chernomor O, Schrempf D, Woodhams MD, von Haeseler A, Lanfear R (2020) IQ-TREE 2: new models and efficient methods for phylogenetic inference in the genomic era. Mol Biol Evol 37(5):1530–1534. https://doi.org/10.1093/molbev/msaa015

Okonechnikov K, Golosova O, Fursov M (2012) Unipro UGENE: a unified bioinformatics toolkit. Bioinformatics 28(8):1166–1167. https://doi.org/10.1093/bioinformatics/bts091

O'Leary NA, Wright MW, Brister JR, Ciufo S, Haddad D, McVeigh R, Rajput B, Robbertse B, Smith-White B, Ako-Adjei D, Astashyn A, Badretdin A, Bao Y, Blinkova O, Brover V, Chetvernin V, Choi J, Cox E, Ermolaeva O, Farrell CM, Goldfarb T, Gupta T, Haft D, Hatcher E, Hlavina W, Joardar VS, Kodali VK, Li W, Maglott D, Masterson P, McGarvey KM, Murphy MR, O'Neill K, Pujar S, Rangwala SH, Rausch D, Riddick LD, Schoch C, Shkeda A, Storz SS, Sun H, Thibaud-Nissen F, Tolstoy I, Tully RE, Vatsan AR, Wallin C, Webb D, Wu W, Landrum MJ, Kimchi A, Tatusova T, DiCuccio M, Kitts P, Murphy TD, Pruitt KD (2016) Reference sequence (RefSeq) database at NCBI: current status, taxonomic expansion, and functional annotation. Nucleic Acids Res, 44(D1):D733–45. https://doi.org/10.1093/nar/gkv1189

Pooggin MM (2018) Small rna-omics for plant virus identification, virome reconstruction, and antiviral defense characterization. Front Microbiol 9. https://doi.org/10.3389/fmicb.2018.02779

Real N, Garcia-Molina A, Stolze SC, Harzen A, Nakagami H, Martín-Hernández AM (2025) Comprehensive proteomic profiling of Cucumber mosaic virus infection: identifying key proteins and pathways involved in resistance and susceptibility in melon. BMC Plant Biol 25(1). https://doi.org/10.1186/s12870-025-06464-3

Ritsch M, Cassman NA, Saghaei S, Marz M (2023) Navigating the landscape: a comprehensive review of current virus databases. Viruses, 15(9):1834. https://doi.org/10.3390/v15091834

Roux S, Matthijnssens J, Dutilh BE (2021) Metagenomics in virology. Encyclopedia Virol, 133–140. https://doi.org/10.1016/B978-0-12-809633-8.20957-6

Samson S, Lord É, Makarenkov V (2022) SimPlot++: a Python application for representing sequence similarity and detecting recombination. Bioinformatics 38(11):3118–3120. https://doi.org/10.1093/bioinformatics/btac287

Sanseverino W, Roma G, De Simone M, Faino L, Melito S, Stupka E, Frusciante L, Ercolano MR (2009) PRGdb: a bioinformatics platform for plant resistance gene analysis. Nucleic Acids Res 38(suppl_1):D814–D821. https://doi.org/10.1093/nar/gkp978

Sarkar M, Gupta D, Singh OW, Paul S, Kumar R, Mandal B, Roy A (2024) Identification of pathogenicity determinants in ToLCNDV and their RNAi-based knockdown for disease management in *Nicotiana benthamiana* and tomato plants. Front Microbiol 15:1481523. https://doi.org/10.3389/fmicb.2024.1481523

Schilbert HM, Rempel A, Pucker B (2020) Comparison of read mapping and variant calling tools for the analysis of plant NGS data. Cold Spring Harbor Laboratory. https://doi.org/10.1101/2020.03.10.986059

Seemann T (2014) Prokka: rapid prokaryotic genome annotation. Bioinform 30(14):2068–9. https://doi.org/10.1093/bioinformatics/btu153

Shean RC, Makhsous N, Stoddard GD, Lin MJ, Greninger AL (2019) VAPiD: a lightweight cross-platform viral annotation pipeline and identification tool to facilitate virus genome submissions to NCBI GenBank. BMC Bioinformatics 20(1). https://doi.org/10.1186/s12859-019-2606-y

Sievers F, Higgins DG (2018) Clustal omega for making accurate alignments of many protein sequences. Protein Sci, 27(1):135–145. https://doi.org/10.1002/pro.3290
Sundaresha S, Jeevalatha A, Kumar R, Sood S, Sharma S, Bhardwaj V, Singh BP, Chakrabarti SK (2022) RNA interference: a versatile tool to augment plant protection strategies in potato. In: Chakrabarti SK, Sharma S, Shah MA (eds) Sustainable management of potato pests and diseases. Springer, Singapore, pp 369–388. https://doi.org/10.1007/978-981-16-7695-6_15
Szklarczyk D, Kirsch R, Koutrouli M, Nastou K, Mehryary F, Hachilif R, Gable AL, Fang T, Doncheva NT, Pyysalo S, Bork P, Jensen LJ, von Mering C (2023) The STRING database in 2023: protein-protein association networks and functional enrichment analyses for any sequenced genome of interest. Nucleic Acids Res, 51(D1):D638–D646. https://doi.org/10.1093/nar/gkac1000
Tamura K, Peterson D, Peterson N, Stecher G, Nei M, Kumar S (2011) MEGA5: molecular evolutionary genetics analysis using maximum likelihood, evolutionary distance, and maximum parsimony methods. Mol Biol Evol 28(10):2731–2739. https://doi.org/10.1093/molbev/msr121
Tirnaz S, Zhang Y, Batley J (2020) Genome-wide mining of disease resistance gene analogs using conserved domains. In: Methods in molecular biology. Springer US, pp 365–375. https://doi.org/10.1007/978-1-0716-0235-5_20
Urban M, Cuzick A, Seager J, Wood V, Rutherford K, Venkatesh SY, De Silva N, Martinez MC, Pedro H, Yates AD, Hassani-Pak K, Hammond-Kosack KE (2019) PHI-base: the pathogen–host interactions database. Nucleic Acids Res. https://doi.org/10.1093/nar/gkz904
Vinodhini J, Rajendran L, Raveendran M, Rajasree V, Karthikeyan G (2020) Characterization of cucumber mosaic virus (CMV) subgroup IB infecting chilli in Tamil Nadu, India. 3 Biotech 10(11). https://doi.org/10.1007/s13205-020-02492-y
Wang J, Watson JL, Lisanza SL (2024) Protein design using structure-prediction networks: AlphaFold and rosettafold as protein structure foundation models. Cold Spring Harb Perspect Biol 16(7):a041472. https://doi.org/10.1101/cshperspect.a041472
Waterhouse A, Bertoni M, Bienert S, Studer G, Tauriello G, Gumienny R, Heer FT, de Beer TAP, Rempfer C, Bordoli L, Lepore R, Schwede T (2018) SWISS-MODEL: homology modelling of protein structures and complexes. Nucleic Acids Res, 46(W1):W296–W303. https://doi.org/10.1093/nar/gky427
Wood DE, Lu J, Langmead B (2019) Improved metagenomic analysis with Kraken 2. Genome Biol, 20(1):257. https://doi.org/10.1186/s13059-019-1891-0
Wylie SJ, Adams M, Chalam C, Kreuze J, López-Moya JJ, Ohshima K, Praveen S, Rabenstein F, Stenger D, Wang A, Zerbini FM (2017) ICTV virus taxonomy profile: Potyviridae. J Gen Virol 98(3):352–354. https://doi.org/10.1099/jgv.0.000740
Yadav RK, Tripathi MK, Tiwari S, Tripathi N, Asati R, Chauhan S, Tiwari PN, Payasi DK (2023) Genome editing and improvement of abiotic stress tolerance in crop plants. Life 13(7):1456. https://doi.org/10.3390/life13071456
Yang Z, Zeng X, Zhao Y, Chen R (2023) AlphaFold2 and its applications in the fields of biology and medicine. Signal Transduct Target Ther 8(1). https://doi.org/10.1038/s41392-023-01381-z
Zaagueri T, Miozzi L, Mnari-Hattab M, Noris E, Accotto GP, Vaira AM (2017) Deep sequencing data and infectivity assays indicate that chickpea chlorotic dwarf virus is the etiological agent of the hard fruit syndrome of watermelon. Viruses 9(11):311. https://doi.org/10.3390/v9110311
Zerbini FM, Simmonds P, Adriaenssens EM, Lefkowitz EJ, Oksanen HM, Alfenas-Zerbini P, Aylward FO, Freitas-Astúa J, Hughes HR, Łobocka M, Krupovic M, Kuhn JH, Mushegian A, Penzes JJ, Reyes A, Robertson DL, Roux S, Rubino L, Sabanadzovic S, Smith DB, Suzuki N, Turner D, Van Doorslaer K, Varsani A (2025) Virus species names have been standardized; virus names remain unchanged. mSphere, 10(5):e0002025. https://doi.org/10.1128/msphere.00020-25
Zheng Y, Gao S, Padmanabhan C, Li R, Galvez M, Gutierrez D, Fuentes S, Ling K-S, Kreuze J, Fei Z (2017) VirusDetect: an automated pipeline for efficient virus discovery using deep sequencing of small RNAs. Virology 500:130–138. https://doi.org/10.1016/j.virol.2016.10.017

# Host Plant Resistance: Breeding Strategies and QTL Mapping for Virus Resistance

# 20

R. L. R. Cabral, C. Gonzalez-Calquin, D. R. S. Jesus,
B. W. S. Lima, and A. S. Xavier

**Abstract**

Genetic resistance in host plants is one of the most effective and sustainable strategies for managing agricultural viruses. This chapter addresses the main mechanisms of antiviral resistance, from vertical defense, characterized by specificity and intensity, to polygenic and more durable resistance. Resistance mechanisms in plants, such as gene silencing by RNAi, susceptibility genes, and hormonal regulation in modulating the immune response, are also discussed. The application of biotechnological and molecular tools has expanded the possibilities for exploring these mechanisms. QTL mapping, molecular markers, and assisted selection have accelerated the identification and use of loci associated with resistance. Recent technologies, including RNA-seq, GWAS, and gene editing, broaden our understanding of the mechanisms underlying antiviral resistance and provide new opportunities for identifying target genes and developing more resilient cultivars. Despite these advances, problems persist, such as rapid viral evolution, the breakdown of resistance in cultivars, the impact of mixed infections, and technical limitations in perennial crops. Prospects include the

R. L. R. Cabral (✉) · B. W. S. Lima · A. S. Xavier
Department of Agronomy, Plant Pathology Graduate Program, Federal Rural University of Pernambuco, Recife, PE, Brazil
e-mail: brendo.silva@ufrpe.br; andre.sxavier@ufrpe.br

C. Gonzalez-Calquin
Center for Plant Molecular Biology, Faculty of Sciences, University of Chile, Ñuñoa, Santiago, Chile
e-mail: christian.gonzalez@usach.cl

D. R. S. Jesus
Department of Agronomy, Genetic and Plant Breeding Graduate Program, Federal University of Espírito Santo, Alegre, ES, Brazil
e-mail: deyse.silvino@ufrpe.brl

 
R. Kumar et al. (eds.), *Vegetable Crops and Viral Pathogens Volume I*, Plant Pathology for Sustainable Agriculture, https://doi.org/10.1007/978-981-95-7265-6_20

integration of omics data, high-throughput phenotyping, and bioinformatic tools, which can increase the accuracy and reduce the costs of breeding programs. The convergence of biotechnology, classical breeding, and digital agriculture promises to strengthen plant resilience and reduce losses associated with viruses. Therefore, understanding and exploring the multiple mechanisms of resistance remains essential to promoting food security and agricultural sustainability in a scenario of increasing pressure from viral pathogens.

**Keywords**

Agronomy · Phytopathology · Genetic mechanisms · Biotechnology · Molecular markers · Assisted selection · Plant responses · Omics

## 20.1 Introduction

World agricultural production faces a series of challenges related to diseases caused by microorganisms that negatively impact the yield and quality of crops. Historically, the study of resistance to these phytopathogens started empirically, observing that certain varieties or germplasm access showed lesser symptoms in field conditions (Kang et al. 2007). Among plant diseases, viruses are responsible for devastating losses, many times irreversible, estimated at more than US$30 billion per year around the world (Sastry and Zitter 2014). These pathogens affect not only basic food crops such as potato and cassava (Kumar et al. 2021) but also fruits and vegetables of high commercial value (Pallas et al. 2012; Zhou and Tzanetakis 2019). Contrary to bacteria and fungi, which can be partially controlled by antibiotics, fungicides, or specific bioinputs, viruses cannot be controlled by direct chemical control, making the intrinsic genetic resistance of the plant host the most efficient, sustainable, and affordable tool to minimize the effects of these infections (Trebicki 2020). In this context, viruses are considered as mandatory intracellular parasites, composed of genetic material (either DNA or RNA) enveloped by a protein layer and dependent on the cellular machinery of the host. Some conditions are needed for a plant virus to cause a negative impact on crops. Seventy-five percent of the dissemination of plant viruses is due to biological vectors such as whitefly (*Bemisia tabaci*), known for transmitting a high variety of plant diseases in different species (Ghosh and Ghanim 2021). These insects depend directly on two climatic aspects for their own propagation (Li et al. 2023). Furthermore, an infected plant needs to coexist with a noninfected plant within the reach of the biological vector, allowing its propagation (Trebicki 2020). Hence, interdisciplinary methods reinforce the fight against plagues and viral diseases that affect diverse crops around the world. After the entry of the virus into the cell, the host plant depends on specific molecular mechanisms to recognize and neutralize the virus. On the other hand, viruses face the effects of plant resistance in response to their infection, which includes physical, biochemical, and molecular changes. Plant resistance to viruses is not a simple nor unidimensional phenomenon. On the contrary, it is the result of a group of interlinked mechanisms that go from physical barriers that hinder the viral entry to highly specific molecular responses. Nevertheless, resistance could be qualitative, conferred by only one gene, or quantitative, when multiple loci contribute to different levels of tolerance or partial resistance. During the twentieth century, classic genetics allowed the identification of the first resistance genes (R), which confer qualitative resistance in some plant-virus interactions. Another discovered mechanism is viral RNA degradation by interference RNA (RNAi) (Das and Sherif 2020; Sundaresha et al. 2022). These discoveries were crucial for the development of crops capable of enduring specific viral infections (Goldbach et al. 2003). At the moment, the high genetic variability of viruses jeopardizes the durability of products from plant breeding programs. Evolutionary mechanisms developed by different strains from the same viral species can result in the loss of the acquired resistance, generating a deficit in crop yields. In this context, mapping of quantitative trait loci (QTLs) emerges as a powerful option, as it allows for identifying genomic regions associated with complex quantitative traits, unravelling a unique architecture of unique characteristics, and identifying candidate genes associated with total or partial virus resistance (Raj and Nadarajah 2022). This strategy is particularly useful for long-cycle crops or perennial crops, where traditional breeding involves substantial effort. An integration of genomic and transcriptomic data has broadened the possibilities, offering a systemic vision of virus-host interaction. This chapter addresses the resistance of host plants to viral infections, the mechanisms of the biological process, recent technological advances in resistance improvement, and breeding strategies focused on QTL mapping. Along the text, examples applied to different agronomic crops are discussed, as well as challenges and perspectives that shape the future of this area of knowledge.

## 20.2 Virus Resistance Mechanisms in Plants

### 20.2.1 Gene Silencing

Gene silencing in eukaryotes is a natural defense system that provides resistance against viral infections. This process is triggered, inside the infected cell, by the presence of viral double-stranded RNA. Dicer-like (DCL) proteins recognize these molecules and process them into virus-derived small interfering RNAs with a length between 21 and 24 nucleotides (Sundaresha et al. 2024). In turn, these small RNAs are loaded into the plant Argonaut (AGO) proteins. Then, the RNA-induced silencing complex (RISC) is assembled, which in turn will locate the targeted viral mRNA and degrade it, resulting in posttranscriptional silencing of the viral genes. A similar pathway is followed by microRNAs (miRNAs), which are transcribed by the host via DNA-dependent RNA polymerase (Fig. 20.1). When transcribed, a double-stranded hairpin is formed, named primary miRNA. These are processed by DCL and loaded into AGO. Interestingly, this complex can both degrade mRNA and also block translation of the targeted genes (Qu 2010; Jin et al. 2021; Akbar et al. 2022). Nevertheless, viruses were found capable of circumventing the described

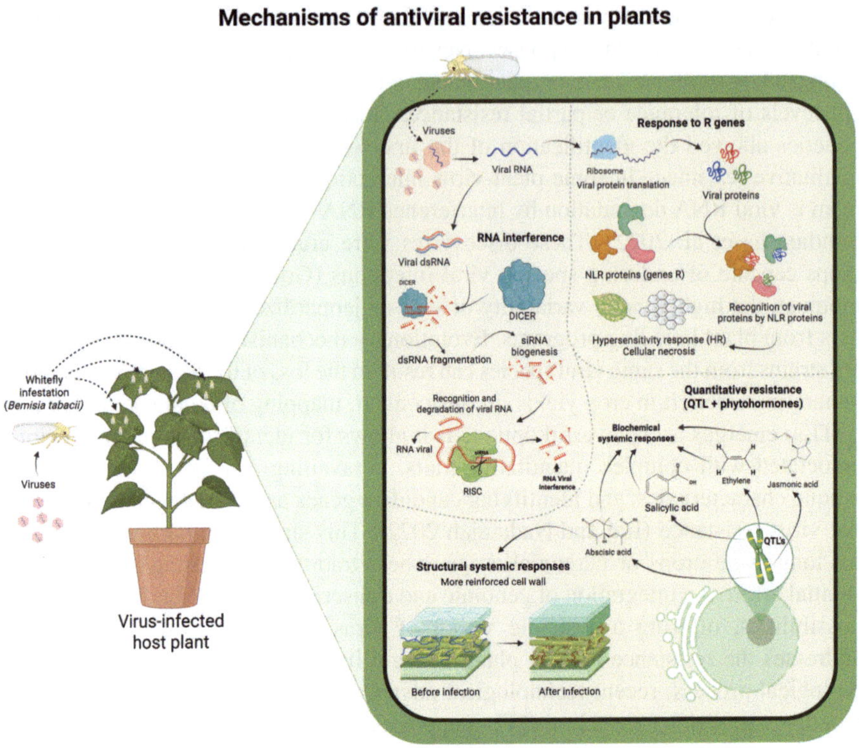


**Fig. 20.1** Illustrative model of the main mechanisms of antiviral resistance in plants. On the left, vector-mediated viral infection (Bemisia tabaci). Within the cell, three defense strategies are highlighted: (i) vertical resistance, mediated by R genes (NLR), which recognize specific viral proteins and trigger the hypersensitive response (HR); (ii) gene silencing by RNAi, in which double-stranded viral RNAs are processed by DCL into small siRNAs, subsequently guided by the RISC complex to degrade viral RNAs; and (iii) quantitative resistance, based on the combined action of multiple QTLs and regulation by phytohormones (SA, JA, ET, ABA), resulting in systemic biochemical responses and structural strengthening of the cell wall. These mechanisms, in a complementary way, contribute to limiting viral replication, reducing dissemination between cells and increasing the durability of resistance

mechanism by encoding viral suppressors of RNA silencing (VSR). These repressors are capable of modulating the expression of AGO/DCL, either by impeding their translation or by direct protein-protein interaction. For example, P19, encoded by tomato bushy stunt virus (TBSV), is able to directly bind siRNAs to prevent viral siRNAs from programming RISC (Lakatos et al. 2004). P21 of beet yellows virus (BYV) and HC-Pro of tobacco etch virus (TEV) impede the formation of the RISC-siRNA complex, similar to TBSV P19 (Chapman et al. 2004). On the other hand, CMV 2b and P0 of beet Western yellows virus (BWYV) repress the function of AGO through protein-protein interaction, changing its affinity and also impeding the formation of the RISC (Bortolamiol et al. 2007). Said mechanisms are only a few of the diverse ways in which plant viruses avoid the host immune system. Some viruses even encode various VSRs, such as citrus tristeza virus (CTV), whose genome is around 20 kilobases, with 11 open reading frames. The group of Lu et al. (2004) determined that the products p20 and the capsid protein prevented intercellular signaling of RNA silencing (Lu et al. 2004). Interestingly, some VSRs are multipurpose proteins: a 130 K protein encoded by the tomato mosaic virus (ToMV) can accumulate in the cytoplasm, showing silencing suppression activity. While in the membrane plasma, it is responsible for viral RNA replication (Hagiwara-Komoda et al. 2008).

### 20.2.2 Resistance Based on Susceptibility Genes

Plants have developed highly sophisticated defense mechanisms against pathogens, so successful colonization requires pathogens to overcome both constitutive barriers and induced defenses. This selective pressure forces pathogens to coevolve, promoting strong specialization in relation to their host (van Schie and Takken 2014). Susceptibility to pathogens is actively promoted by the plant itself, while its immunity can be suppressed by pathogen effectors (proteins that interfere with the host's cellular processes) and intrinsic immunity regulators, facilitating infection and disease development (Lapin and Van den Ackerveken 2013). All genes that favor infection and sustain compatibility with the pathogen are classified as susceptibility genes (S genes). Mutations or loss of function in these genes can restrict a pathogen's ability to cause disease. Unlike resistance genes (R genes), which act dominantly, resistance associated with the inactivation of S genes tends to be recessive (Garcia-Ruiz et al. 2021). Functionally, S genes facilitate viral infection at all stages of the virus cycle: virion disassembly, viral RNA translation, replication complex formation, genome replication, transcription, cell-to-cell movement, systemic spread, and assembly of new virions. A widely conserved example is the *eIF4* gene, which is necessary for potyvirus translation. Viral RNA translation depends on the recognition of VPg, a viral protein linked to the genome, by the eIF4 factor. This factor recognizes the 5′ cap of the viral RNA, a chemical structure at the 5′ end of messenger RNA that protects the RNA from degradation and facilitates the initiation of translation. Alterations in eIF4, including amino acid substitutions or loss-of-function mutations, reduce viral RNA translation, conferring recessive resistance to various viruses (Fig. 20.1). In crops lacking natural resistant alleles, mutated variants of eIF4 can be introduced via genome editing to generate resistant varieties (Garcia-Ruiz et al. 2021). S genes support infection through three main mechanisms: (a) facilitating basic compatibility and host recognition during prepenetration, (b) maintaining compatibility by providing essential nutrients for the pathogen, and (c) suppressing innate immunity. Disrupting an S gene can interrupt these processes and enhance plant resistance to specific diseases (Moniruzzaman et al. 2020). When S genes are mutated, they can confer specific resistance to a pathogen by preventing a compatible interaction, either due to failures in prepenetration requirements (host recognition, penetration) or postpenetration requirements (nutrient provision). Alternatively, mutations can induce broad-spectrum resistance by activating constitutive or prolonged defenses, although this may lead to side effects similar to autoimmunity (van Schie and Takken 2014). The modification of S genes constitutes an innovative approach in genetic breeding, taking advantage of genes that can play roles in both plant susceptibility and defense (Lorang et al. 2007). Genome editing techniques, such as CRISPR-Cas9, oligonucleotide-directed mutagenesis (ODM), TALE-type nucleases, and zinc-finger nucleases (ZF), allow for precise disruption of genes or promoter regions, making them a promising strategy for developing crops with durable resistance to pathogens, including viruses (Zaidi et al. 2018; Rani et al. 2024). The rise of these technologies enables precise manipulation of S genes, with the potential to generate plants resistant to different pathogens and produce transgene-free offspring (Syed et al. 2025; Chen et al. 2019). However, ethical and regulatory issues must be considered to ensure equitable access, environmental sustainability, and social acceptance (Rani et al. 2024).

### 20.2.3 Phytohormones

Phytohormones, traditionally recognized for their roles in plant growth and development, also play a central role in regulating antiviral responses (Islam et al. 2019). Viral infection can alter both the accumulation and signaling of hormones, affecting symptom expression and the efficiency of defense mechanisms. Among the hormones most directly linked to antiviral immunity are salicylic acid (SA), jasmonic acid (JA), ethylene (ET), and abscisic acid (ABA). Salicylic acid (SA) is considered a key regulator of systemic-acquired resistance (SAR), frequently activated by resistance (R) genes. Its signaling stimulates the production of reactive oxygen species, callose deposition, activation of pathogenesis-related proteins, and the initiation of the hypersensitive response (HR), thereby limiting viral multiplication and spread (Gao et al. 2015; Fu and Dong 2013). Consequently, SA forms the foundation of defense against a wide range of pathogens, including viruses. Jasmonic acid (JA), derived from α-linolenic acid, regulates growth, development, and responses to both biotic and abiotic stresses. Traditionally associated with defense against herbivores and necrotrophic pathogens, JA also participates in antiviral responses. In rice, infection by rice stripe virus (RSV) was found to induce the JA pathway via the coat protein (CP), activating defenses against the virus while simultaneously increasing the attractiveness of the plant to its vector, the small brown planthopper (SBPH), thereby facilitating viral spread (Han et al. 2020). This example highlights the dual role of JA in plant-virus interactions: promoting defense while paradoxically enhancing transmission. Ethylene (ET) functions synergistically or antagonistically with JA in various immune responses. In interactions with potyviruses, ET signaling has been shown to attract aphids to infected plants, promoting virus spread. Chemical or genetic inhibition of this pathway reduced viral dissemination by over 80% in laboratory experiments, although field conditions revealed more complex dynamics (Bak et al. 2019). These findings underscore the strategic role of ET in balancing antiviral defense and plant-vector interactions. Abscisic acid (ABA), classically associated with responses to abiotic stresses, also modulates antiviral immunity. Studies indicate that ABA acts at multiple levels: it enhances antiviral gene silencing, influences the microRNA pathway, and promotes callose deposition at plasmodesmata, restricting cell-to-cell viral movement (Alazem and Lin 2017). Unlike its sometimes ambiguous role in interactions with fungi and bacteria, ABA generally exerts protective effects against viral infections. Together, these findings demonstrate that phytohormones play decisive roles in plant-virus interactions, not only regulating resistance but also modulating plant attractiveness to vectors.

## 20.3 Breeding Strategies for Viral Resistance

### 20.3.1 Biotechnology and Assisted Breeding

Molecular markers represent key resources that identify genetic variations associated with virus resistance. These markers, such as RFLPs, SSRs, SNPs and AFLPs, facilitate mapping of resistance (R) genes and QTLs related to antiviral defense, providing accurate support for breeding programs (Şavkan et al. 2025). The advantage lies in its independence from environmental factors and its use in the early stages of the plant cycle, accelerating the selection of promising genotypes. In this field, marker-assisted selection (MAS) emerges, which integrates these resources into breeding practices, allowing the tracking of specific resistance genes or polygenic combinations with greater reliability, corroborating the phenotypic assessment. Thus, MAS reduces the time needed to develop resistant cultivars and increases the efficiency of crossings, being applied in conjunction with modern biotechnology to enhance virus management strategies in cultivated plants.

### 20.3.2 Transgeny and Cisgeny

Traditional breeding is an extensive and laborious process: it requires screening for wild parentals or resistant cultivars among hundreds of individuals. Faster and more efficient alternatives must be considered. The directed introduction of new DNA sequences is one of those alternatives. The study of plant genes and their interactions within their context is an area of knowledge that has been growing intensely since the development of NGS technology. Information about genes and their regulation has risen in the last few years. In the same line, the use of foreign DNA has been used extensively when testing resistant crops. For example, in 1986, Powell et al. generated transgenic tobacco plants harboring a genetic construction for expressing the coat protein from the tobacco mosaic virus (TMV) (Abel et al. 1986). This led to resistance to the TMV in said plants. Since then, similar approaches have been made in squash, potato, tomato, and other agronomically relevant crops (Tepfer 2002; Bhardwaj et al. 2019). Nevertheless, and since the introduction of these transgenic crops in the US market in 1990, rejection from the public, and concerns of genetic flow to wild relatives, which could turn into an ecological problem, other methodologies have to be explored (González-Calquín et al. 2022). An alternative that arose around 2005 was cisgenesis, which implies that the genetic material present in the final modified plant can only have its origin in plant species within the sexual compatibility group. Also, their regulatory sequences (i.e., promoters, terminators, introns) must come from the gene itself. The selection of genes and their regulatory sections is mainly targeted to resistance (R) genes. These genes make up a complex regulatory network that starts with the pathogen infection, which is recognized by membrane-bound receptors. These receptors then start a signaling cascade that ends up regulating the expression of said R genes, activating elicitor-triggered immunity (ETI). Some pathogens express enzymes or toxins that bind directly to these effectors, blocking the ETI response. The hierarchy in which each gene responds needs to be explored by means of NGS, genome-wide studies, RNA sequencing, and metabolomics, mainly targeted to individuals sexually compatible with the candidate crop to improve (Kushalappa et al. 2016). Resistance to virus infection in plants is generally related to SNPs and natural variations within their genome, and these differences are a wide source of functional R genes to evaluate. Some advances have been made in *Prunus*: *P. davidiana* clone P1908, a close relative of peach, presents a complex polygenic plum pox virus (PPV) resistance, involving at least six QTLs, three of which colocalize in genomic regions where gene copies of eukaryotic translation initiation factors 4 (*eIF4E*, *eIF(iso)*) are present (Marandel et al. 2009). The product of these genes interacts with the 5′ cap of mRNA, and experiments carried out by Wittmann et al. (1997) demonstrated that point mutations in said genes diminished symptoms drastically. Hence, the introduction of those genes into economically relevant plum cultivars could be a source of resistance. Nevertheless, many of these options are being considered at the moment and no cisgenic crops have been released yet.

### 20.3.3 Gene Editing and the Use of CRISPR/Cas9

The use of the CRISPR/Cas9 system in plant breeding has improved the specificity and efficiency of the process. This system is based on the design of a guide RNA (gRNA), a specific sequence complementary to the site targeted for editing (Butt and Bastas 2025). Adjacent to this gRNA, a protospacer adjacent motif (PAM), consisting of 2–5 nucleotides, needs to be present for the nuclease activity of Cas9. The sequence of PAMs varies depending on the species from which the Cas9 is obtained (Mojica et al. 2009). When the CRISPR/Cas9 complex is assembled, a cut on the double strand of DNA occurs, frequently three base pairs upstream of the protospacer. This leads to two possible scenarios: (1) the cell will quickly join both ends of the DNA double helix in a process named nonhomologous end joining (NHEJ), which is prone to adding or deleting nucleotides, changing the open reading frame and potentially producing truncated, dysfunctional versions of the protein (Jeyaraj et al. 2024), and (2) on the other hand, the cell can repair the double-stranded break (DSB) by means of homologous-directed repair, where template DNA in proximity to the DSB is used to complete the missing nucleotides (Chen et al. 2022). Both events can be utilized as a tool for accelerating the generation of new varieties that can resist viral infections, but NHEJ has been preferred as HDR is very inefficient in plants. For example, a host protein needed for the viral infection can be knocked out, preventing the virus from replicating and generating symptoms. This has been achieved in various species, such as potato, soybean, and tomato (among others), targeting the *elf4* gene and its isoforms, as this factor plays a crucial role in initiating protein translation by recognizing and interacting with mRNA cap structures and ribosome (Zhang et al. 2020; Kumar et al. 2022; Chavan et al. 2025). The disruption of such a family of genes has conferred resistance to the rice tungro spherical virus (RTSV) in *Oryza sativa* (Macovei et al. 2018), resistance to potato virus Y (PVY) (Kumar et al. 2022), and reduced tomato yellow leaf curl virus (TYLCV) accumulation in *Solanum lycopersicum* (Pramanik et al. 2021). CRISPR/Cas9 technology has advanced further, making use of the inactive Cas9 nuclease domain, which harnesses the specificity of the system to locate specific DNA sequences, but as the nuclease activity is missing, the system stays in place, blocking transcription (Boettcher and McManus 2015). This approach has been used to confer cotton leaf curl virus (CLCuV) resistance in tobacco plants, targeting the promoter region of the *AC1* gene, essential for virus replication (Khan et al. 2019).

## 20.4 QTL Mapping for Virus Resistance

The mapping of QTL (quantitative trait loci) consists of the identification of gene groups associated with the same phenotype. In other words, characteristics are controlled by multiple genes and modulated by the environment. The use of QTL mapping began around the 1980s, when the first genetic maps based on molecular markers were developed (Botstein et al. 1980). The construction of gene linkage maps and the use of markers like RFLP and SSRs allowed the association of phenotypic traits of interest with their corresponding loci. With the development of SNP arrays and NGS, the resolution of mapping improved, allowing to locate of specific genomic regions related to phytopathogen defense responses. These loci explain the variability of resistance observed in different germplasm of the same species, which exhibit partial or total tolerance. These regions are responsible for controlling the reduction of the viral replication, limiting the movement of the virus within the plant tissue and modulating the immune response of the host (Monnot et al. 2021). As the major resistance genes can confer qualitative immune responses, QTLs are fundamental for more subtle traits but as important in breeding programs. The use of segregating populations is an essential component of QTL mapping, like F2 populations or backcrossing (BC) of recombinant inbred lines (RILs). These populations explore genetic segregation at different levels, allowing us to find a correlation between phenotype and genotype. Each one of these populations presents advantages and limitations, and the choice is made upon the main objective of the study, the cultivated species, and the availability of resources. F2 populations are derived from the crossing of two contrasting lines and are one of the simplest ways of studying the heritage of resistance, even though they present high heterogeneity. BC lines are especially useful when the objective is the introgression of resistance genes from wild relatives to agronomically relevant crops, maintaining the maximum of the genetic background of the elite material. RILs, obtained after successive generations of self-crossing, have a major genetic stability, being widely used for high-resolution QTL analyses. These efforts have been used successfully in crops like pepper (Guo et al. 2017), cotton (Zhang et al. 2015), and peanuts (Khera et al. 2016). Subsequently, a precise evaluation of symptom severity, the quantity of virus present, and the pathogen's dissemination rate is carried out, characterizing the phenotype. Genetic data involve molecular markers distributed throughout the genome, such as SSRs or SNPs. Lastly, statistical analysis combines phenotypic and genotypic information to identify genomic regions significantly linked to resistance. Beyond identification, the validation of QTLs is indispensable. Robust QTLs must be consistent across different environments and populations. The integration of transcriptomic and proteomic data within these genomic regions allows for the identification of candidate genes and an understanding of the molecular mechanisms underlying resistance. Large-scale association studies, known as genome-wide association studies (GWAS), complement traditional QTL mapping methods. GWAS uses natural diversity panels, exploring the historical recombination accumulated in natural populations to identify loci associated with the resistance trait at high resolution (He and Gai 2023). Thus, the combination of classical mapping, functional validation, and modern approaches like GWAS bridges genetics and genomics, strengthening the foundation for more precise and sustainable plant breeding programs.

### 20.4.1 Genotyping Technologies

#### 20.4.1.1 SNP Arrays

SNP arrays represented a significant milestone in modern genetics, offering efficient, large-scale genotyping with high reproducibility. The basic principle involves the hybridization of fragmented, fluorescently labeled genomic DNA to a glass or silica chip containing thousands to millions of probes specific for SNPs, previously cataloged in the species' genome. During hybridization, DNA molecules bind to complementary probes, and the intensity of the generated fluorescent signal is detected by optical readers, enabling the identification of which allele is present at each genomic position (Ganal et al. 2012), thereby accelerating marker-assisted selection. In practice, SNP arrays are extremely useful for species with well-characterized reference genomes, such as rice, corn, and soybean (Eckardt 2000; Haberer et al. 2005). In QTL mapping programs for virus resistance, these analyses facilitate the identification of genomic regions related to phenotypes with great statistical robustness. For example, in tomato (*Solanum pimpinellifolium*), Víquez-Zamora et al. (2014) detected high-density arrays, aiding in delimiting loci associated with resistance to tomato yellow leaf curl virus (TYLCV). However, despite their precision, arrays have limitations. Probes need to be designed beforehand, and therefore, SNP arrays do not enable the discovery of new variants; they only recognize already known polymorphisms. Furthermore, their cost per sample can still be relatively high in large-scale studies, and their efficiency strongly depends on the quality of the reference genome. Consequently, in "orphan" or understudied crops, the use of arrays may be limited compared to other approaches.

#### 20.4.1.2 Genotyping by Sequencing (GBS)

Genotyping by sequencing (GBS) is a high-throughput technique that has emerged as a cost-effective and flexible alternative to SNP array platforms. The method involves digesting genomic DNA with specific restriction enzymes, which reduces genome complexity, followed by adapter ligation and sequencing of the fragments on next-generation sequencing (NGS) platforms. This strategy bypasses the need to sequence the entire genome, enabling the identification and genotyping of thousands of broadly distributed SNPs (Deol et al. 2025). Effectively, after digestion and adapter addition, libraries are multiplexed and sequenced in parallel, generating raw data that undergoes bioinformatics pipelines for SNP calling. These SNPs can then be used in QTL mapping studies or GWAS analyses. The significant advantage of GBS lies on its applicability to species with large, complex, or poorly characterized genomes, where well-established SNP arrays do not yet exist. In the context of viral resistance, GBS has been applied to horticultural and fruit crops. A notable example is the mapping of QTLs involved in resistance to yellow mosaic disease in bitter melon (*Momordica charantia* L.), caused by a variety of viral pathogens. In this study, Kaur et al. (2021) demonstrate a genetic linkage map with 3144 SNP markers, validating the identity of three QTLs located on chromosomes 3, 4, and 5, which conferred resistance to the disease in segregating populations. This case exemplifies how GBS can provide a large volume of informative markers even in crops where genetic resources are still limited.

#### 20.4.1.3 Next-Generation Sequencing (NGS)

Unlike the Sanger method, which sequences individual DNA fragments with low throughput, NGS enables the massive and parallel sequencing of millions of fragments in a single assay, exponentially increasing the amount of information obtained and reducing costs per sequenced base. This technological revolution has allowed genomic mapping, previously restricted to a few study models, to be applied on a large scale to cultivated plants, including those with complex and poorly studied genomes. The most well-known platforms, such as Illumina, HiSeq 2000, and PacBio, describe distinct approaches, varying in read length, accuracy, and cost (Shukla and Yadav 2023). In the search for resistance to viral phytopathogens, the use of NGS reveals high-resolution genomic variations, refining the construction of genetic maps and SNPs, which function as markers for QTL mapping. This marker density is essential for dissecting immune response traits, such as viral resistance, which generally involve multiple genes and epistatic interactions. Despite its advantages, the utilization of NGS still faces challenges. The cost per sample, although reduced, can still be a limitation in less-funded breeding programs. Furthermore, data analyses require specialized bioinformatics pipelines capable of handling large data volumes and correctly interpreting genomic variants in highly repetitive regions, commonly found in plants. Another challenge is translating SNP identification into functional information, as not every detected variation is directly related to resistance (Kayihan et al. 2023). In summary, NGS has consolidated itself as the primary technology for genomic studies applied to QTL mapping, as it has boosted significant advances in the understanding of virus resistance in crops. Its integration with other approaches, such as transcriptomics and proteomics, tends to accelerate the identification of genes of interest and their practical application in breeding programs through marker-assisted selection and gene editing (Mushtaq et al. 2021). These technologies are not mutually exclusive but complementary. While SNP arrays offer robustness and standardization, GBS provides flexibility in less-studied species, and NGS delivers very high-resolution data. The choice depends on the study's objective, financial resource availability, and the species under analysis. In all cases, modern genotyping has established itself as an indispensable tool for connecting genetic variation with resistance phenotypes, providing key information for plant breeding.

### 20.4.2 QTL-Seq and GWAS Applied to Viral Resistance

With the advent of genotyping technologies and bioinformatics, innovative methodologies have emerged for identifying loci associated with resistance to viral phytopathogens. Among the most utilized are QTL sequencing (QTL-seq) and genome-wide association studies (GWAS). Both applications rely on NGS data but follow different logics: while QTL-seq is based on biparental populations and the contrast between extreme phenotypes, GWAS explores the natural diversity accumulated in germplasm collections. QTL-seq emerged as a rapid and cost-effective approach for mapping genomic regions of complex traits. The principle is straightforward: individuals from a segregating population (such as $F_2$ or RILs) are evaluated for a trait of interest, and two contrasting groups are formed—resistant and susceptible. The DNA from each group is sequenced together, generating two datasets. Comparing the allelic frequency between these collections highlights genomic regions in disequilibrium, suggesting the presence of a QTL (Wang and Wang 2023). GWAS, on the other hand, uses a different reasoning: it employs natural diversity panels or cultivar collections. The method is based on the statistical association between genetic variants (like SNPs) and the phenotype of interest, considering the historical recombination that has occurred over many generations. This ensures much finer resolution, often narrowing down regions to a few genes. However, GWAS is sensitive to population structure and requires large volumes of well-characterized data (Altaf et al. 2024). The value of integrating QTL-seq and GWAS analyses simultaneously is justified by their complementary qualities. While QTL-seq is agile and direct, GWAS offers higher resolution and applicability in breeding contexts. This integration strengthens the reliability of results and accelerates the incorporation of QTLs into marker-assisted selection programs, reducing the interval between discovery and practical application (Tyagi et al. 2024). Thus, studies including QTL-seq and GWAS are establishing themselves as central methodologies in the search for resistance to viral phytopathogens, enhancing the precision and applicability of results and efficiently connecting genomic research to plant breeding (Table 20.1).

**Table 20.1** Map of QTLs and genes related to virus resistance in different cultivars

| Cultivate | Virus target | Gen/QTL identified | Type of resistance | References |
|---|---|---|---|---|
| Tomato (*Solanum lycopersicum*) | Tomato mottle virus bipartido (ToMoV) | Ty-1 (chr 6) and Ty-3 (chr 6) | Partial to complete | Ji et al. (2007) |
| | Tomato yellow leaf curl virus (TYLCV) | | | |
| Cucumber (*Cucumis sativus*) | Cucumber mosaic virus (CMV) | *RDR1* (chr 5) | Partial | Monnot et al. (2022) |
| | Cucumber vein yellowing virus (CVYV) | | | |
| | Cucumber green mottle mosaic virus (CGMMV) | | | |
| | Watermelon mosaic virus (WMV) | | | |
| | Zucchini yellow mosaic virus (ZYMV) | *VPS4* (chr 6) | | |
| | Papaya ringspot virus (PRSV) | | | |
| Peanut (*Arachis hypogaea*) | Tomato Spotted Wilt Virus (TSWV) | LG'A01' and LG'A09' | Partial | Khera et al. (2016) |
| Maize (*Zea mays*) | Maize streak virus (MSV) | Fourteen meta-QTLs present on chromosomes 1, 3, and 5 were identified, which participate in resistance. | Partial—The number of genes involved justifies this variation in resistance | Gupta et al. (2025) |
| | Mal de río cuarto virus (MRCV) | | | |
| | Sugarcane mosaic virus (SCMV) | | | |
| | Maize stripe virus disease (MSD) | | | |
| | Barley yellow dwarf virus (BYDV) | | | |
| | Foxtail mosaic virus (FoMV) | | | |
| | Maize chlorotic dwarf virus (MCDV) | | | |
| | Maize chlorotic mottle virus (MCMV) | | | |
| | Maize mosaic virus (MMV) | | | |
| | Maize lethal necrosis (MLN) | | | |
| | Maize rayado fino virus (MRFV) | | | |
| Sugarcane (*Saccharum* spp.) | Sugarcane mosaic virus (SCMV) | Seven QTLs were identified | Partial | Dhaliwal et al. (2022) |
| | Sorghum mosaic virus (SrMV) | | | |
| Black gram (*Vigna mungo*) | Mung bean yellow mosaic India virus (MYMIV) | qMYMIV6.1.1 (chr 6) | Complete | Lu et al. (2023) |

## 20.5 Integration of Omic Data in the Study of Resistance

### 20.5.1 Genomics

Genomics, the study of an organism's whole genetic material, is crucial for understanding how plants develop resistance to pathogens. This field studies the structure, function, and interactions of genes and the entire genome. Insights into the organization, diversity, and adaptability of plant genomes form the foundation for comprehending resistance mechanisms. The increasing availability of genomes from cultivated plants and their wild relatives facilitates comparative studies, revealing genetic variations linked to resistance. In fruit and horticultural species, genomic analyses have been vital for mapping chromosomal regions containing resistance genes (R genes) or quantitative trait loci (QTLs) associated with susceptibility factors (Argyris et al. 2015). One example is the comparison between susceptible and resistant cultivars of bell pepper (*Capsicum annuum* L.) to Tobacco mosaic virus (TMV). Şavkan et al. (2025) analyzed 120 plant lines and identified the *L4* allele and *Tsw* gene conferring disease resistance. Similarly, Tomás et al. (2011) observed in wild tomato species (*Solanum habrochaites*) that resistance to begomoviruses stems from gene variants absent in commercial cultivars. These examples highlight the role of comparative genomics in introgressing resistance genes. Furthermore, next-generation sequencing (NGS) technologies have enabled the assembly of high-quality reference genomes and the identification of structural variations. These variations, such as insertions, deletions, and gene duplications, significantly influence the plant's immune response. In fruit crops like papaya, genomic applications have aided in identifying loci associated with resistance to RNA viruses, providing valuable resources for marker-assisted selection and gene editing strategies (Jyotika et al. 2024).

### 20.5.2 Transcriptomics

Transcriptomics is dedicated to studying the complete set of RNAs expressed in a cell, tissue, or organism under specific conditions. Because viruses reprogram gene expression to favor their own replication and spread, this transcriptomic data is particularly crucial for understanding resistance. Providing a dynamic window, transcriptomics reveals how genes are regulated during a plant's physiological processes in response to environmental and biotic stimuli, such as viral infections. It identifies which genes are active, repressed, or differentially expressed in response to viral infection, thereby enriching the genetic potential data provided by genomics. Traditionally, early transcriptomic studies utilized techniques like microarray hybridization, displaying large-scale expression map profiles (Aharoni and Vorst 2002). With advancements in modern genetics, sequencing-based technologies like SAGE and RNA-seq emerged, utilizing Sanger or NGS sequencing. Among these, RNA-seq offers superior data coverage and resolution compared to other approaches (Zanardo et al. 2019). RNA-seq enables massive transcript sequencing, creating unique experimental alternatives. These include analyses comparing infected plants versus uninfected controls or contrasting resistant and susceptible plant genotypes against viral pathogens. This type of analysis has been applied to various crops. For instance, He et al. (2024) identified 2275 differentially expressed genes (DEGs) in zucchini *(Cucurbita pepo)* infected with squash leaf curl China virus (SLCCNV), with 1310 genes showing positive regulation and 965 negative regulation. Many of these studies have identified genes linked to the activation of defense pathways, such as those related to salicylic acid and jasmonate signaling, as well as pathogenesis-related (PR) genes (Li et al. 2018; Song et al. 2022). Furthermore, RNA-seq has facilitated the discovery of nonprotein-coding RNAs, such as microRNAs and siRNAs, which play critical roles in the posttranscriptional regulation of antiviral responses. Another significant aspect is its capacity to identify quantitative resistance genes, which may not involve a single major-effect gene but rather a network of genes modulating the plant's response. Therefore, transcriptomics goes beyond merely describing activated or repressed genes; it can reveal the strategies viruses use to manipulate the host, the plant's vulnerabilities, and robust defense mechanisms that can be leveraged in breeding or gene editing.

### 20.5.3 Proteomics and Metabolomics

In the last 30 years, proteomics and metabolomics have emerged as powerful tools for understanding the complexity of biological systems. While proteomics aims to map and quantify the expression of a total set of proteins in a cell or tissue under a specific state (Van Wijk 2001), metabolomics is dedicated to the characterization of metabolites, the final products of biochemical pathways that reflect cellular activity in real time (Hong et al. 2016). These approaches arose from genome sequencing and the evolution of systems biology, providing in-depth insights into cellular function and adaptive responses to stresses, such as viral diseases. In this context, proteomics and metabolomics become crucial for understanding antiviral responses at a functional level in plants. In the realm of plant proteomics, methodologies like two-dimensional electrophoresis, mass spectrometry, and high-resolution chromatography have facilitated the investigation of thousands of proteins simultaneously. However, proteins that constitute the defense line began to be highlighted through analyses of specific proteins, such as pathogenesis-related (PR) proteins. The direct identification of defense proteins and novel molecules involved in plant-virus interactions is essential for the development of improved crops (Liu et al. 2019). Recent studies show that the integration of proteomics and metabolomics provides unique insights. By evaluating proteomic profiles, Real et al. (2025) found that melon plants resistant and susceptible to cucumber mosaic virus (CMV) remodel their biological response pathways differently during various stages of infection. In parallel, metabolomics contributes by exploring how viral infection alters the profile of primary and secondary metabolites, including sugars, amino acids, and phenolic compounds. These changes are reflected in the plant's metabolic reprogramming during biotic stress. Many of these compounds directly act as antiviral barriers, hindering viral replication or reinforcing cell walls. Techniques such as nuclear magnetic resonance (NMR), liquid chromatography-mass spectrometry (LC-MS), and gas chromatography-mass spectrometry (GC-MS) enable the detection of small variations in the metabolic profile, revealing biomarkers associated with resistance or susceptibility (Sharma et al. 2021). Siriwan et al. (2023) demonstrated that the metabolic profile of cassava plants infected with Sri Lankan cassava mosaic virus (SLCMV) had higher abundance values than those of uninfected samples. Therefore, the practical application of this knowledge in plant breeding is promising. Proteins and metabolites classified as resistance markers represent an excellent resource for marker-assisted selection programs, accelerating the choice of more resistant genotypes. In addition, metabolic manipulation through gene editing or transgenesis opens new perspectives for developing cultivars with a greater capacity for natural defense, reducing reliance on chemical pesticides and increasing the sustainability of agricultural production. By integrating these approaches into plant breeding, researchers can identify new targets for resistance and transform basic knowledge into practical solutions for agriculture.

### 20.5.4 Bioinformatics

Faced with the massive volume of data generated by genomic, transcriptomic, proteomic, and metabolomic studies, bioinformatics has emerged as an interdisciplinary field integrating biology, mathematics, and computer science. Since the first genetic databases were developed in the 1980s and 1990s, bioinformatics has become established, generating robust tools for large-scale analysis, essential for understanding complex biological processes, including plant-virus interactions (Marsh et al. 2021). An emerging aspect is the use of integrated plant resistance databases, which facilitate the comparison of omics data across different plant species and viruses, aiding in the identification of conserved patterns. Furthermore, multiple bioinformatic programs have been developed and applied. Among these, tools for read mapping, genome assembly, functional annotation, gene interaction networks, and metabolic pathway analysis facilitate the transformation of raw data into applicable biological knowledge. For sequence annotation and comparison, programs like *BLAST (NCBI)* and *InterProScan* assist in identifying candidate genes and functional domains associated with resistance (Zainal-Abidin et al. 2021). Alternatively, QTL mapping software, such as MAPtools, correlates molecular markers with resistance phenotypes (Martínez-Guardiola et al. 2024). In gene expression analysis, platforms like Galaxy provide pipelines for RNA-seq (Santos et al. 2024), while resources like *edgeR* and *DESeq2* help identify differentially expressed genes in infected plants (Thawng and Smith 2022). For studying defense networks, databases like *STRING* and *Cytoscape* enable the visualization of protein-protein interactions, revealing critical hubs of antiviral response (Karimipour et al. 2025). In summary, bioinformatics is revolutionizing plant breeding, serving as the link between massive data and practical applications, enabling rapid, reproducible, and large-scale analyses. By integrating information from genomics, transcriptomics, proteomics, and metabolomics, these tools accelerate the discovery of resistance genes and guide breeding strategies. Thus, bioinformatics is establishing itself as an indispensable component for the development of virus-resilient cultivars.

## 20.6 Conclusions

Genetic resistance emerges as the most efficient and sustainable strategy for managing viruses in cultivated plants. The integration of different mechanisms such as R genes, gene silencing, quantitative resistance, and manipulation of susceptibility genes provides a solid basis for developing more resilient cultivars. Advances in genotyping, gene editing, and omics data integration technologies expand the possibilities for breeding programs, making them more precise and faster. However, the effectiveness of these strategies depends on the combination with integrated management practices and constant adaptation to new viral variants. Therefore, virus resistance must be viewed as a dynamic process, which requires continuous innovation to ensure food security and agricultural sustainability.

## References

Abel PP, Nelson RS, De B, Hoffmann N, Rogers SG, Fraley RT, Beachy RN (1986) Delay of disease development in transgenic plants that express the tobacco mosaic virus coat protein gene. Science 232(4751):738–743

Aharoni A, Vorst O (2002) DNA microarrays for functional plant genomics. Plant Mol Biol 48(1):99–118

Akbar S, Wei Y, Zhang MQ (2022) RNA interference: promising approach to combat plant viruses. Int J Mol Sci 23(10):5312

Alazem M, Lin NS (2017) Antiviral roles of abscisic acid in plants. Front Plant Sci 8:1760

Altaf MT, Tatar M, Ali A, Liaqat W, Mortazvi P, Kayihan C et al (2024) Advancements in QTL mapping and GWAS application in plant improvement. Turk J Bot 48(7):376–426

Argyris JM, Pujol M, Martín-Hernández AM, Garcia-Mas J (2015) Combined use of genetic and genomics resources to understand virus resistance and fruit quality traits in melon. Physiol Plant 155(1):4–11

Bak A, Patton MF, Perilla-Henao LM, Aegerter BJ, Casteel CL (2019) Ethylene signaling mediates potyvirus spread by aphid vectors. Oecologia 190(1):139–148

Bhardwaj V, Sood S, Kumar A, Vanishree G, Sharma S, Sundaresha S, Raigond B, Kumar R, Bairwa A, Lal M, Chakrabarti SK (2019) Efficiency and reliability of marker assisted selection for resistance to major biotic stresses in potato. Potato J 46:56–66

Boettcher M, McManus MT (2015) Choosing the right tool for the job: RNAi, TALEN, or CRISPR. Mol Cell 58(4):575–585

Bortolamiol D, Pazhouhandeh M, Marrocco K, Genschik P, Ziegler-Graff V (2007) The Polerovirus F box protein P0 targets ARGONAUTE1 to suppress RNA silencing. Curr Biol 17(18):1615–1621

Botstein D, White RL, Skolnick M, Davis RW (1980) Construction of a genetic linkage map in man using restriction fragment length polymorphisms. Am J Hum Genet 32(3):314

Butt H, Bastas KK (2025) CRISPR/Cas9-based genome engineering in plants for enhancing disease resistance. In: Microbial biocontrol techniques: importance in ensuring food security. Springer Nature Singapore, Singapore, pp 143–154

Chapman EJ, Prokhnevsky AI, Gopinath K, Dolja VV, Carrington JC (2004) Viral RNA silencing suppressors inhibit the microRNA pathway at an intermediate step. Genes Dev 18(10):1179–1186

Chavan DD, Sarkar M, Majumdar A, Mondal F, Babu YM, Lal SK, Mandal B, Kumar R, Roy A (2025) Enhanced expression and interaction of GmRDR1 and GmSGS3 proteins in resistant soybean cultivars synergistically regulate antiviral defense against mungbean yellow mosaic India virus. Plant Biol. https://doi.org/10.1111/plb.70078

Chen K, Wang Y, Zhang R, Zhang H, Gao C (2019) CRISPR/Cas genome editing and precision plant breeding in agriculture. Annu Rev Plant Biol 70(1):667–697

Chen J, Li S, He Y, Li J, Xia L (2022) An update on precision genome editing by homology-directed repair in plants. Plant Physiol 188(4):1780–1794

Das PR, Sherif SM (2020) Application of exogenous dsRNAs-induced RNAi in agriculture: challenges and triumphs. Front Plant Sci 11:946

Deol JK, Ramekar S, Dutt M (2025) The potential value of genotyping by sequencing (GBS) strategies for combating citrus diseases. Euphytica 221(5):1–21

Eckardt NA (2000) Sequencing the rice genome. Plant Cell 12:11

Fu ZQ, Dong X (2013) Systemic acquired resistance: turning local infection into global defense. Annu Rev Plant Biol 64(1):839–863

Ganal MW, Polley A, Graner EM, Plieske J, Wieseke R, Luerssen H, Durstewitz G (2012) Large SNP arrays for genotyping in crop plants. J Biosci 37(5):821–828

Gao QM, Zhu S, Kachroo P, Kachroo A (2015) Signal regulators of systemic acquired resistance. Front Plant Sci 6:228

Garcia-Ruiz H, Szurek B, Van den Ackerveken G (2021) Stop helping pathogens: engineering plant susceptibility genes for durable resistance. Curr Opin Biotechnol 70:187–195

Dhaliwal SK, Gill RK, Sharma A, Kaur A, Bhatia D, & Kaur S (2022) A large-effect QTL introgressed from ricebean imparts resistance to Mungbean yellow mosaic India virus in blackgram (Vigna mungo (L.) Hepper), Theoretical and Applied Genetics, 135(12):4495–4506. https://doi.org/10.1007/s00122-022-04234-5

Ghosh S, Ghanim M (2021) Factors determining transmission of persistent viruses by Bemisia tabaci and emergence of new virus–vector relationships. Viruses 13(9):1808

Goldbach R, Bucher E, Prins M (2003) Resistance mechanisms to plant viruses: an overview. Virus Res 92(2):207–212

González-Calquín C, Univaso L, Stange C (2022) Cisgenesis and Intragenesis as a biotechnological tool to improve abiotic stress tolerance in commercial crops. In: Cisgenic crops: potential and prospects. Springer International Publishing, Cham, pp 169–181

Gupta M, Choudhary M, Singh A, Sheoran S, Kumar H, Singla D, Bohra A, & Rakshit S (2025 Meta-QTL analysis for mining of candidate genes and constitutive gene network development for viral disease resistance in maize (Zea mays L.), Heliyon, 11(1). https://doi.org/10.1016/j.heliyon.2024.e40984

Guo G, Wang S, Liu J, Pan B, Diao W, Ge W et al (2017) Rapid identification of QTLs underlying resistance to cucumber mosaic virus in pepper (Capsicum frutescens). Theor Appl Genet 130(1):41–52

Haberer G, Young S, Bharti AK, Gundlach H, Raymond C, Fuks G et al (2005) Structure and architecture of the maize genome. Plant Physiol 139(4):1612–1624

Hagiwara-Komoda Y, Hirai K, Mochizuki A, Nishiguchi M, Meshi T, Ishikawa M (2008) Overexpression of a host factor TOM1 inhibits tomato mosaic virus propagation and suppression of RNA silencing. Virology 376(1):132–139

Han K, Huang H, Zheng H, Ji M, Yuan Q, Cui W et al (2020) Rice stripe virus coat protein induces the accumulation of jasmonic acid, activating plant defence against the virus while also attracting its vector to feed. Mol Plant Pathol 21(12):1647–1653

He J, Gai J (2023) Genome-wide association studies (GWAS). In: Plant genotyping: methods and protocols. Springer US, New York, pp 123–146

He WZ, Zhao L, Sun K, Feng Z, Zhou G, Rao Q (2024) Transcriptomic profiling reveals the complex interaction between a bipartite begomovirus and a cucurbitaceous host plant. BMC Genomics 25(1):876

Hong J, Yang L, Zhang D, Shi J (2016) Plant metabolomics: an indispensable system biology tool for plant science. Int J Mol Sci 17(6):767

Islam W, Naveed H, Zaynab M, Huang Z, Chen HY (2019) Plant defense against virus diseases; growth hormones in highlights. Plant Signal Behav 14(6):1596719

Jeyaraj G, Alphonse V, Jayanthi P, Angelin N, Govindan G (2024) Harnessing the potential of CRISPR/Cas system for enhancing virus resistance in plants: targets, strategies, and challenges. Physiol Mol Plant Pathol 129:102202

Ji Y, Schuster DJ, & Scott JW (2007) Ty-3, a begomovirus resistance locus near the Tomato yellow leaf curl virus resistance locus Ty-1 on chromosome 6 of tomato, Molecular Breeding, 20(3):271–284. https://doi.org/10.1007/s11032-007-9089-7

Jin Y, Zhao JH, Guo HS (2021) Recent advances in understanding plant antiviral RNAi and viral suppressors of RNAi. Curr Opin Virol 46:65–72

Jyotika RK, Harish S, Karthikeyan G, Kumar KK, Murugan M, Jayakanthan M, Chen TC (2024) Molecular approaches for the management of papaya ringspot virus infecting papaya: a comprehensive review. Mol Biol Rep 51(1):981

Kang MS, Subudhi PK, Baisakh N, Priyadarshan PM (2007) Crop breeding methodologies: classic and modern. In: Breeding major food staples, pp 38–40

Karimipour R, Koolivand D, Ghorbani A, Naderpour M, Rostami M (2025) Integrated gene network analysis and experimental validation identify key hub genes in potato response to Potato Virus Y infection. PLoS One 20(8):e0329747

Kaur G, Pathak M, Singla D, Sharma A, Chhuneja P, Sarao NK (2021) High-density GBS-based genetic linkage map construction and QTL identification associated with yellow mosaic disease resistance in bitter gourd (Momordica charantia L.). Front Plant Sci 12:671620

Kayihan C, Yilmaz H, Özden Çiftçi Y (2023) Next-generation sequencing in plant breeding: challenges and possibilities. In: Advanced crop improvement, volume 1: theory and practice, pp 507–535

Khan Z, Khan SH, Ahmad A, Aslam S, Mubarik MS, Khan S (2019) CRISPR/dCas9-mediated inhibition of replication of begomoviruses. Int J Agric Biol 4:711–718

Khera P, Pandey MK, Wang H, Feng S, Qiao L, Culbreath AK et al (2016) Mapping quantitative trait loci of resistance to tomato spotted wilt virus and leaf spots in a recombinant inbred line population of peanut (Arachis hypogaea L.) from SunOleic 97R and NC94022. PLoS One 11(7):e0158452

Kumar R, Tiwari RK, Jeevalatha A, Siddappa S, Shah MA, Sharma S et al (2021) Potato apical leaf curl disease: current status and perspectives on a disease caused by tomato leaf curl New Delhi virus. J Plant Dis Protect 128(4):897–911

Kumar S, Abebie B, Kumari R, Kravchik M, Shnaider Y, Leibman D et al (2022) Development of PVY resistance in tomato by knockout of host eukaryotic initiation factors by CRISPR-Cas9. Phytoparasitica 50(4):743–756

Kushalappa AC, Yogendra KN, Sarkar K, Kage U, Karre S (2016) Gene discovery and genome editing to develop cisgenic crops with improved resistance against pathogen infection. Can J Plant Pathol 38(3):279–295

Lakatos L, Szittya G, Silhavy D, Burgyán J (2004) Molecular mechanism of RNA silencing suppression mediated by p19 protein of tombusviruses. EMBO J 23(4):876–884

Lapin D, Van den Ackerveken G (2013) Susceptibility to plant disease: more than a failure of host immunity. Trends Plant Sci 18(10):546–554

Li K, Wu G, Li M, Ma M, Du J, Sun M et al (2018) Transcriptome analysis of Nicotiana benthamiana infected by tobacco curly shoot virus. Virol J 15(1):138

Li H, Jiang Z, Zhou J, Liu X, Zhang Y, Chu D (2023) Ecological factors associated with the distribution of Bemisia tabaci cryptic species and their facultative endosymbionts. Insects 14(3):252

Liu Y, Lu S, Liu K, Wang S, Huang L, Guo L (2019) Proteomics: a powerful tool to study plant responses to biotic stress. Plant Methods 15(1):135

Lorang JM, Sweat TA, Wolpert TJ (2007) Plant disease susceptibility conferred by a "resistance" gene. Proc Natl Acad Sci 104(37):14861–14866

Lu R, Folimonov A, Shintaku M, Li WX, Falk BW, Dawson WO, Ding SW (2004) Three distinct suppressors of RNA silencing encoded by a 20-kb viral RNA genome. Proc Natl Acad Sci 101(44):15742–15747

Lu G, Wang Z, Pan YB, Wu Q, Cheng W, Xu F, Dai S, Li B, Que Y, & Xu L (2023) Identification of QTLs and critical genes related to sugarcane mosaic disease resistance. , Frontiers in Plant Science, 14:1107314. https://doi.org/10.3389/fpls.2023.1107314

Macovei A, Sevilla NR, Cantos C, Jonson GB, Slamet-Loedin I, Čermák T et al (2018) Novel alleles of rice eIF4G generated by CRISPR/Cas9-targeted mutagenesis confer resistance to Rice tungro spherical virus. Plant Biotechnol J 16(11):1918–1927

Marandel G, Salava J, Abbott A, Candresse T, Decroocq V (2009) Quantitative trait loci meta-analysis of Plum pox virus resistance in apricot (Prunus armeniaca L.): new insights on the organization and the identification of genomic resistance factors. Mol Plant Pathol 10(3):347–360

Marsh JI, Hu H, Gill M, Batley J, Edwards D (2021) Crop breeding for a changing climate: integrating phenomics and genomics with bioinformatics. Theor Appl Genet 134(6):1677–1690

Martínez-Guardiola C, Parreño R, Candela H (2024) MAPtools: command-line tools for mapping-by-sequencing and QTL-Seq analysis and visualization. Plant Methods 20(1):107

Mojica FJ, Díez-Villaseñor C, García-Martínez J, Almendros C (2009) Short motif sequences determine the targets of the prokaryotic CRISPR defence system. Microbiology 155(3):733–740

Moniruzzaman M, Zhong Y, Yan H, Yuanda L, Jiang B, Zhong G (2020) Exploration of susceptible genes with clustered regularly interspaced short palindromic repeats–tissue-specific knockout (CRISPR-TSKO) to enhance host resistance. Crit Rev Plant Sci 39(5):387–417

Monnot S, Desaint H, Mary-Huard T, Moreau L, Schurdi-Levraud V, Boissot N (2021) Deciphering the genetic architecture of plant virus resistance by GWAS, state of the art and potential advances. Cells 10(11):3080

Monnot S, Cantet M, Mary-Huard T, Moreau L, Lowdon R, Haesendonck MV, Ricard A, & Boissot N (2022) Unravelling cucumber resistance to several viruses via genome-wide association studies highlighted resistance hotspots and new QTLs., Horticulture research, 9, uhac184. https://doi.org/10.1093/hr/uhac184

Mushtaq M, Dar AA, Basu U, Bhat BA, Mir RA, Vats S et al (2021) Integrating CRISPR-Cas and next generation sequencing in plant virology. Front Genet 12:735489

Pallas V, Aparicio F, Herranz MC, Amari K, Sanchez-Pina MA, Myrta A, Sanchez-Navarro JA (2012) Ilarviruses of Prunus spp.: a continued concern for fruit trees. Phytopathology 102(12):1108–1120

Pramanik D, Shelake RM, Park J, Kim MJ, Hwang I, Park Y, Kim JY (2021) CRISPR/Cas9-mediated generation of pathogen-resistant tomato against tomato yellow leaf curl virus and powdery mildew. Int J Mol Sci 22(4):1878

Qu F (2010) Plant viruses versus RNAi: simple pathogens reveal complex insights on plant antimicrobial defense. Wiley Interdiscip Rev: RNA 1(1):22–33

Raj SRG, Nadarajah K (2022) QTL and candidate genes: techniques and advancement in abiotic stress resistance breeding of major cereals. Int J Mol Sci 24(1):6

Rani LU, Shelke M, Sandhya M, Senthilraja G (2024) Targeted editing of susceptibility genes for plant disease resistance: current state and future hopes. Plant Gene 40:100471

Real N, Garcia-Molina A, Stolze SC, Harzen A, Nakagami H, Martín-Hernández AM (2025) Comprehensive proteomic profiling of Cucumber mosaic virus infection: identifying key proteins and pathways involved in resistance and susceptibility in melon. BMC Plant Biol 25(1):1–17

Santos TPFD, Sánchez-Barrantes E, Pereira LF, Gatica-Arias A (2024) Transcriptomic data analysis using the galaxy platform: coffee (Coffea arabica L.) flowers as example. In: Plant functional genomics: methods and protocols, vol 1. Springer US, New York, pp 225–243

Sastry KS, Zitter TA (2014) Management of virus and viroid diseases of crops in the tropics. In: Plant virus and viroid diseases in the tropics: volume 2: epidemiology and management. Springer Netherlands, Dordrecht, pp 149–480
Şavkan AN, Başak H, Uncu AT, Türkmen Ö (2025) Morphological characterization and molecular marker-assisted selection for resistance to tobacco mosaic virus (TMV) and tomato spotted wilt virus (TSWV) in S2 population of capia pepper (Capsicum annuum L.). Genet Resour Crop Evol 72(1):767–782
Sharma V, Gupta P, Kagolla P, Hangargi B, Veershetty A, Ramrao DP et al (2021) Metabolomics intervention towards better understanding of plant traits. Cells 10(2):346
Shukla A, Yadav R (2023) Next-generation sequencing technologies for the development of disease-resistant plants. In: Integrative approaches to biotechnology. CRC Press, pp 231–245
Siriwan W, Vannatim N, Chaowongdee S, Roytrakul S, Charoenlappanit S, Pongpamorn P et al (2023) Integrated proteomic and metabolomic analysis of cassava cv. Kasetsart 50 infected with Sri Lankan cassava mosaic virus. Agronomy 13(3):945
Song L, Wang Y, Zhao L, Zhao T (2022) Transcriptome profiling unravels the involvement of phytohormones in tomato resistance to the tomato yellow leaf curl virus (TYLCV). Horticulturae 8(2):143
Sundaresha S, Jeevalatha A, Kumar R, Sood S, Sharma S, Bhardwaj V, Singh BP, Chakrabarti SK (2022) RNA interference: a versatile tool to augment plant protection strategies in potato. In: Chakrabarti SK, Sharma S, Shah MA (eds) Sustainable management of potato pests and diseases. Springer, Singapore, pp 369–388. https://doi.org/10.1007/978-981-16-7695-6_15
Sundaresha S, Tiwari JK, Bhardwaj V, Sharma S, Jeevalatha A, Kumar R, Sagar V, Singh BP, Chakrabarti SK, Kumar M (2024) Genetic engineering in potato: status and future perspectives. In: Kalia P (ed) Genetic engineering of vegetable crops. CABI, Nosworthy Way/Wallingford, pp 133–140. https://doi.org/10.1079/9781800627123.0008
Syed M, Khan RS, Nazir S, Khan S, Ul Islam Z, Khan S, Ikuo N (2025) Silencing of disease susceptibility genes: an effective disease resistance strategy against fungal pathogens. Plant Cell Rep 44(6):127
Tepfer M (2002) Risk assessment of virus-resistant transgenic plants. Annu Rev Phytopathol 40(1):467–491
Thawng CN, Smith GB (2022) A transcriptome software comparison for the analyses of treatments expected to give subtle gene expression responses. BMC Genomics 23(1):452
Tomás DM, Cañizares MC, Abad J, Fernández-Muñoz R, Moriones E (2011) Resistance to Tomato yellow leaf curl virus accumulation in the tomato wild relative Solanum habrochaites associated with the C4 viral protein. Mol Plant-Microbe Interact 24(7):849–861
Trebicki P (2020) Climate change and plant virus epidemiology. Virus Res 286:198059
Tyagi A, Mir ZA, Almalki MA, Deshmukh R, Ali S (2024) Genomics-assisted breeding: a powerful breeding approach for improving plant growth and stress resilience. Agronomy 14(6):1128
van Schie CC, Takken FL (2014) Susceptibility genes 101: how to be a good host. Annu Rev Phytopathol 52(1):551–581
Van Wijk KJ (2001) Challenges and prospects of plant proteomics. Plant Physiol 126(2):501–508
Víquez-Zamora M, Caro M, Finkers R, Tikunov Y, Bovy A, Visser RG et al (2014) Mapping in the era of sequencing: high density genotyping and its application for mapping TYLCV resistance in Solanum pimpinellifolium. BMC Genomics 15(1):1152
Wang X, Wang G (2023) Application of NGS-BSA and proposal of modified QTL-seq. J Plant Biochem Biotechnol 32(1):31–39
Wittmann S, Chatel H, Fortin MG, Laliberté JF (1997) Interaction of the viral protein genome linked of turnip mosaic potyvirus with the translational Eukaryotic Initiation Factor (iso) 4E of Arabidopsis thaliana using the Yeast two-hybrid system. Virology 234(1):84–92
Zaidi SSEA, Mukhtar MS, Mansoor S (2018) Genome editing: targeting susceptibility genes for plant disease resistance. Trends Biotechnol 36(9):898–906
Zainal-Abidin RA, Ruhaizat-Ooi IH, Harun S (2021) A review of omics technologies and bioinformatics to accelerate improvement of papaya traits. Agronomy 11(7):1356

Zanardo LG, de Souza GB, Alves MS (2019) Transcriptomics of plant–virus interactions: a review. Theor Exp Plant Physiol 31(1):103–125

Zhang J, Yu J, Pei W, Li X, Said J, Song M, Sanogo S (2015) Genetic analysis of Verticillium wilt resistance in a backcross inbred line population and a meta-analysis of quantitative trait loci for disease resistance in cotton. BMC Genomics 16(1):577

Zhang P, Du H, Wang J, Pu Y, Yang C, Yan R et al (2020) Multiplex CRISPR/Cas9-mediated metabolic engineering increases soya bean isoflavone content and resistance to soya bean mosaic virus. Plant Biotechnol J 18(6):1384–1395

Zhou J, Tzanetakis IE (2019) Soybean vein necrosis virus: an emerging virus in North America. Virus Genes 55(1):12–21

# Biotechnological Interventions for Virus-Resistant Vegetable Crops

# 21

P. Jayanthi, F. Neha Angelin, and A. Swapna Geetanjali

**Abstract**

Plant viruses pose a significant threat to global food security, causing major losses in both production and quality of agricultural crops. Managing plant viral disease is very difficult due to emerging new viruses, mixed infections, and viral diversity. While traditional control methods like chemical pesticides, vector control, and classic breeding are limited because of environmental concerns and lack of available natural resistance genes, current management depends on the exclusion, eradication, vector control, and early identification, supported by emerging diagnostics and next-generation sequencing. Biotechnology has enabled new prospects for the sustainable control of viral pathogens. The application of pathogen-derived resistance (PDR), and RNA interference (RNAi) techniques, including hairpin RNAs and synthetic microRNAs, allows long-term and combinational resistance against the viral coat proteins, replicases, or movement genes. The genome editing technology, like CRISPR/Cas, allows the modification of host susceptibility genes. Further approaches, such as virus-induced gene silencing (VIGS), marker-assisted selection (MAS), and next-generation synthetic biology that specifically targets the viral genes, have specifically improved the resistance of plant breeding programmes. This chapter confers the methods for controlling plant viral diseases by using genome engineering, RNA-based technologies, and synthetic biology for high-yielding, resistant crops and thus facilitating sustainable farming and global food security.

P. Jayanthi · F. Neha Angelin · A. Swapna Geetanjali (✉)
Department of Genetic Engineering, SRM Institute of Science and Technology, Kattankulathur, Tamil Nadu, India
e-mail: swapnaga@srmist.edu.in



R. Kumar et al. (eds.), *Vegetable Crops and Viral Pathogens Volume I*, Plant Pathology for Sustainable Agriculture, https://doi.org/10.1007/978-981-95-7265-6_21

**Keywords**

CRISPR/Cas · Genome editing · Marker-assisted selection · Pathogen-derived resistance · RNA interference · Virus-induced gene silencing · Artificial microRNAs · Synthetic biology

## 21.1 Introduction

Vegetable crops are of great importance to humans in supplying vital nutrients, economic returns, and food security. They are cultivated worldwide for economic and medicinal purposes, significantly supporting sustainable agricultural practices. Vegetables, mainly grown in tropical regions, are vital components of the human diet globally (Kumar et al. 2023; Mazumder et al. 2024). In recent decades, demand for tropical fruits and vegetables has increased markedly, especially in the United States and the European Union. Among various tropical fruits and vegetables, banana, mango, citrus, papaya, pineapple, sweet potato, and cassava are the most prominent in international trade (Irfan et al. 2023). However, their production faces challenges from various factors, including biotic and abiotic stresses. For example, drought stress in bananas may result in output losses of up to 65%. Furthermore, several diseases during plant growth and development restrict fruit yield (Sayago-Ayerdi et al. 2021). In some instances, this loss might reach 50% of total output, leading to significant economic loss for these countries.

Plant viruses have genomes (4–20 kb) of DNA or RNA that contain essential conserved proteins, including coat protein, movement protein, and replication-associated protein, with a conserved region (Awasthi et al. 2016). Viral reproduction and transcription depend on the host cellular machinery, making the plant viruses obligate parasites. They can be spread through wounds, seeds, and pollen, as well as via other vectors such as insects, nematodes, soil fungi, or mites (Bragard et al. 2013; Lefeuvre et al. 2019). Compared to other plant pathogens, viruses cannot be directly controlled with chemical treatments on infected materials, so preventive sterile measures are the only way to manage infection. Viral pathogens are particularly challenging because they evolve rapidly, consist of mixed species, and are effectively transmitted by insect vectors. As a result, the incidence of viral diseases is increasing in vegetable crops due to greater diversification. Conventional management methods like vector control, cultural practices, and rotation offer only short-term relief and do not address the core issues. In contrast, methods that include RNA interference (RNAi) are particularly effective; they enable plants to produce short RNAs that inhibit viral gene expression (Taliansky et al. 2021; Sundaresha et al. 2022a). Some RNAi crops are already available on the market. The breakthrough in CRISPR/Cas genome editing has significantly progressed virus resistance in breeding programmes by allowing precise modification of host or viral susceptibility genes, essentially removing the need for transgenic methods (Wang et al. 2022; Sundaresha et al. 2024). Additionally, the use of transgenic techniques, involving viral or host-derived gene introduction, has greatly broadened the scope

**Table 21.1** Viral diseases in vegetable crops and the biotechnological strategies used to develop resistance

| Virus name/genus/family | Host | Strategy used | Target gene | References |
|---|---|---|---|---|
| Tomato yellow leaf curl virus (TYLCV) | Tomato | RNAi and CRISPR | Rep/CP and host S-gene | Khan et al. (2022) |
| Chilli leaf curl virus (ChiLCV) | Chilli | RNAi | AV1 and AC1(Rep) | Nalla et al. (2023) |
| Cucumber mosaic virus (CMV) | Cucumber and tomato | RNAi and CRISPR | CP (coat protein) and MP (movement protein) | Chandrasekaran et al. (2016) |
| Papaya ringspot virus (PRSV) | Papaya, cucurbits | RNAi | Coat protein (CP) | Akhtar et al. (2023) |
| Potato virus Y (PVY) | Potato | RNAi and CRISPR | Nib,CP | Zhan et al. (2019) |
| Badnavirus, Caulimoviridae | Banana | CRISPR | Viral ORF | Tripathi et al. (2019) |
| Mastrevirus, Geminiviridae | Barley | CRISPR | Viral MP/CP, Rep genes, and LIR | Kis et al. (2019) |

of breeding materials for resistance genes, providing greater accuracy and control compared to conventional methods. These biotechnology approaches are currently developing virus-resistant vegetable crops (Table 21.1) and are vital for fostering sustainable yields and safeguarding crops against emerging viruses (Fig. 21.1).

Such methods encourage the establishment of resistance directed towards specific viral proteins or susceptibility factors in hosts, thus improving durability and effectiveness for resistance mechanisms. Through such function, they play a major role in maintaining crop productivity, sustainable crop production methodologies, and food and nutritional security at a global level (Robertson et al. 2022).

## 21.2 Major Viral Risks to Vegetable Production: Major Viruses and Their Impact

Begomoviruses, such as tomato leaf curl virus (ToLCV) and chilli leaf curl virus (ChiLCV), are among the most significant viral diseases infecting vegetable crops worldwide. Whitefly-transmitted viruses are prominent for their high crop yield losses in vegetables like tomato, chilli, and okra. Their huge genetic diversity, high frequency of recombination events, and their association with satellite molecules (i.e. alphasatellites and betasatellites) sustain their pathogenicity and fitness and make them extremely difficult to control; weeds act as reservoirs for such viruses, thereby making the control even more difficult and enabling these pathogens to spread from one crop to another and one geographical location to another (Reddy et al. 2024). Potyviruses, including papaya ringspot virus (PRSV) and zucchini yellow mosaic virus (ZYMV), are a significant group that infects vegetable crops, such as cucurbits and solanaceous plants. Potyviruses are mostly transmitted through

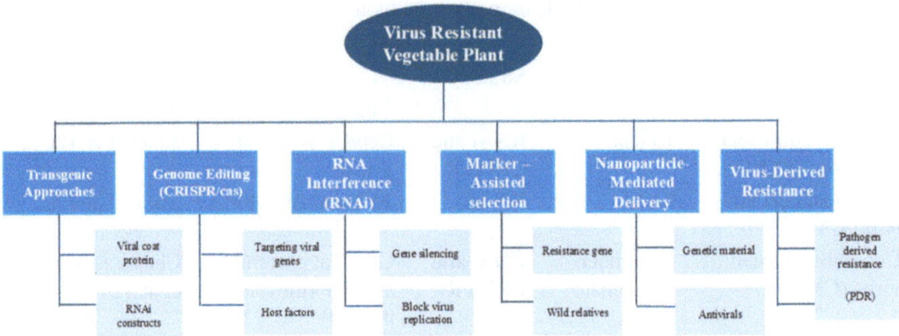


**Fig. 21.1** Overview of biotechnological approaches for virus resistance in vegetable crops

aphids and sometimes through seed, leading to widespread disease and more economic losses. Potyviruses infect more than 14 vegetable hosts in regions such as the Indian subcontinent, leading to symptoms like mosaic patterns, leaf deformation, and poor fruit quality (Tennant et al. 2024). Tospoviruses are transmitted by insect vectors with a broad host range, which includes plants like lettuce, tomato, and pepper. Control of viruses such as this is extremely hard since viruses are capable of breaking resistance in cultivars, thereby eliciting repeated infestations combined with greater intricacy in controls for their populations (Tiberini et al. 2025). Cucumber mosaic virus (CMV) has caused severe damage to more than 1000 plant species, such as beans, cucurbits, peppers, spinach, lettuce, and tomatoes. Transmission of CMV is through aphids and seeds; its broad host range and susceptibility make it a constant risk in vegetable production globally. These viruses cause leaf abnormalities, stunted foliage growth, fruit deformities, and ultimately significant yield losses, thereby threatening food security as well as farmers' financial stability worldwide (Jones and Naidu 2019).

## 21.3 Transgenic Strategies to Increase Viral Resistance

Transgenic methods of virus resistance in plants began with pathogen-derived resistance (PDR), where viral genes are integrated into the plant genome to protect against disease. Coat protein (CP)-mediated resistance is one of the earliest and most common methods, involving the expression of viral coat protein genes in transgenic host plants to disrupt virus uncoating or replication. This approach has shown promising results, such as the production and commercial release of transgenic papaya against papaya ringspot virus in Hawaii and squash lines against ZYMV and watermelon mosaic virus (WMV) in the USA (Bhat and Rao 2020). Another approach involves replicase-mediated resistance, which entails expressing mutant viral replicase genes in plants to inhibit viral RNA replication and infection. Movement protein (MP)-mediated resistance is also utilised to restrict virus spread within the plant by preventing the attachment of proteins responsible for cell-to-cell spread of the virus (Prins et al. 2008). For example, transgenic tomato plants that were expressing the TYLCV (tomato yellow leaf curl virus) coat protein gene have shown delayed symptom development and reduced viral replication, similar to squash plants containing CP genes from ZYMV and WMV. While such transgenic approaches have been found effective for virus resistance and enable virus-resistant crop production on a commercial scale (Tepfer 2002), they have also caused regulatory consistency and consumer acceptance issues due to the introduction of viral transgenes into food crops.

## 21.4 RNA Interference (RNAi) for Virus Resistance

RNA interference is a precise, sequence-specific gene silencing mechanism triggered by double-stranded RNA (dsRNA) and has proven to be an effective method for inducing virus resistance in vegetable crops (Sundaresha et al. 2022b, 2024). The approach involves transforming plants with hairpin RNA constructs that target vital viral genes, encoding the coat protein, replicase, or movement protein. When expressed, these constructs are processed by the plant's cellular machinery into small interfering RNAs (siRNAs), which then guide the RNA-induced silencing complex (RISC) to degrade matching viral RNA, thus inhibiting virus replication and transmission (Kang et al. 2023). The effectiveness of RNA interference (RNAi) in providing resistance to viral diseases has been extensively studied. For example, AV1 (coat protein)-and AC1 (replicase)-mediated ChiLCV (chilli leaf curl virus) silencing through RNAi in chilli plants has conferred high resistance. RNAi-mediated specific disruption of TYLCV Rep gene in tomato has significantly reduced viral accumulation and delayed symptoms. Similarly, transgenic cucumber lines expressing a hairpin RNA-specific CMV have shown marked resistance to CMV in field screenings (Zhang et al. 2018). The advantages of RNA interference (RNAi)-mediated viral resistance include its high specificity, environmental safety due to the absence of the use of chemicals, and stability, especially when targeting multiple viral targets or conserved regions of the genome (Levanova and Poranen 2018). It has some difficulties of potential chance of off-target effects, creation of viral escape mutants, and regulatory concerns regarding transgenic crops (Aman et al. 2018).

## 21.5 CRISPR-Cas-Mediated Genome Editing

CRISPR-Cas genome editing offers efficient and advanced tools for generating virus-resistant vegetable crops (Fig. 21.2). The technology enables two main resistant strategies: directly targeting virus genomes, which is very effective against DNA viruses like begomoviruses and includes editing plant susceptibility (S) genes that limit the virus' activity for replication or transmission. CRISPR-Cas9 has been used to disrupt the SWEET gene in tomato, making it resistant to begomoviruses, while CRISPR-Cas13a, which targets RNA, has been employed to silence RNA

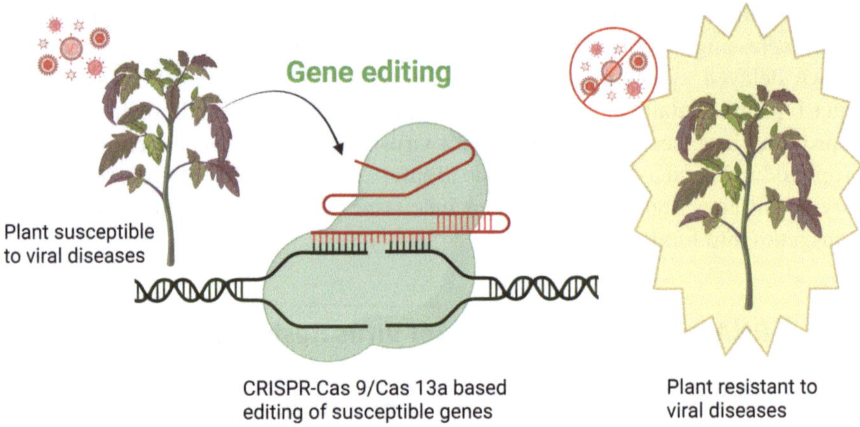


**Fig. 21.2** Virus resistance attained through CRISPR-assisted techniques

viruses such as CMV and potyviruses in cucumber and tobacco. Editing the eIF4E gene in pepper has also conferred resistance to potyviruses (Talakayala et al. 2022). Benefits of CRISPR-based methods include the possibility of nontransgenic, transgene-free editing, meaning no foreign DNA is incorporated into the final plants, as well as the potential of achieving long-term and broad-spectrum resistance. The methods are quicker and more accurate than conventional breeding or transgenic methods, and multiplex genome editing can even minimise the potential for viral escape mutants (Surya Krishna et al. 2023). Overall, CRISPR-Cas systems are transforming the production of virus-resistant vegetables by making targeted, efficient, and potentially affordable solutions.

## 21.6 Marker-Assisted Selection (MAS)

Marker-assisted selection (MAS) is a modern breeding technique that uses closely linked DNA markers associated with disease resistance loci for the rapid and accurate selection of virus-resistant plants (Fig. 21.3). MAS accelerates breeding compared to slower and potentially incorrect phenotypic selection, which is influenced by environmental factors and emerging virus pressure. It has a significant impact on traits governed by major resistance loci or QTLs, reducing the need for constant field screenings. Various marker categories (Table 21.2), such as SSR, SNP, and gene-based markers, have been utilised on key crops like CEDG008 for black gram resistance to mung bean yellow mosaic virus (MYMV) (Roy et al. 2025), SCAR RYSC3 marker against Ryadg gene resistance to potato virus Y (PVY) in potato (Mangal et al. 2021), and indel-based markers against Ty-3-mediated resistance in TYLCV in tomato (Welegama et al. 2021). SSR markers Satt334 and Sct033 enable effective marker-assisted selection for virus resistance in soybean without requiring bulk virus inoculation (Gao and Li 2025). MAS allows pyramiding (stacking) of

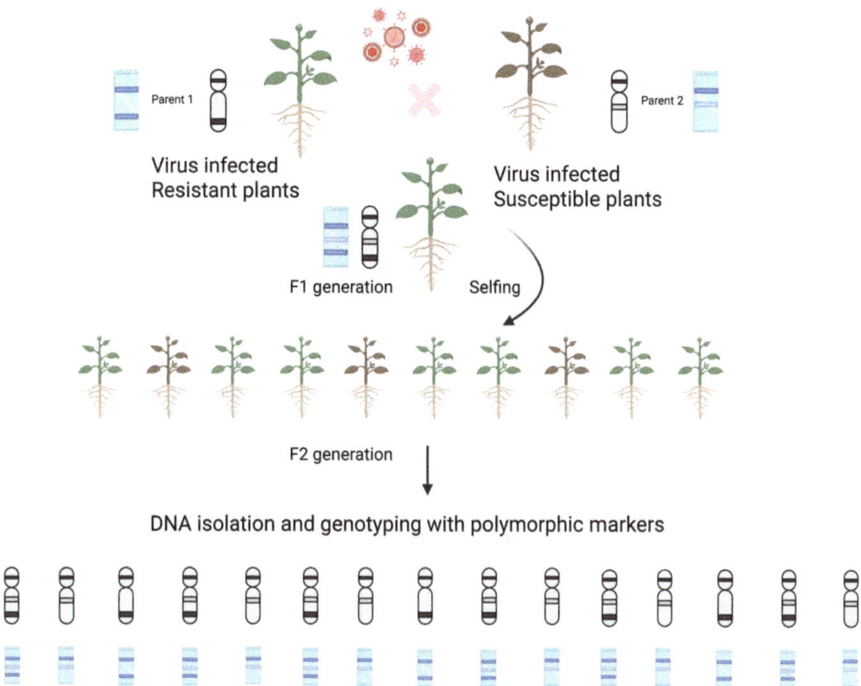


**Fig. 21.3** The figure shows the basic concept of marker-assisted selection

several resistance loci, which enhances the duration and deployment of resistance, as well as reducing the risk of resistance breakdown due to viral mutation or recombination (Roy et al. 2025). Pyramiding of multiple QTLs has led to the development of enhanced resistance to MYMV in black gram.

High-throughput SNP arrays and platforms, such as the application of KASP, have resulted in the enhancement of rapid genotyping at large scales under the limits of breeding populations and thus enhancing marker-assisted selection strategies on intricate viral resistance traits. Further, the incorporation of MAS with recent genomic and transgenic methods speeds up breeding programmes, as observed from rapid systems on citrus tristeza virus resistance (Endo et al. 2022).

## 21.7 Nanoparticle-Mediated Delivery

Nanoparticles are considered for antiviral purposes in which nanomaterials in plants like metal oxides (CuO, ZnO, $TiO_2$, NiO) (Table 21.3) are used for stimulation of the immune system and inactivation of the virions (Liu et al. 2024; Elsahhar et al. 2022), metals (AgNPs) for protein/virion interaction (Elsharkawy 2024; Jiang et al. 2025), biopolymers (chitosan) for the suppression of the virus and regeneration of tissue (Shidore et al. 2021), inorganic carriers (clay nanosheets, cationic polymers)

**Table 21.2** Marker-assisted selection (MAS) for virus resistance in major crops

| Host | Virus | Marker(s)/gene(s) | Key outcome | Citation |
|---|---|---|---|---|
| Black gram | Mung bean yellow mosaic virus (MYMV) | SSR marker CEDG008 | Effectively differentiates resistant and susceptible genotypes | Sathees et al. (2022) |
| Black gram | MYMV | Major QTL on LG 10 | QTL validated, supports MAS/MABC for strong field resistance | Subramaniyan et al. (2022) |
| Potato | Potato virus Y (PVY) | SCAR marker RYSC3 (Ryadg gene) | Achieves ~99.7% selection accuracy in breeding programmes | Kasai et al. (2000) |
| Potato | PVY | Rychc gene and marker MG64–17 | Functional marker enables phenotype-predictive MAS | Li et al. (2022) |
| Tomato | Tomato yellow leaf curl virus (TYLCV) | Indel-based marker (e.g. ACY, Ty-3 linked) | Precise marker-based TYLCV resistance selection | Nevame et al. (2018) |
| Tomato | Tomato spotted wilt virus (TSWV) | Sw-5b gene markers | Gene-specific markers distinguish resistant alleles for MAS | Tong et al. (2023) |
| Soybean | Soybean mosaic virus (SMV) | SSR: Satt334 and Sct_033 | Combined markers enable 95% MAS efficiency for resistance | Ma et al. (2010) |
| Cucumber | Cucumber vein yellowing virus (CVYV) | KASP SNP marker for CsCvy-1 | Rapid MAS in breeding commercial-resistant lines | Pujol et al. (2019) |
| Squash | Papaya ringspot virus W (PRSV-W) | KASP SNP markers (Ch09_2080834, Ch09_5023865–1) | Molecularly validated MAS for field resistance | Shrestha et al. (2022) |

| | | | | |
|---|---|---|---|---|
| Bean | Bean common mosaic virus (BCMV) | KASP/CAPS for I gene | Codominant, tightly linked; high accuracy MAS in diverse lines | Bello et al. (2014) |
| Citrus | Citrus tristeza virus (CTV) | Fast-track MAS and multiple markers | MAS+transgenics cut breeding cycle time by >80% | Endo et al. (2022) |
| Wheat | Wheat curl mite/WSMV vector | KASP SNPs for Cmc4 | Validated markers for robust field MAS against the virus vector | Zhao et al. (2021) |

**Table 21.3** Nanoparticles in plant virus management

| Nanoparticle | Host | Name of the virus | Outcome | References |
|---|---|---|---|---|
| CuO | *Nicotiana benthamiana* | Tobacco mosaic virus (TMV) | CuO NPs (100 mg/L) fragmented TMV in vitro, reduced replication/ accumulation in planta, and upregulated defence genes while promoting growth metrics in treated plants | Liu et al. (2024) |
| Chitosan | Potato | Potato virus Y (PVY) | CS-NP treatment during in vitro culture produced virus-free regenerated plantlets (up to 100% PVY-free at optimal concentration) and improved regeneration parameters | Shidore et al. (2021) |
| Clay nanosheets | *Nicotiana benthamiana* | Various RNA viruses (as dsRNA templates) | Clay nanosheets and other nanocarriers protect and prolong dsRNA stability and release, enabling longer RNAi-based antiviral protection compared with naked dsRNA | Steinmetz and Chariou (2019) |
| Silver | Tomato | Coat protein interactions (PVY, ToMV) | Studies indicate AgNPs can bind viral coat proteins and have antiviral potential in vitro, suggesting applications for inactivation/blocking | Elsharkawy (2024) |
| Silica-phosphonium nanoparticles | Tomato | Tomato brown rugose fruit virus (ToBRFV) | Engineered silica-phosphonium NPs inhibited ToBRFV in laboratory testing; authors note mammalian cytotoxicity concerns for some formulations | Cagno et al. (2018) |
| Titanium dioxide nanoparticles | Pepper | Pepper golden mosaic virus and tobacco mosaic virus (TMV) (single and mixed infections) | Recent experiments evaluated $TiO_2$ NPs for managing single/mixed infections in *Capsicum*, with species-specific symptom and viral load effects reported in controlled studies | Acuña-Fuentes et al. (2022) |

to carry dsRNA, plant viral nanoparticles/VLPs as immune system and delivery platforms (Nissim et al. 2025), and functionalised polymers (silica-phosphonium) to demonstrate surface antiviral activity (Cagno et al. 2018). Nanoparticles present strategies that function through three main mechanisms: direct virucidal action, host immune modulation, and cargo delivery for RNA interference (Masood et al. 2024). Direct virucidal effects include physical disruption, that is, CuO nanoparticles disrupting tobacco mosaic virus particles and reducing infectivity, multivalent binding and deformation by elastic ligand-engineered antiviral nanoparticles, and coat protein binding by metal nanoparticles like AgNPs that can block viral assembly or host attachment. Host immune activation is brought about by systemic acquired resistance (SAR), where CuO nanoparticles induce defence genes (PR1, PR2, ERF1, JAZ3) in *Nicotiana* and increase infection resistance by regulating redox and antioxidant pathways. Cargo delivery platforms incorporate nanocarriers such as clay nanosheets and cationic polymers that protect and deliver dsRNA/siRNA for sustained RNAi-mediated resistance and plant viral nanoparticles (VNPs/VLPs) to deliver agrochemicals or immune modulators to specific tissues. Nanoparticles are also explored for vector and surface control, using coating and formulations that reduce vector viability and sterilise substrates or surfaces to inhibit viral spread.

## 21.8 Pathogen-Derived Resistance (PDR)

Pathogen-derived resistance (PDR) involves using genetic elements from pathogens in plants that may reduce the harmful effects of parasitic organisms (Sanford and Johnston 1985). Building on this concept, a significant breakthrough occurred in 1986 when the Beachy laboratory developed tobacco mosaic virus (TMV) resistance in tobacco species using gene constructs that encode for viral coat protein (CP) (Abel et al. 1986). Since then, many studies have examined viral genes or genomic fragments used to modify plants, ultimately resulting in the successful creation of virus-resistant crops with commercial potential. Transgenic pathogen-derived resistance (PDR) has clearly revolutionised agriculture by producing deployable varieties and field-tested products across various crops. Both protein-based and nucleic acid-based PDR strategies have been implemented across various crops. This includes the control of PRSV in Hawaiian papaya, where coat protein (CP)-mediated transgenic lines provided substantial field protection and enabled commercial production widely recognised as the foundational example of PDR (Gonsalves 2003). Virus-resistant squash containing viral genetic elements was among the first transgenic virus-resistant crops commercialised in the USA. In Brazil, common bean lines featuring the RNAi-based Embrapa 5.1 event have received regulatory approval, demonstrating robust field resistance and compositional similarity to conventional beans (Carvalho et al. 2015). The VIRCA (Virus Resistance cassava for Africa) initiative in East Africa has advanced RNAi-based resistance in cassava, moving from greenhouse trials to field testing (Taylor et al. 2012). In tomatoes, transgenic lines designed with intron hairpin RNA constructs targeting tomato yellow leaf curl virus (TYLCV) showed durable, high-level resistance and stable siRNA expression across multiple field seasons, though new viral recombinants capable of evading resistance were observed (Fuentes et al. 2016). Additional examples include CP-transgenic potatoes with high resistance to PVY and sugarcane with untranslatable CP constructs that improved resistance to sugarcane yellow leaf virus in inoculation trials (Guria and Pandi 2019). Floricultural crops and model systems such as gerbera modified with CMV CP and tobacco with inverted repeat constructs have also demonstrated enhanced resistance under controlled conditions (Stashevski and Il'inskaya 2011). The molecular mechanisms of PDR vary; CP-mediated resistance disrupts viral uncoating or assembly, providing effective protection in some crops, notably papaya (Lindbo and Falk 2017) and squash, although efficacy can be inconsistent and the risk of breakdown increases in mixed-virus infections. RNA interference (RNAi), utilising hairpin or inverted repeat constructs to produce siRNAs that break down viral RNA (Lundstrom 2020), offers high specificity and has demonstrated strong field results, as seen in transgenic tomatoes; however, viral suppressors, emerging recombinants, and possible off-target effects may restrict the longevity of this method. Strategies based on viral replicase or antisense constructs target essential replication genes (such as NIb/Rep in potyviruses and geminiviruses), often providing strong resistance. However, issues may occur due to strain specificity and dependence on expression levels. Movement protein-based approaches, which aim to block systemic viral spread, have shown limited effectiveness and have not gained widespread adoption. Nontransgenic alternatives—such as topical application of dsRNA and the use of transgenic rootstocks (Zhao et al. 2020) offer flexible delivery options while avoiding genetic modification of edible tissues. Nonetheless, challenges related to stability, systemic movement, cost, and regulatory acceptance persist. While each strategy presents unique advantages and limitations, RNAi- and dsRNA-based approaches currently exhibit the greatest promise for achieving durable and effective viral resistance in agricultural fields.

## 21.9 Emerging Biotechnological Tools and Approaches

A range of cutting-edge biotechnological strategies is being designed to instil virus resistance in crops. Synthetic microRNAs (amiRNAs) are specifically designed molecules that mimic the activity of endogenous miRNAs, enabling highly precise and stable silencing of viral genes in plants (Zand Karimi and Innes 2022). The method is more specific than traditional RNA interference (RNAi) techniques and has demonstrated efficacy in imparting resistance to insects and pathogens (Yogindran and Rajam 2021). Host-Induced Gene Silencing (HIGS) is a technique for genetically modifying plants to produce double-stranded RNA (dsRNA) or artificial microRNAs (amiRNAs) that specifically reduce the expression of essential genes in viruses and their insect vectors (Wytinck et al. 2020). These guide RNA molecules hold for their ability to be delivered from the plant into the pathogen or pest for gene silencing, while allowing a possible trans-kingdom RNA interference (RNAi) method for disease control; it may vary depending on a specific target and method of delivery. Delivery of double-stranded RNA (dsRNA) in nanoparticles is a nongenetic, target-specific method of achieving resistance. Encapsulation of dsRNA in nanoparticles such as chitosan, synthetic nanovesicles, or carbon quantum dots stabilises and enhances the absorption of dsRNA, protecting it from environmental degradation and ensuring delivery to target organisms (Qiao et al. 2023). Stacked resistance genes through genome editing integrate multiple mechanisms of resistance, like CRISPR-edited modifications on the susceptibility genes with RNAi transgenes or double editing on multiple S-genes to express long-term and broad-spectrum resistance (Niu et al. 2022).

## 21.10 Virus-Induced Gene Silencing (VIGS)

VIGS has emerged as a powerful biotechnological tool for breeding virus-resistant vegetable crops. It is an efficient, rapid platform that promises to be a premier tool both for functional genomics and crop improvement. RNA-mediated reverse genetics that uses a plant's own posttranscriptional gene silencing machinery to suppress the production of the plant's native genes and thereby inhibit their ability to spread viral diseases, as well as silence the genes of the plant. It is easy to suppress the expression of genes temporarily and does not require stable incorporation of foreign DNA into the genome of the plant. Therefore, this technique has particular utility in species where there are low levels of successful transformation or long breeding cycles (Burch-Smith et al. 2004). There are several different types of engineered viral vectors used in RNA-mediated gene silencing techniques, including tobacco rattle virus (TRV), barley stripe mosaic virus (BSMV), cabbage leaf curl virus (CaLCuV), and wheat dwarf virus (WDV). These vectors are designed to carry pieces of cDNA of the target gene of the plant being studied, and upon infection of the host plant, they activate RNA silencing that produces small interfering RNAs (siRNAs) that degrade both the mRNA of the viral vector and of the endogenously expressed mRNA of the same gene (Lu et al. 2003). Once the gene of interest is silenced, the phenotypic observations are used to infer the biological role of the gene of interest. Gene silencing typically occurs within 2–4 weeks postinfection (Wang et al. 2023). These systems have demonstrated their utility as efficient tools for studying a wide range of vegetable species such as tomato, peppers, eggplant, cucurbitaceae, and brassicas and are also being studied in cereal crops such as wheat, barley, and rice (Li et al. 2024) and were identified as a key resource to identify and assess the importance of genes involved in disease resistance, stress tolerance, and plant growth/development. An example of this is the use of a TRV-based VIGS delivery system using *Agrobacterium tumefaciens* in soybeans to study the effects of silencing the phytoene desaturase (GmPDS) gene, the rust-resistant gene GmRpp6907, and the defence-related gene GmRPT4. Silencing efficiencies ranged from 65% to 95% (Deng et al. 2025). A WDV-based VIGS system has been used to knock down the expression of the rice blast-resistant gene Pi21 in rice, which led to rice plants with a significant increase in resistance to the rice blast pathogen (Zhang et al. 2025). In addition, in wheat and barley, VIGS using BSMV has allowed researchers to identify the genetic pathways related to leaf rust resistance and other protective mechanisms that are induced by the virus (Gunupuru et al. 2018). At this time, virus-induced overexpression (VOX), virus-induced genome editing (VIGE), and host-induced gene silencing (HIGS) can be used to expand our knowledge of plant functional genomics as well as to develop improved crops (Rössner et al. 2022). The RNA-directed DNA methylation (RdDM) pathway allows the induction of stable heritable epigenetic modification, leading to the generation of genotypes that exhibit desired traits. The epigenetic properties of VIGS will open up new avenues for plant breeding to produce crops with an increased level of resistance to viral pathogens. Selection of the appropriate viral vector, identification of the optimal inoculation technique, and experimental conditions to maximise RNA interference efficacy represent key components of this process. Techniques for inoculating plants include *Agrobacterium-mediated* leaf infiltration, vacuum infiltration, and root wounding immersion. Each technique has its own merits and may be most applicable to certain plant species and developmental stages (Xiao et al. 2020). Root wounding-immersion is particularly well suited for inoculating numerous plants that are at an earlier developmental stage; therefore, this technique is generally best for high-throughput functional screening of plants. Phytoene desaturase (PDS) and Mg-chelatase H subunit (ChlH) have been commonly employed as visual marker genes to assess the effectiveness of RNA interference by identifying phenotypically altered plants due to RNA interference through visible signs of the altered phenotype (such as bleached or yellowed leaves) (Feng et al. 2025). Variables that can affect the efficacy and longevity of RNA interference include the type of plant, the targeted gene, and the particular viral vector utilised. Moreover, the off-target effects and the manifestation of symptoms due to the viral vectors used for gene silencing can add complexity to interpreting gene silencing data (Akbar et al. 2022).

VIGS has proven effective for the study of dicotyledons but has had greater difficulty with monocots, although recent advancements have mitigated some of the problems associated with the use of VIGS for monocots. Consequently, researchers continue to focus their efforts to address the limitations of VIGS through improvements to vector design, broadening the host range for VIGS and integrating VIGS with other biotechnology techniques such as CRISPR/Cas9. The development of future generations of viral vectors that will be less restricted to host cells and will have lower levels of pathogenicity will make it easier to utilise VIGS for species such as *Brassica oleracea* and *Atriplex canescens*. VIGS and genome editing technologies provide a unique opportunity to produce crops that are able to transmit long-lasting resistance to viruses and pathogens (Wang et al. 2023). Additionally, the development of VIGS has greatly advanced our understanding of interactions between viruses and plants as well as the molecular mechanisms that regulate plant resistance. Specifically, the rapid repression of the resistance genes of the host plants by VIGS has greatly facilitated the identification of the key participants in plant resistance pathways, that is, nucleotide-binding site leucine-rich repeat (NBS-LRR) proteins, the heat shock protein HSP90, and the various cofactors involved in disease resistance (Scofield et al. 2005). The knowledge gained from this research has contributed to the creation of novel methodologies for developing virus-resistant crops that include the incorporation of multiple resistance genes and the management of resistance responses. With current advancements in genome sequencing research generating a large quantity of genome information rapidly, the validation of gene function via VIGS will be pivotal in bridging fundamental genetic information with effective breeding progression. Furthermore, by integrating VIGS with other technologies (transcriptomics and proteomics), analysis of gene expression for complex traits will also be enhanced and the development of virus-resistant varieties. Due to the nontransgenic nature of virus-induced gene silencing, concerns and regulatory hurdles associated with genetically modified organisms are minimised, and VIGS can be integrated into crop improvement programmes (Wang et al. 2023). As a key aspect of current plant biotechnology, VIGS provides several advantages, including rapid, flexible, and efficient gene functionalisation and the production of virus-resistant vegetable crops.

## 21.11 Conclusion

The potential advancements in biotechnological approaches for cultivating virus-resistant vegetable crops appear very promising due to the synergy between multi-target RNA interference (RNAi), CRISPR-mediated genome editing, and advanced molecular breeding techniques. These technologies enable the development of crops with long-lasting, broad-spectrum resistance against a wide array of viruses, thereby overcoming many of the limitations associated with conventional breeding methods and single-gene strategies. New delivery systems, such as nanotechnology and virus-based vectors, are making gene editing and RNAi technologies more effective and efficient. Evolution of antiviral engineering in plants presents a clear linear progression from the original concept of pathogen-derived resistance and CP-mediated protection and identification of RNA silencing-based methods such as hairpin RNAs and artificial miRNAs in the 1990s and 2000s. These were later superseded by protein-based DNA-binding technologies (AZPs, TALEs, ZFNs). Most recently, CRISPR/Cas technologies have revolutionised the field, delivering precise, flexible, and scalable technologies to produce next-generation virus-resistant crops. The biotechnological methods, including genetically modified (GM) and genome-edited crops, are successfully enhancing the virus resistance of vegetable crops. The safety of genetically modified (GM) crops for food and environmental exposure is established by stringent biosafety evaluations, particularly for those producing double-stranded RNA (dsRNA) to facilitate RNA interference (RNAi)-mediated viral resistance. Public acceptance is the biggest problem and includes concerns related to environmental and health hazards, ethical issues, and the potential for gene flow to nontarget organisms. These are met with strong capacity building, open risk communication, and proactive stakeholder engagement, including farmers, consumers, and policymakers. In the European Union, a stringent risk assessment and licensing process is required for all genetically modified plants, including those having RNA interference characteristics. In some parts of the world, like the USA and Japan, transgene-free genome-edited crops are less. Combining conventional plant breeding strategies with advanced biotechnological techniques (omics) will provide new options for the long-term sustainability of tropical crop development by increasing yields and extending postharvest shelf life. Clearly, biotechnological tools like CRISPR, RNAi, MAS, and PDR are more advantageous than many other alternatives; therefore, the development of tropical crops utilising these technologies will be an important component in providing a sustainable method to decrease losses, thus enabling economic growth in developing countries while improving food availability for the growing world population.

## References


Abel PP, Nelson RS, De B, Hoffmann N, Rogers SG, Fraley RT, Beachy RN (1986) Delay of disease development in transgenic plants that express the tobacco mosaic virus coat protein gene. Science 232(4751):738–743

Acuña-Fuentes NL, Vargas-Hernandez M, Rivero-Montejo SDJ, Rivas-Ramirez LK, Macias-Bobadilla I, Palos-Barba V et al (2022) Antiviral activity of TiO2 NPs against tobacco mosaic virus in chili pepper (Capsicum annuum L.). Agriculture 12(12):2101

Akbar S, Wei Y, Zhang MQ (2022) RNA interference: promising approach to combat plant viruses. Int J Mol Sci 23(10):5312

Akhtar N, Perween S, Ansari AM (2023) Resistance to papaya ringspot virus: a review. https://doi.org/10.58321/AATCCReview.2023.11.03.68

Aman R, Ali Z, Butt H, Mahas A, Aljedaani F, Khan MZ et al (2018) RNA virus interference via CRISPR/Cas13a system in plants. Genome Biol 19(1):1

Awasthi S, Chauhan R, Narayan RP (2016) Plant viruses: history and taxonomy. In: Plant viruses: evolution and management. Springer Singapore, Singapore, pp 1–17

Bello MH, Moghaddam SM, Massoudi M, McClean PE, Cregan PB, Miklas PN (2014) Application of in silico bulked segregant analysis for rapid development of markers linked to Bean common mosaic virus resistance in common bean. BMC Genomics 15(1):903

Bhat AI, Rao GP (2020) Production of virus-resistant plants through transgenic approaches. In: Characterization of plant viruses: methods and protocols. Springer US, New York, pp 491–510

Bragard C, Caciagli P, Lemaire O, Lopez-Moya JJ, MacFarlane S, Peters D et al (2013) Status and prospects of plant virus control through interference with vector transmission. Annu Rev Phytopathol 51(1):177–201

Burch-Smith TM, Anderson JC, Martin GB, Dinesh-Kumar SP (2004) Applications and advantages of virus-induced gene silencing for gene function studies in plants. Plant J 39(5):734–746

Cagno V, Andreozzi P, D'Alicarnasso M, Jacob Silva P, Mueller M, Galloux M et al (2018) Broad-spectrum non-toxic antiviral nanoparticles with a virucidal inhibition mechanism. Nat Mater 17(2):195–203

Carvalho JL, de Oliveira Santos J, Conte C, Pacheco S, Nogueira EO, Souza TL et al (2015) Comparative analysis of nutritional compositions of transgenic RNAi-mediated virus-resistant bean (event EMB-PV051-1) with its non-transgenic counterpart. Transgenic Res 24(5):813–819

Chandrasekaran J, Brumin M, Wolf D, Leibman D, Klap C, Pearlsman M et al (2016) Development of broad virus resistance in non-transgenic cucumber using CRISPR/Cas9 technology. Mol Plant Pathol 17(7):1140–1153

Deng K, Lu Z, Yang H, Chen S, Li C, Cao D et al (2025) Efficient virus-induced gene silencing (VIGS) method for discovery of resistance genes in soybean. Plants 14(10):1547

Elsahhar S, Shahba M, Elsayed T, Mikhail M, Galal A (2022) Effect of chitosan nanoparticles (CS-NPs) on in vitro regeneration response and production of potato virus Y (PVY)-free plants of potato. Agronomy 12(11):2901

Elsharkawy MM (2024) Antiviral activity of nanoparticles in plants. In: Nanotechnology in plant health. CRC Press, pp 285–299

Endo T, Fujii H, Shimada T (2022) Development of cultivar identification system using 12 InDel markers for widely distributed citrus cultivars in Japan. Jpn Agric Res Q: JARQ 56(4):329–340

Feng S, Chen JD, Bao AK (2025) Establishment and optimization of a tobacco rattle virus-based virus-induced gene silencing in Atriplex canescens. Plant Methods 21(1):107

Fuentes A, Carlos N, Ruiz Y, Callard D, Sánchez Y, Ochagavía ME et al (2016) Field trial and molecular characterization of RNAi-transgenic tomato plants that exhibit resistance to tomato yellow leaf curl geminivirus. Mol Plant-Microbe Interact 29(3):197–209

Gao H, Li H (2025) Marker-assisted selection (MAS) in soybean breeding. Mol Plant Breed 16:35

Gonsalves D (2003) The papaya story: a special case or a generic approach? NABC

Gunupuru LR, Perochon A, Ali SS, Scofield SR, Doohan FM (2018) Virus-induced gene silencing (VIGS) for functional characterization of disease resistance genes in barley seedlings. In: Barley: methods and protocols. Springer New York, New York, pp 95–114

Guria A, Pandi G (2019) Commercial applications of transgenic crops in virus management. In: Advances in plant Transgenics: methods and applications. Springer Singapore, Singapore, pp 213–238

Irfan M, Kumar P, Ahmad MF, Siddiqui MW (2023) Biotechnological interventions in reducing losses of tropical fruits and vegetables. Curr Opin Biotechnol 79:102850

Jiang X, Khan S, Dykes A, Stulz E, Zhang X (2025) Biogenic synthesis of silver nanoparticles and their diverse biomedical applications. Molecules 30(15):3104

Jones RA, Naidu RA (2019) Global dimensions of plant virus diseases: current status and future perspectives. Annu Rev Virol 6(1):387–409

Kang H, Ga YJ, Kim SH, Cho YH, Kim JW, Kim C, Yeh JY (2023) Small interfering RNA (siRNA)-based therapeutic applications against viruses: principles, potential, and challenges. J Biomed Sci 30(1):88

Kasai K, Morikawa Y, Sorri VA, Valkonen JPT, Gebhardt C, Watanabe KN (2000) Development of SCAR markers to the PVY resistance gene Ry adg based on a common feature of plant disease resistance genes. Genome 43(1):1–8

Khan ZA, Kumar R, Dasgupta I (2022) CRISPR/Cas-mediated resistance against viruses in plants. Int J Mol Sci 23(4):2303

Kis A, Hamar É, Tholt G, Bán R, Havelda Z (2019) Creating highly efficient resistance against wheat dwarf virus in barley by employing CRISPR/Cas9 system. Plant Biotechnol J 17(6):1004

Kumar R, Lal MK, Tiwari RK, Kumar A, Behera B, Nayak L, Dash GK, Sahoo SK, Jena J, Lal P, Altaf MA (2023) Traditional and emerging climate-resilient agricultural practices for enhancing food production and nutritional quality. In: Aftab T (ed) New frontiers in plant-environment interactions: innovative technologies and developments, Environmental Science and Engineering. Springer, Cham, pp 551–570. https://doi.org/10.1007/978-3-031-43729-8_21

Lefeuvre P, Martin DP, Elena SF, Shepherd DN, Roumagnac P, Varsani A (2019) Evolution and ecology of plant viruses. Nat Rev Microbiol 17(10):632–644

Levanova A, Poranen MM (2018) RNA interference as a prospective tool for the control of human viral infections. Front Microbiol 9:2151

Li G, Shao J, Wang Y, Liu T, Tong Y, Jansky S et al (2022) Rychc confers extreme resistance to potato virus Y in potato. Cells 11(16):2577

Li X, Tao N, Xu B, Xu J, Yang Z, Jiang C et al (2024) Establishment and application of a root wounding–immersion method for efficient virus-induced gene silencing in plants. Front Plant Sci 15:1336726

Lindbo JA, Falk BW (2017) The impact of "coat protein-mediated virus resistance" in applied plant pathology and basic research. Phytopathology 107(6):624–634

Liu S, Tian W, Liu Z, Wei X, Yuan K, Du W et al (2024) Biosynthesis of cupric oxide nanoparticles: its antiviral activities against TMV by directly destroying virion and inducing plant resistance. Phytopathol Res 6(1):30

Lu R, Martin-Hernandez AM, Peart JR, Malcuit I, Baulcombe DC (2003) Virus-induced gene silencing in plants. Methods 30(4):296–303

Lundstrom K (2020) Viral vectors applied for RNAi-based antiviral therapy. Viruses 12(9):924

Ma Y, Li H, Wang D, Liu N, Zhi H (2010) Molecular mapping and marker assisted selection of soybean mosaic virus resistance gene RSC12 in soybean. Legume Genomics Genet 8:41

Mangal V, Sood S, Raigond B, Dalamu D, Thakur AK, Kumar V et al (2021) Updates on molecular markers linked to potato virus Y (PVY) resistance genes in potato. Potato J 48(2):137

Masood HA, Qi Y, Zahid MK, Li Z, Ahmad S, Lv JM et al (2024) Recent advances in nano-enabled immunomodulation for enhancing plant resilience against phytopathogens. Front Plant Sci 15:1445786

Mazumder P, Mondal F, Sarkar M, Majumdar A, Biswas KK, Sharma SK, Lal MK, Tiwari RK, Kumar R, Roy A (2024) Biological and molecular characterization of a new isolate of tomato mottle mosaic virus causing severe shoestring and fruit deformities in tomato plants in India. Plants 13:2811. https://doi.org/10.3390/plants13192811

Nalla MK, Schafleitner R, Pappu HR, Barchenger DW (2023) Current status, breeding strategies and future prospects for managing chilli leaf curl virus disease and associated begomoviruses in Chilli (Capsicum spp.). Frontiers in Plant Sci 14:1223982

Nevame AYM, Xia L, Nchongboh CG, Hasan MM, Alam MA, Yongbo L et al (2018) Development of a new molecular marker for the resistance to tomato yellow leaf curl virus. Biomed Res Int 2018(1):8120281

Nissim M, Shoshani S, Jacobi G, Afrimzon E, Levy L, Banin E et al (2025) Synthesis, characterization, and potential applications of new silica-Phosphonium nanoparticles. ACS Omega 10(31):34548–34558

Niu L, Yan H, Sun Y, Zhang D, Ma W, Lin Y (2022) Nanoparticle facilitated stacked-dsRNA improves suppression of the lepidopteran pest Chilo suppressalis. Pestic Biochem Physiol 187:105183

Prins M, Laimer M, Noris E, Schubert J, Wassenegger M, Tepfer M (2008) Strategies for antiviral resistance in transgenic plants. Mol Plant Pathol 9(1):73–83

Pujol M, Alexiou KG, Fontaine AS, Mayor P, Miras M, Jahrmann T, Aranda MA (2019) Mapping cucumber vein yellowing virus resistance in cucumber (Cucumis sativus L.) by using BSA-seq analysis. Front Plant Sci 10:1583

Qiao L, Niño-Sánchez J, Hamby R, Capriotti L, Chen A, Mezzetti B, Jin H (2023) Artificial nanovesicles for dsRNA delivery in spray-induced gene silencing for crop protection. Plant Biotechnol J 21(4):854–865

Reddy MK, Nagendran K, Kumari S, Pradeep RM, Karthikeyan G (2024) Global scenario of begomovirus diseases in vegetable crops. Veg Sci 51:43–53

Robertson G, Burger J, Campa M (2022) CRISPR/Cas-based tools for the targeted control of plant viruses. Mol Plant Pathol, 23(11):1701-1718

Rössner C, Lotz D, Becker A (2022) VIGS goes viral: how VIGS transforms our understanding of plant science. Annu Rev Plant Biol 73(1):703–728

Roy G, Afandi F, Vishnuvardhannaidu K, Raut RN, Amir M (2025) A comprehensive review of genomics assisted breeding and mutation strategies for yellow vein mosaic virus resistance in mung bean. Int J Res Agron 8(5):176–186

Sanford JC, Johnston SA (1985) The concept of parasite-derived resistance—deriving resistance genes from the parasite's own genome. J Theor Biol 113(2):395–405

Sathees N, Shoba D, Mani N, Saravanan S, Kumari MP, Pillai MA (2022) Tagging of SSR markers associated to yellow mosaic virus resistance in black gram (Vigna mungo (L.) Hepper). Euphytica 218(3):23

Sayago-Ayerdi S, García-Martínez DL, Ramírez-Castillo AC, Ramírez-Concepción HR, Viuda-Martos M (2021) Tropical fruits and their co-products as bioactive compounds and their health effects: a review. Foods 10(8):1952

Scofield SR, Huang L, Brandt AS, Gill BS (2005) Development of a virus-induced gene-silencing system for hexaploid wheat and its use in functional analysis of the Lr21-mediated leaf rust resistance pathway. Plant Physiol 138(4):2165–2173

Shidore T, Zuverza-Mena N, White JC, da Silva W (2021) Nanoenabled delivery of RNA molecules for prolonged antiviral protection in crop plants: a review. ACS Appl Nano Mater 4(12):12891–12904
Shrestha S, Fu Y, Michael VNE, Meru G (2022) Whole genome re-sequencing and bulk segregant analysis reveals chromosomal location for papaya ringspot virus W resistance in squash. Front Plant Sci 13:848631
Stashevski Z, Il'inskaya ON (2011) Virus resistance of transgenic potato plants (Solanum tuberosum L.) bearing PVY NTN-CP gene of potato virus Y coat protein. Russ J Genet Appl Res 1(2):103–111
Steinmetz NF, Chariou PL (2019) U.S. Patent Application No. 16/306,855
Subramaniyan R, Narayana M, Krishnamoorthy I, Natarajan G, Gandhi K (2022) Novel and stable QTL regions conferring resistance to MYMV disease and its inheritance in blackgram (Vigna mungo (L.) Hepper). J Genet 101(1):18
Sundaresha S, Bairwa A, Tomar M, Kumar R, Venkatasalam EP, Sagar V, Bhardwaj V, Sharma S (2022a) In vitro method for synthesis of large-scale dsRNA molecule as a novel plant protection strategy. In: Mysore KS, Senthil-Kumar M (eds) Plant gene silencing, Methods in molecular biology, Vol 2408. Humana, New York, pp 211–226. https://doi.org/10.1007/978-1-0716-1875-2_14
Sundaresha S, Jeevalatha A, Kumar R, Sood S, Sharma S, Bhardwaj V, Singh BP, Chakrabarti SK (2022b) RNA interference: a versatile tool to augment plant protection strategies in potato. In: Chakrabarti SK, Sharma S, Shah MA (eds) Sustainable management of potato pests and diseases. Springer, Singapore, pp 369–388. https://doi.org/10.1007/978-981-16-7695-6_15
Sundaresha S, Tiwari JK, Bhardwaj V, Sharma S, Jeevalatha A, Kumar R, Sagar V, Singh BP, Chakrabarti SK, Kumar M (2024) Genetic engineering in potato: status and future perspectives. In: Kalia P (ed) Genetic engineering of vegetable crops. CABI, Nosworthy Way/Wallingford, pp 133–140. https://doi.org/10.1079/9781800627123.0008
Surya Krishna S, Viswanathan R, Valarmathi R, Lakshmi K, Appunu C (2023) CRISPR/Cas-mediated genome editing approach for improving virus resistance in sugarcane. Sugar Tech 25(4):735–750
Talakayala A, Ankanagari S, Garladinne M (2022) CRISPR-Cas genome editing system: a versatile tool for developing disease resistant crops. Plant Stress 3:100056
Taliansky M, Samarskaya V, Zavriev SK, Fesenko I, Kalinina NO, Love AJ (2021) RNA-based technologies for engineering plant virus resistance. Plants 10(1):82
Taylor NJ, Halsey M, Gaitán-Solís E, Anderson P, Gichuki S, Miano D et al (2012) The VIRCA project: virus resistant cassava for Africa. GM Crops Food 3(2):93–103
Tennant P, Rampersad S, Alleyne A, Johnson L, Tai D, Amarakoon I et al (2024) Viral threats to fruit and vegetable crops in the caribbean. Viruses 16(4):603
Tepfer M (2002) Risk assessment of virus-resistant transgenic plants. Annu Rev Phytopathol 40(1):467–491
Tiberini A, Cillo F, Gentili A, Bertin S (2025) Tomato spotted wilt virus (Orthotospovirus tomatomaculae), a cyclically occurring threat to crop production worldwide. Ann Appl Biol 186(2):93–114
Tong C, Huang S, Shi Y, Wu Q, Shangguan L, Yu H et al (2023) Development of novel specific molecular markers for the Sw-5b gene to assist with tomato spotted wilt virus-resistant tomato breeding. Phytopathol Res 5(1):59
Tripathi JN, Ntui VO, Ron M, Muiruri SK, Britt A, Tripathi L (2019) CRISPR/Cas9 editing of endogenous banana streak virus in the B genome of Musa spp. overcomes a major challenge in banana breeding. Commun Biol 2(1):46
Wang Y, Zafar N, Ali Q, Manghwar H, Wang G, Yu L et al (2022) CRISPR/Cas genome editing technologies for plant improvement against biotic and abiotic stresses: advances, limitations, and future perspectives. Cells 11(23):3928
Wang Z, Cao S, Xu X, He Y, Shou W, Munaiz ED et al (2023) Application and expansion of virus-induced gene silencing for functional studies in vegetables. Horticulturae 9(8):934

Welegama T, Rafii MY, Ahmad K, Ramlee SI, Oladosu Y (2021) Development of high yield and tomato yellow leaf curl virus (TYLCV) resistance using conventional and molecular approaches: a review. Biocell 45(4):1069

Wytinck N, Manchur CL, Li VH, Whyard S, Belmonte MF (2020) dsRNA uptake in plant pests and pathogens: insights into RNAi-based insect and fungal control technology. Plants 9(12):1780

Xiao Z, Xing M, Liu X, Fang Z, Yang L, Zhang Y et al (2020) An efficient virus-induced gene silencing (VIGS) system for functional genomics in brassicas using a cabbage leaf curl virus (CaLCuV)-based vector. Planta 252(3):42

Yogindran S, Rajam MV (2021) Host-derived artificial miRNA-mediated silencing of ecdysone receptor gene provides enhanced resistance to Helicoverpa armigera in tomato. Genomics 113(1):736–747

Zand Karimi H, Innes RW (2022) Molecular mechanisms underlying host-induced gene silencing. Plant Cell 34(9):3183–3199

Zhan X, Zhang F, Zhong Z, Chen R, Wang Y, Chang L et al (2019) Generation of virus-resistant potato plants by RNA genome targeting. Plant Biotechnol J 17(9):1814–1822

Zhang T, Zheng Q, Yi X, An H, Zhao Y, Ma S, Zhou G (2018) Establishing RNA virus resistance in plants by harnessing CRISPR immune system. Plant Biotechnol J 16(8):1415–1423

Zhang Y, Zhang X, Yu L, Yan Y, Zhu S, Huang W et al (2025) An efficient Rice virus-induced gene silencing system mediated by wheat dwarf virus. Appl Sci 15(11):5818

Zhao D, Zhong GY, Song GQ (2020) Transfer of endogenous small RNAs between branches of scions and rootstocks in grafted sweet cherry trees. PLoS One 15(7):e0236376

Zhao L, Liu S, Abdelsalam NR, Carver BF, Bai G (2021) Characterization of wheat curl mite resistance gene Cmc4 in OK05312. Theor Appl Genet 134(4):993–1005

# Advances in Biotechnology for Engineering Virus Resistance in Vegetable Crops

22

Darshana Goswami, Uchakankhi Kashyap, Adrij Pawan Neog, Anindita Saikia, and Nima D. Namsa

**Abstract**

Viral diseases significantly impact the productivity and quality of vegetable crops, posing a major challenge to global food and nutritional security. Traditional control methods, such as chemical treatments, vector management, and crop rotation, offer limited success due to the rapid evolution of viruses and their wide host range. Biotechnological interventions have emerged as effective and sustainable strategies for developing virus-resistant vegetable crops. Genetic engineering, RNA interference (RNAi), CRISPR/Cas genome editing tool, and marker-assisted selection, which enable precise manipulation of host or viral genes to confer resistance, have been discussed in this book chapter. Transgenic approaches involve the introduction of viral coat protein genes, movement protein genes, or other viral sequences into plants to trigger pathogen-derived resistance (PDR). RNAi-based strategies use small interfering RNAs (siRNAs) or hairpin RNAs to silence essential viral genes, effectively reducing viral replication. Recent advancements in CRISPR/Cas systems have opened new avenues for editing plant susceptibility genes or viral genomes, providing a targeted, non-transgenic method for inducing resistance. Several successful applications of biotechnological interventions have been demonstrated in vegetable crops such as tomato, pepper, cucumber, and squash against viruses such as tomato yellow leaf curl virus (TYLCV), cucumber mosaic virus (CMV), and zucchini yellow mosaic virus (ZYMV). Biotechnological intervention allows for the pyramiding of resistance traits, enhancing durability and broad-spectrum protection. Biotechnological tools offer promising, eco-friendly alternatives to traditional

D. Goswami · U. Kashyap · A. P. Neog · A. Saikia · N. D. Namsa (✉)
Department of Molecular Biology and Biotechnology, Tezpur University, Tezpur, Assam, India
e-mail: namsa@tezu.ernet.in





R. Kumar et al. (eds.), *Vegetable Crops and Viral Pathogens Volume I*, Plant Pathology for Sustainable Agriculture, https://doi.org/10.1007/978-981-95-7265-6_22

methods, paving the way for the development of virus-resistant vegetable varieties and sustainable agricultural practices.

**Keywords**

RNA interference · Vegetable crops · Virus-resistant · Susceptibility genes · CRISPR/Cas systems

## 22.1 Introduction

With the increasing demands of livestock and processed food, it is necessary to establish some modern methods to stabilise global food security. With the increasing instability in the climatic conditions, the management of crops is important to prevent the unnatural death of plants. However, one of the major reasons for the decrease in agricultural output is due to microbiological diseases (i.e. viral infections), which is one important jeopardy. Unlike bacterial or fungal infections, which can be effectively managed by fertilisers or agricultural chemicals, plant viruses cannot be managed using such methods. Resistant types crop type was documented to be produced using traditional breeding. The conventional breeding though cost effective has its own drawbacks, like time-ineffective, resource-intensive, and laborious. Furthermore, in many cases, viral resistance in various plants is not available or not possible. Another challenge that occurs while attempting to produce resistance for various viral diseases in plants is particularly preserving the traits like a high-yielding cultivar. For these limitations, alternative methods are required to alleviate the existing and potential viral infections in crops, and scientists have come up with various methods like RNA Inference, CRISPR Cas9 editing, and recombinant DNA technology. A major scientific approach that has acquired necessary attention is the genetic modification of crop plants (Baulcombe 1996). Genetic engineering of plants directly modifies the plants to prevent viral infection using various technologies and biotechnological instruments. This methodology navigates the exact insertion of features across species borders, successfully eliminating prefertilisation barriers. Genetically engineering crops to resist plant viral diseases provides a robust technique for preventing agricultural losses, alleviating economic instability, and contributing to perianal nutritional availability.

## 22.2 Plant-Virus Interaction

Vegetable crops are susceptible to viruses, obligate biotrophic pathogens, which are the most serious threat to their production. Viruses do not have secretion systems like bacteria and fungi, but they exploit numerous viral proteins as effectors, which are represented as movement proteins, coat proteins, replicases, and viral proteins of RNA silencing (VSRs). These factors serve as effectors for the virus to infect the host cells, replicate and spread systematically, and counteract host defences. The plant host system has developed layers of immunity. At the basal layer, the double-stranded RNA (dsRNA) of the virus, which is synthesised during replication, is the PAMP that triggers RNA silencing and multiple PTI-like antiviral defences. In response to this, viruses encode VSRs, which block the silencing machinery and facilitate infection. More specifically, in the case of vegetable crops, effector-triggered immunity (ETI) is activated by resistance R genes that perceive specific viral proteins and defence responses, including the hypersensitive response (HR). As in tomato, the Tm-2 gene can recognise the tomato mosaic virus movement protein, and the tomato Sw-5 gene can activate localised cell death to prevent the movement of the tomato spotted wilt virus (El-Sappah et al. 2022). As viruses are totally dependent on living host tissue, hypersensitive reaction is an effective antiviral strategy, unlike in necrotrophic fungal pathogens, where cell death favours infection. Additionally, several plant viruses, including begomoviruses and tobamoviruses, release effectors that decrease systemic defensive responses by interfering with chloroplast function, plasmodesmata control, and plant hormone signalling. Rapid recombination and mutation rate speed up viral evolution, allowing for the creation of novel strains that can overcome host resistance or spread through other vegetable crop hosts (Beachy 1994). Even while it is less frequent in viruses than in fungi and bacteria, horizontal gene transfer can nonetheless happen in mixed infections and result in synergistic interactions that enhance the severity of the disease. For instance, coinfection of solanaceous crops by potyviruses and cucumovirus frequently causes more severe symptoms and lower yields compared to the impact of either virus alone.

## 22.3 Engineered Resistance to Plant Viruses

### 22.3.1 Pathogen-Derived Resistance

Plant viruses have become favourable models to target genetic engineering techniques due to their relatively short genomes encoding only necessary proteins, the majority required in their replication cycles. Pathogen-derived resistance (PDR), initially introduced by Sanford and Johnson, is one of the most significant theories in this field (Sanford and Johnson 1985). This approach proposed the widespread utilisation of pathogen-derived genes to produce targeted resistance in host plants. The fundamental idea underlying PDR suggests that if the modifications do not impair vital host activities, it is possible to manipulate some viral genes or their modified derivatives to interrupt the pathogen's life cycle and develop resistance. Viruses encode specific genes necessary for the replication of their genome, encapsidation, and subsequent cell-to-cell and systemic dissemination inside the plant system, followed by the transmission to other hosts via various vectors or the environment. These genes have turned out to be great candidates for PDR-induced resistance development. One of the most effective and well-investigated uses of PDR has been the introduction of viral coat protein (CP) genes, which results in CP-mediated resistance. Transgenic expression of CP in host plants has resulted in high levels of operational resistance to various RNA viruses, notably tobacco mosaic virus (TMV), potato virus X (PVX), alfalfa mosaic virus (AIMV), cucumber mosaic virus (CMV), and tobacco rattle virus (TRV) (Beachy 1994; Wilson 1993; Baulcombe 1996). The latter observation substantially supported the theory that CP-mediated resistance affects the disassembly of arriving virions, which is disturbed in the presence of transgenic CP. In addition to coat protein techniques, further PDR strategies have been successfully attempted. These include satellite RNAs present in some plant viruses, which can decrease or inhibit symptoms causing diseases. This satellite RNA is converted into cDNA copies and introduced into the plant genome. Expression of these cDNA transcribes RNA satellite, reducing disease symptoms. Movement proteins (MP) produced by viruses, allowing cell-to-cell mobility through plasmodesmata, can be prevented using PDR. Plant genome is engineered to produce MP that are nonfunctioning or interfere with the movement of viruses, countering the virus from spreading. PDR has also engineered plants to produce mutated or modified replicase proteins, preventing the virus proficiency to replicate within the host cell. Each of these strategies focuses on a distinct step of the life cycle of the virus and has been verified in greenhouse and limited field trials. However, the PDR-developed transgenic crops are not accessible and available throughout the globe due to unknown risks. The transgenic crops might undergo recombination between the viral RNA and the pathogenic viral genome, which has been established in laboratory conditions. Such hazards have prompted discussion over the prospective implementation of PDR techniques in the field, unleashing ecological and evolutionary consequences. However, PDR has additionally brought forth some exceptional achievements, for example, transgenic papaya plants can resist papaya ringspot virus from infecting the plant, successfully preventing the local papaya sector from collapse (Tripathi et al. 2007). This instance serves as a paradigm for the future use of genetically modified crops, especially vegetable crops, and demonstrates the tremendous potential of PDR when used appropriately.

### 22.3.2 Posttranscriptional Gene Silencing as Mean to Achieve Virus Resistance

Posttranscriptional gene silencing (PTGs), also known as RNA silencing, has been used to develop resistance techniques that circumvent one of the main drawbacks of traditional pathogen-derived resistance (PDR), which requires high-level production of viral proteins. Plants use PTGS, an adaptive defence system, to target RNA degradation to defend against invading nucleic acids, such as viruses and transposable elements. First seen in transgenic Petunia lines, this phenomenon was caused by the unknown cosuppression of both transgenes and endogenous genes when extra copies of the chalcon synthase gene were introduced (Napoli et al. 1990). Later, it was suggested as the fundamental cause of several instances where PDR techniques produced virus resistance despite the use of dysfunctional or untranslatable virus gene sequences. These results showed that resistance was, ironically, associated with the sequence-specific destruction of viral transgenic RNA rather than with its substantial accumulation. As a result, PTGs developed naturally as an extension of PDR and were rapidly identified as a powerful, all-purpose antiviral defence mechanism. PTGS has been effectively used in vegetable crops to combat a variety of viruses, regardless of whether they have DNA or RNA genomes. Strong resistance to TMV has been demonstrated by transgenic tomato, pepper, and cucurbit plants that produce viral gene segments as sense, antisense, or inverted-repeat constructions have demonstrated robust resistance to polyviruses, tobacco mosaic virus (TMV), and cucumber mosaic virus (CMV) (Vinodhini et al. 2023; Carr 2024). Viral double-stranded RNA (dsRNA), whether from replication intermediates or transgenic inverted repeats, is detected and converted into 21–23 nucleotide siRNAs, or small interfering RNAs, by plants' Dicer-like RNases, which is the mechanism by which PTGS is initiated (Kuo and Falk 2020). RISC (RNA-inducing silencing complexes) control the sequence-specific destruction of homologous viral RNAs and are subsequently formed from these siRNAs (Fig. 22.1). Furthermore, silencing signals are amplified by RNA-dependent RNA polymerases, which produce secondary dsRNAs from abnormal virus transcripts. The modified siRNAs serve as mobile signals and mediate the local destruction of viral RNA, which helps the plant's antiviral defence propagate across its whole system.

The miRNA pathway begins in the nucleus with transcription, giving the hairpin-folded primary miRNA (pri-miRNA). The transcription hairpin-folded primary miRNA inside the nucleus marks the start of the miRNA pathway. The pri-miRNA is cleaved to form the precursor miRNA by DCL, and then the 22-nucleotide miRNA duplex is formed inside the nucleus. The formed duplex gets transported into the cytoplasm and is loaded into the RISC. The guide strand of the miRNA is retained in the RISC, whereas the passenger strand is discarded

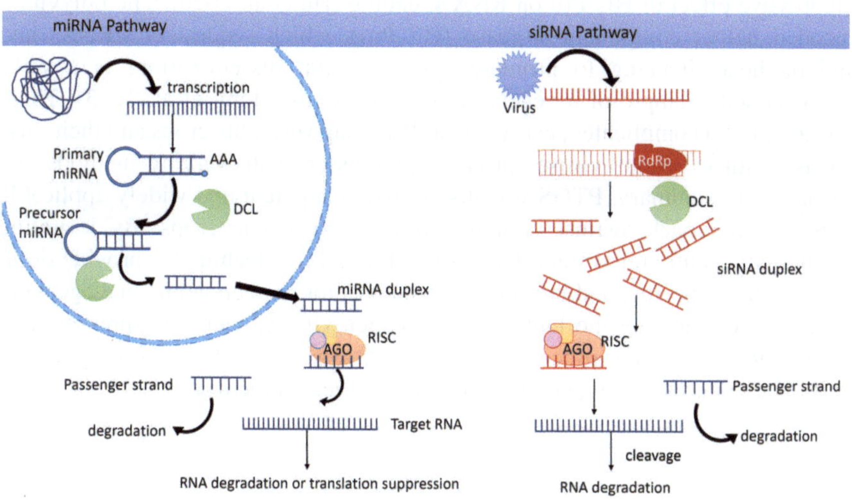


**Fig. 22.1** The RNA interference pathway (miRNA and siRNA) in the plant cell

from the miRNA duplex. The binding of the RISC-bound guide RNA (by Watson-Crick base-pairing) with the target RNA leads to cleavage or translational arrest of the target RNA, whereas the siRNA pathways function in the cytoplasm when the target RNA (which is normally an aberrant RNA or that of an infecting plant virus) gets recognised by the DCL, which results in cleavage of the target RNA into short (21–24 nucleotide) RNA duplexes. Similarly, one strand (passenger strand) is released and degraded while the other strand that binds to RISC (guide strand) is retained. And cleavage occurs when the RISC-bound guide RNA binds to a target RNA. In the cell, the RNAi response depends upon the RNA-dependent RNA polymerase activity. The keywords are AAA, 3′ polyadenylate tail; AGO, argonaute protein; DCL, Dicer-like protein; Pri-miRNA, primary microRNA; RdRP, RNA-dependent RNA polymerase; RISC, RNA-induced silencing complex; and RNAi, RNA interference. One important aspect of PTGS-based resistance is that it eliminates biosafety issues associated with the disadvantages of PDR, by not requiring the production of the entire viral proteins. Rather, resistance results from the quick, sequence-specific breakdown of viral RNA, which is less dangerous and more effective. Because of this, PTGS has become especially desirable in crops, as fast viral development and mixed infections are prevalent. In fact, transgenic vegetable plants that have been modified using dsRNA constructs generated from viruses have shown broad and long-lasting resistance in both greenhouse and natural environments. Since most plant viruses have developed proteins that function as silencing suppressors, the evolutionary relevance of PTGS is highlighted. As a primary line of plant antiviral defence, RNA silencing is highlighted by these viral suppressors, which disrupt multiple stages associated with the PTGS pathway, from siRNA biogenesis to RISC-mediated cleavage. The potyvirus fusion protein P1/HC-Pro was the first viral RNA silencing suppressor to be identified, with early studies on its mechanism showing that P1 enhances the suppressive effect of HC-Pro on RNA silencing (Hu et al. 2020). The polyvirus-encoded helper component-protease (HC-Pro), which sequesters siRNAs and inhibits host silencing to promote systemic viral dissemination, is a well-characterised example of a vegetable-infecting virus (Hong et al. 2023). There appears to be a complicated, evolving battle among vegetable crops and their viral diseases, since various viruses produce suppressors with varying methods and potencies. In summary, PTGS was discovered as a potent and widely applicable resistance approach against viral infections of vegetable crops. By levitating plants' native antiviral system, PTGS-based transgenic techniques provide durable, sequence-specific, and biosafety alternatives for viral epidemic management. Furthermore, knowing the interactions between viral silencing suppressors and the host RNA silencing process opens up new avenues for building long-lasting resistance features using genome editing and targeted breeding.

## 22.4 Functional Genomics and Genome Editing in Virus-Resistant Vegetable Crops

Various RNA silencing precursors have been used to design resistance in vegetable crops, including tomato, pepper, squash, and cucurbits. These transcripts include sense and antisense siRNA, microRNA (miRNA), hairpin RNAs (hpRNAs), and genes encoding for proteins required to survive inside the host plant cell and counter the host plant's defence mechanism such as movement protein (MP), coat protein (CP), RNA-dependent RNA polymerase (RdRP), and replication-associated protein and proteases, used to employed to create hpRNA constructs. Once transcribed, these structures are processed into siRNAs, which serve as the primary antiviral agents that direct RISC to destroy viral RNA. In many plant species, the RNAi techniques have been proven to be an effective approach in controlling the virus diseases in them (Rosa et al. 2018; Lindbo and Falk 2017). Majorly dsRNAs or structured RNAs, like hairpin RNAs (powerful inducer of RNAi), are produced in plants instead of using untranslatable RNAs. It was observed that the benefits of the transgenic RNAi-bound viral resistance that has been used in agricultural systems so far are extremely dramatic. One major achievement is the development of a generation of genetically modified papaya ringspot virus-resistant papaya, used in Hawaii when there were no alternative methods available for controlling the PRSV in the agricultural sector (Gonsalves 2006). Now the transgenic virus-resistant papaya dominates the Hawaiian papaya market, which has essentially saved the Hawaiian papaya industry (Gonsalves 2006). Likewise, the papaya plants were simultaneously engineered towards multivirus resistance using the RNAi technique to reduce susceptibility to more than one virus. Engineered to have robust, simultaneous RNAi resistance against two potyviruses and one cucumovirus (zucchini yellow mosaic virus, watermelon mosaic virus and cucumber mosaic virus, respectively), the first commercially grown Destiny III yellow crookneck squash (*Cucurbita pepo*) was accepted by the consumers (Fuchs et al. 1998). Despite its effectiveness, the hpRNA technique has drawbacks, including fluctuating efficiency and the requirement for sustained transgene expression. To address these limitations, innovative high-throughput techniques based on artificial microRNAs (amiRNAs) and trans-acting siRNAs (ta-siRNAs) are currently developed. These technologies allow for precise viral sequence targeting and may be multiplexed to offer resistance to numerous unrelated viruses at once. For example, amiRNA constructs have been created to target numerous begomoviruses that infect tomatoes, and microRNA-induced gene silencing (MIGS) has been tried in cucurbits to give wide-ranging resistance towards both RNA and DNA viruses (Sharma and Prasad 2020; Al-Roshdi et al. 2023). These enhanced RNA silencing methods increase the possibility of transgenic resistance by boosting distinctiveness, endurance, and adaptability across a wide range of vegetable systems. Aside from viral gene-derived constructions, nonviral genes are also used in GE techniques to develop virus resistance. The integration of resistance-associated (R) transcripts, ribosome-associated protein, and inhibitors for protease has given alternate tactics, frequently affecting host cellular processes that viruses use. For example, manipulating translational control by ribosomal inactivation has reduced viral synthesis of proteins in tomato, while inhibiting protease have been demonstrated to cause disruption with viral maturation in cucurbits. Furthermore, RNA silencing of host genes required for viral replication has resulted in resistance in a variety of vegetable crops. Extending this principle, several treatments have targeted transmitting insects of viruses, decreasing transmission and dissemination in the field. The establishment of genome editing technology has opened the door for new possibilities regarding viral surveillance in vegetable crops. Site-specific nucleases (SSNs), which include meganucleases, zinc finger nucleases (also called ZFNs), transcription activator-like effector nucleases, or TALENs, and, most notably, the clustered commonly interconnected palindromic repeats (CRISPR)-associated 9 (CRISPR/Cas9) system have been used to interfere with viral genomes or host susceptibility genes. In early genome editing efforts, ZFNs and TALENs showed some promise in targeting begomoviruses (Sera 2005). However, the limitation with these technologies is tailoring the DNA-binding protein to target the viral sequence, and extensive protein engineering needs hampered wider acceptance (Sera 2005; Mori et al. 2013). In contrary to popular belief, the CRISPR/Cas9 system offers a simple, adaptable, and extremely effective mechanism for identifying genetic code, making it the primary method in ongoing virus-resistance projects (Zaidi et al. 2016a, b). CRISPR/Cas9 has been used effectively in vegetable crops to develop resistance to gemini-viruses, which include the tomato yellow leaf curl virus (TYLCV), bean yellow dwarf virus (BeYDV), and tobacco curly shoot virus (TbCSV) (Ghorbani Faal et al. 2020; Tashkandi et al. 2018; Cao et al. 2020; Zaidi et al. 2016b). Guide RNAs (gRNAs) have been developed for targeting the virus intergenic region (IR), coat protein, and Rep genes, inhibiting replication and systemic infection. Beyond directly targeting viral genomes, CRISPR/Cas has been used to modify host-responsive proteins such as eukaryotic initiation factor 4E (eIF4E) and its isoform eIFiso4E, both of which are essential for viral RNA translation (Yoon et al. 2020). Knockouts or changes associated with these genes in tomato, cucumber, and pepper have resulted in long-lasting resistance to a variety of RNA-based viruses, especially potyviruses and cucumoviruses, without impacting plant growth or production (Gomez et al. 2019). Collectively, these results demonstrate the promise of silenced RNA and genetic modification as complementary techniques for long-term resistance to viruses in vegetable crops. The suppression of RNA allows adaptability and broad-spectrum protection by degrading viral RNA at a particular sequence, whereas genome editing gives precision and permanence by changing host or viral DNA. The combination of these techniques, together with conventional cultivation and vector management tactics, will most certainly determine the subsequent stage in developing environmentally friendly alternatives for managing viral infections in vegetable production.

## 22.5 Advantages of Genetically Modified Virus-Resistant Plants

Genetic engineering is a vital method for creating virus-resistant cultivars that can protect global food security and enhance the economic viability of vegetable crops because of its adaptability, accuracy, and efficiency. Genetic modification is an important method for producing virus-resistant plants, which can be utilised to manage the economic plant livestock and increase the sustainability of the vegetable crops for the accuracy, efficiency, and adaptability. By overcoming many of the drawbacks of traditional breeding and agronomic techniques, genetic engineering (GE) has become a potent tool in the fight against plant viruses (Niraula and Fondong 2021). Shorter lead times, generally low manufacturing costs, high efficiency, ease of gene transfer through plant barriers, minimal effect on the genetic makeup of elite varieties, and suitability for vegetatively propagated crops are just a few of the clear advantages that genetic engineering (GE) offers over traditional methods. The short lead time of GE is one of its biggest advantages over plant breeding, which sometimes takes a few generations of selective breeding to get resistant lines. On the other hand, even for crops like cassava that are often challenging to work with, the transformation of plants can be finished in a matter of weeks to months. Designing, synthesising, and introducing gene constructs into plant tissue may be completed very fast. The turnaround time has decreased further with the creation of the institutional transformation facilities that specialise in certain crops. This quick development is more economical as well as time-efficient. Reduced labour input, quicker transformation procedures, and the accessibility of centralised facilities with costly equipment and technological know-how all contribute to the comparatively cheap manufacturing cost of GE-based viral resistance (Niraula and Fondong 2021; Dong and Ronald 2019). For commodities like potatoes and tomatoes, the expense of transformation is generally low in nations like the USA when compared to the time and costs of extensive breeding projects. The effectiveness and longevity of GE's designed resistance are another significant benefit. Viral genomes may now be precisely targeted using a variety of techniques thanks to developments in cell biology and genomics. Although somewhat successful, earlier transgenic plant generations that expressed movement proteins, antisense RNAs, or viral coat proteins occasionally lacked endurance. The creation of RNA silencing technologies, such as artificial microRNAs (amiRNAs) and synthetic transacting siRNAs (syn-tasiRNAs), has allowed for the creation of constructs that selectively target conserved viral regions, significantly boosting the stability and resilience of resistance (Hu et al. 2020; Hong et al. 2023; Rosa et al. 2018). Additionally, these short RNA-based tactics can be multiplexed to target several viruses or viral types at once and lower the chance of recombination occurring with infecting viruses. By allowing the immediate targeting of the viral genome or host-associated genes and the potential to multiplex guide RNAs to improve durability, the development of the CRISPR/Cas9 approach has significantly broadened the toolkit (Zaidi et al. 2016a; Ghorbani Faal et al. 2020). Although resistance durability is at risk due to viral mutation, this issue can be lessened by carefully selecting target locations in highly prevalent viral areas. Along with effectiveness, GE offers special chances to get over species barriers that restrict traditional breeding. Traditional gene transfer is almost impossible since virus resistance genes are frequently found in wild cousins, which are sexually incompatible with domesticated species. By permitting the immediate transmission of resistance genes or brief functional gene segments between distantly related species, GE gets over this problem. The creation of genetically engineered papaya resistance to papaya ringspot virus (PRSV) is a prime example; it transformed papaya farming in Hawaii and showed how GE may be used to address issues that would otherwise be unsolvable (Gonsalves 2006). Unlike the conventional breeding used historically that lead to the transfer of humongous, unplanned sets of genes, often transferring undesirable qualities simultaneously, genetic modification (GE) transfers only desirable genetic modifications. This is another significant benefit of GE. Before being released, the added gene in GE is carefully chosen, its function is well understood, and its phenotypic impacts are rigorously assessed. Lastly, GE has special potential for clonally reproduced crops like cassava and potatoes, where variability, polyploidy, and sterility problems pose significant challenges to traditional breeding (Sundaresha et al. 2022a, b). Through direct introduction of engineered viral resistance into top clones, GE circumvents the difficulties associated with cross-breeding, guarantees long-lasting protection, and lowers the possibility of transgene escape through pollen. As a result, GE offers a better platform for controlling plant viral infections from a biological and ecological standpoint, particularly in crops like vegetables, where traditional breeding is either excessively slow or unavailable. Some of the successful case studies of genetically engineered transgenic lines conferring resistance to plant viruses have been described briefly.

## 22.6 Tomato Yellow Leaf Curl Virus (TYLCV) Resistance in Tomato

The tomato yellow leaf curl virus (TYLCV), a virus spread by whiteflies, has a significant negative impact on tomatoes, one of the most commercially significant vegetable crops. Stunted plants, vertical leaf curling, decreased production of fruit, and yield losses of up to 100% during severe epidemics are all symptoms of TYLCV infection. The fast development of the virus and the scarcity of indigenous resistance sources have hampered the use of conventional resistance breeding. In order to overcome this difficulty, scientists created TYLCV-resistant transgenic tomatoes using RNA silencing techniques (El-Sappah et al. 2022; Tashkandi et al. 2018). The plants were modified to generate double-stranded RNA (dsRNA) transcripts by adding hairpin RNA (hpRNA) structures produced by the TYLCV coat protein (CP) and replication-related protein (Rep) genes. These dsRNAs were converted into siRNAs, which directed the complex responsible for RNA-induced silencing (RISC) to destroy TYLCV mRNA during invasion. When exposed to large inoculum pressures, the transgenic plants displayed great resilience, exhibiting low or no symptoms. Subsequent attempts with artificial microRNAs (amiRNAs) improved accuracy by targeting conserved areas of several begomoviruses. The tomato case study is notable because it demonstrated whether each hpRNA, along with amiRNA-based RNA interference techniques, might guard against one of the most destructive viral illnesses affecting crops globally (Ghorbani Faal et al. 2020). It also underlined the need for multiplexing in developing broad-spectrum resistance.

## 22.7 Cucumber Mosaic Virus (CMV) Resistance in Cucurbits

Cucumber mosaic virus (CMV) is one of the most devastating viruses that affects cucurbit crops, including cucumber, melon, and squash. The virus has an extremely broad host variety and can be transmitted irregularly by aphid-like insects, making vector control virtually impossible. CMV infection leads to mosaic patterns, leaf deformation, fruit deformity, and considerable production loss. CRISPR/Cas genome editing has been used successfully to fight CMV in cucurbits (Ruiz et al. 2021). One technique targeted the plant's eukaryotes, initiating factors 4E (eIF4E), the host gene required by CMV to translate viral proteins. Researchers used CRISPR/Cas9 to generate targeted mutations in eIF4E, creating cucumber saplings no longer susceptible to CMV replication. The agronomic performance is notably positive with typical development and fruiting with no negative impact. The major complementary strategy used CRISPR/Cas9 to clearly cut the viral genome, specifically to target the CMV 2b repressor gene, which will suppress the host RNA-silencing methods. When compared with liable controls, the plants that are modified with CRISPR constructs are observed to be less susceptible to the development of the virus. The host susceptibility genes are altered with targeted viral sequences added using this collaborative technique, and the viral control in curcubits is revealed in CRISPR technology's sole adaptability. Not only the CMV disease but also other related cucurbit viruses, such as zucchini yellow mosaic virus (ZYMV), generated resistance in the CMV case study. These findings assure the productivity control in a crop group critical to vegetable production globally, opening the door for the practical application of genetically modified cucurbits with stable viral resistance (Table 22.1).

## 22.8 Potato Virus Y (PVY) Resistance in Potato

Potatoes (*Solanum tuberosum*) are a vital and important food and vegetable crop grown across the world, but they are extremely sensitive to viral infections, notably potato virus Y (PVY) (Kumar et al. 2022). PVY infection causes leaf mosaic discoloration, decay, and loss of yield of up to 80%. Because potatoes are vegetatively reproduced, viral infections build up over multiple generations, making management particularly difficult (Kumar et al. 2020, 2025). Researchers used RNA interference (RNAi) technology to create PVY-resistant potatoes (Grech-Baran et al. 2020). Genetically engineered potato plants were created that produce hairpin RNA structures derived from the PVY coating protein (CP) and assistance factor

**Table 22.1** Virus-resistant genetically modified (GM) crops have been approved for commercialisation in various countries

| Year of approval | Crop | Technology company/ Organisation | Trade name | Target viruses | Gene targeted |
|---|---|---|---|---|---|
| 1996 | Squash | Seminis and Monsanto | CZW3 and ZW20 | Cucumber mosaic virus (CMV), watermelon mosaic virus (WMV), and zucchini yellow mosaic virus (ZYMV) | Coat protein |
| 1997 | Papaya | Cornell University and the University of Hawaii | 55-1 (Rainbow & SunUp),63-1 | Papaya ringspot virus (PRSV) | Coat protein |
| 1998 | Potato | Monsanto | NewLeaf™Plus Russet Burbank | Potato leaf roll virus (PLRV) | Replicase and helicase |
| | | | Shepody NewLeaf™Y potato | Potato virus Y (PVY) | Coat protein |
| 2007 | Plum | USDA-ARS | C-5 | Plum pox virus (PPV) | Coat protein |
| 2009 | Papaya | University of Florida | X17–2 | Papaya ringspot virus (PRSV) | Coat protein |
| 1998 | Sweet pepper | Beijing University | PK-SP01 | Cucumber mosaic virus (CMV) | Coat protein |
| 1999 | Tomato | Beijing University | PK-TM8805R(8805R) | Cucumber mosaic virus (CMV) | Coat protein |
| 2006 | Papaya | Cornell University and the University of Hawaii | 55-1(Rainbow & SunUp),63-1 | Papaya ringspot virus (PRSV) | Coat protein |
| 1999 | Potato | Monsanto | Shepody NewLeaf™Y potato | Potato virus Y (PVY) | Coat protein |
| 2011 | Bean | EMBRAPA | BRSFc401 RMD | Bean golden mosaic virus (BGMV) | RNA of viral replication protein |
| 2011 | Papaya | Cornell University and the University of Hawaii | 55-1 (Rainbow & SunUp),63-1 | Papaya ringspot virus (PRSV) | Coat protein |
| 2021 | Cassava | Danforth Plant Science Center, St. Louis, USA | Cassava Event-4046 | Cassava brown streak virus (CBSV) | RNAi against CP |

proteinase (HC-Pro) genes. The PVY transcripts are silenced when the constructs produce dsRNAs that are subsequently transformed into siRNAs. Resistant plants are often observed to exhibit lower viral accumulation and delayed symptom development, sometimes leading to near-complete or complete immunity to infection. The PVY genome involves a major sophisticated method using artificial microRNAs (amiRNAs). The chances of viral escape via mutations and breakdown of virus RNAs can be lowered by the amiRNAs. An additional alternative to regular pesticide treatments for aphid vector control is the use of field experiments using RNAi-based potato lines. Potatoes are an important vegetable in most of the countries and also a major food security crop. The tuber quality and yield, reducing the pesticide use, can be significantly enhanced using RNAi, leading to engineered viral resistance. In the genetically produced vegetable crops (i.e. viral disease control), the RNA silencing techniques can be used beyond the greenhouse step to the field settings.

## 22.9 Chilli Leaf Curl Virus (ChiLCV) Resistance in Chilli Pepper

The chili peppers (*Capsicum annuum*) are an important crop of vegetables grown across the world, notably in Asia as well as Africa, where it provides substantial money and nourishment. However, it is severely impacted by chili leaf curl virus (ChiLCV) along with associated begomoviruses, which cause leaf curving, yellowing, and yield losses of up to 90% during epidemic years. The variety of viruses makes cultivating long-term resistance exceedingly challenging. Researchers used multiplexed CRISPR/Cas9 systems to create resistance in chili peppers. Multiple guide RNAs (gRNAs) were developed to target critical viral regions, such as the region known as intergenic (IR), the replication-related protein (Rep), and coat protein (CP). Genetically modified chilli plants producing these designs demonstrated considerable resistance, with lower replication of the virus, milder symptoms, and higher yields in both greenhouses and field environments (Verma et al. 2021; Sundaresha et al. 2022b, 2024). Multiplex CRISPR has the benefit of attacking numerous critical sections within the viral genome at once. The major concerns of begomovirus populations are the likelihood of viral escape by single mutations or recombination, which is minimised using this technique. The begomoviruses decimate chilli output, providing hope to producers in the agricultural sector. This indicates the applicability of the CRISPR/Cas9 method. It also demonstrates how genome editing may directly help to promote sustainable farming by lowering synthetic pesticide use while maintaining consistent crop yields. While maintaining the consistent crop yields of the agricultural plants like tomato and cucumber, it also indirectly helps in promoting sustainable farming by lowering the use of synthetic pesticides using genome editing.

## 22.10 Some Threats of Genetically Engineered Crops

Viral infections in vegetable crops, including papaya ringspot virus in papaya, potato virus Y in potatoes, and bean golden mosaic virus in beans, can be managed by genetically modified techniques. However, genetic engineering also has some dangers or risks in the practical world. Though the genetically modified organisms are as reliable as those created by traditional methods, it is equally important to carefully examine the possible or masked hazards or dangers. The consumers, policymakers, farmers, and scientist can provide major educated decisions if these hazards or dangers are properly understood. One of the most prominent hazards is posed by the gene transfer via horizontal gene transfer, which can lead to "escape" from the GM crops onto different plants or even microbes (Sundaresha et al. 2024). In plants, this often occurs via cross-pollination with wild cousins or contact with soil microbes. For example, detractors frequently highlight the prospect that potential resistance to antibiotics marker genes found in some GM veggies might spread to microorganisms, triggering antibiotic resistance. While this is potentially possible, no studies have produced conclusive evidence that it occurs in large quantities. In reality, spontaneous gene transfer among plants and microorganisms happens all the time in nature; the only question that remains is whether the gene that is transmitted will cause any further harm. In the event of virus resistance, the majority of the genes employed are virus coat proteins or replicase segments, which already exist spontaneously in infected plants. As a result, the risk of causing a new disease via gene escape is quite low. Another contentious subject is genetic degradation. When farmers cultivate GM vegetables in large quantities, traditional landraces and heritage kinds may be replaced. The lack of genetic variety may make farming systems more susceptible in the future. For example, if everyone shifts to a particular virus-resistant bean variety and a new illness appears, the loss of diversity might be terrible. On the other hand, GM methods can occasionally conserve kinds that might otherwise become extinct. A classic example is the "San Marzano" tomato, which is favoured in Italy for processing but is very susceptible to the cucumber mosaic virus (Niraula and Fondong 2021). With biotechnology, this variety would have been discarded, but with engineered resistance, it has been conserved. GM crops can thus destroy or aid in maintaining biodiversity, depending on how effectively they are managed. Integrating transgenic viral sequences from genetically modified crops with naturally occurring infectious viruses is a potential, though highly unlikely, biosafety risk. When a GM crop containing a viral gene is coinfected with a different, natural virus, the two viral genomes could potentially exchange genetic material. This recombination process could theoretically generate new viruses with unpredictable characteristics, such as increased virulence, wider host range, or altered transmission (Tepfer et al. 2015). Researchers suspect whether a virus-derived gene in a GM vegetable would recombine with an invading virus, resulting in a new, more hazardous strain. While this sounds scary, data shows that it is uncommon. Viruses normally recombine at high rates, and adding a tiny portion from the viral genes into a crop has no substantial effect on this risk. This shows which modifying genes over viral resistance may be similar to an approach that plants have utilised for several million years. The final big concern is the toxic effects of gene transfers to humans, animals, and even plants themselves. People are concerned about eating vegetables containing viral pieces since viral genes serve as the foundation for most resistance methods. However, it is crucial to realise that we already receive viral proteins when we eat vegetables that are spontaneously infected in the field; the sole distinction is that the viral sequence in GM crops is stable and regulated. Multiple animal feeding tests, particularly long-term trials, have found no indication of neurotoxicity or allergic reaction from virus-resistant genetically modified crops (Su et al. 2020). Furthermore, untended risks or consequences can be prevented by the use of biotechnology, which enables precise editing, frequently employing minuscule segments of viral genomes or RNA molecules. There are also societal and economic implications beyond these scientific issues. Seed sovereignty concerns would be created when the farmers are growing reliant on a few businesses for seeds, particularly when trademarks are involved. Based on certain scepticism or ethical concerns regarding "tampering with nature" in many areas, the GM veggies face consumer opposition not solely on the unpredicted hazards. As shown with Hawaii's GM papaya, which first encountered an export hurdle, there has been an impact on the market acceptability and commerce. It is preferable to handle each one using scientific evidence and prudent legislation instead of ignoring the GM crops. Horizontal transfers of genes, genomic erosion, recombination, and toxicity are all conceivable but uncommon, and with sufficient biosafety precautions, these may be avoided. Ultimately, what counts is not whether an agricultural product is GM or organic but whether it is environmentally friendly, sustainable, and useful to both farmers and consumers. With proper management, GM technology will safeguard crops for vegetables from viruses while also preserving biodiversity and customer trust.

## 22.11 Conclusion

Viral infections in vegetable crops continue to pose a significant danger to the world food supply, consumption, and farmer livelihoods. The current biotechnological technologies, including breeding, demonstrates most promising approaches given that the pathogens lack basic chemical controls. The fast development of viruses frequently destroys the resistance that was produced by the conventional breeding to generate beneficial resistant variants. This led to the investigation and enquiry of novel techniques or methods, including pathogen-induced resistance, RNA silencing, synthetic RNA microRNAs, and CRISPR-based DNA editing. Against damaging viral infection, these approaches have been seen as effective in safeguarding essential vegetable crops such as tomato, potato, cucumber, and chili. However, the questions related to biosafety, regulatory barriers, and consumer acceptability are yet to be tackled regarding genetically modified crops. The open risk assessment and inclusive policies have to integrate with scientific innovations. The combined methods of biotechnology, traditional breeding methods, and sustainable practices are the way forward to a future with innovative and economically effective products.

These methods would opt for the availability of critical vitamins and minerals for mankind consumption and also maintain yields and earnings. Towards a sustainable agricultural scenario and global food security, biotechnology provides the potential to modify the generation of more robust, virus-resistant vegetable, when used responsibly.

## References

Al-Roshdi MR, Ammara U, Khan J, Al-Sadi AM, Shahid MS (2023) Artificial microRNA-mediated resistance against Oman strain of tomato yellow leaf curl virus. Front Plant Sci 14:1164921

Baulcombe DC (1996) Mechanisms of pathogen-derived resistance to viruses in transgenic plants. Plant Cell 8(10):1833–1844

Beachy RN (1994) Mechanisms and applications of pathogen-derived resistance in transgenic plants. Curr Opin Biotechnol 5(2):117–121

Cao Y, Zhou H, Zhou X, Li F (2020) Control of plant viruses by CRISPR/Cas system-mediated adaptive immunity. Front Microbiol 11:593700

Carr JP (2024) Engineered resistance to tobamoviruses. Viruses 16(7):1007

Dong OX, Ronald PC (2019) Genetic engineering for disease resistance in plants: recent progress and future perspectives. Plant Physiol 180(1):26–38

El-Sappah AH, Qi S, Soaud SA, Huang Q, Saleh AM, Abourehab MAS, Wan L, Cheng G, Liu J, Ihtisham M, Noor Z, Mir RR, Zhao X, Yan K, Abbas M, Li J (2022) Natural resistance of tomato plants to tomato yellow leaf curl virus. Front Plant Sci 13:1081549

Fuchs M, Tricoli DM, Carney KJ, Schesser M, McFerson JR, Gonsalves D (1998) Comparative virus resistance and fruit yield of transgenic squash with single and multiple coat protein genes. Plant Dis 82(12):1350–1356

Ghorbani Faal P, Farsi M, Seifi A, Mirshamsi Kakhki A (2020) Virus-induced CRISPR-Cas9 system improved resistance against tomato yellow leaf curl virus. Mol Biol Rep 47(5):3369–3376

Gomez MA, Lin ZD, Moll T, Chauhan RD, Hayden L, Renninger K et al (2019) Simultaneous CRISPR/Cas9-mediated editing of cassava eIF 4E isoforms nCBP-1 and nCBP-2 reduces cassava brown streak disease symptom severity and incidence. Plant Biotechnol J 17:421–434

Gonsalves D (2006) Transgenic papaya: development, release, impact and challenges. Adv Virus Res 67:317–354

Grech-Baran M, Witek K, Szajko K, Witek AI, Morgiewicz K, Wasilewicz-Flis I, Jakuczun H, Marczewski W, Jones JDG, Hennig J (2020) Extreme resistance to Potato Virus Y in potato carrying the Ry sto gene is mediated by a TIR-NLR immune receptor. Plant Biotechnol J 18(3):655–667

Hong SF, Fang RY, Wei WL, Jirawitchalert S, Pan ZJ, Hung YL, Pham TH, Chiu YH, Shen TL, Huang CK, Lin SS (2023) Development of an assay system for the analysis of host RISC activity in the presence of a potyvirus RNA silencing suppressor, HC-Pro. Virol J 20(1):10

Hu SF, Wei WL, Hong SF, Fang RY, Wu HY, Lin PC, Sanobar N, Wang HP, Sulistio M, Wu CT, Lo HF, Lin SS (2020) Investigation of the effects of P1 on HC-pro-mediated gene silencing suppression through genetics and omics approaches. Bot Stud 61(1):22

Kumar R, Jeevalatha A, Baswaraj R, Tiwari RK (2020) Viral and viroid diseases of potato and their management. In: Singh AK, Chakrabarti SK, Singh B, Sharma J, Dua VK (eds) Potato Science & Technology for Sub Tropics. New India Publishing Agency, New Delhi, pp 267–292. (ISBN: 978-93-89571-93-6)

Kumar R, Tiwari RK, Sundaresha S, Kaundal P, Raigond B (2022) Potato viruses and their management. In: Chakrabarti SK, Sharma S, Shah MA (eds) Sustainable management of potato pests and diseases. Springer, Singapore, pp 309–335. https://doi.org/10.1007/978-981-16-7695-6_12

Kumar R, Lal MK, Tiwari RK, Kumar D, Sagar V, Jena R, Kumar R, Singh G, Singh B (2025) Antioxidant defence mechanism of potato (Solanum tuberosum) by regulation of physiological

and biochemical response under potato virus Y necrotic strain (PVYNTN) infection. J Plant Growth Regul. https://doi.org/10.1007/s00344-025-11850-0
Kuo YW, Falk BW (2020) RNA interference approaches for plant disease control. BioTechniques 69(6):469–477
Lindbo JA, Falk BW (2017) The impact of “coat protein-mediated virus resistance” in applied plant pathology and basic research. Phytopathology 107(6):624–634
Mori T, Takenaka K, Domoto F, Aoyama Y, Sera T (2013) Inhibition of binding of tomato yellow leaf curl virus rep to its replication origin by artificial zinc-finger protein. Mol Biotechnol 54(2):198–203
Napoli C, Lemieux C, Jorgensen R (1990) Introduction of a chimeric Chalcone synthase gene into petunia results in reversible co-suppression of homologous genes in trans. Plant Cell 2(4):279–289
Niraula PM, Fondong VN (2021) Development and adoption of genetically engineered plants for virus resistance: advances, opportunities and challenges. Plants (Basel, Switz) 10(11):2339
Rosa C, Kuo YW, Wuriyanghan H, Falk BW (2018) RNA interference mechanisms and applications in plant pathology. Annu Rev Phytopathol 56:581–610
Ruiz L, López C, Picó B, Janssen D (2021) Resistance to cucumber green mottle mosaic virus in Cucumis melo. Plants 10(6):1077
Sanford JC, Johnson SA (1985) The concept of pathogen-derived resistance—deriving resistance genes from the parasite’s own genome. J Theor Biol 113(2):395–405
Sera T (2005) Inhibition of virus DNA replication by artificial zinc finger proteins. J Virol 79(4):2614–2619
Sharma N, Prasad M (2020) Silencing AC1 of Tomato leaf curl virus using artificial microRNA confers resistance to leaf curl disease in transgenic tomato. Plant Cell Rep 39(11):1565–1579
Su S, Ezhuthachan ID, Ponda P (2020) Genetically modified foods and food allergy. J Food Allergy 2(1):111–114
Sundaresha S, Bairwa A, Tomar M, Kumar R, Venkatasalam EP, Sagar V, Bhardwaj V, Sharma S (2022a) In vitro method for synthesis of large-scale dsRNA molecule as a novel plant protection strategy. In: Mysore KS, Senthil-Kumar M (eds) Plant gene silencing. Methods in molecular biology, vol 2408. Humana, New York, pp 211–226. https://doi.org/10.1007/978-1-0716-1875-2_14
Sundaresha S, Jeevalatha A, Kumar R, Sood S, Sharma S, Bhardwaj V, Singh BP, Chakrabarti SK (2022b) RNA interference: a versatile tool to augment plant protection strategies in potato. In: Chakrabarti SK, Sharma S, Shah MA (eds) Sustainable management of potato pests and diseases. Springer, Singapore, pp 369–388. https://doi.org/10.1007/978-981-16-7695-6_15
Sundaresha S, Tiwari JK, Bhardwaj V, Sharma S, Jeevalatha A, Kumar R, Sagar V, Singh BP, Chakrabarti SK, Kumar M (2024) Genetic engineering in potato: status and future perspectives. In: Kalia P (ed) genetic engineering of vegetable crops. CABI, Nosworthy Way, Wallingford, pp 133–140. (Hardback: ISBN: 978-1-80062-710-9; ePDF:eISBN: 978-1-80062-711-6; ePub:eISBN: 978-1-80062-712-3). https://doi.org/10.1079/9781800627123.0008
Tashkandi M, Ali Z, Aljedaani F, Shami A, Mahfouz MM (2018) Engineering resistance against Tomato yellow leaf curl virus via the CRISPR/Cas9 system in tomato. Plant Signal Behav 13(10):e1525996
Tepfer M, Jacquemond M, García-Arenal F (2015) A critical evaluation of whether recombination in virus-resistant transgenic plants will lead to the emergence of novel viral diseases. New Phytol 207(3):536–541
Tripathi S, Suzuki J, Gonsalves D (2007) Development of genetically engineered resistant papaya for papaya ringspot virus in a timely manner: a comprehensive and successful approach. Methods Mol Biol (Clifton, N.J.) 354:197–240
Verma RK, Mishra M, Marwal A, Akhtar A, Petrov NM, Gaur RK (2021) CRISPR/Cas9 protection system for Chilli leaf curl virus. Front Microbiol 12:765536
Vinodhini J, Rajendran L, Karthikeyan G (2023) Engineering resistance against cucumber mosaic virus in *Nicotiana tabacum* through virus derived transgene expressing hairpin RNA. *3.* Biotech 13(5):143

Wilson TMA (1993) Strategies to protect crop plants against viruses: pathogen-derived resistance blossoms. Proc Natl Acad Sci 90(8):3134–3141

Yoon YJ, Venkatesh J, Lee JH, Kim J, Lee HE, Kim DS, Kang BC (2020) Genome editing of *eIF4E1* in tomato confers resistance to pepper mottle virus. Front Plant Sci 11:1098

Zaidi SS, Tashkandi M, Mansoor S, Mahfouz MM (2016a) Engineering plant immunity: using CRISPR/Cas9 to generate virus resistance. Front Plant Sci 7:1673

Zaidi SSEA, Mansoor S, Ali Z, Tashkandi M, Mahfouz MM (2016b) Engineering plants for geminivirus resistance with CRISPR/Cas9 system. Trends Plant Sci 21(4):279–281

# Harnessing RNA Interference for Antiviral Protection in Vegetable Crops

23

P. J. Krishnapriya, K. P. Sindura, Amitha Paul, and Bhumika Prasad Kumar

**Abstract**

Viral diseases significantly affect agriculture, contributing to a 10–15% reduction in crop yield and negatively impacting sustainability and productivity. Conventional mitigation measures like management of insect vectors, field sanitation, antivirals, quarantine measures, and cross protection often fail in tackling the wide array of plant viruses. Genetic engineering and genome editing are advanced tools that offer reliable solutions; however, they face serious biosafety and ethical apprehensions. In this framework, RNA-based solutions are surfacing as a sustainable, encouraging, and non-GMO alternative for opposing plant viral diseases. In this technique, exogenous double-stranded RNA (dsRNA) developed either in vitro or in vivo is introduced into plants to initiate antiviral defence. The dsRNA activates the plant's RNA interference (RNAi) pathway, a natural immune process by targeting and quelling viral genes, thereby inhibiting virus replication and movement. Successful RNAi-based plant vaccination relies on standardisation of critical parameters such as dsRNA dosage, systemic mobility within the plant, and the selection of appropriate viral target regions. The potential of RNA vaccination in reducing viral load and disease severity under laboratory conditions has been exemplified in cases like chilli leaf curl virus, groundnut bud necrosis virus, tomato yellow leaf curl virus, and tomato spotted wilt virus. However, the technology still faces challenges, including the lack of scalable formulation methods, limited stability of dsRNA, low acceptability, and high production costs, which need to be tackled to envisage a commercial-grade RNA biopesticide against plant viruses.

P. J. Krishnapriya (✉) · K. P. Sindura · A. Paul · B. P. Kumar
College of Agriculture, Kerala Agricultural University, Trivandrum, Kerala, India
e-mail: krishnapriya.pj@kau.in



R. Kumar et al. (eds.), *Vegetable Crops and Viral Pathogens Volume I*, Plant Pathology for Sustainable Agriculture, https://doi.org/10.1007/978-981-95-7265-6_23

**Keywords**

Virus · ds-RNA · Resistance · Vector · Biopesticide · RNAi

## 23.1 Introduction

Viral diseases are a major peril to agriculture, shrinking crop yields by 10–15%, leading to a computed perennial global loss of $30 billion. Traditional control strategies, such as controlling vectors, roguing, crop rotation, sanitation, heat therapy, antiviral chemicals, meristem culture, quarantine measures, and cross-protection, often fall short in managing the wide variety of plant viruses because these methods are labour-intensive, nonspecific, and easily affected by environmental changes (Kumar et al. 2020; Jeevalatha et al. 2014). In contemporary years, advances in molecular biology and biotechnology have transformed our understanding of host defense mechanisms. RNA interference (RNAi) is a sturdy natural antiviral pathway seen in crops. RNAi is triggered by the existence of double-stranded RNA (dsRNA), which is modified into small interfering RNAs (siRNAs) that direct the silencing of matching viral RNAs. This sequence-specific and highly flexible mechanism enables the plants defense system to target many viruses. Working on this natural process, scientists have evolved strategies to artificially induce RNAi-mediated resistance, either by creating recombinant plants that express viral dsRNA or, more recently, by externally applying dsRNA molecules as "RNA vaccines". Spray-induced gene silencing (SIGS) is a prominent headway in the field of RNAi. RNA vaccination against plant viruses in combination with traditional control measures and advanced molecular innovations is set to become a key part of next-generation crop protection strategies, providing farmers with a safe, sustainable, and adaptable tool to reduce the devastating incidences of viral diseases. RNA-based vaccination has tremendous potential; however, drawbacks like regulatory uncertainties, challenges of large-scale deployment in different cropping systems, and instability during uptake in field environments comprise the affordable production of dsRNA. Nevertheless, ongoing breakthroughs in dsRNA synthesis techniques, nanotechnology-based carriers, and a deeper understanding of plant uptake mechanisms continue to improve the plausibility of this approach. This chapter offers a detailed overview of plant virus management strategies, with a particular focus on dsRNA-based RNA vaccination. It starts by summarising traditional and eco-friendly methods for controlling viral diseases and then discusses genetic resistance, RNAi mechanisms, dsRNA production platforms, uptake pathways, case studies, advantages, and challenges, ending with future outlooks on integrating RNA vaccination into sustainable agriculture.

## 23.2 Traditional Approaches for Plant Virus Management

Managing plant viral diseases has traditionally relied on methods refined through years of observation, breeding, and integrated pest management. Some traditional techniques are described below.

### 23.2.1 Cultural Practices

Cultural practices form the basis of traditional virus management methods. Plants expressing disease symptoms are visually identified and eliminated from the cropping land to arrest the promotion of the virus and vectors. The method is effective only during localised outbreaks; it is labour-intensive and may be ineffective when infections are latent. Alternating susceptible crops with nonhost species can shatter the infection cycle by depriving viruses and their vectors of ceaseless hosts. However, this approach is less effective against viruses with broad host ranges or those transmitted by persistent insect vectors. Many viruses survive in wild plant reservoirs, which act as alternative hosts. Weed eradication around crop fields reduces opportunities for spillover infections but requires intensive management. Changing the sowing date or plant spacing may lower the coincidence with peak vector populations, thus reducing initial infection pressure. While cultural practices are widely used and budget-friendly, they are often insufficient on their own, especially for highly virulent or quickly spreading viruses.

### 23.2.2 Use of Healthy Planting Material

For clonally multiplicated crops such as potato, banana, cassava, sugarcane, and citrus, using virus-free propagules has been a key part of disease management. Techniques like meristem-tip culture, thermotherapy, and tissue culture indexing help to produce clean planting stock. The number of virus-free onion plants increased significantly when meristem culture and thermotherapy were combined, and the virus frequencies for shallot latent virus, onion yellow dwarf virus, and their combined infection decreased by 10%, 15%, and 10%, respectively (Wang et al. 2021). The development of virus-free plants using apical meristem works well against cassava mosaic virus (Vasanthi et al. 2022), apple mosaic virus, and strawberry latent ringspot in rose (Aljuaid et al. 2022). According to Thanuja et al. (2025), meristem-derived plants of the banana variety CO 3 (ABB) generated progenies that were 92.5% free of banana bunchy top virus, banana streak MY virus, and cucumber mosaic virus (CMV). Despite its advantages, autoclonal variations, somaclonal variations, and recalcitrance are some of the drawbacks associated with meristem culture.

### 23.2.3 Cross-Protection

Cross-protection is a strategy of introducing an avirulent virus strain to protect crops from virulent strains of the same virus. The process is linked to RNA silencing or competition for resources. Shimomoto et al. (2021) demonstrated that an attenuated melon yellow spot virus strain, SA08-8, furnished cross-protection against its virulent isolate, which caused spotted wilt in cucumbers, up to 56 days after the liberation of viruliferous thrips. Chances of mild strains reverting to virulent forms and limited effectiveness against genetically different isolates are some shortcomings of the technique.

### 23.2.4 Antiviral Agents

The use of antiviral chemicals like ribavirin, ningnanmycin, dufulin, vanisulfane, etc. can successfully eliminate plant viruses. Vanisulfane showed protective and therapeutic activities against potato virus Y and tobacco mosaic virus, through the upregulation of ICS1 and PR-1 genes (Sun et al. 2022). Tissue culture medium embellished with 10 mg/l of ribavirin removed *sugarcane streak mosaic virus* and *sugarcane mosaic virus* from the meristem tip of sugarcane without hindering shoot emergence (Neelamathi et al. 2014).

### 23.2.5 Vector Control

Managing insects is essential for controlling viral diseases because insect vectors spread most plant viruses. Targeting insect vectors using techniques such as insecticides, microbial modification, insect transgenics, and integrated pest management measures can assist in reducing the transmission of plant viruses. The insecticide, flupyradifurone, resulted in the highest death rates of nymphs and adults of *Bemisia tabaci*, vector of Tomato leaf curl New Delhi virus, which incited potato apical leaf curl disease (Modgill et al. 2025). However, issues with vector control include ecological harm, exorbitant expenses, insect resistance, and low efficiency.

### 23.2.6 Natural Antivirals

Plant viruses can be effectively combatted with natural antiviral medicines. The antiviral qualities of an aqueous extract made from *Alpinia zerumbet* isolate OG1 were proved by Narusaka et al. (2020) in *Nicotiana benthamiana* against mosaic viruses. *Ganoderma lucidum* filtrate at a concentration of 0.1% significantly reduced groundnut bud necrosis virus titre in tomato plants. The antiviral effect is ascribed to squalene, a triterpenoid compound present in *G. lucidum* (Sangeetha et al. 2020). Restricted availability of basic materials and formulation challenges are key obstacles in harnessing such antiviral compounds.

### 23.2.7 Host Resistance Breeding

Breeding for resistant varieties to combat plant viruses remains one of the most effective and durable procedures for managing these viruses. This method evaluates the genetic defences of different cultivars using marker-assisted selection techniques and develops virus-resistant transgenic variants. EMBRAPA 5.1 is identified as a valuable source of resistance against bean golden mosaic virus (BGMV) in crosses with commercial Brazilian bean varieties. A BGMV-resistant cultivar that was created as a result of this research is currently formally sanctioned and grown under the name “BRSFC401 RMD” (Pinheiro-Lima et al. 2020). Asadi et al. (2025) created segregating tomato populations resistant to tomato mosaic virus and tomato yellow leaf curl virus using marker-aided selection. However, progress in this direction is at a slow pace due to the prolonged breeding process, dearth of suitable host genes, and resistance breakdown.

### 23.2.8 Quarantine Measures

National and international quarantine rules are essential for preventing the ingress of exotic viruses into new areas. However, globalisation of trade and the rise of informal seed exchange networks have increased the risk of transboundary virus spread.

### 23.2.9 Elicitors to Induce Host Plant Resistance

Elicitors can induce systemic resistance against plant viral infections. In potatoes, Bettoni et al. (2022) showed that arachidonic acid has antiviral qualities against the X and M viruses. Steroid glycosides (Poliksenova 2009), polyphenols/flavonoids (Chojnacka et al. 2021), and zircon from *Echinacea purpurea* (Malevannay 2001) are a few examples of elicitors with demonstrated antiviral properties. Problems related to commercialisation, host specificity, and suboptimal field results are some of the limitations of concern.

### 23.2.10 Limitations of Traditional Routes

Main drawbacks include the following:

- Limited effectiveness against mixed infections and transient host resistance
- High labour and finances for cultural practices
- Environmental concerns and regulatory hurdles

These issues emphasise the urgent need for next-generation approaches that are precise, flexible, environmentally friendly, and capable of providing long-term protection.

## 23.3 Molecular Approaches for Plant Virus Management

The molecular approaches leverage advances in plant molecular biology, genomics, and biotechnology to create resistance at the genetic or epigenetic level, with greater accuracy, durability, and potential for broad-spectrum protection.

### 23.3.1 Pathogen-Derived Resistance (PDR)

The pathogen-derived resistance is based on the concept that parts of viral genomes, like coat protein (CP), viral replicase, or movement protein (MP) genes, could be used to save plants from the same or related viruses. Tobacco and tomato expressing the CP gene of TMV showed strong resistance to TMV infection (Pappu et al. 1995). PDR is often strain-specific, and concerns about transgene stability and biosafety have limited its widespread adoption worldwide.

### 23.3.2 RNA Silencing-Based Resistance

The discovery of RNAi in plants revolutionised virus management strategies. A significant benefit of RNAi-based methods is the capability for comprehensive resistance, especially when targeting conserved regions of the viral genome. However, viral suppressors of RNA silencing (VSRs) can reduce the effectiveness of this method.

### 23.3.3 DNA-Based Approaches and Antisense Technologies

One of the earliest molecular approaches involved the use of antisense RNA and ribozymes that bind to and break down viral RNA. Although proof-of-concept studies demonstrated promise, their stability and efficacy under field conditions were limited when compared to RNAi technologies. DNA-based vaccines and the expression of viral defective interfering RNAs (DI RNAs) are also being investigated, limiting virus replication.

### 23.3.4 Genome Engineering for Virus Resistance

The emergence of CRISPR-Cas systems has created new platforms for unfolding virus resistance. Zhang et al. (2019) reported that the CRISPR/Cas13a displayed resistance against RNA viruses, such as the *Southern rice black-streaked dwarf virus* and *rice stripe mosaic virus*, in monocots. CRISPR/Cas9 was harnessed to create resistance against TYLCV in tomato (Faal et al. 2020).

### 23.3.5 Use of Artificial microRNAs (amiRNAs)

amiRNAs are small RNAs specially designed to precisely target viral genomes. Unlike hairpin-based RNAi constructs, amiRNAs reduce off-target effects and need to be combined with multiple constructs to maintain effectiveness against diverse viral strains.

### 23.3.6 Expression of Antiviral Proteins and Antibodies

Plants have also been engineered to express antiviral proteins, including the following:

- Ribonucleases and dsRNA-binding proteins that degrade viral nucleic acids
- Virus-specific antibodies capable of binding viral coat proteins and blocking infection

Though scientifically intriguing, scalability and biosafety considerations have limited their application.

### 23.3.7 Nanophytovirology

Nanophytovirology is an innovative technology that uses nanoparticles (NPs) to detect and control plant viruses; it also employs NPs as carriers of antiviral drugs and biostimulants, to create transgenic crops. Research by Adeel et al. (2021) demonstrated that applying 200 ppm of carbon-based nanoparticles, mainly carbon nanotubes (CNTs) and C60, to *Nicotiana benthamiana* before inoculation with *tobacco mosaic virus* reduced virus multiplication and mobility. Considering the possible long-term effects of NPs on plants and animals, their application rates need to be standardised and experimentally validated.

### 23.3.8 Challenges of Molecular Approaches

Despite the successes, various obstacles to molecular approaches need to be addressed.

*Regulatory hurdles*: Many molecular methods rely on transformation, which faces resistance due to public perception and biosafety concerns.

*Resistance durability*: Host resistance gets weakened due to rapid mutation of viruses.

*Adoption barriers*: High costs involved in accessing such transgenic technologies and intellectual property restrictions curtail smallholder farmers in developing countries from adopting such technologies.

## 23.4 RNAi and Its Role in Plant Virus Management

RNAi is a highly conserved, sequence-specific mechanism of posttranscriptional gene silencing that acts as a natural antiviral defence in plants. This phenomenon, which is present in all eukaryotes and capable of silencing DNA in a sequence-specific manner, was first discovered in plants. The process starts with dsRNA, which can initiate from intermediates of viral replication, structured areas within viral genomes, or artificially developed constructs (Chen and Hur 2021; Mazumder et al. 2025). Upon detection, dsRNA is acted upon by Dicer-like (DCL) endoribonucleases into small interfering RNAs (siRNAs) that are generally 21–24 nucleotides long. Small RNAs are grouped into various classes such as siRNAs, microRNAs (miRNAs), and transacting and phased siRNAs (ta-siRNAs, pha-siRNAs) based on characteristics like size, structural features, and duties in fixed biological pathways (Rosa et al. 2018). DCL proteins (DCL1–DCL4) are responsible for producing siRNAs of contrasting lengths in plants (Fukudome and Fukuhara 2017). siRNAs of 24 nucleotides long are associated with the silencing of recurring DNA sequences (Mishra et al. 2021). RNA-induced silencing complex (RISC) is then loaded with siRNA, where Argonaute (AGO) proteins use these as sequence guides to degrade complementary viral RNAs or to block their translation. Among the various types of AGO proteins, AGO2 is the primary player in antiviral responses (Harvey et al. 2011). AGO proteins consist of three parts, PAZ, MID, and PIWI, with MID and PIWI located at the C-terminal end (Boland et al. 2011). Among the two RNA strands, the sense strand is usually eliminated by Argonaute (AGO) proteins, while the antisense strand serves as a guide, channelling the RISC to its complementary mRNA target. This target mRNA integrates with guide RNA via Watson-Crick base pairing, enabling the PIWI domain of AGO, which shares structural equivalence with RNase H, to split the mRNA (Wang et al. 2008). The resulting cleavage either demolishes or obstructs translation of mRNA, leading to transcriptional silencing and gene inactivation. This RNAi mechanism provides plants with a rapid, flexible, immune-like defence system that helps to subdue viral replication. Furthermore, siRNAs can traverse through plasmodesmata and the phloem, spreading the defence response to unaffected tissues, showcasing the widespread action of RNAi (Uddin and Kim 2013). siRNA-mediated DNA methylation (RdDM) in plants has a pivotal role in the defence against viroids, DNA viruses, and transposable elements. Specifically, 24-nucleotide siRNAs are merged into the Argonaute 4 (AGO4) protein, which facilitates the recoupment of DOMAINS REARRANGED METHYLTRANSFERASE 2 (DRM2) to start de novo DNA methylation at homologous genomic loci (Xie and Yu 2015). The RdDM caters to the function of gene-silencing mediated through the PIWI and PAZ domains of the AGO4 protein, which are indispensable for siRNA binding and target identification (Movahedi et al. 2015). In *Arabidopsis thaliana*, DCL4 processes viral dsRNA into 21-nucleotide siRNAs that specifically target and reduce viral titres of turnip mosaic virus (TuMV) transcripts (Garcia-Ruiz et al. 2010). In *Nicotiana benthamiana*, 24-nt siRNAs play a key role in defending against DNA viruses, namely, tomato yellow leaf curl virus (TYLCV), by mediating transcriptional gene knockdown (Prakash et al. 2020). Likewise, in rice, 22-nt siRNAs are vital for countering rice stripe virus (RSV), illustrating how coevolution between hosts and viruses has fine-tuned RNAi mechanisms (Yang et al. 2018).

### 23.4.1 Transgenic RNAi Strategies Defending Plant Viruses

Initial usages of RNAi in managing plant viral diseases were predominantly transgenic, relying on genetic engineering to confer resistance. In these approaches, plants were modified to express dsRNA targeting viral genes, most commonly those encoding the CP or replicase. Several key examples demonstrate how effective this strategy is: recombinant papaya varieties were developed using CP-derived dsRNA for managing major *papaya ringspot virus* (PRSV) outbursts in Hawaii during the 1990s. This strategy not only facilitated stopping the spread but also contributed to the recovery of Hawaii's papaya industry. Transgenic potato plants having hairpin RNA constructs targeting the viral replicase gene of *potato virus Y* showed strong viral resistance, with notably lower virus titre levels (Tabassum et al. 2016). Although these transgenic RNAi strategies are proven effective, strict regulations and ongoing public scepticism towards genetically modified organisms hinder their widespread promotion. In response to these limitations, current research has increasingly turned to nontransgenic alternatives—most notably, the external usage of dsRNA, often called "RNA vaccines".

## 23.5 RNA Vaccination in Plants

RNA vaccination has become a prominent system in treating plant viruses, using RNAi without requiring genetic modification. Unlike transgenic methods, where dsRNA is always expressed from integrated constructs, exogenous dsRNA delivery offers a transient, nontransgenic preference for triggering antiviral defence. Greater public acknowledgement and adaptability make it a powerful and viable crop shield device. To initiate RNAi, the introduced dsRNA must enter host cells and interact with the cellular silencing mechanisms, ultimately in the specific breakdown of viral RNA (Voloudakis et al. 2022). Therefore, the efficacy of RNA vaccination relies not only on the precise design and synthesis of dsRNA molecules but also on their successful assimilation and processing inside plant cells (Mazumder et al. 2025; Sundaresha et al. 2022b).

### 23.5.1 Design and Construction of Double-Stranded RNA (dsRNA)

Production of dsRNA is made possible through numerous strategies. Conventional approaches included the development of sense and antisense RNA strands, followed by their annealing to form dsRNA. However, this method was relatively incompetent in fuelling antiviral responses. Thereafter, researchers adopted hairpin RNA (hpRNA) constructs, wherein both the sense and antisense sequences are expressed under the same promoter (Waterhouse and Helliwell 2003; Sundaresha et al. 2022a). This method proved more effective in generating small interfering RNA (siRNA). A key advantage of hpRNA constructs is their independence from the plant's RNA-dependent RNA polymerase, making them competent to grant resistance to multiple viruses (Béclin et al. 2002). Currently, dsRNA production approaches can broadly be divided into ex vivo and in vivo categories. Ex vivo methods typically involve PCR amplification of the target viral DNA, cloning into a vector under the power of a T7 promoter and transcription using purified RNA polymerases and nucleotide kits (Beckert and Masquida 2010). In most cases, dsRNA is produced chemically or transcribed using DNA-dependent RNA polymerases with a DNA template. *Escherichia coli* strains like HT115(DE3) are the most widely used host for in vivo dsRNA production, which lack RNase III to prevent dsRNA collapse (Rosa et al. 2018). In *E. coli*, expression vectors encase either single or dual T7 promoters; however, research suggests that vectors with a single promoter generally produce higher amounts of dsRNA (Chen et al. 2019). Genetically modified *Pseudomonas syringae* has been identified as an alternative prokaryotic system for dsRNA synthesis to eliminate RNase III activity (Niehl et al. 2018). Choice of bacterial strain and expression vector significantly controls dsRNA yield and stability. Symbiotic bacteria and yeast are promising in vivo systems for dsRNA production against insect pests. Constructing RNase III-deficient strains for this purpose remains a technical challenge.

### 23.5.2 Exogenous dsRNA Delivery Strategies for RNA Vaccination

The success of RNA-based plant vaccines relies not only on the careful design and synthesis of double-stranded RNA (dsRNA) molecules but also on their effective delivery into plant tissues. After entering the host, the dsRNA must be accurately polished to switch on antiviral RNAi. Exogenous application provides a nontransgenic and adaptable technique to trigger virus resistance, unlike transgenic methods, which require strong integration of dsRNA-synthesising constructs into the crop genome (Huang and Tang 2022; Sundaresha et al. 2024). Thus, exogenously applied dsRNA has improved as a prominent instrument in phytovirus control strategies of the present-day era.

### 23.5.3 Principles of Exogenous Delivery

Extrinsic application of dsRNA and successful antiviral protection can be ensured only when several crucial conditions are satisfied:

1. *Environmental stability*: The dsRNA must stay durable throughout the application process and withstand environmental degradation.
2. *Cellular entry and processing*: The dsRNA must traverse the plant's physical barriers, natural openings, or mechanical injuries and be taken up by plant cells where it is cleaved by Dicer-like (DCL) enzymes into small interfering RNAs.
3. *Systemic spread*: Silencing signal must propagate from the site of application to distal tissues via plasmodesmata and the phloem, ensuring whole-plant protection, promoting effective RNAI (Cagliari et al. 2019).

The present delivery methods differ remarkably in their mechanisms of assimilation, molecular stability, and practical relevance in field conditions. Such strategies vary from simple approaches like foliar spraying to evolved techniques using nanoparticle-based carriers, which upgrade the stability of dsRNA and support accurate delivery to target sites.

## 23.6 dsRNA Conveyance Methods in Plants

Each dsRNA delivery perspective has its own set of merits and demerits, relying on factors like the crop type, the specific pathogen being aimed, and the scale at which the application is intended. The main dsRNA delivery methods are detailed below.

### 23.6.1 Foliar Sprays

Foliar application, commonly recognised as Spray-Induced Gene Silencing (SIGS), remains the widely studied method for delivering dsRNA to plants. This method entails the direct application of dsRNA to leaf surfaces, either as unmodified dsRNA, as a collection of small interfering RNAs (siRNAs), or combined with protective substances that improve stability and absorption (Venu et al. 2024). It is a noninvasive, cost-effective, and easily extensible technique suitable for widespread use in agriculture. However, its challenges include variability in uptake efficiency, which is species-dependent. Additionally, naked dsRNA is rapidly degraded by natural factors such as ultraviolet (UV) radiation, RNases, and rainfall.

### 23.6.2 Trunk Injection and Root Absorption

Delivering systemic treatments through trunk injection or root uptake of woody fruit trees serves as an efficient, optional approach. In these methods, dsRNA is introduced directly into the vascular system of plants, thereby circumventing the physical and physiological blocks connected with foliar uptake (Dalakouras et al. 2018). However, the technique is generally inappropriate for annual or herbaceous crops and is labour-intensive.

### 23.6.3 Nanocarrier-Mediated Delivery

Recent improvements in nanotechnology have led to the generation of several nanocarriers that enhance the stability, uptake, and controlled release of dsRNA. These systems provide potential solutions to many of the challenges posed by traditional delivery techniques. Remarkable examples include layered double hydroxides (LDHs), also known as "BioClay", which bind to dsRNA and enable its slow release on plant surfaces (Mitter et al. 2017). Chitosan nanoparticles are biodegradable and protect dsRNA from enzymatic breakdown while being safe for plants. Carbon nanotubes and graphene oxide enhance the ability of dsRNA to penetrate the plant cuticle and cellular membranes (Xing et al. 2025). Polymers made of liposomes and cations enhanced cellular uptake via electrostatic interactions with the negatively charged dsRNA (Xia et al. 2016). Nevertheless, high production costs, regulatory uncertainties, and restricted validation in field-scale studies are significant barriers to nanocarrier-mediated delivery.

### 23.6.4 Microbial Vectors

Genetically modified viruses can temporarily express dsRNA within host tissues, inducing robust RNAi responses without requiring genomic integration (Kolliopoulou et al. 2017). *Pseudomonas fluorescens* has been estimated as a carrier for dsRNA release due to its ability to colonise plant tissues and release RNA molecules over extended time periods (Goodfellow et al. 2019). Such vector-based methods are still at their preliminary stage but assure sustainable, pragmatic RNAi delivery.

### 23.6.5 High-Pressure Sprays and Mechanical Inoculation

High-pressure spraying methods, often used in parallel to surfactants or abrasive materials, ameliorate dsRNA penetration by breaking through the plant cuticle (Hoang et al. 2022). Likewise, in laboratory environments, mechanical inoculation with compounds such as carborundum has been utilised to aid in dsRNA absorption (Venu et al. 2024).

### 23.6.6 Exogenous dsRNA-Intervened RNAi

The technique is effective mainly under controlled conditions and lacks scalability and practicality for large-scale agricultural use. Continued innovation in nanotechnology, microbial vectors, and formulation chemistry is likely to expand the applicability and efficacy of RNAi-based designs in crop protection. Currently, one of the most effective and widely adopted methods for dsRNA delivery is spray-induced gene silencing (SIGS).

### 23.6.7 Spray-Induced Gene Silencing (SIGS) Approach

As noted earlier, the SIGS methodology necessitates the direct application of small interfering RNA (siRNA) or dsRNA onto plant leaves. After application, the plant engrosses the dsRNA through stomata, minuscule crevices in the cuticle, or other natural entry sites. Within the plant cells, the exogenous dsRNA is refined by Dicer-like (DCL) enzymes, commencing the RNAi pathway. This method can be tailored to aim multiple pathogens concurrently by scheming siRNAs that silence essential genes across different organisms (Chen et al. 2025). Designing siRNAs that bind specifically to nonconserved regions of target genes can minimise off-target effects on beneficial microorganisms (Neumeier and Meister 2021). Foliar application of dsRNA targeting the CP and MP genes reduced the viral load (70–90%) of cucumber green mottle mosaic virus in cucurbits (Delgado-Martín et al. 2022). External usage of dsRNA targeting coding regions of tomato leaf curl New Delhi virus (ToLCNDV) DNA-A component led to delayed manifestation of symptoms in zucchini plants, with a 60–80% decrease in viral load (Frascati et al. 2024). These investigations signify the promise of dsRNA-based approaches in offering effective protection against both RNA and DNA viruses in crops (Table 23.1).

## 23.7 Limitations of RNA Vaccination

RNA-based vaccination is a sustainable, nontransgenic approach to plant virus management; however, its application beyond experimental systems faces significant hurdles. Challenges to widespread adoption may stem from biological, technological, regulatory, and socioeconomic issues that need to be addressed. One major hurdle for field use is the natural vulnerability of externally applied dsRNA. On application, dsRNA is quickly shattered by elements such as UV light, temperature fluctuations, rain, and nucleases present on leaf surfaces (Hoang et al. 2022). Encapsulation methods such as layered double hydroxide (LDH) nanosheets can increase shelf life, but their use is restricted by scalability and cost issues (Imran et al. 2025). Uptake efficacy is another major barrier. While dsRNA can enter through stomata or cuticular pores, absorption varies by species, developmental stage, and environmental conditions. Besides, the constricted mobility of dsRNA and its derived siRNAs often narrows protection to treated tissues, requiring

**Table 23.1** The table below presents various examples of dsRNA used against different viruses

| Virus(es) | Crop | Delivery methods | Viral genes used | Efficacy/protection (%) | References |
|---|---|---|---|---|---|
| CMV, watermelon mosaic virus 2 (WMV 2)/zucchini yellow mosaic virus (ZYMV) | Summer squash (*Cucurbita pepo*) | Agrobacterium-mediated transformation | CP | Multiple genes conferred PDR to the viruses tested | Tricoli et al. (1995) |
| Soybean mosaic virus (SMV) | Soybean (*Glycine max*) | Agrobacterium-mediated transformation | Inverted repeats of HC-Pro | Imparts a high level of virus resistance | Gao et al. (2015) |
| Tobacco mosaic virus (TMV) | Tobacco (*Nicotiana tabacum*) | dsRNA spray | CP | 50–60% | Konakalla et al. (2016) |
| CMV and TMV | Cucumber (*Cucumis sativus*), watermelon (*Citrullus lanatus*), and squash (*Cucurbita pepo*) | dsRNA spray | HC-Pro and CP | 18–82% with HC-Pro and 16–70% with CP | Kaldis et al. (2018) |
| CMV | Bhut Jolokia pepper (*Capsicum chinense* Jacq.) | dsRNA spray | MP | 5–29.4% | Borah et al. (2018) |
| PRSV | Papaya (*Carica papaya)* | dsRNA spray | HC-Pro and CP | 84–100% | Vadlamudi et al. (2020) |
| Pigeonpea sterility mosaic emaraviruses (PPSMVs) | Pigeonpea (*Cajanus cajan*). | dsRNA spray | RdRp, NP, and MP | First field-based demonstration of dSRNA | Patil et al. (2021) |
| Potato virus X (PVX) and potato virus Y (PVY) | Tobacco (*Nicotiana tabacum*) | Artificial microRNAs (amiRNAs) | HC-Pro and MP | 100% | Ai et al. (2011) |
| CMV | Tomato (*Lycopersicon esculentum*) | Artificial microRNAs (amiRNAs) | RdRp, VSR, and UTR | 100% | Zhang et al. (2011) |
| Tomato yellow leaf curl virus (TYLCV) | Tomato (*Lycopersicon esculentum*) | dsRNA spray | Rep | 70% | Melita et al. (2021) |

| | | | | | |
|---|---|---|---|---|---|
| Tomato spotted wilt orthotospovirus (TSWV) | Tobacco (*Nicotiana tabacum*) | dsRNA spray | N | 50–75% | Konakalla et al. (2021) |
| TSWV | Tomato (*Lycopersicon esculentum*) and tobacco (*Nicotiana tabacum*) | dsRNA spray | N and MP | 100% | Tabein et al. (2020) |
| Cymbidium mosaic virus (CymMV) | Orchids (*Brassolaeliocattleya hybrida*) | dsRNA spray | CP | 40-fold lower in virus titre | Lau et al. (2014) |
| Tomato leaf curl virus (ToLCV) and CMV | Tomato (*Lycopersicon esculentum*) and tobacco (*Nicotiana benthamiana*) | dsRNA spray | Rep & viral suppressor | 4–60% | Namgial et al. (2019) |
| CMV | Tobacco (*Nicotiana tabacum*) and *Chenopodium* (*Chenopodium quinoa*) | dsRNA spray | CP and viral suppressor | – | Holeva et al. (2021) |
| PRSV | Papaya (*Carica papaya*) | Mechanical inoculation | CP | 65% | Shen et al. (2014) |
| Sugarcane mosaic virus (SCMV) | Maize (*Zea mays*) | dsRNA spray | Nia and CP | 95% | Gan et al. (2010) |
| Potato virus X (PVX) and potato virus Y (PVY) | Tobacco (*Nicotiana benthamiana*), | dsRNA spray | DCL2 and DCL4 | Absence of systemic symptoms with reduced viral titre | Necira et al. (2021) |
| Cassava brown streak virus (CBSV) and Ugandan cassava brown streak virus (UCBSV) | Cassava | Artificial microRNA (amiRNA) | P1, P3, CI, NIb, and CP | 20–60% | Wagaba et al. (2016) |

(continued)

**Table 23.1** (continued)

| Virus(es) | Crop | Delivery methods | Viral genes used | Efficacy/protection (%) | References |
|---|---|---|---|---|---|
| Tomato leaf curl New Delhi virus (ToLCNDV) | Tomato (*Lycopersicon esculentum*) | amiRNA | AV2 and AV1 | Tolerance to the virus infection | Vu et al. (2013) |
| Cotton leaf curl virus (CLCuV) | Cotton (*Gossypium hirsutum*) | amiRNA | Rep | Stable resistance to the virus | Ali et al. (2013) |
| CLCuV | Cotton (*Gossypium hirsutum*) | Agrobacterium-mediated transformation | hr-miR159a and ghr-miR166a | Overexpression of host-encoded microRNAs could reduce the symptoms and virus accumulation. | Akmal et al. (2017) |
| TMV | Tobacco (*Nicotiana benthamiana*) | dsRNA spray | Rep and CP | Interpolyelectrolyte complexes (IPECs) composed of dsRNA, chitosan, and alginate | Moorlach et al. (2025) |
| CMV | Tobacco (*Nicotiana tabacum*) | siRNAs spray | Multiple sites of the viral genome | 80–90% | Knoblich et al. (2025) |
| Pepper mild mottle virus (PMMoV) and plum pox virus (PPV) | Tobacco (*Nicotiana benthamiana*) | dsRNA spray | Rep | Virus multiplication was blocked | Tenllado et al. (2003) |
| TMV | Tobacco (*Nicotiana benthamiana*) | Graphene oxide nanosheets (GONs)-dsRNA complex spray | Rep | 87.2% and stimulate plant immunity by ROS production | Zhang et al. (2025) |
| Bean common mosaic virus (BCMV) | Common bean *(Phaseolus vulgaris L.)*, tobacco *(Nicotiana tabacum), and* Cowpea *(Vigna unguiculata)* | Multiple dsRNA spray | CP<br>NIb | Reduced the disease incidence and severity | Wani et al. (2025) |

| | | | | | |
|---|---|---|---|---|---|
| Cucumber green mottle mosaic virus (CGMMV) and tomato leaf curl New Delhi virus (ToLCNDV) | Cucumber (*Cucumis sativus*) | dsRNA spray and agroinoculation | CP and MP | dsRNA spray and agroinoculation could effectively reduce the disease symptoms | Delgado-Martín et al. (2022) |
| Beet mosaic virus (BtMV) | Tobacco (*Nicotiana benthamiana*) and sugar beet (*Beta vulgaris*) | dsRNA spray | P1 and Nib | 80% | Rahenbrock et al. (2025) |
| Tomato brown rugose fruit virus (ToBRFV) | Tomato (*Lycopersicon esculentum*) | High-quality (hq)-dsRNA spray | RdRp and MP | (hq)-dsRNA can effectively control viruses and induce plant defence mechanisms | Weiss et al. (2025) |

periodical applications for systemic efficacy. Although numerous studies have successfully described the efficacy of RNA-based vaccines for virus resistance, most of these studies are confined to in vivo conditions; very few have been done at the field level. High manufacturing costs, environmental instability of dsRNA, regulatory challenges, and the limited availability of suitable adjuvants and innovations for large-scale deployment are the major limitations in bringing dsRNA applications to the field. Though advances such as molecular design improvements and delivery systems—like nanoparticles, clay carriers, and seed coatings—have enhanced dsRNA stability, several other challenges still need to be handled. Concerns around environmental persistence, nontarget impacts, and nanoparticle delivery systems add to the complexity. Further detailed inquiries are needed to understand the basic molecular mechanism behind the dsRNA uptake and its efficacy determinants in plants. The application of genomics and bioinformatics can identify target sequences, thereby increasing specificity to a wide range of phytoviruses. The optimisation of RNA size also has to be done to enhance the assimilation of dsRNA into the plant system. Identification of any off-target binding sites of these sequences in the plant epigenome must be verified. All these optimisations and validations have to be done based on the field studies as the inferences from laboratory experiments with a limited number cannot be the same as the field scenario, especially while standardising an RNA-based biopesticide. The carrier or adjuvants used for delivering the dsRNA molecule should efficiently stick to the plant surface and bind promptly with the RNA molecules. Technology must be cost-efficient so that a desirable formulation can be developed, which can be used worldwide without any statutory constraints. Compared to conventional tools, RNA vaccination still lags in durability, affordability, and ease of application. Addressing these limitations will be essential to realising the maximum capacity of RNA-based technologies for viable crop protection.

## 23.8 Future Perspectives

RNA-based vaccination serves as an alternative approach offering more flexibility and precise application. The exterior application of RNA-based vaccines such as small RNAs, dsRNA, and hpRNAs is nontransgenic, easily taken by plants, and effective against an extensive range of viruses. While this method is still in its introductory stages, it holds great assurance for transformation through advances in molecular biology, nanotechnology, and regulatory expansions. RNA secondary structures are nowadays being determined by machine learning algorithms, which enhance siRNA efficacy and reduce off-target effects. Embodying RNA-based approaches with CRISPR-Cas technologies could provide quick and more comprehensive plant immune defences. Genetically engineered microbes like bacteria, fungi, or algae can act as biofactories for extensive dsRNA synthesis, posing a serious challenge in marketable use. Including RNA vaccination in precision agriculture could increase its efficiency and affordability. Technologies such as autonomous drones, electrostatic sprayers, IoT devices, and sensor-driven systems allow for precise and persistent dsRNA application in the field, minimising costs and environmental effects.

## 23.9 Conclusion

RNA vaccination illustrates a pioneering disease management strategy that deviates from broad-spectrum chemical treatments towards accurate, stable, eco-friendly, and scientifically guided approaches for mitigating plant viruses. When blended with nanotechnology, synthetic biology, and AI-steered RNA design, RNA vaccines can be seamlessly incorporated into agricultural operations, enabling the development of robust food production systems in the current era of environmental and biological challenges. Harmonised efforts among policymaking bodies, research institutions, industry stakeholders, and farming communities are essential to streamline innovation, regulation, and user acceptance.

## References


Adeel M, Farooq T, White JC, Hao Y, He Z, Rui Y (2021) Carbon-based nanomaterials suppress tobacco mosaic virus (TMV) infection and induce resistance in *Nicotiana benthamiana*. J Hazard Mater 404:124167

Ai T, Zhang L, Gao Z, Zhu CX, Guo X (2011) Highly efficient virus resistance mediated by artificial microRNAs that target the suppressor of PVX and PVY in plants. Plant Biol (Stuttg) 13(2):304–316

Akmal M, Baig MS, Khan JA (2017) Suppression of cotton leaf curl disease symptoms in *Gossypium hirsutum* through overexpression of host-encoded miRNAs. J Biotechnol 263:21–29

Ali I, Amin I, Briddon RW, Mansoor S (2013) Artificial microRNA-mediated resistance against the monopartite begomovirus cotton leaf curl Burewala virus. Virol J 10:231

Aljuaid BS, Attia OA, Ismail IA, Eldessoky SD, Sadik AS, Khalifa ME (2022) Production of virus-free rose plants using meristem-tip culture and in vitro thermotherapy. Pak J Biol Sci 25(2):160–167

Asadi Z, Savadroudbari NS, Amini F et al (2025) Marker-assisted selection in segregating populations of tomatoes for resistance to TYLCV, ToMV, and Fusarium wilt. Mol Biol Rep 52:107

Beckert B, Masquida B (2010) Synthesis of RNA by in vitro transcription. In: RNA: methods and protocols. Humana Press, Totowa, pp 29–41

Béclin C, Boutet S, Waterhouse P, Vaucheret H (2002) A branched pathway for transgene-induced RNA silencing in plants. Curr Biol 12(8):684–688

Bettoni JC, Mathew L, Pathirana R, Wiedow C, Hunter DA, McLachlan A et al (2022) Eradication of potato virus S, potato virus A, and potato virus M from infected in vitro-grown potato shoots using in vitro therapies. Front Plant Sci 13:878733

Boland A, Huntzinger E, Schmidt S, Izaurralde E, Weichenrieder O (2011) Crystal structure of the MID-PIWI lobe of a eukaryotic Argonaute protein. Proc Natl Acad Sci USA 108(26):10466–10471

Borah M, Berbati M, Reppa C, Holeva M, Nath PD, Voloudakis A (2018) RNA-based vaccination of Bhut Jolokia pepper (*Capsicum chinense* Jacq.) against cucumber mosaic virus. Virus 29(2):207–211

Cagliari D, Dias NP, Galdeano DM, Dos Santos ÉA, Smagghe G, Zotti MJ (2019) Management of pest insects and plant diseases by non-transformative RNAi. Front Plant Sci 10:1319

Chen YG, Hur S (2021) Cellular origins of dsRNA, their recognition and consequences. Nat Rev Mol Cell Biol 23(4):286–301

Chen Z, He J, Luo P, Li X, Gao Y (2019) Production of functional double-stranded RNA using a prokaryotic expression system in *Escherichia coli*. Microbiology 8(7):e00787

Chen C, Imran M, Feng X, Shen X, Sun Z (2025) Spray-induced gene silencing for crop protection: recent advances and emerging trends. Front Plant Sci 16:1527944

Chojnacka K, Skrzypczak D, Izydorczyk G, Mikula K, Szopa D, Witek-Krowiak A (2021) Antiviral properties of polyphenols from plants. Foods 10(10):2277

Dalakouras A, Jarausch W, Buchholz G, Bassler A, Braun M, Manthey T et al (2018) Delivery of hairpin RNAs and small RNAs into woody and herbaceous plants by trunk injection and petiole absorption. Front Plant Sci 9:1253

Delgado-Martín J, Ruiz L, Janssen D, Velasco L (2022) Exogenous application of dsRNA for the control of viruses in cucurbits. Front Plant Sci 13:895953

Faal PG, Farsi M, Seifi A, Mirshamsi Kakhki A (2020) Virus-induced CRISPR-Cas9 system improved resistance against tomato yellow leaf curl virus. Mol Biol Rep 47(5):3369–3376

Frascati F, Rotunno S, Accotto GP, Noris E, Vaira AM, Miozzi L (2024) Exogenous application of dsRNA for protection against tomato leaf curl New Delhi virus. Viruses 16(3):436

Fukudome A, Fukuhara T (2017) Plant dicer-like proteins: double-stranded RNA-cleaving enzymes for small RNA biogenesis. J Plant Res 130(1):33–44

Gan D, Zhang J, Jiang H, Jiang T, Zhu S, Cheng B (2010) Bacterially expressed dsRNA protects maize against SCMV infection. Plant Cell Rep 29(11):1261–1268

Gao L, Ding X, Li K, Liao W, Zhong Y, Ren R, Liu Z, Adhimoolam K, Zhi H (2015) Characterization of soybean mosaic virus resistance derived from inverted repeat-SMV-HC-Pro genes in multiple soybean cultivars. Theor Appl Genet 128(8):1489–1505

Garcia-Ruiz H, Takeda A, Chapman EJ, Sullivan CM, Fahlgren N, Brempelis KJ, Carrington JC (2010) Arabidopsis RNA-dependent RNA polymerases and dicer-like proteins in antiviral defense and small interfering RNA biogenesis during turnip mosaic virus infection. Plant Cell 22(2):481–496

Goodfellow S, Zhang D, Wang MB, Zhang R (2019) Bacterium-mediated RNA interference: potential application in plant protection. Plants 8(12):572

Harvey JJ, Lewsey MG, Patel K, Westwood J, Heimstädt S, Carr JP, Baulcombe DC (2011) An antiviral defense role of AGO2 in plants. PLoS One 6(1):e14639

Hoang BTL, Fletcher SJ, Brosnan CA, Ghodke AB, Manzie N, Mitter N (2022) RNAi as a foliar spray: efficiency and challenges to field applications. Int J Mol Sci 23(12):6639

Holeva MC, Sklavounos A, Rajeswaran R, Pooggin MM, Voloudakis AE (2021) Topical application of double-stranded RNA targeting 2b and CP genes of cucumber mosaic virus protects plants against local and systemic viral infection. Plants 10(5):963

Huang Y, Tang X (2022) RNAi-based pest control: production, application, and the fate of dsRNA. Front Bioeng Biotechnol 10:1080576

Imran M, Feng X, Sun Z, Al Omari H, Zhang G, Zhu J et al (2025) Nanotechnology-driven gene silencing: advancements in SIGS–dsRNA technology for sustainable disease management. Chem Biol Technol Agric 12(1):31

Jeevalatha A, Singh BP, Kaundal P, Kumar R, Raigond B (2014) RCA-PCR: a robust technique for the detection of tomato leaf curl New Delhi virus-[potato] at ultra-low virus titre. Potato J 41:76–80

Kaldis A, Berbati M, Melita O, Reppa C, Holeva M, Otten P, et al (2018) Exogenously applied dsRNA molecules deriving from the Zucchini yellow mosaic virus genome move systemically and protect cucurbits against ZYMV. Mol Plant Pathol, 19(4):883–895

Knoblich M, Gursinsky T, Gago-Zachert S, Weinholdt C, Grau J, Behrens SE (2025) A new level of RNA-based plant protection: dsRNAs designed from functionally characterized siRNAs are highly effective against cucumber mosaic virus. Nucleic Acids Res 53(5):gkaf136

Kolliopoulou A, Taning CN, Smagghe G, Swevers L (2017) Viral delivery of dsRNA for control of insect agricultural pests and vectors of human disease: prospects and challenges. Front Physiol 8:399. https://doi.org/10.3389/fphys.2017.00399

Konakalla NC, Kaldis A, Berbati M, Masarapu H, Voloudakis AE (2016) Exogenous application of double-stranded RNA molecules from TMV p126 and CP genes confers resistance against TMV in tobacco. Planta 244(4):961–969

Konakalla NC, Bag S, Deraniyagala AS, Culbreath AK, Pappu HR (2021) Induction of plant resistance in tobacco (*Nicotiana tabacum*) against tomato spotted wilt orthotospovirus through foliar application of dsRNA. Viruses 13(4):662

Kumar R, Jeevalatha A, Baswaraj R, Tiwari RK (2020) Viral and viroid diseases of potato and their management. In: Singh AK, Chakrabarti SK, Singh B, Sharma J, Dua VK (eds) Potato science & technology for sub tropics. New India Publishing Agency, New Delhi, pp 267–292. (ISBN: 978-93-89571-93-6)

Lau SE, Mazumdar P, Hee TW, Song ALA, Othman RY, Harikrishna JA (2014) Crude extracts of bacterially-expressed dsRNA protect orchid plants against Cymbidium mosaic virus during transplantation from in vitro culture. J Hortic Sci Biotechnol 89(5):569–576

Malevannay N (2001) The explosive temperament of zircon in the service of plants. New Gardener Farmer 1:45

Mazumder P, Majumdar A, Roy B, Mohanbabu Y, Valli NS, Sarkar M, Mondal F, Verma N, Kumar R, Roy A (2025) RNA vaccination in plants: a non-transgenic approach to combat plant viruses with special reference to tomato leaf curl New Delhi virus. In: Tomato Leaf Curl New Delhi Virus (ToLCNDV) insights into virome dynamics and management. Springer Nature Singapore, Singapore, pp 339–352. https://doi.org/10.1007/978-981-95-2852-3_17

Melita O, Kaldis A, Berbati M, Reppa C, Holeva M, Lapidot M, et al (2021) Topical application of double-stranded RNA molecules deriving from Tomato yellow leaf curl virus reduces cognate virus infection in tomato. Biol Plant 65:100–110

Mishra SN, Abdul Kader Jailani A, Mandal B, Mukherjee SK (2021) Secondary siRNAs in plants: biosynthesis, various functions, and applications in virology. Front Plant Sci 12:610283

Mitter N, Worrall EA, Robinson KE, Li P, Jain RG, Taochy C et al (2017) Clay nanosheets for topical delivery of RNAi for sustained protection against plant viruses. Nat Plants 3(2):17008

Modgill N, Naga KC, Firake DM, Tiwari RK, Shivaramu S, Kumar R, Bhatnagar A, Buckseth T, Bairwa A, Maharana C, Sharma S, Singh B (2025) Flupyradifurone application: a promising approach to curtail tomato leaf curl New Delhi virus (ToLCNDV) transmission vectored by *Bemisia tabaci* in potato farming. Crop Prot 187:106951

Moorlach BW, Sede AR, Hermann KM, Levanova AA, Poranen MM, Westphal M, Wortmann M, Stepula E, Jakobs-Schönwandt D, Heinlein M, Keil W (2025) Interpolyelectrolyte complexes of in vivo produced dsRNA with chitosan and alginate for enhanced plant protection against tobacco mosaic virus. Int J Biol Macromol 306:141579

Movahedi A, Sun W, Zhang J, Wu X, Mousavi M, Mohammadi K et al (2015) RNA-directed DNA methylation in plants. Plant Cell Rep 34(11):1857–1862

Namgial T, Kaldis A, Chakraborty S, Voloudakis A (2019) Topical application of double-stranded RNA molecules containing sequences of tomato leaf curl virus and cucumber mosaic virus confers protection against the cognate viruses. Physiol Mol Plant Pathol 108:101432

Narusaka M, Yamaji Y, Uraji M, Hatanaka T, Narusaka Y (2020) Inhibitory effects of *Alpinia zerumbet* extract against plant virus infection in solanaceous plants. Plant Biotechnol 37(1):93–97

Necira K, Makki M, Sanz-Garcia E, Canto T, Djilani-Khouadja F, Tenllado F (2021) Topical application of *Escherichia coli*-encapsulated dsRNA induces resistance in *Nicotiana benthamiana* to potato viruses and involves RDR6 and combined activities of DCL2 and DCL4. Plants 10(4):644

Neelamathi D, Jerold Manuel, Philomena George (2014) Influence of apical meristem and chemotherapy on production of virus-free sugarcane plants. Res J Recent Sci 3(ISC-2013):305–309

Neumeier J, Meister G (2021) siRNA specificity: RNAi mechanisms and strategies to reduce off-target effects. Front Plant Sci 11:526455

Niehl A, Soininen M, Poranen MM, Heinlein M (2018) Synthetic biology approach for plant protection using dsRNA. Plant Biotechnol J 16(9):1679–1687

Pappu HR, Niblett C, Lee RF (1995) Application of recombinant DNA technology to plant protection: molecular approaches to engineering virus resistance in crop plants. World J Microbiol Biotechnol

Patil BL, Raghu R, Dangwal M et al (2021) Exogenous dsRNA-mediated field protection against Pigeonpea sterility mosaic emaravirus. J Plant Biochem Biotechnol 30:400–405

Pinheiro-Lima B, Pereira-Carvalho RC, Alves-Freitas DMT, Kitajima EW, Vidal AH, Lacorte C, Godinho MT, Fontenele RS, Faria JC, Abreu EFM et al (2020) Transmission of the bean-associated Cytorhabdovirus by the Whitefly *Bemisia tabaci* MEAM1. Viruses 12:1028

Poliksenova V (2009) Induced plant stability to pathogens and abiotic stress factors. Bull BSU 2(1):48–60

Prakash V, Singh A, Singh AK, Dalmay T, Chakraborty S (2020) Tobacco RNA-dependent RNA polymerase 1 affects the expression of defence-related genes in *Nicotiana benthamiana* upon tomato leaf curl Gujarat virus infection. Planta 252(1):11

Rahenbrock D, Bode M, Varrelmann M (2025) Differential effects of local dsRNA application on systemic Beet Mosaic Virus (BtMV) resistance in *Nicotiana benthamiana* and *Beta vulgaris*. Mol Plant-Microbe Interact

Rosa C, Kuo YW, Wuriyanghan H, Falk BW (2018) RNA interference mechanisms and applications in plant pathology. Annu Rev Phytopathol 56(1):581–610

Sangeetha B, Krishnamoorthy AS, Renukadevi P, Malathi VG, Sharmila DJ, Amirtham D (2020) Antiviral activity of basidiomycetous fungi against groundnut bud necrosis virus in tomato. Pestic Biochem Physiol 166:104570

Shen W, Yang G, Chen Y, Yan P, Tuo D, Li X, Zhou P (2014) Resistance of non-transgenic papaya plants to papaya ringspot virus (PRSV) mediated by intron-containing hairpin dsRNAs expressed in bacteria. Acta Virol 58(3):261–266

Shimomoto Y, Ishikawa K, Okada T, Oki T, Hayashi K, Yano K, Morita Y, Takeuchi S (2021) Isolation and characterization of an attenuated strain of an orthotospovirus, melon yellow spot virus. J Gen Plant Pathol 87:300–306

Sun ZR, Wei CL, Wu SK, Zhang WB, Song RJ, Hu DY (2022) Synthesis, anti-potato virus Y activities, and interaction mechanisms of novel quinoxaline derivatives bearing dithioacetal moiety. J Agric Food Chem 70:7029

Sundaresha S, Bairwa A, Tomar M, Kumar R, Venkatasalam EP, Sagar V, Bhardwaj V, Sharma S (2022a) In vitro method for synthesis of large-scale dsRNA molecule as a novel plant protection strategy. In: Mysore KS, Senthil-Kumar M (eds) Plant gene silencing, Methods in molecular biology, vol 2408. Humana, New York, pp 211–226. https://doi.org/10.1007/978-1-0716-1875-2_14

Sundaresha S, Jeevalatha A, Kumar R, Sood S, Sharma S, Bhardwaj V, Singh BP, Chakrabarti SK (2022b) RNA interference: a versatile tool to augment plant protection strategies in potato. In: Chakrabarti SK, Sharma S, Shah MA (eds) Sustainable management of potato pests and diseases. Springer, Singapore, pp 369–388. https://doi.org/10.1007/978-981-16-7695-6_15

Sundaresha S, Tiwari JK, Bhardwaj V, Sharma S, Jeevalatha A, Kumar R, Sagar V, Singh BP, Chakrabarti SK, Kumar M (2024) Genetic engineering in potato: status and future perspectives. In: Kalia P (ed) Genetic engineering of vegetable crops. CABI, Nosworthy Way, Wallingford, pp 133–140. https://doi.org/10.1079/9781800627123.0008

Tabassum B, Nasir IA, Khan A, Aslam U, Tariq M, Shahid N, Husnain T (2016) Short hairpin RNA engineering: in planta gene silencing of potato virus Y. Crop Prot 86:1–8

Tabein S, Jansen M, Noris E, Vaira AM, Marian D et al (2020) The induction of an effective dsRNA-mediated resistance against tomato spotted wilt virus by exogenous application of double-stranded RNA largely depends on the selection of the viral RNA target region. Front Plant Sci 11:533338

Tenllado F, Martínez-García B, Vargas M, Díaz-Ruíz JR (2003) Crude extracts of bacterially expressed dsRNA can be used to protect plants against virus infections. BMC Biotechnol 3:3

Thanuja K, Arulmozhiyan R, Saraswathi MS, Selvarajan R, Jegadeeswari V, Babu VR (2025) Production of virus-free quality planting material in banana (*Musa* spp.) variety CO 3 (ABB) through meristem culture. Plant Sci Today 12(1):1–7

Tricoll DM, Carney KJ, Russell PF, McMaster JR, Groff DW, Hadden KC, Himmel PT, Hubbard JP, Boeshore ML, Quemada HD (1995) Field evaluation of transgenic squash containing single or multiple virus coat protein gene constructs for resistance to cucumber mosaic virus, watermelon mosaic virus 2, and zucchini yellow mosaic virus. Nat Biotechnol 13(12):1458–1465

Uddin NM, Kim JY (2013) Intercellular and systemic spread of RNA and RNAi in plants. WIREs RNA 4(3):279–293

Vadlamudi T, Patil BL, Kaldis A, Gopal DVRS, Mishra R, Berbati M, et al (2020) sRNA-mediated protection against two isolates of Papaya ringspot virus through topical application of dsRNA in papaya. J Virol Methods, 270:113750

Vasanthi VJ, Vijai Anandh A, Puvila P, Vaidehi G (2022) Standardization of culture media for meristems and nodal segments in cassava (*Manihot esculenta* Crantz) and virus-indexing using TAS-ELISA. Pharm Innov J 11(5):2413–2418

Venu E, Ramya A, Babu PL, Srinivas B, Kumar S, Reddy NK, Manik S (2024) Exogenous dsRNA-mediated RNAi: mechanisms, applications, delivery methods, and challenges in the induction of viral disease resistance in plants. Viruses 17(1):49

Voloudakis AE, Kaldis A, Patil BL (2022) RNA-based vaccination of plants for control of viruses. Annu Rev Virol 9:521–548

Vu TV, Roy Choudhury N, Mukherjee SK (2013) Transgenic tomato plants expressing artificial microRNAs for silencing the pre-coat and coat proteins of a begomovirus, Tomato leaf curl New Delhi virus, show tolerance to virus infection. Virus Res, 172(1-2):35–45

Wagaba H, Patil BL, Mukasa S, Alicai T, Fauquet CM, Taylor NJ (2016) Artificial microRNA-derived resistance to Cassava brown streak disease. J Virol Methods 231:38–43

Wang Y, Sheng G, Juranek S, Tuschl T, Patel DJ (2008) Structure of the guide-strand-containing argonaute silencing complex. Nature 456(7219):209–213

Wang MR, Hamborg Z, Blystad DR, Wang QC (2021) Combining thermotherapy with meristem culture for improved eradication of onion yellow dwarf virus and shallot latent virus from infected in vitro-cultured shallot shoots. Ann Appl Biol 178(3):442–449

Wani F, Rashid S, Saleem S, Ali G, Mohiddin FA, Hamid A (2025) Topical application of cocktail dsRNA induces plant resistance against Bean Common Mosaic Virus (BCMV). Appl Biochem Biotechnol 197:3431

Waterhouse PM, Helliwell CA (2003) Exploring plant genomes by RNA-induced gene silencing. Nat Rev Genet 4(1):29–38

Weiss D, Sede AR, Levanova AA, Leibman-Markus M, Gupta R, Mishra R, Hak H, Bar M, Poranen MM, Heinlein M, Spiegelman Z (2025) Control of tomato brown rugose fruit virus (ToBRFV) in tomato plants using in vivo-synthesized dsRNA. J Exp Bot

Xia Y, Tian J, Chen X (2016) Effect of surface properties on liposomal siRNA delivery. Biomaterials 79:56–68

Xie M, Yu B (2015) siRNA-directed DNA methylation in plants. Curr Genomics 16(1):23–31

Xing Y, Jiang H, Cai L (2025) Engineered nanotransporters for efficient RNAi delivery in plant protection applications. J Integr Plant Biol 67(5):1223–1245

Yang M, Xu Z, Zhao W, Liu Q, Li Q, Lu L, Cui F (2018) Rice stripe virus-derived siRNAs play different regulatory roles in rice and in the insect vector *Laodelphax striatellus*. BMC Plant Biol 18(1):219

Zhang X, Li H, Zhang J, Zhang C, Gong P, Ziaf K, Xiao F, Ye Z (2011) Expression of artificial microRNAs in tomato confers efficient and stable virus resistance in a cell-autonomous manner. Transgenic Res 20(3):569–581

Zhang T, Zhao Y, Ye J, Cao X, Xu C, Chen B, An H, Jiao Y, Zhang F, Yang X et al (2019) Establishing CRISPR/Cas13a immune system conferring RNA virus resistance in both dicot and monocot plants. Plant Biotechnol J 17:1185–1187

Zhang H, Fan G, Leng D, Liu S, Cai L (2025) Graphene oxide nanosheets for delivery of RNAi and plant immune stimulation for sustained protection against plant viruses. J Agric Food Chem 73(18):11110–11120

# Intervention of Spray-Induced Gene Silencing by dsRNA in Managing Vegetable Plant Viruses

24

Muhammad Imran, Kamal A. M. Abo-Elyousr, Munirah F. Aldayel, and Zhongke Sun

**Abstract**

Viral diseases are among the most destructive constraints to vegetable crop production, causing severe yield losses in tomato, pepper, cucurbits, eggplant, and other major vegetable systems. Conventional management strategies, including resistant cultivars, vector control, and cultural practices, often provide incomplete or unstable protection, especially as new viral variants emerge. Spray-induced gene silencing (SIGS), based on the exogenous application of double-stranded RNA (dsRNA), has recently emerged as a promising nontransgenic strategy for controlling viruses infecting vegetable crops. This chapter outlines the molecular basis of dsRNA-triggered RNA interference (RNAi), including the roles of Dicer enzymes, siRNAs, and RNA-induced silencing complexes (RISC) in degrading viral RNA. We discuss how externally applied dsRNA is absorbed by vegetable plants and generates localized and systemic antiviral protection. Special emphasis is placed on examples relevant to vegetable viruses such as tomato yellow leaf curl virus (TYLCV), cucumber mosaic

M. Imran (✉) · Z. Sun
Department of Biological Engineering, Henan University of Technology, Zhengzhou, Henan, China
e-mail: Imranpathologist@cau.edu.cn

K. A. M. Abo-Elyousr
Department of Plant Pathology, Faculty of Agriculture, University of Assiut, Assiut, Egypt

Department of Agriculture, Faculty of Environmental Sciences, King Abdulaziz University, Jeddah, Saudi Arabia

M. F. Aldayel (✉)
Department of Biological Sciences, College of Science, King Faisal University, Al-Ahsa, Saudi Arabia
e-mail: maldayel@kfu.edu.sa



R. Kumar et al. (eds.), *Vegetable Crops and Viral Pathogens Volume I*, Plant Pathology for Sustainable Agriculture, https://doi.org/10.1007/978-981-95-7265-6_24

virus (CMV), tomato mosaic virus (ToMV), potato virus Y (PVY) in solanaceous vegetables, and key cucurbit-infecting viruses. Advances in nanoformulated dsRNA carriers that enhance stability, uptake, and persistence on vegetable foliage were also summarized. Current challenges, such as environmental degradation, delivery inefficiencies, and regulatory complexity and proposed future innovations, such as multitarget dsRNA constructs and improved nanodelivery systems, were tailored for vegetable production systems. SIGS represents a scalable and sustainable approach with strong potential to strengthen virus management in modern vegetable cropping systems.

**Keywords**

SIGS · RNAi · dsRNA · Virus control · Sustainable agriculture · Climate resilience

## 24.1 Plant Virus Management by dsRNA-Based SIGS

Viral diseases cause substantial economic losses in vegetable crop production, particularly in tomato, pepper, cucurbits, eggplant, okra, and leafy vegetables, where yield and quality are severely compromised. These pathogens collectively contribute to annual global losses exceeding $30 billion, posing a serious threat to food security and sustainable vegetable production (Tatineni and Hein 2023). Their management is especially challenging because viruses in vegetables evolve rapidly; utilize diverse transmission mechanisms mainly insect vectors such as thrips, aphids, and whiteflies; and lack effective postinfection therapeutic options (Soumia et al. 2021; Hamim et al. 2023). Environmental fluctuations, including global warming and altered cultivation practices, further influence virus epidemiology in vegetable production systems by affecting vector abundance, behavior, and dispersal patterns (Ristaino et al. 2021; Tatineni and Hein 2023). Conventional virus management in vegetables relies on resistant cultivars, selective chemical insecticides for vector suppression, and cultural practices. Although resistant varieties remain an essential and eco-friendly strategy, resistance genes are often crop-specific and can be quickly overcome by emerging viral variants capable of evading plant defense responses (Niraula and Fondong 2021; Zhang et al. 2022b). Chemical vector control is still widely practiced in vegetable farming, but its limitations, including ecological contamination, negative impacts on beneficial microbiota, development of insecticide resistance, and human health risks, are well documented (Montes and Pagán 2022; Manikandan et al. 2023). Biological control options have also been explored, although their success remains inconsistent due to ecological variability and the complexity of virus-vector-host interactions (Roudine et al. 2023). Thus, even under integrated pest management programs, viral diseases in vegetable crops remain difficult to suppress effectively, emphasizing the need for innovative, precise, and environmentally benign solutions (Krenz et al. 2024; Akhtar et al. 2024).

RNA-based biotechnologies have emerged as promising approaches for vegetable virus management, particularly those exploiting the mechanism of RNAi. RNAi is a naturally occurring, sequence-specific gene-silencing process enabling plants to degrade viral RNA and suppress viral replication (Akbar et al. 2022; Liu et al. 2023; Sundaresha et al. 2022a). Among RNAi-based approaches, SIGS, which utilizes exogenous application of dsRNA, provides a nontransgenic and highly targeted method to trigger antiviral responses. Unlike host-induced gene silencing, SIGS does not require stable transformation, making it suitable and publicly acceptable for vegetable crops under stringent biosafety regulations (Venu et al. 2024; Chen et al. 2025a; Mazumder et al. 2025). SIGS offers high specificity to viral gene targets, minimizes off-target effects, and reduces environmental toxicity relative to conventional chemical pesticides (Siddique et al. 2025; Chen et al. 2025b). Recent technological advancements have improved dsRNA stability, delivery efficiency, and uptake in vegetable plants through nanoformulations such as liposomes, clay nanosheets, and polymer-based carriers (Qiao et al. 2023; He et al. 2024b; Vatanparast et al. 2024; Islam et al. 2025; Imran et al. 2025a). These innovations significantly enhance the applicability of SIGS under practical vegetable production conditions. Despite these strides, widespread deployment of SIGS in vegetable farming will depend on overcoming challenges related to field stability, delivery efficiency, and regulatory approval. Nonetheless, SIGS represents a dynamic, climate-resilient antiviral strategy with strong potential to manage existing and emerging viral threats in vegetable crops, providing a forward-looking solution to the limitations of traditional virus control methods.

## 24.2 Mechanisms of RNAi

In plants, RNAi constitutes a central antiviral defense mechanism, enabling precise degradation of viral RNA and thereby suppressing infection. The process begins when viral replication generates dsRNA, which is recognized and cleaved by Dicer-like (DCL) enzymes into small interfering RNAs (siRNAs) of approximately 21–25 nucleotides (Muhammad et al. 2019; Gebremichael et al. 2021; Koeppe et al. 2023). These siRNAs are subsequently loaded into the RISC, where Argonaute (AGO) proteins guide the complex to complementary viral RNA sequences, resulting in cleavage or translational repression and ultimately limiting viral replication. Distinct DCL isoforms exhibit specific substrate preferences: DCL4 predominantly generates 21-nt siRNAs associated with posttranscriptional gene silencing and antiviral defense, whereas DCL3 produces 24-nt siRNAs involved in transcriptional gene silencing via chromatin modification (McFarlane et al. 2011; Sundaresha et al. 2022b). Structural features of dsRNA, such as blunt ends or 3′ overhangs, can influence but do not considerably restrict DCL activity, highlighting the versatility of the processing machinery (Leonetti et al. 2023). The PAZ domain of DCL anchors dsRNA ends for accurate cleavage, while the helicase domain facilitates terminal recognition and remodeling to enhance processing efficiency (Paturi and Deshmukh 2021; Leonetti et al. 2023; Zapletal et al. 2023). Through these coordinated steps, viral replication intermediates are efficiently converted into phased siRNAs,

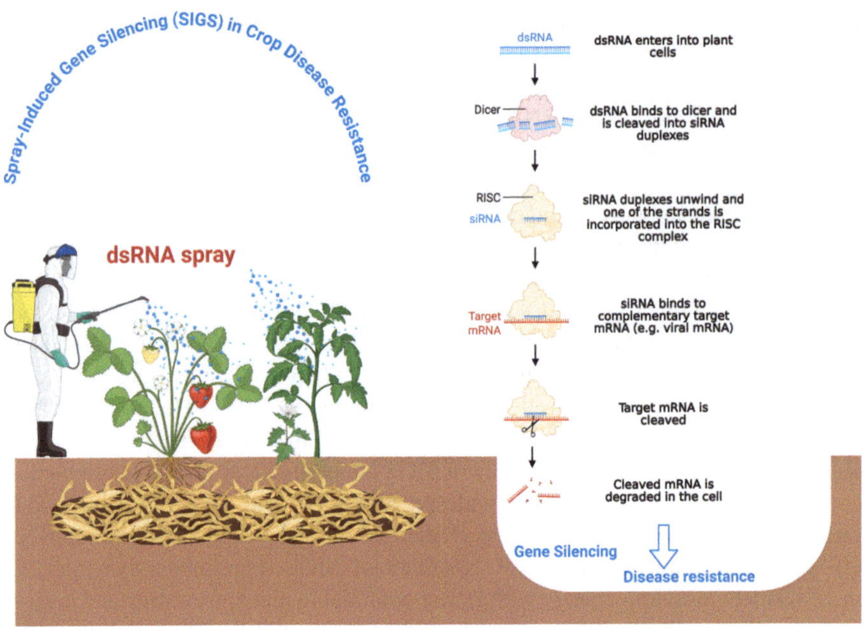


**Fig. 24.1** Mechanism of dsRNA application for effective gene silencing in SIGS technology

forming the molecular basis for subsequent degradation of viral transcripts by RISC (Fig. 24.1).

A key advantage of dsRNA as an RNAi trigger lies in its ability to generate a diverse population of siRNAs upon intracellular processing, broadening silencing coverage across multiple regions of a viral genome and enhancing antiviral durability. In comparison, siRNAs can elicit rapid but often less persistent silencing, as they bypass DCL processing and are more susceptible to degradation (Rego-Machado et al. 2020; Voloudakis et al. 2022; Knoblich et al. 2025). To counter host RNAi defenses, many vegetable-infecting viruses encode viral suppressors of RNAi (VSRs) that interfere with different steps of the silencing pathway. For example, the 2b protein of cucumber mosaic virus binds AGO proteins to inhibit RISC activity, while HcPro of potyviruses sequesters siRNAs and disrupts both local and systemic RNAi signaling (Murray et al. 2013; Li et al. 2013). Similarly, the P0 protein of poleroviruses promotes AGO degradation, dismantling a core effector component (Akbar et al. 2022). Additional VSRs suppress long-distance silencing by interfering with mobile RNAi signals; for instance, the P25 protein of potato virus X and the P30 protein of tobamoviruses hinder RNAi amplification and intercellular movement required for systemic immunity (Murray et al. 2013). Some VSRs act at multiple regulatory points, reflecting the coevolutionary arms race between viruses and host defense pathways.

Understanding VSR-mediated suppression is crucial for improving antiviral strategies. Notably, exogenously applied dsRNA can help overcome certain viral suppression mechanisms by providing an abundant supply of RNAi triggers, enabling the plant to effectively silence viral genes despite VSR activity (Omarov and Bersimbai 2010; Murray et al. 2013; Zhu et al. 2024). Therefore, leveraging the plant's natural RNAi machinery through externally delivered dsRNA, particularly via SIGS, offers a precise and environmentally sustainable approach for managing viral diseases in vegetable crops. These strategies enable broad, sequence-specific targeting without altering the plant genome, supporting practical antiviral applications in modern vegetable production systems (Akbar et al. 2022; Rêgo-Machado et al. 2023; Imran et al. 2025a, b).

## 24.3 dsRNA-Based SIGS

Spray-induced gene silencing using dsRNA represents a transformative strategy for managing viral diseases in vegetable crops, as it activates the plant's endogenous RNAi machinery without requiring genomic modification. In contrast to chemical control measures or transgenic approaches, SIGS enables precise suppression of viral genes through exogenous application of dsRNA designed to target essential viral sequences (Sang and Kim 2020; Beernink et al. 2024; Chen et al. 2025a). By leveraging the plant's natural RNAi pathway, externally delivered dsRNA is recognized and processed into siRNAs that guide sequence-specific degradation of viral RNA, thereby limiting viral accumulation and movement. When applied to foliage commonly through spray formulations, dsRNA must traverse multiple barriers in vegetable crop leaves, including the waxy cuticle, cell wall, and plasma membrane. Uptake is facilitated by cellular processes such as endocytosis, which enable dsRNA to enter the cytoplasm where RNAi is initiated (Beernink et al. 2024). Once inside the cell, Dicer-like enzymes cleave dsRNA into siRNAs that associate with AGO-containing RISC complexes, enabling targeted degradation of viral transcripts. Importantly, dsRNA-derived siRNAs do not remain localized at the application site. Instead, they can move systemically through plasmodesmata and vascular tissues, including both xylem and phloem (Morozov et al. 2019; Pal et al. 2024). Phloem-mediated transport is especially important in rapidly growing vegetable crops, as it enables siRNAs to reach young, meristematic, or highly susceptible tissues, providing broader antiviral protection.

The design of dsRNA constructs for SIGS relies on high sequence specificity, guided by bioinformatic and omics-based tools that identify conserved and functionally essential viral genes such as replication-associated proteins, movement proteins, and coat proteins (Chen et al. 2025a). This precision minimizes off-target effects and avoids disruption of beneficial microbiota associated with vegetable crops. However, dsRNA performance in field-grown vegetables is strongly influenced by environmental conditions. Factors such as UV radiation, temperature fluctuation, and microbial degradation can reduce dsRNA stability before uptake. Among these, UV-induced photodegradation is particularly detrimental, often limiting the persistence of naked dsRNA on exposed leaf surfaces (Chen et al. 2025b). Consequently, formulation improvements such as encapsulation in liposomes, clay nanosheets, or polymeric carriers are essential to enhance stability, retention, and delivery efficiency under open-field vegetable production conditions. After absorption, dsRNA and its siRNA derivatives spread through local and systemic pathways. Symplastic movement via plasmodesmata facilitates lateral spread from treated tissues, while phloem transport enables long-range movement to distant organs (Maizel et al. 2020; Zand Karimi and Innes 2022). The presence of siRNAs in phloem sap has been experimentally confirmed, demonstrating their ability to function as mobile silencing signals across multiple tissues (Morozov et al. 2019). This coordinated mobility allows SIGS to protect tissues that were not directly sprayed, including developing leaves and reproductive structures, an important feature for maintaining yield quality in vegetable crops. SIGS also offers an advantage in countering viral suppressors of RNAi (VSRs), which many vegetable-infecting viruses produce to inhibit silencing pathways. When VSRs sequester endogenous siRNAs or promote AGO degradation, the exogenous supply of dsRNA can saturate these suppressive mechanisms and restore antiviral activity (Murray et al. 2013; Omarov and Bersimbai 2010; Zhu et al. 2024). This capacity to bypass or overwhelm VSR-mediated suppression makes SIGS not only a preventive measure but also a potential tool for suppressing ongoing infections in vegetable crops. Thus, dsRNA-based SIGS represents a biologically rational, highly targeted, and environmentally sustainable approach for managing viral diseases in vegetables. By activating natural RNAi pathways through well-designed dsRNA constructs and ensuring their efficient uptake and systemic movement, SIGS integrates molecular precision with practical field-level applicability. Its ability to circumvent viral counter defenses, coupled with its nontransgenic nature, positions SIGS as a promising component of next-generation viral disease management strategies in vegetable production systems (Rego-Machado et al. 2020; Knoblich et al. 2025).

## 24.4 Design and Development of dsRNA-Based Formulations for Effective Viral Suppression

Effective dsRNA-based SIGS for managing viral diseases in vegetable crops begins with the production of stable, high-quality dsRNA molecules that can withstand environmental exposure and support efficient uptake. Although SIGS provides a precise, nontransgenic RNAi strategy, its performance in open-field vegetable production is often limited by the rapid degradation of naked dsRNA, restricted mobility within plant tissues, nuclease activity, and insufficient internalization. Therefore, optimizing both the synthesis and formulation of dsRNA is essential for achieving consistent antiviral protection under real-world conditions. Three primary methods are employed for dsRNA synthesis: in vitro transcription (IVT), cloning-based microbial production, and chemical synthesis. Among these, IVT is the most widely used because it is scalable and cost-effective and allows flexible sequence design to target viral genes relevant to vegetable-infecting viruses (Li and Zamore 2019; Sundaresha et al. 2022b). IVT utilizes bacteriophage RNA polymerases (T7, T3, SP6) to produce dsRNA fragments typically 150–3000 bp in length, depending on target requirements. High-fidelity template preparation is critical; the process begins with PCR amplification using high-accuracy polymerases, followed by gel electrophoresis to verify template specificity (Liu et al. 2024). After transcription, dsRNA is purified via DNase treatment and column-based removal of shorter RNA species, and quality is confirmed through agarose gel electrophoresis and spectrophotometry (Liu and Chen 2016). Cloning-based dsRNA production uses microbial expression systems to generate large dsRNA volumes, offering a low-cost option for long-term production despite slower turnaround and reduced design flexibility (Voloudakis et al. 2015). Chemical synthesis provides highly pure dsRNA fragments or siRNAs but remains expensive and is generally limited to short sequences, making it suitable mainly for validation assays or mechanistic studies.

Recent enzymatic advances have improved IVT efficiency and purity by reducing undesirable byproducts that compromise stability or generate immunogenic contaminants. Engineering the T7 RNA polymerase, for example, the G753A substitution in the specificity loop suppresses promoter-independent antisense transcription without lowering yield (He et al. 2024a). Combination with mutations such as K389A further enhances the accuracy of RNA synthesis and minimizes unintended dsRNA impurities (He et al. 2024b). These innovations are particularly valuable for producing large volumes of dsRNA required for field-scale SIGS trials in vegetable crops. Because unformulated dsRNA rapidly degrades under field conditions, formulation strategies must enhance stability, uptake, and controlled release. Foundational approaches include the use of surfactants, adjuvants, and pH buffers that increase leaf surface adherence and facilitate movement across lipophilic barriers such as the cuticle. Encapsulation technologies further protect dsRNA from UV degradation and nucleases while enabling gradual release within plant tissues (Yang et al. 2022; Arjunan et al. 2024). Among these, chitosan-based matrices have shown strong potential to extend dsRNA persistence on vegetable foliage and prolong RNAi activity under field conditions (Yang et al. 2022). A rapidly growing area in dsRNA formulation is the application of nanocarriers engineered for efficient delivery. Biocompatible nanoparticles such as chitosan NPs, liposomes, and dendrimers help shield dsRNA from environmental degradation and facilitate translocation across cellular membranes. Chitosan nanoparticles not only protect dsRNA from degradation but also enhance delivery through surface charge interactions (Grewal and Salar 2024; Stefanache et al. 2025). Liposomes, with their lipid bilayer architecture, readily fuse with plant membranes and release dsRNA intracellularly (Venu et al. 2024). Cationic dendrimers, owing to their high charge density and structural uniformity, form stable complexes with dsRNA and promote entry into target cells (Yan et al. 2021). Layered double hydroxides (LDHs) adsorb dsRNA onto their layered surfaces and release it slowly over time, improving the duration of gene silencing and reducing application frequency in vegetable fields (Dalakouras et al. 2024). Advanced polymeric carriers such as star polycation (SPc) nanoparticles condense dsRNA into compact nanostructures that enhance stability, resist environmental degradation, and improve systemic transport within plant tissues (Qiao et al. 2025). Unlike simple encapsulation, these synthetic nanoparticles create controlled-release systems and enable multitarget dsRNA delivery for simultaneous suppression of virus and vector genes.

Chemical modification of dsRNA represents another formulation strategy for improving field applicability. Adjusting backbone or sugar moieties can reduce nuclease sensitivity while maintaining RNAi activity, an approach adapted from therapeutic RNA technologies (Arjunan et al. 2024). Additionally, combining dsRNA with surfactants or adhesives enhances leaf surface retention and improves penetration into internal tissues (Wytinck et al. 2020). Carbon dot-based nanocarriers have recently emerged as a highly promising platform for dsRNA delivery. Synthesized from saccharides and passivated with polyethylenimines, carbon dots exhibit high surface area, excellent biocompatibility, and strong dsRNA-binding capacity. When applied to vegetable plants, carbon-dot-dsRNA complexes significantly enhance cellular uptake and induce robust local and systemic RNAi responses (Delgado-Martín et al. 2022a; Ren et al. 2025). Their intrinsic fluorescence also enables real-time tracking of dsRNA movement in planta, offering valuable insights for optimizing future delivery systems (Gyawali et al. 2024). Together, these formulation and delivery innovations address key challenges limiting SIGS performance in vegetable crops, particularly environmental degradation, inefficient uptake, and short-lived silencing. By integrating optimized dsRNA synthesis, engineered IVT systems, nanotechnology-based carriers, and carefully selected adjuvants, SIGS is rapidly advancing from laboratory validation to practical field deployment. These developments form the technological foundation for the next generation of dsRNA-based antiviral products, supporting their scalability, regulatory acceptance, and long-term adaptability in vegetable crop protection.

## 24.5 Applications in Virus Disease Management in Vegetable Crops

Spray-induced gene silencing using dsRNA has emerged as a powerful strategy for managing viral diseases in vegetable crops, capitalizing on the plant's native RNAi machinery to suppress targeted viral genes (Sundaresha et al. 2022a; Vatanparast et al. 2024; Spada et al. 2025). Unlike conventional genetic modification or chemical antiviral strategies, SIGS provides a transient, nontransgenic, and environmentally friendly option that aligns well with the sustainability goals of modern vegetable production systems. Its efficacy has been validated across multiple virus-host systems, including key vegetable viruses of global economic significance. For tomato yellow leaf curl virus (TYLCV), a major begomovirus threatening tomato production, dsRNA constructs homologous to viral genomic regions effectively reduce viral titers and symptom severity without modifying the host genome (Melita et al. 2021; Frascati et al. 2024). Similarly, for cucumber mosaic virus (CMV), dsRNAs targeting essential genes such as the coat protein have demonstrated strong antiviral effects in cucurbits under both controlled and field conditions (Holeva et al. 2021; Delgado-Martín et al. 2022b; Manamohan et al. 2024). Potato virus Y (PVY), which affects solanaceous vegetables including tomato and pepper, has also been effectively suppressed using dsRNA molecules directed against conserved functional regions, offering protection comparable to chemically based antiviral measures but with significantly lower ecological risks (Samarskaya et al. 2022; Routhu et al. 2023; Chen et al. 2025a). Central to SIGS performance in vegetables is the strategic design of dsRNA constructs that target conserved and crucial viral genes. This not only improves efficacy but also increases the durability of protection (Table 24.1). One challenge is the presence of viral suppressors of RNAi (VSRs), which viruses deploy to counter plant defenses. Current strategies incorporate dsRNAs that simultaneously target essential viral genes and VSR-encoding regions, weakening the virus' capacity to inhibit RNAi (Shao et al. 2019; Hussain et al. 2021). Multigene targeting approaches, such as simultaneous silencing of movement and replication genes, exert strong selective pressure on viral populations and reduce the likelihood of resistance development (Atabekova et al. 2023).

Advancements in dsRNA delivery technologies have greatly improved the utility of SIGS in vegetable systems. Nanocarrier-based delivery protects dsRNA from UV degradation and nuclease activity, enhances cellular uptake, and prolongs antiviral activity (Quilez-Molina et al. 2024). In CMV-infected cucurbits, combinatorial dsRNAs targeting multiple genome segments show synergistic silencing effects, interrupting different stages of the viral life cycle (Holeva et al. 2021; Manamohan et al. 2024). Nevertheless, due to the transient nature of dsRNA protection, repeated applications may be necessary, highlighting the importance of sustained-release formulations and more stable delivery matrices. Although much of the foundational work on SIGS has been demonstrated in diverse virus-host pathosystems, these studies collectively provide essential mechanistic insight relevant to vegetables (Worrall et al. 2019; Rego-Machado et al. 2020; Sundaresha et al. 2024). Multiple experimental evaluations across solanaceous and cucurbit vegetable crops further demonstrate the robustness of SIGS-based antiviral protection (Table 24.1). In these systems, dsRNA constructs targeting essential viral genes such as coat protein (CP), nucleocapsid protein, replication-associated proteins, or multiple genome segments consistently reduced viral accumulation, suppressed symptom development, and enhanced overall plant vigor. These outcomes underscore the versatility of SIGS as a nontransgenic, sequence-specific antiviral strategy and support its potential integration into sustainable vegetable crop protection programs. These investigations, originally conducted in model or nonvegetable systems, have played a crucial role in shaping our understanding of dsRNA processing, antiviral responses, and delivery optimization knowledge directly transferable to vegetable crop protection. Their inclusion establishes the broader scientific foundation upon which vegetable-targeted SIGS applications have been built. Virus-specific design principles also guide successful SIGS deployment in vegetable cropping systems. For TYLCV, effective silencing generally targets conserved intergenic regions or overlapping open reading frames involved in replication (Yang et al. 2019). For CMV, single or combined dsRNAs against the coat protein or replicase genes disrupt both replication and movement. In the case of PVY, which encodes a large polyprotein, dsRNA design focuses on critical protein-coding regions to address strain variability and avoid off-target effects (Samarskaya et al. 2022; Routhu et al. 2023). These approaches are increasingly supported by robust bioinformatic tools and advanced

**Table 24.1** Applications of dsRNA-based SIGS for the management of viral diseases in vegetable crops

| Target virus | Host | Target gene (s) | Effect on virus | Effect on plant | Reference |
|---|---|---|---|---|---|
| Tomato mosaic virus | Tomato | cp (coat protein) | Reduced infection frequency symptom severity; protection in roughly half to two-thirds of treated plants compared with controls | Healthy plants, limited systemic spread; no notable phytotoxic effects | Rego-Machado et al. (2020) |
| Potato virus Y | Tomato | cp (coat protein) | Marked decrease in PVY infections; dsRNA gave about 50–60% protection and did not protect against unrelated ToMV | Fewer PVY symptoms; maintained better than mock-treated control | Rego-Machado et al. (2020) |
| Tomato spotted wilt virus | Tomato | N (nucleocapsid protein) | Strong suppression of accumulation, virus-free and symptomless plants | Normal appearance, heavily diseased control | Tabein et al. (2020) |
| Cucumber mosaic virus | Tomato, cucumber, and pepper | Multiple essential genes in single edsRNA | Very high protection (80–100%), virus titers remained very low | High survival and maintenance; normal growth; stayed vigorous; controls collapsed | Knoblich et al. (2025) |
| Cucumber mosaic virus | Cucumber | cp and 2b | Reduced CMV infection and systemic spread; showed clear sequence-specific antiviral activity | Symptomless/ strongly attenuated mosaic compared with control | Holeva et al. (2021) |
| Cucumber green mottle mosaic virus | Cucumber | cp and mp | Reduced CGMMV accumulation up to tenfold, lowered the number of strongly symptomatic plants | Retained greater leaf area and biomass; appeared close to healthy control in spring cycles | Delgado-Martín et al. (2022b) |

(continued)

**Table 24.1** (continued)

| Target virus | Host | Target gene (s) | Effect on virus | Effect on plant | Reference |
|---|---|---|---|---|---|
| Zucchini yellow mosaic virus | Cucumber | HC-Pro and cp | Provided 70–80% resistance, strong reduction in virus detection | Symptomless/ only mild mosaic, severely stunted and chlorotic controls | Kaldis et al. (2018) |
| Zucchini yellow mosaic virus | Watermelon | HC-Pro and cp | Lowered disease incidence, complete protection with dose/time | Delayed/much milder symptoms; maintained better vigor than control | Kaldis et al. (2018) |
| Zucchini yellow mosaic virus | Squash (zucchini type) | HC-Pro and cp | Weaker and more variable protective effect; Low resistance %; no significant difference from controls | Developed clear symptoms; growth reduction despite, limited benefit in this host | Kaldis et al. (2018) |
| Tomato leaf curl New Delhi virus | Zucchini (courgette) | Chimeric dsRNA covering all DNA-A ORFs | Significant decrease in the proportion of infected plants, delayed symptom onset, and reduced viral titers | Milder leaf curling; less deformation over severely distorted untreated plants | Frascati et al. (2024) |

delivery technologies that consider both viral genomic diversity and host biological traits.

Minimizing off-target effects is essential for the safe deployment of SIGS in vegetable systems. Rigorous bioinformatic screening ensures low homology with host and nontarget genes (Zhang et al. 2022a). Designing shorter, highly specific dsRNAs and combining multiple siRNAs targeting different viral regions further improves precision (Persengiev et al. 2004; Neumeier and Meister 2021). Chemical modifications of dsRNA, including backbone or base alterations, enhance stability and reduce degradation without compromising RNAi activity (Howard et al. 2022; Mendoza and Beal 2023). Asymmetric dsRNA molecules, optimized for guide-strand selection, offer another layer of refinability (Zarrabian and Sherif 2024). Field validation remains critical for confirming both efficacy and biosafety. Compared with conventional chemical pesticides, dsRNA-based SIGS provides substantial ecological and practical advantages. It is biodegradable and nontoxic to beneficial organisms and aligns with the principles of integrated pest management (Delgado-Martín et al. 2022b; Dalakouras et al. 2024; Vatanparast et al. 2024). Its high sequence specificity prevents off-target impacts and slows resistance development. Moreover, the modular nature of dsRNA design enables rapid optimization in response to emerging viral strains, addressing limitations of breeding or transgenic resistance (Frascati et al. 2024). Consequently, dsRNA-based SIGS offers a targeted, adaptable, and environmentally responsible approach to viral disease management in vegetable crops. Continued innovations in delivery matrices, construct optimization, and large-scale field deployment will strengthen its role in sustainable vegetable production and contribute significantly to resilient food systems.

## 24.6 Challenges and Limitations

Although SIGS using dsRNA offers a promising and environmentally friendly strategy for combating viral diseases in vegetable crops, its large-scale implementation remains constrained by several biological, technological, ecological, and regulatory challenges. Overcoming these limitations is essential to achieving consistent field efficacy and successful commercialization of dsRNA-based solutions. A major bottleneck lies in the efficient delivery and uptake of dsRNA by plant tissues and, in some cases, by insect vectors involved in virus transmission. For SIGS to initiate a robust RNAi response, externally applied dsRNA must first traverse multiple structural barriers on the leaf surface, including the waxy cuticle, rigid cell wall, and selectively permeable plasma membrane to reach the cytoplasm, where viral replication occurs. These barriers significantly restrict foliar absorption and limit access to the intracellular RNAi machinery. Even once inside the plant, the systemic movement of dsRNA or siRNA through the phloem and other vascular routes is highly variable across plant species and is not fully understood, making whole-plant protection inconsistent. In vector-borne viruses, the ability of insects to ingest, retain, or transmit dsRNA adds another dimension to delivery complexity, further influencing SIGS outcomes. Beyond plant-internal constraints, the potential dissemination of dsRNA into surrounding ecological compartments introduces both possibilities and uncertainties. Low levels of dsRNA detected in root exudates or insect honeydew indicate that RNA molecules may circulate across trophic levels, raising the theoretical potential for horizontal RNAi effects through herbivory or predator-prey interactions. While such movement could amplify antiviral silencing, it also highlights the need to better understand cross-kingdom RNA transfer and its ecological implications before broad field deployment. Environmental degradation represents another major barrier to SIGS effectiveness. After application, dsRNA is rapidly exposed to abiotic stresses such as UV radiation, fluctuating temperatures, rainfall, and humidity, as well as biotic degradation by plant-surface nucleases and microbial communities (Fig. 24.2). These factors drastically shorten the persistence of dsRNA on foliage and reduce the duration of antiviral protection. Encapsulation using nanocarriers, including clay nanosheets, liposomes, biodegradable polymers, and other protective matrices, has been explored to enhance environmental stability, prolong release, and improve uptake. However, achieving an optimal balance

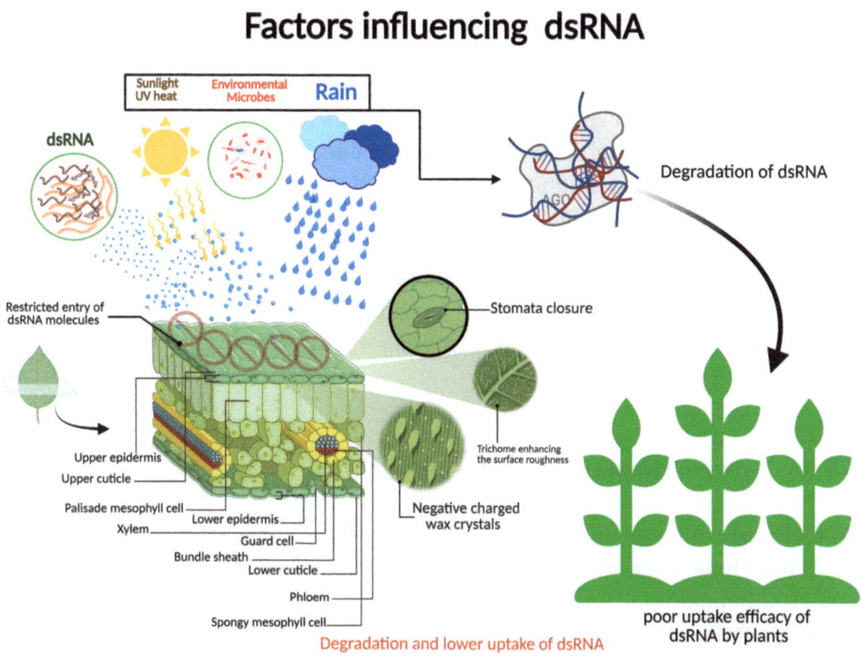


**Fig. 24.2** Factors affecting the dsRNA efficacy in effective gene silencing

between environmental durability and the preservation of RNAi activity remains a formulation challenge yet to be fully resolved.

Despite the sequence specificity inherent to RNAi, off-target gene silencing remains a potential risk. Partial sequence homology or imperfect base pairing can lead to unintended regulation of nontarget genes in beneficial microbes, neutral insects, or even the crop plant itself. Such outcomes could disrupt ecological balance or plant physiology if not carefully controlled. Therefore, stringent in silico analyses, empirical validation, and biosafety profiling are essential components of dsRNA design. Additional concerns relate to the persistence and movement of dsRNA residues in soil or water bodies and the potential impacts on nontarget species across broader agroecosystems. Regulatory bodies increasingly emphasize comprehensive environmental risk assessments to ensure compatibility with ecological and human health standards. The pathway to commercialization is further complicated by regulatory ambiguity and economic considerations. dsRNA-based agricultural products do not fit neatly into existing categories such as pesticides, biologicals, or genetic technologies. As a result, regulatory approval often requires extensive data on efficacy, toxicology, environmental fate, and potential genetic interactions. The lack of harmonized global guidelines delays standardized evaluation and market entry. Although significant progress in microbial fermentation and cell-free synthesis has reduced dsRNA production costs, challenges persist in scaling formulation technologies, maintaining consistent field performance, and meeting quality-control standards. Public perception also plays a key role; because RNA technologies are frequently misunderstood or conflated with transgenic approaches, transparent communication and stakeholder engagement are essential for acceptance.

To establish dsRNA-based SIGS as a next-generation tool for managing viral diseases, particularly in vegetable crops, it is crucial to systematically address these biological, environmental, and regulatory bottlenecks. Enhancing dsRNA uptake and systemic mobility, improving formulation stability under field conditions, ensuring ecological safety, and developing clear regulatory frameworks will require interdisciplinary collaboration across molecular biology, nanotechnology, plant physiology, ecology, and policy. Only through such coordinated efforts can SIGS transition from a promising conceptual technology to a reliable, scalable, and environmentally conscious solution for achieving virus-resilient vegetable production systems.

## 24.7 Future Directions and Innovations

To realize the full potential of dsRNA-based SIGS for managing viral diseases in vegetable crops, future research must strategically address the biological and technological limitations that currently restrict its consistent field performance. Central to these efforts is the development of next-generation delivery platforms capable of enhancing both the uptake and systemic movement of dsRNA within plant tissues and, where applicable, within virus-transmitting insect vectors. Nanocarrier systems, including chitosan nanoparticles, LDH nanosheets, liposomes, and biodegradable polymers, remain among the most promising candidates for improving environmental stability and cellular internalization. Fine-tuning nanocarrier features such as particle size, surface charge, and ligand functionalization will be important for maximizing delivery efficiency in diverse vegetable crop species while simultaneously reducing exposure to non-target organisms. Environmental degradation remains one of the greatest constraints to SIGS performance under field conditions. Naked dsRNA is highly susceptible to breakdown by UV radiation, rainfall, fluctuating temperatures, and microbial nucleases, resulting in very short persistence on leaf surfaces. Encapsulation strategies using LDH nanosheets and chitosan-based carriers have demonstrated significant improvements in dsRNA durability, enabling gradual release and prolonged RNAi activity (Fig. 24.3). LDH formulations, for example, can sustain dsRNA release over multiple weeks, whereas unprotected dsRNA often degrades within days. Chitosan-based systems further enhance uptake due to their biocompatibility and natural affinity for plant membranes. Despite these advances, issues related to large-scale formulation, cost of production, and long-term field stability still require dedicated technological solutions before commercial deployment becomes feasible.

Climatic variability introduces another layer of complexity for SIGS applications in vegetable cropping systems. High humidity, heavy rainfall, and temperature extremes not only accelerate dsRNA degradation but can also influence crop physiology and viral replication dynamics, thereby altering RNAi efficiency. Additionally, virus-vector interactions, especially in begomoviruses and potyviruses affecting

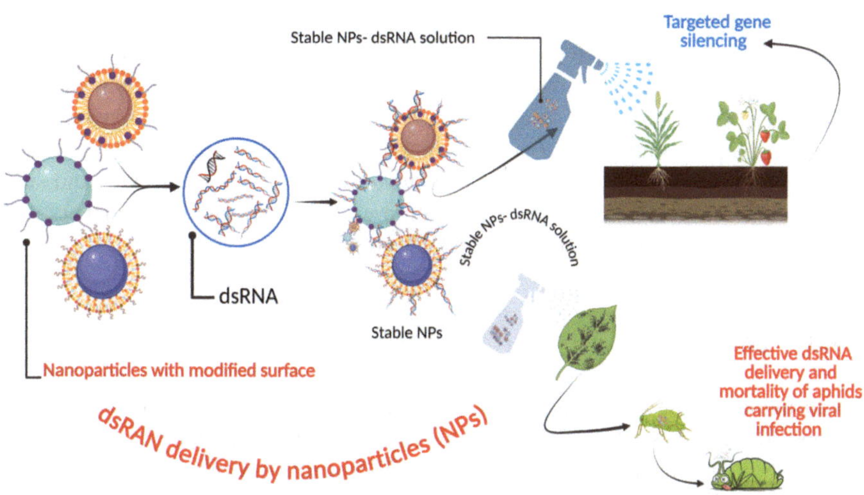


**Fig. 24.3** Effective delivery of dsRNA for gene silencing via encapsulation with NPs

solanaceous and cucurbit vegetable crops, may be affected by environmental factors that influence dsRNA ingestion by insect vectors. Conducting multiseason, multilocation field trials across diverse agroecological regions will be essential for understanding formulation behavior under variable climatic stresses and for refining spray schedules to achieve consistent protection. A deeper mechanistic understanding of dsRNA uptake, processing, and translocation is also crucial. While SIGS success depends on efficient internalization and movement through vascular tissues such as the phloem, the molecular pathways mediating these processes remain poorly characterized and vary widely among vegetable crop species. Future research should therefore prioritize elucidating receptor-mediated uptake pathways, plasmodesmata regulation, and long-distance transport through xylem and phloem. Parallel studies examining how viruses or their insect vectors internalize exogenous dsRNA will enable the development of more targeted delivery strategies tailored to specific virus-vector-vegetable crop systems. Mitigating off-target effects and ensuring biosafety are equally important for the acceptance and regulation of dsRNA-based technologies. Although RNAi is inherently sequence-specific, partial homology with beneficial microorganisms, nontarget insects, or host plant genes could lead to unintended gene silencing. Robust bioinformatic screening, experimentally validated safety assessments, and long-term ecological studies examining dsRNA persistence in soil and water will be essential components of future SIGS development. Such studies will also provide the foundational evidence required to address concerns regarding dsRNA movement across trophic levels and potential impacts on agroecosystem health. Future progress also depends heavily on the development of economically viable dsRNA production systems. Advances in microbial fermentation, yeast-based expression platforms, and cell-free enzymatic synthesis are rapidly reducing the cost of dsRNA manufacturing. Integrating these scalable production platforms with nanoparticle-enabled delivery systems will be key to generating affordable, field-ready formulations for vegetable growers. Progress in formulation science must be complemented by parallel advancements in regulatory science. There is currently no globally harmonized framework governing dsRNA-based agricultural products, resulting in complex and inconsistent approval pathways. Developing standardized criteria for evaluating environmental fate, efficacy, off-target risks, and residue behavior will streamline regulatory processes and support commercialization. Clear communication with growers and policymakers will also be critical for public acceptance and adoption.

## 24.8 Conclusion

Spray-induced gene silencing using exogenous dsRNA represents a highly promising and sustainable strategy for managing viral diseases in vegetable crops. By harnessing the plant's intrinsic RNAi machinery, SIGS enables precise, sequence-specific suppression of viral genes without the need for transgenic modification. This aligns with the global demand for safer, more targeted alternatives to chemical pesticides and offers an adaptable platform that can rapidly respond to emerging viral variants affecting vegetable production systems. Despite its potential, the translation of SIGS from controlled environments to field-level vegetable production remains constrained by several unresolved challenges. Environmental degradation of dsRNA, variable uptake efficiency across different vegetable species, limited systemic mobility, and the need for robust delivery formulations continue to hinder consistent performance. Nanocarrier technologies have significantly improved dsRNA stability and cellular internalization, yet issues related to cost, large-scale formulation, and field durability must still be addressed before commercialization becomes feasible. Knowledge gaps in dsRNA uptake pathways, mechanisms of long-distance movement, and potential off-target interactions also require deeper investigation, particularly within the context of diverse vegetable crop physiology and virus-vector dynamics.

As with most RNA-based technologies, biosafety considerations, including ecological persistence, effects on nontarget organisms, and regulatory acceptance, will play a decisive role in shaping public and commercial confidence. SIGS should therefore be viewed not as a stand-alone intervention but as an integral component of future integrated pest management (IPM) programs. Recent advances in sustainable plant protection have highlighted the potential of integrating biological and chemical strategies to enhance crop resilience. Studies from our group have demonstrated the dual benefits of culture filtrates and nanoparticles in improving plant health under stress conditions, as seen in onion cultivation (Imran et al. 2025b) and in the management of anthracnose in pepper using potash alum and frankincense essential oil (El-Fawy et al. 2025). Moreover, our investigations into biochemical and molecular responses of vegetables, such as cucumber, under pathogen stress provide foundational insights into gene expression modulation, which is critical for the successful application of dsRNA-based SIGS strategies. Collectively, these studies underscore the feasibility of combining molecular and eco-friendly interventions to manage viral and fungal stresses in vegetable crops. Its true strength lies in its complementarity: combining targeted RNAi with optimized delivery systems, predictive bioinformatic design, climate-resilient agronomic practices, and modern breeding efforts aimed at enhancing host responsiveness to RNA-based therapeutics. With continued advancements in formulation science, mechanistic understanding, field validation, and harmonized regulatory frameworks, SIGS has the potential to transition from an emerging research tool to a transformative, scalable, and environmentally responsible platform for managing viral diseases in vegetable crops. As innovation accelerates, SIGS may ultimately redefine antiviral protection, offering a precise, intelligent, and sustainable approach to safeguarding global vegetable production.

**Acknowledgments** The authors extend their appreciation for the support of the Deanship of Scientific Research, Vice Presidency for Graduate Studies and Scientific Research, King Faisal University, Saudi Arabia [Grant No. KFU254309].

## References

Akbar S, Wei Y, Zhang MQ (2022) RNA interference: promising approach to combat plant viruses. Int J Mol Sci 23(10):5312. https://doi.org/10.3390/ijms23105312

Akhtar H, Usman M, Binyamin R, Hameed A, Arshad SF, Aslam HMU, Khan IA, Abbas M, Zaki HE, Ondrasek G, Shahid MS (2024) Traditional strategies and cutting-edge technologies used for plant disease management: a comprehensive overview. Agronomy 14(9):2175. https://doi.org/10.3390/agronomy14092175

Arjunan N, Thiruvengadam V, Sushil SN (2024) Nanoparticle-mediated dsRNA delivery for precision insect pest control: a comprehensive review. Mol Biol Rep 51(355):1–22. https://doi.org/10.1007/s11033-023-09187-6

Atabekova AK, Solovieva AD, Chergintsev DA, Solovyev AG, Morozov SY (2023) Role of plant virus movement proteins in suppression of host RNAi defense. Int J Mol Sci 24(10):9049. https://doi.org/10.3390/ijms24109049

Beernink BM, Amanat N, Li VH, Manchur CL, Whyard S, Belmonte MF (2024) SIGS vs. HIGS: opportunities and challenges of RNAi pest and pathogen control strategies. Can J Plant Pathol 46(6):675–689. https://doi.org/10.1080/07060661.2024.2392610

Chen C, Imran M, Feng X, Shen X, Sun Z (2025a) Spray-induced gene silencing for crop protection: recent advances and emerging trends. Front Plant Sci 16:1527944. https://doi.org/10.3389/fpls.2025.1527944

Chen P, Lan Y, Ding S, Du R, Hu X, Zhang H, Yu H, Xu L, Li C, Lin F, Du L (2025b) RNA interference-based dsRNA application confers prolonged protection against rice blast and viral diseases, offering a scalable solution for enhanced crop disease management. J Integr Plant Biol 67(6):1633–1648. https://doi.org/10.1111/jipb.13896

Dalakouras A, Koidou V, Papadopoulou K (2024) DsRNA-based pesticides: considerations for efficiency and risk assessment. Chemosphere 352:141530. https://doi.org/10.1016/j.chemosphere.2024.141530

Delgado-Martín J, Delgado-Olidén A, Velasco L (2022a) Carbon dots boost dsRNA delivery in plants and increase local and systemic siRNA production. Int J Mol Sci 23(10):5338. https://doi.org/10.3390/ijms23105338

Delgado-Martín J, Ruiz L, Janssen D, Velasco L (2022b) Exogenous application of dsRNA for the control of viruses in cucurbits. Front Plant Sci 13:895953. https://doi.org/10.3389/fpls.2022.895953

El-Fawy MM, Imran M, Ali EF, Alqurashi AS, Issa AA, Emam RA, El-Mageed MFA, Almasoudi NM, Sallam NM, Abo-Elyousr KAM (2025) Managing anthracnose fruit rot disease in pepper caused by *Colletotrichum capsici* via potash alum and frankincense essential oil. J Plant Pathol 107:1471–1482. https://doi.org/10.1007/s42161-025-01924-z

Frascati F, Rotunno S, Accotto GP, Noris E, Vaira AM, Miozzi L (2024) Exogenous application of dsRNA for protection against tomato leaf curl New Delhi virus. Viruses 16(3):436. https://doi.org/10.3390/v16030436

Gebremichael DE, Haile ZM, Negrini F, Sabbadini S, Capriotti L, Mezzetti B, Baraldi E (2021) RNA interference strategies for future management of plant pathogenic fungi: prospects and challenges. Plants 10(4):650. https://doi.org/10.3390/plants10040650

Grewal AK, Salar RK (2024) Chitosan nanoparticle delivery systems: an effective approach to enhancing efficacy and safety of anticancer drugs. Nano TransMed 3:100040. https://doi.org/10.1016/j.ntm.2024.100040

Gyawali B, Rahimi R, Alizadeh H, Mohammadi M (2024) Graphene quantum dots (GQD)-mediated dsRNA delivery for the control of fusarium head blight disease in wheat. ACS Appl Bio Mater 7(3):1526–1535. https://doi.org/10.1021/acsabm.3c00972

Hamim I, Komatsu K, Prager SM, Wu B (2023) Viruses in agricultural systems: interactions with plants, insect pollinators and fungi. Front Microbiol 14:1170402. https://doi.org/10.3389/fmicb.2023.1170402

He L, Zhou Y, Mo Q, Huang Y, Tang X (2024a) Spray-induced gene silencing in phytopathogen: mechanisms, applications, and progress. Adv Agrochem 3(4):289–297. https://doi.org/10.1016/j.aac.2024.06.001

He W, Geng Q, Ji G, Li J, Wang D, He Y, Jin Q, Ye J (2024b) Effective synthesis of mRNA during in vitro transcription with fewer impurities produced. Molecules 29(19):4713. https://doi.org/10.3390/molecules29194713

Holeva MC, Sklavounos A, Rajeswaran R, Pooggin MM, Voloudakis AE (2021) Topical application of double-stranded RNA targeting 2b and CP genes of cucumber mosaic virus protects plants against local and systemic viral infection. Plants 10(5):963. https://doi.org/10.3390/plants10050963

Howard JD, Beghyn M, Dewulf N, De Vos Y, Philips A, Portwood D, Kilby PM, Oliver D, Maddelein W, Brown S, Dickman MJ (2022) Chemically modified dsRNA induces RNAi effects in insects in vitro and in vivo: a potential new tool for improving RNA-based plant protection. J Biol Chem 298(9):102311. https://doi.org/10.1016/j.jbc.2022.102311

Hussain MD, Farooq T, Chen X, Tariqjaveed M, Jiang T, Liu S, Zhou T (2021) Viral suppressors from members of the family Closteroviridae combating antiviral RNA silencing: a tale of a sophisticated arms race in host-pathogen interactions. Phytopathol Res 3(27):1–36. https://doi.org/10.1186/s42483-021-00104-y

Imran M, Feng X, Sun Z, Al Omari H, Zhang G, Zhu J, Aldayel MF, Li C (2025a) Nanotechnology-driven gene silencing: advancements in SIGS-dsRNA technology for sustainable disease management. Chem Biol Technol Agric 12:31. https://doi.org/10.1186/s40538-025-00738-6

Imran M, Alomran MM, Khalifa HA, Bilal MS, Ahmed AF, Ali EF, Alqurashi AS, Aldayel MF, Abo-Elyousr KAM (2025b) Exploring smart solutions for stressed soils: dual-action of culture filtrates and nanoparticles for sustainable onion cultivation. Eur J Plant Pathol:1–22. https://doi.org/10.1007/s10658-025-03113-1

Islam MS, Ahmed MR, Noman M, Zhang Z, Wang J, Lu Z, Cai Y, Ahmed T, Li B, Wang Y, Golam Sarwar AKM, Wang J (2025) Integrating RNA interference and nanotechnology: a transformative approach in plant protection. Plants 14(6):977. https://doi.org/10.3390/plants14060977

Kaldis A, Berbati M, Melita O, Reppa C, Holeva M, Otten P, Voloudakis A (2018) Exogenously applied dsRNA molecules deriving from the Zucchini yellow mosaic virus (ZYMV) genome move systemically and protect cucurbits against ZYMV. Mol Plant Pathol 19(4):883–895. https://doi.org/10.1111/mpp.12572

Knoblich M, Gursinsky T, Gago-Zachert S, Weinholdt C, Grau J, Behrens SE (2025) A new level of RNA-based plant protection: dsRNAs designed from functionally characterized siRNAs highly effective against cucumber mosaic virus. Nucleic Acids Res 53(5):gkaf136. https://doi.org/10.1093/nar/gkaf136

Koeppe S, Kawchuk L, Kalischuk M (2023) RNA interference past and future applications in plants. Int J Mol Sci 24(11):9755. https://doi.org/10.3390/ijms24119755

Krenz B, Niehl A, Krczal G (2024) Emerging strategies in plant virus disease control: insights from the 56th meeting of the DPG working group "Viruskrankheiten der Pflanzen". J Plant Dis Prot 131:1761–1768. https://doi.org/10.1007/s41348-024-00992-0

Leonetti P, Consiglio A, Arendt D, Golbik RP, Rubino L, Gursinsky T, Behrens SE, Pantaleo V (2023) Exogenous and endogenous dsRNAs perceived by plant Dicer-like 4 protein in the RNAi-depleted cellular context. Cell Mol Biol Lett 28:64. https://doi.org/10.1186/s11658-023-00469-2

Li Y, Lu J, Han Y, Fan X, Ding SW (2013) RNA interference functions as an antiviral immunity mechanism in mammals. Sci, 342(6155):231–234. https://doi.org/10.1126/science.12419

Li C, Zamore PD (2019) RNA interference and small RNA analysis. Cold Spring Harb Protoc, 4:247–262. https://doi.org/10.1101/pdb.top097436

Liu L, Chen X (2016) RNA quality control as a key to suppressing RNA silencing of endogenous genes in plants. Mol Plant 9(6):826–836. https://doi.org/10.1016/j.molp.2016.03.011

Liu S, Han Y, Li WX, Ding SW (2023) Infection defects of RNA and DNA viruses induced by antiviral RNA interference. Microbiol Mol Biol Rev 87(2):e00035–e00022. https://doi.org/10.1128/mmbr.00035-22

Liu J, Zheng T, Xu L, Chen Z, Zhang K, Wang X, Xu X, Li Y, Sun Y, Zhu L (2024) An improved method for the detection of double-stranded RNA suitable for quality control of mRNA vaccines. Protein Cell 15(11):791–795. https://doi.org/10.1093/procel/pwae043

Maizel A, Markmann K, Timmermans M, Wachter A (2020) To move or not to move: roles and specificity of plant RNA mobility. Curr Opin Plant Biol 57:52–60. https://doi.org/10.1016/j.pbi.2020.05.005

Manamohan M, Krishna Reddy M, Patil BL (2024) Exogenous application of dsRNAs targeting multiple genes of cucumber mosaic virus in Cucumis sativus significantly reduces the viral titre. J Plant Biochem Biotechnol 34:426–435. https://doi.org/10.1007/s13562-024-00935-8

Manikandan S, Mathivanan A, Bora B, Hemaladkshmi P, Abhisubesh V, Poopathi S (2023) A review on vector borne disease transmission: current strategies of mosquito vector control. Indian J Entomol 85(2):503–513. https://doi.org/10.55446/IJE.2022.593

Mazumder P, Majumdar A, Roy B, Mohanbabu Y, Valli NS, Sarkar M, Mondal F, Verma N, Kumar R, Roy A (2025) RNA vaccination in plants: a non-transgenic approach to combat plant viruses with special reference to tomato leaf curl New Delhi virus. In: Tomato leaf curl New Delhi virus (ToLCNDV) insights into virome dynamics and management. Springer Nature Singapore, Singapore, pp 339–352. https://doi.org/10.1007/978-981-95-2852-3_17

Mcfarlane L, Svingen T, Braasch I, Koopman P, Schartl M, Wilhelm D (2011) Expansion of the Ago gene family in the teleost clade. Dev Genes Evol 221:95–104. https://doi.org/10.1007/s00427-011-0363-7

Melita O, Kaldis A, Berbati M, Reppa C, Holeva M, Lapidot M, Gelbart D, Otten P, Voloudakis A (2021) Topical application of double-stranded RNA molecules deriving from tomato yellow leaf curl virus reduces cognate virus infection in tomato. Biol Plant 65:100–110. https://doi.org/10.32615/bp.2020.172

Mendoza HG, Beal PA (2023) Chemical modifications in RNA: elucidating the chemistry of dsRNA-specific adenosine deaminases (ADARs). Acc Chem Res 56(18):2489–2499. https://doi.org/10.1021/acs.accounts.3c00390

Montes N, Pagán I (2022) Challenges and opportunities for plant viruses under a climate change scenario. Adv Virus Res 114:1–66. https://doi.org/10.1016/bs.aivir.2022.08.001

Morozov SY, Solovyev AG, Kalinina NO, Taliansky ME (2019) Double-stranded RNAs in plant protection against pathogenic organisms and viruses in agriculture. Acta Nat 11(4):13–21. https://doi.org/10.32607/20758251-2019-11-4-13-21

Muhammad T, Zhang F, Zhang Y, Liang Y (2019) RNA interference: a natural immune system of plants to counteract biotic stressors. Cells 8(1):38. https://doi.org/10.3390/cells8010038

Murray GG, Kosakovsky Pond SL, Obbard DJ (2013) Suppressors of RNAi from plant viruses are subject to episodic positive selection. Proc Biol Sci 280(1765):20130965. https://doi.org/10.1098/rspb.2013.0965

Neumeier J, Meister G (2021) siRNA specificity: RNAi mechanisms and strategies to reduce off-target effects. Front Plant Sci 11:526455. https://doi.org/10.3389/fpls.2020.526455

Niraula PM, Fondong VN (2021) Development and adoption of genetically engineered plants for virus resistance: advances, opportunities and challenges. Plants 10(11):2339. https://doi.org/10.3390/plants10112339

Omarov RT, Bersimbai RI (2010) Biochemical mechanisms of suppression of RNA interference by plant viruses. Biochemistry (Mosc) 75:965–970. https://doi.org/10.1134/S0006297910080031

Pal G, Ingole KD, Yavvari PS, Verma P, Kumari A, Chauhan C, Chaudhary D, Srivastava A, Bajaj A, Vemanna RS (2024) Exogenous application of nanocarrier-mediated double-stranded RNA manipulates physiological traits and defence response against bacterial diseases. Mol Plant Pathol 25(1):e13417. https://doi.org/10.1111/mpp.13417

Paturi S, Deshmukh MV (2021) A glimpse of "dicer biology" through the structural and functional perspective. Front Mol Biosci 8:643657. https://doi.org/10.3389/fmolb.2021.643657

Persengiev SP, Zhu X, Green MR (2004) Nonspecific, concentration-dependent stimulation and repression of mammalian gene expression by small interfering RNAs (siRNAs). RNA 10:12–18. https://doi.org/10.1261/rna5160904

Qiao L, Niño-Sánchez J, Hamby R, Capriotti L, Chen A, Mezzetti B, Jin H (2023) Artificial nanovesicles for dsRNA delivery in spray induced gene silencing for crop protection. 21(4):854–865. https://doi.org/10.1101/2023.01.03.522662

Qiao H, Jiang Q, Zhao J, Xiao L, Zhu-Salzman K, Xu D, Xu G, Shen J, Gu A, Hao D, Yan S (2025) Nano-delivery platform with strong protection and efficient delivery: preparation of self-assembled RNA pesticide with dual RNAi targets against Apolygus lucorum. J Nanobiotechnol 23:93. https://doi.org/10.1186/s12951-025-03155-x

Quilez-Molina AI, Niño-Sanchez J, Merino D (2024) The role of polymers in enabling RNAi-based technology for sustainable pest management. Nat Commun 15:9158. https://doi.org/10.1038/s41467-024-53468-y

Rego-Machado CM, Nakasu EY, Silva JM, Lucinda N, Nagata T, Inoue-Nagata AK (2020) siRNA biogenesis and advances in topically applied dsRNA for controlling virus infections in tomato plants. Sci Rep 10:22277. https://doi.org/10.1038/s41598-020-79360-5

Rêgo-Machado CM, Inoue-Nagata AK, Nakasu EY (2023) Topical application of dsRNA for plant virus control: a review. Trop Plant Pathol, 48(1):11–22. https://doi.org/10.1007/s40858-022-00534-9

Ren Q, Zhang Q, Liu Y, Li S, Zhang J, Wang Y, El Wakil A, Moussian B, Zhang J (2025) PEI-SWNT improves RNAi efficiency in Locusta migratoria via dsRNA injection delivery system. Pestic Biochem Physiol 209:106361. https://doi.org/10.1016/j.pestbp.2025.106361

Ristaino JB, Anderson PK, Bebber DP, Brauman KA, Cunniffe NJ, Fedoroff NV, Finegold C, Garrett KA, Gilligan CA, Jones CM, Martin MD (2021) The persistent threat of emerging plant disease pandemics to global food security. Proc Natl Acad Sci USA 118(23):e2022239118. https://doi.org/10.1073/pnas.2022239118

Roudine S, Le Lann C, Bouvaine S, Le Ralec A, Van Baaren J (2023) Can biological control be a strategy to control vector-borne plant viruses? J Pest Sci 96:451–470. https://doi.org/10.1007/s10340-022-01587-0

Routhu GK, Borah M, Siddappa S, Nath PD (2023) Topical application of coat protein specific dsRNA molecules confers resistance against cognate potato virus Y (PVY) infecting potato of North-East India. Gene Rep 31:101776. https://doi.org/10.1016/j.genrep.2023.101776

Samarskaya VO, Spechenkova N, Markin N, Suprunova TP, Zavriev SK, Love AJ, Kalinina NO, Taliansky M (2022) Impact of exogenous application of potato virus Y-specific dsRNA on RNA interference, pattern-triggered immunity and poly (ADP-Ribose) metabolism. Int J Mol Sci 23(14):7915. https://doi.org/10.3390/ijms23147915

Sang H, Kim JI (2020) Advanced strategies to control plant pathogenic fungi by host-induced gene silencing (HIGS) and spray-induced gene silencing (SIGS). Plant Biotechnol Rep 14:1–8. https://doi.org/10.1007/s11816-019-00588-3

Shao ZQ, Xue JY, Wang Q, Wang B, Chen JQ (2019) Revisiting the origin of plant NBS-LRR genes. Trends Plant Sci 24(1):9–12. https://doi.org/10.1016/j.tplants.2018.10.015

Siddique AB, Rahman MZ, Gain N, Rahman MS, Rahman J (2025) Harnessing double-stranded RNA (dsRNA): a sustainable approach to pest management. Physiol Mol Biol Plants 31:1237–1257. https://doi.org/10.1007/s12298-025-01564-8

Soumia PS, Guru Pirasanna Pandi G, Krishna R, Ansari WA, Jaiswal DK, Verma JP, Singh M (2021) Whitefly-transmitted plant viruses and their management. In: Emerging trends in plant pathology, pp 175–195. https://doi.org/10.1007/978-981-15-6275-4_8

Spada M, Pugliesi C, Fambrini M, Palpacelli D, Caneo A, Pecchia S (2025) Spray-induced gene silencing (SIGS): nanocarrier-mediated dsRNA delivery improves RNAi efficiency in the management of lettuce gray mold caused by Botrytis cinerea. Agronomy 15(1):194. https://doi.org/10.3390/agronomy15010194

Stefanache A, Lungu II, Anton N, Damir D, Gutu C, Olaru I, Plesea Condratovici A, Duceac M, Constantin M, Calin G, Duceac LD (2025) Chitosan nanoparticle-based drug delivery systems: advances, challenges, and future perspectives. Polymers 17(11):1453. https://doi.org/10.3390/polym17111453

Sundaresha S, Bairwa A, Tomar M, Kumar R, Venkatasalam EP, Sagar V, Bhardwaj V, Sharma S (2022a) In vitro method for synthesis of large-scale dsRNA molecule as a novel plant protection strategy. In: Mysore KS, Senthil-Kumar M (eds) Plant gene silencing, Methods in molecular biology, vol 2408. Humana, New York, pp 211–226. https://doi.org/10.1007/978-1-0716-1875-2_14

Sundaresha S, Jeevalatha A, Kumar R, Sood S, Sharma S, Bhardwaj V, Singh BP, Chakrabarti SK (2022b) RNA interference: a versatile tool to augment plant protection strategies in potato. In: Chakrabarti SK, Sharma S, Shah MA (eds) Sustainable management of potato pests and diseases. Springer, Singapore, pp 369–388. https://doi.org/10.1007/978-981-16-7695-6_15

Sundaresha S, Tiwari JK, Bhardwaj V, Sharma S, Jeevalatha A, Kumar R, Sagar V, Singh BP, Chakrabarti SK, Kumar M (2024) Genetic engineering in potato: status and future perspectives. In: Kalia P (ed) Genetic engineering of vegetable crops. CABI, Nosworthy Way, Wallingford, pp 133–140. https://doi.org/10.1079/9781800627123.0008

Tabein S, Jansen M, Noris E, Vaira AM, Marian D, Behjatnia SAA, Accotto GP, Miozzi L (2020) The induction of an effective dsRNA-mediated resistance against tomato spotted wilt virus by exogenous application of double-stranded RNA largely depends on the selection of the viral RNA target region. Front Plant Sci 11:533338. https://doi.org/10.3389/fpls.2020.533338

Tatineni S, Hein GL (2023) Plant viruses of agricultural importance: current and future perspectives of virus disease management strategies. Phytopathology 113(2):117–141. https://doi.org/10.1094/PHYTO-05-22-0167-RVW

Vatanparast M, Merkel L, Amari K (2024) Exogenous application of dsRNA in plant protection: efficiency, safety concerns and risk assessment. Int J Mol Sci 25(12):6530. https://doi.org/10.3390/ijms25126530

Venu E, Ramya A, Babu PL, Srinivas B, Kumar S, Reddy NK, Babu YM, Majumdar A, Manik S (2024) Exogenous dsRNA-mediated RNAi: mechanisms, applications, delivery methods and challenges in the induction of viral disease resistance in plants. Viruses 17(1):49. https://doi.org/10.3390/v17010049

Voloudakis AE, Holeva MC, Sarin LP, Bamford DH, Vargas M, Poranen MM, Tenllado F (2015) Efficient double-stranded RNA production methods for utilization in plant virus control. In: Plant virology protocols: new approaches to detect viruses and host responses, pp 255–274. https://doi.org/10.1007/978-1-4939-1743-3_19

Voloudakis AE, Kaldis A, Patil BL (2022) RNA-based vaccination of plants for control of viruses. Annu Rev Virol 9:521–548. https://doi.org/10.1146/annurev-virology-091919-073708

Worrall EA, Bravo-Cazar A, Nilon AT, Fletcher SJ, Robinson KE, Carr JP (2019) Exogenous application of RNAi-inducing double-stranded RNA inhibits aphid-mediated transmission of a plant virus. Front Plant Sci 10:00265. https://doi.org/10.3389/fpls.2019.00265

Wytinck N, Manchur CL, Li VH, Whyard S, Belmonte MF (2020) dsRNA uptake in plant pests and pathogens: insights into RNAi-based insect and fungal control technology. Plants 9(12):1780. https://doi.org/10.3390/plants9121780

Yan S, Ren BY, Shen J (2021) Nanoparticle-mediated double-stranded RNA delivery system: a promising approach for sustainable pest management. Insect Sci 28(1):21–34. https://doi.org/10.1111/1744-7917.12822

Yang Y, Liu T, Shen D, Wang J, Ling X, Hu Z, Chen T, Hu J, Huang J, Yu W, Dou D, Wang MB, Zhang B (2019) Tomato yellow leaf curl virus intergenic siRNAs target a host long noncoding RNA to modulate disease symptoms. PLoS Pathog 15(1):e1007534. https://doi.org/10.1371/journal.ppat.1007534

Yang W, Wang B, Lei G, Chen G, Liu D (2022) Advances in nanocarriers to improve the stability of dsRNA in the environment. Front Bioeng Biotechnol 10:974646. https://doi.org/10.3389/fbioe.2022.974646

Zand Karimi H, Innes RW (2022) Molecular mechanisms underlying host-induced gene silencing. Plant Cell 34(9):3183–3199. https://doi.org/10.1093/plcell/koac165

Zapletal D, Kubicek K, Svoboda P, Stefl R (2023) Dicer structure and function: conserved and evolving features. EMBO Rep 24:e57215. https://doi.org/10.15252/embr.202357215

Zarrabian M, Sherif SM (2024) Silence is not always golden: a closer look at potential environmental and ecotoxicological impacts of large-scale dsRNA application. Sci Total Environ 912:175311. https://doi.org/10.1016/j.scitotenv.2024.175311

Zhang H, Chen J, Gao J, Zhang Q, Liu X, Han Z (2022a) New insights into transmission pathways and possible off-target effects of insecticidal dsRNA released by treated plants. Pestic Biochem Physiol 188:105281. https://doi.org/10.1016/j.pestbp.2022.105281

Zhang S, Griffiths JS, Marchand G, Bernards MA, Wang A (2022b) Tomato brown rugose fruit virus: an emerging and rapidly spreading plant RNA virus that threatens tomato production worldwide. Mol Plant Pathol 23(9):1262–1277. https://doi.org/10.1111/mpp.13229

Zhu LJ, Zhu Y, Zou C, Su LY, Zhang CT, Wang C, Bai YN, Chen B, Li R, Wu Q, Ding SW (2024) New persistent plant RNA virus carries mutations to weaken viral suppression of antiviral RNA interference. Mol Plant Pathol 25(12):e70020. https://doi.org/10.1111/mpp.70020

# Perspectives on Peptide Applications in Vegetable Virus Disease Management

# 25

Jose Isagani B. Janairo, Peter Immanuel Tenido, and Mark Angelo O. Balendres

**Abstract**

Producing more food to feed a rapidly growing population while available land and resources continue to shrink is one of the greatest challenges that humanity faces today. Compounding this problem is productivity loss due to infestations, such as those caused by viruses targeting vegetables. Despite the availability of various methods for managing vegetable viruses, a pressing and ongoing need remains to develop effective strategies for virus management. The ability of viruses to quickly mutate necessitates a rapid response from a diverse arsenal of antivirals. Among the interventions available, using peptides is a promising approach due to the versatility and efficacy of this type of biomolecule. This chapter presents recent innovations in using peptides to manage viral infections in vegetables. Challenges and issues hindering the broader implementation of peptide-based interventions are also discussed. The chapter concludes by highlighting commercially promising examples of peptide-based interventions.

J. I. B. Janairo (✉)
Department of Biology, College of Science, De La Salle University, Manila, Philippines

Institute of Biological Control, College of Science, De La Salle University, Manila, Philippines
e-mail: jose.isagani.janairo@dlsu.edu.ph

P. I. Tenido
DLSU Innovation and Technology Office, De La Salle University, Manila, Philippines

M. A. O. Balendres (✉)
Department of Biology, College of Science, De La Salle University, Manila, Philippines

Plant and Soil Health Research Unit, Center for Natural Sciences and Environmental Research, College of Science, De La Salle University, Manila, Philippines
e-mail: mark.angelo.balendres@dlsu.edu.ph



R. Kumar et al. (eds.), *Vegetable Crops and Viral Pathogens Volume I*, Plant Pathology for Sustainable Agriculture, https://doi.org/10.1007/978-981-95-7265-6_25

**Keywords**

Antiviral strategies · Biocontrol applications · Plant disease management · Tomato virus

## 25.1 Introduction

The agricultural sector plays a major role in achieving food security and many of the Sustainable Development Goals. According to the 2024 report on agricultural production statistics (FAO 2024), global production of key crops reached 9.9 billion tons in 2023, with cereals, sugar crops, and vegetables being the most widely produced commodities. For vegetables, production reached 1.2 billion tons in 2023, with tomato, onion, cucumber, cabbage, and eggplant as key crops with 192, 111, 98, 74, and 61 million tons produced, respectively (FAO 2024). This growth underscores the rising demand for vegetables amid a growing global population. Vegetables play an important role in human nutrition (Ndhlovu et al. 2025) and the livelihoods of growers (Shaheb et al. 2016), particularly in rural areas. In Bangladesh, vegetables produced by some growers are mainly consumed, and the surplus becomes a source of income (Shaheb et al. 2016). Despite production growth and steady income, vegetable production still faces several problems that limit its potential as a high-value cash crop. This calls for a sustainable and practical approach to address vegetable production problems. Among the many constraints in vegetable production are, but are not limited to, lack of knowledge by growers about improved varieties, nonavailability of storage facilities, prevalence of pests and diseases, and the high costs associated with pest and disease management (Noopur et al. 2023; Okoma et al. 2025). The low price of vegetable produce is also a limiting factor. For instance, in India, Noopur et al. (2023) found that the low price of produce discourages growers, consequently reducing the potential yield and income. Pests and diseases also have a significant impact on growers' decision-making. In the Philippines, Dela Cueva and Balendres (2017) reported that some farmers abandoned their pepper farm when severe infestation of anthracnose occurred. The high cost of controlling the disease outweighs the potential profit of selling the healthy fruit harvest. Among the diseases challenging to vegetable growers are those caused by viruses, due to the biology of the pathogen and the nature of the infection. Tomato, the most produced vegetable worldwide, is highly susceptible to many viral diseases.

Viruses are among the most important emerging plant diseases that are easily spread through seeds and mechanical means, and some are vectored by insects (Anderson et al. 2004; Kumar et al. 2022; Raigond et al. 2022; Naga et al. 2020), including whiteflies (Hemiptera: Aleyrodidae) that are known to transmit Begomoviruses (Naga et al. 2021; Modgill et al. 2025) and aphids that are known to transmit potyviruses (Caciagli 2009). *Tomato mosaic virus, Tomato yellow leaf curl virus* (TYLCV), *Tomato spotted wilt virus, Cucumber mosaic virus, and Zucchini yellow mosaic virus* are some of the widely known and important viruses affecting vegetable crops (Tomlinson 1987; Nono-Womdim 2003; Jones 2003; Thompson

2011). More recently, an emerging tobamovirus, *Tomato brown rugose fruit virus*, has threatened tomato and pepper production worldwide (Zhang et al. 2022; Salem et al. 2023). Yield losses due to viruses vary with the host plant, environment, and location (Sastry 2013) but can reach up to 100%. For instance, *Yellow vein mosaic virus* can cause yield loss ranging from 17.09% to 96.49% (Jamir et al. 2020), while the average perceived yield loss of TYLCV in tomato was 100% in Sudan (CIAT). There is no known single effective method against vegetable viruses. For instance, aside from using less susceptible cultivars, controlling the population of insect vectors remains important as plants may succumb to infection. An integrated disease management (IDM) strategy is implemented to control vegetable diseases. This IDM approach may include a combination of cultural (crop rotation, use of resistant varieties), physical (destroying infected plants, insect traps), biological (beneficial insects, biocontrol agents), chemical (use of pesticides), and application of emerging technologies (e.g., nanotechnology, artificial intelligence, and RNA interference). Many reviews have discussed the benefits of utilizing these emerging technologies in managing vegetable virus diseases (Lata et al. 2022; Koeppe et al. 2023; Kaur et al. 2025; Eid et al. 2025; Singh et al. 2025; Sánchez-Tovar et al. 2025).

This chapter offers perspectives on peptide applications in the management of vegetable virus diseases. Peptides are biomolecules made up of a short chain of amino acids. Peptides are much shorter and smaller than proteins, although a definitive threshold between the two has yet to be reached. Peptides are promising alternatives to the conventional agents used against vegetable viruses. The versatility of peptides, as demonstrated by the multiple ways these biomolecules can address viral infection in plants, greatly expands the existing arsenal of antivirals. This is especially important considering that viruses can mutate rapidly, thus necessitating the discovery and development of novel antivirals. The combinatorial nature of peptides can lead to the development of novel and potent antivirals.

## 25.2 Virus Management Using Peptides

The following section briefly discusses the mechanisms of action of antiviral peptides, common strategies on how they are discovering and developing them, and the methods by which they are utilized for vegetable virus management.

## 25.3 Mechanism of Action

One of the advantages of using peptides for vegetable virus management is their ability to address viral infection in multiple ways. Broadly, peptides in this application class can be categorized based on their mechanism of action. Some peptides directly target the virus, and some peptides can enhance plant immunity and thus reduce the recovery period from the infection. Some peptides can reduce viral transmission by interacting with the vector or carrier cells. The first class of peptides that directly eradicate the virus, which are also referred to as antimicrobial peptides, usually disrupts the membrane of the pathogen, interferes with ribosomal protein synthesis, or blocks the pathways related to viral replication. Several antimicrobial peptides have been developed against the *tobacco mosaic virus* (TMV), a virus that infects tobacco and other plants, such as tomatoes, and is spread through direct contact, resulting in leaf mottling (Rifkind and Freeman 2005).

One of the earliest synthetic peptides designed against TMV was based on sequence modifications of the peptide melittin (Marcos et al. 1995). Melittin is found in bee venom and has been reported to possess antibacterial properties (Fennell et al. 1968). It was then determined that the biological activity of melittin emanates from the ability of the peptide to create pores at bacterial membranes (Vogel and Jähnig 1986). These promising characteristics of melittin may have inspired the exploration of its biological activity in other systems, such as in combating viruses. In their paper, sequence homology was observed between melittin and the coat protein of the TMV. This served as the basis for modifying the sequence of melittin to understand the mechanism of action. It was found that the sequence modifications led to alterations in the secondary structures of the peptides, which influenced their interaction with the virus and the viral RNA, from which their antiviral is believed to emanate.

Similar to melittin and its analogs, proteins and peptides known to exhibit antimicrobial activity can be repurposed for managing viruses. In one study, the antiviral activity of the AMP thanatin extracted from the spined soldier bug is explored (Sabokkhiz et al. 2019). In the same study, known AMPs lactoferrin and lactoferrampin were fused together to produce a chimeric protein; thereafter, the antiviral activity was likewise examined. Both recombinant peptides exhibited antiviral activity against TMV, wherein it was postulated that their efficacy is due to their direct interaction with the virus. A specific example of direct interaction with the virus is interfering with its replication. In one study, a peptide derived from the enzymatic digestion of amaranth was reported to be effective against the *Tomato yellow leaf curl virus* (TYLCV) (Mendoza-Figueroa et al. 2018). Further analysis through Raman spectroscopy coupled with molecular simulations revealed the manner in which the active peptide exerts its antiviral activity (Mendoza-Figueroa et al. 2019). It was found that H-bonding, electrostatic, and pi-pi interactions were important noncovalent interactions that drive the binding of the peptide with its target. The extracted peptides exhibit strong affinity toward the replication origin sequence (OriRep) of TYLCV as measured by surface plasmon resonance (SPR). Treatment of infected plants with bioactive peptides showed a lower virus load, indicating that the virus was unable to accumulate and replicate.

Artificially designed short peptide sequences called aptamers represent an important class of peptides in the antiviral arsenal. Aptamers are known to exhibit a strong binding affinity with their target and can therefore be used to disrupt the normal function of viral proteins. Peptide aptamers that bind to the AL1 protein of the *tomato golden mosaic virus* were discovered through combinatorial approaches (Luisa et al. 2006). Using the discovered aptamers resulted in a substantial decline in the ability of the virus to replicate, thus demonstrating the efficacy of the aptamers. In another example, peptide aptamers were successfully developed for TYLCV and the tomato mottle virus (Ines et al. 2013). The cationic peptide, indolicidin, which was derived from bovine neutrophils, was also reported to exhibit antiviral activity against TMV (Bhargava et al. 2007). The charged nature of the peptide is believed to be important for its bioactivity, which is attributed to the peptide's ability to disrupt the membrane.

Peptides can also be used to enhance plant fitness and resistance to pathogens, including viruses. In a recent paper, a short signal peptide was found to enhance the resistance of tobacco against the Chinese wheat mosaic virus (Liu et al. 2025). The study reports that the peptide named CLE7 is involved in the activation of transcription factors that enhance plant resistance to infections. In another example of this mechanism of action, crude peptides derived from the mycelia of *Penicillium chrysogenum* were found to increase the systemic resistance of tobacco against TMV (Zhong et al. 2021). Another class of peptides called peptaibols has also been reported to elicit an immune response against viral pathogens, leading to increased plant fitness. Peptaibols are peptides derived from fungi, which often bear nonstandard amino acids (Ramachander Turaga 2020). The incorporation of nonstandard amino acids in the peptaibol sequence imparts a certain degree of resistance against the proteases of pathogens, thus enhancing and prolonging their efficacy.

Peptaibol extracted from *Trichoderma pseudokoningii* SMF2 exhibited the ability to improve systemic resistance of tobacco against TMV (Luo et al. 2010). The application of the extracted peptaibol, referred to as *trichokonins*, led to increased fitness of tobacco due to the elevated expression of genes related to responding to pathogens, such as those involved in the reactive oxygen intermediate (ROI-mediated pathway. Peptide polysaccharides (PSPs), extracted from fungi, were also found to exhibit relevance in virus management in plants (Zhao et al. 2015). In their study, PSP was able to directly combat TMV and induce the expression of genes related to the immune response. Similar to peptaibol and PSP, biomolecules derived from other microorganisms also induce the plant's immune response against pathogens. Another example is the harpin protein, which is derived from Gram-negative plant-pathogenic bacteria (Choi et al. 2013). The harpin protein Hpa1 was found to be effective in increasing the resistance of *Pinellia ternata* to TMV (Wang et al. 2020). The exogenous application of the protein to the plant also led to elevated expression of genes involved in defense against pathogens, as well as increased enzymatic activity of key defensive enzymes. While the Hpa1 is a protein and not a peptide, it is important to mention this biomolecule since harpin-derived peptides are now commercially available and marketed as plant growth and resistance promoters.

An indirect mechanism in which peptides can control virus infection and transmission involves interaction with the host cell of the vector. In this pathway, bioactive peptides can target vector cells, which will eventually inhibit virus transmission, thereby effectively managing both insect and virus infestations. An example of this method of virus management was presented in a study involving the feeding of thrips synthetic peptides derived from the soybean vein necrosis virus (Zhou and Tzanetakis 2020). Thrips fed with the synthetic peptides demonstrated a substantially lower transmission rate of the orthotospovirus. In another study, a 12–residue gut-binding peptide was discovered that can slow down transmission of *Pea enation mosaic virus* (PEMV) in pea aphids (Liu et al. 2010). The gut-binding peptide was discovered through phage display using dissected gut epithelium from aphids as the substrate. The identified peptide is able to reduce the transmission of PEMV by competing with the virus for binding to the aphid gut, thereby limiting the uptake of PEMV into the hemocoel. In the beet armyworm (*Spodoptera exigua*), a peptidoglycan was discovered that mediates viral infection (Li et al. 2023). Overexpression of the peptidoglycan led to a significant decrease in viral infection, whereas treatment with RNA interference to silence the expression of the peptidoglycan increased the susceptibility of the insect to viral infection. Thus, this discovery can be used to modulate insect immunity, which can be leveraged to reduce viral infection through vectors.

One of the most notorious insect vectors is the whitefly (*Bemisia tabaci*). It is a polyphagous insect that can transmit a variety of viruses, including begomoviruses, criniviruses, and torradoviruses (Fiallo-Olivé et al. 2019). Several studies have examined how viral transmission and accumulation occur in these insects, which can then be used to develop interventions that can minimize the ability of whiteflies to serve as vectors. In one study, a defensin-like antifungal peptide derived from *Bemisia tabaci* was discovered (Wang et al. 2013). The peptide, referred to as Btdef, is composed of 46 amino acids and exhibits antimicrobial activity. A follow-up study aimed to elucidate the role of Btdef in virus transmission mediated by whiteflies revealed that suppressing the expression of this peptide led to decreased transmission of the *tomato yellow leaf curl virus* (Wang et al. 2017). Similar to Btdef, an insulin-like peptide was discovered to be involved in the transmission of the *tomato chlorosis virus* by whiteflies (Xu et al. 2023). Reducing the expression of this peptide also resulted in reduced transmission of the virus, making it a valuable target for the development of antiviral interventions.

## 25.4 Discovery and Development

As demonstrated in the previous section, sequence modifications to create synthetic analogs of biologically active peptides are a viable approach in discovering antiviral peptides. Apart from this, there are various ways to discover promising leads and candidates. Crucial to the discovery of novel sequences is understanding how viruses infect host cells and how viruses are passed on from one organism to another. Thus, understanding virus interactions at both the molecular and ecological levels can lead to the discovery of novel signaling pathways and associated proteins, as well as other biomolecules. This, in turn, may point to the discovery of peptides that can combat plant viruses. Transcriptome analysis is a powerful tool that exemplifies this framework effectively. In one study, the salivary gland of the black-faced leaf hopper (*Graminella nigrifrons*) underwent transcriptome analysis (Rajarapu et al. 2020). This insect is a vector for several viruses, and examining the functional state of cells in the salivary gland through transcriptome analysis can reveal proteins involved in pest control and virus transmission. In a more targeted study, the proteins in the stylet of aphids were characterized (Webster et al. 2018). Aphids are important vectors of plant viruses, wherein the majority of the transmitted viruses through aphids are via a noncirculative route (NG and PERRY 2004). Noncirculative transmission occurs when the virus primarily resides in the stylet of the insect and is transmitted as the insect feeds on multiple plants. The stylet of aphids is therefore highly relevant in understanding virus transmission, which may lead to the identification of potential protein targets that may be inhibited by peptides to combat virus infection. In the study, two highly conserved proteins were identified via immunolabeling. When these two proteins, named stylin-01 and stylin-02, were silenced, a noticeable decrease in transmission of the cauliflower mosaic virus (CamV) was observed. These proteins may therefore serve as targets to which peptides may bind and inhibit. As a follow-up to the study, a proteomics approach was conducted to characterize the proteins present in the stylet of pea aphids (Deshoux et al. 2020). The comprehensive comparative analysis revealed distinct proteins present only in the stylet and not in other insect organs and appendages, such as the wings and legs. The findings shortlist candidate proteins that may be involved in virus transmission, which can be tested in future studies.

Mechanistic analysis of virus infection routes in their hosts is a promising strategy in identifying potential "druggable targets." This approach, which is considered target-based drug discovery, is the common route in which candidates and leads are discovered and developed, including for antimicrobials (Selzer et al. 2000). In this framework, a target is first identified, followed by an exhaustive analysis of the signal transduction pathways the target is involved, in addition to biochemical and structural characterization. Once this information is known and validated, ligands that can bind with the target to disrupt its normal function are designed and developed. Thus, target identification and validation are of fundamental importance in the development of peptide antivirals for plants. One way this type of analysis can be conducted is through the examination of posttranslational modifications (PTM) of peptides and proteins. This offers an opportunity to better understand plant response to infection, which can lead to the production of more diverse antivirals.

Glycosylation is an important form of PTM that influences the interactions of peptide ligands with their corresponding receptors. Additionally, glycosylation is involved in several biological processes, particularly those associated with the endoplasmic reticulum and the Golgi apparatus. In the case of *wheat yellow mosaic virus* (WYMV), glycosylated peptides were discovered to play critical roles in the transmission of the virus (Yang et al. 2025). In the study, comprehensive glycoproteomic analyses of WYMV-infected wheat samples were conducted. In comparison to the control group composed of mock-infected wheat samples, the WYMV group demonstrated distinct glycosylation profiles. In relation to the development of antivirals, studies like this provide important insights since detailed and specific stress responses are identified. For example, one of the findings of the study was that in response to WYMV infection, wheat increases the expression of glycans related to mounting resistance. Thus, peptide promoters can be developed that are involved in these pathways, potentially leading to enhanced plant resistance to infection. In another study related to PTM analysis, it was found that phosphorylation activates the lipid transfer protein, NbLTP1, leading to the accumulation of the bamboo mosaic virus (BaMV) in *N. benthamiana* (Chiu et al. 2020). This is a significant finding, as this pathway can be potentially disrupted by synthetic peptides, which may interfere with viral replication and accumulation. Another protein relevant to membranes and that plays a role in virus-host interaction was identified and is a promising target for inhibiting viral host entry and transmission. The capsid protein P10 of the rice black-streaked dwarf virus (RBSDV) was found to exhibit high affinity toward the phosphatidylinositol 3,5-bisphosphate (PI(3,5)$P_2$)of its host (Liu et al. 2022). The interaction between the two biomolecules is believed to be part of the uncoating of the virus during entry and transmission and may therefore be explored as a means of disrupting and preventing viral entry.

Another method of gaining fundamental insights regarding target proteins and those with which they interact is through site-directed mutagenesis. In this method, residues in the binding site or catalytic domain of proteins or enzymes that are believed to be critical for the function are mutated. The activity of the mutant protein is then compared with the wild type. This particular method was applied to identify key residues in a movement protein of the *Citrus psorosis virus*, which is involved in the production of tubules in the plasmodesmata of the host cells (Gabriel et al. 2018). This is part of the process of virus transmission. In the study, site-directed mutagenesis revealed that the movement protein contains an aspartic protease sequence motif wherein mutation of the aspartic residue changed the activity of the protein. This is important for designing ligands that can modulate the activity of this protein, which can be exploited to inhibit virus transmission.

## 25.5 Methods of Deployment

One of the advantages of using peptides for virus management in plants is the versatility of application. In the studies mentioned in the previous sections, the common route of application is through spraying and direct application. This usually involves dissolving the peptides in an aqueous medium and then spraying the solution on the plants. Perhaps one of the most promising ways to utilize peptides in controlling and managing viral infections in plants is through the systemic integration of the peptide into the plant by creating transgenic plants. In the case of the cationic peptide indolicidin, transgenic tobacco plants were prepared. These transgenic specimens exhibited enhanced resistance to various infectious agents, including viruses, as demonstrated by leaf assays. Another example of preparing transgenic plants that express antiviral proteins was demonstrated for the use against various tospoviruses (Rudolph et al. 2003). In the study, transgenic tobacco plants were created that express peptide aptamers, which are 29 amino acids in length. The transgenic plants exhibited increased resistance to a variety of viruses, including *tomato spotted wilt virus*, *tomato chlorotic spot virus*, *groundnut ring spot virus*, and *chrysanthemum stem necrosis virus*. A promising approach to altering the gene expression of plants is the use of biomolecular delivery systems that target specific cells and organelles. This approach was demonstrated by creating hybrid systems composed of cell-penetrating peptides (CPP) or organelle-targeting peptides (OTP) that have been tagged by nucleic acids (Thagun et al. 2022). The sequence of the CPP or OTP confers specificity in cargo delivery, wherein DNA can be delivered and eventually expressed by the plant. Although the paper did not demonstrate the use of this approach in expressing antiviral peptides, this method presents a promising way to integrate and express antiviral peptides through foliar application. This can greatly expand the scope and utility of peptides for managing plant viruses.

## 25.6 Challenges in Peptide Application

Despite the benefits offered by using peptides for managing vegetable viruses, persistent challenges remain that hinder their full potential. One of the key challenges in the full-scale application of peptides for virus management is the reproducible and consistent efficacy in field tests and trials. For example, the application of the Harpin protein Hpa1 to *Pinellia ternata* was effective in greenhouse experiments; however, different results were obtained in field tests and experiments (Wang et al. 2020). Challenges in translating laboratory success to the field can be attributed to several fundamental issues inherent in the nature of peptides, such as stability. The following section presents a SWOT (strengths, weaknesses, opportunities, threats) analysis of the application of peptides for plant virus management.

## 25.7 Strengths

### 25.7.1 Versatility

The numerous modes of action of peptides against vegetable viruses are a great advantage. As outlined above, peptides used for virus management in plants can exert their efficacy by directly inhibiting the virus, enhancing plant resistance to the infection, and reducing transmission with the vector. It is possible that a single peptide can address viral infection in multiple ways, which increases efficacy and likelihood of successful recovery.

### 25.7.2 Environment-Friendly

One of the components of virus management is preventing the spread of the virus, and since the virus can be spread by insect vectors, pesticide application may help in reducing virus transmission. However, pesticides pose a great risk to the environment (Beaumelle et al. 2023) and human health (Shekhar et al. 2024). In contrast, peptides can be considered more ecological since they do not persist in the environment, they are effective at low concentrations, and they pose minimal risk to humans and nontarget organisms. However, studies that have examined the safety of peptides in this kind of application are scarce. This is a promising area of research that can be pursued.

## 25.8 Weaknesses

### 25.8.1 Cost and Scalability

Producing peptides is expensive and could be a severely limiting factor in the application of peptides for plant virus management. Peptides are usually produced using solid-phase peptide synthesis (Coin et al. 2007), which involves anchoring an initial protected amino acid onto a solid resin followed by a series of deprotection, activation, and coupling steps until the desired sequence is reached. Leveraging biotechnology, such as using cell factories (Baeshen et al. 2014), similar to how insulin is commercially produced, may address problems of scaling, and may potentially decrease costs. Another possible method that is also based on biotechnology is molecular farming (Bharathi et al. 2024). In this method, plants are used to produce the target biomolecule, which can be extracted and purified later on.

### 25.8.2 Stability

Compared to other pest control strategies, peptides can be considered delicate since this biomolecule is prone to denaturation and degradation. The structure of peptides is crucial to their function, and exposure to the natural environment often leads to denaturation. Bacteria, for instance, can also metabolize peptides, which may lead to their premature degradation (Decker et al. 2023). Thus, improving the stability of peptides is an active field of research for peptide pharmaceuticals (Ghosh and Tal-Gan 2025), which should also be conducted for this application domain.

## 25.9 Opportunities

### 25.9.1 Growing Awareness of the Environment

As mentioned above, using peptides for pest management offers a "greener" alternative. The growing consciousness about the environmental impacts of farming, which has become an issue in selecting produce and commodities (Ornelas Herrera et al. 2025), is a great opportunity for peptides.

### 25.9.2 The Potential of Artificial Intelligence and Machine Learning

The widespread development and utilization of artificial intelligence/machine learning present opportunities for the further development of peptides for vegetable viruses. AI/ML models can accelerate the discovery of bioactive peptides as well as the development of short but potent sequences. For instance, an ML classification model that utilized the random forest algorithm was recently created and can predict whether a peptide sequence has antiviral properties or not (Wang et al. 2025b). The classifier was trained and tested on 1334 specimens and exhibits 85.8% accuracy. Aside from using AI/ML tools for screening, these tools can also be used in elucidating the structure of potential targets. AlphaFold was used in characterizing and visualizing the structure of a virus protein that played an important role in its entry to the host (Liu et al. 2022). Through the help of AlphaFold and other methods, putative binding sites were identified. AlphaFold is an AI model that can accurately predict the structure of proteins based only on the amino acid sequence (Jumper et al. 2021). This example clearly demonstrates how AI/ML can speed up the discovery not only of bioactive peptides but also in the discovery of targets in which peptide ligands can be developed. In addition, synergistic applications of AI tools such as AlphaFold and other computational tools, such as molecular dynamics simulations, have tremendous potential in exploring the expansive bioactive peptide space (Chang et al. 2024), including those related to the development of antivirals for plants.

## 25.10 Threats

### 25.10.1 Conservative Views

One method of using peptides for virus management involves the preparation of transgenic plants that express the antiviral peptide. However, conservative views and limited understanding of the science behind transgenic crops present a significant impediment to the wider applications of peptides for virus management. For example, a systematic review on the perception of the population from the Middle East, North Africa, and Turkey (MENAT) region showed negative sentiments for genetically modified crops (Medani et al. 2024). In the Philippines, genetically modified (GM) crops face legal challenges, as the Supreme Court has banned the field tests and commercial release of golden rice and BT eggplant. Golden rice is genetically modified to express beta-carotene, whereas BT eggplant was also genetically modified to produce peptides that are toxic to insect pests. The SC issued a writ of *kalikasan* (environment) (Philippines 2023), which is a legal remedy under Philippine law that seeks to uphold the right to a healthy environment. Hopefully, as more evidence emerges regarding the safety of GM crops, greater acceptability will follow. For example, a seven-year study on the effects of GM corn consumption in monkeys showed no adverse impact on their health (Wang et al. 2025a).

## 25.11 Patent Analysis of Peptides for Virus Management

Peptides for pathogen management are part of the broad repertoire of biopesticides. This nature-derived pest management technique is expected to reach $ 13.4 billion in market value by 2030, a sharp increase from its 2024 market value of $ 7.5 billion (Lucintel 2024). The significant increase in the market value of biopesticides reflects a growing preference for green innovations. However, data on the market share of peptides cannot be found, but it can be inferred that peptide-based products form a minority. Searching for commercial peptide-based biopesticides has been challenging and has only returned a handful of entries. Among these is KeyPlex, which is a US-based company specializing in the formulation of nutrition and biopesticides for crops. Some of their products that rely on peptides for their efficacy are KeyPlex350 and KeyPlex 350 OR. According to the company, the application of both KeyPlex 350 and KeyPlex 350 OR can stimulate the plant to produce defensive proteins, which can increase the tolerance of plants to pathogens.

Despite the apparent absence of commercially available peptides for managing vegetable viruses, the patent literature is rich in information about peptides for this application domain. A patent serves two primary functions. First, a patent protects the rights of inventors over the fruits of their inventive ability. Second, the patent document, once published, teaches the invention to the public, enabling members of the public to learn about the invention, design around it to develop alternative or better inventions, or use the invention when the patent is no longer enforceable. This second "disclosure" function of patents makes patent documents useful sources of information for analyzing technological trends, gathering business intelligence, navigating infringement risks, measuring innovation capabilities, and identifying research and commercial opportunities. For example, patent information has been used to review advances in peptides to treat fungal infections (Zambrano et al. 2022), to comprehensively analyze patents on the therapeutic use of antimicrobial peptides (Annunziato and Costantino 2020), and to explain some of the progress in antiviral peptides (Mousavi Maleki et al. 2022). In the USA, drugs granted patents by the USPTO, along with FDA approval, demonstrate a higher likelihood of commercialization (Moaniba et al. 2020). Moreover, other studies have also found patents to be a good indicator of corporate technological strength (Narin et al. 1987) and have used granted patents as an indicator of innovation (Burhan et al. 2017). Thus, examining patents related to peptides for managing vegetable viruses can provide insights into commercial viability and related emerging technologies.

This section considers patent activity relating to the use of peptides in managing viruses that affect plants or vegetables. A selection of patent documents is examined to observe the technology development directions suggested by the inventions disclosed in them. The purpose is to gain a preliminary sense of areas that may hold commercial potential, assuming that applicants typically seek patents for technologies with at least some potential for commercial value. Patent searches were conducted using a combination of keywords and International Patent Classification (IPC) codes related to three main concepts: (1) peptides, (2) vegetables, and (3) antivirals. Keywords related to peptides include "peptide," "polypeptide," "oligopeptide," "dipeptide," "tripeptide," and "tetrapeptide", while the IPC code C07K relates to "peptides." Keywords related to vegetables include "vegetable," "vegetation," "crop," "fruit," "horticulture," and "botany." Keywords for specific vegetables and crops, particularly those commonly affected by viral diseases (Angira et al. 2024), were also considered. Keywords related to antivirals include "antivirus" and "antiviral." Although there appear to be no IPC codes specifically covering the use of peptides in managing viruses that target vegetables, there are applicable codes relating to "Biocides; Pest repellents or attractants; Plant growth regulators" (A01N25/00 to A01N65/00) and to "Antivirals" (A61P31/12). The Espacenet database of the European Patent Office was used. The search was limited to patent documents with the earliest publication dates from 1 January 2010 to 31 July 2025. The query was intentionally designed with a narrow scope, as the aim was to identify a focused set of patents most likely to be substantively relevant to the subject matter of interest, allowing for closer qualitative analysis.

The invention taught in US10856546B2 (US'546) relates to harpin peptides, which trigger a plant's systemic-acquired resistance (SAR). SAR is a defense mechanism against various pathogens (including viruses) and enhances a plant's resistance to pathogens and other stresses. Improving a plant's defenses would be beneficial and desirable for growers, as it would improve plant health, boost crop yield and quality, and reduce reliance on pesticides. In its background section, US'546 states that a key limitation of harpin peptides has been their typical formulation as powdered solids coated on starch; liquid suspensions of these forms tend to have short shelf lives, low stability, and poor solubility. The invention taught in US'546 addresses these issues by providing isolated harpin peptides with specific sequences. These peptides lack cysteine and methionine and exhibit average hydropathy indices of less than 0.3. The peptides claimed in US'546 are reportedly water-soluble, stable, and resistant to degradation. Example 18, presented in the description of US'546, describes the effectiveness of the claimed peptides in improving tobacco's resistance to the tobacco mosaic virus, demonstrating the potential for SAR-inducing peptides to play a role in viral disease management in crops. On Google Patents, a search of patents listing "Plant Health Care Inc." (the applicant for US'546) generates thirty-seven (37) documents (by default, search results on Google Patents are deduplicated by patent family), many of which relate to isolated molecules (e.g., DNA, proteins, polypeptides) that function as hypersensitive response elicitors. According to its website (Care.), Plant Health Care Inc.'s core patented products act as "vaccines for plants." Furthermore, the company's business is driven partly by the sales of PHC279, a harpin-derived peptide that amplifies a plant's natural defense. Viewing together the technical information in US'546, consistent patenting activity around hypersensitive response elicitors, the company's own description of its products as "vaccines for plants," and the reported importance of PHC279 as a key sales driver for the company suggests a commercial strategy centered on the use of immune-modulating peptides to enhance plant resistance, including to viral pathogens.

CN110526960 (CN'960) teaches antiviral polypeptides derived from *Arabidopsis thaliana* and rice. The background of CN'960 explains that plants produce many secreted peptides involved in growth, development, and stress responses when exposed to pathogens. However, few secretory peptides are known to confer virus resistance, as most documented examples involve subjecting plants to stress from bacteria, fungi, or insects, thereby inducing the secretion of polypeptides that protect against those specific stresses. Only a small number of reports describe polypeptides with antiviral activity. This gap exists in part because the genes encoding secretory peptides are short, making it difficult to obtain function-deficient mutants for study. CN'960 addresses this gap, offering several advantages. First, the polypeptides exhibit strong antiviral effects. The infection rate of various viruses in plants transformed with the inventive polypeptide gene is significantly reduced (up to 99% in some examples). Second, the invention provides a straightforward preparation method for the antiviral polypeptides. Third, the inventive antiviral polypeptides can be directly applied to the surface of a plant by spraying and still produce significant antiviral effects: up to 89% reduction in the infectious load of various viruses where the polypeptide has been applied.

CN119798379A (CN'379) discloses short peptides that, when applied externally, can enhance a plant's antiviral ability. The background explains that plant virus diseases have a significant agricultural impact, spread rapidly, mutate frequently, and are difficult to control, with no specific antiviral treatments available. In the disclosed examples of CN'379, eight short peptides were extracted from naturally infected cucumber, pepper, tomato, and tobacco plants with viruses in the field. When these peptides were applied exogenously by spraying onto the leaves of *Nicotiana benthamiana* or cucumber plants, followed by inoculation with *tobacco mosaic virus* (TMV) or *cucumber mosaic virus* (CMV), virus accumulation and disease symptoms were significantly reduced. The invention therefore provides nontoxic, pollution-free, and efficient peptides that can be used conveniently either directly or formulated into virus inhibitors to control plant viral diseases.

CN115850373A (CN'373) discloses a short peptide, CLE7, and nucleic acid molecules encoding it, for use in inhibiting plant virus infection or in preparing antiviral agents. The background explains that plant RNA viruses are among the most harmful and diverse plant pathogens, spreading through soil, insects, water, and other routes and that existing control measures (e.g., chemical pesticides, crop rotation, and adjustments to planting schedules) are either environmentally damaging or only partially effective. While the CLE (CLAVATA3/Embryo surrounding region-related) family of plant peptides is well studied for developmental regulation, the function of short CLE peptides in activating plant immunity remains poorly understood. CN'373 reports that, when injected into or sprayed onto the leaves of *Nicotiana benthamiana*, CLE7 inhibits infection by the tobacco mosaic virus, the tomato mosaic virus, potato virus X, and the turnip mosaic virus. The findings indicate CLE7's potential as a targeted, environmentally friendly antiviral treatment for crops, which can be delivered via injection or foliar spray.

CN117441722A (CN'722) discloses the use of a plant peptide regulator, phytosulfokine (PSK), in preparations for resisting *turnip mosaic virus* (TuMV) in cruciferous crops. The background section of CN'722 explains the global spread of TuMV and the virus' severe economic impact on vegetables and brassicas. It further discusses the limitations of current control measures, including the negative environmental impact of chemical pesticides, the limited availability of virus-resistant plant varieties, and—in the case of varieties that are resistant—the degradation of resistance over time. PSK is a disulfated pentapeptide known to regulate plant growth and development. However, the antiviral effects of PSK against TuMV have not been reported. CN'722 describes the synthesis of PSK and its application via spraying on the leaves of *Nicotiana benthamiana*. PSK-treated plants exhibited visibly reduced systemic infection and significantly lower viral protein accumulation compared to controls. CN'722 highlights PSK's potential as a green and effective immune-inducing antiviral agent with broad application potential in improving crop resistance to viral diseases.

The teachings of KR102751628B1 (KR'628) are of a composition for preventing or controlling plant viral diseases. These compositions contain an antiviral protein derived from *Pseudomonas chlororaphis* O6. The background section of the patent document explains that effective antiviral agents for plant viruses remain underdeveloped. Present interventions rely mainly on physically removing infected plants from other plants or on indirect vector control. The inventors of KR'628 found that suspensions of viable, heat-treated, or ultrasonicated O6 cells exhibited strong inhibitory effects against *tobacco mosaic virus* (ToMV) in *Nicotiana tabacum* compared to controls, observed as a reduction in the formation of systemic lesions typical of ToMV symptoms. The composition can be applied directly to plants or soil and may include additives, bulking agents, nutrients, or sealing agents. The compositions of KR'628 could be applied to other viruses, such as cucumber mosaic virus, making them interesting candidates for sustainable, protein-based management of plant viruses.

## 25.12 Conclusion

Vegetable production has increased over the last decade. However, viral diseases remain one of the key constraints in global vegetable production. Management strategies are available and focus on controlling insect vectors and enhancing the immunity and resistance of host plants. Emerging technologies for managing vegetable virus diseases have also been introduced over the last decade. This chapter discusses the potential for peptides as novel and potent antivirals. Recent research has highlighted the support for using peptides in the management of viral diseases. However, reproducibility and consistency in field trials remain a challenge. Nevertheless, this chapter provides a SWOT analysis that will further improve the application of peptides in managing viral diseases in vegetables. Patent analysis also reveals themes across some or all of the patents. The peptide inventions activate innate plant defenses, rather than directly attacking the virus. Ecological friendliness and sustainability are important advantages; the backgrounds often cite the environmental damage caused by the continued use of chemical pesticides as a motivation for finding new ways to manage plant viruses. The disclosed peptides are potentially effective against multiple viruses, suggesting potential for activating mechanisms that strengthen broad resistance. Finally, ease of application (often by spraying or injection) and nontoxicity are consistently highlighted as commercial advantages. Ease of preparation is also important. Overall, this chapter highlights peptides as emerging technologies in plant virus management, potentially functioning as plant "vaccines." Their appeal lies in their efficacy, alignment with sustainability goals, and societal and regulatory trends toward pesticide reduction.

## References


Anderson PK, Cunningham AA, Patel NG, Morales FJ, Epstein PR, Daszak P (2004) Emerging infectious diseases of plants: pathogen pollution, climate change and agrotechnology drivers. Trends Ecol Evol 19:535–544

Angira A, Sharma T, Choudhary N et al (2024) Viral diseases of field and horticultural crops. In: Awasthi LPBT-VD of F and HC. Academic Press, pp 15–21

Annunziato G, Costantino G (2020) Antimicrobial peptides (AMPs): a patent review (2015–2020). Expert Opin Ther Pat 30:931–947. https://doi.org/10.1080/13543776.2020.1851679

Baeshen NA, Baeshen MN, Sheikh A et al (2014) Cell factories for insulin production. Microb Cell Fact 13:141. https://doi.org/10.1186/s12934-014-0141-0

Beaumelle L, Tison L, Eisenhauer N et al (2023) Pesticide effects on soil fauna communities—a meta-analysis. J Appl Ecol 60:1239–1253. https://doi.org/10.1111/1365-2664.14437

Bharathi JK, Suresh P, Prakash MAS, Muneer S (2024) Exploring recent progress of molecular farming for therapeutic and recombinant molecules in plant systems. Heliyon 10:e37634. https://doi.org/10.1016/j.heliyon.2024.e37634

Bhargava A, Osusky M, Hancock RE et al (2007) Antiviral indolicidin variant peptides: evaluation for broad-spectrum disease resistance in transgenic Nicotiana tabacum. Plant Sci 172:515–523. https://doi.org/10.1016/j.plantsci.2006.10.016

Burhan M, Singh AK, Jain SK (2017) Patents as proxy for measuring innovations: a case of changing patent filing behavior in Indian public funded research organizations. Technol Forecast Soc Change 123:181–190. https://doi.org/10.1016/j.techfore.2016.04.002

Caciagli P (2009) Vegetable viruses. In: Desk encyclopedia of plant and fungal virology, pp 479–487

Chang L, Mondal A, Singh B et al (2024) Revolutionizing peptide-based drug discovery: advances in the post-AlphaFold era. WIREs Comput Mol Sci 14:e1693. https://doi.org/10.1002/wcms.1693

Chiu L-Y, Chen I-H, Hsu Y-H, Tsai C-H (2020) The lipid transfer protein 1 from Nicotiana benthamiana assists bamboo mosaic virus accumulation. Viruses 12

Choi M-S, Kim W, Lee C, Oh C-S (2013) Harpins, multifunctional proteins secreted by gram-negative plant-pathogenic bacteria. Mol Plant-Microbe Interact 26:1115–1122. https://doi.org/10.1094/MPMI-02-13-0050-CR

Coin I, Beyermann M, Bienert M (2007) Solid-phase peptide synthesis: from standard procedures to the synthesis of difficult sequences. Nat Protoc 2:3247–3256. https://doi.org/10.1038/nprot.2007.454

Decker AP, Su Y, Mishra B et al (2023) Peptide stability is important but not a general requirement for antimicrobial and antibiofilm activity in vitro and in vivo. Mol Pharm 20:738–749. https://doi.org/10.1021/acs.molpharmaceut.2c00918

Dela Cueva FM, Balendres MA (2017) Elucidating the cause and impact of a chilli anthracnose outbreak in Batangas, Philippines. J Trop Plant Pathol 53:37–51

Deshoux M, Masson V, Arafah K et al (2020) Cuticular structure proteomics in the pea aphid Acyrthosiphon pisum reveals new plant virus receptor candidates at the tip of maxillary stylets. J Proteome Res 19:1319–1337. https://doi.org/10.1021/acs.jproteome.9b00851

Eid NA, Ibrahim AM, Elsharawy AA, Salem KF (2025) Nanofertilizers for plant viral disease management. In: Nanofertilizers for sustainable agriculture: assessing impacts on health, environment, and economy. Springer Nature Switzerland, Cham, pp 149–187

FAO (2024) Agricultural production statistics 2010–2023. FAOSTAT Analytical Briefs, No. 96. Rome

Fennell JF, Shipman WH, Cole LJ (1968) Antibacterial action of melittin, a polypeptide from bee venom. Proc Soc Exp Biol Med 127:707–710. https://doi.org/10.3181/00379727-127-32779

Fiallo-Olivé E, Pan L-L, Liu S-S, Navas-Castillo J (2019) Transmission of begomoviruses and other whitefly-borne viruses: dependence on the vector species. Phytopathology 110:10–17. https://doi.org/10.1094/PHYTO-07-19-0273-FI

Gabriel RL, José PE, Belén BM et al (2018) Citrus psorosis virus movement protein contains an aspartic protease required for autocleavage and the formation of tubule-like structures at plasmodesmata. J Virol 92. https://doi.org/10.1128/jvi.00355-18

Ghosh U, Tal-Gan Y (2025) Designing highly potent side-chain lactam-bridged cyclic competence-stimulating peptide-based quorum-sensing modulators in streptococcus oligofermentans. ACS Infect Dis 11:197–203. https://doi.org/10.1021/acsinfecdis.4c00773

Ines RM, Nash TE, Dallas MM et al (2013) Peptide aptamers that bind to geminivirus replication proteins confer a resistance phenotype to tomato yellow leaf curl virus and tomato mottle virus infection in tomato. J Virol 87:9691–9706. https://doi.org/10.1128/jvi.01095-13

Jamir I, Mandal AK, Devi AP, Bhattacharjee T, Maurya PK, Dutta S et al (2020) Screening of genotypes against viral diseases and assessment of yield loss due to yellow vein mosaic virus in okra grown in the eastern part of India. Indian Phytopathol 73(1):125–133

Jones DR (2003) Plant viruses transmitted by whiteflies. Eur J Plant Pathol 109:195–219

Jumper J, Evans R, Pritzel A et al (2021) Highly accurate protein structure prediction with AlphaFold. Nature 596:583–589. https://doi.org/10.1038/s41586-021-03819-2

Kaur J, Aakriti, Thakur N, Harish BM (2025) Revolutionizing vegetable production: the promising role of nanotechnology. Commun Soil Sci Plant Anal:1–19

Koeppe S, Kawchuk L, Kalischuk M (2023) RNA interference past and future applications in plants. Int J Mol Sci 24(11):9755

Kumar R, Tiwari RK, Sundaresha S, Kaundal P, Raigond B (2022) Potato viruses and their management. In: Chakrabarti SK, Sharma S, Shah MA (eds) Sustainable management of potato pests and diseases. Springer, Singapore, p 309. https://doi.org/10.1007/978-981-16-7695-6_12

Lata H, Sharma A, Chadha S, Kaur M, Kumar P (2022) RNA interference (RNAi) mechanism and application in vegetable crops. J Hortic Sci Biotechnol 97(2):160–170

Li J, Li J, Jing Z et al (2023) Antiviral function of peptidoglycan recognition protein in Spodoptera exigua (Lepidoptera: Noctuidae). Insect Sci 30:1092–1104. https://doi.org/10.1111/1744-7917.13158

Liu S, Sivakumar S, Sparks WO et al (2010) A peptide that binds the pea aphid gut impedes entry of pea enation mosaic virus into the aphid hemocoel. Virology 401:107–116. https://doi.org/10.1016/j.virol.2010.02.009

Liu H, Peck XY, Choong YK et al (2022) Identification of putative binding interface of PI(3,5)P2 lipid on rice black-streaked dwarf virus (RBSDV) P10 protein. Virology 570:81–95. https://doi.org/10.1016/j.virol.2022.03.010

Liu P, Zhang J, Liu S et al (2025) The plant signal peptide CLE7 induces plant defense response against viral infection in Nicotiana benthamiana. Dev Cell 60:934–948.e5. https://doi.org/10.1016/j.devcel.2024.11.020

Lucintel (2024) Bio-pesticides market size worldwide in 2024, with a forecast for 2030. In: Statista. https://www-statista-com.dlsu.idm.oclc.org/statistics/1350343/bio-pesticides-market-size-globally/. Accessed 30 May 2025

Luisa L-O, Jorge R-P, Linda H-B (2006) Peptide aptamers that bind to a geminivirus replication protein interfere with viral replication in plant cells. J Virol 80:5841–5853. https://doi.org/10.1128/jvi.02698-05

Luo Y, Zhang D-D, Dong X-W et al (2010) Antimicrobial peptaibols induce defense responses and systemic resistance in tobacco against tobacco mosaic virus. FEMS Microbiol Lett 313:120–126. https://doi.org/10.1111/j.1574-6968.2010.02135.x

Marcos JF, Beachy RN, Houghten RA et al (1995) Inhibition of a plant virus infection by analogs of melittin. Proc Natl Acad Sci 92:12466–12469. https://doi.org/10.1073/pnas.92.26.12466

Medani KR, Neill A, Garrod G et al (2024) Societal perceptions and attitudes towards genetically modified (GM) crops, feed, and food products in the Middle East, North Africa, and Turkey (MENAT) region: a systematic literature review. Food Qual Prefer 117:105148. https://doi.org/10.1016/j.foodqual.2024.105148

Mendoza-Figueroa JS, Kvarnheden A, Méndez-Lozano J et al (2018) A peptide derived from enzymatic digestion of globulins from amaranth shows strong affinity binding to the replication origin of tomato yellow leaf curl virus reducing viral replication in Nicotiana benthamiana. Pestic Biochem Physiol 145:56–65. https://doi.org/10.1016/j.pestbp.2018.01.005

Mendoza-Figueroa JS, Badillo-Ramírez I, Kvarnheden A et al (2019) The peptide AmPep1 derived from amaranth recognizes the replication hairpin of TYLCV disturbing its replication process in host plants. J Agric Food Chem 67:9241–9253. https://doi.org/10.1021/acs.jafc.9b02526

Moaniba IM, Lee P-C, Su H-N (2020) How does external knowledge sourcing enhance product development? Evidence from drug commercialization. Technol Soc 63:101414. https://doi.org/10.1016/j.techsoc.2020.101414

Modgill N, Naga KC, Firake DM, Tiwari RK, Shivaramu S, Kumar R, Bhatnagar A, Buckseth T, Bairwa A, Maharana C, Sharma S, Singh B (2025) Flupyradifurone application: a promising approach to curtail Tomato Leaf Curl New Delhi Virus (ToLCNDV) transmission vectored by Bemisia tabaci in potato farming. Crop Prot 187:106951. https://doi.org/10.1016/j.cropro.2024.106951

Mousavi Maleki MS, Sardari S, Ghandehari Alavijeh A, Madanchi H (2022) Recent patents and FDA-approved drugs based on antiviral peptides and other peptide-related antivirals. Int J Pept Res Ther 29:5. https://doi.org/10.1007/s10989-022-10477-z

Naga KC, Buckseth T, Subhash S, Bairwa A, Verma G, Kumar R, Malik K, Sharma S, Chakrabarti SK (2020) Transmission efficiency of potato leafroll virus (PLRV) by potato aphid Aulacorthum solani and green peach aphid Myzus persicae. Indian J Entomol 71:68–71. https://doi.org/10.5958/0974-8172.2020.00017.6

Naga KC, Sundaresha S, Kumar R, Tiwari RK, Subhash S, Verma G, Buckseth T, Bairwa A, Sharma S, Katare S, Srivastava RM (2021) A new record of Asia II 5 genetic group of Bemisia tabaci (Gennadius) in the major potato growing areas of India and its relationship with tomato leaf curl New Delhi virus infecting potato. 3 Biotech 11:421. https://doi.org/10.1007/s13205-021-02966-7

Narin F, Noma E, Perry R (1987) Patents as indicators of corporate technological strength. Res Policy 16:143–155. https://doi.org/10.1016/0048-7333(87)90028-X

Ndhlovu PT, Mokgehle SN, Kola ME, Chauke S, Falemara BC, Otang-Mbeng W (2025) The contribution of underutilized fruits and vegetables to enhanced food and nutrition security: a review. In: Food security and nutrition, pp 4–33

Ng JCK, Perry KL (2004) Transmission of plant viruses by aphid vectors. Mol Plant Pathol 5:505–511. https://doi.org/10.1111/j.1364-3703.2004.00240.x

Nono-Womdim R (2003) An overview of major virus diseases of vegetable crops in Africa and some aspects of their control. In: Hughes J d'A, Odu J (eds) Plant virology in sub-Saharan Africa, Proc. Conf. Organized by IITA. International Institute of Tropical Agriculture, Ibadan, pp 213–232

Noopur K, Chauhan JK, Walia SS, Verma MR, Dhar U, Choudhary S, Chikkeri SS (2023) Constraints in vegetable production in India: a review. Indian Res J Ext Edu 23(3):14–19

Okoma RN, Omuse ER, Mutyambai DM, Beesigamukama D, Murongo MF, Subramanian S, Chidawanyika F (2025) An assessment of vegetable production constraints, trait preferences and willingness to adopt sustainable intensification options in Kenya and Uganda. Front Sustain Food Syst 9:1471333

Ornelas Herrera SI, Baba Y, Kallas Z et al (2025) The role of environmental attitudes and consumption patterns in consumers' preferences for sustainable food from circular farming system: a six EU case studies. Agric Food Econ 13:7. https://doi.org/10.1186/s40100-025-00350-0

Philippines SC of the (2023) SC Issues Writ of Kalikasan on genetically modified rice and eggplant products. https://sc.judiciary.gov.ph/sc-issues-writ-of-kalikasan-on-genetically-modified-rice-and-eggplant-products/. Accessed 7 Aug 2025

Raigond B, Verma G, Kumar R, Tiwari RK (2022) Serological and molecular diagnosis of potato viruses: an overview. In: Chakrabarti SK, Sharma S, Shah MA (eds) Sustainable management of potato pests and diseases. Springer, Singapore, pp 337–352. (Hardcover ISBN 978-981-16-7694-9; Softcover ISBN978-981-16-7697-0; eBook ISBN978-981-16-7695-6). https://doi.org/10.1007/978-981-16-7695-6_13

Rajarapu SP, Bansal R, Mittapelly P, Michel A (2020) Transcriptome analysis reveals functional diversity in salivary glands of plant virus vector, Graminella nigrifrons. Genes (Basel). 11

Ramachander Turaga VN (2020) Peptaibols: antimicrobial peptides from fungi. In: Singh J, Meshram V, Gupta M (eds) BT - bioactive natural products in drug discovery. Springer Singapore, Singapore, pp 713–730

Rifkind D, Freeman GL (2005) 11 - Tobacco Mosaic Virus. In: Rifkind D (ed) Freeman GLBT-TNPWD in ID. Academic Press, London, pp 81–84. https://doi.org/10.1016/B978-012369353-2/50018-7

Rudolph C, Schreier PH, Uhrig JF (2003) Peptide-mediated broad-spectrum plant resistance to tospoviruses. Proc Natl Acad Sci 100:4429–4434. https://doi.org/10.1073/pnas.0730832100

Sabokkhiz MA, Tanhaeian A, Mamarabadi M (2019) Study on antiviral activity of two recombinant antimicrobial peptides against tobacco mosaic virus. Probiotics Antimicrob Proteins 11:1370–1378. https://doi.org/10.1007/s12602-019-09539-4

Salem NM, Jewehan A, Aranda MA, Fox A (2023) Tomato brown rugose fruit virus pandemic. Annu Rev Phytopathol 61(1):137–164

Sánchez-Tovar MR, Rivera-Bustamante RF, Saavedra-Trejo DL, Guevara-González RG, Torres-Pacheco I (2025) Mixed plant viral infections: complementation, interference and their effects, a review. Agronomy 15(3):620

Sastry KS (2013) Impact of virus and viroid diseases on crop yields. In: Plant virus and viroid diseases in the tropics: volume 1: introduction of plant viruses and sub-viral agents, classification, assessment of loss, transmission and diagnosis. Springer Netherlands, Dordrecht, pp 99–159

Selzer PM, Brutsche S, Wiesner P et al (2000) Target-based drug discovery for the development of novel antiinfectives. Int J Med Microbiol 290:191–201. https://doi.org/10.1016/S1438-4221(00)80090-9

Shaheb M, Nazrul M, Sarker A (2016) Improvement of livelihood, food and nutrition security through homestead vegetables production and fruit tree management in Bangladesh. J Bangladesh Agric Univ 12(2):377–387. https://doi.org/10.3329/jbau.v12i2.28698

Shekhar C, Khosya R, Thakur K et al (2024) A systematic review of pesticide exposure, associated risks, and long-term human health impacts. Toxicol Rep 13:101840. https://doi.org/10.1016/j.toxrep.2024.101840

Singh J, Kashyap R, Bansal K, Das R, Sandhu K, Singh G, Saini DK (2025) Transition from conventional to AI-based methods for detection of foliar disease symptoms in vegetable crops: a comprehensive review. J Plant Pathol:1–24

Thagun C, Horii Y, Mori M et al (2022) Non-transgenic gene modulation via spray delivery of nucleic acid/peptide complexes into plant nuclei and chloroplasts. ACS Nano 16:3506–3521. https://doi.org/10.1021/acsnano.1c07723

Thompson WMO (ed) (2011) The whitefly, Bemisia tabaci (Homoptera: Aleyrodidae) interaction with Geminivirus-infected host plants. Springer, New York

Tomlinson JA (1987) Epidemiology and control of virus diseases of vegetables. Ann Appl Biol 110(3):661–681

Vogel H, Jähnig F (1986) The structure of melittin in membranes. Biophys J 50:573–582. https://doi.org/10.1016/S0006-3495(86)83497-X

Wang Z-Z, Shi M, Ye X-Q et al (2013) Identification, characterization and expression of a defensin-like antifungal peptide from the whitefly Bemisia tabaci (Gennadius) (Hemiptera: Aleyrodidae). Insect Mol Biol 22:297–305. https://doi.org/10.1111/imb.12021

Wang Z, Bing X, Liu S, Chen X (2017) RNA interference of an antimicrobial peptide, Btdef, reduces tomato yellow leaf curl China virus accumulation in the whitefly Bemisia tabaci. Pest Manag Sci 73:1421–1427. https://doi.org/10.1002/ps.4472

Wang D, Wang B, Wang J et al (2020) Exogenous application of Harpin protein Hpa1 onto Pinellia ternata induces systemic resistance against tobacco mosaic virus. Phytopathology 110:1189–1198. https://doi.org/10.1094/PHYTO-12-19-0463-R

Wang C, Fan S, Li M et al (2025a) A 7-year feed study on the long-term effects of genetically modified maize containing cry1Ab/cry2Aj and EPSPS genes on gut microbiota and metabolite profiles across two generations of cynomolgus macaques. Food Chem Toxicol 200:115419. https://doi.org/10.1016/j.fct.2025.115419

Wang H, Li R, Yu Q et al (2025b) iAVP-RFVOT: identify antiviral peptides by random forest voting machine learning with unified manifold learning embedded features. Biochemistry. https://doi.org/10.1021/acs.biochem.5c00237

Webster CG, Elodie P, van Manuella M et al (2018) Identification of plant virus receptor candidates in the stylets of their aphid vectors. J Virol 92. https://doi.org/10.1128/jvi.00432-18

Xu H, Zhang Z, Zhang Z et al (2023) Effects of insulin-like peptide 7 in Bemisia tabaci MED on tomato chlorosis virus transmission. Pest Manag Sci 79:1508–1517. https://doi.org/10.1002/ps.7329

Yang J, Liu J, Liu Y et al (2025) In-depth site-specific glycoproteomic analysis reveals ER-resident protein PDI regulating wheat yellow mosaic virus infection. Int J Biol Macromol 293:139303. https://doi.org/10.1016/j.ijbiomac.2024.139303

Zambrano P, Xavier LC, Santos AM et al (2022) What do we have that is new in antifungal peptides? A patent review. Future Microbiol 17:1421–1432. https://doi.org/10.2217/fmb-2022-0105

Zhang S, Griffiths JS, Marchand G, Bernards MA, Wang A (2022) Tomato brown rugose fruit virus: an emerging and rapidly spreading plant RNA virus that threatens tomato production worldwide. Mol Plant Pathol 23(9):1262–1277

Zhao L, Hao X, Wu Y (2015) Inhibitory effect of polysaccharide peptide (PSP) against Tobacco mosaic virus (TMV). Int J Biol Macromol 75:474–478. https://doi.org/10.1016/j.ijbiomac.2015.01.058

Zhong Y, Li Y, Huang K et al (2021) Crude peptides extracted from dry mycelium of Penicillium chrysogenum serve as a micro-associated molecular pattern to induce systemic resistance against tobacco mosaic virus in tobacco. Physiol Mol Plant Pathol 115:101677. https://doi.org/10.1016/j.pmpp.2021.101677

Zhou J, Tzanetakis IE (2020) Transmission blockage of an orthotospovirus using synthetic peptides. J Gen Virol 101:112–121. https://doi.org/10.1099/jgv.0.001352

MIX
Papier aus verantwortungsvollen Quellen
Paper from responsible sources
FSC® C105338
www.fsc.org

If you have any concerns about our products,
you can contact us on
**ProductSafety@springernature.com**

In case Publisher is established outside the EU,
the EU authorized representative is:
**Springer Nature Customer Service Center GmbH**
**Europaplatz 3, 69115 Heidelberg, Germany**

Printed by Libri Plureos GmbH
in Hamburg, Germany